Communications in Computer and Information Science   **575**

Commenced Publication in 2007
Founding and Former Series Editors:
Alfredo Cuzzocrea, Dominik Ślęzak, and Xiaokang Yang

More information about this series at http://www.springer.com/series/7899

Kangshun Li · Jin Li
Yong Liu · Aniello Castiglione (Eds.)

Computational Intelligence and Intelligent Systems

7th International Symposium, ISICA 2015
Guangzhou, China, November 21–22, 2015
Revised Selected Papers

 Springer

Editors

Kangshun Li
College of Mathematics and Informatics
The South China Agricultural University
Guangzhou
China

Jin Li
School of Computer Science
Guangzhou University
Guangzhou
China

Yong Liu
School of Computer Science and
  Engineering
The University of Aizu
Aizu-Wakamatsu, Fukushima
Japan

Aniello Castiglione
Department of Informatics
University of Salerno
Fisciano
Italy

ISSN 1865-0929   ISSN 1865-0937 (electronic)
Communications in Computer and Information Science
ISBN 978-981-10-0355-4   ISBN 978-981-10-0356-1   (eBook)
DOI 10.1007/978-981-10-0356-1

Library of Congress Control Number: 2015958854

Printed on acid-free paper

This Springer imprint is published by SpringerNature
The registered company is Springer Science+Business Media Singapore Pte Ltd.

Preface

The present volume contains the proceedings of the 7th International Symposium on Intelligence Computation and Applications (ISICA 2015) held in Guangzhou, China, November 21–22, 2015. ISICA 2015 successfully attracted over 189 submissions. Through rigorous reviews, 77 high-quality papers were selected for this volume of *Communications in Computer and Information Science* (CCIS 575). ISICA conferences are one of the first series of international conferences on computational intelligence that combine elements of learning, adaptation, evolution, and fuzzy logic to create programs as alternative solutions to artificial intelligence. The past ISICA proceedings including three volumes of CCIS and four volumes of LNCS have been accepted in both the Index to Scientific and Technical Proceedings (ISTP) and Engineering Information (EI).

Following the success of the past six ISICA events, ISICA 2015 persisted in exploring new problems emerging in the fields of computational intelligence. In recent years, a number of intelligent driving systems for driverless cars have been developed. For example, at least ten of Google's self-driving cars, including six Toyota Prius, an Audi TT, and three Lexus RX450h, have undergone road safety testing. Such impressive progress makes people think that current techniques have solved all issues in the design of an intelligent driving system in the sense of overall human performance. However, it is simply not the case. There are still many unsolved problems. For example, Google's cars are not able to spot a police officer who is waving for traffic to stop on the side of road. The car's sensors cannot tell whether a road obstacle is a rock or a crumpled piece of paper. It is expected that these unsolved problems in such intelligent systems will become increasingly difficult. While it is difficult to create intelligence directly, an intelligent system should inherit the simple mechanism of evolution in which the simple models could produce the evolution of complex morphologies.

ISICA 2015 featured the most up-to-date research in the analysis and theory of evolutionary computation, neural network architectures and learning, neuro-dynamics and neuro-engineering, fuzzy logic and control, collective intelligence and hybrid systems, deep learning, knowledge discovery, learning, and reasoning. It provided a venue for fostering technical exchanges, renewing everlasting friendships, and establishing new connections.

On behalf of the Organizing Committee, we would like to thank warmly the sponsors, South China Agricultural University, Guangzhou University, Wuhan University, and China University of Geosciences, who helped in one way or another to achieve our goals for the conference. We wish to express our appreciation to Springer for publishing the proceedings of ISICA 2015. We also wish to acknowledge the dedication and commitment of both the staff at the Springer Beijing office and the CCIS editorial staff. We would like to thank the authors for submitting their work, as well as the Program Committee members and reviewers for their enthusiasm, time, and expertise. The invaluable help of active members of the Organizing Committee, including Wei Li, Lei Yang, Lixia Zhang, Yan Chen, Lu Xiong, Lei Zuo, Liang Zhong, Weiguang Chen, and Luyan Guo, in setting up and maintaining the online submission systems, assigning the papers to the reviewers, and preparing the camera-ready version of the proceedings is highly appreciated. We would like to thank them for helping to make ISICA 2015 a success.

November 2015

Kangshun Li
Jin Li
Yong Liu
Aniello Castiglione

Organization

Program Committee

Shaowei Cai	Chinese Academy of Sciences, China
Jiannong Cao	The Hong Kong University of Science and Technology, China
Aniello Castiglione	University of Salerno, Italy
Weineng Chen	Sun Yat-sen University, China
Yan Chen	South China Agricultural University, China
Debiao He	Wuhan University, China
Jun He	Aberystwyth University, UK
Shuqiang Huang	Jinan University, China
Xinyi Huang	Fujian Normal University, China
Ying Huang	Gannan Normal University, China
Chunfu Jia	Nankai University, China
Dazhi Jiang	Shantou University, China
Nan Jiang	East China Jiao Tong University, China
Jin Li	Guangzhou University, China
Kangshun Li	South China Agrictural University, China
Ping Li	Sun Yat-sen University, China
Wei Li	Jiangxi University of Science and Technology, China
Xuan Li	Fujian Normal University, China
Zhiqiang Lin	Chinese Academy of Sciences, China
Xiaozhang Liu	Hainan University, China
Zheli Liu	Nankai Universisy, China
Xu Ma	Qufu Normal University, China
Wen Sheng	Deakin University, Australia
Ke Tang	University of Science and Technology of China, China
Ming Tao	Dongguan University of Technology, China
Xiang Tao	Chongqing University, China
Cong Wang	City University of Hong Kong, China
Jiahai Wang	Sun Yat-sen University, China
Jianfeng Wang	Xidian University, China
Yilei Wang	Ludong University, China
Yong Wang	Central South University, China
Xianglin Wei	Nanjing Telecommunication Technology Research Institute, China
Di Wu	Sun Yat-sen University, China
Lu Xiong	Jiangxi University of Science and Technology, China
Honyang Yan	Guangzhou University, China

Lei Yang	South China Agricultural University, China
Shuling Yang	South China Agricultural University, China
Dongbo Zhang	South China University of Technology, China
Lixia Zhang	South China Agricultural University, China
Kuo Zhao	Jilin University, China
Wei-Shi Zheng	Sun Yat-sen University, China
Liang Zhong	South China Agricultural University, China

Additional Reviewers

Chen, Peng
Chen, Yan
Li, Wei
Ye, Jun
Zeng, Ling

Contents

Intelligent Simulation Algorithms

Data Mining and Cloud Computing

Applications and Security

Evolutionary Algorithms

A Hybrid Group Search Optimizer with Opposition-Based Learning and Differential Evolution

Chengwang Xie[✉], Wenjing Chen, and Weiwei Yu

School of Software, East China Jiaotong University, Nanchang 330013, China
chengwangxie@163.com

Abstract. Group search optimizer (GSO) is a recently developed heuristic inspired by biological group search resources behavior. However, it still has some defects such as slow convergence speed and poor accuracy of solution. In order to improve the performance of GSO in solving complex optimization problems, an opposition-based learning approach (OBL) and a differential evolution method (DE) are integrated into GSO to form a hybrid GSO. In this paper, the strategy of OBL is used to enlarge the search region, and the operator of DE is utilized to enhance local search to improve. Comparison experiments have demonstrated that our hybrid GSO algorithm performed advantages over previous GSO and DE approaches in convergence speed and accuracy of solution.

Keywords: Group search optimizer · Opposition-based learning · Differential evolution · Hybrid group search optimizer

1 Introduction

In recent years, some swarm intelligence optimization algorithms have emerged, such as genetic algorithm (GA) [1], simulated annealing algorithm (SA) [2], particle swarm optimization algorithm (PSO) [3], ant colony algorithm (ACO) [4], differential evolution algorithm (DE) [5] and so on. These optimization models are all based on group search, and they have made great progress on the theoretical and applied aspects recently.

In 2006, Sheldon *et al.* [6] proposed a new algorithm of group intelligence, that is group search optimization (GSO). The GSO is a novel random group search algorithm based on producer-scrounger model, simulated the behavior of searching resources for animals. Some investigations have showed that the GSO has better accuracy and convergence speed compared to PSO and some other evolutionary algorithms (EAs). Moreover, the GSO not only employs the simplicity and easy implementation compared to other EAs, but also it uses a particular search model. These characteristics enable the GSO to avoid some unnecessary structure analysis, thereby saving a lot of time, and it is especially suitable for the optimization design. However, there exists no a general optimizer being suitable to solve all kinds of optimization problems due to the theorem of No Free Lunch. That is to say, optimizers may obtain good performance on some optimization problems and may behave poor for others.

© Springer Science+Business Media Singapore 2016
K. Li et al. (Eds.): ISICA 2015, CCIS 575, pp. 3–12, 2016.
DOI: 10.1007/978-981-10-0356-1_1

The standard GSO is suitable for solving the complex structure design problems with multimodal, high-dimension and varying dimensions, but it still employs some shortcomings such as premature convergence and poor accuracy of solution. In order to improve the GSO, a hybrid GSO with opposition-based learning and differential evolution (OBDGSO) is proposed in the paper. The opposition-based learning method travels the current solution space and the opposition-based solution space simultaneously, and the better solution is found to view as the new solution, which can increase the probability of finding the global optimum. In general, DE has a strong ability of global search at the early stage of the algorithm. The reason is that the early behavior of mutation in DE can obtain remarkable differences between individuals of the population. However, the differences between the individuals become insignificant gradually at the later stage because the population tends to converge with the evolution process, correspondingly, the ability of local search of DE become stronger and stronger. So, it can be found that some advantages of DE, such as fast convergence and not easy to fall into local optimum, are precisely what the GSO does not have. In this paper, the OBGGSO on the one hand improves the diversity at the later stage of algorithm, and enhances the information exchange of the above optimizers (for example, DE, OBL and GSO), which facilitates the OBDGSO to jump out of the local optimum area and accelerate the convergence speed.

2 Background

2.1 Group Search Optimizer

Group search optimizer is a new algorithm derived from the nature of social animals foraging behavior, the process of finding the optimal solution can be viewed as the behavior of animals foraging. The population of GSO is called group and each individual in the population is a member, the number of the individuals in the population is population size. In GSO, a group consists of three kinds of members: producers, scroungers and rangers [7]. The behaviors of producers and scroungers follow the Producer-Scrounger (PS) model, and the rangers perform random walk motions.

In a n-dimensional search space, the ith member at the tth iteration has a current position $X_i^t \in R^n$, the head angle $\phi_i^t = (\phi_{i1}^t, \ldots, \phi_{i(n-1)}^t) \in R^{n-1}$ and the head direction $D_i^t(\phi_i^t) = (d_{i1}^t, \ldots, d_{in}^t) \in R^n$ can be calculated from ϕ_i^t via a polar to Cartesian coordinates transformation as follows:

$$\begin{cases} d_{i1}^t = \prod_{p=1}^{n-1} \cos(\phi_{ip}^t) \\ d_{ij}^t = \sin(\phi_{i(j-1)}^t) \cdot \prod_{p=i}^{n-1} \cos(\phi_{ip}^t) \\ \ldots\ldots \\ d_{in}^t = \sin(\phi_{i(n-1)}^t) \end{cases} \tag{1}$$

In GSO, at the kth iteration, the steps of producer X_p behaves as follows:

(1) the X_p will scan at zero degree and then scan laterally by randomly sampling three points in the scanning area. The way of calculation for the point at zero degree is as follow:

$$X_z = X_p^t + r_1 l_{max} D_p^t(\phi_p^t)$$ (2)

The point in the right hand side hypercube follows formula (3):

$$X_r = X_p^t + r_1 l_{max} D_p^t(\phi_p^t + r_2 \theta_{max}/2)$$ (3)

and the point in the left hand side hypercube abides by Eq. (4):

$$X_l = X_p^t + r_1 l_{max} D_p^t(\phi_p^t - r_2 \theta_{max}/2)$$ (4)

where $r_1 \in R$ is a normally distributed random number with mean 0 and standard deviation 1 and $r_2 \in R^{n-1}$ is a random sequence in the range (0,1).

(2) The producer X_p will then find the best point with the best resource (fitness value). If the better point has a better resource than its current position, then it will move to this point. Or it will still stay in its current position and turn its head to a new angle according to Eq. (5).

$$\phi_p^{k-1} = \phi_p^k + r_2 a_{max}$$ (5)

where a_{max} represents the maximum turning angle.

(3) If the producer X_p cannot find a better area after α iterations, it will turn its head back to zero degree such as formula (6).

$$\phi^{t+\alpha} = \phi^t$$ (6)

where α is a constant.

At each iteration, a number of group members are selected as scroungers. The scroungers will keep searching for opportunities to join the resources found by the producer. At the tth iteration, the area copying behavior of the ith scrounger can be modeled as a random walk towards the producer followed the Eq. (7)

$$X_i^{t+1} = X_i^t + r_3(X_p^t - X_i^t)$$ (7)

where $r_3 \in R^n$ is a uniform random sequence in the range (0,1).

Besides the producer and the scroungers, a small number of rangers have been also introduced into GSO. In nature, group members often have different searching and competitive abilities; subordinates, who are less efficient foragers than the dominant ones, will be dispersed from the group. In GSO, random walks, which are viewed as the most efficient searching method for randomly distributed resources, are adopted by the rangers. If the ith group member is selected as a ranger, at the tth iteration, it will generate a random head angle ϕ_i, denoted by formula (8).

$$\phi_i^{t+1} = \phi_i^t + r_2 \alpha_{max} \tag{8}$$

where α_{max} is the maximum turning angle; and the ith member will choose a random distance as Eq. (9)

$$l_i = ar_1 l_{max} \tag{9}$$

and it will move to the new point denoted by formula (10).

$$X_i^{k+1} = X_i^k + l_i \cdot D_i^k(\phi_i^{k+1}) \tag{10}$$

In order to improve the probability of finding resources, animals use several strategies to confine their search to a profitable area, and the important action is to turn back to the patch where its edge is detected. This strategy is adopted by GSO to handle the boundary search space. That is to say when a member is outside the search space, it will turn back to its previous position inside the search space.

2.2   Opposition-Based Learning

As a new model of machine intelligence, the method of opposition-based learning (OBL) proposed by Rahnamayan et al. [8] has been successfully integrated into many optimizers. Because OBL can increase the diversity of the population to a certain extent, then reduce the possibility of falling into local optimum, OBL is often used to improve the ability of global exploration. Furthermore, Wang *et al.* [9] presented a generalized opposition-based learning (GOBL) based on opposition-based learning by introducing the randomized parameter k, and the approach of GOBL was integrated into the PSO to design an efficient optimizer. Some basic concepts concerning OBL are presented as follows.

Definition 1. Let $x \in R$ be a real number defined on a certain interval: $x \in [a, b]$. The *opposite number* $\tilde{x}$ is defined as follows:

$$\tilde{x} = a + b - x \tag{11}$$

For $a = 0$ and $b = 1$ we receive

$$\tilde{x} = 1 - x \tag{12}$$

Analogously, the opposite number in a multidimensional case can be stated as Definition 2.

Definition 2. Let $P = (x_1, x_2, \ldots, x_n)$ be a point in a n-dimensional coordinate system with $x_1, x_2, \ldots, x_n \in R$, and $x_i \in [a_i, b_i]$. The opposite point $\tilde{P}$ is completely defined by its coordinates $\tilde{x}_1, \tilde{x}_2, \ldots, \tilde{x}_n$, where

$$\tilde{x} = a_i + b_i - x_i, i = 1, 2, \ldots, n \tag{13}$$

2.3 Differential Evolution

As a global optimizer based on group differences, the differential evolution (DE) has been widely used in various engineering applications. The advantages of DE can be stated as follows. (1) DE employs good property of stabilization for some optimization problems with non-convex, multimodal and non-linear functions. (2) the convergence speed of DE is faster compared to other optimizers in the same accuracy of solution. (3) DE is especially suitable to solve multi-variable function optimization problems. (4) the operation of DE is simple and easy to implement [10].

The main operator of DE is mutation, and DE uses a perturbation of two members as the vector to add to the third member, which will generate a new vector. This operation is called mutation. Then, the new vector is mixed with the predefined parameters according to certain rules to generate trial vector, where the process is called crossover. If the trial vector of function is inferior to the target function, the test vector will be replaced in the next generation.

The basic steps of DE [11, 12] are presented as follows.

1) Mutation operation

   Based on the generating ways of mutating individuals, we can classify the mutation into some types. In general, there are three kinds of mutation operators, such as DE/rand/1/bin, DE/best/1/bin and DE/current-to-best/2/bin. And the first type of mutation is adopted in this paper.

$$\begin{cases} v_{i,j} = x_{best,j} + F * (x_{r1,j} - x_{r2,j}) \\ v_{i,j} = x_{best,j} + F * (x_{r1,j} - x_{r2,j}) \\ v_{i,j} = x_{i,j} + \lambda * (x_{best,j} - x_{i,j}) + F * (x_{r1,j} - x_{r2,j}) \end{cases} \tag{14}$$

   where $r_1 \neq r_2 \neq r_3 \neq i$, $r_1, r_2, r_3 \in \{1, 2, \ldots, N\}$, and $(x_{r1,j} - x_{r2,j})$ is differential vector, F is scaling factor.

2) Crossover operation

   In order to improve the diversity of population, the crossover operator is used as follows.

$$u_{i,j} = \begin{cases} v_{i,j}, (rand(0, 1) \leq CR) \, or \, (j = rand(1, D)) \\ x_{i,j}, else \end{cases} \tag{15}$$

   where $rand(0,1)$ is a uniform random number within [0,1], and CR is the crossover constant with outcome $\in$ [0,1].

3) Selection operation

   DE uses a "greedy" selection strategy to ensure the better individual having better fitness to enter into the next generation. The trial individual will be compared to parent individuals after the process of mutation and crossover. If the fitness of the trial individual is better than the parent, it will replace the parent and join into the next population. Otherwise, the parent remains unchanged and enters into the next iteration directly.

3 Hybrid GSO

A lot of investigations have showed that the shortcomings of GSO, such as poor convergence speed and easily trapping into local optimum, could attribute to the loss of diversity of group gradually. So, it is crucial to GSO to improve the diversity to enhance the efficiency of GSO to solve the complex optimization problems.

As mentioned in Sect. 1, the main merit of DE contains the strong local search ability, and the method of opposition-based learning is good at global exploration. Considering the above factors, we combine DE and OBL into GSO to design a hybrid GSO optimizer.

Let the population size be N, and select 30 percents of the population randomly to carry out OBL operator to generate a new opposition-based population. Combining the new population and the original population (whose size is $0.3*N$) to sort in descending order based on the fitness, then the better half of the mixed population would be selected. Afterwards, the number of $0.4*N$ of individuals from the rest of the population are selected to perform DE operation. At last, the remainder population, whose size is $0.3*N$, will be carried out GSO optimization. The hybrid GSO integrates the advantages of DE and OBL to better balance the global exploration and local exploitation to solve the complicated optimization problems effectively.

The flowchart of the hybrid GSO in the paper is presented as follows.

Step 1. Initialize the population P randomly, let the population size be N and the maximum iteration number be T_{max}. Calculate the fitness of each individual, and set the counter of iteration $t = 0$.

Step 2. Random select $0.3*N$ individuals to form a subpopulation SP_1, applying OBL to SP_1 to generate an opposition-based population OBP. Combined SP_1 and OBP to select the better half of the mixing population to form subpopulation P_1 based on the fitness values.

Step 3. Random select $0.4*N$ individuals to construct a subpopulation SP_2, and apply DE to SP_2 to generate a differential evolution population P_2, whose size is also $0.4*N$.

Step 4. Apply GSO to the remainder population whose size is $0.3*N$ to generate a population P_3, and $| P_3 | = 0.3*N$.

Step 5. Combine the subpopulation P_1, P_2 and P_3 to form the next population, and $t = t+1$.

Step 6. If $t > T_{max}$, stop; otherwise, goto Step 2.

4 Experimental Results and Analysis

In order to test the validity of the hybrid GSO, we select two representative optimizers as peer comparison algorithms, the one is GSO proposed by *He S* [13]. in 2009, and the other is the original DE proposed by *Storn et al*. The comparable experiments in the paper are all based on 13 benchmark single objective optimization problems [14]. And the 13 test problems are listed in Table 1. These 13 benchmark functions can be classify into three kinds as follows. (1) unimodal functions. (2) simple multi-peak functions.

Table 1. The 13 classical test functions

Function	Name	Feasible solution space	f_{min}
f_1	Sphere	[−100,100]	0
f_2	Schwefel's Problem 2.22	[−10,10]	0
f_3	Schwefel's Problem 1.2	[−100,100]	0
f_4	Schwefel's Problem 2.21	[−100,100]	0
f_5	Rosenbrock	[−30,30]	0
f_6	Step	[−5.12,5.12]	0
f_7	Quartic	[−1.28,1.28]	0
f_8	Schwefel's Problem 2.6	[−500,500]	0
f_9	Rastrigin	[−5.12,5.12]	0
f_{10}	Ackley	[−32,32]	0
f_{11}	Griewank	[−600,600]	0
f_{12}	Penalized 1	[−50,50]	0
f_{13}	Penalized 2	[−50,50]	0

Table 2. The average results for each algorithm on each test function

Function	GSO	DE	OBDGSO	f_{min}
f_1	21.2789	0.1803	**9.12E-07**	0
f_2	0.1779	1.0614	**7.20E-08**	0
f_3	4.63E+04	**2.19E+03**	1.41E+04	0
f_4	73.1417	**4.5009**	62.4728	0
$f_{5'}$	4.48E+07	6.14E+06	**3.61E+06**	0
$f_{6'}$	1.31E+03	77.4618	**22.3586**	0
f_7	3.23E-08	1.40E-06	**1.49E-11**	0
f_8	3.91E+03	7.64E+03	**3.71E+03**	0
f_9	1.05E+02	2.18E+02	**18.3557**	0
f_{10}	11.4299	10.7654	**6.0519**	0
f_{11}	1.2249	0.3278	**3.08E-06**	0
$f_{12'}$	1.79E+06	6.1801	**4.2264**	0
$f_{13'}$	3.90E+06	18.4162	**8.4528**	0

The above two types of functions are mainly used to test the optimization accuracy. (3) non-rotating multi-peak functions. This kind of functions have many local extreme points, which is hard for the average optimizers to find the global optimum. So, the third type of test functions are often used to test the global optimization performance and the ability of avoiding premature convergence.

All experiments in the paper are based on Windows 7 operating system, dual-core 2.50 GHz Intel processor and 4G memory, and Matlab 2010 programming platform.

In order to compare the performances of peer algorithms, we set some identical running parameters, such as the group (population) size is 100, the dimension of decision variable is 30, and the maximum number of iterations is set to 1000.

Table 3. Statistical mean and standard deviation of obtained by OBDGSO, GSO and DE on 13 test functions over 30 independent runs

Test instance		OBDGSO	GSO	DE
f_1	Mean	**1.00E-09**	8.1584	0.1432
	Std.	1.70786E-06	18.99110904	0.032472599
	t-test		+	+
f_2	Mean	**7.40E-09**	0.0408	0.8419
	Std.	4.39996E-08	0.118550676	0.211006374
	t-test		+	+
f_3	Mean	9.44E+03	2.96E+04	**1.61E+03**
	Std.	3997.819643	10703.80471	671.7363074
	t-test		+	=
f_4	Mean	51.6286	61.9726	**3.9978**
	Std.	8.146587239	7.455841639	0.349443413
	t-test		+	=
f_5	Mean	**9.41E+05**	3.04E+07	5.21E+06
	Std.	3270285.171	14123670.35	900927.2766
	t-test		+	+
f_6	Mean	**5.7405**	3.85E+02	50.5079
	Std.	11.23317659	964.833203	19.58850474
	t-test		+	+
f_7	Mean	**8.93E-11**	1.00E-10	8.00E-09
	Std.	4.36417E-12	5.54543E-08	2.56192E-06
	t-test		+	+
f_8	Mean	**3.08E+03**	3.33E+03	7.33E+03
	Std.	387.9434117	387.4388644	555.615874
	t-test		=	+
f_9	Mean	**0.0029**	6.15E+01	2.10E+02
	Std.	24.75970942	47.18212305	6.143087229
	t-test		=	=
f_{10}	Mean	**2.9365**	10.2191	10.3556
	Std.	2.376233926	0.664516162	0.203911203
	t-test		=	=
f_{11}	Mean	**1.60E-07**	1.0465	0.2912
	Std.	4.22386E-06	0.229185544	0.066716038
	t-test		+	+
$f_{12'}$	Mean	**2.2658**	2.42E+05	5.3495
	Std.	1.654364408	1861291.275	1.137839227
	t-test		+	=
$f_{13'}$	Mean	1.6379	6.09E+05	10.8051
	Std.	7.899360368	3198753.277	5.047741332

(*Continued*)

Table 3. (*Continued*)

Test instance	OBDGSO	GSO	DE
t-test		+	=
Better(+)		10	7
Same(=)		3	6
Worse(−)		0	0
Score		10	7

In addition, the algorithmic parameters of GSO are used followed [15], and the algorithmic parameters of DE are followed [16] and [17].

Our OBDGSO is compared with the other two optimizers, such as GSO and DE, all three algorithms are carried out 30 times repeatedly, and we can obtain the statistical average data as the experimental results, which are listed in Table 2. We can observe that the OBDGSO has the best accuracy of solution in all 13 test functions among the three peer algorithms.

Table 3 lists the statistical mean and standard deviation results obtained by OBDGSO, GSO and DE on 13 test functions over 30 independent runs. It can be seen that the OBDGSO has significant performance advantages over GSO and DE. So we can conclude that OBDGSO is promising optimizer in solving multi-modal, high-dimensional functions.

5 Conclusion

The paper proposed a hybrid GSO with opposition-based learning and differential evolution, called OBDGSO. The hybrid GSO utilizes the method of opposition-based learning to enhance the ability of global exploration and uses differential evolution to improve the local search ability. Three peer comparison algorithms (GSO, DE and OBDGSO) are performed to comparing experiments on 13test functions, the experimental results show that OBDGSO has significant advantages over GSO and DE in accuracy of solution and convergence speed. So, the conclusion is that the OBDGSO in the paper is a promising optimizer in solving multi-modal, high-dimensional functions.

References

1. Holland, J.H.: Adaptation in Natural and Artificial Systems. University of Michigan Press, AnnArbor (1975)
2. Kirkpatrick, S., Gelatt, C.D., Vecchi, P.M.: Optimization by simulated annealing. Science **220**, 671–680 (1983)
3. Kennedy, J., Eberhart, R.C.: Particle swarm optimization. In: Proceedings of IEEE International Conference on Neural Networks, pp. 1942–1948 (1995)

4. Dorigo, M., Maniezzo, V., Colorni, A.: The ant system: optimization by a colony of cooperating agents. IEEE Trans. Syst. Man Cybern. Part B Cybern. **26**, 29–41 (1996)
5. Storn, R., Price, K.: Differential evolution-a simple efficient adaptive scheme for global optimization. J. Global Optim. **11**, 341–359 (1997)
6. He, S., Wu, Q.H., Saunders, J.R.: A novel group search optimizer inspired by animal behavioural ecology. In: 2006 IEEE Congress on Evolutionary Computation (CEC), pp. 1272–1278. IEEE Xplore, Vancouver, BC, Canada. New York, 16–21 July 2006
7. He, S., Wu, Q.H., Saunders, J.R.: Group search optimizer:an optimization algorithm inspired by animal searching behavior. IEEE Trans. Evol. Comput. **13**(5), 973–990 (2009)
8. Rahnamayan, S., Tizhoosh, H.R., Salama, M.M.A.: Opposition-based differential evolution algorithms. In: IEEE Congress on Evolutionary Computation Canada (2006)
9. Giraldeau, L.-A., Lefebvre, L.: Exchangeable producer and scrounger roles in a captive flock of feral pigeons - a case for the skill pool effect. Anim. Behav. **34**(3), 797–803 (1986)
10. Storn, R., Price, K.: Differential evolution-a simple and efficient heuristic for global optimization over continuous spaces. J. Glob. Optim. **11**, 341–359 (1997)
11. Yuan, J., Sun, Z., Qu, G.: Simulation study of differential evolution. J. Syst. Simul. **20**, 4646–4647 (2007)
12. Wang, F.-S., Jang, H.-J.: Parameter estimation of a bioreaction model by hybrid differential evolution. Evol. Comput. **1**, 16–19 (2000)
13. He, S., Wu, Q.H., Saunders, J.R.: Group search optimizer: an opimization algorithm inspired by animal searching behavior. IEEE Trans. Evol. Comput. **13**(5), 973–990 (2009)
14. Yao, X., Liu, Y., Liu, G.: Evolutionary programming made faster. IEEE Trans. Evol. Comput. **3**(2), 82–102 (1999)
15. Karaboga, D., Akay, B.: A comparative study of artificial bee colony algorithm. Appl. Math. Comput. **214**(1), 108–132 (2009)
16. Qin, A.K., Huang, V.L., Suganthan, P.N.: Differential evolution algorithm with strategy adaptation for global numerical optimization. IEEE Trans. Evol. Comput. **13**(2), 398–417 (2009)
17. Liu, G., Li, Y., Zhang, Q.: Enhancing the search ability of differential evolution through orthogonal crossover. Inf. Sci. **185**(1), 153–177 (2012)

A New Firefly Algorithm with Local Search for Numerical Optimization

Hui Wang[1,2](✉), Wenjun Wang[3], Hui Sun[1,2], Jia Zhao[1,2],
Hai Zhang[1,2], Jin Liu[4], and Xinyu Zhou[5]

[1] School of Information Engineering, Nanchang Institute of Technology,
Nanchang 330099, China
huiwang@whu.edu.cn
[2] Jiangxi Province Key Laboratory of Water Information Cooperative Sensing
and Intelligent Processing, Nanchang 330099, China
[3] School of Business Administration, Nanchang Institute of Technology,
Nanchang 330099, China
[4] School of Computer, Wuhan University, Wuhan 430072, China
[5] College of Computer and Information Engineering, Jiangxi Normal University,
Nanchang 330022, China

Abstract. Firefly algorithm (FA) is a recently proposed swarm intelligence optimization technique, which has shown good performance on many optimization problems. In the standard FA and its most variants, a firefly moves to other brighter fireflies. If the current firefly is brighter than another one, the current one will not be conducted any search. In this paper, we propose a new firefly algorithm (called NFA) to address this issue. In NFA, brighter fireflies can move to other positions based on local search. To verify the performance of NFA, thirteen classical benchmark functions are tested. Experimental results show that our NFA outperforms the standard FA and two other modified FAs.

Keywords: Firefly algorithm (FA) · Swarm intelligence · Local search · Numerical optimization

1 Introduction

Firefly algorithm (FA) is a new swarm intelligence algorithm developed by Yang in 2010 [1]. It is inspired by the social behavior of fireflies based on the flashing and attraction characteristics of fireflies. In the past five years, the research of FA has attracted much attention. Different versions of FA has been designed to solve benchmark or real-world optimization problems [2–6].

To enhance the performance of FA, Farahani et al. [7] proposed a Gaussian distributed FA (GDFA). Computational results on five benchmark functions show that GDFA outperforms PSO and the standard FA. Tilahun and Ong [8] modified the random movement of the brighter firefly by generating random directions in order to determine the best direction. If such a direction is not generated, it will remain its current position. Moreover, the assignment of attractiveness is modified in such a way that the effect of the objective function is magnified. Simulation results show that the

© Springer Science+Business Media Singapore 2016
K. Li et al. (Eds.): ISICA 2015, CCIS 575, pp. 13–22, 2016.
DOI: 10.1007/978-981-10-0356-1_2

modified FA performs better than the standard FA in finding the best solution with smaller CPU time. Fister et al. [9] proposed a memetic FA (MFA) to solve combinatorial optimization problems. In MFA, the parameter α is dynamically adjusted, and the parameter β is changed in the range [0.2, 1.0] based on the distance between two fireflies. Additionally, the random part $\alpha\varepsilon$ for the movement of the attraction is scaled by the size of the search range. Experimental results show that the MFA is significantly better than the standard FA. In our previous work [10], the MFA is used as the standard FA and combined with other strategies. Gandomi et al. [11] introduced chaos into FA to increase its global search ability for robust global optimization. Different chaotic maps are utilized to tune the attractive movement of fireflies. Results show that the chaotic FA (CFA) outperforms the standard FA. In [12], quaternion is used for the representation of individuals in FA so as to enhance the performance of the firefly algorithm and to avoid any stagnation. Yu et al. [13] designed a new FA with a wise step strategy (WSSFA), which considers the information of firefly's personal and the global best positions. Results show that the modified algorithm outperforms the standard FA on twenty benchmark functions. In [14], a variable step size FA (VSSFA) is proposed, where a dynamical method is used to update the parameter α. Computational results show that WSSFA and VSSFA achieve better solutions than the standard FA on a set of low-dimensional benchmark functions ($D = 2$). However, our experiments demonstrate that both of them can hardly obtain reasonable solutions for some high-dimensional problems ($D = 30$). Compared to WSSFA and VSSFA, MFA can achieve promising solutions.

In the FA, the fitness function for a given problem is associated with the light intensity. The brighter the firefly is, the better the firefly is. That means a brighter firefly has a better fitness value. The search process of FA depends on the attractions between fireflies. Based on these attractions, a firefly tends to move other brighter fireflies. If a firefly is brighter than another one, the brighter firefly will not be conducted any search. In this paper, we propose a new FA (called NFA) to avoid this case. When the current firefly is brighter than another one, a local search operation is conducted on the current one to provide more chances of finding more accurate solutions. It is noted that the proposed NFA is implemented based on the MFA. Therefore, the NFA is a hybrid algorithm by combining the MFA and the proposed local strategy. To verify the performance of NFA, a set of well-known benchmark function with $D = 30$ are tested. Experimental results show that NFA performs better than the standard FA, MFA, and VSSFA.

The rest paper is organized as follows. In Sect. 2, the standard FA is briefly introduced. In Sect. 3, the proposed NFA is described. Experimental results are presented in Sect. 4. Finally, the work is concluded in Sect. 5.

2 Firefly Algorithm

As mentioned before, the FA mimics the behavior of the social behavior of the flashing characteristics of fireflies. To simply the behavior of fireflies and construct the search mode of FA, three rules are used as follows [1]:

- All fireflies are unisex so that one firefly is attracted to other fireflies regardless of their sex;
- Attractiveness is proportional to their brightness. For any two fireflies, the less bright one is attracted by the brighter one. The attractiveness is proportional to the brightness and they both decrease as their distance increases. If no one is brighter than a particular firefly, it moves randomly;
- The brightness or light intensity of a firefly is affected or determined by the landscape of the objective function to be optimized. For a minimization problem, the brightness can be proportional to the objective function. It means that the brighter firefly has smaller objective function value.

As light intensity and thus attractiveness decreases as the distance from the source increases, the variations of light intensity and attractiveness should be monotonically decreasing functions. This can be approximated by the following Eq. [1]:

$$I(r) = I_0 e^{-\gamma r^2}. \tag{1}$$

where I is the light intensity, I_0 is the original light intensity, and γ is the light absorption coefficient. The attractiveness of a firefly is proportional to the light intensity. The attractiveness β of a firefly can be defined by [1]:

$$\beta(r) = \beta_0 e^{-\gamma r^2}. \tag{2}$$

where β_0 is a constant and presents the attractiveness at $r = 0$. The distance between r_{ij} between any two fireflies i and j can be calculated by [1]:

$$r_{ij} = \left\| X_i - X_j \right\| = \sqrt{\sum_{d=1}^{D} (x_{id} - x_{jd})^2}. \tag{3}$$

where D is the dimensional size of the given problem.

Based on the above definitions, the movement of this attraction is defined by [1]:

$$x_{id}(t+1) = x_{id}(t) + \beta_0 e^{-\gamma r_{ij}^2} (x_{jd}(t) - x_{id}(t)) + \alpha \varepsilon_{id}(t). \tag{4}$$

where x_{id} and x_{jd} is the dth dimension of firefly i and j, respectively, α is a random value with the range of [0,1], ε_{id} is a Gaussian random number for the dth dimension, and t indicates the index of generation.

3   Proposed Approach

Recently, some new FA variants were proposed to enhance the performance of FA. However, these algorithms only work well on some low-dimensional problems. For high-dimensional problems (such as $D = 30$), they can hardly find reasonable solutions. In [9], Fister et al. designed a memetic FA (MFA) by introducing multiple strategies. Experimental results show that MFA performs well when $D = 30$.

In MFA, a new updating search equation is defined as follows [9].

$$x_{id}(t+1) = x_{id}(t)(1-\beta) + x_{jd}(t)\beta + \alpha(r-0.5). \tag{5}$$

$$\alpha = \alpha \left(\frac{1}{9000}\right)^{\frac{1}{t}} (up - low). \tag{6}$$

$$\beta = \beta_{min} + (\beta_0 - \beta_{min})e^{-\gamma r_{ij}^2}. \tag{7}$$

where α is the generation index, r is a random number between 0 and 1, and β_{min} is a constant value. In [9], β_0 and β_{min} are set to 1.0 and 0.2, respectively. The initial a is set to 0.2. In the proposed NFA, we use the MFA as the basic algorithm. Then, we embed a local search strategy into the MFA.

Algorithm 1: The Proposed NFA

```
 1: Begin
 2:     Randomly initialize all fireflies in the swarm;
 3:     while FEs <= MaxFEs do
 4:         for i=1 to N do
 5:             for j=1 to i do
 6:                 if firefly j is better than firefly i then
 7:                     Generate a new firefly according to Eq. (5);
 8:                     Evaluate the new solution;
 9:                 end if
10:                 else
11:                     Conduct the local search according to Eq. (8);
12:                 end else
13:             end for
14:         end for
15:     end while
16: End
```

In the standard FA and its most variants, a firefly can move to other brighter fireflies based on the attraction operations. However, if the current firefly is brighter than another one, the current one will not be conducted any search. To avoid this case, we design a new solution updating model. When the above case occurs, a local search is conducted on the brighter firefly as follows.

$$x_{id}^*(t) = Best_d(t) + (x_{id}(t) - x_{kd}(t))(2r - 1). \tag{8}$$

where r is a random number within [0, 1], $Best$ is the global best solution found so far, and X_k is a randomly selected solution from the current population ($i \neq j$). If X_i^* is better than X_i, then replace X_i with X_i^*; otherwise keep the X_i.

The main steps of the proposed NFA are presented in Algorithm 1, where N is the population size, and FEs is the number of fitness evaluations, and Max_FEs is the maximum number of fitness evaluations.

4 Experimental Results

4.1 Test Functions

In order to verify the performance of the proposed NFA, there are thirteen classical benchmark functions used in the following experiments [15, 16]. According to their properties, they are divided into two groups: unimodal functions (f_1-f_7) and multimodal functions (f_8-f_{13}). All test functions are minimization problems. In this paper, we only consider the problems with $D = 30$. The mathematical descriptions of these functions are listed as follows.

(1) Sphere

$$f_1(x) = \sum\nolimits_{i=1}^{D} x_i^2$$

where $x_i \in [-100, 100]$, and the global optimum is 0.

(2) Schwefel 2.22

$$f_2(x) = \sum\nolimits_{i=1}^{D} |x_i| + \prod\nolimits_{i=1}^{D} x_i$$

where $x_i \in [-10, 10]$, and the global optimum is 0.

(3) Schwefel 1.2

$$f_3(x) = \sum\nolimits_{i=1}^{D} \left(\sum\nolimits_{j=1}^{i} x_j \right)^2$$

where $x_i \in [-100, 100]$, and the global optimum is 0.

(4) Schwefel 2.21

$$f_4(x) = \max_i(|x_i|, 1 \le i \le D)$$

where $x_i \in [-100, 100]$, and the global optimum is 0.

(5) Rosenbrock

$$f_5(x) = \sum\nolimits_{i=1}^{D-1} \left[100(x_i^2 - x_{i+1})^2 + (x_i - 1)^2 \right]$$

where $x_i \in [-30, 30]$, and the global optimum is 0.

(6) Step

$$f_6(x) = \sum_{i=1}^{D} (\lfloor x_i + 0.5 \rfloor)^2$$

where $x_i \in [-100, 100]$, and the global optimum is 0.

(7) Quartic with noise

$$f_7(x) = \sum_{i=1}^{D} i x_i^4 + rand[0, 1)$$

where $x_i \in [-1.28, 1.28]$, and the global optimum is 0.

(8) Schwefel 2.26

$$f_8(x) = \sum_{i=1}^{D} -x_i \sin\left(\sqrt{|x_i|}\right)$$

where $x_i \in [-500, 500]$, and the global optimum is -12569.5.

(9) Rastrigin

$$f_9(x) = \sum_{i=1}^{D} [x_i^2 - 10\cos(2\pi x_i) + 10]$$

where $x_i \in [-5.12, 5.12]$, and the global optimum is 0.

(10) Ackley

$$f_{10}(x) = -20 \cdot \exp\left(-0.2 \cdot \sqrt{\frac{1}{D}\sum_{i=1}^{D} x_i^2}\right) - \exp\left(\frac{1}{D}\sum_{i=1}^{D} \cos(2\pi x_i)\right) + 20 + e$$

where $x_i \in [-32, 32]$, and the global optimum is 0.

(11) Griewank

$$f_{11}(x) = \frac{1}{4000}\sum_{i=1}^{D} x_i^2 - \prod_{i=1}^{D} \cos\left(\frac{x_i}{\sqrt{i}}\right) + 1$$

where $x_i \in [-600, 600]$, and the global optimum is 0.

(12) Penalized 1

$$f_{12}(x) = 0.1\{\sin^2(3\pi x_1) + \sum_{i=1}^{D-1} (x_i - 1)^2[1 + \sin^2(3\pi x_{i+1})]$$
$$+ (x_D - 1)^2[1 + \sin^2(2\pi x_D)]\} + \sum_{i=1}^{D} u(x_i, 10, 100, 4)$$

where $x_i \in [-50, 50]$, and the global optimum is 0.

(13) Penalized 2

$$f_{13}(x) = \frac{\pi}{D}\{10\sin^2(3\pi y_1) + \sum_{i=1}^{D-1}(y_i - 1)^2[1 + \sin^2(3\pi y_{i+1})]$$
$$+ (y_D - 1)^2[1 + \sin^2(2\pi x_D)]\} + \sum_{i=1}^{D}u(x_i, 5, 100, 4)$$

where $x_i \in [-50, 50]$, and the global optimum is 0.

4.2 Involved Algorithms and Parameter Settings

In this section, the proposed NFA is compared with the standard FA and two other FA variants. The involved algorithms are listed as follows:

- The Standard FA.
- Memetic FA (MFA) [9].
- Variable step size FA (VSSFA) [14].
- The proposed NFA.

The parameter settings of the above four algorithms are listed as follows. To have a fair comparison, all algorithms use the same Max_FEs and N as the termination condition, and the Max_FEs and N are set to 5.0E + 05 and 20, respectively. For the standard FA and VSSFA, $\alpha = 0.2$, $\beta_0 = 1$, and $\gamma = 1$ are used [14]. For MFA and NFA, $\beta_0 = 1.0$, $\beta_{min} = 0.2$, $\alpha = 0.2$, and $\gamma = 1$ are used. For each test function, each algorithm is run 30 times and the mean best fitness values are reported.

4.3 Results

Table 1 presents the computational results of FA, VSSFA, MFA, and NFA on the test set, where "*Mean*" indicates the mean best fitness values. As shown, both FA and

Table 1. Results achieved by FA, VSSFA, MFA, and NFA on the test suite.

Functions	FA	VSSFA	MFA	NFA
	Mean	*Mean*	*Mean*	*Mean*
f_1	6.67E + 04	5.84E + 04	1.56E − 05	**6.59E − 09**
f_2	5.19E + 02	1.13E + 02	1.85E − 03	**3.21E − 05**
f_3	2.43E + 05	1.16E + 05	5.89E − 05	**7.29E − 07**
f_4	8.35E + 01	8.18E + 01	1.73E − 03	**2.29E − 04**
f_5	2.69E + 08	2.16E + 08	2.29E + 01	**1.57E − 03**
f_6	7.69E + 04	5.48E + 04	**0.00E + 00**	**0.00E + 00**
f_7	5.16E + 01	4.43E + 01	1.30E − 01	**7.68E − 04**
f_8	−1563.4	−1854.6	**−7634.35**	−7160.3
f_9	3.33E + 02	3.12E + 02	6.47E + 01	**4.97E + 01**
f_{10}	2.03E + 01	2.03E + 01	4.23E − 04	**1.68E − 05**
f_{11}	6.54E + 02	5.47E + 02	9.86E − 03	**7.39E − 03**
f_{12}	7.16E + 08	3.99E + 08	5.04E − 08	**3.43E − 11**
f_{13}	1.31E + 09	8.12E + 08	6.06E − 07	**5.02E − 10**

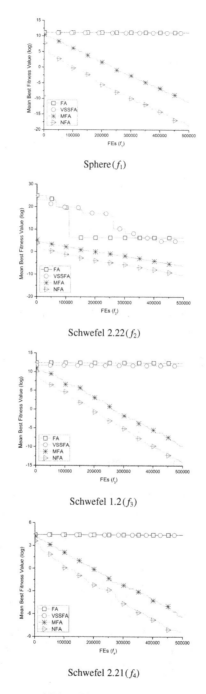

Fig. 1. The search processes of FA, VSSFA, MFA, and NFA on four selected functions.

VSSFA could hardly achieve reasonable solutions on all test functions. Even on simple unimodal function Sphere (f_1), they still cannot converge to promising solutions. Compared to FA and VSSFA, MFA and NFA obtains much better solutions on all test functions. NFA outperforms MFA on 11 functions, while MFA only performs better than MFA on f_8. Both of them can find the global optimum on f_6.

Figure 1 lists the search processes of FA, VSSFA, MFA, and NFA on four selected functions. As seen, NFA shows the fastest convergence speed among all four algorithm. FA and VSSFA cannot improve the fitness value during the whole search process.

5 Conclusion

In this paper, we propose a new firefly algorithm (NFA) to improve the performance of the standard FA. Unlike the standard FA and its most modifications, the NFA defines a new operation for brighter fireflies. When a firefly is brighter than another one, the brighter firefly will be conducted on a local search. This is helpful to enhance the local search and improve the accuracy of solutions. Moreover, the NFA employs the modifications of MFA. By the hybridization of MFA and the proposed local search, NFA achieves much better results than the standard FA, VSSFA, and MFA on the majority of test functions. The proposed local search can be embedded in other FAs. This will be investigated in the future work.

Acknowledgement. This work is supported by the Humanity and Social Science Foundation of Ministry of Education of China (No. 13YJCZH174), the National Natural Science Foundation of China (Nos. 61305150 and 61261039), the Science and Technology Plan Project of Jiangxi Provincial Education Department (Nos. GJJ14747 and GJJ13762), and the Natural Science Foundation of Jiangxi Province (No. 20142BAB217020).

References

1. Yang, X.S.: Engineering Optimization: An Introduction with Metaheuristic Applications. Wiley, Hoboken (2010)
2. Chandrasekaran, K., Simon, S.P., Padhy, N.P.: Binary real coded firefly algorithm for solving unit commitment problem. Inf. Sci. **249**, 67–84 (2013)
3. Coelho, L.S., Mariani, V.C.: Improved firefly algorithm approach applied to chiller loading for energy conservation. Energy Build. **59**, 273–278 (2013)
4. Gandomi, A.H., Yang, X.S., Alavi, A.H.: Mixed variable structural optimization using firefly algorithm. Comput. Struct. **89**(23–24), 2325–2336 (2013)
5. Miguel, L.F.F., Lopez, R.H., Miguel, L.F.F.: Multimodal size, shape, and topology optimisation of truss structures using the firefly algorithm. Adv. Eng. Softw. **56**, 23–37 (2013)
6. Marichelvam, M.K., Prabaharan, T., Yang, X.S.: A discrete firefly algorithm for the multi-objective hybrid flowshop scheduling problems. IEEE Trans. Evol. Comput. **18**(2), 301–305 (2014)

7. Farahani, S.M., Abshouri, A.A., Nasiri, B., Meybodi, M.R.: A gaussian firefly algorithm. Int. J. Mach. Learn. Comput. **1**(5), 448–453 (2011)

8. Tilahun, S.L., Ong, H.C.: Modified firefly algorithm. J. Appl. Math. Article ID 467631 (2012). doi:10.1155/2012/467631

9. Fister Jr., I., Yang, X.S., Fister, I., Brest, J.: Memetic firefly algorithm for combinatorial optimization. In: Filipic, B., Silc, J. (eds.) Bioinspired Optimization Methods and their Applications (BIOMA 2012). Jozef Stefan Institute, Ljubljana, Slovenia (2012)

10. Wang, H., Wang, W.J., Sun, H., Rahnamayan, S.: Firefly Algorithm with Random Attraction. Int. J. Bio-Inspired Comput. (2016, to be published)

11. Gandomi, A.H., Yang, X.S., Talatahari, S., Alavi, A.H.: Firefly algorithm with chaos. Commun. Nonlinear Sci. Numer. Simul. **18**(1), 89–98 (2013)

12. Fister, I., Yang, X.S., Brest, J., Fister Jr., I.: Modified firefly algorithm using quaternion representation. Expert Syst. Appl. **40**(18), 7220–7230 (2013)

13. Yu, S.H., Su, S.B., Lu, Q.P., Huang, L.: A novel wise step strategy for firefly algorithm. Int. J. Comput. Math. **91**(12), 2507–2513 (2014)

14. Yu, S.H., Zhu, S.L., Ma, Y., Mao, D.M.: A variable step size firefly algorithm for numerical optimization. Appl. Math. Comput. **263**, 214–220 (2015)

15. Wang, H., Wu, Z.J., Rahnamayan, S., Liu, Y., Ventresca, M.: Enhancing particle swarm optimization using generalized opposition-based learning. Inf. Sci. **181**(20), 4699–4714 (2011)

16. Wang, H., Rahnamayan, S., Sun, H., Omran, M.G.H.: Gaussian bare-bones differential evolution. IEEE Trans. Cybern. **43**(2), 634–647 (2013)

A New Trend Peak Algorithm with X-ray Image for Wheel Hubs Detection and Recognition

Wei Li[1], Kangshun Li[1(✉)], Ying Huang[2], and Xiaoyang Deng[3]

[1] College of Mathematics and Informatics, South China Agricultural University, Guangzhou 510642, Guangdong, People's Republic of China
likangshun@sina.com
[2] School of Information Engineering, Jiangxi University of Science and Technology, Ganzhou 341000, Jiangxi, People's Republic of China
[3] Institute of Mathematical and Computer Sciences, Gannan Normal University, Ganzhou 341000, Jiangxi, People's Republic of China

Abstract. The automatic detection and recognition of automotive wheel hubs defects has important significance to improve the quality and efficiency of automotive wheel production and vehicle safety. In order to improve accuracy of detection and recognition of automotive wheel hub defect images, an improved peak location algorithm - trend peak algorithm is proposed to extract region of wheel hub defect, combined with BP neural network to classify and recognize wheel hub defect. Firstly, initial defect positions are extracted using peak locations of vertical and horizontal directions. Then mathematical morphology is used to remove pseudo defects, and the exact locations of the defects are obtained. Finally, the wheel hub defect features are classified to reach the target of defect recognition by BP neural network. In actual industrial conditions, the algorithm is found to obtain good recognition results and reach real-time detection request in low contrast, high noise, uneven illumination, and complex structure of the products, by experiments of X-ray images of four common defects of the actual wheel hubs.

Keywords: X-ray image · Trend peak algorithm · Defects detection and recognition

1 Introduction

Currently, X-ray detection is one of the most common methods that used in the wheel hubs detection [1]. Despite the fact that the detection of wheel hubs is gradually transforming to digitized graphic method, in the industry, people still depend on X-ray images to detect the defects. Because of the progress in automotive manufacturing process, and the drawbacks in human detection (Heavy workload, affected by human subjective judgment, and low efficiency), the demand of automotive wheel hubs detection is increasing. And in recent years, with the development in the field of computer image technology and pattern recognition,

© Springer Science+Business Media Singapore 2016
K. Li et al. (Eds.): ISICA 2015, CCIS 575, pp. 23–31, 2016.
DOI: 10.1007/978-981-10-0356-1_3

these two technologies have made many achievements in X-ray image application [2,3]. Automatic detection of X-way images, which based on the pattern recognition and image processing technology, can effectively increase the productive efficiency and capacity, and make it easier to implement the integration of embedded device and meet the requirement of automatic production.

The method we used today to automatically detect the X-ray image includes two types: template image method and non-template image method. Specifically, the most common way that used in X-ray template image is standard reference image method [4]. And in terms of the non-template image, it has wavelet analysis [5] and defects tracking and matching method [6]. But these algorithms all have their own defects. When dealing with the actual industrial conditions, like low contrast ratio, loud noises, complicated background, and bad lighting condition of the X-ray image. In terms of the normal four types of defects of wheel hubs (air hole, shrinkage porosity, shrinkage cavity and crack), the accuracy, efficiency, and accuracy of classification of image recognition are still not ideal enough.

Aiming at the defects of current methods, in order to implement the automatic detection of automotive wheel hubs, this thesis presents the trend peak algorithm which based on the improvement of peak location algorithm [7] to locate the defects of wheel hubs in images. And then the mathematical morphology is used to eliminate the pseudo-flaw. Finally using BP neural network classification analysis to classify the selected defects and obtain the defect type. After testing the four common defects on real wheel hubs under actual industrial condition, this thesis comes up with the conclusion that the trend peak algorithm meets the industrial requirement which could work under the conditions of low contrast ratio, loud noises, complicated background, and bad illumination condition of the X-ray image.

2   The Main Procedure of the Automatic Detection of Automotive Wheel Hubs Defects

The X-ray automatic detection of automotive wheel hubs defects system have four common types: air hole, shrinkage porosity, shrinkage cavity and crack. So that the intensity, contrast ratio and size of illumination are all different. Besides, the X-ray images are also affected by the camera that has the voltage oscillation. The Fig. 1 shows the four types of common defects, and they were labeled by red rectangle frames.

In general, the recognition and classification of the X-ray automatic detection of automotive wheel hubs defects has three phases: segmentation, feature detection and classification.

(1) Segmentation: segment the image into pieces and the valid regions are the ROI (regions of interesting) [8]. It can separate the defects from other regions. In this thesis, the algorithm used in segmentation is the trend peak algorithm.

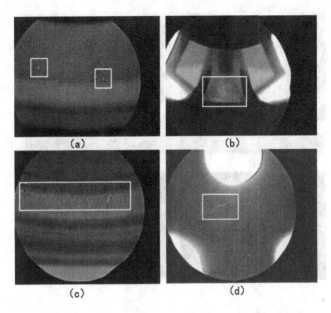

Fig. 1. The four types of defects of automotive wheel hubs ((a) air hole, (b) shrinkage porosity, (c) shrinkage cavity, (d) crack) (Color figure online)

(2) Feature detection: after phase 1, in order to increase the accuracy of recognition, the image still needs the feature detection. For example, extract the grey-scale feature or geometrical characteristic. In our research, after implementing peaks location analysis, the normalized characteristics are extracted.

(3) Defects classification: when we get the defects, they still need to be classification. In this thesis the BP neural network classification were used to analysis.

3   The Trend Peak Algorithm

3.1   The Theory of Peak Location Algorithm

The defects of wheel hubs in X-ray images have the features of complicated structure, unbalanced brightness and numerous varieties. The exist defects extract algorithms are all using good image preprocessing to decrease difficulties, and implementing the global or local algorithm to extract the defects [9]. However, when deal with the X-ray images which has the features of complicated structure, unbalanced brightness and numerous varieties, this global detection method and ROI extracting algorithm could not come up with ideal results. In this paper, we improve peak location algorithm and using trend peak algorithm to extract the defects region.

In Fig. 2, the grey curve graph, the horizontal axis means width, and vertical axis means the grey value of each pixel on the red line. The values are between

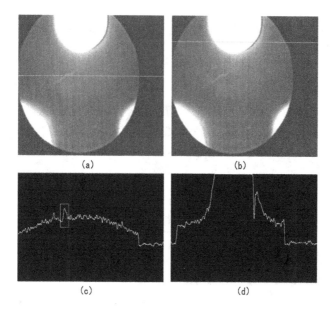

Fig. 2. Images of wheel hubs and their gray curve graphs (Color figure online)

0–255 (in order to have a more clear observation, the graph firstly transformed to the grey image and then implement Gaussian filter). In terms of the figure c, the grey curve graph of defect region has the features that are mountain shaped and the grey value was first increasing and then decreasing. However, sometimes the non-defect region also has the same shape. The height, width and gradient of the defect region are within a specific range. So this algorithm can be used when extract the defect region: firstly, scanning each row of the image horizontally and obtain its binary image; then scanning each column and get its vertical binary image; combine these two binary images to get the general one; using mathematical morphological method to analysis the binary image and get the small pieces of defects region. These is the basic principle of peaks location algorithm.

3.2 Trend Peak Algorithm

The original peak location algorithm only considers about three factors: width, height, and gradient. During the experiment, we found out that in the real industrial manufacturing, images are always complicated and has tiny peaks which have the same feature as the peaks we actually need, showing in Fig. 3 (These images were not processed by Gaussian filter). Observing the Fig. 3(b), it can be seen that in non-defect regions, there are some small peaks. If we still use the original algorithm, these peaks would be unrecognized as defects.

But if we finish the Gaussian filter analysis first, some details of images would be lost. It may eliminate the defects at the same time. So in this thesis, it came up

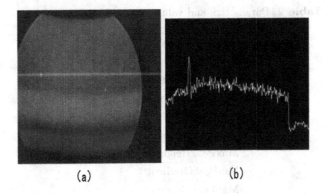

(a)   (b)

Fig. 3. Images of Wheel hubs and their gray curve graphs

Table 1. Trend peak algorithm

Input :	gray value I_i, trend value t_i, presupposed parameter p
Output:	No. i pixel of the Image I
1	if $v_{i+1} = v_i$
2	$t_i = FLAT$
3	else if $v_{i+1} > v_i$
4	if $|v_{i+1} - v_i| \geq p$
5	$t_i = UP$
6	else
7	$t_i = t_{i-1}$
8	end if
9	else
10	if $|v_{i+1} - v_i| \geq p$
11	$t_i = DOWN$
12	else
13	$t_i = t_{i-1}$
14	end if
15	end if
16	end for

with an improved algorithm-trend peaks algorithm, which set a specific reference value to decide whether the trend is changed. The definition of the trend is:

$$t_i = \begin{cases} UP & \text{if } v_i < v_{i+1} \\ FLAT & \text{if } v_i = v_{i+1} \\ DOWN & \text{if } v_i > v_{i+1} \end{cases}$$

In the definition, the t_i represents the trend of the No. i elements, v_i represents the grey value of the No. i pixel. UP, FLAT and DOWN are means increasing, gentle and decreasing respectively. The algorithm is indicated in the Table 1.

Table 2. Parameters and values of trend peak algorithm

Parameter	Value
Increment	5
MinPeakHeight	10
MaxPeakHeight	40
MinPeakWidth	5
MinPeakHeight	9
MaxPeakHeight	32
MinPeakGradient	1
MaxPeakGradient	6
MinPeakTop	140
MaxPeakTop	178

Another improvement is that the new algorithm adds the judging condition of the grey value at the top of peaks. Only when the grey value is within a specific range, it can be recognized as a peak. As it shown in the Table 2, the trend peaks algorithm needs 9 parameters.

The Fig. 4 shows the comparing result that between the original algorithm and the improved one. It can be seen that the trend peaks algorithm have more accurate result when extract defects, which decreases the pseudo defects.

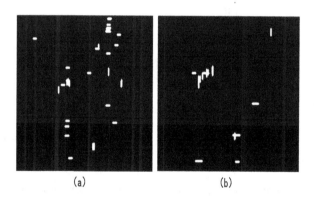

(a)   (b)

Fig. 4. The algorithm result:(a) peak location algorithm result of Fig. 1(d), (b) trend peak algorithm result of Fig. 1(d)

4   Experimental Results

All the images in this thesis were taken from the actual wheel buds factories and under the X-ray condition of 160kv. These images all include the bad condition that in actual factories may happen and the four common types of defect

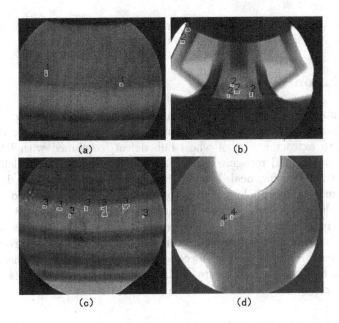

Fig. 5. Defect recognition results of Fig. 1 (Figure (x) is the defect image of Fig. 1(x))

Table 3. Experiment results of detection of wheel hubs defects

A	B	C	D	E
Air hole	25	100 %	0	11.20
Shrinkage porosity	25	92 %	4 %	11.86
Shrinkage cavity	25	96 %	0	12.20
Crack	25	96 %	0	10.72

features. So the methods used in detection, recognition and classification are representative.

The algorithm is implemented by Visual Studio 2013, based on OpenCV 2.4.9. The operating system is 64-bit Windows 8.1, and CPU is Intel Core i7, and RAM is 4G. Figure 5 shows the results of images of 4 types of defect. It can be seen that most defects are extracted and classified successfully. (The meaning of the numbers in Fig. 5: 1 is air hole, 2 is shrinkage porosity, 3 is shrinkage cavity and 4 is crack). 100 images of wheel hubs that have all kinds of defects are tested. The number of images of each type of defects is 25. The results of the experiments show in Table 3. In Table 3, A denotes the defect type, B denotes the number of images, C denotes the defect recognize rate, D denotes the defect recognize error rate, E denotes the average elapsed time of processing one image(ms), each defect's and pseudo defect's recognition rate is high and low respectively, which means this algorithm is applicable and meets the retirement of industrial standard. Furthermore, in terms of the average execution time, this algorithm has high efficiency and can be used in real-time detection.

5   Conclusion

In order to improve accuracy of detection and recognition of automotive wheel hub defect images, an improved peak location algorithm - trend peak algorithm is proposed to extract region of wheel hub defect, combined with BP neural network to classify and recognize wheel hub defect. Firstly, initial defect positions are extracted using peak locations of vertical and horizontal directions. Then mathematical morphology is used to remove pseudo defects, and the exact locations of the defects are obtained. Finally, the wheel hub defect features are classified to reach the target of defect recognition by BP neural network. In actual industrial conditions, the algorithm is found to obtain good recognition results and reach real-time detection request in low contrast, high noise, uneven illumination, and complex structure of the products, by experiments of X-ray images of four common defects of the actual wheel hubs.

Acknowledgements. This work was supported by the National Natural Science Foundation of China under Grant No. 61573157, by the Fund of the Natural Science Foundation of Guangdong Province of China under Grant No. 2014A030313454, and by the Natural Science Foundation of Jiangxi, China (No. 20142BAB217028).

References

1. Liu, M., Xie, J., Yu, G., et al.: Online nondestructive inspection of automotive aluminum wheel hubs. Nondestructive Inspection (2), pp. 41–43 (2000)
2. Kyle, J.R., Ketcham, R.A.: Application of high resolution X-ray computed tomography to mineral deposit origin, evaluation, and processing. Ore Geol. Rev. **65**(4), 821–839 (2015)
3. Zhou, T., Lu, H., Chen, Z., et al.: Research progress of multi-model medical image fusion and recognition. J. Biomed. Eng. **30**(5), 1117–1122 (2013)
4. Rebuffel, V., Sood, S.: Defect detection method in digital radiography for porosity in magnesium castings. In: 9th European Conference on NDT, Berlin, pp. 1–10 (2006)
5. Li, X.L., Tso, S.K., Guan, X.P., et al.: Improving automatic detection of defects in castings by applying wavelet technique. IEEE Trans. Ind. Electron. **6**(53), 1927–1934 (2006)
6. Mery, D., Filbert, D.: Automated flaw detection in aluminum castings based on the tracking of potential defects in a radioscopic image sequence. IEEE Trans. Robot. Autom. **18**(6), 890–900 (2002)
7. He, Z., Zhang, S., Huang, C.: Approach to detecting defects in wheels based on flaw's characteristics and seeded region growing method. J. Zhejiang Univ. (Eng. Sci.) **43**(7), 1230–1237 (2009)
8. Tiwari, K., Gupta, P.: An efficient technique for automatic segmentation of fingerprint ROI from digital slap image. Neurocomputing **151**(3), 1163–1170 (2015)

9. Veenman, C., Reinders, M., Backer, E.: A Cellular Coevolutionary Algorithm for Image Segmentation. IEEE Trans. Image Process. **12**(3), 304–316 (2003)
10. Li, G.: Research on automatic detection and recognition technology of automobile hub defect based on X-ray images. North University of China, Shanxi (2013)

Community Detection Based on an Improved Genetic Algorithm

Kangshun Li[(✉)] and Lu Xiong

College of Mathematics and Informatics, South China Agricultural University,
Guangdong 510642, P.R. China
likangshun@sina.com

Abstract. When the traditional genetic algorithm was used to solve the community detection problem, it was not easy to avoid the problems of low efficiency and slow convergent speed. To be aim at these problems, a improved genetic algorithm which is based on the immune mechanism was proposed in this paper. In this new algorithm, the immune mechanism was used to ensure the diversity of population. Meanwhile, a improved character encoding was adopted to further reduce the search space. The results shows that the shortcomings of slow convergent speed and low efficiency could be overcome by using the improved genetic algorithm to solve these problems, compared with the traditional genetic algorithm.

Keywords: Genetic algorithm · Community discovery · Data mining

1 Introduction

The discovery of community structure in networks has become a hot issue in recent years, which has been widely concerned by researchers in the fields of computer, mathematics, biology and sociology.

Communities in networks are a set of nodes that are similar to each other and have different nodes in the network. The nodes in the same community are connected with each other, and the nodes in the different community are sparse [1]. At present, the community structure can be found in the real networks, such as biological network, science and technology and social network. Complex network community structure discovery has important theoretical significance and practical value for topology analysis, function analysis and behavior prediction in complex network.

Because the discovery of community structure is very important for the research of networks, many researchers have proposed many different community structure discovery algorithms. Girven and Newman [2] proposed the most famous community mining method GN (Girven-Newman) in 2002. The GN algorithm plays a very important role in the research of complex network community mining. The important significance of work of Girven and Newman is: They discovered the community structure in

© Springer Science+Business Media Singapore 2016
K. Li et al. (Eds.): ISICA 2015, CCIS 575, pp. 32–39, 2016.
DOI: 10.1007/978-981-10-0356-1_4

networks in the first time, which inspired other researchers to make a further study on this problem, and then set off a boom of complex network community mining research. The researchers put forward some improved methods for the deficiency of GN algorithm. The biggest drawback of the algorithm is that the computation speed is slow. Because of the computational overhead ($O(mn)$) of the edge number is too large, it has a high time complexity ($O(m^2n)$), where, m represents the number of links, n represents the number of points. Therefore, the proposed algorithm is suitable for processing small and medium scale networks (The knot points are usually less than 10^{-3}). In 2004, Newman and Girven [3] proposed a quantitative standard used to describe the quality of the network community structure, which is called the modular function Q. Function Q gives a clear definition of community structure, and get a great success in practical application. As a result, the function Q has been widely accepted by researchers in related fields. At the same time, the optimization method of function Q as the objective function is one of the main methods in the complex network community mining.

In 2004, Newman [4] proposed the first method of community mining (fast Newman, FN) based on module optimization. The search strategy of the candidate solution is: select and merge two existing communities. At the time of initialization, the candidate solution contains only one node in each community. At each iteration, the algorithm FN merging the communities which make the function value increase (or decrease the least). When the candidate solution only corresponds to one community, the algorithm ends. The algorithm FN outputs a hierarchical clustering tree through this bottom-up hierarchical clustering process. And then Maximum community division of the corresponding function values as the final clustering results. The time complexity of the algorithm is $O(mn)$.

The researchers first noticed the character of the encoding way for the problem of community mining.

String encoding is simple and intuitive, but the disadvantage is that it is not suitable for the traditional cross operator. In 2007, Tasgin et al. [5] proposed first genetic algorithms for community mining, and its main contribution is to give a single path crossover operation for a string of encoding. They verified the algorithm by using small network. In 2010, Dongxiao He et al. [6] proposed a genetic algorithm based on Clustering Fusion for community mining. The algorithm introduced the clustering merging into the crossover operator, and it generate a new individual by using the clustering information of the parent individual and the local information of the network topology structure, which avoids the problems caused by the traditional crossover operator to simply exchange character blocks and neglect the clustering problem. The algorithm can obtain high quality clustering results, but the disadvantage is that the operation efficiency is low. In 2011, Dayou Liu et al. [7] proposed a fast and effective local search mutation strategy based on the analysis of the local monotonic property of the module function Q. With the combination of single path crossover operator and selection operator, a genetic algorithm for community mining is presented. Experiments show that the proposed algorithm can effectively analyze large-scale networks. In 2011, Jack et al. [8] proposed a new dense mother algorithm for community detection combined the genetic algorithm and local search of mountain climbing method, but the convergence speed of the algorithm is slow. This paper solve proposed a improved genetic algorithm based on the traditional algorithm to s the problem of community detection. Experimental results show that this method can effectively suppress the degradation of the population, improve the convergence speed.

2   Description of Algorithm

2.1   Encoded Mode

At present, the community detection algorithm based on genetic algorithm is mainly used encoding based on string [6–10] and encoding based on the graph [11–14].

The string encoding is more intuitive and efficient in the network community structure compared with the encoding based on the graph, so the genetic algorithm in this paper adopt the string encoding mode. If given a complex network N, any division of the network can be represented as a string encoding (chromosome) $R = \{r(1), r(2), \ldots r(n) \mid k\}$, where n is the number of nodes in network N, $r(i)$ represents the tags of node i, which can be expressed in an integer represent k communities into which is divided by the network.

If $r(i) = r(j)$ for any node pairs i and j in R, then the two nodes are located in the same community; or they are in different communities. Obviously, the string encoding can be very convenient to represent the any candidate solutions of the network community detection problem (network clustering results).

Supposing a complex network packet contains 5 nodes $\{v1, v2, v3, v4, v5 \mid k\}$, if the string encoding of one chromosome is $\{1, 2, 1, 3 \mid 3\}$, then it means that the nodes $\{v1, v3, v5\}$ in a complex network belongs to the same community.

2.2   The Selection Operator Based on Immune

The selection operator is a global search operator in genetic algorithm. In this paper, we reference on the idea of antibody concentration inhibition in immune algorithm to increase the concentration of regulatory factors, and to ensure the diversity of the population according to the choice of fitness. And it can improve the convergence of the genetic algorithm. The similarity index $q_{i,j}$ of individual i and j are:

$$q_{i,j} = \left| f_i - f_j \right| \big/ f_{\max} \tag{1}$$

Where, f_i, f_j represent the values of fitness function for individual i and individual j respectively, $f_{\max}$ indicates the larger value of f_i and f_j

$$S_{i,j} = \begin{cases} 1 & if \quad q_{i,j} \le \varepsilon \\ 0 & if \quad q_{i,j} > \varepsilon \end{cases} \tag{2}$$

Where, ε refers to the individual's similarity threshold, we assign 0.001 to ε in this paper. The proportion of the same or similar individuals in the group (Also known as concentration c_i):

$$c_i = \frac{\sum_{j=1}^{m} S_{i,j}}{m} \tag{3}$$

Where, m refers to the population scale.

For a specific population of M, the selection probability p_i of individual i $i = 1, 2, \ldots, m$ is defined as follows:

$$p_i = \frac{f_i}{\sum_{j=1}^{m} f_j} \cdot e^{-\beta * c_i} \tag{4}$$

Where, f_i refers to the fitness of the individual i and $f_i \geq 0$, The transformation is defined as $f = (1 + Q)/2$ according to the objective function Q. β refers to the parameters that can be adjusted to the concentration and fitness, $\beta \in [0, 1]$ and take $\beta = 1$ in this paper.

2.3 Algorithm Description

In summary, the algorithm described as follows:

Procedure IGA

Input N, L, μ, m, c

//N represents a complex network, L represents the evolution algebra, μ represents the size of the population, m represents the mutation probability, c indicates the cross probability.

Output O, Q

//O represents the optimal individual, Q represents the value of function Q of the optimal individual O.

```
Begin
1. P = initpopulation(N, μ)        //Initial population
2.     for g=1:L
3.         for i =1:μ/2      //interlace operation
4.             p=rand()
5.         end for
6.         for i =1:μ              //Mutation operation
7.             P(i)=mutation(P(i),m)   // mutation operator
8.         end for        //selecting operation
9.         P=select(P) //selection operator based on immune
10.    end for
11. End
```

3   Simulation Experiments and Results Analysis

In order to verify the performance of the IGA, we use the benchmark test network and the real large-scale network to verify the performance, and then give the parameters analysis.

Algorithm experiment environment:

Processor Intel(R) Core(R) i5-4200CPU@1.60 GHz 2.30 GHz;

memory 4.00 GB;

hard disk 500 G;

operating system Microsoft Windows Server 2003,

programming environment: Matlab 7.3.

The parameters and values of IGA in the experiment are shown in Table 1.

Table 1. IGA parameters and setting

Name of parameter	Parameter meaning	Parameter values
μ	Population size	200
L	Evolutionary algebra	200
c	Crossover probability	0.8
m	Variation probability	0.02
ε	Computing and judging the threshold of 2 individual similarity when individual select probability	0.001
α	Parameters of individual optimization times	0.4

3.1 Artificial Network

Girven and Newman [4] proposed a benchmark artificial network, which can be defined as $GN(C, s, d, z_{out})$, Where C represents the number of communities in the network, s means the number of nodes in each community, d represents the degree of each node in the network, z_{out} represents the number of links each node and the node outside the community.

In order to test the clustering accuracy of different algorithms, we adopt the random network $GN(4, 32, 16, z_{out})$ widely here. Each node is an independent node at the beginning, and the network is constructed according to the following method:

Connecting to the other nodes in the same community in the probability $p1$ randomly for each node, and connecting to the other nodes in the different community in the probability $p2$ randomly, and $p2 < p1$. By setting the appropriate $p1$ and $p2$, the number of connections between each node in the network and its community internal nodes is z_{in}. The number of connections with other community nodes is z_{out}, and satisfy $z_{in} + z_{out} = 16$, With the increase of z_{out}, the community structure of the network is becoming more and more fuzzy. When $z_{out} > 8$, it is generally considered that the network has no obvious community structure or the result of the community is vague. The performance of IGA was tested by using computer generated z_{out} which was 0 ~ 8.

The algorithm measure the performance by the correct partition rate. This method can measure the performance by comparing the difference of the structure of the community structure and the structure of the network community structure. The proportion of correct partition node number in the total number of nodes). Figure 1 is IGA and algorithm FN [4], TGA (Tasgin's genetic algorithm). A comparison of the results of community division of computer generated network community. The data used in the graph are the average values of the correct partitioning rate for 50 random networks with the same value z_{out}.

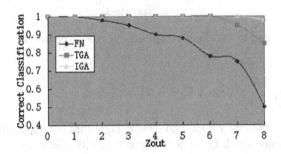

Fig. 1. Comparing IGA with FN, TGA in correct classification

X axis in Fig. 1 indicates the value of z_{out}, y axis represents the correct ratio of community partition for the network z_{out} by 3 different algorithms. It can be seen that the clustering accuracy of the algorithm IGA is significantly higher than that of FN and TGA, and with the increasing of z_{out}, the network community structure is more fuzzy, the effect of IGA algorithm is more obvious. Even though the $z_{out} = 8$, the algorithm IGA can still correctly divide the higher network nodes, the results show that the algorithm IGA has a considerable clustering accuracy.

Computing speed is an important index for evaluating the performance of network community detection algorithm. From the experimental point of view to further evaluate the efficiency of the algorithm.

Figure 1 shows the trend of the actual running time of the algorithm with the network size. In the experiment, the random network $GN(a, 100, 16, 5)$ is used to test, the community structure of the network is determined, but the number of the network community can be adjusted by the value a, which contains 100 a nodes of network, 1600 a network connections.

In Fig. 2, the x axis indicates the network size (node number + connection number), y axis represents the actual running time of the algorithm (s). It can be seen that the running time of the IGA algorithm is proportional to the size of the complex network in the premise of the average community size of the network. Therefore, this experiment verifies the correctness of the time complexity $O(n^2)$ of the IGA algorithm.

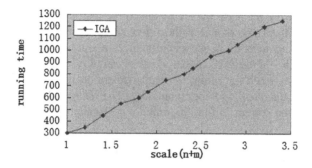

Fig. 2. The actual running time of IGA on networks

4   Conclusion

In this paper, we presented a improved genetic algorithm of community discovery to solving the problems which bring about by community discovery method based on traditional genetic algorithm. Using the improved string encoding method and the method of immune algorithm to divide the community. Based on the immune principle, an immune based selection operator is proposed. Through the analysis of the experimental results, it shows that the clustering quality and the efficiency of the IGA algorithm are obviously higher than other algorithms.

In this paper, the next work is to set the parameters of the algorithm more reasonable, improve the stability of the algorithm, and try to use the algorithm for Web community mining and other research areas.

Acknowledgment. This work is supported by the National Natural Science Foundation of China with the Grant No. 61573157, the Fund of Natural Science Foundation of Guangdong Province of China with the Grant No. 2014A030313454.

References

1. Luo, J., Yuan, C., Hu, H., Yuan, H.: Community structure division in complex networks based on gene expression programming algorithm. J. Comput. Appl. **32**(2), 317–321 (2012)
2. Girven, M., Newman, M.E.J.: Community structure in social and biological networks. Proc. Natl. Acad. Sci. **9**(12), 7821–7826 (2002)
3. Newman, M.E.J., Girven, M.: Finding and evaluating community structure in networks. Phys. Rev. E **69**(2), 026113 (2004)
4. Newman, M.E.J.: Fast algorithm for detecting community structure in networks. Phys. Rev. E **69**(6), 066133 (2004)
5. Tasgin, M., Herdagdelen, A., Bingol, H.: Community detection in complex networks using genetic algorithms [EB/OL] (2007). http://arxiv.org/abs/0711.0491v1
6. He, D., Zhou, X., Wang, Z., et al.: Community mining in complex networks-Clustering combination based genetic algorithm. Acta Automatica Sinica **36**(8), 1160–1170 (2010)

7. Jin, D., Liu, J., Bo, Y.: Genetic algorithm with local search for community detection in large-scale complex networks. Acta Automatica Sin. **37**(7), 873–882 (2011)
8. Gong, M., Fu, B., Jiao, L.: Memetic algorithm for Community detection in networks. Phys. Rev. E **84**(5), 056101 (2011)
9. Wu, F., Huberman, B.A.: Finding communities in linear time: a physics approach. Eur. Phys. J. B **38**(2), 331–338 (2003)
10. Li, S., Chen, Y., Du, H., Feldman, M.W.: A genetic algorithm with local search strategy for improved detection of community structure. Complexity **15**(4), 53–60 (2010)
11. Pizzuti C.: Community detection in social networks with genetic algorithms. In: Proceedings of the 10th Annual Conference on Genetic and Evolutionary Computation, NewYork, USA, pp. 1137–1138. ACM (2008)
12. Pizzuti, C.: A multi-objective genetic algorithm for community detection in networks. In: Proceedings of the 21st IEEE International Conference on Tools with Artificial Intelligence, New Jersey, USA, pp. 379–386. IEEE (2009)
13. Shi, C., Yan, Z., Wang, Y., Cai, Y., Wu, B.: A genetic algorithm for detecting communities in large-scale complex networks. Adv. Complex Syst. **13**(1), 3–17 (2010)
14. Jin, D., He, D., Liu, D., Baquero, C.: Genetic algorithm with local search for community mining in complex networks. In: Proceedings of the 22nd IEEE International Conference on Tools with Artificial Intelligence, Arras, France, pp. 105–112. IEEE (2010)
15. Zhou, S., Xu, Z., Tang, X.: New method for determining optimal number of clusters in k-means clustering algorithm. Comput. Eng. Appl. **46**(16), 27–31 (2010)
16. Guo. S., Lu, Z.: Basic theory of complex networks. pp. 270–271. Science Press, Beijing (2012)
17. Zachary, W.W.: An information flow model for conflict and fission in small groups. J. Anthropol. Res. **33**(4), 452–473 (1977)
18. Lusseau, D., Schneider, K., Boisseau, O.J., et al.: The bottlenose dolphin community of doubtful sound features a large proportion of long-lasting associations-can geographic isolation explain this unique trait. Behav. Ecol. Sociobiol. **54**(4), 396–405 (2003)

Selecting Training Samples from Large-Scale Remote-Sensing Samples Using an Active Learning Algorithm

Yan Guo[1]([⊠]), Li Ma[2], Fei Zhu[2], and Fujiang Liu[3]

[1] College of Computer Science, China University of Geosciences,
Wuhan, Hubei, China
323110966@qq.com
[2] Faculty of Mechanical and Electronic Information,
China University of Geosciences, Wuhan, Hubei, China
maryparisster@gmail.com, 582952469@qq.com
[3] Faculty of Information Engineering, China University of Geosciences,
Wuhan, Hubei, China
felixwuhan@163.com

Abstract. Based on margin sampling (MS) strategy, an active learning approach was introduced for proposed sample selection from large quantities of labeled samples using a Landsat-7 ETM+ image to solve remote sensing image classification problems for large number of training samples. As a breakthrough from conventional random sampling and stratified systematic sampling methods, this approach ensures classification of only using a few hundred training samples to be as effective as that of using several thousand and even tens of thousands of samples by conventional methods, thereby avoiding enormous calculations, substantially reducing operating time and improving training efficiency. The test results of the proposed approach was compared with those of random sampling and stratified systematic sampling, and the effects of training samples on classification under optimized and non-optimized selection conditions was analyzed.

Keywords: Training sample selection · Active learning · Margin sampling · Remote sensing image classification

1 Introduction

As a key factor affecting remote sensing (RS) images' classification precision, training data plays a significant role. Constructing classifiers and their performance depends directly on training data quality. Conventionally, training data are collected through manual labeling, which is usually expensive, time consuming and is too randomized. Therefore, scholars are exploring automatic database construction for acquiring training data to realize low-cost, efficient learning from RS images. As a result, many automatic interpretation algorithms based on spectral knowledge have been developed to acquire large-scale training samples.

© Springer Science+Business Media Singapore 2016
K. Li et al. (Eds.): ISICA 2015, CCIS 575, pp. 40–51, 2016.
DOI: 10.1007/978-981-10-0356-1_5

This research was performed under the background of the Global Forest Cover Monitoring Program, a research program led by Prof. John Townshend and Prof. Chengquan Huang from the University of Maryland. It is aimed at monitoring global forest cover changes by using global Landsat ETM+ RS images in two different periods, and classifying them into the following four groups through automatic training sample acquisition and classification: forest to non-forest (forest areas changed to non-forest areas), forest to forest (forest areas without change), non-forest to forest (non-forest areas changed to forest areas), and non-forest to non-forest (non-forest areas without change) [1–6].

Automatic training data acquisition is crucial to the program's success; therefore the team introduced spectral knowledge into computer interpretation and developed an algorithm for automatic training sample acquisition. This algorithm can automatically obtain nearly 10 million forest/non-forest samples from one Landsat ETM+ image.

Automatic training sample acquisition can solve sampling quantity problem and save manual labeling costs; however the large amounts of high-dimensional RS data creates a hindrance on follow-up research. Excessive training data is usually acquired in the classification system, resulting in serious redundant information. Therefore, collecting effective training samples from enormous amounts of training data has become a critical task. Only by collecting high-quality training samples can we avoid unnecessary data processing and operations, and achieve satisfactory classification results.

However, there will be numerous problems if large-scale sample sets (usually more than 500,000) are processed by conventional means, such as high memory requirement, slow execution speed and inability to execute [7]. Additionally, large-scale data sets may result in serious data redundancy, and redundant information is not only useless, but also increases processing difficulty and cost. Therefore, the most effective way to solve large sample set problem is to select samples from a sample set to construct the classification model [8].

Based on existing research, conventionally random sampling or stratified systematic sampling is used to select appropriate-sized samples from "10 million-level" sample sets. Selected samples are submitted to a SVM for learning purposes in a passive manner [9], and finally RS images are classified using the SVM obtained from training. Such approaches are easy to operate, but their operational blindness often leads to unsatisfactory results because they do not consider characteristics of different data. Thus, it is absolutely necessary to propose an effective method for training data selection.

Based on the above idea, this paper proposes an active learning algorithm to replace the original random sampling or stratified systematic sampling method for optimized large-scale training samples selection to improve large-scale data classification precision.

First, a small number of initial training samples are established using "big data", and then remaining samples are selected based on proposed active learning strategy until a desired training samples size or classification precision is achieved. At the same time, the test results of proposed approach are compared with those of random sampling and stratified systematic sampling, and the effects of training samples on classification under optimized and non-optimized selection conditions are analyzed, and time complexities of various algorithms are performed. In the end, the effectiveness of proposed active learning algorithm in solving a sub-problem (i.e., classification of forest and non-forest areas) in Global Forest Cover Monitoring Program is validated.

2   Margin Sampling Algorithm

Margin sampling (MS) algorithm is an active learning algorithm that uses geometric attributes of support vector machines (SVM) [10] to label samples based on prior selection of samples that are closest to the SVM hyper-plane. Based on the MS algorithm, samples in the margin have more uncertainties and contain more information than samples in other regions; hence they are the most difficult to classify.

As shown in Fig. 1, ○ and □ denote labeled samples of Class A and Class B respectively; △ denotes unlabeled samples. Given the condition of linearly separable binary classification, initial hyper-plane classification is shown by a solid line, as shown in Fig. 1(a), the points lie on the two parallel hyper-planes and are called current support vectors (as shown by the dash lines). Based on the MS algorithm, points with maximal amount of information should be in the margin (bounded by the dash lines in the figure). As these points are most likely to become support vectors of new classifiers, so unlabeled samples within maximal margin in relation to the hyper-plane are selected; then samples with minimal distance from the hyper-plane (i.e., unlabeled samples marked △ as shown in (b)) are selected.

These samples may improve decision surface for classification within the shortest possible time. Samples marked △ in (b) are manually labeled and are added into the training data, as shown by the dash lines in (c). They are likely to become support vectors of new classifiers, resulting in a change to the hyper-plane. As shown in (c), the improved hyper-plane has a better classification effect. Figure 1 shows changes to the SVM hyper-plane brought about by the MS algorithm.

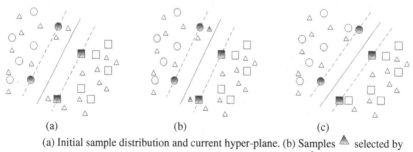

(a)   (b)   (c)

(a) Initial sample distribution and current hyper-plane. (b) Samples ▲ selected by using MS algorithm. (c) Hyper-plane after samples ▲ are added.

Fig. 1. Diagrams illustrating changes to hyper-plane due to MS algorithm

So candidate samples selected by using MS algorithm for binary classification x_{query} is as follows:

$$\mathbf{x}_{query} = \text{argmin}_{\mathbf{x}_i \in \mathbf{U}} |f(\mathbf{x}_i)| \tag{1}$$

Where, U is unlabeled sample set; $|f(\mathbf{x}_i)|$ is the distance between unlabeled sample $\mathbf{x}_i$ and current hyper-plane.

For multi-class classification, several one-versus-rest SVM classification models need to be built by classifying each class and other classes of data, in order to obtain a number of decision surfaces for classification. MS strategy-based criterion is as follows:

$$\mathbf{x}_{query} = \text{argmin}_{\mathbf{x}_i \in \mathbf{U}} (\text{min}_c |f(\mathbf{x}_i|c)|) \tag{2}$$

Similar to (1), $|f(\mathbf{x}_i|c)|$ is the distance between unlabeled sample $\mathbf{x}_i$ and hyper-plane corresponding to Class c.

The iteration process of MS algorithm is listed as follows:

Input: Training data: $\mathbf{L} = \{(\mathbf{x}_i, y_i)\}_{i=1}^l$ Unlabeled data: $\mathbf{U} = \{(\mathbf{x}_j)\}_{j=1}^u$ The number of classes: c Labeled sample number in each iteration: q
Step 1: Learn current training set **L** with an SVM classifier and obtain a classification model.   **Step 2**: Calculate the distance from each candidate point to c hyper-planes in **U** and obtain corresponding distance matrix.   **Step 3**: Select q nearest points to hyper-plane according to Formula (2), i.e., a point set satisfying given conditions $\mathbf{Q} = \{(\mathbf{x}_k)\}_{k=1}^q$.   **Step 4**: Manually label **Q**, add it to training set **L** and remove it from unlabeled set **U**. Update **L** and **U**, i.e., $\mathbf{L} = \mathbf{L} \cup \mathbf{Q}$, $\mathbf{U} = \mathbf{U}/\mathbf{Q}$.   **Step 5**: Repeat steps 1-4 until given conditions are met.
Output: Current classifier.

3 Selecting Large-Scale Training Samples Based on an Active Learning Algorithm

With the emergence of 10-million-level RS training sample sets, a new problem is encountered in RS data processing, that is, how to select subsets from such super-sized training sample sets for automatic RS image classification. Problems such as slow operation, insufficient memory and even "over-fitting" encountered in classifying large quantity of training samples often result in failure to classify; hence the need for optimized training samples selection from super-sized RS training sample sets has drawn the attention of researchers to automatic super-sized RS data sets classification.

The most direct and effective way to solve large-scale training sample sets' problem is to select typical samples from original data for developing classification models through training. Thereby, operations are simplified by effectively reducing the original data size, learning process is accelerated, "over-fitting" is prevented and the generalization capacities of classification algorithm are improved [11]. However, blind reduction of training samples will undoubtedly affect classification algorithm performance, so an ideal way to achieve optimized sample selection is to meet storage requirements and reduce operating time, while maintaining classification algorithm performance. In other words, priority should be given to selecting most useful samples for large-scale RS training samples classification.

The Global Forest Cover Monitoring System is a typical case for processing super-sized RS training sample sets, and addresses large-scale data classification problems through stratified systematic sampling. It is an easy-to-operate approach, but has problems, such as blind operation and poor generalization capacity. In the machine learning field, active learning algorithms are aimed to quickly improve classification surfaces and achieve stable performance by searching for most significant samples that contain most useful information, so as to improve classification precision with as few samples as possible. The ability to select high-quality samples from a large variety of samples is crucial to an active learning algorithm. Although active learning algorithms are used to select unlabeled samples, their selection concepts are the same as those for optimized large-scale training samples' selection, that is, they extract the most effective samples by simplifying original data sets to achieve "higher precision with fewer samples". Therefore, this strategy is of great research significance in selecting active learning samples in large-scale training samples' classification.

A lot of research has been conducted on selection strategies for active learning samples [12], but most research was based on a small number of training samples, and none has dealt with validating optimized large-scale training samples' selection. The active learning concept for optimized large-scale training samples' selection is proposed with the hope to achieve better classification results for large-scale data compared to random sampling and stratified systematic sampling.

To solve global forest cover information classification problem, an active learning algorithm was used in optimized large-scale training samples' selection, and selected samples were used in classifying forest and non-forest areas on a Landsat RS image. An MS-based algorithm was used to select a small quantity of effective samples from large-scale training samples to train SVM classifiers in classifying RS images with the hope that training samples selected using the active learning algorithm would have better classification results and faster operation, compared to those obtained using conventional methods.

Figure 2 shows the steps in the proposed learning strategy for selecting large-scale sample sets:

First, obtain a small-scale initial training sample subset through random selection from original large-scale RS data, and take the rest as candidate training sample set.

Then, input initial training sample subset and candidate training sample set, and run active learning system; repeatedly prune and sieve candidate sample set by using the classifier resulting from initial training sample subset and in accordance with active learning sample selection constraints, and select training samples with maximal amount

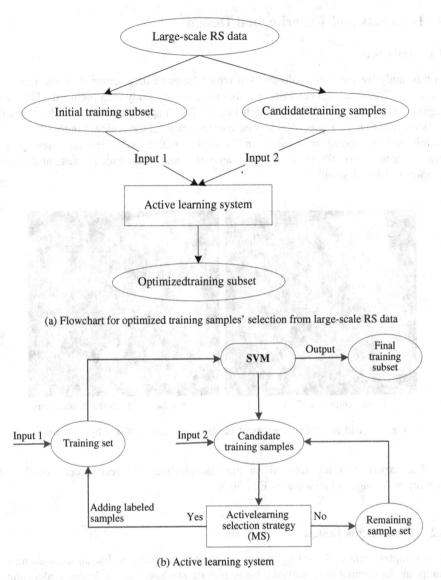

(a) Flowchart for optimized training samples' selection from large-scale RS data

(b) Active learning system

Fig. 2. Flowchart for optimized training samples' selection from large-scale RS data based on active learning

of information from candidate sample set; finally obtain an optimized training subset formed by the initial training subset and high-quality candidate training samples.

We believe that such a small-scale training subset can not only represent the original large-scale RS data, but also effectively reduce classification complexity. Test results prove that active learning strategy can substantially reduce learning cost, significantly increase classification speed, and achieve high-precision classification with a small number of training samples.

4 Data Sets and Experimental Design

4.1 Data Sets

In this study we used the multi-spectral remote sensing data acquired by the Landsat Enhanced Thematic Mapper Plus (ETM+) sensor. The study area included Myanmar region. The image was taken on March 1st, 1984 with spatial resolution of 30×30 m^2 (as shown in Fig. 3). Figure 3(a) is a false color image of the region. Figure 3(b) is the distribution of labeled information in the region where green denotes forest areas, orange denotes non-forest areas, blue denotes water and shaded areas, and white denotes unlabeled samples.

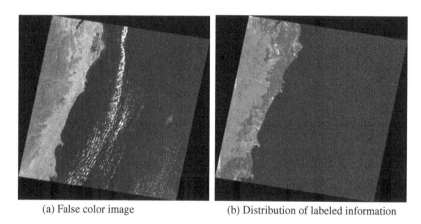

(a) False color image   (b) Distribution of labeled information

Fig. 3. Landsat-7 ETM+ images of the Myanmar region (Color figure online)

The experiments are focused on the classification of forest (green color) and non-forest (orange color) areas in Fig. 3(b).

4.2 Experiments Design

Test samples were collected by means of random sampling and stratified systematic sampling for comparative analysis, using uncertainty-based active learning algorithm. Tests mainly involve the following:

1. In view of SVM classifiers' high computing complexity, a maximum of 10,000 training samples were used in the test, aimed to determine classification results with varying numbers of training samples (X = [20 60 100 200 300 400 500 800 1000 1500 2000 3000 5000 10000]) by using the following algorithms:

   - MS algorithm: Default initial number of training samples = 20 and number of samples for each iterative sampling operation = 20.
   - Random sampling algorithm: Randomly obtain X training samples.

- Stratified systematic sampling algorithm: Obtain X training samples at equal spatial intervals.

2. To carefully observe classification precision changes in active learning and random sampling algorithms with small number of samples, classification tests were carried out using 20–620 training samples, with an interval of 20 samples.

   - MS algorithm: Default initial number of training samples = 20 and number of samples for each iterative sampling operation = 20.
   - Random sampling algorithm: Randomly obtain same number of training samples as MS algorithm.

3. SVM classifier's time complexity changes with varying sample sets [20–10000] were analyzed.
   All the above tests used SVM classifier models, and average value of 10 test results were used for each test.

4.3   Experiments Results and Analysis

Results of the above tests are as follows:

1. Total classification precision of the three algorithms with number of samples X = [20 60 100 200 300 400 500 800 1000 1500 2000 3000 5000 10000] are shown in Fig. 4 and Table 1.

   As shown in the figure and table, random sampling and stratified systematic sampling produced different results as the number of samples changed, but the difference was insignificant. The reason is that stratified systematic sampling is aimed to obtain general and typical pixels from a spatial perspective in order to ensure sample generality and typicality in geographic information, and retain most information of original images. However, sampling interval may have an impact on sampling efficiency. Improper intervals may result in non-uniform sampling and incomplete information. Therefore, stratified systematic sampling method is unstable. However, the proposed active learning strategy can significantly improve classifier performance and optimize classification efficiency. When the number of samples reaches 200, it can produce same results as other algorithms produce with 3000-5000 samples. When number of samples reaches 300, it can produce same results as other algorithms produce with 10,000 samples. So this algorithm can greatly reduce calculations required for constructing a trainer. As the MS algorithm is obviously better than the other two when number of training samples is relatively small, hence only classification precision below 620 are listed.

2. To observe precision changes at specific points, total classification precision curves of MS and random sampling algorithms with number of training samples between 20 and 620 with an interval of 20 samples are given in Fig. 5. As shown in the figure, the MS algorithm exhibited a significant advantage over the random sampling algorithm, with a precision difference of up to 1.5863 %. The MS algorithm curve began to rise abruptly when sample numbers reached 40, and quickly reached

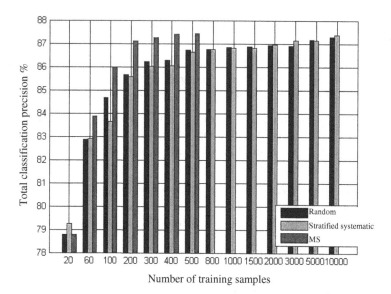

Fig. 4. Diagram illustrating total classification precision of three sampling algorithms with varying number of training samples

Table 1. Total classification precision of three algorithms

Number of training samples	Random sampling	Stratified systematic sampling	Margin sampling
20	78.78 %	79.26 %	78.78 %
60	82.88 %	82.90 %	83.87 %
100	84.64 %	83.61 %	86.00 %
200	85.66 %	85.57 %	87.11 %
300	86.21 %	86.00 %	87.26 %
400	86.29 %	86.04 %	87.39 %
500	86.70 %	86.63 %	87.43 %
800	86.76 %	86.74 %	–
1000	86.84 %	86.79 %	–
1500	86.87 %	86.81 %	–
2000	86.90 %	86.97 %	–
3000	86.91 %	87.12 %	–
5000	87.17 %	87.12 %	–
10000	87.26 %	87.35 %	–

convergence. When training sample numbers reached 160, it produced a better classification result than that of random sampling algorithm using 620 samples. It exhibited excellent performances both horizontally and vertically.

3. Figure 6 shows time complexity changes (in seconds) with varying number of SVM training samples X = [20 60 100 200 300 400 500 800 1000 1500 2000 3000 5000

10000]. As shown in the figure, as sample numbers increase, operating time increases exponentially. The more samples that are used, heavier the burden on the classifier.

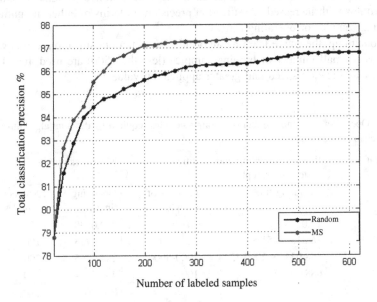

Fig. 5. Comparison of total classification precision curves of two sampling algorithms with relatively small number of training samples

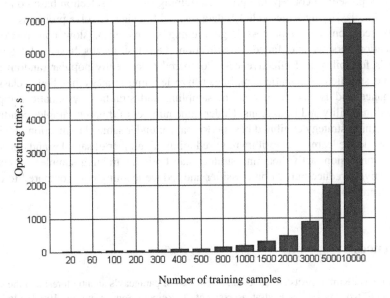

Fig. 6. Diagram illustrating changes in operating time with varying number of SVM training samples

4. Table 2 shows operating time and number of training samples required by random sampling and stratified systematic sampling and MS algorithms under same precision levels. As shown in the table, the active learning algorithm exhibited an increasing advantage over random sampling and stratified systematic sampling algorithms with increased classification precision, in terms of either the number of training samples or operating time.

To sum up, the active learning algorithm has better classification results when a relatively small number of training samples (less than 500) are used, thus having faster classification speed and greater application value.

Table 2. Operating time and number of training samples required by different algorithms under same precision conditions

Classification precision	Random sampling and stratified systematic sampling		Margin sampling	
	Operating time (s)	Number of samples	Operating time (s)	Number of samples
78.78 %	12.0138	20	12.0138	20
86.00 %	36.8152 ~ 51.5007	200-300	92.1723	100
87.11 %	888.9244 ~ 2000.3394	3000-5000	244.3265	200
87.26 %	6888.5975	10000	440.2104	300

5   Conclusion

This paper presents a concept of optimized training samples' selection based on active learning for classifying large-scale RS data sets, with a view to solve problems such as memory requirements to process large training sample data, slow construction of classification models and failed classification. Our test results have validated the concept's feasibility and effectiveness. Compared to currently popular random sampling and stratified systematic sampling, active learning strategy has better classifier performance and is free from random sampling and stratified systematic sampling methods' instability and limitations. Under same number of training samples condition, active learning strategy exhibited best performance; under same classification precision condition, active learning algorithm required minimal number of samples and operating time. In conclusion, active learning strategy can optimize training samples' selection, and effectively reduce burden on classifier and reduce learning cost compared to other sampling methods.

References

1. Huang, C., Kim, S., Altstatt, A., et al.: Rapid loss of paraguay's atlantic forest and the status of protected areas — a landsat assessment. J. Remote Sens. Environ. **106**(4), 460–466 (2007)

2. Huang, C., Song, K., Kim, S., et al.: Use of a dark object concept and support vector machines to automate forest cover change analysis. J. Remote Sens. Environ. **112**(3), 970–985 (2008)
3. Huang, C., Kim, S., Song, K., et al.: Assessment of paraguay's forest cover change using landsat observations. J. Global Planet. Change **67**(1), 1–12 (2009)
4. Huang, C., Thomas, N., Goward, S.N., et al.: Automated masking of cloud and cloud shadow for forest change analysis using landsat images. J. Int. J. Remote Sens. **31**(20), 5449–5464 (2010)
5. Sexton, J.O., Song, X., Feng, M., et al.: Global, 30-m resolution continuous fields of tree cover: landsat-based rescaling of MODIS vegetation continuous fields with lidar-based estimates of error. J. Int. J. Digit. Earth **6**(5), 427–448 (2013)
6. Townshend, J.R., Masek, J.G., Huang, C., et al.: Global characterization and monitoring of forest cover using landsat data: opportunities and challenges. J. Int. J. Digit. Earth **5**(5), 373–397 (2012)
7. Li, H., Wang, C., Yuan, B., et al.: A learning strategy for SVM-based large-scale training sets. J. Chin. J. Comput. **27**(5), 715–719 (2004)
8. Zhai, J., Li, S., Wang, X.: Comparative research of condense nearest rules based on fuzzy rough sets. J. Comput. Sci. **39**(2), 236–239 (2012)
9. Vapnik, V.N.: Statistical Learning Theory, pp. 231–244. Wiley, New York (1998)
10. Schohn, G., Cohn, D.: Less is more: active learning with support vector machines. In: International Conference on Machine Learning, pp. 839–846. Morgan Kaufmann Publishers Inc. (2000)
11. Jiang, W.: Research on the selection of samples for pattern recognition and its application. D. Nanjing University of Science and Technology (2008)
12. Tuia, D., Ratle, F., Pacifici, F., et al.: Active learning methods for remote sensing image classification. J. IEEE Trans. Geosci. Remote Sens. **47**(7), 2218–2232 (2009)

Coverage Optimization for Wireless Sensor Networks by Evolutionary Algorithm

Kangshun Li, Zhichao Wen[(✉)], and Shen Li

College of Mathematics and Information, South China Agricultural University,
Guangzhou 510642, China
283072731@qq.com

Abstract. Wireless sensor network consists of a large number of tiny sensor nodes owned capable of perception in monitoring region by self-organized wireless communication, has been widely applied in military and civil fields. From the perspective of resource- saving, under the condition of the network's connectivity and specific coverage, the number of sensor nodes is assumed to be opened as few as possible. So, computing the sensor nodes collection which meeting the requirements is called the problem of network coverage optimization for Wireless Sensor Network; also called the problem of minimum connected covering node set. The innovation point of the article is: Firstly, it analyzed the deficiencies of traditional evolution algorithm fitness function, put forward an improved fitness function design scheme, and has been proved that it has advantage of solving problem on wireless sensor networks coverage optimization; Secondly, it applied the method of control variables, comparison and analysis of the influence on the various operations and parameters selection in evolution algorithm on the optimization results and performance, and then point out how to design algorithm to manage to the best optimize effect and performance.

Keywords: Wireless sensor network · Evolutionary algorithm · Coverage optimization

1 Introduction

With the rapid development of embedded system, wireless communication, network and micro electro mechanical system, a new measurement and control network appeared—the Wireless Sensor Network (WSN for short). This technology, arose in 1990s, has developed unprecedentedly in the first decade of 20 century, and now has played an important role in human society and all aspects of modern life.

WSN is a self-organizing intelligent network for exclusive use, composed by a lot of micro sensor with the capability of sensing, calculating and wireless network communicating. Through wireless communication it can synergistically achieve the specific function. It synthesizes sensor technology, embedded computing, communication technology, distributed information processing, microelectronics manufacture technology and software programming technology [1]. It can also monitor, sense and collect the

© Springer Science+Business Media Singapore 2016
K. Li et al. (Eds.): ISICA 2015, CCIS 575, pp. 52–63, 2016.
DOI: 10.1007/978-981-10-0356-1_6

information that comes from the monitored objects or circumstance in sensor field, and then transmits the collected information to terminal user.

2 The Evaluation Standard of WSN Coverage

After deploying wireless sensor network in a certain area, how to evaluate the effect of the wireless sensor network on detecting the designated area is the first thing that we should give consideration to when we are to set up a wireless sensor network model. The effect of wireless sensor network' detection to the designated area is known as Sensing Coverage, corresponding to different application requirements, we can define it as different sensing coverage.

Currently, based on specific application requirements and specific sensor model, there are three types of coverage definitions as follows: Barrier Coverage, Area Coverage and Point Coverage.

2.1 Barrier Coverage

Barrier coverage is based on the question: "how can we deploy a wireless sensor network to make the probability the lowest when a spy goes through the monitoring area successfully?" A specific model definition is given by Meguerdichian et al. [5]: that is to set a deployment of wireless sensor network in the monitoring area and the spy's starting and ending positions. We need to figure out the Maximal Breach Path (MBP) and the Maximal Support Path (MSP), which each corresponds to the worst and the best coverage. However in exact study, the most common things are the Area Coverage [6–8] and the Point Coverage [9–11]. Then, these will be analyzed in the following essay.

2.2 Area Coverage

Assuming that all sensors included in wireless sensor networks are set as $S = \{S_i, i = 1, 2, \ldots, n\}$; the detection range of each sensor S_i as c_i; the target detection area as D, and the expected effect of detection as $\bigcup_{i=1}^{n} c_i \supseteq A$. Assuming that $S_1 = |\bigcup_{i=1}^{n} c_i \supseteq A|$ is the area that the sensor can cover, $S_0 = |A|$ is the target detecting area, and then $\rho = S_1/S_0$ is called the coverage of wireless sensor networks.

2.3 Point Coverage

If the detection target isn't area, but a limited number of target detection points $D = \{d_i, i = 1, 2, \ldots, m\}$, then the desired probing effect transforms into $\bigcup_{i=1}^{n} d_i \supseteq D$. Suppose there are n_1 points in D that can be detected by sensor network and a total number n_0 of target detection points in D, similarly Point Coverage can be defined as $\rho = n_1/n_0$.

In certain study, in order to simplify the problem, we usually divide the target area into discrete grid, which will transform an Area Coverage problem into a Point Coverage problem. Chen and his partners [11], who have carefully discussed this subject, proposed a method called the sampling of the coverage area. The area will be divided by some equal space and each vertical lines. The intersection of the straight line (grid) is called sample points, which will replace the detected area. In this case, the density of the sample points is an issue needed to be considered because it determines the discrete accuracy directly.

3   Energy-Saving of Wireless Sensor Network

In practical applications, depending on the actual situation, the position of the sensors can sometimes be deployed by the deployers, but sometimes can't. In this case, the deployment location of the wireless sensors cannot be determined in advance and the number of deployed wireless sensors is often much more than the minimum number needed to cover the target area. As a result, what we need to consider is how to find out the set of nodes to meet the minimum number of the coverage, and design the sensor activation programs to extend the life of the network as long as possible at the same time. Corresponding to both the actual situation above, currently there are two researching directions: pre-deployment decisions (Deterministic Deployment) and random deployment (Random Deployment).

3.1   Deterministic Deployment

Refers to the so-called Deterministic Deployment, according to the topography of the detection area's specific characteristics and monitoring requirements and determine the number and position, what is the most classic of the deployment of wireless sensor is the art gallery theorem [12]. Besides, Coverage opportunities for global ocean color in a multimission era, discussed by Gregg et al. [13], is also a case of pre-determined deployment. In general two-dimensional cases, it is easier for a pre-determined deployment, but in 3 d, the complexity of the problem increases greatly. In fact, as Hoffmann notes, three-dimensional predetermination deployment is a NP-hard problem. Therefore, people often use heuristic algorithm to calculate the approximate optimal solution of the problem such as Yong et al. [14], they used genetic algorithm to solve a pre-determined deployment issues.

Aim at the problems mentioned above, Dhillon [10, 15] puts forward two kinds of algorithms based on probabilistic sensor model to calculate the optimal layout. Yi et al. [16] adopted Dhillon's first algorithm in applications, however, in the selection of lattice, the author not only consider the priority of placing sensor in the smallest lattice among **M (k)**, but also put forward another algorithm opposite to it, namely consider the priority of placing sensors in the biggest point among **M (k)**.

3.2   Random Deployment

In the region of the enemy or dangerous areas, especially in need of a wide range of areas such as forest fire to monitor the situation, people often use aircraft to throw down a large number of wireless sensors in it in order to achieve detection, and in this case, the exact location of deployed wireless sensor cannot be controlled. According to the existing technical level, monitors are basically battery power monitors. Due to the deployment of wireless sensor network is in inaccessible areas, it is impossible to replace the battery. Therefore, how to achieve the purpose of energy conservation through rational planning becomes a problem needed to be discussed. We call it a random deployment issues, also refers to as NSP (Node Scheduling Problem). To this end, this thesis is studied.

According to the actual needs, the NSPs have tiny difference as well. In some problems, for example, we hope that the target area is covered absolutely [10, 15, 16] in every moment, while in some problems, the coverage just needs to reach a certain level [6, 9]. At that time, we need to take a balance between coverage and system's lifetime. This balance, sometimes is achieved through adding the optimization algorithm of penalty term to the objective function value, sometimes by multi-objective optimization algorithm just as multi-objective genetic algorithm Jourdan [6] has used.

4   The Evolution of the WSN Coverage Optimization Algorithm Design

4.1   Evolutionary Algorithm

Evolutionary Algorithm (EA) is a kind of bionic Algorithm which solves global complex optimization problem [17]. It based on Darwin's Natural Evolution and Mendel's Genetic Mutation Theory and was first come forward by Professor Holland j. h of the United States in the mid - 1960s. Evolutionary Algorithm exists mainly in the form of the Genetic Algorithm in its early time. Evolutionary Algorithm searches the optimization groups for the best individuals by global parallel searching technology, in order to meet the requirements of the optimal solution or near optimal solution. Zhang Jun et al. [18] points out the meaning and scope of application of Evolutionary Algorithm, believing that Evolutionary Algorithm can be widely used in combinatorial optimization, machine learning, self-adaptive control, planning and design, artificial life, artificial neural network training and image processing, and it is one of the key technology of intelligent calculation of the 21st century.

4.2   Coding

Because there are only two states of the sensor nodes in Problem Model, the "sleeping" and the "working" states, naturally we think of using 0/1 string to encode it. The number 0 means "sleeping" state, number1 indicates "working" state.

Each gene is described as a 0/1 matrix $\theta = [|T|][|S|]$ and $\theta = [|T|][|S|]$ is the matrix consists of $|T|$ rows and $|S|$ columns; T is the amount of time and $|S|$ is the number of sensor nodes; each one of each line represents a sensor and the corresponding number represents the state of sensor; For a certain one of the line t (t∈1,..., $|T|$), if the node is active in the period t, the value of this node is 1, otherwise it is 0.

A wireless sensor network node activation solution was defined as a chromosome, or an individual.

4.3 Generation of Initial Population

We define the generation of initial population scheme as generating a number of individuals randomly. In the experiment of this article, set the population size as POPULATION_SIZE = 100. At the same time, each individual in the initial population is also randomly generated, namely each of the individual's random value is 0 or 1. When reflected on the decoding, it means the sensor nodes in the area are randomly in "working" or "sleeping" state.

4.4 The Improved Fitness Function

The goal for optimization is to make less nodes open but cover areas as much as possible. For the argument's sake, we make up the following definitions:

WorkingNodes: it means the total number of Nodes in a Working state;

Network coverage (point coverage) is defined as:

$$COV\_RATE = \frac{coverPoints}{amount} \tag{1}$$

Among them, coverPoints is the number of the covered points and amount is the total number of Points.

The goal of this paper is to make the workingNodes as less as possible, the Cover Points as many as possible, and COV_RATE meets the requirement. In the experiment of this paper, we stipulate COV_RATE must be bigger than 0.9.

The design goal of fitness function: on the premise of meet the coverage requirements, reduce the workingNodes, and increase the coverPoints at the same time. Utilizing intuitive analysis is to make the overlapping area of the circle of the rendering smaller, and require more lattices are covered by the circle. This would involve the issue that whether a sensor node is worth to open or not because once is opened, workingNodes will plus 1, but coverPoints not always increase more. This is because the lattices node A can cover after it is opened probably have already been covered by other opened nodes. In this case, the opening of A seems not valuable.

In the following, a geometric method will be given to explore that, if a node is opened, the number of lattices can be covered are as shown:

What can be seen from Fig. 1 above is that an opened node can detect the number of lattice points is related to its detection radius as well as the density of lattice points.

The radius of the circle is RADIUS, the length of the side of the square inscribed in the circle is $RADIUS\sqrt{2}$. Take factors such as the boundary into consideration, we can figure out the area of the square inscribed in the circle is about $(RADIUS\sqrt{2} + 1)^2$. Because all the lattices are arranged 1 meter vertically and horizontally from each other, the number of lattice points of a square inscribed in a circle can be equal to the square's area. This can also be regarded as the lower bound of the number of lattice points the circle covers.

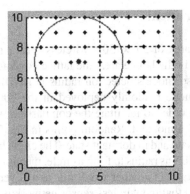

Fig. 1. Schematic diagram of the number of lattice points that nodes covered effectively

For further analysis, we believe that if an additional node's open can make the value of **coverPoints** reach up for $(RADIUS\sqrt{2} + 1)^2$, we certainly think it's worth it. But for most of the time, an additional node's open cannot make the value of **coverPoints** increase like this $(RADIUS\sqrt{2} + 1)^2$. That is because some lattice points are likely to have already been covered by other opened nodes.

Therefore a parameter **CIRCLE_GRID** is introduced, and **CIRCLE_GRID** ε (0, 1) is called the lattice parameter inside the circle. We believe that if an additional node's open can make the value of **coverPoints** increase **CIRCLE_GRID** $\times (RADIUS\sqrt{2} + 1)^2$, it's worthy.

With the analysis above, the improved fitness function can be:

$$f(individual) = coverPoints - CIRCLE_{GRID} \times \left(RADIUS\sqrt{2} + 1\right)^2 \times workingNodes \qquad (2)$$

Obviously, if **workingNodes** adds 1, the increment **coverPoints** can bring is more than $CIRCLE\_GRID \times (RADIUS\sqrt{2} + 1)^2$, then the fitness function f (individual) increase to the positive direction, which indicates the optimization is worthy.

This paper adopts the probability method of the fitness value which is proportional for selecting operation, also called Roulette Wheel Selection. The higher the fitness of individual fitness is, the more likely it would be selected as the parent. Here set $P_i = \frac{f_i}{\sum_i f_i}$ as the individual i's selection probability and f_i is the individual i's fitness function value.

4.5   The Mutation Probability

Mutation occurred when two individuals cross and produce offspring. It can be considered as a situation that the offspring's random bit 0/1 turn into the other. In the genetic evolution process of nature, the gene mutation probability of higher plants and animals ranges from one hundred-thousandth to one hundred-millionth. Because the number of the organism's genes is large, there is always gene mutation. In this study, the number of gene is less, so we study the mutation probability between 0 and 1.

5   Simulation Experiment

Given a detection lattice area of 30 × 30's wireless sensor network and 200 sensor nodes randomly deployed in the lattice area; the lattice coverage requires 0.9.

Assume that each sensor node's detection radius is **RADIUS** = 3, the number of wireless sensor network node randomly deployed in the 30 × 30 detection area is **NODE_NUM** = 200. In order to facilitate the comparison, we suppose each node's state is random in the initial moment. Random points in Fig. 2 above represent nodes which are in the working state and in dormant state. The circle uses working state points as the center and the **RADIUS** as radius to form a circle, meaning the detection boundary of the nodes in working state. In another word, the area covered by the pink circle in Fig. 2 is right now the area that the wireless sensor network is able to cover while it is working. Obviously, the black points indicate the probes that have not been covered, also indicate the area that the network does not cover.

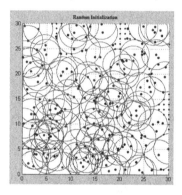

Fig. 2. A diagram of wireless sensor network

5.1   Optimization Experiment of Fitness Function

To test the improved fitness function in this paper, we have carried out an experiment, the results of this experiment are as shown in Fig. 3 (set **CIRCLE_GRID** = 0.3):

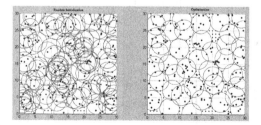

Fig. 3. Effect picture of the experiment of fitness function optimization

The Fig. 4 shows that when using the improved fitness function, the number of covering lattice points only cuts down from the original number—about 840 to more than 820, and the number of sensors in working state is from nearly 90 to less than 40, and the point coverage also remains above 0.9.

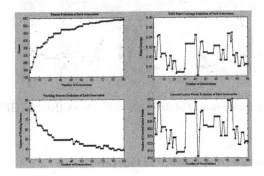

Fig. 4. Performance picture of fitness function optimization

5.2   Contrast Experiment of the Lattice Parameters Inside the Circle

In the following paper we will discuss the size of lattice parameters **CIRCLE_GRID**'s effect on optimization.

Based on the definition of the improved fitness function, we can see that the bigger the **CIRCLE_GRID**, the smaller the overlapping area of the circle after optimized; the smaller the **CIRCLE_GRID**, the larger the overlapping area of the circle after optimized. Figure 5 is the result of the experiment when **CIRCLE_GRID** = 0.7:

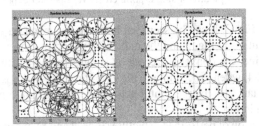

Fig. 5. Effect picture of the experiment when **CIRCLE_GRID** = 0.7

From Fig. 5 it can be seen clearly that the overlap of circle is smaller as well as the total covering areas compared with the experiment of Fig. 3 (**CIRCLE_GRID** = 0.3). Seen from the Fig. 6, point coverage is only 0.75, which does not conform to the actual detecting demand of the above analysis this paper–0.9 coverage requirements.

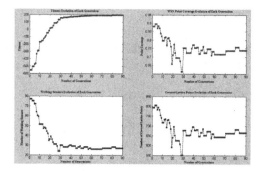

Fig. 6. Performance picture of the experiment when CIRCLE_GRID = 0.7

5.3 Contrast Experiment of Crossover Operator

When using single-point crossover operator, the other parameters are consistent with the parameters of the fitness function of optimization experiment. Its result is as shown in Fig. 7:

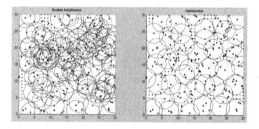

Fig. 7. Effect picture of single-point crossover operator experiment

Compared with Fig. 3, it can be seen that both the optimization effect is the same. But the fitness evolution of Fig. 8 and the number of working sensors' evolution graph is almost straight line, the convergence rate is slow compared with experiment two. Specifically, in Fig. 4, when proceeding to the 70th iteration, it almost has obtained the optimal value and start to converge. But it gets the optimal solution only it proceeding to the 90th iteration in Fig. 8.

From the comparison of performance picture (Figs. 4 and 8), the experiments' points coverage are all above 0.9, which meets the requirements. The number of optimized working sensors in Fig. 8 is 52, more than that of Fig. 3 which is 39. The number of the covering lattice points in Fig. 8 after optimized is below 820 while is higher than 820 in Fig. 4.

Therefore, compared with the single-point crossover operator, multi-point crossover operator is characterized by faster convergence rate, high efficiency and better optimization effect.

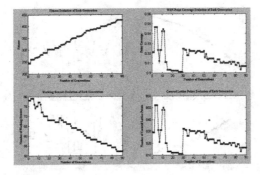

Fig. 8. Performance picture of single-point crossover operator experiment

5.4 The Mutation Probability

In the previous experiments, we set the mutation probability as 0.1. Now, we will set the mutation probability in experiment of Fig. 9 as 1, but the other related parameters keep unchanged.

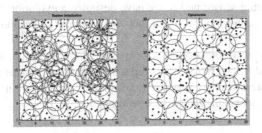

Fig. 9. Effect picture of the experiment when mutation probability is 1

Compared Figs. 9 and 3, it can be seen that the optimization effect of each is roughly the same. But the contrast between the performance pictures indicates that: the fitness evolution in Fig. 10 is a straight line and its convergence rate is slower than that in Fig. 4. The reason for this kind of phenomenon is that the mutation probability is 1, while the mutation probability of experiment 5.1 is 0.1; the mutation probability is too big, causing the searching space decentralized and slowing down convergence. The number of working nodes in Fig. 10 is 43 after optimizing, and it is 39 in Fig. 4. Consequently, selecting the small mutation probability is better than the big one based on the optimization results.

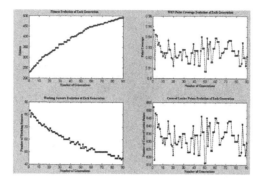

Fig. 10. Performance picture of the experiment when mutation probability is 1

6   Conclusion

In this article, for the problem of optimization of the wireless sensor network coverage in the high density nodes environment, we set up a mathematical model with minimizing the working nodes and maximizing the nodes coverage. In the problem of optimization of the wireless sensor network coverage, aiming at meeting the coverage and minimizing the number of working sensor nodes, we have designed a evolutionary algorithm, and reduced the energy consumption of the network in order to prolong the lifetime of network.

Aiming at traditional fitness function of evolutionary algorithm, we have analyzed its shortcomings and figured out an improved scheme of fitness function, which has been proved to have more advantages by the experimental data in solving the problem of optimization of wireless sensor network coverage.

Acknowledgements. This work is supported by the National Natural Science Foundation of China with the Grant No. 61573157, the Fund of Natural Science Foundation of Guangdong Province of China with the Grant No. 2014A030313454.

This work was jointly supported by Natural Science Foundation of Guangdong Province of China (#2015A030313408).

References

1. Limin, Sun, et al.: Wireless Sensor Networks. Tsinghua University Press, Beijing (2005)
2. Lu, Y., Zhu, G., Liu, A.: Intelligent network system research and construction program. In: International Conference on Multimedia Technology (ICMT), pp. 4954–4957 (2011)
3. Jamali, M.A., Bakhshivand, N., Easmaeilpour, M., et al.: An energy-efficient algorithm for connected target coverage problem in wireless sensor networks. In: 3rd IEEE International Conference on Computer Science and Information Technology (ICCSIT), pp. 249–254 (2010)
4. Mao, Y.-C., Chen, L.-J., Chen, D.-X.: A survey on coverage control techniques for wireless sensor networks. Comput. Sci. **21**(5), 20–26 (2007)

5. Meguerdichian, S., Koushanfar, F., Potkonjak, M., et al.: Coverage problems in wireless ad-hoc sensor networks: INFOCOM. In: Proceedings of Twentieth Annual Joint Conference of the IEEE Computer and Communications Societies, vol. 3, pp. 1380–1387. IEEE (2001)

6. Jourdan, D., de Weck, O.L.: Layout optimization for a wireless sensor network using a multi-objective genetic algorithm. In: 2004 IEEE 59th Vehicular Technology Conference, VTC 2004-Spring, vol. 5, pp. 2466–2470 (2004)

7. Cardei, M., Jie, W., Mingming, L., et al.: Maximum network lifetime in wireless sensor networks with adjustable sensing ranges. In: IEEE International Conference on Wireless and Mobile Computing, Networking and Communications, WiMob 2005, vol. 3, pp. 438–455 (2005)

8. Erfu, Y., Erdogan, A.T., Arslan, T., et al.: Multi-objective evolutionary optimizations of a space-based reconfigurable sensor network under hard constraints. In: ECSIS Symposium on Bio-inspired, Learning, and Intelligent Systems for Security, BLISS 2007, pp. 72–75 (2007)

9. Quintao, F.P., Nakamura, F.G., Mateus, G.R.: Evolutionary algorithm for the dynamic coverage problem applied to wireless sensor networks design. In: The 2005 IEEE Congress on Evolutionary Computation, vol. 2, pp. 1589–1596 (2005)

10. Dhillon, S.S., Chakrabarty, K., Iyengar, S.S.: Sensor placement for grid coverage under imprecise detections. In: Proceedings of the Fifth International Conference on Information Fusion, vol. 2, pp. 1581–1587 (2002)

11. Chen, H., Wu, H., Tzeng, N.F: Grid-based approach for working node selection in wireless sensor networks. In: IEEE International Conference on Communications, vol. 6, pp. 3673–3678 (2004)

12. Hoffmann, F., Kaufmann, M., Kriegel, K.: The art gallery theorem for polygons with holes. In: Proceedings of 32nd Annual Symposium on Foundations of Computer Science, pp. 39–48 (1991)

13. Gregg, W.W., Esaias, W.E., Feldman, G.C., et al.: Coverage opportunities for global ocean color in a multimission era. IEEE Trans. Geosci. Remote Sens. **36**(5), 1620–1627 (1998)

14. Yong, X., Xin, Y.: A GA approach to the optimal placement of sensors in wireless sensor networks with obstacles and preferences. In: 2006 3rd IEEE on Consumer Communications and Networking Conference, CCNC 2006, pp. 127–131 (2006)

15. Dhillon, S.S., Chakrabarty, K.: Sensor placement for effective coverage and surveillance in distributed sensor networks. In: 2003 IEEE Conference on Wireless Communications and Networking, WCNC 2003, vol. 3, pp. 1609–1614 (2003)

16. Yi, Z., Chakrabarty, K.: Uncertainty-aware sensor deployment algorithms for surveillance applications. In: 2003 Global Telecommunications Conference, GLOBECOM 2003, vol. 5, pp. 2972–2976. IEEE (2003)

17. Zhengjun, Pan, Lishan, Kang, Yuping, Chen: Evolutionary Computation. Tsinghua University Press, Beijing (2009)

18. Zhang, J., et al.: Computational Intelligence. Tsinghua University Press, Beijing (2009)

Combining Dynamic Constrained Many-Objective Optimization with DE to Solve Constrained Optimization Problems

Xi Li[1,2], Sanyou Zeng[1(✉)], Liting Zhang[1], and Guilin Zhang[1]

[1] School of Computer Science, China University of Geosciences, Wuhan 430074, Hubei, People's Republic of China
1589441554@qq.com, sanyouzeng@gmail.com
[2] School of Information Engineering, Shijiazhuang University of Economics, Shijiazhuang 050031, Hebei, People's Republic of China

Abstract. This paper proposes a dynamic constrained many-objective optimization method for solving constrained optimization problems. We first convert a constrained optimization problem (COP) into an equivalent dynamic constrained many-objective optimization problem (DCMOP), then present many-objective optimization evolutionary algorithm with dynamic constraint handling mechanism, called MaDC, to solve the DCMOP, thus the COP is addressed. MaDC uses DE as the search engine, and reference-point-based nondominated sorting approach to select individuals to construct next population. The effectiveness of MaDC has been verified by comparing with peer algorithms.

Keywords: Constrained optimization problem · Many-objective optimization · Dynamic constraint · DE · Reference points

1 Introduction

In science and engineering disciplines, it is common to encounter a large number of constrained optimization problems (COPs). During the past decades, researchers have widely used evolutionary algorithms (EAs) to deal with COPs [1–3], and made considerable achievements. In recent years, with the development of the multi-objective and adaptive evolutionary theories and methodologies, more and more works are managed to add these fruits to solving constrained problems.

Coello first used dominance-based selection strategy to deal with constraints [4]. In [5] Coello and Mezura proposed a new version of the Niched-Pareto Genetic Algorithm (NPGA). This approach uses dominance-based selection scheme to assign fitness function value, and adopts an additional parameter called S_r to control the diversity of the population. Venkatraman and Yen [6] proposed genetic algorithm-based two-phase framework for solving COPs. In the first phase the objective function is completely disregarded, and only the constraints of the problem are focused on. In the second phase, the objective

© Springer Science+Business Media Singapore 2016
K. Li et al. (Eds.): ISICA 2015, CCIS 575, pp. 64–73, 2016.
DOI: 10.1007/978-981-10-0356-1_7

function and satisfaction of the constraints are treated as two objectives to be simultaneously optimized. Hsieh [7] proposed an algorithm based on well-known multi-objective evolutionary algorithm, NSGA-II. The procedure, used as a hybrid constraint handling mechanism, combines ϵ-comparison method of multi-objective optimization and penalty method of constraints-handling. Yong Wang [8] presented hybrid constrained optimization EA (HCOEA), which effectively combines multi-objective optimization with global and local search models. In global model, Pareto-dominance-based tournament selection among parent and offspring and similarity measuring by Euclidean distance among individuals are used to promote population diversity; in the local model, a parallel search in subpopulations is implemented to accelerate convergence. Penalty function is a classical method used for solving COP, but the determination of penalty parameters is a difficulty. Deb [9] proposed a hybrid algorithm which combines bi-objective evolutionary approach with the penalty function methodology. The bi-objective approach provides a good estimate of the penalty parameter, and the unconstrained penalty function approach being constructed using provided penalty parameter generates the optimal solutions of overall hybrid algorithm. Zeng and Li [10,11] used not only multi-objective optimization technology but also dynamic constraint mechanism for COPs. They first convert COP to a dynamic constrained multi-objective optimization problem, then adopt a dynamic constrained multi-objective optimization algorithm to solve the problem.

In this paper, we convert the COP into the many-objective optimization problem, there are m+1 objectives (m is the number of constraints) in a problem, in other words, each constraint function is converted into a violation objective function. So we can introduced many-objective optimization technique into our method. Besides, we adopt dynamic constraint handling mechanism to deal with constraints. The proposed many-objective optimization evolutionary algorithm with dynamic constrained handling, MaDC, uses DE to generate offspring and reference-point-based nondominated sorting approach to create next parent population.

The rest of this paper is organized as follows. Section 2 introduces process of converting a COP into an equivalent dynamic constrained many-objective optimization problem (DCMOP). Section 3 describes implementation of MaDC algorithm. Experiments and results are shown in Sect. 4 to test whether the methodology is effective. Section 5 gives the conclusion.

2   Convert COP to DCMOP

This section first converts a COP into a constrained many-objective optimization problem (CMOP) which is equivalent to the COP. Then the CMOP is converted into a dynamic constrained many-objective optimization problem (DCMOP), a series of CMOPs. In this way, the COP can be solved by solving the equivalent DCMOP.

2.1   Convert COP to CMOP

Without loss of generality, minimization optimization is assumed unless specified otherwise in this paper. A constrained optimization problem (COP) can be stated as follows:

$$
\begin{aligned}
min \quad & y = f(\boldsymbol{x}) \\
st: \quad & \boldsymbol{g}(\boldsymbol{x}) = (g_1(\boldsymbol{x}), g_2(\boldsymbol{x}), ..., g_m(\boldsymbol{x})) \leq \boldsymbol{0} \\
where \quad & \boldsymbol{x} = (x_1, x_2, ..., x_n) \in \mathbf{X} \\
& \mathbf{X} = \{\boldsymbol{x} | \boldsymbol{l} \leq \boldsymbol{x} \leq \boldsymbol{u}\} \\
& \boldsymbol{l} = (l_1, l_2, ..., l_n), \boldsymbol{u} = (u_1, u_2, ..., u_n)
\end{aligned} \tag{1}
$$

where $\boldsymbol{x}$ is the solution vector and $\mathbf{X}$ is the whole search space, $\boldsymbol{l}$ and $\boldsymbol{u}$ are the lower bound and upper bound of the solution space, respectively, $\boldsymbol{g}(\boldsymbol{x}) \leq \boldsymbol{0}$ is the constraint and $\boldsymbol{0}$ is the constrained boundary. When an equality constraint $h(\boldsymbol{x}) = 0$ is involved in the COP, it is usually transformed into an inequality constraint $|h(\boldsymbol{x})| - \epsilon \leq 0$, ϵ is a positive close-to-zero number, $\epsilon = 0.0001$ in this paper.

A solution $\boldsymbol{x} = (x_1, x_2, ..., x_n)$ is **feasible** if it satisfies the constraints conditions $\boldsymbol{g}(\boldsymbol{x}) \leq \boldsymbol{0}$, otherwise it is **infeasible**. A feasible set $\mathbf{S_F}$ of a COP is defined as $\mathbf{S_F} = \{\boldsymbol{x} | \boldsymbol{x} \in \mathbf{X} \ and \ \boldsymbol{x} \ is feasible\}$.

Now we would like to construct a constrained many-objective optimization problem equivalent to the COP discussed above. This can be implemented by converting the constraint function $\boldsymbol{g}(\boldsymbol{x}) = (g_1(\boldsymbol{x}), g_2(\boldsymbol{x}), ..., g_m(\boldsymbol{x}))$ to violation objective function $\boldsymbol{\varphi}(\boldsymbol{x}) = (\varphi_1(\boldsymbol{x}), \varphi_2(\boldsymbol{x}), ..., \varphi_m(\boldsymbol{x}))$ and inserting $\boldsymbol{\varphi}(\boldsymbol{x})$ into the COP as additional objectives without deleting the constraints $\boldsymbol{g}(\boldsymbol{x}) \leq \boldsymbol{0}$, i.e., a constrained many-objective optimization problem (CMOP) is constructed as follow:

$$
\begin{aligned}
min \ \boldsymbol{y} = & (f(\boldsymbol{x}), \varphi_1(\boldsymbol{x}), \varphi_2(\boldsymbol{x}), ..., \varphi_m(\boldsymbol{x})) \\
st: \ \boldsymbol{g} = & \boldsymbol{g}(\boldsymbol{x}) = (g_1(\boldsymbol{x}), g_2(\boldsymbol{x}), ..., g_m(\boldsymbol{x})) \leq \boldsymbol{0}
\end{aligned} \tag{2}
$$

where $\boldsymbol{x} = (x_1, x_2, ..., x_n)$ is n dimension search vector, $\boldsymbol{y}$, $\boldsymbol{g}$ are functions of vector $\boldsymbol{x}$, $\varphi_i(\boldsymbol{x})$ $(i = 1, 2, \cdots, m)$ is a violation objective function converted from $g_i(\boldsymbol{x})$, the conversion is stated as:

$$
\varphi_i(\boldsymbol{x}) = max\{g_i(\boldsymbol{x}), 0\}, i = 1, 2, \cdots, m \tag{3}
$$

so the COP is transformed CMOP with m+1 evolution objectives and m constraint conditions.

Obviously, the CMOP in Eq. (2) has the same feasible set and the same optimal solutions as the COP in Eq. (1). Then the CMOP is equivalent to the COP, and therefore, we could solve the COP by the way of solving the CMOP by using a constrained many-objective optimization algorithm.

In multi-objective optimization, Pareto dominance is an essential relation in comparing two solution individuals. Given two solutions $\boldsymbol{x}_1$ and $\boldsymbol{x}_2$, $\boldsymbol{x}_1$ is called **Pareto dominates** $\boldsymbol{x}_2$ if and only if $f_i(\boldsymbol{x}_1) \leq f_i(\boldsymbol{x}_2)$ for every objective index i, and $f_j(\boldsymbol{x}_1) < f_j(\boldsymbol{x}_2)$ for at least one index j. A solution $\boldsymbol{x}^*$ is **Pareto optimal (non-dominated)** solution if there is no solution $\boldsymbol{x}$ such that $f(\boldsymbol{x})$ Pareto dominates $f(\boldsymbol{x}^*)$.

2.2   Convert CMOP to DCMOP

Many-objective evolutionary algorithm (MOEA) in solving CMOP will face the same difficulty of handling constraints as that of EA in solving COP. We know that multi-objective evolutionary algorithm in solving a multi-objective optimization problem (MOP) without constraints performs very well, if we can make the CMOP look MOP without constraints and use MOEA to overcome, we will obtain the optimal resolution. So, the key issue is to achieve a feasible population all the time, which can be addressed by adopting dynamic constraint handling technique.

First, the original constrained boundary $\mathbf{0}$ of the CMOP in Eq. (2) is largely broadened to $e^{(0)}$ at the beginning. Then the broadened boundary $e^{(0)}$ shrinks gradually back to $\mathbf{0}$. Each change of boundary is small enough so that the whole population is always near feasible.

This process constructs a sequence of CMOPs $\{CMOP^{(s)}\}, s = 0, 1, 2, \cdots, S$, i.e., a dynamic constrained many-objective optimization problem (DCMOP) as follows:

$$
\begin{aligned}
COMP^0 &\begin{cases} min\ \mathbf{y} = (f(\mathbf{x}), \varphi_1(\mathbf{x}), \varphi_2(\mathbf{x}), ..., \varphi_m(\mathbf{x})) \\ st : g(\mathbf{x}) \leq e^{(0)} \end{cases} \\
COMP^1 &\begin{cases} min\ \mathbf{y} = (f(\mathbf{x}), \varphi_1(\mathbf{x}), \varphi_2(\mathbf{x}), ..., \varphi_m(\mathbf{x})) \\ st : g(\mathbf{x}) \leq e^{(1)} \end{cases} \\
&............ \\
COMP^S &\begin{cases} min\ \mathbf{y} = (f(\mathbf{x}), \varphi_1(\mathbf{x}), \varphi_2(\mathbf{x}), ..., \varphi_m(\mathbf{x})) \\ st : g(\mathbf{x}) \leq e^{(S)} = \mathbf{0} \end{cases}
\end{aligned} \tag{4}
$$

where $e^{(s)} = (e_1^{(s)}, e_2^{(s)}, ..., e_m^{(s)}), s \in \{0, 1, 2, \cdots, S\}$, $e^{(0)} \geq e^{(1)} \geq \cdots \geq e^{(S)} = \mathbf{0}$.

$e^{(s)}$ is called **elastic constrained boundary**, and s is called **environment state**.

The initial boundary $e^{(0)}$ on the initial state $s = 0$ needs to enable initial population $\mathbf{P}(0)$ feasible, It is set as $e_i^{(0)} = \max_{\mathbf{x} \in \mathbf{P}(0)} \{g_i(\mathbf{x})\}$, $g_i(\mathbf{x})$ is the function value of the ith constraint, $i = 1, 2, ..., m$. On the final state $s = S$, the boundary goes back to $\mathbf{0}$, i.e., $e^{(S)} = \mathbf{0}$. the boundary change on every environment state is modelled as follow:

$$
e_i^{(s)} = A_i e^{-(\frac{s}{B_i})^2} - \varepsilon, i = 1, 2, \cdots, m \tag{5}
$$

Regarding to elastic constrained boundary, if a solution satisfies inequality $g \leq \mathbf{x}$, it is said to be **e-feasible**, otherwise, it is said to be **e-infeasible**. Obviously, a feasible solution is e-feasible, while an e-feasible solution might by infeasible or feasible.

Pareto-domination is defined without considering the constraints, see Subsect. 2.2. An e-constrained Pareto-domination for the DCMOP Eq. (4) is stated as follows:

Given two solutions:

- if both are e-feasible, the one which dominates the other at all objectives (involve the original objective and the violation objectives) wins;
- if one is e-feasible and the other is e-infeasible, the e-feasible solution wins;
- if both are e-infeasible, the one which dominates the other at violation objectives wins.

3   Algorithm Description

This section gives the implementation of many-objective optimization algorithm with dynamic constraints (MaDC) for solving DCMOP.

Algorithm 1. Framework of MaDC

---

step 1 : Initiation

   1.1 Initialize parent population $\mathbf{P}(0) = \{\boldsymbol{x}_1, \boldsymbol{x}_2, ...\boldsymbol{x}_N\}$. Set global generation counter $t = 0$.

   1.2 Initialize elastic constrained boundary $e = e^{(0)}$. Set environment state $s = 0$.

   1.3 Determine reference points $\mathbf{Z}$.

step 2 : Change state

   IF population is e-feasible THEN reduce boundary $e = e^{(s+1)}$, $s = s + 1$.

step 3 : Generate offspring population

   Use DE to generate offspring population $\mathbf{S}(t)$ from $\mathbf{P}(t)$ and evaluate $\mathbf{S}(t)$.

step 4 : Generate next population

   Use reference-point-based nondominated sorting approach to select individuals from combined $\mathbf{S}(t)$ and $\mathbf{P}(t)$ to create next population $\mathbf{P}(t + 1)$.

step 5 : $t = t + 1$, IF s achieves final state S or t achieves $MaxG$, THEN goto *Step 6*, ELSE goto *Step2*.

step 6 : Output results.

MaDC use DE to generate offspring, and reference-point-based nondominated sorting approach to create next population. Reference-point-based nondominated sorting approach is proposed in literature [12], it is an evolutionary many-objective optimization technique of combining nondominated sorting and reference-point-based selection strategy.

Note if the algorithm could not evolve to achieve an e-feasible population on a certain state s, then it would iterate infinitely on this state. A maximal run generation $MaxG$ is set to abort the run.

The generation of offspring population in step 3 of Algorithm 1 is a combination of some genetic operators: affine mutation, crossover and uniform mutation. The detail of genetic operators is as Algorithm 2:

Algorithm 2. Generate offspring procedure

---

input : $\mathbf{P}(t)$, F, CR, P_m

output : Offspring population $\mathbf{S}(t)$

step 1 : $\mathbf{S}(t) = \Phi$.

step 2 : For every individual $\boldsymbol{x}_i \in \mathbf{P}(t), i = 1, 2, 3, ..., N$ do:

    2.1 Affine Mutation:

    $\boldsymbol{v}_i = \boldsymbol{x}_a + F(\boldsymbol{x}_b - \boldsymbol{x}_c)$

    $\boldsymbol{x}_a, \boldsymbol{x}_b, \boldsymbol{x}_c \in \mathbf{P}(t), a \neq b \neq c \neq i$, are selected randomly three individuals.

    2.2 Crossover on $\boldsymbol{x}_i = (x_{i1}, x_{i2}, \cdots, x_{in})$ and $\boldsymbol{v}_i = (v_{i1}, v_{i2}, \cdots, v_{in})$:

$$u_{ij} = \begin{cases} v_{ij} & if \ r_{rnd} < CR \ or \ j = j_{rnd} \\ x_{ij} & if \ r_{rnd} \geq CR \ or \ j \neq j_{rnd} \end{cases}$$

$$j = 1, 2, ..., n$$

$$j_{rnd} = rndInt(1, n), r_{rnd} = rndReal(0, 1)$$

    2.3 Uniform Mutation on $\boldsymbol{u}_i = (u_{i1}, u_{i2}, \cdots, u_{in})$:

    Change $u_{ij} = rndReal(0, 1)$ with probability P_m for $j = 1, 2, ..., n$.

step 3 : Add each $\boldsymbol{u}_i(i = 1, 2, 3, ..., N)$ into $\mathbf{S}(t)$.

step 4 : Return $\mathbf{S}(t)$.

Algorithm 3 was given as the details of creating next parent population. It uses reference-point-based nondominated sorting method to select individuals to construct the next population.

Algorithm 3. Generate next population

---

input : $\mathbf{Q}(t)$, $\mathbf{Z}$, N

output : $\mathbf{N}(t)$

step 1 : $\mathbf{N}(t) = \Phi$. $(\mathbf{F}_1, \mathbf{F}_2, \cdots)$ = Non-dominated-sort($\mathbf{Q}(t)$). $i = 1$.

step 2 : IF $|\mathbf{N}(t)| + |\mathbf{F}_i| < N$, THEN $\mathbf{N}(t) = \mathbf{N}(t) \cup \mathbf{F}_i$, $i = i + 1$, goto *Step 2*;

    IF $|\mathbf{N}(t)| + |\mathbf{F}_i| = N$, THEN $\mathbf{N}(t) = \mathbf{N}(t) \cup \mathbf{F}_i$, goto *Step 4*;

    IF $|\mathbf{N}(t)| + |\mathbf{F}_i| > N$, THEN goto *Step 3*.

step 3 : Select $N - |\mathbf{N}(t)|$ individuals from $\mathbf{F}_i$ and add them into $\mathbf{N}(t)$:

    3.1 Associate each solution of $\mathbf{N}(t)$ with closest reference point by the perpendicular distance, compute the niche count of each reference point.

    3.2 Select randomly a point $\mathbf{r}$ which have smallest niche count.

    3.3 Let $\mathbf{I_r}$ be a set of individuals associated with $\mathbf{r}$, $\mathbf{I_r} \subseteq \mathbf{F}_i$. IF $|\mathbf{I_r}| = \Phi$, THEN remove $\mathbf{r}$ from $\mathbf{Z}$ temporarily at this generation, goto *3.2*;

    ELSE: IF nichecount($\mathbf{r}$) = 0, THEN select the member s which has smallest perpendicular distance to $\mathbf{r}$; IF nichecount($\mathbf{r}$) $\neq$ 0, THEN select a member s randomly from $\mathbf{I_r}$

    3.4 Add the selected member s into $\mathbf{N}(t)$, nichecount($\mathbf{r}$)= nichecount($\mathbf{r}$)+1, $\mathbf{F}_i = \mathbf{F}_i \setminus s$.

    3.5 Repeat the selection above, until all $N - |\mathbf{N}(t)|$ individuals are chosen.

step 4 : Return $\mathbf{N}(t)$.

---

4   Experiments and Results

In this section, we apply our proposed methodology to a number of benchmark problems proposed in Problem Definitions and Evaluation Criteria for the CEC 2006 Special Session on Constrained Real-Parameter Optimization [13], online available: http://www.ntu.edu.sg/home/epnsugan/. The 24 test instances are minimization problems. The detail of the test problems refers to [13].

4.1   Determination of Reference Points and Algorithm Parameters

The proposed algorithm uses a predefined set of reference points to ensure diversity of many-objective optimization. We use determination method presented in [12] that places points on a normalized hyper-planean $(M-1)$-dimensional unit simplex—which is equally inclined to all objective axes and has an intercept of one on each axis. If p divisions are considered along each objective, the total number of reference points (H) in an M-objective problem is given by:

$$H = C^p_{M+p-1} \tag{6}$$

For example, in a three-objective problem $(M = 3)$, if six divisions $(p = 6)$ are chosen for each objective axis, $H = 28$ reference points will be created [12]. When there are many objectives $(M \geq 5)$, one layer of reference points is not appropriate. For eight-objective problems, even if we use $p = 8$ (to have exactly one intermediate reference point), it requires 5040 reference points. To avoid such a situation, we use two layers of reference points (boundary layer and inside layer) in many-objective problems.

The population size N is set the number of reference points. Table 1 shows the number of chosen reference points (H) and corresponding population sizes.

Table 1. Number of reference points and corresponding population size

M	p of boun.	p of insi.	H	popsize	M	p of boun.	p of insi.	H	popsize
2	90	0	91	91	8	2	2	72	72
3	12	0	91	91	9	2	2	90	90
4	6	0	84	84	10	2	2	110	110
5	4	2	85	85	14	2	1	119	119
6	3	2	77	77	39	1	1	78	78
7	3	1	91	91	–	–	–	–	–

Other parameters are as follow:
Number of repeats: 25.

In the offspring generation procedure (Algorithm 2), the scaling factor $F = 0.5$, crossover rate $CR = 0.9$, uniform mutation probability $P_m = 0.01$.

Table 2. Function values obtained by MADC, SAMO-DE, ECHT-EP2, DE-DPS, HCOEA and DCMOEA

Pro.	Crit.	MaDC	SAMO-EA	ECHT-EP2	DE-DPS	HCOEA	DCMOEA
g01	best	**−15.0000**	−15.0000	−15.0000	−15.0000	−15.0000	−15.0000
	Avg.	**−15.0000**	−15.0000	−15.0000	−15.0000	−15.0000	−15.0000
g02	best	**−0.8036191**	−0.8036191	−0.8036191	−0.8036190	−0.803241	−0.8036191
	Avg.	−0.8010908	−0.7987352	−0.7998220	−0.8036189	−0.801258	−0.7969470
g03	best	**−1.0005**	−1.0005	−1.0005	−1.0005	−1.0005	−1.0005
	Avg.	**−1.0005**	−1.0005	−1.0005	−1.0005	−1.0005	−1.0005
g04	best	**−30665.5386**	−30665.5386	−30665.5386	−30665.5386	−30665.5386	−30665.5386
	Avg.	**−30665.5386**	−30665.5386	−30665.5386	−30665.5386	−30665.5386	−30665.5386
g05	best	**5126.4967**	5126.497	5126.497	5126.497	5126.498	5126.498
	Avg.	**5126.4967**	5126.497	5126.497	5126.497	5148.960	5126.498
g06	best	**−6961.8138**	−6961.8138	−6961.8138	−6961.8138	−6961.8138	−6961.8138
	Avg.	**−6961.8138**	−6961.8138	−6961.8138	−6961.8126	−6961.8138	−6961.8138
g07	best	**24.3062**	24.3062	24.3062	24.3062	24.3062	24.3062
	Avg.	24.3063	24.3096	24.3063	24.3062	24.307	24.3064
g08	best	**−0.095825**	−0.095825	−0.095825	−0.095825	−0.095825	−0.095825
	Avg.	**−0.095825**	−0.095825	−0.095825	−0.095825	−0.095825	−0.093491
g09	best	**680.630**	680.630	680.630	680.630	680.630	680.630
	Avg.	**680.630**	680.630	680.630	680.630	680.630	680.630
g10	best	**7049.249**	7049.249	7049.249	7049.248	7049.287	7049.248
	Avg.	7049.304	7059.813	7049.249	7059.248	7049.525	7049.248
g11	best	**0.7499**	0.7499	0.7499	0.7499	0.750	0.75
	Avg.	**0.7499**	0.7499	0.7499	0.7499	0.750	0.75
g12	best	**−1.000**	−1.000	−1.000	−1.000	−1.000	−1.000
	Avg.	**−1.000**	−1.000	−1.000	−1.000	−1.000	−1.000
g13	best	**0.05394**	0.05394	0.05394	0.05394	0.05395	0.05395
	Avg.	**0.05394**	0.05394	0.05394	0.81702	0.05395	0.05395
g14	best	**−47.76488**	−47.76488	−47.7649	−47.76488	–	–
	Avg.	−47.76395	−47.68115	−47.7648	−47.76488	–	–
g15	best	**961.71502**	961.71502	961.71502	961.71502	–	–
	Avg.	**961.71502**	961.71502	961.71502	962.13142	–	–
g16	best	**−1.905155**	−1.905155	−1.905155	−1.905155	–	–
	Avg.	**−1.905155**	−1.905155	−1.905155	−1.905155	–	–
g17	best	**8853.5338**	8853.5397	8853.5397	8862.6287	–	–
	Avg.	**8853.5338**	8853.5397	8853.5397	8934.8675	–	–
g18	best	**−0.866025**	−0.866025	−0.866025	−0.866025	–	–
	Avg.	−0.866024	−0.866024	−0.866025	−0.866025	–	–
g19	best	**32.65559**	32.65559	32.6591	32.65559	–	–
	Avg.	32.65564	32.75734	32.6623	32.65559	–	–
g21	best	**193.72451**	193.72451	193.7246	193.72451	–	–
	Avg.	**193.72451**	193.77137	193.7348	193.72451	–	–
g23	best	−400.0451	−396.1657	−398.9731	−400.0551	–	–
	Avg.	−395.8492	−360.8176	−373.2178	−395.6745	–	–
g24	best	**−5.508013**	−5.508013	−5.508013	−5.508013	–	–
	Avg.	**−5.508013**	−5.508013	−5.508013	−5.508013	–	–

The ε in Eq. 5 was set to 0.000 000 1.

The number of environment changes was set $S = 240000/N$.

The maximal run generation $MaxG = 10000$, if a problem has no feasible solutions or the algorithm could not find feasible solutions, the algorithm would abort after evolving 10 000 generations. Problems g20 and g22 could not find feasible solution.

4.2   Results and Comparison

The detailed results of MaDC are provided in Table 2, along with that of the state-of-the-art algorithms such as: (1) self-adaptive multioperator differential evolution (SAMO-DE) [14]; (2) ensemble of constraint handling techniques based on evolutionary programming (ECHT-EP2) [15]; (3) differential evolution with dynamic parameters selection (DE-DPS) [3]; (4) hybrid constrained optimization evolutionary algorithm (HCOEA) [8]; (5) dynamic constrained multi-objective evolutionary algorithm (DCMOEA) [10]. All algorithms solved 22 test problems, except HCOEA and DCMOEA, in which only the 13 test problems were solved.

From Table 2, MaDC was able to obtain the optimal solutions for all problems except g23. The algorithm SAMO-DE, ECHT-EP2, DE-DPS were able to obtain the optimal solutions for 20, 19, 20 problems, respectively. The algorithm HCOEA and DEMOEA obtained the optimal solutions for 9, 12 out of 13 problems. In regard to the average results, MaDC is superior to SAMO-DE, ECHT-EP2, DE-DPS, HCOEA and DEMOEA for eight, four, three, three, two text problems, respectively. It can be seen that our proposed method performs better than or is competitive to state-of-the-art algorithms.

5   Conclusion

In this paper, we have suggested a many-objective optimization algorithm with dynamic constraint mechanism for solving constrained optimization problem. We first construct an equivalent dynamic constrained many-objective optimization problem to the COP, then adopt MaDC algorithm to solve the DCMOP, thus the COP is solved. Dynamic technology is implemented by setting an elastic boundary for the constrained problem, and the trade-off between the population diversity and accuracy is mainly handled by reference-point-based nondominated sorting method. The proposed algorithm is tested by a number of benchmark problems. Experimental results show that it is competitive to state-of-the-art algorithms referred in this paper. The future work should be: (1) Retaining more feasible solutions to improve the performance of the algorithm in each evolutionary generation by adopting other selection strategy; (2) Introducing other better many-objective optimization technique in the algorithm; (3) Using dynamic parameters selection mechanism in DE to speed up the convergence of the algorithm; (4) To explore other candidates of the dynamic environment.

Acknowledgment. This work was supported by the National Natural Science Foundation of China and other foundations(No.s: 61271140, 61203306, 2012001202, 61305086).

References

1. Mezura-Montes, E., Coello, C.A.C.: Constraint handling in nature-inspired numerical optimization: Past, present and future. Swarm Evol. Comput. **1**(4), 173–194 (2011)
2. Kramer, O.: A review of constraint-handling techniques for evolution strategies. In: Applied Computational Intelligence and Soft Computing, vol. 2010 (2010)
3. Sarker, R.A., Elsayed, S.M., Ray, T.: Differential evolution with dynamic parameters selection for optimization problem. IEEE Trans. Evol. Comput. **18**(5), 689–707 (2014)
4. Coello, C.A.C.: Constraint-handling using an evolutionary multi-objective optimization technique. Civil Eng. Environ. Syst. **17**, 319–346 (2000)
5. Coello, C.A.C., Mezura-Montes, E.: Constraint-handling in genetic algorithms through the use of dominance-based tournament election. Adv. Eng. Inform. **16**(3), 193–203 (2002)
6. Venkatraman, S., Yen, G.G.: A generic framework for constrained optimization using genetic algorithms'. IEEE Trans. Evol. Comput. **9**(4), 424–435 (2005)
7. Hsieh, M., Chiang, T., Fu, L.: A hybrid constraint handling mechanism with differential evolution for constrained multiobjective optimization. In: IEEE Congress on Evolutionary Computation, pp. 1785–1792 (2011)
8. Wang, Y., Cai, Z., Guo, G., Zhou, Y.: Multiobjective optimization and hybrid evolutionary algorithm to solve constrained optimization problems. IEEE Trans. Syst. Man Cybern. **37**(3), 560–575 (2007)
9. Deb, K., Datta, R.: A fast and accurate solution of constrained optimization problems using a hybrid bi-objective and penalty function approach. In: IEEE Congress on Evolutionary Computation, pp. 165–172 (2010)
10. Zeng, S., Chen, S., Zhao, J., Zhou, A., Li, Z., Jing, H.: Dynamic constrained multi-objective model for solving constrained optimization problem. In: IEEE Congress on Evolutionary Computation, pp. 2041–2046 (2011)
11. Li, X., Zeng, S., Qin, S., Liu, K.: Constrained optimization problem solved by dynamic constrained NSGA-III multiobjective optimizational techniques. In: IEEE Congress on Evolutionary Computation, pp. 2923–2928 (2015)
12. Deb, K., Jain, H.: An evolutionary many-objective optimization algorithm using reference-point-based nondominated sorting approach, part i: solving problems with box constraints. IEEE Trans. Evol. Comput. **18**(4), 577–601 (2014)
13. Liang, J.J., Runarsson, T.P., Mezura-Montes, E., Clerc, M., Suganthan, P.N., Coello, C.A.C., Deb, K.: Problem definitions and evaluation criteria for the CEC2006 special session on constrained real-parameter optimization (2006). http://www.ntu.edu.sg/home/epnsugan/
14. Elsayed, S.M., Sarker, R.A., Essam, D.L.: Multi-operator based evolutionary algorithms for solving constrained optimization problems. Comput. Oper. Res. **38**(12), 1877–1896 (2011)
15. Mallipeddi, R., Suganthan, P.N.: Ensemble of constraint handling techniques. IEEE Trans. Evol. Comput. **14**(4), 561–579 (2010)

Executing Time and Cost-Aware Task Scheduling in Hybrid Cloud Using a Modified DE Algorithm

Yuanyuan Fan, Qingzhong Liang$^{(\boxtimes)}$, Yunsong Chen, Xuesong Yan,
Chengyu Hu, Hong Yao, Chao Liu, and Deze Zeng

School of Computer Science, China University of Geosciences, Wuhan 430074, China
{yyfan,qzliang,yunschen,yanxs,huchengyu,yaohong,
liuchao,dzzeng}@cug.edu.cn

Abstract. Task scheduling is one of the basic problem on cloud computing. In hybrid cloud, tasks scheduling faces new challenges. In order to better deal the multi-objective task scheduling optimization in hybrid clouds, on the basis of the GaDE and Pareto optimum of quick sorting method, we present a multi-objective algorithm, named NSjDE. This algorithm also makes considerations to reduce the frequency of evaluation Comparing with experiment of Min-Min algorithm, GaDE algorithm and NSjDE algorithm, results show that for the single object task scheduling, GaDE and NsjDE algorithms perform better in getting the approximate optimal solution. The optimization speed of multi-objective NSjDE algorithm is faster than the single-objective jDE algorithm, and NSjDE can produce more than one non-dominated solution meeting the requirements, in order to provide more options to the user.

Keywords: Hybrid cloud · Task scheduling · Executing time-aware · Cost-aware

1 Introduction

Hybrid is the composition of private cloud and public cloud. It allows organizations who maintain their private computing infrastructure (private cloud) has a higher utilization rate Achieve the goal of optimization between computing needs and cost. In hybrid cloud, tasks scheduling faces new challenges. First of all, hybrid cloud often contains a huge number of cloud computing node, so that task scheduling is NP for a large-scale optimization problems. Secondly, the differences between lots of users in hybrid cloud, lead to variety of tasks. Different tasks are real-time bandwidth-aware, computing cost-aware, QoS-aware, or both, or above of all. In order to better deal the multi-objective task scheduling optimization in hybrid clouds, on the basis of the GaDE, combining with the non-dominated sorting algorithm, NSGA-II, based on Pareto optimum of quick sorting method, we present a multi-objective algorithm, named NSjDE algorithm. This algorithm also makes considerations to reduce the frequency of

© Springer Science+Business Media Singapore 2016
K. Li et al. (Eds.): ISICA 2015, CCIS 575, pp. 74–83, 2016.
DOI: 10.1007/978-981-10-0356-1_8

evaluation Comparing with experiment of Min-Min algorithm, GaDE algorithm and NSjDE algorithm, results show that for the single object task scheduling, GaDE and NSjDE algorithms perform better in getting the approximate optimal solution. The optimization speed of multi-objective NSjDE algorithm is faster than the single-objective jDE algorithm, and NSjDE can produce more than one non-dominated solution meeting the requirements, in order to provide more options to the user.

2 Related Work

In the hybrid cloud system, task scheduling has new challenges. First of all, there is a lot of computing nodes involved in hybrid, so that scheduling is a large-scale NP optimization problem. In addition, due to the large differences between users on hybrid cloud, tasks they submitted have a wide variety of performance demands, such as one or more of bandwidth, executing time, and computing cost, etc. [1, 2].

Some scholars solute the scheduling problem as a kind of knapsack problem where computing and storage resources are regarded as receiving space and tasks as cargoes. They use greedy algorithm to find the optimization scheduling goals equivalent to achieving a maximum cargo capacity in the limited knapsack space. In fact, since task scheduling problem has been proved to be a NP hard problem, it is usually converted to a Multiple choice Multi-dimensional Knapsack ProblemMMKP, and finds a approximate optimal solution based on heuristic algorithms, such as Particle Swarm Optimization algorithm, PSO [3], Simulated Annealing algorithm, SA [4] and Genetic algorithm, GA [5]. In a recent work, some scholars solve the tasks scheduling problem as a hard constrained multi-objective problems [6], and using genetic algorithms to solve it. They get a global approximate optimal solution by GA considering the various demands and hard constraint conditions of tasks.

Wei Liu etc. propose a novel two-phase Adaptive Energy-efficient Scheduling (AES) [7], which combines the Dynamic Voltage Scaling (DVS) technique with the adaptive task duplication strategy. The AES algorithm justifies threshold automatically, thus improving the system flexibility. In the first phase, they propose an adaptive threshold-based task duplication strategy, which can obtain an optimal threshold. It then leverages the optimal threshold to balance schedule lengths and energy savings by selectively replicating predecessor of a task. Therefore, the proposed task duplication strategy can get the suboptimal task groups that not only meet the performance requirement but also optimize the energy efficiency. In the second phase, it schedules the groups on DVS-enabled processors to reduce processor energy whenever tasks have slack time.

To improve the overall performance of cloud computing, with the deadline constraint, a task scheduling model [8] is established for reducing the system power consumption of cloud computing and improving the profit of service providers. For the scheduling model, a solving method based on multi-objective genetic algorithm (MO-GA) is designed and the research is focused on encoding rules, crossover operators, selection operators and the method of sorting Pareto solutions.

3 Improved DE Algorithm for Hybrid Cloud Scheduling

3.1 Scheduling Modeling

In the proposed model, the users submit jobs to the hybrid cloud, and each job consists of several tasks. Several tasks can be handled by a resource host, which is usually constructed as a virtual machine. The definitions for a host and a task are introduced as follows.

Host j is denoted as H_j, represents a computing resource slot in hybrid cloud. In our model, a host may consist of five attributions, such as $H_j = \{Mips_j, Mem_j, Stg_j, Band_j, Cost_j\}$.

$Mips_j$ represents the computing power of host j which is measured by MIPS (Million Instructions per Second).

Mem_j represents the memory size of host j which is available to executing tasks.

Stg_j represents the storage capacity of host j which is the available disk space for data and tasks storing during executing tasks.

$Band_j$ represents the available network bandwidth for host j.

$Cost_j$ represents the computing cost of host j. It is the sum of data storing cost, I/O cost and device renting cost.

Task i is denoted as T_i. It is a basic unit of scheduling and can be submitted to any host in hybrid cloud. T_i is a six tuple, $T = \{tD_i, tW_i, tL_i, tM_i, tB_i, tC_i\}$, where:

tD_i represents the deadline of task i, that is, the latest completion time when the task is completed. The argument as task scheduling to meet the constraints parameters of QoS.

tW_i represents the workload of task i, that is, the data size that produced in task execution process. You can simply expressed as the number of instructions contained in the task. This parameter also indicates that the task on the host required storage space.

tL_i represents the work length of task i, that is, the size of the task assignments, you can simply expressed as the number of instructions included in the task.

tM_i represents the data cache requirements of task i, that is, the maximum amount of memory should consumed during the execution of tasks.

tB_i represents the bandwidth requirements of task i, that is, the maximum network bandwidth need to consume during the execution of the tasks.

tC_i represents the pricing calculation of task i, that is, the maximum acceptable cost ceiling that tasks required in the implementation process. The argument as task scheduling parameters to meet the constraints of cost control user needs.

Suppose there are N tasks required on hybrid cloud environments M hosts scheduling. Five tuple, $\{Mips_j, Mem_j, Stg_j, Band_j, Cost_j\}$ represents the jth host with five dimensions of property, such as computing power, the capacity of computing buffer, data storage capacity, network bandwidth and computational cost.

Six tuple, $\{tD_i, tW_i, tL_i, tM_i, tB_i, tC_i\}$, represents the ith task with six dimensions of property, such as deadline, workload, work length, data cache requirements, bandwidth requirements, pricing calculation. The mathematical model about task scheduling problem can be expressed as formulas 1 and 2:

$$\max \sum_{i=1}^{N} \sum_{j=1}^{M} v_j x_{ij} \tag{1}$$

$$s.t = \begin{cases} \sum_{i=1}^{N} a_i x_{ij} \leq Mem_j \\ \sum_{i=1}^{N} b_i x_{ij} \leq Stg_j \\ \sum_{i=1}^{N} tB_i x_{ij} \leq Band_j \\ x_{ij} = 1 \ or \ 0 \end{cases} \tag{2}$$

where:

$i = 1, 2, ..., n$, and $j = 1, 2, ..., m$. Parameter a_i and b_i is defined as formulas 3 and 4:

$$a_i = tW_i + tM_i \tag{3}$$

$$b_i = tW_i + tL_i \tag{4}$$

a_i represents consuming memory during execution b_i represents machine disk space during execution. v_{ij} Represents scheme benefits of scheduling about assign task i to host j. Depending on the user's target, which is defined as a single target sensitive and the smallest total bandwidth consumption or the shortest total execution time, also can be defined as multiple targets simultaneously met. x_{ij} Represents the ith task whether can be assigned to the host j, you can use the formula expressed as formula 5:

$$x_{ij} = \begin{cases} 0, \text{task } i \text{ is not assigned to host } j \\ 1, \text{task } i \text{ is assigned to host } j \end{cases} \tag{5}$$

In addition, in the task assignment process, if you need to consider the latest finish time and calculate the cost of the case, each of examples of task assignments also exist two constraints that execution time and computing costs. Specific expressed as: Time of Executing (ToE) as the sum of Trans Time (time cost that tasks transfer to a host) and Runtime (time cost that host execute all task instruction). ToE must be limited to the task within the deadline, namely as formula 6:

$$ToE_{ij} = TransTime_{ij} + RunTime3 - 10_{ij} = \frac{tL_i}{Band_j} + \frac{tL_i}{Mips_j} \leq tD_i \tag{6}$$

Cost of Computing (CoC) is calculated in accordance with the length of time it takes up on a host, It must be less than the task of computing the expected pricing, namely as formula 7:

$$CoC_{ij} = RunTime_{ij} \times Cost_j = \frac{tL_i \times Cost_j}{Mips_j} \leq tC_i \qquad (7)$$

Thus, task scheduling problem can be transformed into a multi-dimensional knapsack problem, and this problem is a NP-complete problem in combinatorial optimization. So for solving this problem, especially when relatively large scale of the problem, it becomes difficult to get the optimal solution in limit time.

3.2 Modifying jDE Algorithm - GaDE

Because hybrid cloud complex task scheduling problem, computationally intensive, time consuming very long, which is the speed and parameter selection algorithm presented a great challenge. In this regard, the article attempts to improve the work from two aspects: the observed differential evolution algorithm basic crossover is able to generate two offspring, and work a discarded child individual also added to produce a new individual tournament selection, with the least number of simulations to retain the most effective genetic information; the same time, because the time to evaluate a scheduling scheme is very long, you want to rely on a lot of experiments to get the experience of control parameters is very difficult, so in this study, to borrow the jDE algorithm automatically adjusts the control parameters thinking, and its improvements.

As well known, those evolution algorithm like jDE are based on randomly adapt strategy, but uncertainty on such strategy may lead to great impacts on parameters. In this paper, we propose a new adaptive DE algorithm GaDE base on deterministic mechanism.

Modification on Cross Operation and Selection Strategy. In DE, only one of two offspring successfully enters the next generation, which definitely making convergence slow. To fix it, we add the discarded one into competition of new offspring creation.

Formula of creating two offspring $v_i(t+1)$ and $u_i(t+1)$ as formulas 8 and 9:

$$v_{ij}(t+1) = \begin{cases} h_{ij}(t+1), \; if(rand1_{ij} \leq CR)or(j == rand(i)) \\ x_{ij}(t), \quad otherwise \end{cases} \qquad (8)$$

$$u_{ij}(t+1) = \begin{cases} x_{ij}(t), \quad if(rand1_{ij} \leq CR)or(j == rand(i)) \\ h_{ij}(t+1), \, otherwise \end{cases} \qquad (9)$$

To speed up algorithm evaluation, new individual selection adopted as following, at first, offspring $v_i(t+1)$ competes with target individual $x_i(t)$, if $v_i(t+1)$ perform better, than it will be saved. Otherwise, calculate $u_i(t+1)$ and then compete with $x_i(t)$. In this way, more genetic information will enter the next generation. For an example, using minimize function f, formula of selection algorithm is described as formula 10:

$$x_i(t+1) = \begin{cases} v_i(t+1), \; if(f(v_i(t+1)) < f(x_i(t))) \\ u_i(t+1), \; elseif(f(u_i(t+1)) < f(x_i(t))) \\ x_i(t), \quad otherwise \end{cases} \qquad (10)$$

Modification of Control Parameters. jDE adopt random control parameters adjustment. As a matter of fact, a good offspring generated from father generation means that the current scale factor and crossover rate is perfect, and they has no need to change. Meanwhile, because of adjustment of crossover operation and selection strategy, scale factor F and crossover CR has to change its specific modification way, as formula 11:

$$F_i(t+1) = \begin{cases} F_l + rand_1 \times F_u, \ if(x_i(t) == x_i(t+1) \ and \ rand_2 < \tau_1) \\ F_i(t), \qquad\qquad\qquad otherwise \end{cases} \tag{11}$$

where $rand_j$, $j \in \{1, 2, 3, 4\}$, is a random decimal in $[0, 1]$; τ_1 τ_2 stands for probability of F and CR. Generally, $F_l = 0.1$, $F_u = 0.9$, $F \in [0.1, 1.0]$, $CR \in [0, 1]$, and $\tau_1 = \tau_2 = 0.1$.

Only when the two offspring behave better than its father objective, control parameters will adjust according to reserved probability. Otherwise, the old parameters enter into next generation without changing.

3.3 Multi-objective jDE Algorithm Based on Non-dominated Sorting - NSjDE

Introducing the non-dominated sorting concept of NSGA-II to improved jDE algorithm, it will be converted to multi-objective evolutionary algorithm, which is mainly changed in some mutation and selection policies. Fast non-dominated sorting method NSGA-II algorithm uses the partial order. In order to make NSjDE be able to handle multiple targets with constrained optimization problem, we show a partial order $\prec_{pBetter}$ aiming at constrained optimization problem for multi-objective, and replace $\prec$ order fast in non-dominated sorting method.

Constraint Pareto dominant in NSjDE is defined as follow:

If $(V(x_u) < V(x_v) or (V(x_u) == V(x_v)$ and $x_u \prec x_v)$, then decision vector $x_u \in \Omega$ constraints dominant decision vector $x_v \in \Omega$, denoted as $x_u \prec_{pBetter} x_v$.

4   Experimental Performance Results

4.1   Design of the Experiments

In order to verify the improved GaDE algorithm with NSjDE algorithm effectiveness of task scheduling in the hybrid cloud, they are compared with the classic Min - Min algorithm and standard evolution algorithm. Experiment on the CloudSim respectively simulates the 40 cloud computing resource nodes. Ten of the nodes are to simulate the private cloud, which computation cost is zero. The other 30 nodes, are respectively to simulate different bandwidth of public cloud nodes. There are 10 nodes with the bandwidth of 5 Mbps, 10 nodes with 10 Mbps, and 10 nodes with 100 Mbps. The specific parameter configuration is shown in Table 1.

Table 1. Hybrid cloud simulation environment configuration tables

Cloud type	Hosts	MIPS	Mem	Stg	Band	Cost
Private clouds	10	$rand(300, 500)$	$rand(300, 500)$	$rand(3k, 5k)$	100	0
Public clouds(L)	10	$rand(300, 500)$	$rand(300, 500)$	$rand(3k, 5k)$	5	0.1
Public clouds(M)	10	$rand(1k, 2k)$	$rand(1k, 2k)$	$rand(10k, 20k)$	10	0.4
Public clouds(H)	10	$rand(2k, 4k)$	$rand(2k, 4k)$	$rand(20k, 40k)$	100	1

4.2 Experimental Results

The number of arrived task respectively is 50, 100, 150, 100, 150, and the goal of scheduling is the shortest completion time and minimum computational cost. Comparing with the Min - Min algorithm, the standard evolutionary algorithm, GaDE algorithm and NSjDE algorithm, the scheduling experiment result is shown in Figs. 1 and 2:

As we can see from Fig. 1, since the Min - Min algorithm schedule the task aiming to the shortest completion time, so, as a result, there is a little difference in the performance of completion time for the four algorithm. However, heuristic algorithm can get a better solution. In theory, the GA algorithm and GaDE, NSjDE can both get the optimal solution (can be seen from the diagram, the optimal results of GaDE and NSjDE are basically identical), but in the actual performance, the classical GA convergence speed is slow, so the result with a limited algebra is worse than the result in GaDE and NSjDE algorithm, if increase the number of iterations, may be able to get closer solution comparing with the GaDE and NSjDE algorithm.

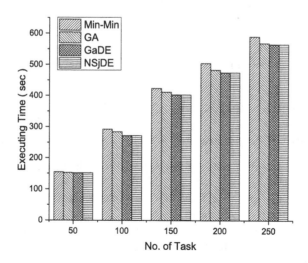

Fig. 1. The completion time scheduled by four algorithms.

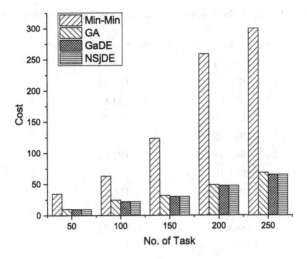

Fig. 2. The cost scheduled by four algorithms.

As we can see from Fig. 2, since the Min - Min algorithm don't care about cost in the process of task scheduling, only concern on the shortest complete of a single task. Therefore, in the process of scheduling, Min-Min algorithm may be more inclined to make task assigned to the high performance of resource host, which cause high cost. When use the classical GA, GaDE and NSjDE algorithm, under the constraints of deadline, tasks (if do not set the deadline constraints, all tasks will be assigned to the private cloud, so the computation cost is 0) are assigned as far as possible to the low cost of the machine, so can effectively maintain cost at a lower level. In the same way, the scheduling results of GaDE close to NSjDE algorithm, and is better than classical GA algorithm. However, the completion time of task scheduling is worse than Min - Min algorithm.

As we can see from the above experiments, the Min-Min algorithm, the classical GA algorithm and GaDE algorithm is aimed at a optimization of single objective problem, cannot solve the multi-objective optimization problem. Although can through the method of combine or aggregate all subgoals, the multi-objective optimization problem is converted into a single objective optimization problem, but in such practices, in order to get better result, you need to do further optimization in the accumulation function of weights, thus makes the problem more complicated. Aiming at this problem, using multi-objective method NSjDE to make multi-objective task scheduling experiment. In experiment, we choose 500 random tasks, with the goal of minimizing the completion time and minimization calculation cost. Repeating experiments after 10 times, the corresponding statistical data is shown in Table 2.

The results can be seen from the Table 2, the use of multi-objective algorithm NSjDE based on non-dominated sorting, can greatly speed up the solving multi-objective optimization problem speed, which for task scheduling based hybrid cloud is very important; at the same time, since the algorithm is based Pareto optimal, so an operation may simultaneously obtain a plurality of non-inferior solutions, to provide users with more choices.

Table 2. NSjDE best result of task scheduling evaluation data

No.	Task completion time	Computing costs
01	1757.21	176.14
02	1876.83	155.98
03	1525.50	124.81
04	1910.99	143.06
05	1916.97	141.234
06	1564.18	175.87
07	1991.19	172.01
08	1946.08	116.17
	1910.75	129.61
09	1863.58	160.71
	1779.96	172.96
10	1918.91	136.51
	1916.07	138.85

5   Conclusions

This paper propose evolutionary algorithms to solve the hybrid cloud task scheduling problems. At the first, a differential evolution algorithm for single target task scheduling –GaDE algorithm is proposed with the crossover operation, select the policy and controls adjustment parameters on jDE algorithm have been improved. In order to better deal with multi-objective scheduling, the second work of this paper is to propose a multi-objective scheduling algorithm, name NSjDE algorithm, based on the combination of non-dominated sorting algorithm according to Pareto optimal from NSGA-II. Experimental results show that, the multi-objective NSjDE algorithm is obviously superior to the single objective jDE algorithm in the optimization speed, and may generate a plurality of non-inferior solutions to meet the requirements in a single run.

Acknowledgment. The work was partially supported by Project 61501412 supported by National Natural Science Foundation of China.

References

1. Goudarzi, H., Ghasemazar, M., Pedram, M.: Sla-based optimization of power and migration cost in cloud computing. In: 2012 Conference Proceedings on 12th IEEE/ACM International Symposium on Cluster, Cloud and Grid Computing (CCGrid), pp. 172–179. IEEE (2012)

2. Kumar, B.A., Ravichandran, T.: Time and cost optimization algorithm for scheduling multiple workflows in hybrid clouds. Eur. J. Sci. Res. **89**(2), 265–275 (2012)
3. Xue, S.-J., Wu, W.: Scheduling workflow in cloud computing based on hybrid particle swarm algorithm. TELKOMNIKA Indonesian J. Electr. Eng. **10**(7), 1560–1566 (2012)
4. Xu, X., Hu, N., Ying, W.Q.: Cloud task and virtual machine allocation strategy based on simulated annealing-genetic algorithm. Appl. Mech. Mater. **513**, 391–394 (2014)
5. Sellami, K., Ahmed-Nacer, M., Tiako, P.F., Chelouah, R.: Immune genetic algorithm for scheduling service workflows with qos constraints in cloud computing. S. Afr. J. Ind. Eng. **24**(3), 68–82 (2013)
6. Yassa, S., Sublime, J., Chelouah, R., Kadima, H., Jo, G., Granado, B.: A genetic algorithm for multicobjective optimisation in workflow scheduling with hard constraints. Int. J. Metaheuristics **2**(4), 415–433 (2013)
7. Liu, W., Du, W., Chen, J., Wang, W., Zeng, G.: Adaptive energy-efficient scheduling algorithm for parallel tasks on homogeneous clusters. J. Netw. Comput. Appl. **41**, 101–113 (2013)
8. Liu, J., Luo, X.-G., Zhang, X.-M., Zhang, F., Li, B.-N.: Job scheduling model for cloud computing based on multi-objective genetic algorithm. Int. J. Comput. Sci. Issues (IJCSI) **10**(1), 134–139 (2013)

A Novel Differential Evolution Algorithm Based on JADE for Constrained Optimization

Kangshun Li[1,2], Lei Zuo[1(✉)], Wei Li[1,2], and Lei Yang[1]

[1] College of Mathematics and Informatics, South China Agricultural University,
Guangzhou 510642, China
853483687@qq.com
[2] School of Information Engineering, JiangXi University of Science
and Technology, Ganzhou 341000, China

Abstract. To overcome the problem of slow convergence and easy to be plunged to premature when the traditional differential evolution algorithm for solving constrained optimization problems, a novel differential evolution algorithm (CO-JADE) based on adaptive differential evolution (JADE) for constrained optimization was proposed. The algorithm used skew tent chaotic mapping to initialize the population, generated the crossover probability of each individual according to the normal distribution and the Cauchy distribution and the mutation factor according to the normal distribution. CO-JADE used improved adaptive tradeoff model to evaluate the individuals of population. The improved adaptive tradeoff model used different treatment scheme for different stages of population, which aimed to effectively weigh the relationship between the value of the objective function and the degree of constraint violation. Simulation experiments were conducted on the night standard test functions. CO-JADE was much better than COEA/ODE and HCOEA in terms of the accuracy and standard variance of final solution. The experimental results demonstrate that the CO-JADE has better accuracy and stability.

Keywords: Adaptive differential evolution · Adaptive tradeoff model · Constrained optimization · Chaotic mapping

1 Introduction

Constrained optimization problem is often appear in the application of the science and engineering a mathematical programming problem, the first based on numerical linear and nonlinear programming method to solve, but the algorithm has significant limitations. In order to avoid and solve the problems of the existing numerical methods, the researchers focused on the study of the heuristic algorithm. Heuristic algorithms are usually based on the combination of rules and randomness to imitate the natural phenomenon [1]. These natural phenomena include the evolution of biological processes, the behavior of animals, the physical processes of annealing and so on. According to these phenomena, a series of related algorithms are proposed. For example, these phenomena include genetic algorithms proposed by Holland [2] and Goldberg [3], difference evolution (DE) algorithm proposed by Storn and Price [4], particles swarm optimization proposed by Kennedy and Eberhart [5] and simulated

© Springer Science+Business Media Singapore 2016
K. Li et al. (Eds.): ISICA 2015, CCIS 575, pp. 84–94, 2016.
DOI: 10.1007/978-981-10-0356-1_9

annealing proposed by Kirkpatrick et al. and so on. In these optimization algorithms, differential evolution algorithm is generally considered to be one of the general method to solve constrained optimization problems has very successfully applied to solving a variety of constraints optimization problems.

In Ref. [6], a hybrid evolutionary algorithm (HEAA) which used to solve constrained optimization problems pointed out by Wang Yong et al. This algorithm uses simplex crossover and a mutation operator to produce offspring, then according to the state of the current offspring individual to deal the constraint conditions different constraint processing technology. There is a new hybrid algorithm called PSO-DE which based on the combination of particle swarm optimization and differential evolution algorithm proposed by Liu Hui et al. [7]. PSO-DE not only use the powerful search capability of the differential evolution algorithm to help the particles of particle swarm jump out of stagnation, and also employ particle swarm algorithm to strengthen the local search capability of the different evolution algorithm. The research shows that PSO-DE has faster convergence speed and higher performance. It proposed a new evolutionary algorithm based on orthogonal design constraints in the literature [8]. This algorithm uses orthogonal experimental design to arrange the cross operation of multiple parents. A new orthogonal crossover operator is proposed. According to the problem of constrained optimization problem is transformed into a bi-objective optimization problem, it presents a partial multi-objective evolutionary algorithm at paper [9].

Although researchers have proposed many algorithms and strategies for constrained optimization problems, but these algorithms have their own shortcomings and limitations, so this paper proposes a solution for a class of constrained optimization problems, which aims to solve this kind of constrained optimization problems.

2 Previous Work

2.1 Constrained Optimization Problems

The general constrained optimization problem can be described as

$$
\begin{aligned}
&\text{min: } f(\mathbf{x}), x = (x_1, x_2, \cdots, x_N) \in \mathrm{R}^N \\
&\text{s.t.: } g_j(\mathbf{x}) \le 0, j = 1, 2, \cdots, l \\
&\qquad h_j(\mathbf{x}) = 0, j = l+1, \cdots, m
\end{aligned}
\tag{1}
$$

where $x \in \Omega \in S$ is an decision vector, $f(\mathbf{x})$ is the objective function, $g_j(\mathbf{x})$ is the jth inequality constraint, $h_j(\mathbf{x})$ is the jth equality constraint, l is the number of inequality constraints, $m-l$ is the number of equality constrains and S is decision space. Generally, S is an n-dimensional rectangle of R^N:

$$
l(i) \le x_i \le u(i), i = 1, 2, \cdots, N
\tag{2}
$$

where $l(i)$ and $u(i)$ is constant. Ω is the feasible region only if

$$
\Omega = \{\mathbf{x} \in \mathbf{S} | g_j(\mathbf{x}) \le 0, j = 1, 2, \cdots, l; h_j(\bar{x}) = 0, j = l+1, \cdots, m\}
\tag{3}
$$

The complement set in S with Ω is infeasible domain. The solution in the feasible domain is called feasible solution and the solution in the infeasible domain is called infeasible solution.

2.2   Constraint Handling

When the optimization problem is constrained, the objective function is very difficult to be transformed into the fitness function. This is because the fitness function is not only to evaluate the quality of a solution, but also to describe the extent of the feasible region in the search space.

When the optimization problem has a lot of linear and nonlinear, equality and inequality constraints, the process will become more complex. It is worth noting that the evolutionary algorithm is an unconstrained search technology, because it lacks a clear constraint handling mechanism, which prompted the researchers to develop different methods to deal with constraints. In general, the combination of constraint handling techniques and the evolutionary algorithm will bring some additional parameters, and the selection of these parameters is usually determined by the user. Because of this, the design of better performance constraint processing technology is particularly important.

In general, the constraint handling technique based on the evolutionary algorithm converts the equality constraints into the following inequality constraints like Eq. (4):

$$|h(x)| - \sigma \leq 0 \tag{4}$$

where σ is tolerance value for equality constraints. When the constraints of equality constraints are transformed to the inequality constraints, the constrained optimization problems will only contain the inequality constraints.

In general, the degree of constraint violation of a solution x on the jth constraint is expressed as:

$$G_j(x) = \begin{cases} max\{0, g_j(x)\}, 1 \leq j \leq l \\ max\{0, |h(x)| - \sigma\}, l+1 \leq j \leq m \end{cases} \tag{5}$$

Then, the degree of constraint violation about a solution $\vec{x}$ on all constraints can be expressed as

$$G(\mathbf{x}) = \sum_{j=1}^{m} G_j(\mathbf{x}) \tag{6}$$

2.3   Adaptive Differential Evolution

In Ref. [10], an adaptive differential evolution (JADE) was proposed by Zhang et al. At each generation, the crossover probability of each individual is independently generated according to a normal distribution of mean CR and the mutation factor F of each individual is independently generated according to a Cauchy distribution. One records CR and

F of the individual that successfully participates in the differential variation, and then calculates their means and generates new CR and F according to a specialized equation.

Firstly, for each individual i at generation G, generate CR_i and F_i by Eqs. (7), (8):

$$CR_i = randni(\mu CR, 0.1) \tag{7}$$

$$F_i = randci(\mu F, 0.1) \tag{8}$$

where μCR is the crossover probability and μF is the mutation factor.

Then, for each individual i perform the three operations: selection, crossover and mutation according to formulas (9)–(11):

$$v_{i,G+1} = x_{i,G} + F_i \cdot \left(x_{best,G}^p - x_{i,G}\right) + F_i \cdot \left(x_{r1,G} - x_{r2,G}\right) \tag{9}$$

$$u_{ji,G+1} = \begin{cases} v_{ji,G+1} & if(randb(j) \leq CR_i \text{ or } j = rnbr(i)) \\ x_{ji,G} & if(randb(j) > CR_i \text{ or } j = rnbr(i)) \end{cases} \tag{10}$$

$$x_{i,G+1} = \begin{cases} x_{i,G} & if \ f(x_{i,G}) \leq f(u_{i,G+1}) \\ u_{i,G+1}; x_{i,G} \to A; CR_i \to S_{CR}; F_i \to S_F; otherwise \end{cases} \tag{11}$$

where $x_{best,G}^p$ is randomly chosen as one of the top $100p$ % individuals in the current population, $x_{r1,G}$ is randomly chosen from the current population, $x_{r2,G}$ is randomly chosen from the current population P and the external population A, and satisfy $x_{r2,G} \neq x_{r1,G} \neq x_{i,G}$.

Meanwhile, update μCR and μF by Eqs. (12) and (13):

$$\mu CR = (1 - c) \cdot \mu CR + c \cdot mean_A(S_{CR}) \tag{12}$$

$$\mu F = (1 - c) \cdot \mu F + c \cdot mean_L(S_F) \tag{13}$$

where c is a positive constant between 0 and 1, $mean_A(S_{CR})$ is the usual arithmetic mean, and $mean_L(S_F)$ is the Lehmer mean calculated by Eq. (14).

$$mean_L(S_F) = \frac{\sum_{F \in S_F} F^2}{\sum_{F \in S_F} F} \tag{14}$$

3 Improved dE Algorithm

3.1 Initialization

In this paper, the chaotic mapping function is used to initialize the population. Chaos theory is a method that combines qualitative thinking and quantitative analysis. It can be used to investigate the behavior of the dynamic system which cannot use in a single data relationship to express. Chaotic motion has its unique properties such as randomness, ergodicity and regularity [11]. Chaotic characteristics are the starting point of the initial population iteration function.

There are lots of chaotic mapping functions. It used one of them which called Skew tent map. Skew tent map function compared to some other simple mapping function (Logistic mapping) has some advantages such as uniformity and neat. Skew tent map function is defined as:

$$F(x) = \begin{cases} \frac{x}{p}, x \in [0, p) \\ \frac{1-x}{1-p}, x \in [p, 1] \end{cases} \tag{15}$$

where $p \in (0, 1)$ is the control parameter.

3.2   An Improved Adaptive Tradeoff Model

In this paper, we proposed an improved adaptive tradeoff model based on archiving-based adaptive tradeoff model in the literature [12]. In constrained optimization problems, there are two states of individual in the population: one is the infeasible solution which satisfy $G(\mathbf{x}_i) > 0$, the other one is the feasible solution which satisfy $G(\mathbf{x}_i) = 0$. However, there are three states for the whole population: one is that all individuals in the population are the infeasible solution, the second is that all individuals in the population are the feasible solution, three is that some individuals are the infeasible solution and the others are the feasible solution. Therefore, it will take different treatment schemes to handle these three different states:

The Infeasible Solution. In the infeasible solution, all the individuals in the population are infeasible. In order to make the individual in the population quickly into the feasible region, it should be possible to select the small amount of constraint violations into the next generation population. But in order to keep the diversity of the population, the individuals that which have big degree of constraint violations and small values of the objective function are also appropriate to enter the next generation. In Ref. [12], it proposed a constraint handling strategy for solve this problem. But this strategy did not reflect the influence of degree of constraint violations. So an improved constraint handling strategy is proposed in this paper.

Firstly, the objective function is converted to a normalized objective function $f_{nor}(\mathbf{x}_i)$ by Eq. (16):

$$f_{nor}(\mathbf{x}_i) = \frac{f(\mathbf{x}_i) - f(\mathbf{x}_{best})}{f(\mathbf{x}_{best}) - f(\mathbf{x}_{worst})} \times \frac{G(\mathbf{x}_i)}{G(\mathbf{x}_{worst})}, i = 1, \cdots, num \tag{16}$$

where num is the number of the individuals in the population, $f(\mathbf{x}_{best})$ is the best objective function value for the individual in the population, $f(\mathbf{x}_{worst})$ is the worst objective function value for the individual in the population, $G(\mathbf{x}_{worst})$ is the maximum degree of constraint violations:

$$f(\mathbf{x}_{best}) = \min_{i \in \{1, \cdots, num\}} f(\mathbf{x}_i) \tag{17}$$

$$f(\mathbf{x}_{worst}) = \max_{i \in \{1,\cdots,num\}} f(\mathbf{x}_i) \tag{18}$$

$$G(\mathbf{x}_{worst}) = \max_{i \in \{1,\cdots,num\}} G(\mathbf{x}_i) \tag{19}$$

Finally, a final fitness function $f_{final}(\mathbf{x}_i)$ is obtained by adding the normalized objective function value $f_{nor}(\mathbf{x}_i)$ and the degree of constraint violation $G(\mathbf{x}_i)$:

$$f_{final}(\mathbf{x}_i) = f_{nor}(\mathbf{x}_i) + G(\mathbf{x}_i), i = 1, \cdots, num \tag{20}$$

The Semi-Feasible Solution. In the semi-feasible solution, some individuals in the population are the infeasible solution and the others are the feasible solution. In general, the feasible solution is superior to the infeasible solution. But it is not to say that all infeasible solutions are useless, because some good infeasible solutions have excellent genes and those genes have great help for the evolution of the population. So it is necessary to retain the infeasible solution which has excellent genes and copy to the next generation. The processing method is as follows:

Firstly, the feasible individuals and the infeasible individuals in the population are identified and their subscripts are recorded into two sets Z_1 and Z_2.

$$\mathbf{Z}_1 = \{i | G(\mathbf{x}_i) = 0, i \in \{1, \cdots, num\}\} \tag{21}$$

$$\mathbf{Z}_2 = \{i | G(\mathbf{x}_i) > 0, i \in \{1, \cdots, num\}\} \tag{22}$$

Then, the best objective function value of the feasible individual, the worst objective function value of the feasible individual and the average objective function value of the feasible individual are found by Eqs. (23)–(25):

$$f(\mathbf{x}_{best}) = \min_{i \in \mathbf{Z}_1} f(\mathbf{x}_i) \tag{23}$$

$$f(\mathbf{x}_{worst}) = \max_{i \in \mathbf{Z}_1} f(\mathbf{x}_i) \tag{24}$$

Subsequently, the objective function is converted to a converted objective function $f'(\mathbf{x}_i)$ by Eq. (25).

Meanwhile, a normalized objective function $f_{nor}(\mathbf{x}_i)$ is obtained by normalizing the converted objective function $f'(\mathbf{x}_i)$ by Eq. (26).

Finally, the fitness function $f_{final}(\mathbf{x}_i)$ by Eq. (20).

$$f'(\mathbf{x}_i) = \begin{cases} f(\mathbf{x}_i), & i \in \mathbf{Z}_1 \\ \max\{f(\mathbf{x}_i), \varphi \times f(\mathbf{x}_{best}) + (1 - \varphi) \times f(\mathbf{x}_{worst})\}, & i \in \mathbf{Z}_2 \end{cases} \tag{25}$$

$$f_{nor}(\mathbf{x}_i) = \frac{f'(\mathbf{x}_i) - f(\mathbf{x}_{best})}{f(\mathbf{x}_{best}) - f(\mathbf{x}_{worst})}, i = 1, \cdots, num \tag{26}$$

The Feasible Solution. In the feasible solution, all individuals are the feasible solution and the degree of constraint violation is zero. So the fitness functions $f_{final}(\mathbf{x}_i)$ satisfy $f_{final}(\mathbf{x}_i) = f(\mathbf{x}_i), i = 1, \ldots, num$.

3.3   Algorithmic Framework

01	Begin
02	Set $\mu CR = 0.5; \mu F = 0.5; A = \varnothing$;
03	Create an initial population $\{x_{i,0} \mid i = 1, 2, ..., NP\}$ by Skew tent map;
04	For g=1 to G
05	$S_F = \varnothing; S_{CR} = \varnothing; NP' = 0$;
06	For i=1 to NP
07	Set $CR_i = randni(\mu CR, 0.1), CR_i^2 = randci(\mu CR, 0.1), F_i = randni(\mu F, 0.1)$;
08	Randomly choose $x_{best,g}^p$ as one of the 5% best vectors
09	Randomly choose $x_{r1,g} \neq x_{i,g}$ from current population P
10	Randomly choose $x_{r2,g} \neq x_{r1,g} \neq x_{i,g}$ from $P \cup A$
11	$v_{i,g+1} = x_{i,g} + F_i \cdot \left(x_{best,g}^p - x_{i,g} \right) + F_i \cdot \left(x_{r1,g} - x_{r2,g} \right)$
12	Generate $j_{rand} = rand\mathrm{int}(1, D), j_{rand}^2 = rand\mathrm{int}(1, D)$
13	For j=1 to D
14	If $rand(0,1) < CR_i$ or $j = j_{rand}$
15	$u_{ij,g} = v_{ij,g}$
16	Else
17	$u_{ij,g} = x_{ij,g}$
18	End If
19	If $rand(0,1) < CR_i^2$ or $j = j_{rand}^2$
20	$u_{ij,g}^2 = v_{ij,g}$
21	Else
22	$u_{ij,g}^2 = x_{ij,g}$
23	End If
24	End For
25	End For
26	Generate fitness $f_{final}(\bullet)$ by an improved individual comparative criterion
27	For i=1 to NP
28	If $f_{final}(u_{i,g}^2) \leq f_{final}(u_{i,g})$
29	$u_{i,g} = u_{i,g}^2; CR_{i,g} = CR_{i,g}^2; u_{i,g} = u_{i,g}^2; f(u_{i,g}) = f(u_{i,g}^2)$
30	End If
31	If $f_{final}(x_{i,g}) \leq f_{final}(u_{i,g})$
32	$x_{i,g+1} = x_{i,g}; CR_i' = 0$
33	Else $x_{i,g+1} = u_{i,g}$; $x_{i,g} \to A; CR_i \to S_{CR}; F_i \to S_F; NP' = NP' + 1$
34	For j=1 to D
35	If $x_{i,g+1} == x_{i,g}; diff_{ij} = 0$
36	Else $diff_{ij} = 1$
37	End If
38	End For
39	$CR_i' = \sum_{j=1}^{D} diff_{ij} / D;$
40	End If
41	End For
42	$statmean = \sum_{i=1}^{NP} CR_i' / NP'$
43	Randomly remove solutions from A so that $|A| \leq NP$
44	If $NP' / NP < Threshold$
45	$\mu CR = (1 - c) \cdot \mu CR + c \cdot mean_A(S_{CR})$
46	Else
47	$\mu CR = (1 - c) \cdot \mu CR + c \cdot statmean$
48	End If
49	$\mu F = (1 - c) \cdot \mu F + c \cdot mean_L(S_F)$
50	End For
51	End

4 Experimental Study

In this section, the performance of the proposed algorithm is tested on 9 benchmark test function [13]. 9 benchmark test functions have a variety of objective functions and different constraints as shown in Table 1. In this table, n is the number of function arguments, ρ is the estimated ratio between the feasible region and the search space, LI is the number of linear inequality constraints, NI is the number of nonlinear inequality constraints, LE is the number of linear equality constraints, NE is the number of nonlinear equality constraints, and $f(\mathbf{x}^*)$ is the objective function value of the best known solution.

Table 1. 9 benchmark test functions

Prob.	n	Type of objective function	$\rho(\%)$	LI	NI	LE	NE	$f(\mathbf{x}^*)$
g03	10	Polynomial	0.0000	0	0	0	1	−1.0005001000
g04	5	Quadratic	51.1230	0	6	0	0	−30665.5386717834
g06	2	Cubic	0.0066	0	2	0	0	−6961.8138755802
g07	10	Quadratic	0.0003	3	5	0	0	24.3062090681
g08	2	Nonlinear	0.8560	0	2	0	0	−0.0958250415
g09	7	Polynomial	0.5121	0	4	0	0	680.6300573745
g10	8	Linear	0.0010	3	3	0	0	7049.2480205286
g11	2	Quadratic	0.0000	0	0	0	1	0.7499000000
g12	3	Quadratic	4.7713	0	1	0	0	−1.0000000000

According to the data in Table 1, we can know these test functions have various kinds of objective functions and different kinds of constraints. Because of the complexity of the function, the 9 benchmark test functions are used to test the performance of the algorithm.

For each test function, the algorithm runs 25 times independently and compare with HEAA [6], COEA/OED [8], HCOEA [14], ATMES [15], M-ABC [16] and SAFF [17]. The results show in Table 2. In the table, Best is the best of the achieved best solution, Worst is the worst of the achieved best solution, Mean is the mean of the achieved best solution and St.dev is the standard deviation of the achieved best solution.

Table 2. Comparison of experimental results of 9 benchmark test functions

Fcn	Status	HEAA	COEA/OED	HCOEA	ATMES	M-ABC	SAFF	CO-JADE
g03	Best	−1.000	−1.000	−1.00000	−1.000	−1.000	−1.000	−1.0005
	Mean	−1.000	−1.000	−1.00000	−1.000	−1.000	−1.000	−1.0005
	Worst	−1.000	−1.000	−1.00000	−1.000	−1.000	−1.000	−1.0005
	St.dev	2.09E − 04	5.8E − 09	1.30E − 12	5.90E − 05	4.68E − 05	7.5E − 05	1.99E − 07
g04	Best	−30665.539	−30665.539	−30665.539	−30665.539	−30665.539	−30665.50	−30665.5386
	Mean	−30665.539	−30665.539	−30665.539	−30665.539	−30665.539	−30665.20	−30665.5386
	Worst	−30665.539	−30665.539	−30665.539	−30665.539	−30665.539	−30663.30	−30665.5386
	St.dev	0.00E + 00	1.2E − 11	5.40E − 07	7.40E − 12	2.22E − 11	4.9E − 01	3.71E − 12

(Continued)

Table 2. (*Continued*)

Fcn	Status	HEAA	COEA/OED	HCOEA	ATMES	M-ABC	SAFF	CO-JADE
g06	Best	−6961.814	−6961. 814	−6,961.814	−6961.814	−6961.814	−6961.800	−6961.8138
	Mean	−6961.284	−6961. 814	−6,961.814	−6961.814	−6961.814	−6961.800	−6961.8138
	Worst	−6952.482	−6961. 814	−6,961.814	−6961.814	−6961.814	−6961.800	−6961.8138
	St.dev	1.85E + 00	3.9E − 11	8.51E − 12	4.60E − 12	0.00E + 00	0.0E + 00	0.00E + 00
g07	Best	24.327	24.306	24.306	24.306	24.315	24.48	24.3062
	Mean	24.475	24.306	24.307	24.316	24.415	26.58	24.3062
	Worst	24.843	24.306	24.309	24.359	24.854	28.40	24.3062
	St.dev	1.32E − 01	1.1E − 05	7.12E − 04	1.10E − 02	1.24E − 01	1.1E + 00	2.65E − 10
g08	Best	−0.095825	−0.095825	−0.095825	−0.095825	−0.095825	−0.095825	−0.095825
	Mean	−0.095825	−0.095825	−0.095825	−0.095825	−0.095825	−0.095825	−0.095825
	Worst	−0.095825	−0.095825	−0.095825	−0.095825	−0.095825	−0.095825	−0.095825
	St.dev	2.8E − 17	1. 6E- 17	2.42E − 17	2.80E − 17	4.23E − 17	0.0E + 00	1.41E − 17
g09	Best	680.632	680.630	680.630	680.630	680.632	680.64	680.6300
	Mean	680.643	680.630	680.630	680.639	680.647	680.72	680.6300
	Worst	680.719	680.630	680.630	680.673	680.691	680.87	680.6300
	St.dev	1.55E − 02	3.5E − 12	9.41E − 08	1.00E − 02	1.55E − 02	5.9E − 02	2.93E − 13
g10	Best	7051.903	7049.521	7049.287	7052.253	7051.706	7061.34	7049.248
	Mean	7253.047	7072.167	7049.525	7250.437	7233.882	7627.89	7049.249
	Worst	7638.366	7155.754	7049.984	7560.224	7473.109	8288.79	7049.283
	St.dev	1.36E + 02	3.7E + 01	1.50E − 01	1.20E + 02	1.10E + 02	3.7E + 02	7.04E − 03
g11	Best	0.75	0.75	0.75	0.75	0.75	0.750	0.7499
	Mean	0.75	0.75	0.75	0.75	0.75	0.750	0.7499
	Worst	0.75	0.75	0.75	0.75	0.75	0.750	0.7499
	St.dev	1.52E − 04	2. 8E − 07	1.55E − 12	3.40E − 04	2.30E − 05	0.0E + 00	1.13E − 16
g12	Best	−1.000	−1.000	−1.000	−1	−1.000	−1.000	−1.0000
	Mean	−1.000	−1.000	−1.000	−1	−1.000	−1.000	−1.0000
	Worst	−1.000	−1.000	−1.000	−1	−1.000	−1.000	−1.0000
	St.dev	0.00E + 00	0.0E + 00	0.00E + 00	1.00E − 03	0.0E + 00	0.0E + 00	0.0E + 00

In the Table 2, it shows CO-JADE can obtain the optimal solution more accurately and the standard deviation also is the first three of the seven algorithms. For function g03, all algorithms can get the optimal value and only then standard deviation of HCOEA is better than CO-JADE. For function g05 and g12, CO-JADE can get the optimal value and the standard deviation is zero. For function g08, all algorithms can get the optimal value and CO-JADE is better than others expect SAFF. For other 5 test function (i.e., g04, g07, g08, g09, g10 and g11), CO-JADE can get the optimal value more accurately and smaller standard deviation.

In terms of computational cost, for the nine standard test functions, the number of function evaluations of SAFF is 1400000, HEAA is 200000, and COEA/OED, HCOEA, ATMES, M-ABC and CO-JADE only needs 240000.

The experimental results show that CO-JADE is superior to the other seven kinds of algorithms, shows better optimization performance and better stability.

5 Conclusion and Future Work

CO-JADE is improves JADE to solve constrained optimization problems. It used chaotic mapping function to initialize population and the mutation probability which conforms to normal distribution and Cauchy distribution to mutate the population individual. At the same time, the fitness value is solved by using an improved adaptive tradeoff model. The experimental simulations show that CO-JADE has better stability and higher accuracy and is better than HEAA, COEA/OED and HCOEA. This paper is use CO-JADE to solve the single objective optimization problem, but the multi-objective optimization problem is not involved. So the next research direction is use CO-JADE to solve the multi-objective optimization problem.

Acknowledgments. This work is supported by the National Natural Science Foundation of China with the Grant No. 61573157, the Fund of Natural Science Foundation of Guangdong Province of China with the Grant No. 2014A030313454, the Natural Science Foundation of Guangdong Province of China with the Grant No. 2015A030313408.

References

1. Lee, G.: A new meta-heuristic algorithm for continuous engineering optimization: harmony search theory and practice. Comput. Methods Appl. Mech. Eng. **194**, 3902–3933 (2005)
2. Holland, J.H.: Adaptation in Natural and Artificial Systems. Mich: University of Michigan Press, Ann Arbor (1975)
3. Goldberg, D.E.: Genetic Algorithms in Search, Optimization and Machine Learning. Addison-Wesley Publishing Company, Massachusetts (1989)
4. Storn, R., Price, K.: Differential evolution-a simple and efficient heuristic for global optimization over continuous spaces. J. Global Optim. **11**(4), 341–359 (1997)
5. Kennedy, J., Eberhart, R.: Particle swarm optimization. In: Proceedings of IEEE International Conference on Neural Networks. IEEE, pp. 1942–1948 (1995)
6. Wang, Y., Cai, Z., Zhou, Y., et al.: Constrained optimization based on hybrid evolutionary algorithm and adaptive constraint-handling technique. Struct. Multi. Optim. **37**(4), 395–413 (2009)
7. Liu, H., Cai, Z., Wang, Y.: Hybridizing particle swarm optimization with differential evolution for constrained numerical and engineering optimization. Appl. Soft Comput. **10**(2), 629–640 (2010)
8. Zi-Xing, C., Zhong-Yang, J., Yong, W., et al.: A novel constarained optimization evolutionary algorithm based on orthogonal experimental design. J. Comput. Sci. **33**(5), 855–864 (2010)
9. Ning, D., Yuping, W.: Multi-objective evolutionary algorithm based on preference for constrained optimization problems. J. Xidian Univ. **41**(1), 98–104 (2014)
10. Zhang, J., Sanderson, A.C.: JADE: adaptive differential evolution with optional external archive. IEEE Trans. Evol. Comput. **13**(5), 945–958 (2009)
11. Zhenyu, G., Bo, C., Min, Y., et al.: Parallel chaos differential evolution algorithm. J. Xi'an Jiaotong Univ. **41**(3), 299–302 (2007)
12. Jia, G., Wang, Y., Cai, Z., Jin, Y.: An improved $(\mu + \lambda)$-constrained differential evolution for constrained optimization. Inf. Sci. **222**, 302–322 (2013)

13. Liang, J.J., Runarsson, T.P., Mezura-Montes, E., et al.: Problem Definitions and Evaluation Criteria for the CEC 2006 Special Session on Constrained Real-Parameter Optimization. Nanyang Technological University, Singapore (2006)
14. Wang, Y., Cai, Z., Guo, G., et al.: Multiobjective optimization and hybrid evolutionary algorithm to solve constrained optimization problems. IEEE Trans. Syst. Man Cybern. Part B: Cybern. **37**(3), 560–575 (2007)
15. Wang, Y., Cai, Z., Zhou, Y., et al.: An adaptive tradeoff model for constrained evolutionary optimization. IEEE Trans. Evol. Comput. **12**(1), 80–92 (2008)
16. Mezura-Montes, E., Cetina-Domínguez, O.: Empirical analysis of a modified artificial bee colony for constrained numerical optimization. Appl. Math. Comput. **218**(22), 10943–10973 (2012)
17. Farmani, R., Wright, J.A.: Self-adaptive fitness formulation for constrained optimization. IEEE Trans. Evol. Comput. **7**(5), 445–455 (2003)

A New Ant Colony Classification Mining Algorithm

Lei Yang[1], Kangshun Li[1(⊠)], Wensheng Zhang[2], Yan Chen[1],
Wei Li[1], and Xinghao Bi[1]

[1] College of Mathematics and Informatics, South China Agricultural University,
Guangzhou 510642, China
yanglei_s@scau.edu.cn, likangshun@sina.com
[2] Institute of Automation, Chinese Academy of Sciences, Beijing 100190, China
wensheng.zhang@ia.ac.cn

Abstract. Ant colony optimization algorithms have been successfully applied in classification rule mining, but in general, the basic ant colony classification mining algorithms have the problems of premature convergence, easily falling into local optimum, and etc. In this paper, a new ant colony classification mining algorithm based on pheromone attraction and exclusion (Ant-Mine$_{rPAE}$) is proposed, where the new pheromone calculation method is designed and the search is guided by the new probability transfer formula. Our experiments using 12 publicly available data sets show that the predictive accuracy obtained by the Ant-Miner$_{PAE}$ algorithm is statistically significantly higher than the predictive accuracy of other rule induction classification algorithms, such as CN2, C4.5rules, PSO/ACO2, Ant-Miner, $_c$Ant-Miner$_{PB}$, and the rules discovered by the Ant-Miner$_{PAE}$ algorithm are considerably simpler than those discovered by the counterparts.

Keywords: Data mining · Ant colony algorithm · Classification rule · Pheromone

1 Introduction

Data mining [1] is a method to extract valuable knowledge from a large amount of data. It combines traditional data analysis method and the complex algorithm of processing large amount of data. Data classification is an important task in the field of data mining, which describes important data mining model and predicts the future trend of the data. Although in some applications, such as credit approval, medical diagnostics and protein structure prediction, the comprehensibility of the model plays an important role [2, 3], but one of the main objectives of classification is to create a classification model, to classify and maximize the prediction accuracy.

Ant colony algorithm proposed by Parpinelli et al. has been applied in classification rule mining; it simulates ant foraging to construct classification rules. Ant colony is a self-organizing system, which can find the shortest path between the nest and food. This cooperation behaviors caused by autocatalytic form a positive feedback mechanism, which makes more and more ants choose the optimal foraging path [4, 5]. Ultimately,

© Springer Science+Business Media Singapore 2016
K. Li et al. (Eds.): ISICA 2015, CCIS 575, pp. 95–106, 2016.
DOI: 10.1007/978-981-10-0356-1_10

the majority of ants will follow the same path, most likely the shortest path between the food source and the nest.

The definition of Ant colony optimization (ACO) in the book written by Dorigo and Stützle [6] in 2004 gives a broader description. The ACO algorithm uses artificial ant colony. The ants build a solution for the optimization problem, which is selected by iteration based on their associated pheromone and heuristic information, corresponding to the assessment of how well the solution in the current [7]. The candidate solution is established after iteration, according to the pheromone and heuristic information. Eventually ant colony converges to optimal solution or near optimal solution.

This paper proposes a new ant colony classification mining algorithm based on pheromone attraction and exclusion (Ant-Miner$_{PAE}$). It can effectively improve the rules list search quality and accuracy. We use JAVA to implement the algorithm. Our experiments using 12 publicly available data sets show that the predictive accuracy obtained by our Ant-Miner$_{PAE}$ algorithm is statistically significantly higher than the predictive accuracy of other rule induction classification algorithms, such as CN2, C4.5rules, PSO/ACO2, Ant-Miner, $_c$Ant-Miner$_{PB}$, and the rule lists discovered by the Ant-Miner$_{PAE}$ algorithm are considerably simpler.

The remainder of this paper is organized as follows. Section 2 presents a discussion of the traditional ant colony classification algorithms. Section 3 presents our new ant colony classification mining algorithm based on pheromone attraction and exclusion. The experimental results and discussion are presented in Sect. 4. Finally, Sect. 5 concludes this paper and presents future research directions.

2   The Traditional Ant Colony Classification Algorithms

Classification refers to a given data object, and the class of the object is determined by a specific classification model. Data samples are usually required to determine the classification model, which is called the training data set. And it is needed to evaluate the accuracy of the model by testing data sets. The general form of classification rules is:

$$\text{IF} < conditions > \text{THEN} < class >$$

where <conditions> is called rule antecedent. The rule antecedent (IF part) contains a set of conditions, usually connected by a logical conjunction operator (AND). We refer to each rule condition as a term, so that the rule antecedent is a logical conjunction of terms in the form IF $< term_1 \ AND \ term_2 \ AND \cdots AND \ term_n >$. Each term is a triple <attribute, operator, value>, where value belongs to the domain of attributes, and operator is a relational operator. The current version of classification algorithm copes only with categorical attributes, so that the operator is always "=".

In 2002, Parpinelli firstly used the ant colony algorithm to obtain the classification rules [8], and proposed a famous ant colony classification algorithm, which is called Ant-Miner. The basic idea of Ant-Miner [9] is as follows: At first, the list of discovered rules is empty and the training set consists of all the training cases. Each search of the algorithm completes three basic tasks, i.e. rules construction, rule pruning and pheromone updating, and then discovers one classification rule. This rule is added to the list of discovered rules and the training cases that are covered correctly by this rule are removed from the training set. This process is performed iteratively whenever the number of uncovered training cases is greater than a user-specified threshold, called *Maximum uncovered*. The algorithm will eventually get a set of classification rules. At the same time, we need to carry out the convergence test in the whole process of the discovery of the rules. The convergence threshold can be set as desired. If the rules found by ants meet the formula:

$$R_t = R_{t+1}(t > 1) \tag{1}$$

The convergence threshold, named j, is set to $j + 1$ if the rule just discovered is the same as the last one, or as to 1 otherwise, and then the next ant begins to rebuild the rules.

Studies have shown [8, 10, 11] that compared with the classical classification algorithms C4.5 [12] and CN2 [13, 14], the Ant-Miner algorithm simplifies the discovered rules, improves the comprehensibility and prediction accuracy of classification. In recent years, there are a large number of researchers working on ant colony classification algorithm. In the construction of ant colony, Meyer and Parpinelli [15] proposed construction of multi ant colony method, Jin et al. [16] presents multi swarm ant colony construction method. Regarding the state transition rule, Liu et al. [17] proposed the new state transfer rule, which extends the exploration ability for Ant-Miner state transition rules and reduces the discovered rule preferences. Wang et al. [18] simplified the state transfer probability calculation method based on Liu's research foundation. Regarding heuristic strategy, Liu et al. [19] put forward the evaluation method based on density, Martens et al. [20] made more effective ant-based assessment of the selected class of equations, Wang and Feng [18] proposed a simple and effective heuristic function for the problem of premature. In terms of rule pruning, in order to improve the pruning speed of Ant-Miner in dealing with large data set of attributes, Chan and Freitas [21] proposed a hybrid pruning approach. Reference[17, 18] discussed pheromone updating in order to control the pheromone evaporation rate of old trail, Otero et al. [22, 23] put forward the processing method of the discretization of continuous attributes in the construction rules. In addition, Chan and Freitas [24] proposed an improved algorithm for solving multi label classification tasks, Smaldon and Freitas [25] proposed an improved algorithm for generating a random classification rule.

Based on the existing research of Ant-Miner algorithms, we aim to find an algorithm in this paper, which can improve the prediction precision and stability of the algorithm, and simplify the rules. So as to improve the performance of classification mining algorithm, and make the improved algorithm have more advantages in massive data classification.

3   A New Ant Colony Classification Mining Algorithm

In this section, we discuss in details our proposed the new algorithm for the discovery of classification rules, called Ant-Miner$_{PAE}$.

3.1   General Description of Ant-Miner$_{PAE}$

Ant-Miner aims at discovering a list of classification rules by applying a sequential covering strategy, which creates one-rule-at-a-time until all training examples are covered by one of the rules in the discovered list. In 2013, Fernando et al. proposed a new sequential covering strategy with ant colony mining algorithm [7], called $_c$Ant-Miner$_{PB}$, which do not need to set the number of rules of candidate list constructed by ants, and the length difference of the rule list is flexible. Based on the basic structure of Fernando's approach, we put forward a new pheromone attraction and exclusion method in the rule construction by ants.

3.2   Rule Construction

When the first ant starts its search, all nodes have the same amount of pheromone. The initial amount of pheromone deposited at each position is inversely proportional to the number of values of all attributes and is defined by

$$\tau_{ij}(t = 0) = \frac{1}{\sum_{i=1}^{a} b_i} \tag{2}$$

where a is the total number of attributes and b_i is the number of possible values that can be taken by attribute A_i.

3.2.1   The Method of Rule Construction

In the process of constructing the rules, if attribute nodes are random, the calculation time is very large, which is not allowed to happen [26]. In order to avoid the blind choice of ants, the choice of the $term_{ij}$ to be added to the current partial rule depends on both a problem-dependent heuristic function (η) and on the amount of pheromone (τ) associated with each $term_{ij}$. Each ant starts with an empty rule, i.e., a rule with no term in its antecedent, and adds one term at a time to its current partial rule until the stopping criteria is met. Let $term_{ij}$ be a rule condition of the form $A_i = V_{ij}$, where A_i is the ith attribute and V_{ij} is the jth value of the domain of A_i. The probability that $term_{ij}$ is chosen to be added to the current partial rule is

$$P_{ij}(t) = \frac{\tau_{ij}(t)\eta_{ij}}{\sum_{i=1}^{a} \sum_{j=1}^{b_i} \tau_{ij}(t)\eta_{ij}}, \ \forall i \in I \tag{3}$$

Where the variables are defined as below.

η_{ij} value of a problem-dependent heuristic function for $term_{ij}$. The higher the value of η_{ij}, the more relevant for classification the $term_{ij}$ is and so the higher its probability of being chosen.

$\tau_{ij}(t)$ amount of pheromone associated with $term_{ij}$ at iteration t, corresponding to the amount of pheromone currently available in the position (i,j) of the rule followed by the current ant.

a total number of attributes.

b_i number of values in the domain of the *i*th attribute.

I the set of attributes that have not been visited by the ants.

3.2.2   New Method for Calculating the Pheromone

In this part we propose a new calculation method of pheromone based on the pheromone attraction and exclusion, and apply it to the new transition probability, we define a representation to attract the weight of Λ_{ij}^k and a representation of the exclusion of the weight of Φ_{ij}^k for each attribute node *term*$_{ij}$ of the ant *k*. The weight of representation to attract is defined as follows:

$$\Lambda_{ij}^k(t) = \frac{\tau_{ij}^k(t)}{\sum_{u \in N_i^k(t)} \tau_{iu}^k(t)} \tag{4}$$

where $\tau_{ij}^k(t)$ is the pheromone of *term*$_{ij}$ for ant *k*, standardized with the sum of pheromones left on the other nodes by the ant. The weight of representation to exclude is defined as:

$$\Phi_{ij}^k(t) = \frac{\Pi_{ij}^k(t)}{\sum_{u \in N_i^k(t)} \Pi_{iu}^k(t)} \tag{5}$$

where

$$\Pi_{ij}^k(t) = \sum_{m=1,\cdots,n_k(m \neq k)} \tau_{ij}^m(t) \tag{6}$$

The formula (6) represents the amount of exclusion pheromone of *term*$_{ij}$ for ant *k*, namely the sum of pheromone quantity released by other ants. According to the above definition, the method for calculating the pheromone of *term*$_{ij}$ is:

$$\tau_{ij}(t) = \frac{\left(\Lambda_{ij}^k(t) \big/ \Phi_{ij}^k(t)\right)^\alpha}{\sum_{u \in N_i^k(t)} \left(\Lambda_{iu}^k(t) \big/ \Phi_{iu}^k(t)\right)^\alpha} \tag{7}$$

The ratio of attraction and exclusion in the formula balances the relationship between exploration and development.

3.2.3   Heuristic Function

Ant-Miner$_{PAE}$ computes the value η_{ij} of a heuristic function which is an estimate of the quality of *term*$_{ij}$, with respect to its ability to improve the predictive accuracy of the rule. The heuristic function is defined by

$$\eta_{ij} = \frac{\log_2(k) - InfoT_{ij}}{\sum_{i=1}^{a} \sum_{j=1}^{b_i} (\log_2(k) - InfoT_{ij})} \qquad (8)$$

$$InfoT_{ij} = -\sum_{w=1}^{k} \left[\frac{freqT_{ij}^w}{|T_{ij}|} \right] \times \log_2 \left[\frac{freqT_{ij}^w}{|T_{ij}|} \right] \qquad (9)$$

In $InfoT_{ij}$, k is the number of category, $|T_{ij}|$ is the number of $term_{ij}$'s of the form $A_i = V_{ij}$, A_i is the ith attribute and V_{ij} is the jth value belonging to the domain of A_i. $freqT_{ij}^w$ is the number of observing class w conditioned on $A_i = V_{ij}$. From (8) and (9), it can be known that the higher the value of $InfoT_{ij}$, the smaller the probability that the current ant chooses to add $term_{ij}$ to its partial rule.

3.3　Rule Pruning

Rule pruning potentially increases the predictive power of the rule, helping to avoid its overfitting to the training data. The quality of a rule in this paper is defined by the following formula:

$$Q = \begin{cases} 1 & (FP + FN + TN = 0) \\ \frac{TP}{TP+FN} & (FP + TN = 0) \\ \frac{TP}{TP+FP} & (TN + FN = 0) \\ \frac{TP}{TP+FN} \times \frac{TN}{FP+TN} & (ELSE) \end{cases} \qquad (10)$$

Where, TP is the number of cases covered by the rule that have the class predicted by the rule; FP is the number of cases covered by the rule that have a class different from the class predicted by the rule; TN is the number of cases that are not covered by the rule and that do not have the class predicted by the rule; FN is the number of cases that are not covered by the rule but that have the class predicted by the rule. The larger the value of Q, the higher the quality of the rule.

3.4　The Improved Method of Pheromone Updating

3.4.1　The Pheromone Updating on the Attribute Node in the Rules

Pheromone updating for a $term_{ij}$ is performed according to (11), for all terms $term_{ij}$ that occur in the rule.

$$\tau_{ij}(t+1) = \tau_{ij}(t) + \tau_{ij}(t) \times Q, \forall i, j \in Rule \qquad (11)$$

The pheromone will be increased for the terms that occur in the rule, and be reduced for the terms which do not occur in the rule due to evaporation. The following two basic characteristics of pheromone updating conditions must be met:

(1) For all *term$_{ij}$* occurring in the rule found by the current ant, the amount of pheromone is increased by a fraction of the current amount of pheromone which is given by *Q*, The greater the *Q*, the greater the increasement of the pheromone.

(2) The amount of pheromone associated with each *term$_{ij}$* that does not occur in the rule found by the current ant has to be decreased in order to simulate pheromone evaporation in real ant colonies according to (12).

$$\tau_{ij}(t+1) = (1 - \rho)\tau_{ij}(t) \tag{12}$$

3.4.2 The Pheromone Updating on the Attribute Node Outside the Rules

Pheromone evaporation on the attribute node *term$_{ij}$* outside the rules is implemented in a way of standardization of each attribute node pheromone value τ_{ij}. The standardization of the pheromone values for an attribute node *term$_{ij}$* is calculated by the following formula (13).

$$\tau_{ij}(t+1) = \frac{\tau_{ij}(t)}{\sum_{i=1}^{a} \sum_{j=1}^{b_i} \tau_{ij}(t)}, \ \forall i, j \notin Rule \tag{13}$$

where $\sum_{i=1}^{a} \sum_{j=1}^{b_i} \tau_{ij}(t)$ is the sum of pheromones on all attribute nodes, which also include the attribute nodes *term$_{ij}$* in the rules found by the current *Ant$_t$*.

4 Experiment and Result Analysis

4.1 Data Sets and Preprocessing

In order to evaluate the proposed Ant-Miner$_{PAE}$ algorithm, we carried out experiments using 12 publicly available data sets from the UCI [27] machine learning repository.

Table 1 presents a summary of the data sets used in the experiments. The first column gives the data sets name, while the other columns indicate, respectively, the number of cases, the number of attributes, and the number of classes of the data sets.

This algorithm discovers rules referring only to categorical attributes. Therefore, continuous attributes have to be discretized in a preprocessing step. The discretization is performed by the C4.5-Disc discretization method, which simply uses the well-known C4.5 algorithm [12] for discretizing continuous attributes. The reduced data set containing only two attributes: the attribute to be discretized and the goal (class) attribute. To compare with the results obtained by Ant-Miner$_{PAE}$, we have selected some commonly classification mining algorithms as follows: (1) CN2 [13, 14]: a well-known classification rule mining algorithm based on sequential covering strategy, which uses an iterative method to produce a rule list; (2) C4.5 algorithm [12]: a rule induction algorithm that extracts a set of classification rules from an unpruned decision tree created by the well-known C4.5 algorithm; (3) PSO/ACO2: a hybrid particle swarm optimization/ant colony optimization (PSO/ACO) algorithm for the

Table 1. Data sets used in the experiments

Data set	Size	Attributes	Classes
Anneal	898	38	6
Balance-scale	625	4	3
Breast-cancer	286	9	2
Credit-a	690	14	2
Credit-g	1000	20	2
Dermatology	366	34	6
Glass	214	9	7
Heart-c	303	13	5
Heart-h	294	13	5
Ionosphere	351	34	2
Liver-disorders	345	6	2
Pima	768	8	2

discovery of classification rules, which follows a sequential covering strategy and directly deals with both continuous and nominal attribute; (4) Ant-Miner [8]: proposed by Parpinelli et al. in 2002, aims at discovering a list of classification rules by applying a sequential covering strategy, which creates one-rule-at-a-time until the number of uncovered training cases is less than or equal to a user-specified threshold; (5) $_c$Ant-Miner$_{PB}$ [7]: proposes a new sequential covering strategy, and considers the interaction between rules.

4.2 Ant-Miner$_{PAE}$'s Parameter Setting

Ant-Miner$_{PAE}$ has the following six user-defined parameters, they are set as follows: $m = 100$, *colony-size* $= 5$, $\rho = 0.9$, *min_cases_per_rule* $= 5$, *Maximum uncovered* $= 10$, *no_rules_converge* $= 10$.

4.3 Comparing Ant-Miner$_{PAE}$ with Other Several Algorithms

We have evaluated the performance of Ant-Miner$_{PAE}$ that have been implemented by comparing it with the commonly rule induction classification algorithms. All the results of the comparison are obtained on 12 sets of public data sets. We use an Inter(R) Core (TM) i5-4460 CPU with clock rate of 3.20 GHz and 4G main memory. Ant-Miner$_{PAE}$ is developed in Java.

We have performed a tenfold cross-validation procedure to test the algorithm's accuracy and seek its average as the accuracy of the estimation algorithm. The data format of this algorithm is based on WEKA program, and the format of WEKA is ARFF (File Format Attribute-Relation) which is a kind of ASCII text file.

The comparison is done by two criteria: the predictive accuracy of the discovered rule lists and the simplicity of the discovered rule list.

The results comparing the predictive accuracy of Ant-Miner$_{PAE}$ and other classification algorithms are reported in Table 2. The numbers right after the "±"symbol are the standard deviations of the corresponding predictive accuracies rates. As shown in the table, Ant-Miner$_{PAE}$ discovered rules with a better predictive accuracy than other classification algorithms in eight data sets, namely breast-cancer, Credit-a, Credit-g, dermatology, Heart-c, Heart-h, ionosphere and pima. In the anneal data set, the difference in predictive accuracy between Ant-Miner$_{PAE}$ and other classification algorithms is relatively quite small. In four data sets, breast-cancer, Credit-a, Heart-c and Heart-h, Ant-Miner$_{PAE}$ is significantly more accurate than other classification algorithms. In these data sets, the difference is large. On the other hand, CN2 discovered rules with a better predictive accuracy than Ant-Miner$_{PAE}$ and other classification algorithms in balance-scale data set, with the predictive accuracy of 88.47 ± 1.81. In the glass data set, $_c$Ant-Miner$_{PB}$ is significantly more accurate than other classification algorithms with the predictive accuracy of 73.94 ± 0.49. Experiments show that Ant-Miner$_{PAE}$ obtained results better than other classification algorithms in eight of the twelve data sets.

We now turn to the results concerning the simplicity of the discovered rule list, measured by the average number of terms per rule and the number of discovered rules. The results comparing the simplicity of the rule lists discovered by Ant-Miner$_{PAE}$ and other classification algorithms are reported in Tables 3 and 4. In the anneal data set, Ant-Miner$_{PAE}$ discovers a compact rule list with 8.6 ± 0.37 rules and 12.00 ± 0.45 terms per rule, whereas the other classification algorithms discovered a rule list larger than Ant-Miner$_{PAE}$. In the balance-scale data sets, the rule list discovered by Ant-Miner$_{PAE}$ is simpler, i.e., it has a smaller number of rules with 13.20 ± 0.20 and the terms with 12.30 ± 0.26 per rule, than the rule list discovered by the other classification algorithms. However, the difference in the simplicity of the discovered rule list by Ant-Miner$_{PAE}$ and Ant-Miner and $_c$Ant-Miner$_{PB}$ is small. In the breast-cancer data set, Ant-Miner$_{PAE}$ discovers a compact rule list with 6.10 ± 0.28 rules and 6.70 ± 0.58 terms per rule, similar to that of Ant-Miner. In the Credit-a data set, Ant-Miner discovers a compact rule list with 7.00 ± 0.49 rules and Ant-Miner$_{PAE}$ discovers the simplicity of the average number of terms with 9.40 ± 1.14 terms per rule. In the Credit-g data set, Ant-Miner discovers a compact rule list with 8.40 ± 0.27 rules and 11.30 ± 0.47 terms per rule, and in dermatology data set, Ant-Miner$_{PAE}$ discovers a compact rule list with 7.90 ± 0.53 rules, and Ant-Miner discovered the simplicity of the average number of terms with 19.20 ± 0.83 terms per rule. In the other two data sets, namely, glass and heart-c, Ant-Miner$_{PAE}$ discovered rule lists simpler than other classification algorithms. In the ionosphere data set, PSO/ACO2 discovers a compact rule list with 3.36 ± 0.07 rules and Ant-Miner discovered the simplicity of the average number of terms with 8.20 ± 0.73 terms per rule. In the Liver-disorders data set, Ant-Miner$_{PAE}$ discovers a compact rule list with 9.00 ± 1.43 rules and Ant-Miner discovers the simplicity of the average number of terms with 9.20 ± 0.57 terms per rule. In the pima data set, Ant-Miner discovers the optimal average rule number and the average number of terms per rule, which are 9.00 ± 0.37 and 8.20 ± 0.42 respectively.

Experimental results show that among 12 publicly available data sets, there are 7 groups of data sets Ant-Miner$_{PAE}$ can get the shortest condition number, and at the

Table 2. Average predictive accuracy (average ± standard error) in % after the tenfold cross-validation procedure

Data set	CN2	C4.5rules	PSO/ACO2	Ant-Miner	$_c$Ant-Miner$_{PB}$	Ant-Miner$_{PAE}$
Anneal	94.99 ± 0.62	94.22 ± 0.62	97.25 ± 0.08	92.88 ± 1.16	**97.60 ± 0.10**	97.44 ± 0.55
Balance-scale	**88.47 ± 1.81**	74.87 ± 1.16	79.16 ± 0.19	73.90 ± 2.08	76.83 ± 0.24	84.47 ± 1.48
Breast-cancer	66.77 ± 1.69	68.56 ± 1.93	71.29 ± 0.31	75.95 ± 3.14	72.32 ± 0.31	**85.06 ± 1.88**
Credit-a	81.31 ± 1.36	85.53 ± 1.53	84.66 ± 0.24	85.51 ± 1.65	85.68 ± 0.15	**90.29 ± 1.56**
Credit-g	69.50 ± 1.33	71.60 ± 0.92	69.99 ± 0.23	69.70 ± 1.04	73.63 ± 0.23	**77.80 ± 1.02**
Dermatology	91.79 ± 1.37	93.45 ± 1.22	91.80 ± 0.22	86.95 ± 2.56	92.46 ± 0.31	**94.73 ± 1.03**
Glass	68.14 ± 2.39	68.63 ± 1.70	70.24 ± 0.51	52.08 ± 2.77	**73.94 ± 0.49**	68.74 ± 1.81
Heart-c	48.85 ± 3.14	53.12 ± 1.92	55.20 ± 0.51	79.11 ± 3.32	55.50 ± 0.37	**90.78 ± 1.24**
Heart-h	52.81 ± 3.38	63.31 ± 1.40	63.18 ± 0.45	81.37 ± 3.07	64.75 ± 0.27	**91.50 ± 1.04**
Ionosphere	88.03 ± 2.68	90.85 ± 2.59	86.55 ± 0.43	81.70 ± 2.23	89.65 ± 0.31	**92.37 ± 1.53**
Liver-disorders	62.27 ± 2.84	64.90 ± 3.21	**68.78 ± 0.44**	64.66 ± 2.87	66.72 ± 0.40	67.03 ± 1.52
Pima	72.37 ± 1.26	74.32 ± 1.73	73.10 ± 0.33	69.39 ± 2.05	74.81 ± 0.18	**78.52 ± 1.44**

Table 3. Average number of terms (rule condition) in the discovered list (average ± standard error) measured by tenfold cross-validation

Data set	CN2	C4.5rules	PSO/ACO2	Ant-Miner	$_c$Ant-Miner$_{PB}$	Ant-Miner$_{PAE}$
Anneal	52.10 ± 2.72	43.00 ± 3.05	28.60 ± 0.91	14.40 ± 0.56	22.11 ± 0.33	**12.00 ± 0.45**
Balance-scale	90.20 ± 3.51	72.60 ± 3.27	39.87 ± 0.53	12.50 ± 0.17	12.64 ± 0.03	**12.30 ± 0.26**
Breast-cancer	141.6 ± 1.98	16.50 ± 2.60	23.20 ± 1.80	8.50 ± 0.34	19.15 ± 0.40	**6.70 ± 0.58**
Credit-a	96.30 ± 2.57	37.10 ± 3.06	80.27 ± 2.10	9.70 ± 1.10	17.54 ± 0.32	**9.40 ± 1.14**
Credit-g	222.5 ± 2.19	76.90 ± 4.99	227.7 ± 3.51	**11.30 ± 0.47**	64.75 ± 1.50	12.10 ± 0.64
Dermatology	48.60 ± 0.93	45.00 ± 1.89	30.80 ± 0.31	**19.20 ± 0.83**	44.47 ± 0.63	20.20 ± 1.47
Glass	43.90 ± 1.08	48.30 ± 2.05	60.87 ± 1.12	11.50 ± 1.16	10.73 ± 0.14	**9.20 ± 1.10**
Heart-c	110.1 ± 2.48	50.80 ± 4.61	62.20 ± 1.16	7.20 ± 0.53	27.65 ± 0.58	**5.90 ± 0.35**
Heart-h	108.3 ± 2.00	47.80 ± 2.80	48.67 ± 2.43	8.70 ± 0.86	21.49 ± 0.41	**6.70 ± 0.40**
Ionosphere	23.30 ± 0.87	19.50 ± 1.74	11.73 ± 1.34	**8.20 ± 0.73**	11.04 ± 0.17	10.80 ± 0.81
Liver-disorders	76.50 ± 1.37	33.50 ± 3.42	10.33 ± 0.45	**9.20 ± 0.57**	11.78 ± 0.08	11.76 ± 2.34
Pima	135.5 ± 2.05	36.60 ± 4.09	126.9 ± 2.56	**8.20 ± 0.42**	15.93 ± 0.14	10.20 ± 0.51

Table 4. Average number of rules (average ± standard error) measured by tenfold cross-validation

Data set	CN2	C4.5rules	PSO/ACO2	Ant-Miner	$_c$Ant-Miner$_{PB}$	Ant-Miner$_{PAE}$
Anneal	19.80 ± 0.47	20.03 ± 0.89	15.51 ± 0.09	9.00 ± 0.30	16.12 ± 0.13	**8.6 ± 0.37**
Balance-scale	50.80 ± 1.57	35.70 ± 1.43	25.23 ± 0.16	13.50 ± 0.17	13.64 ± 0.03	**13.20 ± 0.20**
Breast-cancer	60.20 ± 0.77	8.60 ± 1.08	12.97 ± 0.13	6.30 ± 0.21	10.36 ± 0.11	**6.10 ± 0.28**
Credit-a	35.50 ± 0.70	14.00 ± 0.98	25.30 ± 0.11	**7.00 ± 0.49**	12.31 ± 0.10	7.70 ± 0.58
Credit-g	74.40 ± 0.93	27.50 ± 1.27	54.15 ± 0.12	**8.40 ± 0.27**	28.57 ± 0.27	9.30 ± 0.30
Dermatology	18.80 ± 0.44	19.60 ± 0.34	10.29 ± 0.08	8.10 ± 0.35	19.17 ± 0.11	**7.90 ± 0.53**
Glass	16.70 ± 0.26	14.90 ± 0.50	20.34 ± 0.08	7.40 ± 0.45	9.41 ± 0.08	**7.30 ± 0.62**
Heart-c	34.40 ± 0.40	14.90 ± 1.08	15.87 ± 0.05	6.70 ± 0.33	12.85 ± 0.14	**5.50 ± 0.22**
Heart-h	35.10 ± 0.75	12.90 ± 0.59	14.75 ± 0.08	**5.30 ± 0.15**	10.96 ± 0.12	7.00 ± 0.37
Ionosphere	10.40 ± 0.31	10.00 ± 0.36	**3.36 ± 0.07**	8.40 ± 0.64	9.18 ± 0.09	11.10 ± 0.80
Liver-disorders	29.10 ± 0.46	12.20 ± 0.93	21.53 ± 0.07	9.20 ± 0.44	10.39 ± 0.07	**9.00 ± 1.43**
Pima	49.50 ± 0.48	12.50 ± 0.92	33.55 ± 0.13	**9.00 ± 0.37**	14.89 ± 0.09	10.90 ± 0.41

same time, there are 7 groups of data sets Ant-Miner$_{PAE}$ can get the shortest number of rules, but they are not completely consistent.

5 Conclusion

In this paper, we proposed a new improved ant colony classification mining algorithm based on the principle of attraction and exclusion of pheromone. It can effectively improve the rule list search quality and accuracy. Our experiments using 12 publicly available data sets show that the predictive accuracy obtained by the Ant-Miner$_{PAE}$ algorithm with the proposed pheromone attraction and exclusion strategy is statistically significantly higher than the predictive accuracy of other rule induction classification algorithms, and the rule lists discovered by the Ant-Miner$_{PAE}$ algorithm are considerably simpler. There are several interesting directions for future research. First, the difference of prediction accuracy, simplicity of rules and time from different data sets by the same algorithm for classification rules mining is big. It would be interesting to study the correlation between them. Second, we want to further improve the prediction accuracy on different data sets. In addition, it would be interesting to study how to obtain a large speed up by running a parallel version of Ant-Miner$_{PAE}$.

Acknowledgments. This work was partially supported by Science and Technology Project of Guangdong Province of China (Grant Nos. 2015A020209119 and 2014A020208087), National Natural Science Foundation of China (Grant No. 61573157), and Science and Technology Project of Guangdong Province of China (Grant No. 2012BM0500054), and Fund of Natural Science Foundation of Guangdong Province of China (Grant No. S2013040015755) and the Foundation for Distinguished Young Talents in Higher Education of Guangdong Province (Grant 2013LYM_0119). The authors also gratefully acknowledge the reviewers for their helpful comments and suggestions that helped to improve the presentation.

References

1. Han, J., Micheline, K.: Data Mining: Concepts and Techniques. Morgan Kaufmann Publishing, Burlington (2010)
2. Vasant, D., Dashin, C., Foster, P.: Discovering interesting patterns for investment decision making with GLOWER: a genetic learner overlaid with entropy reduction. Data Min. Knowl. Discov. **4**(4), 251–280 (2000)
3. Freitas, A.A., Wieser, D.C., Apweiler, R.: On the importance of comprehensible classification models for protein function prediction. IEEE/ACM Trans. Comput. Biol. Bioinform. **7**(1), 172–182 (2010)
4. Dorigo, M.: Learning by probabilistic Boolean networks. In: Proceedings of the IEEE International Conference on Neural Networks, pp. 887–891 (1994)
5. Dorigo, M., Maniezzo, V., Colorni, A.: Ant system: optimization by a colony of cooperating agents. IEEE Trans. Syst. Man Cybern. B Cybern. **26**(1), 29–41 (1996)
6. Dorigo, M., Stützle, T.: Ant Colony Optimization. MIT Press, Cambridge (2004)
7. Otero, F.E., Freitas, A., Johnson, C.G.: A new sequential covering strategy for inducing classification rules with ant colony algorithms. IEEE Trans. Evol. Comput. **17**(1), 64–76 (2013)

8. Parpinelli, R.S., Lopes, H.S., Freitas, A.: Data mining with an ant colony optimization algorithm. IEEE Trans. Evol. Comput. **6**(4), 321–332 (2002)
9. Liu, B., Pan, J.: Research on classification algorithm based on ant colony optimization. Comput. Appl. Softw. **24**(4), 50–53 (2007)
10. Parpinelli, R.S., Lopes, H.S., Freitas, A.: An ant colony algorithm for classification rule discovery. Data Mining: A Heuristic Approach, pp. 191–208. Springer, Heidelberg (2002)
11. Parpinelli, R.S., Lopes, H.S., Freitas, A.: An ant colony based system for data mining: applications to medical data. In: Proceedings of the Genetic and Evolutionary Computation Conference (GECCO-2001), pp. 791–797 (2001)
12. Quinlan, J.R.: C4.5: Programs for Machine Learning. Morgan Kaufmann Publishing, San Mateo (1993)
13. Clark, P., Niblett, T.: The CN2 induction algorithm. Mach. Learn. **3**(4), 261–283 (1989)
14. Clark, P., Boswell, R.: Rule induction with CN2: some recent improvements. In: Proceedings of the European Working Session on Learning (EWSL-91), pp. 151–163. Springer, Heidelberg (1991)
15. Meyer, F., Parpinelli, R.S.: ACO: Public Software (2012). http://www.aco-metaheuristic. org/aco-code/public-software.html
16. Jin, P., Zhu, Y., Hu, K., et al.: Classification rule mining based on ant colony optimization algorithm. Intell. Control Autom. **344**, 654–663 (2006)
17. Liu, B., Abbass, H.A., McKay, B.: Classification rule discovery with ant colony optimization. In: Proceedings of the IEEE/WIC International Conference on Intelligent Agent Technology, pp. 83–88 (2003)
18. Wang, Z., Feng, B.: Classification rule mining with an improved ant colony algorithm. In: AI 2004: Advances in Artificial Intelligence, pp. 177–203 (2005)
19. Liu, B., Abbass, H.A., McKay, B.: Density-based heuristic for rule discovery with Ant-Miner. In: Proceedings of the 6th Australasia-Japan Joint Workshop on Intelligent and Evolutionary System, Canberra, Australia, pp. 180–184 (2002)
20. Martens, D., De Backer, M., Haesen, R., et al.: Classification with ant colony optimization. IEEE Trans. Evol. Comput. **11**(5), 651–665 (2007)
21. Chan, A., Freitas, A.A.: A new classification-rule pruning procedure for an ant colony algorithm. In: Talbi, E.-G., Liardet, P., Collet, P., Lutton, E., Schoenauer, M. (eds.) EA 2005. LNCS, vol. 3871, pp. 25–36. Springer, Heidelberg (2006)
22. Otero, F., Freitas, A., Johnson, C.: An ant colony classification algorithm to cope with continuous attributes. In: Dorigo, M., Birattari, M., Blum, C., Clerc, M., Stützle, T., Winfield, A.F.T. (eds.) ANTS 2008. LNCS, vol. 5217, pp. 48–59. Springer, Heidelberg (2008)
23. Otero, F., Freitas, A., Johnson, C.: Handling continuous attributes in ant colony classification algorithms. In: Proceedings of the Computational Intelligence and Data Mining, pp. 225–231 (2009)
24. Chan, A., Freitas, A.: A new ant colony algorithm for multi-label classification with applications in bioinformatics. In: Proceedings of the 8th Annual Conference on Genetic and Evolutionary Computation, pp. 27–34 (2006)
25. Smaldon, J., Freitas, A.: A new version of the Ant-Miner algorithm discovering unordered rule sets. In: Proceedings of the 8th Annual Conference on Genetic and Evolutionary Computation, pp. 43–50 (2006)
26. De Mántaras, R.L.: A distance-based attribute selection measure for decision tree induction. Mach. Learn. **6**(1), 81–92 (1991)
27. Asuncion, A., Newman, D.: UCI Machine Learning Repository, School of Information and Computer Science, University of California, Irvine (2010). http://archive.ics.uci.edu/ml/

A Dynamic Search Space Strategy for Swarm Intelligence

Shuiping Zhang, Wang Bi[(✉)], and Xuejiao Wang

Faculty of Information Engineering, Jiangxi University of Science and Technology,
Ganzhou, Jiangxi, China
happyeveryday_386@163.com
http://www.springer.com/lncs

Abstract. As an appendix which is designed to embed in one of the complete swarm intelligence algorithms, the novel strategy, named dynamic-search-spaces (DS) is proposed to deal with the premature convergence of those algorithms. For realizing the decrement of search space, the differences or the distances between individual sites and the site of the global performance are to form the threshold of the self-adaption system. Once the value reached by calculating the quotient of sum of those sitting near the global performance and others over a stated-percentage, the system is working to readjust the borders of search space by the site of the global performance. After each readjustment, the re-initialize to distribute individuals in the whole search space should be achieved to enhance individuals' vitality which prove away from the premature convergence. Meanwhile, the simpler verifications are provided. The improvements of results are exhibited embedding in the genetic algorithm, the particle swarm optimization and the differential evolution. This dynamic search space scheme can be embedded in most of swarm intelligence algorithms easily *abstract* environment.

Keywords: Swarm intelligence · Self-adaption · Search space · Particle swarm optimization

1 Introduction

In the past three decades, swarm intelligence algorithms as portion of Bionic algorithm, such as genetic algorithms (GA), particle swarm optimization algorithms (PSO), are applied in many fields [1,2,4] to optimize performance when solving complex problems. Thousands of strategies have been provided to advance the suitability and robustness, improving the convergence speed for those algorithms. Unfortunately, the premature convergence, as a nightmare, still exists and detracts the performance with a low degree of accuracy of the solution in most of swarm intelligence algorithms. Although many of researches which were collected by this study can be applied to one or a few kinds of swarm intelligence algorithms perfectly, researchers have to discover some novel methods for emerging swarm intelligence algorithms to deal with the same problem–the premature convergence.

© Springer Science+Business Media Singapore 2016
K. Li et al. (Eds.): ISICA 2015, CCIS 575, pp. 107–115, 2016.
DOI: 10.1007/978-981-10-0356-1_11

As a frequently used parameter to solve the global optimization, the search space, the necessary factor in any unconstrained global optimization algorithms, is discussed. The utilization of reducing the search space has been opted in many fields such as data mining [5], SVM [6], engineering design [3,7]. These applications sustain the hypothesis laterally. And meanwhile, there are many researches based the bound handling strategies [12–15] to improve PSOs. But improvements which are supported by the only bound handing are senseless. Mixing with the issue that no all of the swarm intelligence algorithms will meet individuals out of the search space frequently, the control of search space is not usual to optimize singly. And moreover, the threshold for handing the ripe for reconstructing the search space is difficult to set to apply in practical. Meeting this problem, the research should be discussed from the timing of regularizing search space, the center of compressed space and the strategy of re-initialization. In this paper, a hypothesis is to ease even more solve the premature problem through reducing the search space as versatilely as Opposition-Based Learning [8] for most of swarm intelligence algorithms. An additional self-adaption system is designed to reinforce robustness and the ability of real-time readjustment.

Abstracting some traits as the necessary conditions for applying the dynamic search space strategy to all swarm intelligence algorithms are mentioned in section two. Moreover, the simpler verifications are shown to explicate the feasibility of the increase in solution's accuracy degree with dynamic search space strategy. In section three, specific measures and pseudo-codes about how to toggle the trigger of self-adaption system are exhibited. Final part is algorithm implementation based on the simulation experiment of benchmark functions.

2   Basic Ideal

2.1   Conditions

For continuous function, the essence of dynamic search space strategy is that search space is compressed when the individual difference is below the fixed threshold. Ideally, the accuracy of solution is increasing with the size of search space reducing. First of all, the conditions and their mathematical expressions which are regarded as the prerequisites of applying this strategy to one swarm intelligence algorithm are defined.

1. Some individuals in swarm are necessary to reckon the degree of swarm's convergence. Normally, the number of individuals is not less than ten in most of swarm intelligence algorithms.

$$s_i \in S, i \geq K.$$

where
s_i is an individual, S is the swarm, K is the number of individuals.

2. The individuals in swarm are convergent.

$$if \ \forall m \geq M \ then \ \| s_i^m - s_i^{m-1} \| \to 0 \tag{1}$$

where

m is the iteration times;

$\| s_i^m - s_i^{m-1} \|$ is the difference of an individual between twice adjacent positions.

The difference approximates zero meaning that individuals are convergent.

Hitherto, the algorithm which can be used with dynamic search space strategy is definable. Then, its rationality should be discussed immediately.

3   Specification

The premature is caused by the reducing of individual differences with the increasing of iteration times. In this study, the size of search space will be decreased and the individual informations will be reset but the social informations like global solutions if individual differences are tiny. Moreover, the discussion of expanding the search space when meeting a unprecise space is insufficient because lack of effective strategy.

3.1   Decreasing and Expanding Space

Primarily, it is when or which iteration time will start to adjust the search space that need to define very clearly to protect the flair of each algorithm. An early adjustment would result in an unprecise or erroneous scope of the new space. For accounting for how does this research work, Eqs. 2a and 2b is pointed out as a threshold for decreasing the search space.

Definition 1. *This kind of individuals could be called approximate convergence, if the difference between one individual and the global optimal solution is smaller than a predetermined value. The time of iteration could be called approximate convergence point, if almost individuals are approximate convergence at that time.*

$$f\_thold = \|ub, lb\| * \alpha \tag{2a}$$

$$c\_rate = \frac{\sum_{i=1}^{i<N} (\|s_i, s_g\| <= f\_thold)}{N} \tag{2b}$$

Where

ub, lb are the borders of search space;

f_thold is the threshold of approximate convergence;

α is the compression ratio;

$c\_rate \in (0, 1)$ is the ratio of those individuals which near s_g to all.

In Eq. 2a, the threshold which correlated positively with the scope of current search space is the standard to decide which individual is approximate convergence and which is not. In Eq. 2b, an approximate convergence individual is counted as 1. Once c_rate is greater than a key ratio presented key_rate, meeting the approximate convergence point, decrement of search space begin. Generally,

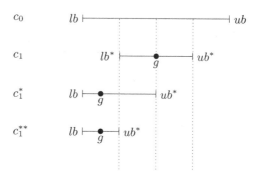

Fig. 1. The scheme of decreasing/expanding the search space.

α is smaller than $1e - 3$ and *key_rate* is greater than 0.8, but not invariable. If individuals' convergence are fast, *key_rate* can be set more closer to 1. And also, the different methods of reckoning $\|lb, ub\|$ lead to different α. If the value is reckoned in each dimension respectively, α will smaller than which reckoned with the integral space.

After setting those thresholds, determining how to decrease the search space is to be concerned in Fig. 1. The primal interval $[lb, ub]$ is decreased based on the global optimal solution g. The length of new search space in each dimension, as the case c_1, c_1^* or c_1^{**}, is flexible. Formula representation for the consider interval is Eq. 3.

$$\begin{cases} l = \|lb, ub\| * ratio \\ lb^* = g - max\,(lb, g - l/2) \\ ub^* = g + min\,(g + l/2, ub) \end{cases} \qquad (3)$$

where
l is the remaining length of each interval;
$[lb^*, ub^*]$ is the new interval;
ratio is a percentage to represent the degree of decrement.

$$l = max\,(l, f\,(\|lb, ub\|)) \qquad (4)$$

3.2   Re-initializing

At the begin of each algorithm, the apace changing of global optimal solution is easy to discover. But, when beginning the adjustment, the individual differences dive to a tiny value. That means if there is nothing else to do except decrease the search space, there will be no advance for final solution. Thereby, with each readjusting the search space, the initialization is required. The unusual initialization need to reserve the information at least including the global optimal solution and distribute individuals to whole search space uniformly using Eq. 5. Those reserved information are heuristic factors applied in the further solving which are to offer great help.

Algorithm 1. Dynamic search space strategy

1: Initialize the algorithm as normal.
2: **while** iteration time smaller than Max-Time **do**
3:    **for** traverse each individual x_k **do**
4:       adjust the information of x_k
5:    **end for**
6:    calculate values in Eqs. 2a and 2b.
7:    **if** $c\_rate > key\_rate$ **then**
8:       //discussed in the Definition 1.
9:      **if** allow to expand **then**
10:        expanding the search space using Eq. 3
11:      **else**
12:        decreasing the search space using Eqs. 3 and 4
13:      **end if**
14:      re-initializing with Eq. 5 and reserving informations.
15:    **end if**
16: **end while**

$$s_i = rand()\frac{ub - lb}{L} + lb + \frac{ub - lb}{L}(i - 1) \qquad (5)$$

Illustrating how dose this strategy work in practice is shown in Algorithm 1. The additional judgement system completes the whole works independently with litter change of primal algorithms. This is what the study seeks. In the following sections, the implementations of algorithm based GA will be more clear expositions.

4   Algorithm Implementation

In order to verify the advantages of dynamic search space strategy, there are six benchmark functions which are selected from the literature [9] in Table 1. Four of benchmark functions' solutions are at Origin. Functions f_1, f_2 and their shifts f_3, f_4 are compared to demonstrate that the origin collapse will not occur.

Table 1. The list of benchmark functions

Name	Search space	Solution
f_1:Ackley	$[-32, 32]^D$	$f_{min}\left(0.0^D\right) = 0$
f_2:Rastrigin	$[-5, 5]^D$	$f_{min}\left(0.0^D\right) = 0$
f_3:SftAckley	$[-22, 45]^D$	$f_{min}\left(10.0^D\right) = 0$
f_4:SftRastrigin	$[5, 15]^D$	$f_{min}\left(10.0^D\right) = 0$
f_5:Rosenbrock	$[-100, 200]^D$	$f_{min}\left(0.0^D\right) = 0$
f_6:Schwefel	$[-100, 110]^D$	$f_{min}\left(0.0^D\right) = 0$

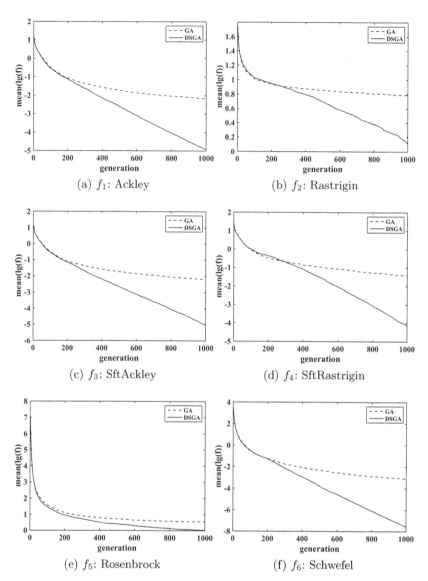

Fig. 2. The convergence curves of GA

Functions f_5, f_6 are set with large search space to exhibit the robust of this study. The solutions of each iteration resulted by one hundred independence experiments were kept during respectively.

$$y_i = \sum \lg(s_i)/100 \tag{6}$$

Both of Fig. 2 represents the average accuracy of solutions (y axis) which calculated by Eq. 6 over the iteration times (x axis). In Table 2, the numerical characteristic features of solutions are displayed. In Table 3, the confidence intervals are estimated in confidence level $\beta = 0.95$ based on the Bernoulli distribution, like Eq. 7, to elucidate the advances of algorithms' stability. To have a better persuasion about the universality of this study, the implements are some general versions of swarm algorithms which have no startling performance. If not, there will be a lot of time to invest for waiting the approximate convergence point.

$$k = \begin{cases} 1 & \lg(s) \le \lg(m); \\ 0 & \lg(s) > \lg(m); \end{cases} \tag{7}$$

4.1 Dynamic Search Space GA

Genetic algorithms [10,11], one of the stochastic search methods, is proposed from the natural evolution law. In this swarm system, chromosomes of individuals are encoding solutions. Evolutionary chromosomes are generated by hybridization to survive of the fittest eliminating the worst. The small differences of each chromosome give promise of convergence.

Notoriously, the solutions of some of GAs are expressed as the binary system or some sequences of 0-1. For those algorithms, an encoding system should be supplied, marching the decoding system, to change solutions' form with decimal system to sequences of 0-1 after each re-initialize.

In this experiment, $f(\|lb, ub\|)$ in Eq. 4 is $f(\|lb, ub\|) = 0$. That means there is no minimum scope. The steps show in Algorithm 2. The choosing of reserving the

---

Algorithm 2. Dynamic search space strategy for GA

1: **while** iteration time smaller than Max-Time **do**
2:     decreasing search space.
3:     re-initializing
4:     calculating fitness
5:     sorting fitness as temp
6:     reserving the middle class of original swarm instead of the top of temp.
7: **end while**

Table 2. The analysis of benchmark functions for GA

Test func	Mean		Std		Best		Worst	
	GA	DSSGA	GA	DSSGA	GA	DSSGA	GA	DSSGA
f_1	7e−03	**2e−05**	3e−03	**4e−05**	1e−03	**3e−07**	1e−02	**3e−04**
f_2	6e+00	**3e+00**	3e+00	**2e+00**	1e+00	**4e−08**	1e+01	**1e+01**
f_3	7e−03	**2e−05**	3e−03	**2e−05**	1e−03	**2e−07**	1e−02	**1e−04**
f_4	9e−01	**4e−01**	9e−01	**5e−01**	9e−07	**3e−14**	3e+00	**2e+00**
f_5	1e+01	**3e+00**	3e+01	**1e+01**	9e−03	**8e−04**	2e+02	**9e+01**
f_6	1e−03	**7e−07**	1e−03	**2e−06**	2e−05	**4e−12**	5e−03	**1e−05**

Table 3. The interval estimation for GA

Precision	Strategy	Interval estimation					
		f_1	f_2	f_3	f_4	f_5	f_6
$m = 1e - 0$	GA	[0.96,1.00]	[0.00,0.04]	[0.96,1.00]	[0.64,0.82]	[0.02,0.11]	[0.96,1.00]
	DSGA	[0.96,1.00]	[0.09,0.24]	[0.96,1.00]	[0.73,0.89]	[0.38,0.58]	[0.96,1.00]
$m = 1e - 2$	GA	[0.70,0.87]	[0.00,0.04]	[0.74,0.90]	[0.27,0.46]	[0.00,0.05]	[0.96,1.00]
	DSGA	[0.96,1.00]	[0.02,0.13]	[0.96,1.00]	[0.41,0.61]	[0.00,0.05]	[0.96,1.00]
$m = 1e - 4$	GA	[0.00,0.04]	[0.00,0.04]	[0.00,0.04]	[0.15,0.32]	[0.00,0.04]	[0.02,0.13]
	DSGA	[0.89,0.98]	[0.00,0.07]	[0.90,0.99]	[0.36,0.56]	[0.00,0.04]	[0.96,1.00]
$m = 1e - 6$	GA	[0.00,0.04]	[0.00,0.04]	[0.00,0.04]	[0.00,0.05]	[0.00,0.04]	[0.00,0.04]
	DSGA	[0.02,0.13]	[0.00,0.05]	[0.04,0.15]	[0.28,0.48]	[0.00,0.04]	[0.79,0.93]
$m = 1e - 8$	GA	[0.00,0.04]	[0.00,0.04]	[0.00,0.04]	[0.00,0.04]	[0.00,0.04]	[0.00,0.04]
	DSGA	[0.00,0.04]	[0.00,0.04]	[0.00,0.04]	[0.19,0.38]	[0.00,0.04]	[0.27,0.46]

top of temp instead with the middle class of original swarm is to reserve the social informations and promote the individual differences between the swarm. Figure 2 illustrates compares of the convergence curves. The source algorithm is GA and the extended algorithm is DSGA. From Fig. 2, at the begin of generations, the curves of GA and DSGA are overlapping. Meantime, the downtrend of those curves are clear. With the gentle curves of GA, the curves of DSGA branch off them keeping the good trends except Fig. 2e and b. Fortunately, Fig. 2b has a good search space, so the curve of DSGA is more steep than Fig. 2e's. That is why the assumption of expanding search space wants to be presented. From Table 2, the advantages of DSGA can be found in all fields, especially the best soultions. From Table 3, the DSGA is more stable than the GA. For example, the probability of getting a more precise solution raises markedly with $m = 1e - 4$ in f_1, f_3 and f_6. Because the bad performances of GA in f_2 and f_5, the DSGA fails too.

5   Conclusion

Discussing how and when to readjustment search space and introducing how to achieve to embed swarm intelligence algorithms are the main objects of the dynamic search space strategy. Necessary proofs and preliminary results were provided. Both of them can present that the rationality of primal hypothesis–decreasing the search space can ease the premature convergence–is sufficient. The results of experiments based on GA also imply that the premature convergence can be prevented. Albeit unable to enumerate, the other swarm intelligence algorithms with those mentioned conditions in Sect. 2.1 will be embedded inductively. This strategy can be applied to almost algorithms with no alteration of the primal design. One could conclude that the cooperation between readjusting search space and re-initialize individuals should enhance the global performance of algorithms. However, this research is still required finding the vicinity of the real solution as early as possible, proposing more effective self-adaption system and readjustment scheme, especially the scheme of expanding and dealing with the further improvement.

References

1. Zhang, J., Xin, B., Chen, J.: Hybridizing differential evolution and particle swarm optimization to design powerful optimizers: a review and taxonomy. IEEE Trans. Syst. Man Cybern. Part C Appl. Rev. **42**(5), 744–767 (2012)
2. Poli, R.: Analysis of the publications on the applications of particle swarm optimisation. J. Artif. Evol. Appl. **2008**, 1–10 (2008)
3. Khare, A., Rangnekar, S.: A review of particle swarm optimization and its applications in solar photovoltaic system. Appl. Soft Comput. **13**(5), 2997–3006 (2013)
4. Ghaemi, R., Sulaiman, N., Ibrahim, H., Mustapha, N.: A review: accuracy optimization in clustering ensembles using genetic algorithms. Artif. Intell. Rev. **35**(4), 287–318 (2011)
5. Huang, J.H., Chen, T.Y.: Application of data mining in a global optimization algorithm. Adv. Eng. Softw. **66**(12), 24–33 (2013)
6. Ortiz, E.: Improving the training time of support vector regression algorithms through novel hyper-parameters search space reductions. Neurocomputing **72**, 3683–3691 (2009)
7. Bland, J.A., Nolle, L.: Self-adaptive stepsize search for automatic optimal design. Knowl.-Based Syst. **29**(3), 75–82 (2012)
8. Tizhoosh, H.R.: Opposition-based learning: a new scheme for machine intelligence. In: International Conference on Computational Intelligence for Modelling, Control and Automation, 2005 and International Conference on Intelligent Agents, Web Technologies and Internet Commerce, pp. 695–701 (2005)
9. Tang, K., Li, X., Suganthan, P.N., Yang, Z., Weise, T.: Benchmark functions for the cec2010 special session and competition on large-scale global optimization. Nature Inspired Computation and Applications Laboratory (2010)
10. Goldberg, D.E., Sastry, K.: Genetic algorithms in Search, Optimization and Machine Learning. Addison-Wesley Longman, Boston (1989)
11. Zhang, W.S., Li, K., Yu, X.: Improved evolutionary algorithm and its application to solving complex optimization problems. Appl. Res. Comput. **29**(4), 1223–1226 (2012)
12. Zhang, W.J., Xie, X.F., Bi, D.C.: Handling boundary constraints for numerical optimization by particle swarm flying in periodic search space. In: Congress on Evolutionary Computation, CEC2004, vol. 2. IEEE (2004)
13. Helwig, S., Branke, J., Mostaghim, S.: Experimental analysis of bound handling techniques in particle swarm optimization. IEEE Trans. Evol. Comput. **17**(2), 259–271 (2013)
14. Gandomi, A.H., Yang, X.-S.: Evolutionary boundary constraint handling scheme. Neural Comput. Appl. **21**(6), 1449–1462 (2012)
15. Chu, W., Gao, X., Sorooshian, S.: Handling boundary constraints for particle swarm optimization in high-dimensional search space. Inf. Sci. **181**(20), 4569–4581 (2011)

Adaptive Mutation Opposition-Based Particle Swarm Optimization

Lanlan Kang[1,2(✉)], Wenyong Dong[1(✉)], and Kangshun Li[3(✉)]

[1] Computer School, Wuhan University, Wuhan 430072, China
victorykll@163.com, dwy@whu.edu.cn
[2] School of Apply Science, Jiangxi University of Science and Technology,
Ganzhou 341000, China
[3] College of Mathematics and Information, South China Agricultural University,
Guangzhou 510641, China
likangsun@sina.com

Abstract. To solve the problem of premature convergence in traditional particle swarm optimization (PSO), This paper proposed a adaptive mutation opposition-based particle swarm optimization (AMOPSO). The new algorithm applies adaptive mutation selection strategy (AMS) on the basis of generalized opposition-based learning method (GOBL) and a nonlinear inertia weight (AW). GOBL strategy can provide more chances to find solutions by space transformation search and thus enhance the global exploitation ability of PSO. However, it will increase likelihood of being trapped into local optimum. In order to avoid above problem, AMS is presented to disturb the current global optimal particle and adaptively gain mutation position. This strategy is helpful to improve the exploration ability of PSO and make the algorithm more smoothly fast convergence to the global optimal solution. In order to further balance the contradiction between exploration and exploitation during its iteration process, AW strategy is introduced. Through compared with several opposition-based PSOs on 14 benchmark functions, the experimental results show that AMOPSO greatly enhance the performance of PSO in terms of solution accuracy, convergence speed and algorithm reliability.

Keywords: Particle swarm optimization · Adaptive mutation · Generalized opposition-based learning · Adaptive inertia weight

1 Introduction

Particle swarm optimization (PSO), firstly proposed by Kennedy and Eberhart in 1995, is a kind of swarm intelligence optimization algorithm [1, 2]. Inspired by the simulation the social behavior of birds flocks and schooling, the swarm intelligence in PSO could lead the particles to searching for global optimal solution through collective collaboration between swarms. As a result of the characteristics it possesses such as simplicity and flexibility, comparatively fewer parameters, as well as easy to implement and understand, PSO has been discussed and improved by lots of researchers in recent years, and widely applied in various scientific and engineering problems [3, 4].

© Springer Science+Business Media Singapore 2016
K. Li et al. (Eds.): ISICA 2015, CCIS 575, pp. 116–128, 2016.
DOI: 10.1007/978-981-10-0356-1_12

Whereas, similar to other evolutionary computation technique, PSO is also a population-based iterative algorithm. The basic PSO exist defects like poor local search ability, lower search accuracy, easy to be trapped into local optima and excessive reliance on parameters set and so on, in which particularly salient in solving complex multimodal problems.

Hence, around the above questions, many researchers have proposed variant methods to improve the traditional PSO from the points of parameters set, algorithm convergence and combination with auxiliary search operators [5, 6]. In the basic PSO, each particle represents a possible solution, whose flight direction in the next generation is influenced by its velocity vector, current global best position (*gbest*) and the best position of each particle (*pbest*). Therefore, it is key to improve the basic PSO that how to control size of velocity when entering into next generation which can better control the scope of the search and how to make full use of the value of gbest and pbest which not only lead particles to fast into the global optimal location but also avoid falling into local optimal.

In the last decade, numerous empirical studies have been conducted in order to get a group of best parameter values aiming to a specific problem, but it is impractical. Preferably, a set of control parameters should be adaptively adjusted for different problem [7]. In literature [8], Zhang et al. presented APSO to adapt the swarm size, inertia weight and the neighborhood size of each particle automatically based on the fitness of particles. M. Hu et al. further put forward a parameter control mechanism to adaptively change the inertia weight etc. related parameters to pick one randomly selected particle close to the global best position and thus improve the robustness of PSO-MAM [9]. In a more general, parameter control in evolutionary algorithms and its trends and challenges are surveyed in literature [10]. Opposition-based learning (OBL) was firstly introduced into PSO for enhance local search ability. Meanwhile, Cauchy mutation operator was used to disturb the current global best position in literature [11]. Wang et al. proposed an adaptive mutation strategy for PSO to dynamically choose different mutation operators during the evolution [12], in which each mutation operator has an independent selection ratio to choose the best mutation operator employing roulette wheel selection mechanism among Gaussian mutation (GPSO), Cauchy mutation(CPSO), Lévy mutation (LPSO). Multifrequency vibrational PSO is proposed in literature [13] to avoid premature convergence by periodic mutation application strategy and diversity variety based on an artificial neural network. In a more general,

For purpose of performance, a novel adaptive mutation opposition-based particle swarm optimization called AMOPSO is proposed in this paper through analyzing relationship of distance between pbest and gbest. AMOPSO is composed of AMS and AW. Thereinto, AMS which means global optimal adaptive mutation selection strategy is first presented in this paper. It is integrated into generalized opposition-based learning (GOBL) [14] and makes the gbest generate mutation position when the pbest of all particles tends to converge to gbest according to the standard deviation between pbest and gbest. AMS helps effectively trapped particles to escape and thus enhance the ability of global search of algorithm. Meanwhile, this paper applied a nonlinear adaptive inertia weight strategy called AW to further balance the global search and local explorative ability of the algorithm which make the inertia weight adaptive change with the objective function value of the particle. Through compared with several opposition-based PSOs on 14 benchmark functions, a series of experimental trials confirm that AMOPSO substantially enhance the performance of PSO in aspect of solution accuracy, convergence speed and algorithm reliability.

The rest of this paper is organized as follows. Section 2 briefly introduces the standard PSO algorithm and GOBL strategy. Section 3 presents the AMS and AW strategy in detail. The AMOPSO is proposed including the detailed procedure of algorithm in Sect. 4. Section 5 experimentally compares the AMOPSO with various opposition-based learning PSO algorithms on 14 benchmark problems, of which the parameter sensitivity study is given simultaneously. Finally, conclusions and discussions on the future work are given in Sect. 7.

2 Related Work

2.1 Particle Swarm Optimization

Particle swarm optimization (PSO) is a population-based stochastic search algorithm. A particle represents a candidate solution, and has two attributes of velocity and position. Each individual in colony is considered as a particle without quality and volume in D dimensional search space and moves in a certain speed in the solution space towards their own best position of each particle (pbest) and historical best position of all particles (gbest) that has realized to the evolution of the candidate solution. It is supposed that the size of population is N, the velocity $v_{i,j}$ and position $x_{i,j}$ of the jth dimension of the ith particle are updated according to the following equations [1, 2]:

$$v_{i,j}(t+1) = w \cdot v_{i,j}(t) + c_1 rand_1 (pbest_{i,j} - x_{i,j}(t)) + c_2 rand_2 (gbest - x_{i,j}(t)) \quad (1)$$

$$x_{i,j}(t+1) = x_{i,j}(t) + v_{i,j}(t+1) \quad (2)$$

Where, $i = 1, 2, \cdots, N$, $j = 1, 2, \cdots, D$, c_1 and c_2 are two acceleration coefficients which control the influence of the social and cognitive components and guide each particle toward the pbest and the gbest, respectively. The inertial factor w was proposed by Shi and Eberhart [15], In general, $c_1, c_2 \in [0, 2]$. $rand_1$ and $rand_2$ are two uniformly distributed random variables in the interval [0,1].

2.2 A Generalized Opposition-Based Learning (GOBL)

Opposition-based learning (OBL) is firstly proposed by H.R.Tizhoosh in 2005 [16]. A number of study shows that OBL is effective concept introduced to various optimization algorithm. In 2007, OBL was applied to PSO and achieved good performance. Let $X = (x_1, x_2, \cdots, x_D)$ is a particle in a D-dimension space. The opposite point $\breve{X} = (\breve{x}_1, \breve{x}_2, \cdots, \breve{x}_D)$ is defined by:

$$\breve{x}_j = a_j + b_j - x_j \quad (3)$$

Where, $x_j \in [a_j, b_j], j = 1, 2, \cdots D, a_j$ and b_j are dynamic boundaries of x_j. Because the chance of fitness of opposite point being better than the original particle is accounted for 50 %, the probability of obtaining optimal solution will be greatly increased in the combined search space between opposite space and original space.

A generalized OBL, called GOBL [17], is an improvement of OBL which can provide more chance of finding candidate solutions closer to the global optimum. GOBL applied to PSO is defined as follows:

Let $X_i = (x_{i,1}, x_{i,2}, \cdots, x_{i,D})$ is the ith particle in a D-dimension space. The opposite particle $\breve{X}_i = (\breve{x}_{i,1}, \breve{x}_{i,2}, \cdots, \breve{x}_{i,D})$ is defined by:

$$\breve{x}_{i,j} = k \cdot (da_j + db_j) - x_{i,j} \tag{4}$$

Where, $j = 1, 2, \cdots D$, k is a random number drawn from the uniform distribution in the interval [0,1]. da_j and db_j are the minimum and maximum values of the jth dimension of the ith particle, respectively:

$$da_j = \min(x_{i,j}), \quad db_j = \max(x_{i,j}) \tag{5}$$

The generated opposite position of each iteration will always located in the diminishing search space in the setting of dynamic boundaries $[da_j, db_j]$. If it jumps out of the feasible solution boundary, the generated opposite position is assigned to a random value as follows:

$$\breve{x}_{i,j} = randn(da_j, db_j) \quad if \ \breve{x}_{i,j} < da_j \ or \ db_j > \breve{x}_{i,j} \tag{6}$$

3   Adaptive Mutation Opposition-Based PSO

This section presents a novel adaptive mutation operator to help current global best particle jump out local optimum and expend global search space. Meanwhile, adaptive inertia weight operator is introduced. Two kinds of adaptive operation integrated into GOBL strategy forms a new algorithm, i.e. adaptive mutation opposition-based PSO.

3.1   Global Adaptive Mutation Selection Operator

Some theoretical analyses have shown that the particle in PSO will oscillate between their history best particle and the current global best particle before it converges [18, 19]. Therefore, it will effectively avoid being trapped into local optimum and occurrence of premature convergence if the search neighborhood of global best particle is expended in each generation. For this purpose, a global adaptive mutation selection operator (AMS) is proposed in this paper. The relations of evolutionary generation and distance between the best position of each particle (*pbest*) and current global best position (*gbest*) are comprehensively considered in AMS, and then calculate the distance in each dimension, the nearer the distance is, which means higher similarity between the particles, the further mutation position could be generated in corresponding dimension. Finally, the new particle *gbest** will be formulated after D time mutation operators in each generation. If the fitness of *gbest** is superior to the fitness of original gbest, *gbest* will be replaced by *gbest**.

In AMS, generated factor *xm* for mutation in the *i*th dimension is defined by:

$$xm(i) = \exp(-\lambda \cdot t/t_{\max}) \cdot (1 - r(i)/r_{\max}) \tag{7}$$

Where, λ is a constant, t is iteration times, $t_{\max}$ is the maximum of t, $r(i)$ is the distance from particles to *gbest* in ith dimension in a generation. $r_{\max}$ is the maximum of r vector. $r(i)$ is defined as follows:

$$r(i) = |gbest(i) - avg\_pbest(i)| \tag{8}$$

Where, $i = 1, 2, \cdots, D, avg\_pbest(i)$, a mean value of the *pbest* of N particles, is be formulated as follows:

$$avg\_pbest(i) = (\sum_{j=1}^{N} pbest[j][i])/N \tag{9}$$

Where, *pbest*[j][i] is the value of pbest in the *i*th dimension of the *j*th population. N is the size of population.

xm as a generated factor insert into mutation function (10).

$$F(xm) = \frac{1}{\pi}\arctan(xm) + \frac{1}{2} \tag{10}$$

By function (10), adaptive mutation value is generated. Figure 1 shows value of *F*(*xm*) when *xm* inserts into function (10).

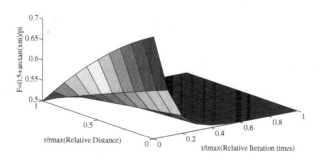

Fig. 1. Expected value of adaptive mutation selection (AMS) in three-dimensional space

The results have been drawn from Fig. 1. At the beginning of iteration, the performance of individual is worse, larger mutation value can cause enough disturbances to *gbest*, thus exploring all the promising regions. With the increasing of iteration times, the mutation value is decreasing, thereby ensuring the optimization problems smooth convergence to the optimal value. Meanwhile, AMS will get larger mutation value when all individuals converge to *gbest* that enhance the global searching ability, whereas, when search space is enough large, AMS will reduce mutation value to avoid fluctuation of optimal value and accelerate convergence speed of algorithm.

Global adaptive mutation operation (AMS) is defined as follows in the *i*th dimension:

$$gbest^*(i) = gbest(i) + F(xm) \tag{11}$$

In each generation, recorded the position of the *i*th dimension after mutation operation using AMS, the global best mutation position is be generated, that is *gbest**.

3.2 Adaptive Inertia Weight Operation

To further balance the global search and local explorative ability of the algorithm, this paper applied a nonlinear adaptive inertia weight (AW). It is defined as follows [20].

$$w = \begin{cases} w_{\min} + \dfrac{(w_{\max} - w_{\min}) * (f_i - f_{\min})}{(f_{avg} - f_{\min})} & , f_i \leq f_{avg} \\ w_{\max} & , f_i > f_{avg} \end{cases} \tag{12}$$

Where, $w_{\max}$ and $w_{\min}$ are the maximum and minimum of inertia weight, respectively; f_i is the fitness of the *i*th individual; f_{avg} and $f_{\min}$ is the mean fitness and the minimal fitness of current all individuals, respectively. One the one hand, Inertia weight w will be increased when the fitness of all individuals is approximate because of convergence or falling into local optimum, whereas w is adaptively decreased. On the other hand, the inertia weight w is less to maintain the advantaged particle if its fitness is superior to mean fitness of all individuals, whereas w is increased to close up the advantaged search space.

3.3 Adaptive Mutation Opposition-Based PSO

Adaptive mutation opposition-based PSO, called AMOPSO, is composed of AMS applied to opposition-based learning PSO and adaptive inertia weight operator (AW). The main steps of AMOPSO algorithm are shown as Table 1. Where *jr* is the probability of using GOBL operator.

Through analysis to Table 1, it can be known that AMOPSO algorithm mainly includes five parts: initial population, GOBL strategy, update operations of speed and position (the basic PSO), AW operation and AMS operation. It is easy to know that the time complexity of update operations of speed and position and AW operation are $O(N)$ and $O(1)$, respectively. Initial population consists of random individuals, creation of opposite points, group selection mechanism. GOBL strategy includes dynamic boundary update operation besides creation of opposite points and group selection mechanism. Therein, the time complexity of random individuals, creation of opposite points and dynamic boundary update operation are all $O(N \cdot D)$, while the time complexity of group selection mechanism is $O(N^2)$. For GOBL strategy, if the dimension D is lesser, the size of population N can be similar to D, otherwise, D is larger. For example, N is usually less than D when $D \geq 100$. Therefore, the time complexity of GOBL strategy is $O(N \cdot D)$. Based on the analysis of the above, the computational time complexity for AMOPSO is $O(N \cdot D)$.

Table 1. The main steps of AMOPSO algorithm

randomly initialize N particles in the population P;
generate the opposite solutions of N particles   to construct opposite population OP according to Eq.(5);
calculate the fitness of P and OP;
select N best fittest solutions from $P \cup OP$ as an initial population P;
update the fitness of P;
while the stopping criterion is not meet do
 If rand(0,1)<jr then
  update the dynamic interval boundaries $[da_j, db_j]$ according to Eq.(6);
  generate the new opposite population OP of N particles according to Eq.(5);
  calculate the fitness of OP ;
  select N best fittest solutions from $P \cup OP$ as a new current population P;
  update *pbest* vector and *gbest* if needed;
 else
  for i=1 to N do
   calculate   w according to Eq.(15);
   calculate the velocity of the ith particle according to Eq.(3);
   update the position of the ith particle according to Eq.(2);
   update the fitness the ith particle;
   update *pbest_i* if needed;
  update gbest if needed;
for j=1 to D do
 generate *gbest** via mutation operation Eq.(14);
if the fitness of   *gbest** is better than gbest
  gbest = gbest*

4 Experiments

In this section, two groups of experiment are carried out based on a set of benchmark problems. The objective of the first group of experiments is to investigate the performance of AMOPSO via compared with a set of OBL-based PSO, analyze the sensitivity of key parameters, and suggest some methods to set up the parameters. In second group of experiments, the performance of AMOPSO is compared with well-known opposition-based PSO.

4.1 Benchmark Problems

All experiments are conducted by MATLAB2012 on 14 benchmark problems with dimension $D = 30$ and the size of population $N = 40$. According to their properties, the problems are divided into two categories, which are unimodal problems $f_1 \sim f_5$ and multimodal problems $f_6 \sim f_{14}$. Multimodal problem is difficult since the number of local optima grows exponentially with the increase of dimensionality. In order to further verify the performance of AMOPSO, the section of multimodal problems are composed of three groups, which are basic functions $f_6 \sim f_8$, rotated functions $f_9 \sim f_{11}$ and shifted functions $f_{12} \sim f_{14}$ [19]. All functions used in this paper are minimization problems. A brief description of these benchmark problems is listed in Table 2.

Noted, in multimodal functions $f_9 \sim f_{11}$, M is an $D \times D$ orthogonal matrix, and $x = (x_1, x_2, \cdots, x_D)$ is a D-dimensional row vector (i.e., a $1 \times D$ matrix). Multimodal functions $f_{12} \sim f_{14}$ are shifted by a random point $o = (o_1, o_2, \cdots, o_D)$ in D-dimensional search space.

Table 2. The 14 Benchmark Functions in the experiments, where D is the dimension, f_{min} is the global minimum value of the test function

	Test Function	D	Search Space	f_{min}	Name of function
Unimodal	$f_1(x) = \sum_{i=1}^{D} x_i^2$	30	$[-100, 100]^D$	0	Sphere
	$f_2(x) = \sum_{i=1}^{D} (\lfloor x_i + 0.5 \rfloor)^2$	30	$[-100, 100]^D$	0	Step
	$f_3(x) = \sum_{i=1}^{D-1} [100 \cdot (x_{i+1} - x_i^2)^2 + (1 - x_i)^2]$	30	$[-30, 30]^D$	0	Rosenbrock
	$f_4(x) = \sum_{i=1}^{D} (\sum_{j=1}^{i} x_j)^2$	30	$[-100, 100]^D$	0	Quadric
	$f_5(x) = \sum_{i=1}^{D} |x_i| + \prod_{i=1}^{D} |x_i|$	30	$[-10, 10]^D$	0	Schwefel2.22
Multimodal	$f_6(x) = \sum_{i=1}^{D} [x_i^2 - 10\cos(2\pi x_i) + 10]$	30	$[-5.12, 5.12]^D$	0	Rastrigin
	$f_7(x) = -20 \cdot \exp(-0.2 \cdot \sqrt{(\sum_{i=1}^{D} x_i^2)/D})$ $- \exp((\sum_{i=1}^{D} \cos(2\pi x_i))/D) + 20 + e$	30	$[-32, 32]^D$	0	Ackley
	$f_8(x) = \frac{1}{4000} \sum_{i=1}^{D} x_i^2 - \prod_{i=1}^{D} \cos(\frac{x_i}{\sqrt{i}}) + 1$	30	$[-600, 600]^D$	0	Griewank
	$f_9(x) = f_6(z), \quad z = x * M$	30	$[-5.12, 5.12]^D$	0	Rotated Rastrigin
	$f_{10}(x) = f_7(z), \quad z = x * M$	30	$[-32, 32]^D$	0	Rotated Ackley
	$f_{11}(x) = f_8(z), \quad z = x * M$	30	$[-600, 600]^D$	0	Rotated Griewank
	$f_{12}(x) = f_6(z), \quad z = x - o$	30	$[-5.12, 5.12]^D$	0	Shifted Rastrigin
	$f_{13}(x) = f_7(z), \quad z = x - o$	30	$[-32, 32]^D$	0	Shifted Ackley
	$f_{14}(x) = f_8(z), \quad z = x - o$	30	$[-600, 600]^D$	0	Shifted Griewank

4.2 Parameter Settings

The selection of the parameters which is very important to bionic algorithm can greatly influence the performance of PSO. To better compare the performance of AMOPSO with other similar algorithms, the parameter settings in this paper is consistent with the original algorithms as far as possible. The velocity v is limited to the half range of the search space on each dimension. The specific parameter settings in this paper are listed as Table 3.

Table 3. The Specific Parameter Settings of AMOPSO

c_1	c_2	w_{max}	w_{min}	jr	λ
1.49618	1.49618	0.4	0.2	0.3	30

Table 4. The mean value of the global optimum in run 30 times algorithm among five OBL-based PSO aiming to the 14 benchmark functions

	Funs.	PSO		OPSO		GOPSO		EOPSO		AMOPSO
	f_1	1.56E-03	+	4.59E-36	+	0.00E+00	~	0.00E+00	~	**0.00E+00**
	f_2	7.18E-02	+	2.09E-35	+	3.02E-321	+	1.97E-323	+	**0.00E+00**
	f_3	1.26E+00	~	7.18E+00	~	2.82E+01	+	**1.57E-24**	-	6.94E+00
	f_4	5.15E-02	+	4.13E+04	+	1.32E+04	+	5.01E-03	+	**0.00E+00**
The Mean Value of Fitness	f_5	1.48E-03	+	6.51E-12	+	6.98E-162	+	3.01E-162	+	**9.88E-324**
	f_6	2.06E+00	+	1.51E+01	+	0.00E+00	~	2.09E+00	+	**0.00E+00**
	f_7	4.45E-02	+	1.85E+00	+	0.00E+00	~	1.86E-01	+	**0.00E+00**
	f_8	1.14E-03	+	3.83E-01	+	0.00E+00	~	1.26E-02	+	**0.00E+00**
	f_9	2.99E+00	+	1.51E+01	+	0.00E+00	~	4.11E+00	+	**0.00E+00**
	f_{10}	2.81E-02	+	2.98E+00	+	**0.00E+00**	-	3.41E-01	+	6.63E-15
	f_{11}	7.95E-04	+	2.33E-02	+	0.00E+00	~	1.22E-02	+	**0.00E+00**
	f_{12}	4.68E+00	+	1.35E+01	+	0.00E+00	~	1.76E+00	+	**0.00E+00**
	f_{13}	4.25E-01	+	2.98E+00	+	**0.00E+00**	-	1.34E-01	+	3.55E-15
	f_{14}	7.69E+00	+	1.95E-02	+	0.00E+00	~	1.44E-02	+	**0.00E+00**
	+	13		13		4		12		--
	~	1		1		8		1		--
	-	0		0		2		1		--

4.3 Comparison Between OBL-Based PSO

OBL-based PSO involved in the comparison with AMOPSO includes EOPSO [21], GOPSO [14] and OPSO [11] in this paper. At the same time, basic PSO is also joined in the algorithms comparison. Every algorithm runs 10000 generations every time, all the experiments were conducted 30 times and then recorded the mean value of the global best optimum of each algorithm as Table 4. Where Symbol "+" denotes the performance of AMOPSO is better than other algorithm, symbol "-" denotes worse and "~" denotes equal to other algorithm. The best results among five algorithms aim to the fourteen benchmark problems are shown in bold.

It can be concluded from Table 4, AMOPSO has achieved good results in all benchmark problems except for f_3. Experimental results show that AMOPSO is obviously superior to OPSO and PSO in almost all problems though only achieved similar effect in f_3. Conversely, optimal value of f_3 is obtained by EOPSO. It is worth noting that AMOPSO is better than GOPSO in four problems and worse only in two problems, i.e. rotated ackley's function, shifted ackley's function although AMOPSO has achieved optimum value together with GOPSO in eight problems.

Table 5. The comparison between GOPSO and AMOPSO aiming at the 14 benchmark functions in a run, where fval is the global optimum, FEs is number of function evaluations.

Funs.	GOPSO		AMOPSO	
	Fval	FEs	Fval	FEs
f_1	9.31E−17	47556	4.27E−17	**10377**
f_2	0.00E + 00	11466	0.00E + 00	**2846**
f_3	2.38E + 01	400080	1.74E + 00	400080
f_4	9.97E−17	254778	3.18E−17	**14201**
f_5	9.87E−17	85230	5.96E−17	**18634**
f_6	0.00E + 00	98481	0.00E + 00	**10195**
f_7	0.00E + 00	93746	0.00E + 00	**15019**
f_8	0.00E + 00	43388	0.00E + 00	**10827**
f_9	0.00E + 00	125458	0.00E + 00	**9850**
f_{10}	0.00E + 00	**79714**	3.55E−15	817901
f_{11}	0.00E + 00	47401	0.00E + 00	**10965**
f_{12}	0.00E + 00	51078	0.00E + 00	**8993**
f_{13}	**0.00E + 00**	89212	3.55E−15	822750
f_{14}	0.00E + 00	47886	0.00E + 00	**10083**
Less FES	–	1	–	**11**

A series of experiment were conducted to further compare the comprehensive performance between AMOPSO and GOPSO and the results was listed in Tables 4 and 5. On the one hand, through comparison of their mean value, best value, worst value and standard deviation after running 30 times aiming to every test function, the results in Table 5 show the worst value of AMOPSO is better than GOPSO in almost all test function, which reflects the execution of AMOPSO is more stable than GOPSO. On the other hand, this paper records the average number of function evaluations (FEs) of every test function during 30 run of AMOPSO and GOPSO on the condition that precision is $1.0e − 16$ and maximum number of iterations is 10000, the experimental results show as Table 5 that FEs of AMOPSO is significant less than GOPSO except for f_{10} and f_{13}.

4.4   Parameter Sensitivity Study

In this section, the two parameters including λ, w in AMOPSO are investigated separately, they have greatly impact to the performance of the algorithm. Besides, it has been indicated in literature [11] that AMOPSO will get best performance when $jr = 0.3$. As mentioned, AMOPSO is composed of AMS operator, AW operator and GOBL strategy, the value of λ in AMS operator is critical factor about whether the algorithm can be smoother and faster convergence to the global optimal value. Due to limited space, Fig. 2 demonstrates the trend of the algorithm convergence to the global optimum only aiming at four test functions through λ gets different values. From Fig. 2, it can be concluded that the test function will get better value as a general rule when $\lambda = 30$.

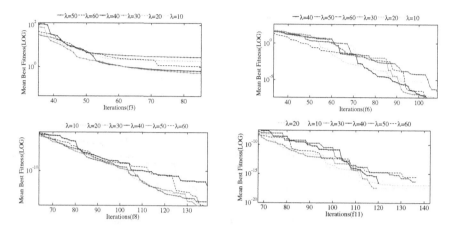

Fig. 2. The process of global convergence when λ gets different value

5   Conclusion and Future Work

PSO is efficient evolutionary computation. In order to solve the problem of premature convergence and enhance speeding of convergence, this paper proposed a new adaptive mutation opposition-based particle swarm optimization shortly AMOPSO. The new algorithm was composed of GOBL strategy and two adaptive operations, i.e. global adaptive mutation selection operation (AMS) and adaptive inertia weight operation (AW). Therein, GOBL strategy is used to generate opposite solutions of individuals, which can provide more chances to find better solution and thus enhance the global exploitation ability of algorithm. AMS operation can adaptively generate mutation value according to the clustering degree of individuals considering the distance between individual best value (*pbest*) and current global optimum (*gbest*) as well as relationship between evolution generations, which can generate enough disturbance to *gbest* when fitness of individuals tend to identical and thus avoid be trapped into local optimum. This strategy is helpful to improve the exploration ability of PSO and make the algorithm more smoothly fast convergence to the global optimal solution. w can be adaptively update by AW operation, which can further balance the global search and local explorative ability of algorithm and significantly accelerate convergence speed of algorithm.

The new algorithm compared with several opposition-based PSO on 14 benchmark functions, the experimental results show AMOPSO greatly improve accuracy and convergence speed of algorithm. However, new algorithm is still inevitable trapped into local optimum in f_3. How to further improve the performance of algorithm in more problems and apply to practical problems are future works.

Acknowledgment. We would like to thank the editors and the anonymous reviewers for their valuable comment and suggestions. This work was supported by the National Natural Science Foundation of China (No. 61170305, No. 61573157).

References

1. Kennedy, J., Eberhart, R.C.: Particle swarm optimization. In: Proceedings of IEEE International Conference on Neural Networks, vol. 4, Perth, Australia, pp. 1942–1948 (1995)
2. Eberhart, R.C., Kennedy, J.: A new optimizer using particle swarm theory. In: Proceedings of 6th International Symposium on Micro-machine Human Science, Nagoya, Japan, pp. 39–43 (1995)
3. Dehuri, S., Nanda, B.K., Cho, S.-B.: A hybrid APSO-aided learnable Bayesian classifier. In: IICAI, pp. 695–706 (2009)
4. Dehuri, S., Roy, R., Cho, S.-B.: An adaptive binary PSO to learn Bayesian classifier for prognostic modeling of metabolic syndrome. In: Genetic and Evolutionary Computation Conference (GECCO), pp. 495–501, 12–16 July 2011
5. Ismail, A., Engelbrecht, A.P.: Self-adaptive particle swarm optimization. In: Bui, L.T., Ong, Y.S., Hoai, N.X., Ishibuchi, H., Suganthan, P.N. (eds.) SEAL 2012. LNCS, vol. 7673, pp. 228–237. Springer, Heidelberg (2012)
6. Zhan, Z.H., Zhang, J., Li, Y., Chung, H.S.H.: Adaptive particle swarm optimization. IEEE Trans. Syst. Man Cybern. Part B **39**, 1369–1381 (2009)
7. Shi, Y., Eberhart, R.C.: Parameter selection in particle swarm optimization. In: Evolutionary Programming VII: Proceedings of the Seventh Annual Conference on Evolutionary Programming, New York, pp. 591–600 (1998)
8. Zhang, W., Liu, Y., Clerc, M.: An adaptive PSO algorithm for reactive power optimization In: Proceedings of 6th International Conference Advances in Power System Control, Operation and Management, pp. 302–307, November 2003
9. Mengqi, H., Teresa, W., Weir, J.D.: An adaptive particle swarm optimization with multiple adaptive methods. IEEE Trans. Evol. Comput. **17**, 705–720 (2013)
10. Karafotias, G., Hoogendoorn, M., Eiben, A.E.: Parameter control in evolutionary algorithms: trends and challenges. IEEE Trans. Evol. Comput. **19**(2), 167–187 (2015)
11. Wang H., Li, H., Liu, Y., et al.: Opposition-based particle swarm algorithm with Cauchy mutation. In: Proceedings of IEEE Congress on Evolutionary Computation, Tokyo, pp. 356–360 (2007)
12. Wang, H., Wang, W.J., Wu, Z.J.: Particle swarm optimization with adaptive mutation for multimodal optimization. Appl. Math. Comput. **221**, 296–305 (2013)
13. Pehlivanoglu, Y.V.: A new particle swarm optimization method enhanced with a periodic mutation strategy and neural networks. IEEE Trans. Evol. Comput. **17**(3), 436–452 (2013)
14. Wang Hui, H., Zhijian, W., Rahnamayan, S., et al.: Enhancing particle swarm optimization using generalized opposition-based learning. Inf. Sci. **181**, 4699–4714 (2011)
15. Shi, Y., Eberhart, R.C.: A modified particle swarm optimizer. In: Proceedings of IEEE World Congress Computational Intelligence, pp. 69–73 (1998)
16. Tizhoosh, H.R.: Opposition-based learning: a new scheme for machine intelligence. In: Proceedings of IEEE International Conference of Intelligent for Modeling, Control and Automation. PiscatAWay: Institute of Electrical and Electronics Engineers Computer Society, pp. 695–701 (2005)

17. Wang, H., Wu, Z., Liu, Y., Wang, J., Jiang, D., Chen, L.: Space transformation search: a new evolutionary technique. In: Proceedings of World Summit Genetic Evolution Computer, pp. 537–544 (2009)
18. Ozcan, E., Mohan, C.K.: Particle swarm optimization: surfing and waves. In: Proceedings of Congress on Evolutionary Computation (CEC 1999), Washington DC, pp. 1939–1944 (1999)
19. van den Bergh, F., Engelbrecht, A.P.: Effect of swarm size on cooperative particle swarm optimizers. In: Genetic and Evolutionary Computation Conference, San Francisco, USA, pp. 892–899 (2001)
20. Gong, C., Wang, Z.: Proficient optimization calculation in MATLAB. Electronic Industry Press, Beijing (2012)
21. Zhou, X.Y., Wu, Z.J., Wang, H., et al.: Elite opposition based particle swarm optimization. Acta Electron. Sin. **41**(8), 1647–1652 (2013)

Quick Convergence Algorithm of ACO Based on Convergence Grads Expectation

Zhongming Yang[1,2(✉)], Yong Qin[3], Huang Han[4], and Yunfu Jia[5]

[1] Computer Engineering Technical College, Guangdong Institute of Science and Technology, Zhuhai, Guangdong, China
yzm8008@126.com
[2] Guangdong Provincial Key Laboratory of Petrochemical Equipment Fault Diagnosis, Maoming, Guangdong, China
[3] College of Computer Science, Dongguan University of Technology, Dongguan, Guangdong, China
mmcqinyong@126.com
[4] College of Computer Science and Engineering, South China University of Technology, Guangzhou, Guangdong, China
hhan@scut.edu.cn
[5] Department of Information Technology, Zhuhai Technician College, Zhuhai, Guangdong, China

Abstract. While the ACO can find the optimal path of network, there are too many iterative times and too slow the convergence speed is also very slow. This paper proposes the Q-ACO QoSR based on convergence expectation with the real-time and the high efficiency of network. This algorithm defines index expectation function of link, and proposes convergence expectation and convergence grads. This algorithm can find the optimal path by comparing the convergence grads in a faster and bigger probability. This algorithm improves the ability of routing and convergence speed.

Keywords: ACO · QoS · QoSR · CG expectation · Q-ACO

1 Introduction

ACO is a heuristic searching algorithm method frequently used in the solution of combinatorial optimization problem based on the simulation of ant's foraging behavior of ants. ACO was firstly proposed by Italian scholar Dorigo [1] in the early 1990s. Accounting for the independence on mathematics description of specified problems, it was successfully used in the solution of traveling salesman problem (TSP) [2], quadratic assignment problem (QAP) [3], graph coloring problem (GCP) [4], fleet scheduling problem (FSP) [5] and other NP-complete problems.

Traditional Qos routing algorithm mainly includes distance vector algorithm and routing state algorithm [6], and there still exist some weaknesses or limitations. One is that these algorithms have only considered either the biggest bandwidth or smallest that are obtainable at the current service flows, but not whether such a QoS service flow path will lead to the barrages [7]. The other problem is that these algorithms cannot

© Springer Science+Business Media Singapore 2016
K. Li et al. (Eds.): ISICA 2015, CCIS 575, pp. 129–138, 2016.
DOI: 10.1007/978-981-10-0356-1_13

efficiently improve the speed of convergence and calculation. These disadvantages need to be addressed in applying QoSR in solving problems in the future. If we set QoS parameter as the optimization condition, multi-constrained-based QoS routing decision-making problem is a NP-complete problem [8]. As for the algorithm's characteristics of positive feedback, self organization and distributive computation, many scholars applies the ACO to the solution of routing optimization and load balance. Some progress has been made in the research and the application of routing optimization. In terms of load balance, Yuan Li's work has yielded a better result compared with the former algorithms for the solution on problem of loss rate, average delay and load balance. It has also effectively the muti-constrain QoS routing optimization problem. To address the disadvantages of network barrages and low network efficiency, Yuan Li and Zhengxin Ma proposed MS-ACO algorithm [9], which mainly solves the problem of pheromone delay. Regarding the routing optimization, Wang et al. [10] applied the updating rule of partial pheromone and whole pheromone in multi-path routing. The result shows that it could quickly find the routing and improve the convergence speed and network efficiency. Based on the thought of parallel ACO algorithm, Lian and Xiang [11] expanded the multi-path routing protocol in the Ad hoc network, which largely enhanced the transmission speed of data packet increases greatly and the parallel ACO has increased the routing forecast probability and convergence speed. Some scholars [12, 13] obtain the optimal convergence speed and optimal result through adjusting the expectation of the partial optimization of pheromone density, pheromone evaporation factors, the pheromone updating rule and the state shifting rule.

Convergence analysis theory only informed us that there is probability [14] in finding the final optimal result in the ACO algorithm, and it is difficult to be applied in the comparative judgment the performance of actual alglorithm. Only by analyzing the convergence speed of ACO algorithm can we know the time spent the optimal result by using ACO algorithm. However, there has been little research on this aspect. Dorigo regarded the problem of research on convergence speed of ACO as the first public problem in 2005, and proposed that scholars should try to analyze the convergence speed problem in some simple ACO algorithm, to fill this research gap. The paper [15] published in 2006 addressed this public problem for the first time. Nevertheless, the conclusion is only limited to the solution on linear function of binary system single ant ACO algorithm.

Han Huang proposed the qualitative analysis [16] on the ACO algorithm's convergence based on the mathematical modeling of the Markov process, which is at absorption state in the ant colony algorithm. This paper defined the convergence expectation and convergence grads, proposed the thought of Q-ACO (quick searching ant colony algorithm) and discussed the QoSR computation. As for the multi-constrain QoS routing model, the algorithm controls the iteration and searches the optimal path that meets the QoS restriction condition under the condition of the quicker convergence speed. This quick searching algorithm can be applied in routing algorithm, as well as the ant colony algorithm optimization in other fields, such as the solution of task scheduling and load balance problems [17, 18] in cloud computing. We can also form the expectation function, with the function decided by restriction conditions such as the execution speed, scheduling time and cost calculation in the process of task scheduling.

We can generate the optimal result of quick convergence through controlling the iteration.

2 Explanation of Symbols

$Bw(i,j)$: path bandwidth;

$D(i,j)$: path delay;

$Pl(i,j)$: path loss ratio;

$Cost(i,j)$: path cost from node i and node j;

f_{ij}: the expectation function of path from node i to node j;

d_{ij}: path length between node i and node j;

ΔT_{ij}^{t}: the amount of pheromones that ants release at the t time of iteration on the path from node i to node j;

$T_{ij}(t)$: the amount of pheromones at the t time of iteration on the path from node i to node j;

$\sum\limits_{k=1}^{m} T_{ij}^{k}$: the amount of newly added pheromones at the t time of iteration on the path from node i to node j;

$P_{ij}^{k}(t)$: the ratio of the ant k on the node i select the node j;

A: a set of optional nodes which is jointed with node i; α: pheromones value of ACO;

β: heuristic value of ACO;

a: convergence grads;

Fi: convergence grads function;

3 Description of ACO-Based QoS Routing Problem

Definition 1 (quality of service). QoS is the collection of service flow's need for network service in the network's transmitting service flow. Also service flow is the specified QoS-related packet flow from source to aim.

Definition 2 (quality of service routing). QoS Routing allows the network to determine a path that supports the QoS needs of one or more flows in the network. Otherwise, QoSR can be regarded as a dynamic routing protocol of which the selection of path may include optional bandwidth, delay, jump times, loss ratio, shake, and other QoS parameters. QoSR seeks to find a feasible path for QoS traffic with two objectives: (a) providing the QoS guarantee for QoS traffic; (b) maximizing the utilization of the whole network.

3.1 QoS Routing Constraint Mode

Based on the characteristics of the path itself and the constraint conditions of the request, the network is regarded as a weighted undirected connected graph. We suppose

$G <V, E>$ as network, V as a set of nodes in the network, E as a set of edges in the network, and $E = \{(i,j)|i,j \in P\}$. with i and j refering to node and P refereing to a set of nodes.

Here we suppose that the path with a length of n from source node S to aim node D as $Path(i_1, \ldots, i_n)$. The general expressions of QoS's index which is correspondent with path $Path$ are as follows. x, y and z mean bandwidth, delay and loss ratio, respectively. $Bw(i, j)$, $D(i, j)$, $Pl(i, j)$ means path bandwidth, path delay and path loss ratio, respectively. Particularly, i, j mean nodes, k means constant.

Definition 3 (path bandwidth). It is the amount of data that can be transmitted in a fixed amount of time from node i to node j on the path of network, or the capability that the data can be transmitted in the transmission pipelines.

$$Bw(i,j) = \min_{j=2,3,\ldots n} \{Bw(i,j)\} \tag{1}$$

Here the formula of $Bw(i, j)$ means the bandwidth index of path(i,j).

Definition 4 (path delay). It means the iterative time of that from data packet's first byte' inputting into routing to the last byte' outputting of routing in the path transmission from node i to node j.

$$D(i,j) = \sum_{j=2}^{n} D[(i,j)] + \sum_{k=1}^{n} D(i_k) \tag{2}$$

Here the formula of $D(i,j)$ and $D(i)$ means the delay indexes of path(i,j) and node i respectively.

Definition 5 (path loss ratio). It means the ratio that the amount of data packet that is lost ranks in the data packet sending in the path test from node i to node j.

$$Pl(i,j) = 1 - \prod_{j=1}^{n} (1 - Pl(i,j)) \tag{3}$$

Here $PL(i)$ means the loss ratio index of node i. QoS routing is to search for the path that meets the constraint conditions as follows. $x \le Bw(i)$, $y \ge D(i)$, $z \le PL(i)$.

As for the multiple constraint routing problem, the value of indexes will show a great difference accounting for the inconsistent index dimension. Given that different service have different requests for service quality and the importance of all indexes differs, we should make an optimization of all the indexes. The mode of multiple QoS index constraint routing is presented below.

$$\min(y, z) or \ \max(x) => S.T. \begin{cases} Bw(i) \le x \\ D(i) \le y \\ PL(i) \le z \end{cases} \tag{4}$$

3.2   The ACO-Based QoS Routing Optimization Algorithm

3.2.1   Path Cost

Definition 6. Path cost is the relation function of bandwidth, delay and loss ratio that data are transmitted for a length of i. $Cost(i, j)$ is the path cost from node i to node j.
The path cost can be expressed as follows.

$$Cost(i,j) = \rho(\frac{Bw(i,j)}{D(i,j) * Pl(i,j)}) \tag{5}$$

ρ is a constant.

3.2.2   Pheromone Update

Definition 7. Pheromone update is the variable relationship of the presence of increase and the disappearance between the amount of path pheromone and ants over time. The shorter the path, the more ants that going by. The quicker the pheromone trace increases, the faster the evaporation, disappearance, and time elapses go.

4   CG Mode

4.1   Expectation Function

Definition 8. The function which is decided by the delay, bandwidth and other iterative conditions is the expectation function. f_{ij} is the expectation function on the path from node i to node j. ℓ is a constant. d_{ij} is the length of path that is from node i to node j.

$$f_{ij} = \frac{\ell * cost(i,j)}{d_{ij}} \tag{6}$$

4.2   The Probability of Selecting Path

Pheromone on the path is one of determinative factors that the ant selects the path. Pheromone is decided by the path cost. So the more the iterative times are, the stronger the pheromone in the path with low costs becomes. $T_{ij}(t)$ is the pheromone at t time iterative from node i to node j. ΔT_{ij}^t is the pheromone that the ant releases at t time iterative from node i to node j. Then the pheromone update formula is shown below:

$$T_{ij}(t) = T_{ij} * (1 - \rho/f_{ij}) + \Delta T_{ij} \tag{7}$$

$$\Delta T_{ij} = \sum_{k=1}^{m} T_{ij}^k \tag{8}$$

$$P_{ij}^k(t) = \begin{cases} \dfrac{T_{ij}^\alpha(t)*f_{ij}^\beta(t)}{\sum\limits_{s\in A} T_{is}^\alpha(t)*f_{is}^\beta(t)} \\ 0 \quad others \end{cases} \tag{9}$$

$\sum\limits_{k=1}^{m} T_{ij}^{k}$ is the increased pheromone on the path which is from node i to node j by t time iterative. ρ is pheromone evaporation parameter. k is the ant on the path. $P_{ij}^{k}(t)$ is probability that the ant k on the node i select the node j. The formula is presented above.

A is a set of optional nodes which is jointed with node i. Ant $k(k = 1, 2, \cdots, m)$ decides its shift direction in the movement according to the pheromone on the path. From formula (9), $\sum\limits_{s \in A} T_{is}^{\alpha}(t) * f_{is}^{\beta}(t)$ is the sum of all optional pheromone on the path of ant on the node i select node j. Each path is equivalent. We can know that divert probability $P_{ij}^{k}(t)$ increases along with the increase of $T_{ij}^{\alpha}(t) * f_{ij}^{\beta}(t)$. α and β are two weight parameters that determine the relative importance of pheromone and heuristic pheromone the ants accumulated in the movement on the decision of selecting path.

Theorem 1. At the beginning of iteration, we suppose that the original pheromone and path cost is equivalent. When $\alpha \geq 0, \beta \geq 0$, then the bigger $P_{ij}^{k}(t)$, the shorter the path length d_{ij}.

Proof: Suppose $d_1, d_2, \ldots, d_m$ as the optional path length from node i to node j and $d_1 < d_2 < \ldots < d_m$. Because Formula 6 So $f_1 > f_2 > f_3 > \ldots > f_m$ Because Formula 7.

There we suppose that every ant releases the same quantity of pheromone, because $T_{1.1} > T_{2.1} > T_{3.1} > \ldots > T_{m.1}$, so, $P_{1.1} > P_{2.1} > P_{3.1} > \ldots > P_{m.1}$.

Theorem 2. If $\alpha \geq 0, \beta \geq 0$ and $d_1 < d_2 < \ldots < d_m$, (d_i is optional path length), d_1 is the path length. The more the iterative times are, the bigger the probability of that the path with the shortest length is selected.

Proof: We suppose Formula 9 as the probability of the number i path by k time iteration as follows:

$$\frac{1}{1 + \left| \frac{T_{2.k} * f_2}{T_{1.k} * f_1} + \frac{T_{3.k} * f_3}{T_{1.k} * f_1} + \ldots + \frac{T_{m.k} * f_m}{T_{1.k} * f_1} \right.} = \frac{1}{1 + \frac{f_2}{f_1} * \frac{T_{2.k}}{T_{1.k}} + \frac{f_3}{f_1} * \frac{T_{3.k}}{T_{1.k}} + \ldots + \frac{f_m}{f_1} * \frac{T_{m.k}}{T_{1.k}}}$$

$$\frac{1}{P_{1,k-1}} - \frac{1}{P_{1,k-1}} = \frac{f_2}{f_1} * \left(\frac{T_{2.k}}{T_{1.k}} - \frac{T_{2.k-1}}{T_{1.k-1}} \right) + \frac{f_3}{f_1} * \left(\frac{T_{3.k}}{T_{1.k}} - \frac{T_{3.k-1}}{T_{1.k-1}} \right) + \ldots + \frac{f_m}{f_1} * \left(\frac{T_{m.k}}{T_{1.k}} - \frac{T_{m.k-1}}{T_{1.k-1}} \right)$$

So put T_{ik}, T_{1k} into $\frac{T_{ik}}{T_{1k}} - \frac{T_{ik-1}}{T_{1k-1}} = \frac{T_{ik} * (1 - \rho/f_i) + \Delta T_{ik}}{T_{1k} * (1 - \rho/f_i) + \Delta T_{1k}} - \frac{T_{ik-1}}{T_{1k-1}}$, So

$$\frac{[T_{ik} * (1 - \rho/f_i) + \Delta T_{ik}] * T_{1k-1} - [T_{1k} * (1 - \rho/f_i) + \Delta T_{1k}] * T_{ik-1}}{[T_{1k} * (1 - \rho/f_i) + \Delta T_{1k}] * T_{1k-1}}.$$

There we suppose every ant releases the same values of pheromone as the original values of pheromone on each path.

$$\frac{[T_{ik} * (1 - \rho/f_i) + \Delta T_{ik}] * T_{1k-1} - [T_{1k} * (1 - \rho/f_i) + \Delta T_{1k}] * T_{ik-1}}{[T_{1k} * (1 - \rho/f_i) + \Delta T_{1k}] * T_{1k-1}} \Rightarrow \rho/f_1 - \rho/f_1$$

Because $f_1 > f_i$ (f_i is the expectation function of selected path on node i), so $\rho/f_1 < \rho/f_1$, then $\frac{T_{ik}}{T_{1k}} - \frac{T_{ik-1}}{T_{1k-1}} < 0$, so $\frac{1}{P_{1k}} - \frac{1}{P_{1k-1}} < 0$, so $P_{1k} > P_{1k-1}$.

The more the iteration times, the bigger the probability of the path with the longest length is selected.

Theorem 3. If $f_i > 0$ and $\alpha \geq 0, \beta \geq 0$, the probability of the path of which the expectation function is the biggest will approach 1 when the iteration times approach ∞. $P_{i.k}$ is the probability of the path i selected at the path k iteration time when the f_i is the biggest expectation function.

Proof: We can get the relation from Theorem 2 as follows: because $d_1 < d_2 < \ldots < d_m$, so $P_{1k} \geq P_{1k-1}$. When f_i is the biggest function, then $\lim\limits_{k \to \infty} \frac{P_{1.k-1}}{P_{1.k}} = 0$, so $\lim\limits_{k \to \infty} P_{1.k} = 1$.

Only by analyzing the convergence speed of ACO algorithm can we know the time spent working out the optimal result of optimization algorithm. However, the research on the convergence speed is scarce. This paper discusses the convergence speed problem based on the definition of convergence grad function. The convergence speed increases through changing judgment conditions of convergence.

5   CG Convergence

Definition 9 (convergence grads function). It refers to the relation function that the pheromone on the path becomes bigger along with the iterative times increasing. When F is the convergence grads function, the formula is as follows:

$$Fi = [T_i * (1 - \rho/f_i) + \Delta T_i^t] * t \tag{10}$$

We suppose that convergence grads are the ratio between differential coefficient of convergence grads function and iteration times.

$$a = \Delta F / \Delta t \tag{11}$$

Theorem 4. The Bigger the expectation function fi is, the bigger the convergence grads is.

Proof: Because Formula 11, so $a_i = T_i * (1 - \rho/f_i) + \Delta T_i^t$. From the Theorems 1 and 3, because f_i is the biggest expectation function, P_i is the biggest probability, so a_i is the biggest convergence grads.

From the theorem above, the convergence grads is the biggest on the path of the biggest expectation function. We can stop the iteration by comparing the convergence speed of every path when we find that convergence grads are bigger than the other convergence grads. We can select the best path of which the convergence grads is the biggest and greatly reduce the time of selecting path and thus improve the network's efficiency.

Definition 10 (convergence grads expectation). The function that reflects the relation between the change of convergence grads and the increase of iteration times

$$y = m^{\frac{t-1}{t}} T_i * (1 - \rho/f_i) \tag{12}$$

m is the number of ants, and T_i is the pheromone on the ith path.

Theorem 5. The convergence grads expectation is along with iteration time.

Proof: From Theorems 2 and 3, we know that $y = m^{\frac{t-1}{t}}T_i * (1 - \rho/f_i)$ is increased along with the iteration times' increasing. a_i will be close to a specified value when iteration time k approach N.

$$a_i = \lim_{k \to N} m^{\frac{t-1}{t}}T_i * (1 - \rho/f_i) = m * T_i * (1 - \rho/f_i) \tag{13}$$

Theorem 6. If the iteration is close to ∞, then the speed of convergence grads expectation is close to 0: $\lambda \to 0$. Here, λ is the speed of convergence grads expectation.

Proof: From Theorem 5, 有 $\lambda = \frac{\partial y}{\partial t} = \frac{\Delta u}{\Delta t} = (m^{\frac{t}{t+1}} - m^{\frac{t-1}{t}}) * T_i(1 - \rho/f_i)$

Because $T_i(1 - \rho/f_i)$ is invariable value, so $\lambda \to 0$.

From Theorems 5 and 6, when iteration $t \to \infty$, the convergence grads expectation reach an invariable value. The speed of convergence is optimized and the speed of convergence grads expectation approach 0. Then ants always select the best path at this time.

6 Value Experiment and Result Analysis

We have simulated the Q-ACO algorithm in the environment of Visual C++. We take a totally undirected graph coordinate of 10 nodes as the data set and calculate the optimal path length by different iteration time using the standard ACO algorithm and Q-ACO algorithm stated in this paper respectively. The parameter value of algorithm is as follows: m = 20, $\alpha = 0.999$, $\beta = 5$, $\rho = 1$. In Table 1, I_{ACO} refers to the iteration times of ACO algorithm. $I_{Q\text{-}ACO}$ refers to the iteration times of Q-ÁCO algorithm. CG is the biggest grads value of Q-ACO algorithm at current time iteration. λ refers to the current error ratio when we obtain the optimal result by using Q-ACO algorithm and standard ACO algorithm.

Table 1. Analysis on the convergence expectation value of ACO and Q-ACO

I_{ACO}	ACO current optimal result	$I_{Q\text{-}ACO}$	Q-ACO current optimal result	CG	λ
50	3.312916	12	3.547081	0.1049	7.07 %
100	3.074772	25	3.338388	0.6867	8.57 %
200	2.918033	42	3.185981	0.3725	9.18 %
400	2.830737	51	3.067977	0.1570	8.38 %

The experiment result shows that Q-ACO quickly reduces the iteration times x under the condition that it is closest to the optimal result. Q-ACO will get the better iteration result at the same iteration times. In addition, the error ratio of Q-ACO compared with standard ACO is around 10 % at the obvious reduction of iteration times.

7 Conclusion

This paper has analyzed some primary factors that constraint the QOS routing problem and stated an ACO-based optimized algorithm of multiple constraint QOS routing problem. This algorithm stops the iteration through comparing the path convergence grads in Table 1 and gets the optimal path in a faster and bigger ratio. It improves the ACO-based QoSR speed of calculation and of searching for path in the limited node network as well as algorithm efficiency of QoS routing service of searching for path.

Acknowledgment. This study was supported by Open project of Guangdong Provincial Key Laboratory of Petrochemical Equipment Fault Diagnosis (GDUPTKLAB201322), a Science and Technology Project of Special fund for High-tech development by Guangdong Provincial Department of Finance in 2013(2013B010401036). Guangdong Provincial Department of Education Science and Technology Innovation Project (2013KJCX0178).

References

1. Colorni, A., Dorigo, M., Maniezzo, V., et al.: Distributed optimization by ant colonies. In: Proceedings of the 1st European Conference on Artificial Life, pp. 134–142(1991)
2. Dorigo, M., Maniezzo, V., Colorni, A.: Ant system: optimization by a colony of cooperating agent. IEEE Trans. Syst. Man Cybern. Part B 26(1), 29–41 (1996)
3. Gambardella, L.M., Taillard, E.D., Dorigo, M.: Ant colonies for the quadratic assignment problem. J. Oper. Res. Soc. 50(2), 167–176 (1999)
4. Costa, D., Hertz, A.: Ants can color graph. J. Oper. Res. Soc. 48(3), 295–305 (1997)
5. Colorni, A., Dorigo, M., Maniezzo, V., et al.: Ant system for job-shop scheduling. Belg. J. Oper. Res. Statist. Comput. Sci. 34, 39–53 (1994)
6. Cui, X., Lin, C.: A Constrained quality of service routing algorithm with multiple objectives. J. Comput. Res. Develop. 41(8), 1368–1375 (2004)
7. Cui, X., Lin, C.: Multicast QoS routing optimization based on multi-objective genetic algorithm. Chin. J. Comput. 41(7), 1144–1150 (2004)
8. Cui, Y., Wu, J., Xu, K., Xu, M.: Research on internetwork QoS routing algorithms: a survey. J. Softw. 13(11), 2065–2073 (2002)
9. Li, Y., Ma, Z.: A mitigating stagnation-based ant colony optimization routing algorithm. In: Proceedings of ISCIT (2005), pp. 34–37 (2005)
10. Wang, Z., Zhang, D., A Qos multicast routing algorithm based on ant colony algorithm. In: IEEE 1007(2005), pp. 1007–1009 (2005)
11. Li, L., Yang, X., et al.: Research of multi-path routing protocol based on parallel ant colony algorithm optimization in mobile ad hoc networks. In: Fifth International Conference on Information Technology: New Generations (2008), pp. 1006–1010 (2008)
12. Qi, J., Zhang, S., Sun, Y., Lei, Y.: Cognitive networks multi-constraint QoS routing algorithm based on autonomic ant colony algorithm. J. Nanjing University Posts Telecommun. (Nat. Sci.) 32(6), 86–91 (2012)
13. Wang, H., Li, Y.: Quantum ant colony algorithm for QoS best routing problem. Comput. Simul.31(3), 295–298 (2014)
14. Dorigo, M., Blum, C.: Ant colony optimization theory: A survey. Theoret. Comput. Sci. 344, 243–278 (2005)

15. Hao, Z.-F., Huang, H., Zhang, X., Tu, K.: A time complexity analysis of ACO for linear functions. In: Wang, T.-D., Li, X., Chen, S.-H., Wang, X., Abbass, H.A., Iba, H., Chen, G.-L., Yao, X. (eds.) SEAL 2006. LNCS, vol. 4247, pp. 513–520. Springer, Heidelberg (2006)
16. Han, H., Hao, Z., et al.: The convergence speed of ant colony optimization. Chin. J. Comput. **8**, 1345–1353 (2007)
17. Zhang, M.: Research of virtual machine load balancing based on ant colony optimization in cloud computing and muiti-dimensional QoS. Comput. Sci. **40**, 60–62 (2013)
18. Duan, W., Fu, X., et al.: QoS constraints task scheduling based on genetic algorithm and ant colony algorithm under cloud computing environment. J. Comput. Appl. **34**, 66–69 (2014)

A New GEP Algorithm and Its Applications in Vegetable Price Forecasting Modeling Problems

Lei Yang[1], Kangshun Li[1(✉)], Wensheng Zhang[2], and Yaolang Kong[1]

[1] College of Mathematics and Informatics, South China Agricultural University,
Guangzhou, China
`yanglei_s@scau.edu.cn`, `likangshun@sina.com`
[2] Institute of Automation, Chinese Academy of Sciences, Beijing, China
`wensheng.zhang@ia.ac.cn`

Abstract. In this paper, a new Gene Expression Programming (GEP) algorithm is proposed, which increase "inverted series" and "extract" operator. The new algorithm can effectively increase the rate of utilization of genes, with convergence speed and solution precision is higher. Taking the Chinese vegetables price change trend of mooli, scallion as example, and discuss the way to solve the forecasting modeling problem by adopting GEP. The experimental results show that the new GEP Algorithm can not only increase the diversity of population but overcome the shortage of primitive GEP. In addition, it can improve convergence accuracy compared to original GEP.

Keywords: Gene expression programming · Vegetable prices prediction · Utilization of gene · Gene extraction

1 Introduction

Vegetable price forecast can improve the forecast of the price of vegetables, take measures to slow the price fluctuation and keep the market price stable. From the point of domestic and foreign research dynamic, short-term prediction in the social economy, the power load [1] and other fields has made great progress. In agricultural products market price prediction, Henry had carried on the regression prediction on American cotton yield, the prediction results are more accurate than the United States department of agriculture forecast [2]. Sarle studied the relationship of market price influence factor of live pig, and established the prediction equation of hog price by using sample data, and the goodness of fit was up to 0.75 [3]. Jarrett uses the exponential smoothing method to estimate the price of wool in Australia [4]. Schmitz and Watts predict the price of live pigs with exponential smoothing and Box - Jenkins method [5]. Cui Guoli uses the chaotic neural network model and ARIMA model to build model and forecast the price trend of Chinese cabbage in the next 10 days. The results show that the average relative error between the chaotic neural network model and the actual price is relatively small [6]. Zhu Xiaoxia verified the vegetable price fluctuation cycle of predictability by using markov chain simulation analysis [7]. In summary, the research on vegetable price forecasting is less, and the existing forecasting methods mainly adopt the econometric methods, and

© Springer Science+Business Media Singapore 2016
K. Li et al. (Eds.): ISICA 2015, CCIS 575, pp. 139–149, 2016.
DOI: 10.1007/978-981-10-0356-1_14

domestic researches mainly use neural networks, the prediction based on intelligent algorithm is not much.

GEP is a new evolutionary algorithm based on genotype and phenotype, it automatically creates a function expression by using the function set and the terminator, and the evolution process is more easy to operate, and has the stronger ability to solve problems. Gene expression programming method is widely used to solve the problem of factor decomposition, function parameter optimization, evolutionary modeling, association rule mining, classification and clustering, and sunspot prediction, etc. and obtained good results [8–11]. In this paper, the program design of gene expression is improved and applied to the prediction of vegetable prices.

The remainder of this paper is organized as follows. Section 2 presents a discussion of the traditional Gene expression programming. Section 3 presents our new Gene expression programming algorithm based on "inverted series" and "extract" operator. The experimental results and discussion are presented in Sect. 4. Finally, Sect. 5 concludes this paper and presents future research directions.

2   Gene Expression Programming (GEP)

2.1   Overview of Gene Expression Programming

Gene expression programming is a new type of evolutionary algorithm, which was proposed by a Portuguese scientist, Candida Ferreira. It has absorbed the advantages of Genetic Programming (GP) [13] and Genetic Algorithm (GA) [14], which is an efficient evolutionary algorithm. GEP uses the same performance method as GP, and uses GA to replace the GP with the long linear genome to replace the indefinite nonlinear entities, thus overcoming the limitations of the phenotype. In addition, Ferreira invented the Karva language to read and express the information of encoding in the GEP chromosome, which makes the chromosome allow the existence of the multiple genomes, and simplifies the process of building a powerful genotype and phenotype system. These advantages make GEP faster than other traditional genetic algorithms 2–4 orders of magnitude.

2.2   Gene Expression Programming

2.2.1   Chromosome Structure

(1)   Gene

Genes are composed of a head and a hail in GEP. The head contains symbols that represent both functions and terminals, whereas the tail contains only terminals. For each problem, the length of the head h is chosen according the corresponding problem, whereas the length of the tail t is a function of h and n (n is the number of arguments of the function with the most arguments), for example, n is 1 in square root function and trigonometric function, and n is 2 in add function. Then t is calculated by the (1), and the length of the gene is calculated by the (2).

$$t = h * (n - 1) + 1 \tag{1}$$

$$l = h + t = n * h + 1 \tag{2}$$

(2) Expression Tree

The expression can be transformed into the expression tree using the regulation from top to bottom and from left to right.

(3) Chromosome

In GEP, multiple genomes are often used to form chromosomes when solving complex problems, then each genome using a connector. Two genes of chromosomes are as follow (dark blue for the genome tail).

The chromosome has 2 open reading frame (ORF); each ORF corresponding to the sub expression tree (sub_ET), which is connected with the "+", the expression tree is as follows.

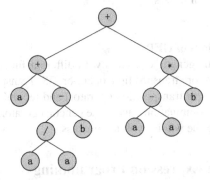

The starting code for each ORF is the position 0 of the above expression, which is related to the corresponding sub expression tree.

2.2.2   Fitness Function of GEP

The ultimate goal of gene expression programming is to find out the optimal solution, and to evaluate whether the chromosome is the optimal solution is the fitness function, and different fitness function may be chosen for different problems. In GEP, Ferreira proposes two kinds fitness function: the model based on the absolute error (3) and the relative error of the model (4).

$$f_i = \sum_{j=1}^{c_r} (M - |C_{(i,j)} - T_j|) \tag{3}$$

$$f_i = \sum_{j=1}^{c_r} \left(M - \left| \frac{C_{(i,j)} - T_j}{T_j} * 100 \right| \right) \tag{4}$$

where M is a constant, the range of f_i, $C_{(i,j)}$ is the value returned by the individual chromosome i for fitness case j, and T_j is the target value for fitness case j, C_t is the size of test samples.

2.2.3   Selection Function

After the fitness function evaluation, each chromosome has a corresponding degree of adaptation. Then need to select the parent chromosome. The father is chosen according to the fitness of the GEP. The selection probability formula is as follows:

$$P_i = \frac{f_i}{\sum_{i=1}^{m} f_i} \tag{5}$$

P_i indicates the selected probability of individual i, and f_i for the individual's fitness, $\sum_{i=1}^{m} f_i$ for the total population fitness.

2.2.4   Genetic Operators of GEP

In GEP, there are various genetic operators with different function, such as selection operator, mutation operator, transposition operator (IS Transposition operator, RIS transposition operator, Gene transposition operator) and recombination operator (one-point, two-point, gene recombination), and these genetic operators can ensure the diversity of population during the whole evolving process.

3   Improved Gene Expression Programming

Many researchers have spent a lot of effort to improve and develop Gene expression program since it has been proposed in 2001 [15]. Some scholars have proposed the strategy of generating the population. Some scholars have proposed the dynamic changes of the chromosome, and some scholars have improved the probability of genetic operators [16]. In order to improve the effective utilization of gene, this paper proposed a new method of "inverted series" and "gene extraction".

3.1   The Methods of Improved GEP

In this paper, a new method is presented, which increase two operators, named "inverted series" and "extract" operator.

3.1.1   Inverse String Operator

Inverted list operator acting on the head of a genome. Randomly selected a piece of string in the head of genes, and then using the string as the center of symmetry, the position sequence of each character swap. The following chromosome second genome head inverted list (green for the occurrence of an inverted string of genes).

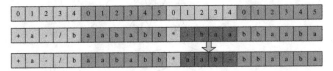

3.1.2   Gene Extraction Operator

In the genome of chromosome, when terminator appears in the first position of genome, it is necessary to retain this gene, but also to express the behind hidden gene, solving method is to extract the gene and connect it to the rest of the genome, the head of the original genome moved forward, and finally fill a head gene. So it can be achieved by retaining the single gene, and the opportunity to get the performance of the genes behind. In this paper, this operation is named "gene extraction", which is a new genetic operator, the operator acts on the gene recombination, before the formation of the next generation of population.

Algorithm for gene extraction operator:

Input: single chromosome

Output: the extracted chromosome

(1) Check the genome one by one, and if encounter the genome that terminator is in its first position, marking the terminator and go to 2. The end of the algorithm is over if the genome is not found.

(2) To search for the gene which the first position is not a terminator, the effective gene is shortest and the length of the gene is at least two genes shorter than the head of the gene, if found then turned to 3, or the algorithm is end.

(3) The head of the genome have been found move two place behind, and then fill the connector in the first position, the second position was filled in the terminator which had been found in step 1, and then go to 1.

The core of the algorithm is to extract the gene that the first position is a terminator, and put it to the genome which the first position is not a terminator and the effective gene is shortest, so that each gene is expressed, so that the first for the end of the gene to the end of the gene is expressed, but also to the effective length of the genome tend to average.

3.2   Improvement Effect

For example, there is a chromosome which has many genomes (connector for plus):

The corresponding expression tree is

After gene extraction, the end of the genome head being extracted was made up for the character "a", the head of the genome which being inserted genes was intercepted the last "/b". After extraction of the gene and the expression tree is as follows:

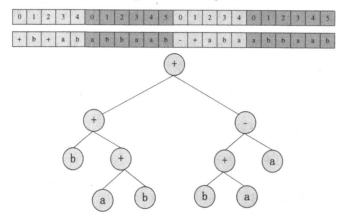

It can be seen from above figure that the utilization of second genomes of the chromosome have been greatly improved by gene extraction, which is very useful for evolution.

4   Examples and Assessment

One of the most common applications of GEP is to predict [17], such as grain yield forecasting [18], stock forecasting, [19], etc. The reason is that the GEP can be used only to evolve according to the fitness function, without human intervention can get accurate results, it has a strong ability to find the solution in function regression [20]. Prediction is often a function of regression problems.

4.1   Forecasting Method

Because vegetables are seasonal, A vegetable market to maintain a relatively short cycle, so the single from the quarter, due to the accuracy of the consideration, not to make a far more distant forecast. This paper mainly studies the application of IGEP in short-term vegetable price forecasting. Short term vegetable price is a weak change time series, and the sliding window method is used to forecast the short-term price of vegetables. In this paper, we use C# language to make a short-term forecasting tool for vegetable prices.

The price data of Chinese vegetables price change trend of mooli, scallion are from Ministry of Agriculture. There are a total of 87 data from 11 May 2015 to 5 August 2015. As shown in the following table (Price unit: yuan/kg) (Tables 1 and 2).

Table 1. Date format of Chinese mooli

Serial	Date	Price	Serial	Date	Price
1	5-11-2015	1.3	5	5-15-2015	1.27
2	5-12-2015	1.29	...	...	...
3	5-13-2015	1.28	86	8-4-2015	1.51
4	5-14-2015	1.27	87	8-5-2015	1.51

Table 2. Date format of Chinese white scallion

Serial	Date	Price	Serial	Date	Price
1	5-11-2015	1.73	5	5-15-2015	1.76
2	5-12-2015	1.72	...	...	...
3	5-13-2015	1.75	86	8-4-2015	3.4
4	5-14-2015	1.76	87	8-5-2015	3.39

4.2 Time Aeries Analysis

Time aeries analysis is one of the most typical and traditional tasks for historical data analysis, and widely used in decision support systems. Researchers have proposed many models for prediction in time series. Expression of time series analysis model is as follows.

$$Y_t = f(Y_{(t-d)}, Y_{(t-2d)}, \ldots \ldots Y_{(t-\tau d)})$$

Y_t indicates the value of the t sample.

4.3 GEP Evolution

Two different GEP-based models were developed for vegetables price prediction. In order to find the optimal prediction model, selecting appropriate parameters of the GEP evolution is necessary. Various parameters involved in the GEP predictive algorithm are shown in Table 3. The parameter selection will affect the model generalization capability of GEP. They were selected based on some previously suggested values [21] and also after a trial and error approach.

Table 3. Parameter settings for the GEP algorithm

Parameter	Value
Number of generations	5000
Population size	50
Number of fitness cases	1000
Function set	+ − */ ~ Q S C L E
Terminal set	a b c d e f g h i j
Head length	8
Number of genes	10
Linking function	+
Mutation rate	0.044
One-point recombination rate	0.3
Two-point recombination rate	0.3
Gene recombination rate	0.1
IS transposition rate	0.1
RIS transposition rate	0.1
Gene transposition rate	0.1
Gene refine rate	0.3

Import data, and set the parameters, GEP began to evolve. The roulette wheel selection method is used to choose the father, and then gene groups evolve through variation, transposition, recombination and other genetic operators constantly toward the evolutionary fitness of maximum degree direction.

4.4 Comparative Analysis of Forecast Results

The following figure is the price forecast of mooli and scallion, the algorithm improves the forecast accuracy, and can be seen, the IGEP training forecast is more close to the true value than the previous forecast, such as linear regression, parabola regression and GEP. This also shows that the improved algorithm can evolve to a more perfect solution, and the fitness is better than before (Figs. 1 and 2).

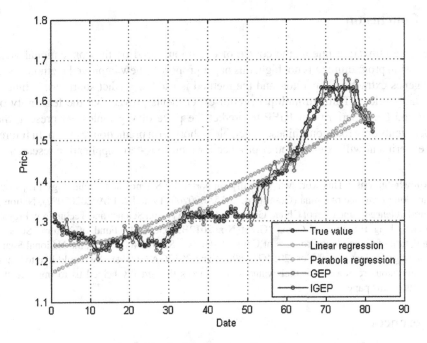

Fig. 1. The forecast of mooli

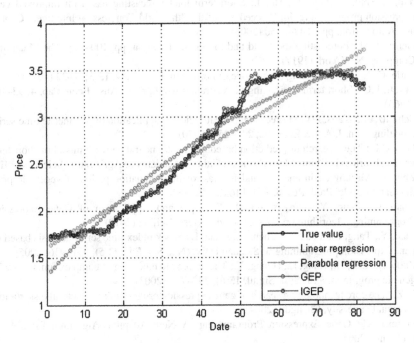

Fig. 2. The forecast of scallion

5 Conclusion

The gene utilization rate of the course of evolution based on the conventional gene expression programming is not high, this paper proposes a new improved method based on "genes extraction" operator, and the method is used to predict short term Chinese vegetable prices. Through multiple sets of experiments proved that the feasibility of GEP and the superiority of IGEP to predict the price of vegetables. At present, the experiments only focus on Chinese vegetable's short-term prediction research. In future, the experiment will extend to other produce for their long-term application research.

Acknowledgment. This work was partially supported by Science and Technology Project of Guangdong Province of China (Grant Nos. 2015A020209119 and 2014A020208087), National Natural Science Foundation of China (Grant No. 61573157), and Science and Technology Project of Guangdong Province of China (Grant No. 2012BM0500054), Fund of Natural Science Foundation of Guangdong Province of China (Grant No. S2013040015755), the National Spark Technology Project (Grant Nos. 2014GA780051 and 2013GA780044). We would like to thank the anonymous reviewers for their valuable comments that greatly helped us to improve the contents of this paper.

References

1. Yin, J., Huo, L., Guo, L., Hu, J.: Short-term load forecasting based on improved gene expression programming. In: Proceedings of the 7th World Congress on Intelligent Control and Automation, pp. 5647–5650 (2008)
2. Henry, L.M.: Forecasting the Yield and the Price of Cotton, pp. 100–113. The Macmillan Company, New York (1917)
3. Sarle, C.F.: The forecasting of the price of hogs. Am. Econ. Rev. **15**(3), 1–22 (1925)
4. Jarrett, F.G.: Short term forecasting of Australian wool prices. Aust. Econ. Pap. **4**, 93–102 (1965)
5. Schmitz, A., Watts, D.G.: Forecasting wheat yield: an application of parametric time series modeling. Am. J. Agric. Econ. **52**, 247–254 (1970)
6. Guo, C.L.: The research of prediction based on chaotic neural network model of short-term vegetable prices in China. Master's thesis, Chinese Academy of Agricultural Sciences (2013)
7. Zhu, X.: Markov chain analysis and forecast of the fluctuation period of vegetable price fluctuation cycle. Prod. Res. **8**, 143–146 (2012)
8. Yuan, C., Peng, Y., Qin, X., et al.: Principle and Application of Gene Expression Programming Algorithm, pp. 37–40. Science Press, Beijing (2010)
9. Liao, Y., Tang, C., Yuan, C., Chen, A., Duan, L.: Stock index time series analysis based on gene expression programming. J. Sichuan Univ. (Nat. Sci. Ed.) **42**(5), 931–936 (2005)
10. Zhou, A., Cao, H., Kang, L., Huang, Y.: The automatic modeling of complex functions with genetic programming. J. Syst. Simul. **15**(6), 797–799 (2003)
11. Ji, Z.: The core technology research of gene expression programming. Doctoral dissertation: Sichuan University, Chengdu (2004)
12. Ferreira, C.: Gene Expression Programming: A New Adaptive Algorithm for Solving Problems (2001)
13. Zhengjun, P., Lishan, K., Yuping, C.: Evolutionary Computation, pp. 112–120. Tsinghua University Press, Beijing (1998)

14. Mitch, M.: An Introduction to Genetic Algorithms. MIT Press, Cambridge (1996)
15. Ryana, N., Hiblerb, D.: Robust Gene Expression Programming, pp. 165–170. Christopher Newport University, Newport News (2011)
16. Kun, L.: Metal fatigue time prediction model based on GEP. Dissertation Doctoral, Wuhan: Wuhan University of Technology (2010)
17. Jingguang, Z.: Research on combination forecasting method based on gene expression programming. Master's thesis, Huazhong Normal University (2011)
18. Yin, L.: Grain yield forecasting based on gene expression programming. Master's thesis, Northwest Agriculture and Forestry University (2010)
19. Zhang, Y., Xiao, J., Sun, S.: BS-GEP algorithm for prediction of software failure series **7**, 243–248 (2012)
20. Lanli, H.: The application of gene expression programming in symbolic regression. Master's thesis, Hunan Normal University (2007)
21. Canakcı, H., Baykasoğlu, A., Güllü, H.: Prediction of compressive and tensile strength of Gaziantep basalts via neural networks and gene expression programming. Neural Comput. Appl. **18**(8), 1031–1041 (2009)

An Optimized Clustering Algorithm Using Improved Gene Expression Programming

Shuling Yang[1], Kangshun Li[1(✉)], Wei Li[1,2], and Weiguang Chen[1]

[1] College of Mathematics and Informatics, South China Agricultural University,
Guangzhou 510642, China
likangshun@sina.com
[2] School of Information Engineering,
Jiangxi University of Science and Technology, Ganzhou 341000, China

Abstract. How to find the better initial center points plays an important role in many clustering applications. In our paper, we propose the novel chromosome representation according to extended traditional gene expression programming used in GEP-ADF. It is aimed at improving the performance of GEP to obtain center points more accurately. Experimental results show that our new algorithm has good performance in clustering and the three real world datasets compared with the other two algorithms.

Keywords: Center points · Novel chromosome representation · Gene expression programming · GEP-ADF

1 Introduction

Nowadays, many applications in various fields require dividing a set of data points into some group. There is no doubt that the classification and clustering can do so. If some examples (training data points) are provided, the partition process is called 'classification'; otherwise, the process is named as 'clustering' [1].

In recent years, many novel clustering methods both hard clustering and soft clustering are proposed in different fields in order to improve the accuracy of clustering results. Liu et al. [2] proposed an automatic clustering method based on evolutionary optimization. Their method can successfully detect the number of clusters for some separable data sets and two real world problems. However, the method can not deal with irregular-shape data sets and study multiple objectives. As pointed out in Agustin-Blas et al. [3], using the grouping genetic algorithm improves classical approaches such as K-means and DBSCAN algorithms to solve clustering problems, but they fail to automatically produce the value of κ which is the number of clustering. Yu et al. [4] used semi-supervised clustering ensemble for tumor clustering from gene expression profiles. The experiments on cancer gene expression profiles show that the proposed framework works well on bio-molecular data, and outperforms most of the state-of-the-art tumor clustering approaches. A novel genetic clustering algorithm is proposed with an encoding schema allowing for variable density based clusters within the same clustering solution by Sabau [5]. Zhou et al. [6] use Gene Expression Programming to evolve accurate and compact classification rules, which is more efficient

K. Li et al. (Eds.): ISICA 2015, CCIS 575, pp. 150–160, 2016.
DOI: 10.1007/978-981-10-0356-1_15

and tends to generate shorter solutions compared with canonical tree-based GP classifiers. Ni et al. [7] proposed a multi-objective cluster algorithm based on GEP. As we all know, it is very hot to apply the concept of multi-object recently. Using a multi-objective clustering algorithm solves the difficult problem that it is hard to find the high-quality solution in the limited search space. But it still has some problems like the parallelization. What is more, some have even tried to estimate the age using facial features based on Gene Expression Programming in the literature [8] and an optimized algorithm has been combined BP neural network and improved GEP by Zha et al. [9]. The new algorithm is more user friendly not requiring any a priori knowledge about the target datasets. In this paper, we propose an optimized clustering method by using improved gene expression programming to generate initial center points which are similar to real center points. Our major contributions are as follows:

- A novel chromosome representation is based on GEP-ADF produces medoid-based points in order to solve a problem of fuzzy C-means which can not automatically provide medoid-based points.
- An optimized morphology similarity distance can consider the particular contributions of different features and enhance the accuracy.

The rest of this paper is organized as follows: In Sect. 2, some related work including basic knowledge about traditional gene expression programming and fuzzy C-mean algorithm are outlined. The basic knowledge helps us understand the rest of the paper. Section 3 presents our new algorithm including the novel chromosome representation on traditional gene expression programming. Then we give a completed pseudocode of the new algorithm. In Sect. 4, the proposed method is empirically studied on a real world dataset. And we compare our approach with other algorithms. The advantages and the limitation of our approach are summarized in Sect. 5, which also concludes some possible extensions in the future.

2 Related Work

2.1 Extended Traditional Gene Expression Programming Used in GEP-ADF

Automatically Defined Functions, briefly named as ADF, was proposed by Ferreira [10], who combined GEP into GEP-ADF. In GEP-ADF, every chromosome is composed by a number of conventional genes and a homeotic gene, as illustrated in Fig. 1. All of the conventional genes and homeotic gene are expressed using gene expression. Some child nodes of the homeotic gene are represented by ADF. The conventional genes encode sub expression trees, and become an ADF.

For instance, the non-terminal symbols of ADF are $\{*, +, -, /\}$, the terminal symbols are $\{x, y\}$, and the length of the head and tail are respectively 3 and 4. The functions of a homeotic gene are $\{*, /, +, -, \sin, \cos\}$, and the terminal symbols are $\{1, 2\}$, and the length of the head and tail are respectively 6 and 7. Consider the following chromosome:

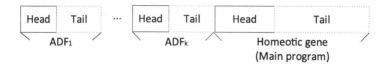

Fig. 1. The structure of the gene expression chromosome in GEP-ADF

$$\begin{bmatrix} +,-,*,x,x,y,x,*,-,+,x,y,x,y \\ *,-,\cos,1,+,\sin,2,1,2,2,1,1,2 \end{bmatrix} \quad (1)$$

The first line encodes two ADFs and the second line encodes a homeotic gene into sub ETs, as illustrated in Fig. 2. We can notice that both ADF1 and ADF2 are used twice in the homeotic gene. Furthermore, ADFs can only be used as terminal of the homeotic gene and contain no input argument.

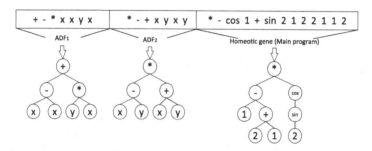

Fig. 2. An example chromosome of GEP-ADF and the corresponding ET

2.2 Standard Fuzzy C-Means Algorithm (FCM)

Fuzzy C-means algorithm based on the concept of fuzzy C-partition, which was presented by variety of researchers in this field Ruspini [11], developed by Dunn [12] and generalized by Bezdek [13]. FCM, which is different with K-Means algorithm, allows one piece of data to belong to two or more clusters. $X = \{x_1, x_2, \ldots, x_n\}$ is a set of n data points where $x_i = (x_{i,1}, x_{i,2}, \ldots, x_{i,d}) \in R^d$ is a d-dimensional point. FCM partitions X into C clusters $v = \{v_1, v_2, \ldots, v_C\}$ which satisfy

- $v_1 \cup v_2 \cup \cdots \cup v_C = X$;
- $v_i \neq \phi$, $i = 1, 2, \ldots, C$;
- $2 \leq C \leq \sqrt{n}$.

The function of FCM algorithm is defined as:

$$J_m(U, V, X) = \sum_{i=1}^{C} \sum_{k=1}^{n} (u_{ik})^m d_{ik}^2(x_k, v_i) \quad (2)$$

Here, the result of cluster is represented as an affiliation matrix $U = [u_{ik}]$ that satisfies: $u_{ik} \in [0, 1]$, $\sum_{i=1}^{C} u_{ik} = 1 \ \forall k = 1, 2, \cdots, n$ and $0 < \sum_{k=1}^{n} u_{ik} < n \ \forall i = 1, 2, \cdots, C$. u_{ik} is on behalf of the x_k affiliation value to cluster i. In addition, $m \in [1, \infty)$ is a scalar parameter that determines the fuzziness of the resulting clusters. If $m = 1$, the formula (2) is represented the hard C-means algorithm. $d_{ik}(x_k, v_i)$ is calculated by the Euclidean distance of the sample data k to the cluster center i:

$$d_{ik}(x_k, v_i) = \|x_k - v_i\| \tag{3}$$

Table 1 shows a description of fuzzy C-means algorithm with details.

Table 1. Original fuzzy C-means clustering algorithm

FunctionFCM(U,V)

Initialize the number of clusters C, medoid-based cluster points $v = \{v_1, v_2, \ldots, v_C\}$, an affiliation matrix $U = [u_{ik}]$, the value of ε and the weighted fuzzy exponent m.

Repeat

Calculate the fuzzy cluster centers V with U:

$$v_i = \frac{\sum_{k=1}^{n} (u_{ik})^m x_k}{\sum_{k=1}^{n} (u_{ik})^m}$$

Update the affiliation matrix U:

$$u_{ik} = \frac{1}{\sum_{j=1}^{C} \left(\frac{d_{ik}(x_k, v_i)}{d_{jk}(x_k, v_j)} \right)^{\frac{2}{m-1}}}$$

Until the end condition satisfies $\begin{cases} E(t) = \|J^t - J^{(t-1)}\| < \varepsilon \\ E(t) = \|V^t - V^{(t-1)}\| < \varepsilon \end{cases}$

---

3 The Optimized Clustering Algorithm

This section describes proposed the improved GEP-ADF in the first part, which is aimed at enhancing version of the traditional GEP as well as the GEP-ADF. And in the second part, we will give a pseudocode for completed NGEP-ADF algorithm.

3.1 The Novel Chromosome Representation

We design the new rule to change the chromosome representation while it meets some conditions. And we improve our algorithm simply named as NGEP-ADF based on GEP-ADF, which was proposed by Ferreira. In the literature [10], we can find out when the number of ADF is one, the experimental results are best. Therefore, we choose one ADF to add traditional gene expression program. Each chromosome comprises of a homeotic gene (main program) and an ADF in our proposed NGEP-ADF, as illustrated in Fig. 3.

In our paper, we aim at both the valid length (VL) and the head length of the chromosome to design the new GEP-ADF algorithm. To start with, we present a novel

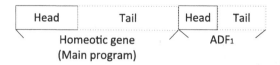

Fig. 3. The structure of the proposed chromosome representation

symbol to enhance the performance of GEP-ADF, which is defined as \$. This symbol \$ has special function in establishing the chromosome structure. When \$ appears in the first place of the head, it becomes a binary operator. But here has a question, which binary operators will it use? It depends on what kind of non-terminal symbols we define. \$ can choose any binary operator as long as it exists. Furthermore, ADF will replace \$ while \$ serves as a terminal symbol in Chromosome representation. Four situations follow as (h means the length of the head):

(1) While VL = 1, it indicates that the root is a terminal symbol, let it enforce to transform a binary operator;
(2) While $1 < VL < h$

- If the terminal symbols contain \$ in the coding region, cast \$ as a binary operator, and add two sub expression trees of ADF;
- If there are no \$, we cut out and place upside down non-coding symbols of the gene in the head, and insert the head root. In the meantime, the rest symbols of the head can be moved after this inverted segment.

(3) While VL = h, do not do any processing;
(4) While VL > h,

- If the terminal symbols contain \$ in the coding region, cast \$ as an ADF;
- If there are no \$, we will cut out and place upside down all non-coding symbols of the whole gene, and insert the tail root. The rest of coding symbol behind the insertion point should move backward in sequence.

Suppose the value of n is 2, and h and t of the homeotic gene are set to be: $h = 5$, $t = 6$, and both the non-terminal symbols and the terminal symbols of the main program are set to be: $F = \{+, -, *, \sin, \cos\}$, $T = \{x, y, \$\}$. The head length ($h'$) and tail length ($t'$) of the ADF are set to be: $h' = 3$, $t' = 4$, the non-terminal symbols and the terminal symbols of the ADF are respectively set to be: $F' = \{+, -, *\}$, $T' = \{x, y\}$. We suppose a generated chromosome with one ADF randomly, expressed as:

$$\begin{bmatrix} \$, -, \sin, x, +, \$, x, y, x, y, x \\ +, *, -, x, y, x, y \end{bmatrix} \tag{4}$$

Figure 4 shows an example chromosome of the proposed chromosome representation in NGEP-ADF. There are two parts of the whole chromosome. One is a homeotic gene and another is an ADF. As we define above, the symbol of \$ not only is a terminal symbol but also has some special function. In Fig. 4, (a) shows the first step after the

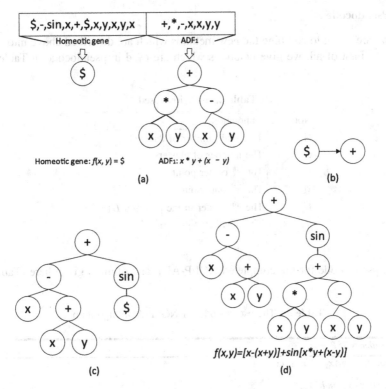

Fig. 4. An example chromosome of a homeotic gene and a related ADF. (a) An example chromosome of a homeotic gene and the corresponding ET using ADF. (b) $ is in the root. (c) The ET has change after (b). (d) The final ET.

generated gene. The main program only has one symbol $, and the ADF_1 gets an equation: $x * y + (x - y)$. According to the picture (b), when $ is in the root, we need to cast $ as a binary operator. Here we randomly choose one +. Picture (c) illustrates the homeotic gene generates an expression tree after (b). Picture (d) casts $ as an ADF in terms of the above forth situation.

In the reproduction of traditional Gene Expression Programming, there are a variety of method to make chromosomes change such as mutation, IS transposition and gene recombination and so on. In our method, we slightly or strongly modulate the structure of the chromosome by the mutation of the root, placing some segments upside down and insertion some specified places. We aim at making the shortcomings of RIS transposition up and slightly modulating in order to keep the consistency of genes.

We are aimed at generating initial medoid-based points by using this new chromosome representation. As we all know, normal clustering methods face two difficult problems, the first one is hard to correctly initialize the medoid-based points. The second problem is difficult to determine the number of clustering groups. Hence, we propose chromosome representation in order to generate accurately the medoid-based points instead of generating without any reason.

3.2   Pseudocode

Now, we are going to combine the new methods which are described above into a new algorithm. First of all, we give notations which are used in pseudocode in Table 2.

Table 2. Notations used

Notation	Annotation
N	The total number of data set
c	The number of clusters
v_i	The ith center point
x_k	The kth data point
U_i	The ith cluster in the partition $U(c, X)$
u_{ik}	The membership of x_k in U_i

The pseudocode for the completed NGEP-ADF algorithm is given here (Table 3).

Table 3. The pseudocode for NGEP-ADF algorithm

Algorithm NGEP-ADF

Input: $X = \{x_1, x_2, ..., x_N\}$; *c;*

Output: $V(\sigma)$ *and U(c,X).*

1. *Initialize parameters and normalize data;*
2. *Use improved Gene Expression Programming algorithm to randomly initialize clusters;*
3. *Execute each program;*
4. *Evaluate fitness;*
5. *do*
6.   *selection and reproduction;*
7.   *New chromosomes of next generation;*
8. *While (the result is not satisfied terminal condition)*

9. *Based on the medoid-based points, use Fuzzy C-means algorithm to cluster and mark the clustering results as U(c,X);*
10. *Output c center points and compare with real center points, If $E(c) = \|V^c - V^{real}\| < \varepsilon$, we can obtain the best clustering result, and the initial center points generated by the proposed method are what we want.*

4   Experiments and Results

In this section, in order to verify the performance and the effectiveness of the proposed NGEP-ADF algorithm, we tested on a real world dataset: Iris. We briefly describe the real world datasets below:

- Iris: This dataset has 150 data points divided into three classes (Setosa, Versicolor and Virginia). Each class has 50 data points. And every data point is in four-dimensional space (the length and the width of sepal, the length and the width of petal).
- Wine: The dataset concludes 178 data points along with 13 attributes (Alcohol, Malic, Acid, Ash, Alcalinity of Ash, Magnesium, Total phenols, Flavanoids, Nonflavanoid phenols, Proanthocyanins, Color, Intensity, Hue, OD280/OD315 of diluted wines, Proline) and it has three clusters.
- Glass: It has 214 data points which are divided into 6 classes. Each data point has 10 attributes such as Id number, Rrefractive index, Sodium, Magnesium, Aluminum, Silicon, Potassium, Calcium, Barium and Iron (Table 4).

Table 4. Basic information of the real world datasets

Datasets	Data size	Data dimension	The number of clusters
Iris	150	4	3
Wine	178	13	3
Glass	214	10	6

Table 5. The average accuracy of algorithms on the real world dataset Iris

Cluster algorithm	Correct number	Error number	Accuracy rate (%)
FCM	134	16	89.33
Paper [14]	140	10	93.33
FCM + MSD	141	9	94.00
Paper [15]	143	7	95.33
The proposed algorithm	144.77	5.23	96.51

Table 6. Results of clustering of algorithms on the Iris dataset (l-e-n = least error number, b-a-r = best accuracy rate)

Cluster algorithm	l-e-n	b-a-r (%)	Cluster center of the best case
FCM	16	89.33	$v_1 = (5.0036, 3.4030, 1.4850, 0.2515)$
			$v_2 = (5.8892, 2.7612, 4.3643, 1.3974)$
			$v_3 = (6.7751, 3.0524, 5.6469, 2.0536)$
FCM + MSD	9	94.00	$v_1 = (5.0086, 3.4170, 1.4749, 0.2480)$
			$v_2 = (6.6887, 2.9972, 5.5705, 2.0070)$
			$v_3 = (5.8930, 2.7816, 4.3225, 1.3828)$
Paper [16]	4	97.33	$v_1 = (5.0422, 3.4264, 1.4688, 0.2498)$
			$v_2 = (5.9489, 2.7635, 4.3052, 1.3465)$
			$v_3 = (6.6022, 2.9584, 5.5209, 1.9972)$
The proposed algorithm	3	98.00	$v_1 = (5.0001, 3.4210, 1.4601, 0.2440)$
			$v_2 = (5.9244, 2.7188, 4.2552, 1.3305)$
			$v_3 = (6.6002, 2.9900, 5.5209, 1.9862)$

We verify our main algorithm named as NGEP-ADF. The parameters for NGEP-ADF in all the experiments are set: the population size is 50, the maximum iteration is 150.

We take the Iris dataset as a particular test set in order to test the effectiveness of the proposed algorithm. We summarize the average accuracy and some results clustering compared with the other algorithms in Table 5. And Paper [17] provides the real clustering center of the Iris dataset: $v_1 = (5.00, 3.42, 1.46, 0.24), v_2 = (5.93, 2.77, 4.26, 1.32), v_3 = (6.58, 2.97, 5.55, 2.02)$. In Table 5, the proposed algorithm is compared with the other four algorithms. It is very clearly that the classical FCM algorithm has the lowest accuracy rate. And our new algorithm can produce a much better result than the others. Table 6 illustrates results of clustering of algorithms on the real world dataset Iris including least error number and best accuracy rate. We can see the proposed algorithm only obtain three errors while clustering center is closer to the real clustering center of the Iris. By analyzing the results of experiments, the data points belonged to Virginica are frequently clustered by mistake. Hence, the third experimental center is more different from the third real center.

At last, three real world datasets are tested in producing clustering solutions with both lower intra-cluster distance and higher inter-cluster distance after comparison of the traditional GEP, MOGEP [7] and our new algorithm. Table 7 shows the new algorithm obtain the better intra-cluster distance and inter-cluster distance. It is obvious that our new algorithm has best performance.

Table 7. Results for the three real world datasets compared with the other two algorithms (l-g-h = the length of gene head, A-GEP = average intra-cluster distance of GEP-Cluster, A-MOGEP = average intra-cluster distance of MOGEP-Cluster, A-N = average intra-cluster distance of NGEP-ADF Cluster, B-GEP = average inter-cluster distance of GEP-Cluster, B-MOGEP = Average inter-cluster distance of MOGEP-Cluster, B-N = average inter-cluster distance of NGEP-ADF Cluster)

Dataset	l-g-h	A-GEP	A-MOGEP	A-N	B-GEP	B-MOGEP	B-N
Iris	4	1.48×10^8	9.19×10^7	7.89×10^6	5.02×10^6	1.91×10^7	4.61×10^7
Wine	4	8.54×10^6	7.91×10^6	5.92×10^6	5.42×10^4	2.98×10^5	4.59×10^5
Glass	7	3.76×10^2	1.85×10^2	1.46×10^2	3.622	3.14×10	5.22×10

5   Conclusion and Future Direction

In our paper, we proposed a novel chromosome representation based on GEP-ADF in order to produce the medoid-based points of the data. The new algorithm was applied to a variety of three real world datasets and it also compares with different algorithms. The experimental results provide strong evidences to prove that the proposed algorithm has good performance and effectiveness.

However, there are some possible methods to improve the effectiveness and the performance of our new approach in the future work. To start with, How to improve the efficiency of program is still a problem. Because the proposed algorithm needs many parameters to calculate the results, it must lead to take too much time to run the procedure. It is worth considering more effective heuristic optimization ways. Furthermore, it is not steady while using improved GEP produces the great central points of datasets especially high-dimensional data. Therefore, we might aim at disposing the high-dimensional data to optimize the new algorithm. Finally, we can consider to use validity index to select the best the number of center points instead of deciding by people. In conclusion, there is still much work to do and we are going to optimize this algorithm on account of existing problems.

Acknowledgements. This work is supported by the National Natural Science Foundation of China with the Grant No. 61573157, the Fund of Natural Science Foundation of Guangdong Province of China with the Grant No. 2014A030313454.

References

1. Theodoridis, S., Koutroumbas, K.: Pattern Recognition, 4th edn. Academic Press, Waltham (2008)
2. Liu, C., Zhou, A., Zhang, G.: Automatic clustering method based on evolutionary optimization. Inst. Eng. Technol. **7**, 258–271 (2013)
3. Agustin-Blas, L.E., Salcedo-Sanz, S., Jimenez-Fernandez, S., Carro-Calvo, L., Del Ser, J., Portilla-Figueras, J.A.: A new grouping genetic algorithm for clustering problems. Expert Syst. Appl. **39**, 9695–9703 (2012)
4. Yu, Z., Chen, H., You, J., Wong, H.-S., Liu, J., Li, L., Han, G.: Double selection based semi-supervised clustering ensemble for tumor clustering from gene expression profiles. IEEE/ACM Trans. Comput. Biol. Bioinform. **11**, 727–740 (2014)
5. Sabau, A.S.: Variable density based genetic clustering. In: 14th International Symposium on Symbolic and Numeric Algorithms for Scientific Computing, pp. 200–206 (2012)
6. Zhou, C., Xiao, W., Tirpak, T.M.: Evolving accurate and compact classification rules with gene expression programming. IEEE Trans. Evol. Comput. **7**, 519–531 (2003)
7. Ni, Y., Du, X., Xie, D.: A multi-objective cluster algorithm based on GEP. In: 2014 International Conference on Cloud Computing and Big Data, pp. 33–38 (2014)
8. Laskar, A.B.Z., Kumar, S., Majumder, S.: Gene expression programming based age estimation using facial features. In: 2013 IEEE 2nd International Conference on Image Information Processing, pp. 442–446 (2013)
9. Zha, B.-B., Wang, R.-L., Sun, H.-L., Wang, L.: A study of the design and parameters optimization of BP neural network using improved GEP. In: 2014 10th International Conference on Computational Intelligence and Security, pp. 714–719 (2014)
10. Ferreira, C.: Automatically Defined Functions in Gene Expression Programming, pp. 21–56. Springer, Heidelberg (2006)
11. Ruspini, E.: Numerical methods for fuzzy clustering. Inf. Sci. **2**, 319–350 (1970)
12. Dunn, J.C.: A fuzzy relative of the ISODATA process and its use in detecting compact, well separated clusters. J. Cybern. **3**, 32–51 (1973)
13. Bezdek, J.C.: Cluster validity with fuzzy sets. J. Cybern. **3**, 58–73 (1974)

14. Zhu, C., Zhang, Y.: Research of improved fuzzy C-mean clustering algorithm. J. Henan Univ. (Nat. Sci.) **1**, 92–95 (2012)
15. Wen, Zhongwei, Li, Rongjun: Fuzzy C-means clustering algorithm based on improved PSO. Appl. Res. Comput. **27**, 2520–2522 (2010)
16. Li, K., Zhang, C., Chen, Z., Chen, Y.: Development of a weighted fuzzy C-means clustering algorithm based on JADE. Int. J. Numer. Anal. Model. Ser. B **5**, 113–122 (2014)
17. Li, J., Gao, X., Jiao, L.: A new feature weighted fuzzy clustering algorithm. Acta Electron. Sin. **34**, 89–92 (2006)

Predicting Acute Hypotensive Episodes Based on Multi GP

Dazhi Jiang[1], Bo Hu[1], and Zhijian Wu[2(✉)]

[1] Department of Computer Science, Shantou University, Shantou 515063, China
jiangdazhi111007@sina.com, 14bhu@stu.edu.cn
[2] State Key Lab of Software Engineering, Computer School, Wuhan University,
Wuhan 430072, China
zhijianwu@whu.edu.cn

Abstract. Acute Hypotensive Episodes (AHE) is one of the hemodynamic insta-bilities with high mortality rate that is common among patients. Timely and rapid intervention is necessary to save patient's life. This paper presents a methodology to predict AHE for ICU patients based on the Multi Genetic Programming (Multi GP). The methodology is applied to the dataset obtained from Multi-parameter Intelligent Monitoring for Intensive Care (MIMIC-II). The achieved accuracy of the proposed methodology is 79.07 % in the training set and 77.98 % in the testing set with the five-fold cross-validation.

Keywords: Time series · Acute hypotensive episodes · Empirical mode decomposition · Multi GP

1 Introduction

Acute Hypotensive Episodes (AHE) is defined as any period of 30 min or more during which at least 90 % of the mean arterial pressure (MAP) measurements were at or below 60 mmHg. The occurrence of AHE is a critical event in the ICU as it can cause multiple organ failure and thus has severe implications on the mortality risk in the ICU [1]. Because the blood flow through the arteries is too low to deliver enough oxygen and nutrients to vital organs such as the brain, heart, and kidney. If not treated promptly and properly, it may cause an irreversible organ damage, and eventually death. As a consequence, the characterization of such episodes is of fundamental importance for the management of critical decision concerning which intervention is more appropriate for each specific condition. In fact, early detection of AHE will provide professionals enough time to select a more effective treatment, without exposing the patient to additional risks because of delaying therapy. Therefore, the development of methodologies is able to not only detect the presence of this condi-tion but also predict its occurrence by computer, which is extremely important concerning appropriated clinical interventions.

Actually, for every patient in the ICU, massive amounts of clinical information are collected in real-time. Such as arterial blood pressure (ABP), heart rate (HR) and oxygen saturation (SO2), etc. And most of the data belong to the big data time series [2].

© Springer Science+Business Media Singapore 2016
K. Li et al. (Eds.): ISICA 2015, CCIS 575, pp. 161–170, 2016.
DOI: 10.1007/978-981-10-0356-1_16

Big data time series in the ICU is now touted as the solution to help clinicians to diagnose the case of the physiological disorder and select proper treatment based on this diagnosis. It's well known that the time series prediction is a hard problem and it plays an important role in various fields, such as in economics, in engineering and in biomedicine. The prediction and classification of AHE time series most initiated from the beginning of 21st century. Moreover, a challenge from PhysioNet/Computers in 2009 held a research competition about predicting the AHE which put this research into the climax. Generally, most research in monitoring ABP time series can be categorized into two aspects, which are pure ABP time series analysis and ABP time series with other parameters analysis methods.

For pure ABP data analysis, in 2001, Bassale J [3] proposed to generate the statistical summaries of Arterial Blood Pressure (ABP) signals to predict hypotension before hypotension episodes, including the mean, standard deviation, variance, skewness and the quantile-quantile. Saeed M [4] introduced a temporal similarity metric, which applied a wavelet decomposition to characterize time series dynamics at multiple time scales to utilize classical information retrieval algorithms based on a vector-space model. This algorithm was used to identify similar physiologic patterns in hemodynamic time series from ICU patients by the detection of imminent hemodynamic deterioration in 2007. In 2011, Teresa Rocha [5] proposed the application of neural network multi-models to predict the AHE which worked under two phases. First phase mainly trained the models according to analyze between the current blood pressure time signal and a collection of historical blood pressure templates, and in another phase, the multi-model structure was employed to detect the occurrence of AHE on the basis of predicting the future evolution of current blood pressure signal. Vaibhav Awandekar, A.N. Cheeran presented an algorithm consists of probability distributions of MAP and information divergence methods for calculating the statistical distance between tow probability distributions. The Bhattacharyya Distance is found out to be most accurate method for calculating such statistical distance [6]. X Chen et al. explored six basic indices derived from ABP data near the forecast window including mean ABP and diastolic ABP [7]. MA Mneimneh [8] presented a rule-based approach for the prediction of the AHE, and this approach could achieve testing accuracy for 87.5 %. PA Foumier [9] used Kullback-Leibler (KL) divergence between two distributions to identify the discriminative features, and then utilized these features to train the classification model based on a nearest neighbor algorithm. TCT Ho and X Chen proposed to process the MAP recordings like a signal but yet do analysis like an image in order to differentiate recordings from patients. They used a band pass filter in processing the MAP, and after processing, a divide and conquer strategy is adopted in evaluating the data sets using histograms [10].

For ABP data with other parameters analysis method, it is clinically accepted if there exists enough patient's clinical information, then a prediction system for hypotensive episodes, over a specific time period, can be developed. Typically, this information is based on the medical record, such as clinical history, laboratory tests and medications, as well as on information besides being extracted from physiologic vital signals, such as electrocardiogram, blood pressure and respiration. In this context, Lin et al. [11] studied the association of specific variables with the increasing risk of hypotensive episodes, namely weight, height, American Society of Anesthesiologist physical status, surgical category (orthopedics, plastic surgery, general surgery, obstetrics, and urology) and systolic blood pressure. Based on these variables, Lin et al. proposed a logistic regression model to assess the risk of developing a hypotensive episode. MA Frolich [12] discovered the higher baseline heart rate, possible reflecting a higher sympathetic tone, might be a useful parameter to predict hypotension in 2002. A Ghaffari [13] aimed to detect AHE and MAP dropping regimes using ECG signal and ABP waveforms in 2010, this method was based on adaptive network fuzzy inference system to incorporate the influences between heart rate, systolic blood pressure, diastolic blood pressure, age, gender, weight and some miscellaneous factors to the calculation of shock occurrence probability.

This work presents a generic methodology for time series prediction, which extracts features using statistics methods. Genetic programming (GP) is an effective method to select features and constructs a classifier simultaneously [14–16]. In this work, Multi GP is used to classify the AHE and NOAHE patients (In particular, AHE means there is an episode of acute hypotension beginning within the forecast window). The proposed methodology has a general applicability on the time series prediction and classification problem for its feature extraction and selection method. The validation set is comprised of 2866 records which are obtained from MIMIC-II database [17]. The achieved accuracy of the proposed methodology is 79.07 % in the training set and 77.98 % in the testing set after the five-fold cross-validation.

The rest of this paper is organized as follows: in Sect. 2, the detailed description of dataset is introduced. In Sect. 3, it introduces the methodology for the EMD and Multi GP. In the Sect. 4, experiment verification is presented and discussed. Finally, the last section is conclusion.

2　Data Set

The Multi-Intelligent Monitoring in Intensive Care (MIMIC) II project has collected data from about 30,000 ICU patients up to now [18]. In MIMIC-II database, the patient records contain most of the information that would appear in a medical record (such as results of laboratory tests, medications and hourly vital signs). About 5,000 of the records also include physiologic waveforms typically including ECG, blood pressure, and respiration etc., and time series that can be observed by the ICU staff. The intent is that a MIMIC-II record should be sufficiently detailed to allow its use in studies that would otherwise require access to an ICU, e.g., for basic research in intensive care medicine, or for development and evaluation of diagnostic and predictive algorithms for medical decision support.

In MIMIC-II database, the blood pressure signal includes systolic arterial blood pressure (SABP), diastolic arterial blood pressure (DABP), heart rate, SpO2, pulse, respiration and so on. The SABP is the maximum pressure (Fig. 1 red box) when the heart contracts and blood begins to flow. The DABP is the minimum pressure occurring (Fig. 1 green circle) between heartbeats (see Fig. 1).

In this experiment, we focus on MAP signal analysis, which is calculated as followed:

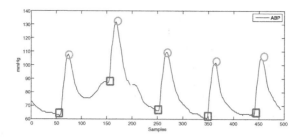

Fig. 1. ABP, SABP and DABP diagram (Color figure online)

$$MAP = DABP + \frac{SABP - DABP}{3} \tag{1}$$

Each data record contains 3 h data, including 2 h data before T_0 and 1 h data after T_0. In addition, the 1 h after the T_0 point is the forecast window. If the record contains AHE, the T_0 is always set at the beginning of the first AHE, else the T_0 is set casually in the case of sufficient data.

In this paper, we will put forward a methodology to predict AHE in the forecast window with the MAP data. The validation set is a big data set which contains 2866 records downloaded from MIMIC-II 3.0 dataset. For all records, the sampling frequency is 1 Hz.

3 Methodology

3.1 Empirical Mode Decomposition

Hilbert-Huang Transform (HHT) is an adaptive method for time series signal analysis, which is proposed by N.E. Huang [18]. HHT is composed of Empirical Mode Decomposition (EMD) and Hilbert Spectrum Analysis (HSA) method. As a nonlinear and nonstationary signal processing tool, HHT has been used in a large variety of applications. In this work, the EMD method is applied to data decomposition of patients' MAP signals.

EMD decomposes a complicated signal into several components called IMFs. As a consequence, each IMF component contains the local characteristics of original signals in different time scales. Each IMF has the following two specifications:

(1) In the whole data sequence, the number of extreme values and the number of zero crossing points must be same or not more than one at most.
(2) At any time, the envelope mean, defined by the signal of local maximum and minimum, is zeros.

For a fixed length time series signal $x(t)$ (for MAP time series, the length of time series is 2 h), the EMD process can be summarized as follows:

Step 1: Finding out all the local maximums and minimums of the signal $x_i(t)$, and getting the upper envelopes ($e_{max}(t)$) and lower envelopes ($e_{min}(t)$) by connecting the maximums and minimums respectively with cubic spline. Then, the average curve of envelopes $(m(t))$ can be calculated by:

$$m(t) = \frac{e_{max}(t) + e_{min}(t)}{2} \tag{2}$$

Step 2: Defining the intermediate variable $h(t) = x_i(t) - m(t)$, and detecting whether the $h(t)$ is an IMF or not on above conditions (1) (2).

Step 3: When $h(t)$ is an IMF, assigning the $c_i(t)$ to be an basic IMF by $c_i(t) = h(t)$.

Step 4: Repeat the process with the residual signal $x_{(i+1)}(t) = x_i(t) - h(t)$, and the Steps 1–3, until residual signal $x_{(i+1)}$ can't be decomposed.

At the end of the decomposition, the original signal $x(t)$ is defined as the sum of N IMFs and residual term $r(t) = x_{(i+1)}$:

$$x(t) = \sum_{t=1}^{N} c_i(t) + r(t) \tag{3}$$

3.2 Multi GP

After the IMFs of each record are obtained by EMD, some statistical features from the original signal and the IMFs are extracted, including min, mean, max, median, variance, the maximum instantaneous frequency, the high frequency energy to low frequency energy ration. Thus, the size of the feature set extracted from original signal and the IMFs produced by EMD is 77. After that, training the best classifier assists us to predict whether the patients suffered from AHE.

GP is an automatic programming technique for evolving computer programs, which is able to solve problems in a wider range of disciplines (may be more powerful than neural networks and other machine learning techniques) [15]. GP is applied in the classifiers design and feature selection frequently [16, 17]. For example, for the attributes A, B, C and the target attributes, *Yes* and *No*, with the mathematical operator set $\{+, -, *\}$, one example of GP classifier model can be showed as Fig. 2.

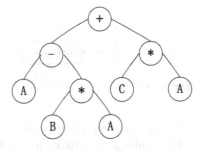

Fig. 2. Basic GP classifier model

The classifier is a discriminant function which is defined as:

$$f(x) = (A - B * A) + C * A \tag{4}$$

In this work, binary classifier algorithm based on GP is used for classifier and the fitness function is defined as follows:

$$fitness = consig * \exp(compl - 1)$$

$$consig = \left(\frac{p}{p+n} - \frac{P}{P+N}\right) * \frac{P+N}{N} \tag{5}$$

$$compl = \frac{p}{P}$$

Where, P and N are respective the total numbers of "*AHE*" and "*NOAHE*" class. The p and n is the correct number of P and N in the obtained discriminant function $f(x)$.

In order to avoid model over-fitting, the dataset is split for training models. The process is showed in Fig. 3. In the beginning ①, the dataset is split into $N + 1$ partitioned sub-datasets (N is odd), which are the $S_1, S_2, ..., S_{N+1}$. Among $N + 1$ sub-datasets, we supposed the S_i is the testing dataset, and the rest datasets are used to train N GP classifiers ②. N GP classifiers are combined together for voting for unlabeled MAP records ③, and the voting combination method is defined as follows:

$$\begin{cases} class = AHE, & \text{if more than half of } N \text{ output results are AHE} \\ class = NOAHE, & \text{otherwise} \end{cases} \tag{6}$$

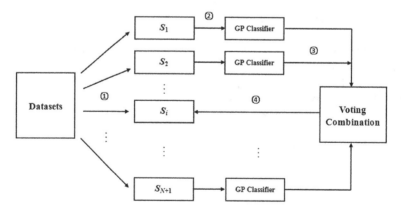

Fig. 3. Multi GP classifier method.

By voting combination, the classification model consists of several discriminant functions (named as Multi GP classifiers). Then for each record in testing data, S_i, the voting combination method is used to predict the "*AHE*" or "*NOAHE*" in one hour after the T_0 point ④.

4   Experiment Verification and Discussion

The Multi GP classifier parameters setting are as follows: The function sets are $\{+, -,$ $*, /, sqrt, exp, ln, x^2, x^3, x^{1/3}, sin, cos, atan\}$, the population size is 50, the mutation and crossover rate are 0.15 and 0.8 respectively, the stop criteria is 10000 generations and the Roulette method is used as the selection method.

The obtained results can be described in terms of accuracy (AC), sensitivity (SE) and specificity (SP). AC, SE and SP are given by the following equations.

$$AC = \frac{TP + TN}{TP + TN + FP + FN}$$

$$SE = \frac{TP}{TP + FN} \tag{7}$$

$$SP = \frac{TN}{TN + FP}$$

Where TP, FP, TN and FN are defined as true positive, false positive, true negative and false negative events detected respectively.

The dataset is randomly divided into 5 sub-datasets (Datasets $S_{i = 1, 2, 3, 4, 5}$). For each sub-dataset (S_j), we train 4 models to test it. For example, suppose the testing sub-dataset S_1, we train 4 different models ($M_{i = 2, 3, 4, 5}$) to predict the S_1, and the results are given in Table 1. Each training accuracy is the value of a classification model M_i in classifying the training sub-datasets S_j, and the testing accuracy is the value of classifier in predicting the testing sub-dataset S_1. Each records in S_1 is voted by 4 classification models ($M_{i = 2, 3, 4, 5}$). Moreover, the training and testing results of dataset in 5-fold cross validation by Multi GP are summarized in the following tables.

Table 1. The 4 models $M_{i= 2, 3, 4, 5}$ and predict result for S_1

No.	Training accuracy	Testing accuracy
M_2	80.35 %	78.19 %
M_3	79.48 %	78.32 %
M_4	80.71 %	79.02 %
M_5	78.42 %	77.39 %

The results of AC, SE and SP in the training set when 5-fold cross validation is used are shown in Table 2.

Table 2. The result of training set

	Training set					
Time Precision	First validation	Second validation	Third validation	Forth validation	Fifth validation	Average validation
ACC	78.39%	77.85%	80.46%	80.21%	78.43%	79.07%
SE	79.18%	76.49%	80.04%	78.63%	77.54%	78.38%
SP	78.67%	77.96%	80.05%	79.28%	78.41%	78.87%

The results of AC, SE and SP in the testing set when 5-fold cross validation is used are shown in Table 3.

Table 3. The result of testing set

	Testing set					
Time Precision	First validation	Second validation	Third validation	Forth validation	Fifth validation	Average validation
ACC	76.49%	81.23%	77.35%	76.68%	78.17%	77.98%
SE	81.67%	80.23%	80.26%	79.19%	80.94%	80.46%
SP	75.39%	80.39%	77.42%	77.52%	77.65%	77.67%

The accuracy of the Multi GP classifier are 79.07 % and 77.98 % with the proposed features in the training data and testing data respectively. Furthermore, within the same training and testing set, the SVM method with radial basis function kernel ($\sigma = 2.4$) is used to compare with Multi GP classification method. For accuracy, Multi GP obtains better performance than SVM in ACC, SE and SP. The results of experiments confirm that the Multi GP method improves the prediction of AHE with higher accuracy compared with the SVM (see Table 4).

Table 4. The results obtained by Multi GP and SVM

	Multi GP		SVM	
	Training	Testing	Training	Testing
ACC	79.07%	77.98%	76.22%	75.49%
SE	78.38%	80.46%	77.11%	74.09%
SP	78.87%	77.67%	75.27%	72.63%

5   Conclusion

Time series data is pervasive across almost all human endeavors, including medicine, finance, science, and entertainment. As such, it is hardly surprising that time series data mining has attracted significant attention. As a typical medical time series data, MAP signals are analyzed tentatively in this work. As a nonlinear and non-stationary signal processing tool, EMD method is used to decompose the MAP time series into a number of IMFs. The complex and unordered MAP data become regular and ordered by the decomposition. After features extraction, Multi GP method is used to establish the classifier for AHE prediction. The result shows that the classification model can provide the medical guidance for predicting, which is significant for the care and cure of AHE in ICU.

In spite of the success achieved by the proposed approach, more effort should be done for future work. For example, one Future work should consider other sources of information from AHE patients. In this study, only the arterial blood pressure signal is used. Further studies maybe take into account the heart rate variability to improve the capability of the model in forecasting MAP time series. This detection has done well in this PhysioNet database. However, it will table more research before the viability can be considered for clinical trials and usage.

Acknowledgements. The authors would like to thank anonymous reviewers for their very detailed and helpful review. This work was supported by National Natural Science Foundation of China (No.: 61502291), Natural Science Foundation of Guangdong Province (No.: S2013010013974), in part by the Shantou University National Foundation Cultivation Project (No.: NFC13003).

References

1. Moody, G.B, Lehman, L.H.: Predicting acute hypotensive episodes: the 10th annual physioNet/computers in cardiology challenge. In: Proceedings of IEEE Computers in Cardiology, pp. 541–544 (2009)
2. Gang, D., Shi-Sheng, Z., Yang, L.: Time series prediction using wavelet process neural network. Chin. Phys. B **17**(6), 1998 (2008)
3. Bassale, J.: Hypotension prediction arterial blood pressure variability. Tecnhical report (2001)
4. Saeed, M.: Temporal pattern recognition in multiparameter ICU data. Ph.D. thesis, Massachusetts Institute of Technology, USA (2007)
5. Rocha, T., Paredes, S., De Carvalho, P., et al.: Prediction of acute hypotensive episodes by means of neural network multi-models. Comput. Biol. Med. **41**(10), 881–890 (2011)
6. Awandekar, V., Cheeran, A.N.: Predicting acute hypotensive episode by bhattacharyya distance. Int. J. Eng. Res. Appl. **3**(2), 370–372 (2013)
7. Chen, X., Xu, D., Zhang, G., et al.: Forecasting acute hypotensive episodes in intensive care patients based on a peripheral arterial blood pressure waveform. In: Proceedings of IEEE Computers in Cardiology 2009, pp. 545–548 (2009)
8. Mneimneh, M.A., Povinelli, R.J.: A rule-based approach for the prediction of acute hypotensive episodes. In: Proceedings of IEEE Computers in Cardiology 2009, pp. 557–560 (2009)
9. Fournier, P.A., Roy, J.F.: Acute hypotension episode prediction using information divergence for feature selection, and non-parametric methods for classification. In: Proceedings of IEEE Computers in Cardiology 2009, pp. 625–628 (2009)
10. Ho, T.C.T., Chen, X.: Utilizing histogram to identify patients using pressors for acute hypotension. In: Proceedings of IEEE Computers in Cardiology 2009, pp. 797–800 (2009)
11. Lin, C.S., Chiu, J.S., Hsieh, M.H., et al.: Predicting hypotensive episodes during spinal anesthesia with the application of artificial neural networks. Comput. Methods Programs Biomed. **92**(2), 193–197 (2008)
12. Frölich, M.A., Caton, D.: Baseline heart rate may predict hypotension after spinal anesthesia in prehydrated obstetrical patients. Can. J. Anesth. **49**(2), 185–189 (2002)

13. Ghaffari, A., Homaeinezhad, M.R., Atarod, M., et al.: Detection of acute hypotensive episodes via a trained adaptive network-based fuzzy inference system (ANFIS). J. Electr. Electron. Eng. Res. 2(2), 025–047 (2010)
14. Koza, J.R.: Genetic Programming: on the Programming of Computers by Means of Natural Selection. MIT Press, Cambridge (1992)
15. Gray, H.F., Maxwell, R.J., Martínez-Pérez, I., et al.: Genetic programming for classification and feature selection: analysis of 1H nuclear magnetic resonance spectra from human brain tumour biopsies. NMR Biomed. 11(4–5), 217–224 (1998)
16. Muni, D.P., Pal, N.R., Das, J.: Genetic programming for simultaneous feature selection and classifier design. IEEE Trans. Syst. Man Cybern. Part B Cybern. 36(1), 106–117 (2006)
17. MIMICII. http://physionet.org/physiobank/database/mimicdb/
18. Huang, N.E., Shen, Z., Long, S.R., et al.: The empirical mode decomposition and the Hilbert spectrum for nonlinear and non-stationary time series analysis. Proc. R. Soc. London A Math. Phys. Eng. Sci. 454(1971), 903–995 (1998). The Royal Society

Research on Evolution Mechanism in Different-Structure Module Redundancy Fault-Tolerant System

Xiaoyan Yang[1,2(✉)], Yuanxiang Li[1], Cheng Fang[3], Cong Nie[3], and Fuchuan Ni[1]

[1] State Key Laboratory of Software Engineering, Computer School, Wuhan University,
Wuhan 430072, People's Republic of China
yxy0197@126.com, {yxli,fcni_cn}@whu.edu.cn
[2] Hubei Collaborative Innovation Center for High-Efficient Utilization of Solar Energy,
Hubei University of Technology, Wuhan, People's Republic of China
[3] School of Science, Hubei University of Technology,
Wuhan 430068, People's Republic of China
fcyoux@sina.com, 15527481340@163.com

Abstract. With the dramatic increase of circuit scale and the harsh environment, the reliability of the system has become the great hidden danger. Triple different-structure modular redundant system based on evolution mechanism shows good fault tolerant ability. How to enhance the efficiency and diversity of the evolution generation module has become the key issue which can ensure the system fault tolerant. This article puts forward two-stage mutation evolution strategy (TMES) and interactive two-stage mutation evolution strategy (ITMES) based on improving virtual reconfigurable architecture platform to evolve combination logical circuit on the fault-tolerant system with different-structure redundancy module. The efficiency of the proposed methodology is tested with the evolutions of a 2-bit multipliers, and a 3-bit multipliers, and a 3-bit full adders. The obtained results demonstrate the effectiveness of the scheme on generation circuit diversity and evolution efficiency.

Keywords: Evolvable hardware · Triple different-structure modular redundancy system · CGP · VRA · Evolution strategies

1 Introduction

With the micromation and intellectualized development of electron device, it has been extensively used in the extreme environment such as aviation, military affairs, the exploration of deep sea. However, with the dramatic increase of circuit scale and the harsh environment, the reliability of the system has become the great hidden danger. Hence, the research on how to enhance the stability of the electron device in extreme environment has been concerned.

Foundation item: Projects (HBSKFMS2014009) supported by the Hubei Collaborative Innovation Center for High-efficient Utilization of Solar Energy.

© Springer Science+Business Media Singapore 2016
K. Li et al. (Eds.): ISICA 2015, CCIS 575, pp. 171–180, 2016.
DOI: 10.1007/978-981-10-0356-1_17

TMR technology has become the widely-used fault tolerant technology due to its high reliability and easy implementation recently. The article [1, 2] points out TMR has greatly increased the reliability of FPGA based on the influence of SEU. However, due to its problems in the extreme environment such as SEU occurrence in the same gate of redundancy module and the high consumption of hardware resource, the development of FPGA partial reconfigurable architecture, different-structure modular redundancy system [3] and small granularity TMR fault tolerant technology [4] have been the research hotspot recently.

The article puts forward the modified virtual reconfigurable platform, two-stage mutation (TMES) and interactive two-stage mutation ES (ITMES) based on CGP coding to evolve combinational logic circuit. The result of the experiment shows that these two schemes respectively enhance the diversity and efficiency of producing successful individual based on the fault tolerant system with different-structure redundant modular, hence promote the stability of fault tolerant and real-time self-repairing capability.

2 Fault Tolerant System Architecture Based on Evolution Mechanism

Since Moore Shannon and Von Neumann et al. put forward the thought of enhancing the system reliability by using the redundancy, it has been the main scheme to solve the problem of fault tolerant by redundant resources to ensure the system stability. Triple different-structure Modular Redundancy system combines voter and three individuals which have the same functions and different circuit configuration. When one individual breaks down, the voter will select the output of multitude from same individual to shield the malfunctioned circuit. This scheme reduces the possibility of individual simultaneous error in severe environment due to the diversity of individual structure, and improves fault-intolerant ability of the system [5]. But the system stability is supposed to be destroyed when the appearance of multi-individual malfunction.

In 1993, Higuchi put forward new thought for the development of fault-tolerant technology, taking advantage of the dynamic reconfigurability of the programmable device itself and achieving system self-organizing, self-reconfiguration and self-repairing by evolutionary algorithm to drive the programmable device [6]. The Triple different-structure Modular Redundancy system based on the evolution mechanism does the error detection to the output of three different-structure modules by control module. When certain module goes wrong, it will select one individual from the successful individual pool to do the repairing. When the successful individual number in the successful individual pool is less than the threshold value, the control module motivate evolution algorithm module to regenerate the new circuit which can achieve objective function in the successful individual pool, and to ensure individual effectiveness by the multi-objective evaluation of heterogeneous degree and circuit power consumption, and finally to achieve the real-time repairing of the error, as the system architecture shown in the Fig. 1. The article [7, 8] proves the good fault-tolerant ability of Triple different-structure Modular Redundancy system based on evolution hardware. However, how to enhance the diversity of the evolution circuit and evolution efficiency has become the key issue of improving fault-tolerant ability of the system. Hence, the design of the evolution platform and evolution algorithm has become utmost important.

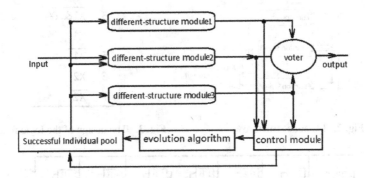

Fig. 1. Triple different-structure Modular Redundancy system based on evolution mechanism

3   Evolution Platform Design

Virtual Reconfigurable (VRC) sees the entire circuit as function element matrix of a r-line an c-row [9–11]. Each FE has two inputs, one output and certain function Fi (as shown in Fig. 2). The inputs stands for the interconnection which can originate from the original inputs or the former column outputs, it's depending on the joint series k. The function Fi stands for the specific operation to the input signal. The output stands for the result after the function operation.

Because VRC architecture restricts the entire circuit output's selection from the last column Fes [9], the generation circuit architecture is restricted. This article adopts the function virtual reconfigurable technology to evolve combinational logical circuit, which increases the diversity of the circuit by Multiplexer (MUX) module array to select all FE's output as the system output. The reference [5] proves the more complex the logical function of the digital circuits is, the lower average connection depth of the node is. Hence, the article adopt that each FE in column 1 can be connected to any one of the system inputs and negated system inputs, each FE in other columns is able to be connected to any one output of FE in previous one column and the system inputs. It can increase the flexibility of the circuit and the solving ability of the complicated circuit. The logic function Fi is shown by Fig. 3. The Fig. 4 shows a 8*8 virtual reconfigurable circuit architecture platform.

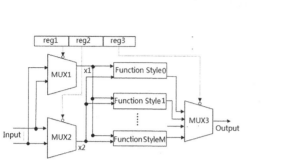

0	X1&&X2
1	X1||X2
2	X1^X2
3	$\overline{X1\&\&X2}$
4	X1
5	$\overline{X1}$
6	X2
7	$\overline{X2}$

Fig. 2. FE node structure   **Fig. 3.** The design of the logic function

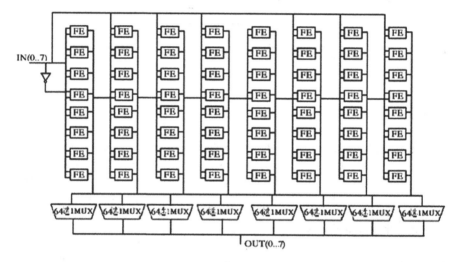

Fig. 4. Function virtual reconfigurable architecture platform

4 Evolution Algorithm

4.1 Coding

During the evolution process of the digital circuit, the mainstream coding scheme include GEP [12], CGP [13] and MEP [14], but the coding scheme CGP has been applied extensively due to the compatibility between the structure and the physical property of FPGA. Based on the above-mentioned VRA platform, this article adopts CGP coding which every FE coding is consist of triad (input1, input2, function). The input1 and input2 that express connections of each FE are selected by its two equipped multiplexers, and the logic function is configured with {000,001,010,011,100,101,110,111} to perform one of 8 logic function. For example, when constructing a VRA platform is 8*8 matrix, a FE node chromosome is {00010110010}, the 0001 represents the input1 of the

FE selects the No.1 among the 16 inputs, and the 0110 represents input2 selecting No. 6 among the 16 inputs, the 010 stands for XOR operation of the two inputs. Circuit output Oi selects certain node from 8*8 matrix as the output by 64-1 MUXs, each MUX need 6 bits to determine the system output. Hence, the configuration memory processes $64*11 + 8*6 = 752$ bits.

4.2 Two-Stage Mutation Evolution Strategy (TMES)

CGP coding simplifies the evolution of the complicated circuit, but there are many redundant nodes. Miller pointed out that the redundant nodes are beneficial to the evolution process [15]. How to enhance the utilization of the redundant node in the basis of maintaining the neutral effect [16]. This article focuses on $(1 + \lambda)$ ES based on the above-mentioned VRA and put forward a two-stage mutation evolution strategy. The basic idea is that as for the evolution stage, the mutation conducts two-stages which is respectively in the matrix connecting chromosome and in the circuit output selecting chromosome with different probability. To compare the adaptive values after two mutations, the better individual will be selected to next generation evolution. Hence, the redundant node's utilization will be enhanced by moderately increasing the assessment ratio of the adaptive value.

TMES Algorithm description:

Step1 Evolution generation initialization :gen = 0;
Step2 Randomly generate λ individual as the initial population x(t);
Step3 Respectively calculate all individual's fitness in the population x(t), select the fittest individual as a parent x1 among the population x(t);
Step4 Collection genes in individual x1 are selected with a given probability of Pm1, the selected genes are changed randomly and the new individual y1 is generated, calculate individual adaptive value f1; MUXs array output genes in individual y1 are selected with a given probability of Pm2, the selected genes are changed randomly and the new individual y2 is generated, calculate individual adaptive value f2. If f1 < f2, select y2 individual into the new population y (t), if f1 > f2, select y1 individual into new population x(t + 1). Repeat the step, until generate λ new individual into population x(t + 1);
Step5 Judge the end condition, if not satisfied t = t + 1, turn to step 3, or output the optimal result.

The basic idea is that as for the evolution stage, the redundant node's utilization will be enhanced by moderately increasing the assessment ratio of the adaptive value.

4.3 Interactive Two-Stage Mutation Evolution Strategy (ITMES)

EA influence the evolution design efficiency. But with the increase of the circuit complexity, the matrix size of virtual reconfigurable platform increase which results in the increase of the chromosome length, so the whole code space increases dramatically. While the increase of the input and output of the circuit, the searching space of the evolution algorithm exponentially increase [17–20].

Hence, based on the above-mentioned TMES, this article puts forward an interactive two-stage variation evolution strategy ITMES. In the process of evolution, when the population adaptive value stays in m generation, users put a individual which is no less that the present optimal individual's fitness or the successful individual as the selected individual in the step 3, and then interactively produce λ new individuals into the population x(t + 1) by the step 4 of TMES. Repeat the two steps until generate the optimal result or satisfy the end condition.

5   Analysis of the Experimental Results

To investigate the effectiveness of the proposed two modified algorithms and VRA platform, three benchmarks are chosen is 2-bit multipliers, 3-bit multipliers and 3-bit full adders. In all ITMES's experiment, the interactive individuals are the successful individuals. Table 1 lists the other's parameters and the settings of all experiments presented in this work.

Table 1. Experimental parameters and settings

	VAR Matrix		EA setting				
	Row	Col	λ	Pm1	Pm2	M	Maxgen
2-bit multipliers	4	4	15	0.015	0.5	50	10000
3-bit multipliers	8	8	4	0.015	0.5	1000	3000000
3-bit full adders	8	8	4	0.015	0.5	1000	1000000

5.1   Two-Bit Multipliers

The first experiment is the evolution of a 2-bit multipliers, which has 4 inputs and 4 outputs. The experimental results produced by ES, TMES, ITMES are summarized in Table 2. It show: considering the evolution generation, the ITMES is the best, ES is followed, and TMES is the worst; considering the circuit types, the TMES and ES is more, and ITMES is the least. An evolved 2-bit multipliers is shown in Fig. 5.

Table 2. Comparison of results for evolving 2-bit multipliers with various schemes

	Succeed times	Optimal generation	Average generation	Circuit types
ES	10	728	4803	10
TMES	10	789	5329	10
ITMES	10	622	2704	4

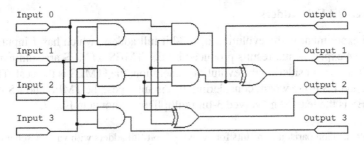

Fig. 5. An evolved 2-bit multipliers

5.2 Three-Bit Multipliers

The second experiment is the evolution of a 3-bit multipliers, which has 6 inputs and 6 outputs. The experimental results produced by ES, TMES, ITMES are summarized in Table 3. It show: considering the evolution generation, the ITMES is the best, TMES is followed, and ES is the worst; considering the circuit types, the TMES and ES is more, and ITMES is the least. An evolved 3-bit multipliers is shown in Fig. 6.

Table 3. Comparison of results for evolving 3-bit multipliers with various schemes

	Succeed times	Optimal generation	Average generation	Circuit types
ES	7	550832	1346724	10
TMES	10	315340	910739	10
ITMES	10	64910	202409	5

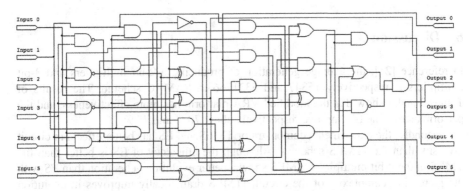

Fig. 6. An evolved 3-bit multipliers

5.3   Three-Bit Full Adders

The third experiment is the evolution of a 3-bit full adders, which has 7 inputs and 4 outputs. The experimental results produced by ES, TMES, ITMES are summarized in Table 4. It show: considering the evolution generation, the ITMES is the best, TMES is followed, and ES is the worst; considering the circuit types, the TMES and ES is more, and ITMES is the least. An evolved 3-bit multipliers is shown in Fig. 7.

Table 4. Comparison of results for evolving 3-bit full adders with various schemes

	Succeed times	Optimal generation	Average generation	Circuit types
ES	9	58335	266417	10
TMES	10	44823	228361	10
ITMES	10	39458	67708	4

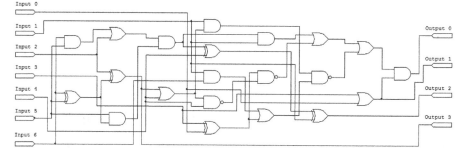

Fig. 7. An evolved 3-bit full adders

6   Discussion

In reference [21], the average generation of producing a 2-bit multipliers and 3-bit multipliers is respectively 15847 and 5321766 in SAMRC scheme. The results of Tables 2 and 3 show that the modified VRA platform has more effectiveness, and the performance of the ES improve respectively about 70 % and 75 %.

Secondly, the experiments demonstrator TMES and ES has better effects in circuit diversity than the ITMES schemes. Although the efficiency of ES is better than the TMES in the 2-bit multipliers, TMES show a better evolution generation than ES with the increasing complexity of the circuit. ITMES dramatically improves in evolution efficiency, but circuit diversity is the worst. We will add evaluation of the heterogeneous degree in ITMES to enhance the diversity of the evolution circuit in the follow-up research to ensure the stability and instantaneity of the large-scale circuit fault-tolerant system.

References

1. Rollins, N., Wirthlin, M., Caffrey, M., et al.: Evaluating TMR techniques in the presence of single event upsets. In: Proceedings of the Conference on Military and Aerospace Programmable Logic Devices (MAPLD), Washington, DC, p. 63 (2003)
2. Carmichael, C.: Triple module redundancy design techniques for virtex FPGAs. xAPP197(v1.0). Xilinx Corp., San Jose (2001)
3. Ping, J., Wang, Y., Kong, D., Yao, R.: Research on technology of different-structure system based on evolvable hardware. J. Chin. Compute Syst. 30(11), 2290–2293 (2009)
4. Pratt, B., Caffrey, M., Carroll, J.F., et al.: Fine-grain SEU mitigation for FPGAs using partial TMR. IEEE Trans. Nucl. Sci. 55(4), 2274–2280 (2008)
5. Lin, Y., Luo, W., Wang, X.: The selective evolution redundancy of hardware circuit. J. USTC 36(5), 523–529 (2006)
6. Higuchi, T., Iwata, M., Keymeulen, D., Sakanashi, H., et al.: Real-world applications of analog and digital evolvable hardware. IEEE Trans. Evol. Comput. 3(3), 220–235 (1999)
7. Gao, G.J., Wang, Y.R., Yao, R.: Research on redundancy and tolerance of system with different structures. Transducer Microsyst. Technol. 26(10), 25–28 (2007)
8. Yao, R., Wang, Y., Yu, S., Cheng, Z.: Design and experiments of enhanced fault–tolerant triple-module redundancy systems capable of online self-repairing. Acta Electro. Sin. 38(1), 177–183 (2010)
9. Sekanina, L.: Virtual reconfigurable circuits for real-world applications of evolvable hardware. In: Tyrrell, A.M., Haddow, P.C., Torresen, J. (eds.) ICES 2003. LNCS, vol. 2606, pp. 186–197. Springer, Heidelberg (2003)
10. Glette, K., Torresen, J.: A flexible on-chip evolution system implemented on a Xilinx virtex-II pro device. In: Moreno, J., Madrenas, J., Cosp, J. (eds.) ICES 2005. LNCS, vol. 3637, pp. 66–75. Springer, Heidelberg (2005)
11. Wang, J., Chen, Q.S., Lee, C.H.: Design and implementation of a virtual reconfigurable architecture for different applications of intrinsic evolvable hardware. IET Comput. Digital Tech. 2(5), 386–400 (2008)
12. Li, K., Liang, J., Zhang, W., et al.: Optimization algorithm for complicated circuit based on GEP. Comput. Eng. Appl. 44(18), 83–86 (2008)
13. Miller, J.F., Thomson, P.: Cartesian genetic programming. In: Banzhaf, W., Fogarty, T.C., Langdon, W.B., Miller, J., Nordin, P., Poli, R. (eds.) EuroGP 2000. LNCS, vol. 1802, pp. 121–132. Springer, Heidelberg (2000)
14. Oltean, M., Grosan, C.: Evolving digital circuits using multi expression programming. In: Proceedings of the 2004 NASA/DoD Conference on Evolution Hardware (EH2004) (2004)
15. Miller, J.F., Smith, S.L.: Redundancy and computational efficiency in Cartesian genetic programming. IEEE Trans. Evol. Comput. 10(2), 167–174 (2006)
16. Vassilev, V.K., Miller, J.F.: The advantages of landscape neutrality in digital circuit evolution. In: Miller, J.F., Thompson, A., Thompson, P., Fogarty, T.C. (eds.) ICES 2000. LNCS, vol. 1801, pp. 252–263. Springer, Heidelberg (2000)
17. Vassilev, V.K., Miller, J.F.: Scalability problems of digital circuit evolution evolvability and efficient designs. In: Proceedings of the Second Conference on Evolvable Hardware (2000)
18. Gordon, T.G.W., Bentley, P.J.: Towards development in evolvable hardware. In: Proceedings of the 2002 NASA/DOD Conference on Evolvable Hardware (EH 2002), pp. 241–250 (2002)
19. Lee, J., Sitte, J.: Issues in the scalability of gate-level morphogenetic evolvable hardware. In: Recent Advances in Artificial Life, Advances in Natural Computation, vol. 3, pp. 145–158. World Scientific, Singapore (2005)

20. Tufte, G.: Discovery and investigation of inherent scalability in developmental genomes. In: Hornby, G.S., Sekanina, L., Haddow, P.C. (eds.) ICES 2008. LNCS, vol. 5216, pp. 189–200. Springer, Heidelberg (2008)
21. Wang, J., Lee, C.H.: Virtual reconfigurable architecture for evolving combinational logic circuits. J. Cent. South. Univ. **21**, 1862–1870 (2014)

Intelligent Simulation Algorithms

Application of Neural Network for Human Actions Recognition

Tomasz Hachaj[1] and Marek R. Ogiela[2(✉)]

[1] Institute of Computer Science and Computer Methods, Pedagogical University of Krakow,
2 Podchorazych Ave, 30-084 Krakow, Poland
tomekhachaj@o2.pl
[2] Cryptography and Cognitive Informatics Research Group,
AGH University of Science and Technology, 30 Mickiewicza Ave, 30-059 Krakow, Poland
mogiela@agh.edu.pl

Abstract. In this paper we have proposed human actions recognition methodology. The main novelty of this paper is application of neural network (NN) trained with the parallel stochastic gradient descent to perform classification task on multi-dimensional time-varying signal. The original motion-capture data consisted of 20 time-varying three-dimensional body joint coordinates acquired with Kinect controller is preprocessed to 9-dimensional angle-based time-varying features set. The data is resampled to the uniform length with cubic spline interpolation after which each action is represented by 60 samples and eventually 540 (60×9) variables are presented to input layer of NN. The dataset we used in our experiment consists of recordings for 14 participants that perform nine types of popular gym exercises (totally 770 actions samples). The averaged recognition rate in k-fold cross validation for different actions classes were between 95.6 % ± 9.5 % to even 100 %.

Keywords: Human actions recognition · Neural network · Classification · Gym exercises · Kinect

1 Introduction

Human actions recognition is among new and challenging tasks for pattern recognition. Due to the fact that in everyday life human behavior is observed by cameras connected to computer systems (for example public security monitoring or digital controllers of games consoles) there is a growing demand on reliable computer methods of human actions classification. In literature we can find many state-of-the-art methods that was applied to solve this task. Most of them uses well-established classifiers like neural networks (NN), support vector machines (SVM), random forests (RF), Hidden Markov Models (HMM) and others.

In [1] authors propose a heterogeneous multi-task learning framework for human pose estimation from monocular images using a deep convolutional neural network. Authors simultaneously learn a human pose regressor and sliding-window body-part

© Springer Science+Business Media Singapore 2016
K. Li et al. (Eds.): ISICA 2015, CCIS 575, pp. 183–191, 2016.
DOI: 10.1007/978-981-10-0356-1_18

and joint-point detectors in a deep network architecture. Paper [2] presents a novel approach for supervised codebook learning and optimization for bag-of words (BoW) models. In presented application, space-time interest points are calculated on each video, and discriminant and invariant features are calculated on a space-time cuboid around each interest point location. Initially, a video is therefore described as a collection of feature vectors. In traditional ways to translate this description into a BoW model, codebook creation and learning of the BoW models of the training set are treated as two different phases addressed with two different methods. Authors present a novel formulation as a single artificial NN. Study [3] proposes human action recognition method using regularized multi-task learning. First authors propose the part Bag-of-Words (PBoW) representation that completely represents the local visual characteristics of the human body structure. Each part can be viewed as a single task in a multi-task learning formulation. Further, they formulate the task of multi-view human action recognition as a learning problem penalized by a graph structure that is built according to the human body structure. Work [4] proposes the volume integral as a new descriptor for three-dimensional action recognition. The descriptor transforms the actor's volumetric information into a two-dimensional representation by projecting the voxel data to a set of planes that maximize the discrimination of actions. Paper tests the volume integral using several Dimensionality Reduction techniques (namely PCA, 2D-PCA, LDA) and different Machine Learning approaches (namely Clustering, SVM and HMM) so as to determine the best combination of these for the action recognition task. In paper [5] authors propose a novel method for human action recognition based on boosted key-frame selection and correlated pyramidal motion feature representations. Instead of using an unsupervised method to detect interest points, a Pyramidal Motion Feature (PMF), which combines optical flow with a biologically inspired feature, is extracted from each frame of a video sequence. The AdaBoost learning algorithm is then applied to select the most discriminative frames from a large feature pool. In the classification phase, a SVM is adopted as the final classifier for human action recognition. Paper [6] addresses the multi-view action recognition problem with a local segment similarity voting scheme, upon which we build a novel multi-sensor fusion method. The random forests classifier is used to map the local segment features to their corresponding prediction histograms. In [7] BoW gives a first estimate of action classification from video sequences, by performing an image feature analysis. Those results are afterward passed to a common-sense reasoning system, which analyses, selects and corrects the initial estimation yielded by the machine learning algorithm. This second stage resorts to the knowledge implicit in the rationality that motivates human behavior. In paper [8] authors present feature descriptor for action recognition based on differences of skeleton joints, i.e., EigenJoints which combine action information including static posture, motion property, and overall dynamics. Accumulated Motion Energy (AME) is then proposed to perform informative frame selection, which is able to remove noisy frames and reduce computational cost. Authors employ non-parametric Naive-Bayes-Nearest-Neighbor (NBNN) to classify multiple actions. In work [9] authors propose an ensemble approach using a discriminative learning algorithm, where each base learner is a discriminative multi-kernel-learning classifier, trained to learn an optimal combination of joint-based features. In [10] authors propose a unsupervised learning method for automatic generation of knowledge base for syntactic Gesture Description Language (GDL) classifier [11] by analyzing unsegmented data recordings of gestures.

The up-to-date implementation of parallel stochastic gradient descent training method [12] allows to relatively quickly train NN that is dependent on hundreds of thousands synaptic weights. This enables faster development of action recognition methods that also requires less pre-processing of incoming signal taking nearly raw information that comes from motion capture hardware. The main novelty of this paper is application of NN trained with the parallel stochastic gradient descent to perform classification task on multi-dimensional time-varying signal. The original motion-capture data consisted of 20 time-varying three-dimensional body joint coordinates acquired with Kinect controller is preprocessed to 9-dimensional angle-based time-varying features set. The data is resampled to the uniform length with cubic spline interpolation after which each action is represented by 60 samples and eventually $540\,(60 \times 9)$ variables are presented to input layer of NN. The dataset we used in our experiment consists of recordings for 14 participants that perform nine types of popular gym exercises (totally 770 actions samples), the same large dataset as we used in [10]. In the following sections we will present the dataset we have used in our experiment, feature selection methodology and architecture of NN. We will also discuss the obtained results and present goals for future researches.

2 Material and Methods

In this section we will present the dataset we have used in our experiment, features selection procedure and architecture of NN we have used in our experiment.

2.1 Dataset and Features Selection

To gather the dataset for evaluation of proposed methodology we have utilized Microsoft Kinect. Despite the fact that Kinect was initially designed to be a game controller its potential as cheap general purpose depth camera was quickly noticed [13]. We have utilized Kinect SDK software library to segment and track 20 joints on human body with acquisition frequency of 30 Hz. The tracking was marker-less. Than we have changed original representation of motion capture data that is three-dimensional coordinates of 20 joints to angle-based representation. We did this because the original representation has two main drawbacks: it is dependent of relative position of user to camera leans and it is 60-dimensional (3 dimensions of Cartesian frame * 20 joints). The dependence from the camera position virtually prevents method from being usable in real-world scenario. In our angle-based representation (see Fig. 1 - left) the vertices of angles are positioned either in some important for movements analysis body joints (like elbows – angle 1 and 2, shoulders – angle 3 and 4, knees – angle 6 and 7) or angles measure position of limbs relatively to each other or relatively to torso. The second type of angles we utilized are angle defined between forearms (angle 5), angle between vector defined by joint between shoulders - joint between hips and thighs (angle 8 and 9). In the next step data is resampled to the uniform length with cubic spline interpolation after which each action is represented by 60 samples and eventually 540 (60 × 9) variables describe each action exemplar in our database. We have chosen 60-sample representation arbitrary.

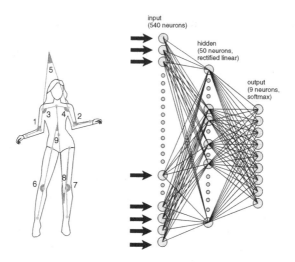

Fig. 1. In this figure on the left we present the positions of angles we used for angle-based representation of actions. In the right we present the schema of NN we used for classification task

We have used dataset that was previously used in our earlier work [10]. It consists of recordings for 14 participants, 4 women (W1-W4) and 10 men (M1-M10) – W means a woman, M – man, numbers defines id of a participant. The exercises that were performed were: body weight lunge left (bwll), body weight lunge right (bwlr), body weight squat (bws), dumbbell bicep curl (dbc), jumping jacks (jj), side lunges left (sll), side lunges right (slr), standing dumbbell upright row (sdur), tricep dumbbell kickback (tdk). In Table 1 we have presented quantities of gestures of a given type that was performed on a given SKL recording by each person. As can be seen not every person have performed each gesture, also the numbers of gestures are not equal. That is because that recordings were made in a certain period of time and not all users were asked to perform all gestures (for example in four recordings bws was skipped). Those lacks were then completed by recordings from other four persons in order to complete the dataset. Each person was asked to perform those exercises how many times he is capable to, but not more than 10 (in order not to get too tired for next exercises). There were some people who made those exercises more than 10 times (for example M1). For the other hand many participants, were getting tired more quickly and it was decided to reduce number of repetitions to 5 of each type. The participant M4 after performing slr was not capable to perform sll correctly.

Table 1. This table presents the test dataset we used in our experiment

Dataset	Bwll	Bwlr	Bws	Dbc	Jj	Sll	Slr	Sdur	Tdk
W1	10	10	10	10	10	10	10	10	10
W2	5	5	5	5	11	5	5	5	6
M1	13	10	12	11	9	10	7	12	10
M2	10	10	10	10	12	12	9	10	10
M3	10	10		10	10	8	10	10	10
W3			10						
M4	7	5		5	5		5	5	5
W4			5						
M5	10	10		10	10	5	10	10	10
M6			10						
M7	10	10	10	10	10	10	10	10	10
M8	5	5		5	5	6	5	5	5
M9			5						
M10	10	10	10	10	10	10	10	10	10
Sum	90	85	87	86	92	76	81	87	86

In Fig. 2 we have presented visualization of 9-dimensional representation of exemplar body weight lunge left exercise before resampling.

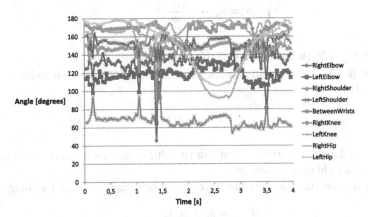

Fig. 2. This figure presents visualization of 9-dimensional representation of exemplar body weight lunge left exercise before resampling

2.2   Classification with NN

Multi-layer, feedforward neural networks consist of many layers of interconnected neuron units: beginning with an input layer to match the feature space followed by multiple layers of nonlinearity and terminating with a linear regression or classification layer to match the output space [14]. Each training example j the objective is to minimize a loss function $L(W, B|j)$.

Here W is the collection $\{w_i\}_{1:N-1}$, where W_i denotes the weight matrix connecting layers i and i + 1 for a network of N layers; similarly B is the collection $\{b\}_{1:N-1}$, where bi denotes the column vector of biases for layer i + 1. This basic framework of multi-layer neural networks can be used to accomplish deep learning tasks. Deep learning architectures are models of hierarchical feature extraction, typically involving multiple levels of nonlinearity. Such models are able to learn useful representations of raw data, and have exhibited high performance on complex data such as images, speech, and text [15]. The training of NN for classification task is based on minimization of cross-entropy loss function [14]:

$$(W, B|j) = - \sum_{y \in o} \left(\ln \left(o_y^{(j)} \right) \cdot t_y^{(j)} + \ln \left(1 - o_y^{(j)} \right) \cdot \left(1 - t_y^{(j)} \right) \right) \tag{1}$$

Where $o_y^{(j)}$ and $t_y^{(j)}$ are the predicted (target) output and actual output, respectively, for training example j, and y denote the output units and O the output layer.

For minimization of (1) stochastic gradient descent (SGD) method can be used which is an iteration procedure for each training example i [16]:

$$\begin{cases} w_{jk} := w_{jk} - \alpha \frac{\partial L(W,B|j)}{\partial w_{jk}} \\ b := b_{jk} - \alpha \frac{\partial L(W,B|j)}{\partial b_{jk}} \end{cases} \tag{2}$$

Where $w_{jk} \in W$ (weights), $b_{jk} \in B$ (biases).

Lately the lock-free parallelization scheme for SGD called Hogwild has been published [12].

$$\alpha = \sum_i w_i x_i + b \tag{3}$$

x_i and w_i denote the firing neuron's input values and their weights, respectively; α denotes the weighted combination.

The activation function in hidden layer might be a rectified linear function:

$$f(\alpha) = max(0, \alpha) \tag{4}$$

In our experiment we have utilized fully connected NN. Input layer had 540 neurons, hidden layer 50 neurons with rectified linear activation function (4) (number of neurons was arbitrary chosen) and output softmax layer with 9 neurons (the same as class number). The input data for network is standardize to $N(0, 1)$.

3 Results

We have implemented our approach in R language using "H2O" package [17] for neural network implementation and "signal" package for spline interpolation. Number of training epochs of NN was arbitrary set to 50. To validate our approach we used cross validation excluding actions of particular persons from training and making them target of NN prediction. The averaged values of recognition results of NN classifier from cross validation test plus/minus standard deviation are presented in Table 2.

Table 2. The averaged values of recognition results of NN classifier from cross validation test plus/minus standard deviation. Rows state for actual condition, columns are obtained recognition results

	bwll	bwlr	bws	dbc	jj	sdur	sll	slr	tdk
bwll	100.0%	0.0%	0.0%	0.0%	0.0%	0.0%	0.0%	0.0%	0.0%
bwlr	0.0%	100.0%	0.0%	0.0%	0.0%	0.0%	0.0%	0.0%	0.0%
bws	0.0%	0.0%	100.0%	0.0%	0.0%	0.0%	0.0%	0.0%	0.0%
dbc	0.0%	0.0%	0.0%	100.0%	0.0%	0.0%	0.0%	0.0%	0.0%
jj	0.0%	0.0%	0.0%	0.0%	100.0%	0.0%	0.0%	0.0%	0.0%
sdur	0.0%	0.0%	0.0%	1.0±3.0%	0.0%	99.0±3.0%	0.0%	0.0%	0.0%
sll	0.0%	0.0%	0.0%	0.0%	0.0%	0.0%	98.0±3.8%	2.0±3.8%	0.0%
slr	0.0%	0.0%	0.0%	1.4±4.3%	0.0%	0.0%	3.0±9.0%	95.6±9.5%	0.0%
tdk	0.0%	0.0%	0.0%	0.0%	0.0%	0.0%	0.0%	0.0%	100.0%

In Fig. 3, we present visualization of results from Table 2.

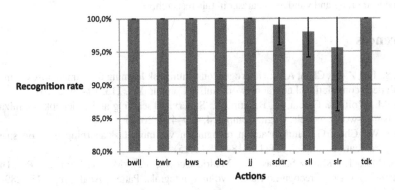

Fig. 3. This figure presents visualization of results from Table 2. Each bar represents averaged recognition rate of NN for single class. Black bars are plus/minus standard deviation

4 Discussion

As can be seen in Table 2 the initial experiment with NN that had most of parameters arbitrary fixed resulted in very good results. Most of the classes were 100 % correctly recognized. Two largest recognition errors were misclassification of sll as slr (2.0 ± 3.8 %) and slr as sll (3.0 ± 9.0 %). This situation happens because our representation of movements uses only information about values of angles between body joints without knowledge of movement direction. However the presence of error on such low level indicates that this is not very serious issue of our angle-based representation.

5 Conclusions

The representation of action to be recognized with angle-based features and resampling the multi-dimensional signal to the same length seems to be very promising approach for classification. Up-to-date implementation of state-of-the-art pattern recognition methods like H2O package for NN deals very well with large number of input features. There are however several open problems that have to be addressed in future research. The convenient method for selection of appropriate angles for features representation has to be established. Also the length of resampled data has to be related to characteristic of movement consisted in recording. We have also to evaluate our method on other actions datasets. What is more the evaluation should not be limited to proposed in this paper architecture of NN. The way to find optimal number of neurons and activation function would be valuable achievement. Also application of other classifiers like SVM and DF can result in even better classification results.

Acknowledgments. We kindly thank company NatuMed Sp. z o.o (Targowa 17a, 42-244 Wancerzow, Poland) for supplying us with SKL dataset that together with our own SKL recordings was used as training and validation dataset in this research.

References

1. Li, S., Liu, Z.-Q., Chan, A.B.: Heterogeneous multi-task learning for human pose estimation with deep convolutional neural network. Int. J. Comput. Vis. **113**, 19–36 (2015)
2. Jiu, M., Wolf, C., Garcia, C., Baskurt, A.: Supervised learning and codebook optimization for bag-of-words models. Cogn. Comput. **4**, 409–419 (2012)
3. Guo, W., Chen, G.: Human action recognition via multi-task learning base on spatial–temporal feature. Inf. Sci. **320**(1), 418–428 (2015)
4. Díaz-Más, L., Muñoz-Salinas, R., Madrid-Cuevas, F.J., Medina-Carnicer, R.: Three-dimensional action recognition using volume integrals. Pattern Anal. Appl. **15**, 289–298 (2012)
5. Liu, L., Shao, L., Rockett, P.: Boosted key-frame selection and correlated pyramidal motion-feature representation for human action recognition. Pattern Recogn. **46**, 1810–1818 (2013)
6. Zhu, F., Shao, L., Lin, M.: Multi-view action recognition using local similarity random forests and sensor fusion. Pattern Recogn. Lett. **34**, 20–24 (2013)

7. del Rincón, J.M., Santofimia, M.J., Nebel, J.-C.: Common-sense reasoning for human action recognition. Pattern Recogn. Lett. **34**, 1849–1860 (2013)

8. Yang, X., Tian, Y.: Effective 3D action recognition using EigenJoints. J. Vis. Commun. Image Represent. **25**, 2–11 (2014)

9. Chen, G., Clarke, D., Giuliani, M., Gaschler, A., Knoll, A.: Combining unsupervised learning and discrimination for 3D action recognition. Sig. Process. **110**, 67–81 (2015)

10. Hachaj, T., Ogiela, M.R.: Full body movements recognition – unsupervised learning approach with heuristic R-GDL method. Digit. Sig. Process. **46**, 239–252 (2015)

11. Hachaj, T., Ogiela, M.R.: Rule-based approach to recognizing human body poses and gestures in real time. Multimedia Syst. **20**, 81–99 (2014)

12. Recht, B., Re, C., Wright, S., Niu, F.: Hogwild: a lock-free approach to parallelizing stochastic gradient descent. In: Shawe-Taylor, J., Zemel, R.S., Bartlett, P., Pereira, F.C.N., Weinberger, K.Q. (eds.) Advances in Neural Information Processing Systems, vol. 24, pp. 693–701 (2011)

13. Hachaj, T., Ogiela, M.R., Koptyra, K.: Effectiveness comparison of Kinect and Kinect 2 for recognition of Oyama karate techniques. NBiS 2015 - The 18-th International Conference on Network-Based Information Systems (NBiS 2015), September 2–4, Taipei, Taiwan, pp. 332–337 (2015). doi:10.1109/NBiS.2015.51

14. Candel, A., Parmer, V.: Deep Learning with H2O, Published by H2O, (2015). http://leanpub.com/deeplearning. Accessed 8 August 2015

15. Bengio, Y.: Learning deep architectures for AI. Found. Trends® Mach. Learn. **2**, 1–127 (2009). doi:10.1561/2200000006

16. LeCun, Y.A., Bottou, L., Orr, G.B., Müller, K.-R.: Efficient BackProp. In: Orr, G.B., Müller, K.-R. (eds.) NIPS-WS 1996. LNCS, vol. 1524, pp. 9–50. Springer, Heidelberg (1998)

17. Official website of H2O machine learning programming library. http://h2o.ai/. Accessed 8 August 2015

The Improved Evaluation of Virtual Resources' Performance Algorithm Based on Computer Clusters

Suping Liu[(⊠)]

Department of Computer Science,
Guangdong University of Science and Technology,
Dongguan, Guangdong, China
457789090@qq.com

Abstract. After analyzing the four aspects of cloud computing, storage, network, infrastructure, and combining the fuzzy evaluation method in fuzzy mathematics theory, the quantitative index and qualitative index are combined to realize the virtual resource performance evaluation. This method breaks through the limitation of the previous evaluation system, and establishes a performance evaluation system based on multi-level fuzzy evaluation. Through the analysis of a case, it is proved that the proposed method can evaluate the resource performance more comprehensively, This will have a certain application value and significance for the research of resource management and scheduling based on cloud computing.

Keywords: Cloud computing · Resource scheduling · Fuzzy assessment · Performance assessment

1 Introduction

With the rapid development of science and technology both here and abroad, Cloud computing and storage solutions provide users and enterprises with various capabilities to store and process their data in third-party data centers. It relies on sharing of resources to achieve coherence and economies of scale, similar to a utility (like the electricity grid) over a network. At the foundation of cloud computing is the broader concept of converged infrastructure and shared services. Cloud computing needs to provide dynamic and flexible virtual resource services through the demand of end users of the Internet. It can also be said that cloud computing is the further development of distributed computing, parallel computing and grid computing [1]. Resource scheduling plays an important role in Cloud computing, its main mechanism is to map the tasks requested by users to appropriate resource nodes to execute. However Cloud Computing has characteristics of virtual, heterogeneous, complex, virtualization makes users have no need to know the requested tasks running on any physical host, which effectively shields the actual physical resource information, but meanwhile it increases the management of virtual resource, so the research of the optimal use of virtual resource under an complicated cloud computing has become a key technology.

© Springer Science+Business Media Singapore 2016
K. Li et al. (Eds.): ISICA 2015, CCIS 575, pp. 192–201, 2016.
DOI: 10.1007/978-981-10-0356-1_19

At present, the typical resource management of virtual system adopts the static resource scheduling, which is not the performance evaluation of virtual resources, which leads to the waste of a large amount of resources, and cannot meet the application of huge data center, cloud computing, etc. then, it will be applied to the allocation of virtual resources based on the market mechanism of the model, improve エthe efficiency of the resources, but the evaluation of the performance of the resources is more simple and single.

At present, there are some literature about on evaluation of Virtual Resources' Performance. The paper[*] proposed the approach of fuzzy assessment can takes a complete account of diverse influential factors and makes full use of the quality information, and respects the fuzzy attribute of the quality, so this approach has a more exquisite procedure and can get a more reliable result.

Hadoop is a distributed system infrastructure developed by the Apache foundation. However, in above case research, no valid assessment result can be obtained no matter which fuzzy assessment model is used. In common, the above case is looked on as an extreme case. However, it may be true because the chosen assessment factors are irrelative. Even if such an extreme case couldn't exist, an approximate case would lead to an invalid and unreliable assessment result.

2   Related Work

2.1   Definition of Cloud Computing

Up to now, there is still no unified definition on Cloud Computing, every huge IT giant enterprises put forward each Cloud computing strategies related to their own enterprise model with their own business, however, we must clear that the life cycle and development of cloud computing comes from users, only the broad users are the decision makers who decide the future of cloud computing, so based on specific technology of cloud computing definition will appear certain biased, some emphasizing on parallel cluster, some on virtualization, some on server, some on client. Regardless of any type of technical architecture, the ultimate purpose of Cloud Computing is to serve end-users, any XaaS (X can represent software, platform, infrastructure, hardware, etc.) should be considered as cloud computing, so the definition of Cloud Computing with user-centered will probably become truly accurated definition [2].

Cloud computing refers to ten thousands or more servers cluster distributed on Internet, it uses high-speed transmission network to deal with private data by calculating from a personal computer or server to cloud computing cluster. Users no longer need to buy high performance hardware and software, but turn to use any enabled Internet device connect to "cloud", to calculate and process data using "cloud" directly that provided software or service. Cloud computing cluster provides on-demand computing resource of cloud to users, it not only saves users' cost but also achieves the same effect of super computer [3, 4].

2.2   Architecture of Cloud Computing

Analyzing the current cloud computing, we take the internal cloud computing system as a collection of a set of services, separating the architecture of cloud computing into three layers, which are application layer, platform layer, infrastructure layer [5], Fig. 1 shows the architecture of cloud computing Fig. 2 shows the service of Cloud Computing

The application layer is faced to users, which realizes the interactive mechanism between users and service providers supported by the infrastructure layer, also called SaaS. During the process of task schedule, users submit tasks and receive results by application layer. Platform layer offer all platform resources needed by application program of running and maintaining. Infrastructure layer is a set of virtualization hardware resource and related management function, and also provides the optimized dynamic infrastructure services.

Cloud computing can be considered to include the following several levels of service: infrastructure as a service (IaaS), platform as a service (PaaS), and software as a service (SaaS).

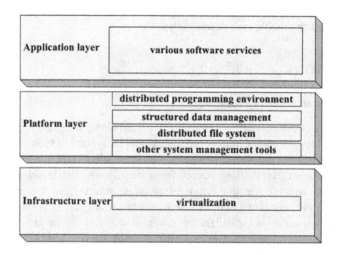

Fig. 1. Architecture of cloud computing

IaaS (Service Infrastructure-as-a-): infrastructure services. Consumers through the Internet can get service from the sound of the computer infrastructure. For example: hardware server rental.

PaaS (Service Platform-as-a-): platform that serves. PaaS is actually refers to the software development of the platform as a service, SaaS mode to the user. Therefore, PaaS is also a kind of SaaS mode application. However, the emergence of PaaS can accelerate the development of SaaS, especially to speed up the development of SaaS application. For example, the development of software customization.

SaaS (Service Software-as-a-): software as a service. It is a model that provides software through the Internet, users do not need to buy the software, but to the provider rented Web based software to manage business activities. For example: the sun cloud server.

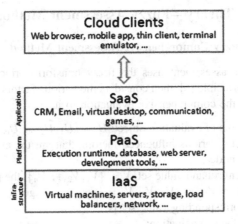

Fig. 2. The service of Cloud Computing

2.3 Virtual Resource Nodes Index System

Cloud Computing takes use of the mature virtualization technology to map actual physical hosts to the virtualization layer, which forms the standard virtualization resource pool. End-users just submit their requests to the virtualization layer to handle without caring where the tasks running on. So one important core of Cloud Computing is to realize the optimized management of virtual resource on virtual resource pool, which to meet all requests by users. Generally, we consider several main facts such as CPU, memory, storage, to evaluate the virtual resource performance. Aforementioned, such facts are still over single and simple. Therefore, in order to be evaluated roundly, the paper would consider from four aspects like computing, storage, network, infrastructure, and making further subdivide on the four aspects. Fig. 3 shows evaluation index system of virtual resource performance.

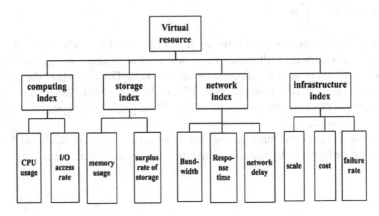

Fig. 3. Evaluation index system of virtual resource performance

3 Maths Fuzzy Theory—Fuzzy Assessment Method

3.1 Principle of Fuzzy Comprehensive Assessment Method

Fuzzy comprehensive assessment uses the fuzzy transform principle and maximum membership degree principle [6], then considers the various factors related to evaluated things, at last makes the comprehensive evaluation to it.

(1) establish the factors of evaluation object: $U = \{U_1, U_2, \ldots U_m\}$, the factors are the evaluation object's various influenced factors, that are the evaluation indexes of virtual resource nodes.
(2) establish comment set and value set: $V = \{V_1, V_2, \ldots V_n\}$, the paper divided into 5 grades, that is $V = V_1, V_2, V_3, V_4, V_5 = \{$very good, good, common, bad, very bad$\}$, and the corresponding value set is;
(3) establish single factor evaluation:

It is to define every factor's evaluation index in comment set, then judging a fuzzy result according to the evaluation grade in comment set which means how much the membership degree to each evaluation grade in comment set for each factor.

Membership degree can be represented by the membership degree function, for convenience, we took the linear function:

$$Y = \begin{cases} \frac{X-X_{\text{worst}}}{X_{\text{best}}-X_{\text{worst}}} \text{ or } \frac{X_{\text{best}}-X}{X_{\text{best}}-X_{\text{worst}}}, (X_{\text{worst}} < X < X_{\text{best}}) \\ 1, (X = X_{\text{best}}) \\ 0, (X = X_{\text{worst}}) \end{cases} \tag{1}$$

In Eq. (1), Y—the regulated membership degree for X_{worst} or X_{best}
X—actual value
X_{worst}, X_{best}—the best index and the worst index
If each single factor U_i had been judged, we could get the fuzzy matrix R.

(4) Due to each facto evaluation index has different emphasis on evaluation, it needs to give different weight on each factor, which presents a fuzzy subset on U, assuming that: $\sum_{i=1}^{m} a_i = 1$.

The model of fuzzy comprehensive evaluation is $B = A \circ R$, where B is the fuzzy subset on V, also called result vector of comprehensive evaluation; "$\circ$" is synthetic computation, which is similar with common matrix multiplication, where the difference are that changing the two multiplication "." to "∧" and getting the smaller one as "multiplication"; changing the two plus "+" to "∨" and getting the bigger one as "plus". We assumed it as (∧, ∨).

After making uniformization on the comprehensive evaluation B, we could make evaluation according to "the maximum membership degree" principle.

3.2 Multi-level Comprehensive Evaluation Model

In complicated systems, there are always many factors to be considered and among various factors usually have different levels. In this case, we should consider to divide the factors sets U into several categories according to certain attributes. Firstly evaluating comprehensively each category, then making further comprehensive evaluation among the categories of the evaluation result, that is to construct muti-level index system. This paper categorized ten indexes about the virtual resource node performance evaluation system, and assumed them for 4 secondary evaluation system (Fig. 2).

(1) devide factor set(evaluation system) into N subsets according to each attribute, $U = \{U_1, U_2, \ldots U_N\}$, where $U_i = \{u_{i1}, u_{i2}, \ldots u_{ik}\}$, = 1, 2,...N, it means that U_i had k_i factors, $\sum_{i=1}^{N} k_i = n$, and meeting the following request: $\bigcup_{i=1}^{N} U_i = U$.

(2) make comprehensive evaluation to each U_N with first model separately.

Assuming U_i's fuzzy subset of factor importance level as A_i (in this paper $A_i = \{0.15, 0.2, 0, 0.25, 0.4\}$), and U_i's total evaluation matrix of k factors as R_i, then we got:

$$A_i \circ R_i = B_i = (b_{i1}, b_{i2}, \ldots b_{in}), i = 1, 2, \ldots, N$$

Where B_i——single factor evaluation of U_i.

(3) regard each subset U_n as an element, taking b_i as their single factors evaluation, that is:

$$\tilde{R} = \begin{pmatrix} \tilde{B}_1 \\ \tilde{B}_2 \\ \cdots \\ \tilde{B}_N \end{pmatrix} = \begin{pmatrix} \tilde{A}_1 \circ \tilde{R}_1 \\ \tilde{A}_2 \circ \tilde{R}_2 \\ \vdots \\ \tilde{A}_N \circ \tilde{R}_N \end{pmatrix} \tag{2}$$

Each U_n is a part of U, representing a certain attribute, so it could give weight allocation with their importance rate: $a = \{a_1^*, a_2^*, \ldots, a_n^*\}$, the second comprehensive evaluation: $b = a^* \circ R$.

(4) finally, calculate the fuzzy subset b with the maximum membership degree principle, then we got the final evaluation result, but in order to get the optimized result, we could also calculate the grade parameter W taking use of the method that using membership degree b_j and weight p as power to calculate the weighted average, we could make comparability with it:

$$W = \sum_{j=1}^{n} b_j^p \bullet w_j / \sum_{j=1}^{n} b_j^p \tag{3}$$

In Eq. (3), w_j is the v_j's regulated value in W, p can be decided by specific problem, generally p = 2.

4 Examples and Assessment

Supposing there were ten virtual resource nodes that all meet the request of users, the weight allocation took Expert judgment method to determine with the quantitative indices and qualitative indices. Using above second-level fuzzy comprehensive assessment method to choose the optimized virtual resource node. Tables 1 and 2 (at the bottom) shows the index of ten virtual resource nodes.

Now taking the R1 resource node for example, we used the second-level fuzzy comprehensive assessment method to measure its performance:

(1) Assuming evaluation factor set as U = {u_i}: As shown in Table 1 we could know the first-level index U = {Computing-u_1, Storage-u_2, Network-u_3, Infrastructure-u_4}, the second-level index u_1 = {CPU usage-u_{11}, I/O access rate-u_{12}}, the second-level index u_2 = {memory usage-u_{21}, storage surplus rate-u_{22}}, the second-level index u_3 = {bandwidth-u_{31}, response time-u_{32}, network delay-u_{33}}, the second-level index u_4 = {scale-u_{41}, cost-u_{42}, failure rate-u_{43}}.

(2) Assuming comment set V = {very good v_1 = 1.0, good v_2 = 0.8, common v_3 = 0.6, bad v_4 = 0.4, very bad v_5 = 0.2}.

(3) Figuring out the first-level evaluation's single factor evaluation matrix R_1: the computing index is composed of two indexes that are CPU usage (u_{11}) and I/O access rate (u_{12}), putting the prepared data into membership degree function to calculate the single factor evaluation matrix of second-level index, then calculating it with prepared second-level evaluation weight matrix to make synthetic computation to get the first-level evaluation's single factor evaluation matrix R_1:

$$R_1 = \begin{bmatrix} 0.9 & 0.1 & 0 & 0 & 0 \\ 0.36 & 0.04 & 0 & 0.12 & 0.48 \\ 0.88 & 0.12 & 0 & 0 & 0 \\ 0.46 & 0.14 & 0 & 0 & 0.4 \end{bmatrix}$$

(4) making the first-level fuzzy comprehensive evaluation $b = a^* \circ R$. Where each b_j used the fuzzy mathematics method of M($\wedge$,$\vee$), which is: $b_j = \sum_{j=1}^{n} a_i^* \bullet r_{ij}$. So, we got R_1's evaluation result: B_j = [0.4 0.14 0 0.12 0.4]. Then, we handled the result with above Eq. (3). Finally, we got the mark of performance evaluation of R_1 resource node: W_1 = 0.603.

Other 9 resource nodes' performance followed the same method as R_1, so the paper didn't describe any more. We got the performance of all virtual resource nodes according to the aforementioned fuzzy comprehensive evaluation method:

$$W_1 = 0.603 \quad W_2 = 0.703 \quad W_3 = 0.682 \quad W_4 = 0.851 \quad W_5 = 0.788$$
$$W_6 = 0.727 \quad W_7 = 0.794 \quad W8 = 0.809 \quad W_9 = 0.760 \quad W_{10} = 0.756$$

Figure 4 showed the fuzzy assessment method (line 1) and cost-center as standard (line 2) to evaluate the mark of ten virtual resource nodes performance:

In Fig. 4, it could be seen from line 1 that R4 was the optimized virtual resource node, while from line 2 that R10 was the optimized virtual resource node considering cost-center as standard. After evaluating objectively, selecting R4 could improve the resource utilization more effectively.

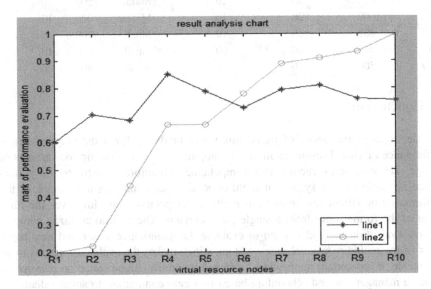

Fig. 4. Mark of performance evaluation

Table 1. The assummed ten virtual resource nodes indexes about computing and storage

Resource	Computing (0.15)		Storage (0.2)	
	CPU usage (0.6)	I/O Access rate (0.4)	Memory usage (0.4)	Storage surplus rate (0.6)
R1	10	10	10	20
R2	20	20	20	80
R3	5	10	20	50
R4	10	20	20	60
R5	20	60	10	50
R6	30	30	30	80
R7	10	40	10	20
R8	10	20	10	30
R9	20	10	5	95
R10	40	40	20	80

Table 2. The assumed ten virtual resource nodes indexes about network and infrastructure

Resource	Network (0.25)			Infrastructure (0.4)		
	Band-width (0.4)	Response time (0.3)	Network delay (0.3)	Scale (0.2)	Cost (0.4)	Failure rate (0.4)
R1	1000	20	20	Big	5000	10
R2	800	50	10	Big	4000	15
R3	1000	20	30	Big	3000	25
R4	900	10	10	Middle	2000	10
R5	1000	30	50	Middle	2000	5
R6	800	30	20	Middle	1500	20
R7	1000	20	10	Middle	1000	10
R8	800	10	10	Small	900	5
R9	700	60	60	Small	800	10
R10	500	10	30	Small	500	20

5 Conclusion

In order to meet the needs of users, this paper firstly analyzes the comprehensive performance of virtual resources in cloud computing after considering four aspects of storage, network, infrastructure and computation. Therefore, constructs the performance evaluation index system of multi-dimension, and analyzes and compares the performance of virtual resources from multiple perspectives, and improves the limitation of the former only from a single point of view. The traditional fuzzy comprehensive evaluation method is used to evaluate the quantitative index and qualitative index of the multidimensional index system, which makes the evaluation results more reasonable. It has certain application value and theoretical significance on research of resource management and scheduling based on cloud computing, to some extent.

Acknowledgment. This project is supported by Guangdong Province's the teaching reform project of higher education in 2015.

References

1. Foster, I., Zhao, Y., Raicu, I., Lu, S.: Cloud computing and grid computing 360-degree compared. In: Proceedings of IEEE Grid Computing Environments Workshop, pp. 1–10 (2008)
2. Peng, W.: The key technology of cloud computing and application examples, pp. 13–14. Posts and Telecom Press, Beijing (2010)
3. Zhu, S.C., Xu, Y., Jin, M.Y., Sheng, L.: Cloud computing security risk assessment based on level protection strategy. Comput. Secur. **5**, 39–42 (2013)
4. Yuan, H.C., Jia, W.G., Hu, J.F.: The research work of information security risk assessment in military colleges and universities network. In: Network security technology and Application. pp. 59–60, May 2009

5. Jian-xun, Z., Zhi-min, G., Chao, Z.: The reviewer of cloud computing research and development. Comput. Appl. **27**(2), 429–432 (2010)
6. Yu-jie, M., Jian, C., Shen-sheng, Z.: The research of extended WebService service quality model. Comput. Sci. **33**(1), 4–9 (2006)
7. Fang, Y., Wang, F., Ge, J.: A task scheduling algorithm based on load balancing in cloud computing. In: Wang, F.L., Gong, Z., Luo, X., Lei, J. (eds.) Web Information Systems and Mining. LNCS, vol. 6318, pp. 271–277. Springer, Heidelberg (2010)
8. Ting-bin, C.,Yan-xia, L., Guo-jun, S.: The application of the fuzzy assessment method based on webservices. J. DaLian Jiaotong Univ. **30**(1), 49–52 (2009)
9. Shi-bin, Z., Da-ke, H.: The research of establishment strategies on fuzzy independent trust. J. Electron. Inf. **28**(8), 1493–1497 (2006)

Bayesian Optimization Algorithm Based on Incremental Model Building

Jintao Yao[1(✉)], Yuyan Kong[2], and Lei Yang[1]

[1] College of Mathematics and Informatics, South China Agricultural University,
Guangzhou 510642, China
{justin_yjt,yanglei_s}@scau.edu.cn
[2] Guangzhou College of Commerce, Guangzhou 511363, China
yuyankong@sina.com

Abstract. In Bayesian Optimization Algorithm (BOA), to accurately build the best Bayesian network with respect to most metrics is NP-complete. This paper proposes an improved BOA based on incremental model building, which learns Bayesian network structure using PBIL instead of greedy algorithm in BOA. The PBIL is effective to learn better Bayesian network. The simulation results also show that the improved BOA has the better performance than BOA.

Keywords: Bayesian optimization algorithm · PBIL · Model building · Incremental learning

1 Introduction

Estimation of distribution algorithms (EDAs) [1–3] have become a new trend of stochastic optimizers in the last decade. Generally, EDAs build probabilistic models of promising solutions from population and sample from the corresponding probability distributions to generate new solutions, which replace the standard crossover and mutation operators in traditional evolutionary algorithm. Therefore, these algorithms are typically classified according to the complexity of the probabilistic models they rely on. Bayesian optimization algorithm (BOA) [4, 5] is one of high order EDAs, which uses Bayesian networks (BNs) [6] to model complex multivariate interactions in solution, and is able to solve a broad class of nearly decomposable and hierarchical problems in a reliable and scalable manner. However, the number of conditional probabilities that must be specified in Bayesian network grows exponentially with the order of interactions between variables encoded by the model, which also can become a bottleneck on more complex problems. In order to reduce the number of stored conditional probabilities of each variable without losing any information at all, BOA uses default tables as local structures. But the limitation of default tables is that they are capable of encoding the similarities among the members of only one subset of probabilities. So decision graphs which are more sophisticated local structures is incorporated into BOA, named DBOA [5, 7]. Decision graphs can find a use for more regularities in the conditional probabilities associated with Bayesian network, so by using decision graphs, BOA could reduce the number of conditional probabilities that must be stored by more than a half comparing to default tables. Besides saving memory

K. Li et al. (Eds.): ISICA 2015, CCIS 575, pp. 202–209, 2016.
DOI: 10.1007/978-981-10-0356-1_20

and time for both more complex model construction and utilization, decision graphs also help BOA to speed up the construction of Bayesian network. In order further to extend BOA to solve difficult hierarchical problems, hBOA [5] incorporate the approach of chucking into BOA with decision graphs without compromising its capability of decomposing the problem properly. At the same time, hBOA also uses restricted tournament replacement to effectively maintain diversity and preserve alternative partial solutions. It was empirically and theoretically shown that hBOA can solve nearly decomposable and hierarchical optimization problems in a quadratic number f evaluation of faster.

However, to learn a near-optimal Bayesian network structure from a population is required to search an exponentially large solution space, which has been proved to be a NP-complete problem [5, 8]. In BOA, the computational complexity of learning the structure is much higher than the computational complexity of learning the parameters and also dominates the overall complexity of BOA. In this paper, we use an incremental technique using PBIL [9, 12] to learn incrementally Bayesian networks structure and replace the greedy algorithm in BOA. Through the simulation test, we show that our new approach is efficient to find the optimal Bayesian networks quickly, which makes BOA more likely converge to the optimal solution.

The paper is organized as follows. Section 2 briefly reviews the BOA and discusses its computational complexity. Section 3 describes incremental Bayesian network structure building for efficiency enhancement of BOA. Section 4 describes the procedure of proposed Bayesian optimization algorithm based on incremental model building. Section 5 presents and discusses simulation results. The paper ends with a brief summery and conclusions.

2 Bayesian Optimization Algorithm

Basically, BOA evolves a population of candidate solutions which are represented by fixed-length binary strings. By using the Bayesian networks, the Bayesian optimization algorithm tries its best to find the dependencies or independencies among the decision variables of the optimization problems. Specially, BOA replaces the crossover and mutation operators used in traditional evolutionary algorithms by iteratively building and sampling a Bayesian network.

As the powerful probabilistic graphical models (PGM), Bayesian networks combine graph theory with probability theory to discover and encode the linkage relationships between variables. Basically, a Bayesian network is comprised of two parts: the structure and corresponding parameters. The structure is represented by a directed acyclic graph, where the nodes correspond to the variables of the optimization problem and the edges correspond to the conditional dependencies. The parameters are represented by the conditional probabilities for each variable. In essence, a fully connected Bayesian network can model and encode any given joint probability distribution as Eq. 1.

$$p(X) = \prod_{i=1}^{n} p(X_i|\Pi_i) \tag{1}$$

Where, $X = (X_1, X_2, \ldots, X_i, \ldots, X_n)$ represents the vector with all variables of the problem, Π_i represents the set of variable X_i's parents, and $p(X_i|\Pi_i)$ represents the conditional probability of X_i given Π_i. The parameters of a Bayesian network can be represented by a set of conditional probability tables or local structures such as decision trees and decision graphs. Because local structures can provide the more efficient and flexible representation, this paper focuses on Bayesian networks with decision graphs.

The procedure of constructing structure in BOA is very important and difficult. BOA uses a greedy algorithm to build the Bayesian network for a good compromise between search efficiency and network structure quality, during which BOA quantifies the quality of a built network structure by BD (Bayesian-Dirichet) metric or BIC (Bayesian Information Criterion) metric shown as Eqs. 2 and 3 respectively [10].

$$BD(BN) = p(BN) \prod_{i=1}^{n} \prod_{l \in L_i} \frac{\Gamma(m_i'(l))}{\Gamma(m_i'(l) + m_i(l))} \prod_{x_i} \frac{\Gamma(m_i'(x_i,l) + m_i(x_i,l))}{\Gamma(m_i'(x_i,l))} \tag{2}$$

Where, $p(BN)$ is the prior probability of Bayesian network structure BN, which can be set $p(BN) = 1$ or $p(BN) = 2^{-0.5 \log_2 N \sum_{i=1}^{n} |L_i|}$ [5], N is the population size and n is the problem size (length of variables); L_i is the set of leaves in the decision graph G_i corresponding to X_i; $m_i(l)$ is the number of instances in the population that contain the traversal path in G_i ending in leaf; $m_i(x_i, l)$ is the number of instances in the population that satisfy $X_i = x_i$ and contain the traversal path in G_i ending in leaf l; $m'(l)$ and $m'(x_i, l)$ represent prior knowledge of $m(l)$ and $m(x_i, l)$ respectively.

$$BIC(BN) = \sum_{i=1}^{n} \left(\sum_{l \in L_i} \sum_{x_i} (m_i(x_i,l) \log_2 \frac{m_i(x_i,l)}{m_i(l)}) - |L_i| \frac{\log_2 N}{2} \right) \tag{3}$$

The BIC metric is based on MDL (Minimum Description Length) principle. In this paper, we only consider and discuss the K2 variant of the BD metric assigning $m'(x_i, l) = 1$ for all i. The reason is that the K2 metric has been proved of more robustness than BIC metric when using decision graphs.

When the building of a Bayesian network is completed, new solutions can be generated by sampling the built Bayesian network with decision graphs according to PLS (Probabilistic Logic Sampling) algorithm [11]. Briefly, PLS firstly computes an ancestral ordering of nodes which are preceded by their parents individually, and then generate the values for each variable according to the ancestral ordering and the conditional probabilities.

3 Incremental Construction of Bayesian Network Structure Using PBIL

Fukuda *et al.* proposed a new method for constructing the Bayesian network based on PBIL [12]. In their algorithm, individuals correspond to each Bayesian network, which is represented as $s = (v_{11}, v_{12}, ..., v_{1N}, v_{21}, v_{22}, ..., v_{N1}, v_{N2}, ..., v_{NN})$, where v_{ij} corresponds to the edge from events X_i to X_j, i.e., if $v_{ij} = 1$ the edge from X_i to X_j exists in s, otherwise $v_{ij} = 0$. Similarly, the algorithm has the probability vector P to generate individuals as $P = (p_{11}, p_{12}, ..., p_{1N}, p_{21}, p_{22}, ..., p_{N1}, p_{N2}, ..., p_{NN})$, where p_{ij} is the probability that the edge from X_i to X_j exists. A probability vector can be regarded as a table. Note that, because Bayesian networks do not allow self-edges, p_{ij} is always 0 if $i = j$. The process of the proposed algorithm is described as Fig. 1.

Step 1. Initialize the probability vector P as $p_{ij} = 0$ if $i = j$ and $p_{ij} = 0.5$.

Step 2. Generate a set S that consists of C individuals, according to P.

Step 3. Compute values of the evaluation criterion for each individual $s \in S$.

Step 4. Select a subset of individuals S' whose members have evaluation values within top C' in S, and update the probability vector according to Eq.4.

$$p_i^{new} = dec(i) \times \alpha + p_i \times (1 - \alpha) \qquad (4)$$

Where, p_i^{new} is the updated value of the new probability vector P^{new}; $dec(i)$ is a function that takes 1 if more than C'/2 individuals have their evaluation values more than 0.5, and takes 0 otherwise; α is the parameter called learning ratio.

Step 5. Apply mutations on the new probability vector P according to Eq.5.

$$p_i^{new} = rand() \times \beta + p_i \times (1 - \beta) \qquad (5)$$

Step 6. Repeat step 2- step 5.

Fig. 1. The procedure of incremental construction Bayesian network using PBIL

The above algorithm requires that Bayesian network is not allowed to have cycles in it. Step 2 can meet the requirement, and please refer to the paper [12] for understanding the detailed step 2.

4 Bayesian Optimization Algorithm Based on Incremental Model Building

The basic steps of improved BOA are the same with those of BOA, and the difference is that improved BOA uses PBIL to incrementally constructing Bayesian network instead of greedy algorithm. The basic procedure of improved BOA is described as Fig. 2.

Step 1. Set t=0 and generate initial population $P(t)$ comprising N candidate solutions at random.

Step 2. Evaluate each candidate solution in population $P(t)$.

Step 3. elect N' promising candidate solutions from P to form a sub-population S(t).

Step 4. Build a Bayesian network $BN(t)$ for S(t) using PBIL(This step is described in Figure.2).

Step 5. Sample the joint probability distribution of $BN(t)$ to generate offspring $O(t)$.

Step 6. Combine O(t) with P(t) to yield P(t+1).

Step 7. Set t=t+1.

Step 8. Stop if some terminate criteria are satisfied. Or return to step 2.

Fig. 2. The procedure of improved BOA

5 Simulation and Results

5.1 Test Problem and Parameter Setting

Test problem: DEC-3 (3-order deceptive function) [13]. In DEC-3, the input string is partitioned into independent groups of 3 bits each. This partitioning is unknown to the algorithm and it does not change during the run. A 3-bit deceptive function is applied to each group and the contributions of all deceptive functions are added together to form the fitness. Each 3-bit deceptive function is defined as follows:

$$dec(u) = \begin{cases} 1.0, & if \quad u = 3 \\ 0.0, & if \quad u = 2 \\ 0.8, & if \quad u = 1 \\ 0.9, & if \quad u = 0 \end{cases} \quad (6)$$

Where, u is the number of 1 in the input string of 3 bits. The task is to maximize the function. An n-bit dec-3 function has one global optimum in the string of all 1 s and $(2n/3 - 1)$ other local optima. To solve DEC-3, it is necessary to consider interactions among the positions in each partition because when each bit is considered independently, the optimization is misled away from the optimum (Table 1).

Table 1. Parameter setting

BOA/Improved BOA			
Para-Name	*Para-Value*	*Para-Name*	*Para-Value*
populationSize	1000	α	0.1 ~ 0.8
problemSize	10	β	0.5
fitnessFunction	DEC-3	*Generation*	10-200
OffspringPercentage	50 %	C	20 ~ 110
maxNumberofGenerations	40	C'	1
randSeed	123	P_{mut}	0.001 ~ 0.008

5.2 Simulation Results

Table 2 shows that improved BOA needs much more running time than BOA when simulations stop in 40th generation. However, improved BOA gains the better Optima Percentage of 40th Generation. The reason is that PBIL can build the better Bayesian network than greedy algorithm in BOA, but the computational complexity is higher.

Table 2. The comparison of statistical result in 40th generation

Measure-Metrics	BOA	Improved BOA
Time(μs)	587	791
Optima percentage of 40th generation	52.9 %	59.8 %
Fitness of 40th generation (Best/Average/Worst)	3.0/3.0/3.0	3.0/3.0/3.0

Figure 3 indicates the BIC score of the best Bayesian network found by improved BOA and BOA respectively with the growth of generations. The improved BOA can converge to the optimal score, while BOA can only converge to the sub-optimal score. The higher BIC score indicates the algorithm can converge to the better solution. Therefore, the performance of the improved BOA is better than BOA.

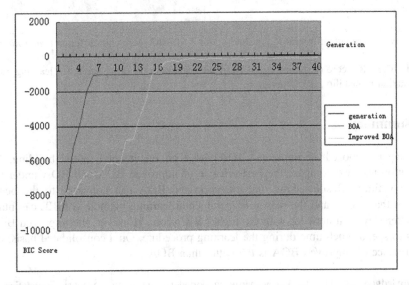

Fig. 3. The BIC score of the best Bayesian network with the growth of generations (X-generation, Y-BIC score)

Figure 4 shows the performance of the improved BOA with variation of C value, learning ratio, mutation probability and generation. These results show the performance of the proposed method depends on the above parameter.

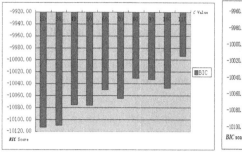
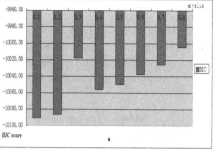

(a) BIC score with different **C** Value   (b) BIC score with different **α** value

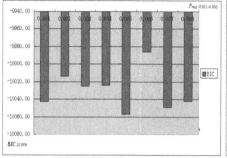
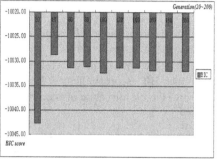

(c) BIC score with different P_{mut} value   (d) BIC score with different **Generation**

Fig. 4. The BIC score of the improved BOA with variation of: (a) C value; (b) learning ratio; (c) mutation probability; (d) *Generation*.

6 Summary

This paper proposed an improved BOA based on incremental model building. By simulation test, we compare the performance of improved BOA and BOA under the same condition. Results show that the improved BOA can converge to the better solution than BOA, and the reason is incremental learning method in PBIL can build the better Bayesian network with the higher BIC score. However, the improved BOA needs to spend much time during the learning procedure. On a consolidated basis, the performance of improved BOA is the better than BOA.

Acknowledgement. This work was partly supported by the Natural Science Foundation of Guangdong Province under Grant No. S2013040015755, the Foundation for Distinguished Young Talents in Higher Education of Guangdong Province under Grant No. 2013LYM_0119, and the Special Foundation for Public Welfare Research and Capacity Building of Guangdong Province under Grant No. 2014A020208087.

References

1. Pelikan, M., Sastry, K., Cantú-Paz, E.: Scalable Optimization Via Probabilistic Modeling: From Algorithms to Applications. Springer, Heidelberg (2006)
2. Lozano, J.A., Larranaga, P., Inza, I., Bengoetxea, E.: Towards a New Evolutionary Computation: Advances on Estimation of Distribution Algorithms. Springer, Heidelberg (2006)
3. Pelikan, M., Goldberg, D.E., Lobo, F.: A survey of optimization by building and using probabilistic models. Comput. Optim. Appl. **21**(1), 5–20 (2002)
4. Pelikan, M., Goldberg, D.E., Cantú-Paz, E.: BOA: the Bayesian optimization algorithm. In: Banzhaf, W., et al. (eds.) Proceedings of the Genetic and Evolutionary Computation Conference GECCO 1999, pp. 525–532. Morgan Kaufmann, San Francisco (1999)
5. Pelikan, M.: Hierarchical Bayesian Optimization Algorithm: Toward a New Generation of Evolutionary Algorithms. Springer, Heidelberg (2005)
6. Pearl, J.: Probabilistic Reasoning in Intelligent Systems: Networks of Plausible Inference. Morgan Kaufmann, San Mateo (1988)
7. Pelikan, M., Goldberg, D.E., Sastry, K.: Bayesian optimization algorithm, decision graphs, and Occam's razor. In: Genetic and Evolutionary Computation Conference (GECCO 2001), pp. 519–529. Morgan Kaufmann: San Francisco, California (2001)
8. Chickering, D.M., Heckerman, D., Meek, C.: Large-sample learning of Bayesian networks is NP-hard. J. Mach. Learn. Res. **5**, 1287–1330 (2004)
9. Baluja, S.: Population-based incremental learning: a method for integrating genetic search based function optimization and competitive learning. Technical report no. CMU-CS-94-163, Carf Michigan, Ann Arbor (1994)
10. Lima, C.F., Lobo, F.G., Pelikan, M., Goldberg, D.E.: Model accuracy in the Bayesian optimization algorithm. Soft. Comput. **15**, 1351–1371 (2011)
11. Henrion, M.: Propagation of uncertainty in Bayesian networks by logic sampling. In: Lemmer, J.F., Kanal, L.N. (eds.) Uncertainty in Artificial Intelligence, pp. 149–163. Elsevier, Amsterdam (1988)
12. Fukuda, S., Yoshihiro, T.: Learning Bayesian networks using probability vectors. In: Omatu, S., Bersini, H., Corchado Rodríguez, J.M., González, S.R., Pawlewski, P., Bucciarelli, E. (eds.) Distributed Computing and Artificial Intelligence 11th International Conference. AISC, vol. 290, pp. 503–510. Springer, Heidelberg (2014)
13. Goldberg, D.E.: Simple genetic algorithms and the minimal, deceptive problem (Chap. 6). In: Davis, L. (ed.) Genetic Algorithms and Simulated Annealing, pp. 74–88. Morgan Kaufmann, Los Altos (1987)

An Improved DBOA Based on Estimation
of Model Similarity

Yuyan Kong[1], Jintao Yao[2(✉)], and Lei Yang[2]

[1] Guangzhou College of Commerce, Guangzhou 511363, China
yuyankong@sina.com
[2] College of Mathematics and Informatics, South China Agricultural University,
Guangzhou 510642, China
{justin_yjt,yanglei_s}@scau.edu.cn

Abstract. In DBOA, to build accurately the best Bayesian network with respect to most metrics is NP-complete and the high time complexity of learning the model structure becomes a bottleneck of DBOA for real application. Consequently, in order to decrease the asymptotic time complexity of model building and make the algorithm more practical even for extremely large and complex problem, this paper presents adaptive sporadic model building based on estimation of model similarity as an efficiency enhancement technique of DBOA. The results show that performing the adaptive model building in DBOA can reduce the number of building model under no increasing on the number of generation and population size necessary to converge to optimal solutions, and achieve a better trade-off between the convergence speed and convergence results.

Keywords: Estimation of distribution algorithms · Bayesian optimization algorithm · Bayesian networks · Model similarity · Adaptive model building

1 Introduction

Genetic algorithms (GAs) and evolutionary algorithms (EAs) has been successfully and broadly applied to solving difficult and complex problems in real-world application since they were proposed by Holland [1]. As the sale and complexity of problems handled by genetic algorithms and evolutionary algorithms constantly increase, the ability of learning linkage relationship between variables becomes a crucial mechanism to have to be integrated into algorithms [2]. However, due to the lack of effective linkage learning ability, the standard crossover and mutation operators of genetic algorithms and evolutionary algorithms inevitably destroy good building blocks in solution during evolution procedure. From this standpoint, estimation of distribution algorithms (EDAs), which guide the search for optimum by building and sampling probabilistic graphical models (PGMs) of promising candidate solutions to replace the standard crossover and mutation operators, have brought a new trend of evolutionary algorithm [3–5]. EDAs have been successfully applied to numerous challenging optimization problems in the past decade and a half [6, 7].

© Springer Science+Business Media Singapore 2016
K. Li et al. (Eds.): ISICA 2015, CCIS 575, pp. 210–218, 2016.
DOI: 10.1007/978-981-10-0356-1_21

As one of the most prominent PGMs, Bayesian networks can model complex multivariate interactions for efficient and scalable optimization in the Bayesian optimization algorithm (BOA) [8, 9]. The results of related experiments indicate that BOA is capable of solving problems of bounded difficulty in a sub-quadratic number of fitness evaluations. However, the number of conditional probabilities that must be specified in Bayesian network grows exponentially with the order of interactions between variables encoded by the model, which also can become a bottleneck on more complex problems. Therefore, decision graphs is incorporated into BOA, named DBOA [9, 10]. In order further to extend DBOA to solve difficult hierarchical problems, hBOA [9] incorporates the approach of chucking into DBOA without compromising its capability of decomposing the problem properly. It was empirically and theoretically shown that hBOA can solve nearly decomposable and hierarchical optimization problems in a quadratic number f evaluation of faster.

However, to find the best Bayesian network has been proved to be a NP-complete problem [8, 9]. In DBOA, learning Bayesian network with decision graphs consists of learning the structure and learning the parameters. The computational complexity of learning the structure is much higher than the computational complexity of learning the parameters and also dominates the overall complexity of DBOA. Therefore, a number of approaches have been used to further enhance efficiency of DBOA [9].

The paper is organized as follows. Section 2 briefly reviews the BOA and discusses its computational complexity. Section 3 describes simple sporadic model building for efficiency enhancement of BOA and analyzes its advantage and disadvantage. Section 4 presents adaptive sporadic model building based on estimation of model similarity. Section 5 presents experimental results. The paper ends with a brief conclusions.

2 Bayesian Optimization Algorithms

Basically, BOA evolves a population of candidate solutions which are represented by fixed-length binary strings. By using the Bayesian networks, the Bayesian optimization algorithm tries its best to find the dependencies or independencies among the decision variables of the optimization problems. Specially, BOA replaces the crossover and mutation operators used in traditional evolutionary algorithms by iteratively building and sampling a Bayesian network.

As the powerful probabilistic graphical models (PGM), Bayesian networks combine graph theory with probability theory to discover and encode the linkage relationships between variables. Basically, a Bayesian network is comprised of two parts: the structure and corresponding parameters. The structure is represented by a directed acyclic graph, where the nodes correspond to the variables of the optimization problem and the edges correspond to the conditional dependencies. The parameters are represented by the conditional probabilities for each variable. In essence, a fully connected Bayesian network can model and encode any given joint probability distribution as following:

$$p(X) = \prod_{i=1}^{n} p(X_i | \Pi_i) \tag{1}$$

Where, $X = (X_1, X_2, \ldots, X_i, \ldots, X_n)$ represents the vector with all variables of the problem, Π_i represents the set of variable X_i's parents, and $p(X_i|\Pi_i)$ represents the conditional probability of X_i given Π_i. The parameters of a Bayesian network can be represented by a set of conditional probability tables or local structures such as decision trees and decision graphs. Because local structures can provide the more efficient and flexible representation, this paper focuses on Bayesian networks with decision graphs.

The procedure of constructing Bayesian network during BOA's running is very important and difficult. BOA uses a greedy algorithm to build the Bayesian network for a good compromise between search efficiency and network structure quality, during which BOA quantifies the quality of a built network structure by BD (Bayesian-Dirichet) metric or BIC (Bayesian Information Criterion) metric shown as formulas (2) and (3) respectively [11].

$$BD(BN) = p(BN) \prod_{i=1}^{n} \prod_{l \in L_i} \frac{\Gamma(m_i'(l))}{\Gamma(m_i'(l) + m_i(l))} \prod_{x_i} \frac{\Gamma(m_i'(x_i, l) + m_i(x_i, l))}{\Gamma(m_i'(x_i, l))} \tag{2}$$

Where, $p(BN)$ is the prior probability of Bayesian network structure BN, which can be set $p(BN) = 1$ or $p(BN) = 2^{-0.5 \log_2 N \sum_{i=1}^{n} |L_i|}$ [9], N is the population size and n is the problem size (length of variables); L_i is the set of leaves in the decision graph G_i corresponding to X_i; $m_i(l)$ is the number of instances in the population that contain the traversal path in G_i ending in leaf; $m_i(x_i, l)$ is the number of instances in the population that satisfy $X_i = x_i$ and contain the traversal path in G_i ending in leaf l; $m'(l)$ and $m'(x_i, l)$ represent prior knowledge of $m(l)$ and $m(x_i, l)$ respectively.

$$BIC(BN) = \sum_{i=1}^{n} \left(\sum_{l \in L_i} \sum_{x_i} (m_i(x_i, l) \log_2 \frac{m_i(x_i, l)}{m_i(l)}) - |L_i| \frac{\log_2 N}{2} \right) \tag{3}$$

3   Sporadic Model Building for Efficiency Enhancement of BOA

Pelikan proposed and introduced sporadic model building (SMB) technique into hBOA [13]. With SMB, hBOA learns the structure of Bayesian network only once in every few generations, during which only the parameters (conditional and marginal probabilities) of Bayesian network need to be updated using the selected solutions. The above method is based on the fact that the models learned closely correspond to problem structure and do not change much over consequent iterations [14]. In addition, the time complexity of learning the structure is much higher than the time complexity of updating the parameters, so SMB largely decreases the overall complexity of model building and increases the efficiency of BOA partly.

However, one of the most important factors that influence the effectiveness of SMB is when to only update the parameters for the previous structure. To solve this question, a very direct and simple method is used in [14], in which a parameter $t_{ab} \propto n^{-2}$ called the structure-building period decides the period to update the structure, where n is the problem size. For example, $t_{ab} = 2$ denotes the scenario in which the structure is built in every other iteration. But t_{ab} is fixed during the running of algorithm, which can not guarantee model having adequate similarity in the period of t_{ab}. Therefore, the lack of accuracy in SMB is obvious and this shortcoming also will seriously limit the actual effect of SMB. Extremely, if we set t_{ab} too large, it is inevitable that the accuracy of model will be badly reduced. In order to achieve a good convergence, the algorithm has to enlarge the population size and convergence time quickly with the increasing of t_{ab}. Instead of speeding up the algorithm, SMB would slow the convergence speed of algorithm down and increases the computational complexity. In contrast, if we set t_{ab} too small (for example, $t_{ab} = 1$), SMB will be of no effect. Anyhow, for each particular optimization problem, it also can be expected that there exists an optimal value of t_{ab} leading to the maximum overall speedup of BOA on this problem. Nonetheless, the research of SMB shows that how to set the value of t_{ab} for all kind of problems with different complexity is independent from selecting the problems for solving. Therefore, the research of SMB would still bias to problems for solving and heavily depend on the theoretical analysis of problem structure. In many practical situations to solve high order problems, although we can use the mathematical statistics and inductive method to obtain the optimal value of t_{ab}, but the lack of adaptability still becomes a disadvantage of SMB, which inevitably limits BOA to play its potential best.

4 Adaptive Model Building Based on Estimation of Model Similarity

In order to increase the adaptability of SMB in DBOA, we introduce a parameter called model similarity s_{ab} into SMB, which can evaluate the difference (distance) between two adjacent populations and the corresponding Bayesian networks. The larger is the distance, the smaller is the model similarity. Therefore, using the parameter s_{ab} during the running of DBOA, the structure-building period t_{ab} can be adaptively decided by DBOA itself. We call this method the adaptive model building (AMB) in this paper. Assuming each individual in population has n variables and each variable lies in $\{0, 1\}$, the joint probability distribution described by population can be presented as n-dimensional matrices with 2^n elements. Since a matrix of n elements can be represented geometrically as a point in the $\Re^n$ space, the joint probability distributions described by adjacent populations can be represented geometrically as points in the $\Re^{2^n}$ space [15]. Specially, AMB technique maps the joint probability distribution of population to $\Re^{2^n}$ space at first, and then calculates the Euclidean Distance (ED) of the corresponding mapping nodes on the $\Re^{2^n}$ space, which can reflect (estimate) the similarity s_{ab} of two Bayesian networks with respect to adjacent populations more accurately. In fact, ED between joint probability distributions of adjacent populations also reflects the uncertainty distribution (diversity) of evolutionary population, which shows the distribution characteristic of corresponding Bayesian networks. Figure 2 shows a sample of using ED between adjacent populations to estimate the similarity o corresponding Bayesian networks (Fig. 1).

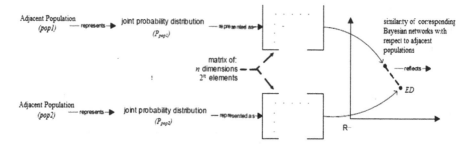

Fig. 1. *ED* between joint probability distribution matrixes of adjacent populations

The calculation formulas of *ED* and s_{ab} are as follows:

$$ED = \sqrt{\sum_{n_{ij}} \sum_{i=0}^{n} \sum_{j=0}^{n} (point_{ij} - point'_{ij})^2} \tag{4}$$

$$S_{ab} = ED + \sqrt{Num_{MB}} \tag{5}$$

Where, n is the problem size and Num_{MB} is the number of constructing structure.

Step 1. Set $t=0$ and generate initial population P(t) comprising N candidate solutions at random; set $s_{ab}=0$ and $\mu =0$.

Step 2. Evaluate each candidate solution in population P(t).

Step 3. Select N' promising candidate solutions from P to form a sub-population S(t); calculate the joint probability distribution $P_{s(t)}$ of S(t) and form a n-dimensional matrix $Ma_{s(t)}$. When $t>0$, calculate the *ED* between S(t) and S(t-1) , s_{ab} and μ. If $s_{ab} \geq \mu$, the procedure skips to step5.

Step 4. Build a Bayesian network BN(t) for S(t) and store it as $BN_{archive}$.

Step 5. Sample the joint probability distribution of $BN_{archive}$ to generate offspring O(t).

Step 6. Combine O(t) with P(t) to yield P(t+1).

Step 7. Set $t=t+1$

Step 8. Stop if some terminate criteria are satisfied. Or return to step 2.

Fig. 2. The procedure of DBOA with AMB

Specially, we can set a dynamic s_{ab} threshold and calculation formula is as follows:

$$\mu = \frac{1}{m+1} \sum_{i=n-m}^{n} ED \tag{6}$$

Where, $m = (\log_{10} n)$.

The experimental results confirm that the average structure-building *speedup* and the average evaluation *slowdown* for AMB are significantly better than the SMB method. Figure 2 shows the procedure of improved DBOA with AMB.

5   Experimental Results

In the section, we will test AMB method and find whether AMB can help DBOA to achieve a high structure-building *speedup* and the average evaluation *slowdown* comparing with SMB. In three algorithms, each individual of the population is represented by an *n*-bit binary string, where *n* is the test problem size. For each problem instance, 72 independent runs are performed. The experimental environment and parameters are set as Table 1, where we set $t_{ab} = 2$ for SMB.

Table 1. Parameter setting

Name of parameter	Parameters value
Population size	1200
Offspring percentage	50
Tournament size	4
Max number of generations	200
Max incoming edges in graph	20
Allow merge operator	0 (DO NOT Allow)

The test functions used in the experiments are *Onemax* function, *3-Deceptive* function, and *Trap-5* function.

Firstly, we respectively used DBOA, DBOA$_{SMB}$ (DBOA with SMB) and DBOA$_{AMB}$ (DBOA with AMB) to test the *3-Deceptive* function for different problem size. Table 2 shows the comparing results of convergence generation ($G_{covergence}$), minimal fitness ($Fitness_{min}$), average fitness ($Fitness_{avg}$) and maximal fitness ($Fitness_{max}$). The convergence generation of DBOA$_{AMB}$ is higher than DBOA. However, DBOA$_{AMB}$ achieves the better results on other aspects than DBOA and DBOA$_{SMB}$. The results also confirm that for difficult problems that AMB can help DBOA to improve its performance for real application.

Figures 3, 4 and 5 respectively shows the number of constructing network structure for different test functions until the convergence of DBOA, DBOA with SMB and DBOA with AMB. With the problem size growing, AMB helps DBOA to achieve the lower structure-building speedup than other two algorithms, which means that overall complexity of DBOA can be decreased significantly under AMB's assistance. This observation agrees with the presented results.

Table 2. Statistical results of three algorithms to solve 3-order Deceptive problem

Problem-Size	$G_{covergence}$ DBOA/ DBOA$_{SMB}$/ DBOA$_{AMB}$	$Fitness_{min}$ DBOA/DBOA$_{SMB}$/ DBOA$_{AMB}$	$Fitness_{avg}$ DBOA/DBOA$_{SMB}$/ DBOA$_{AMB}$	$Fitness_{max}$ DBOA/DBOA$_{SMB}$/ DBOA$_{AMB}$
40	21/31/24	13.79/13.90/13.90	13.89/13.90/13.90	13.90/13.90/13.90
60	26/33/32	19.70/19.90/19.90	19.79/19.99/19.99	19.79/20.00/20.00
80	34/47/39	25.50/25.59/26.69	26.49/26.59/26.70	26.50/26.59/26.79
100	36/49/45	33.00/33.40/33.19	33.00/33.40/33.19	33.00/33.40/33.19
120	37/55/53	38.70/39.10/39.00	38.70/39.10/39.00	38.70/39.10/39.00
140	49/65/55	45.40/44.70/45.20	45.49/44.70/45.20	45.50/44.70/45.20
160	50/63/60	50.50/51.30/52.10	51.49/51.49/52.10	51.50/51.50/52.10
180	56/67/65	55.80/56.70/57.00	55.80/56.70/57.00	55.80/56.70/57.00

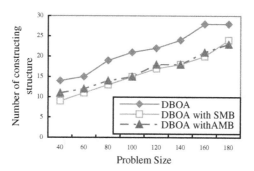

Fig. 3. Num_{MB} comparison of three algorithms to test *Onemax* function

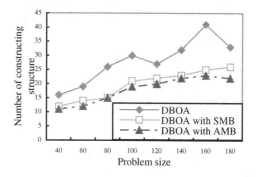

Fig. 4. NumMB comparison of three algorithms to test *3- Deceptive* function

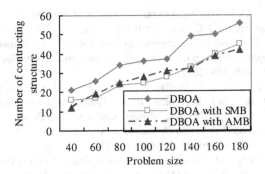

Fig. 5. NumMB comparison of three algorithms to test *Trap-5* function

6   Conclusions

In DBOA, building the model structure is the most computationally expensive part, which becomes a computational bottleneck for DBOA to solve large and complex problems. In order to extend the application of DBOA in real world, this paper presented and analyzed an efficiency enhancement technique called Adaptive model building (AMB), which can be used to speed up model building and decrease the overall complexity. The experimental results presented in this paper also confirm that AMB technique is more efficient than SMB.

Acknowledgement. This work was partly supported by the Natural Science Foundation of Guangdong Province under Grant No. S2013040015755, the Foundation for Distinguished Young Talents in Higher Education of Guangdong Province under Grant No. 2013LYM_0119, and the Special Foundation for Public Welfare Research and Capacity Building of Guangdong Province under Grant No. 2014A020208087.

References

1. Holland, J.H.: Genetic algorithms and the optimal allocation of trials. SIAM J. Comput. **2**(2), 88–105 (1973)
2. Goldberg, D.E.: The Design of Innovation: Lessons form and for Competent Genetic Algorithms. Genetic Algorithms and Evolutionary Computation. Kluwer Academic Publishers, Boston (2002)
3. Lozano, J.A., Larrañaga, P., Inza, I., Bengoetxea, E.: Towards a New Evolutionary Computation: Advances on Estimation of Distribution Algorithms. Springer, Heidelberg (2006)
4. Larrañaga, P., Lozano, J.A.: Estimation of Distribution Algorithms: a New Tool for Evolutionary Computation. Kluwer Academic Publishers, Boston (2002)
5. Hauschild, M.W., Pelikan, M.: An introduction and survey of estimation of distribution algorithms. Swarm Evol. Comput. **1**(3), 111–128 (2011)

6. Sun, J., Zhang, Q., Li, J., Yao, X.: A hybrid estimation of distribution algorithm for CDMA cellular system design. Int. J. Comput. Intell. Appl. **7**(2), 187–200 (2007)

7. Shah, R., Reed, P.: Comparative analysis of multiobjective evolutionary algorithms for random and correlated instances of multiobjective d-dimensional knapsack problems. Eur. J. Oper. Res. **211**(3), 466–479 (2011)

8. Pelikan, M., Goldberg, D.E., Cantú-Paz, E.: BOA: the Bayesian optimization algorithm. In: Banzhaf, W., et al. (eds.) Proceedings of the Genetic and Evolutionary Computation Conference GECCO-99, pp. 525–532. Morgan Kaufmann, San Francisco (1999)

9. Pelikan, M.: Hierarchical Bayesian Optimization Algorithm: Toward a New Generation of Evolutionary Algorithms. Springer, Heidelberg (2005)

10. Pelikan, M., Goldberg, D.E., Sastry, K.: Bayesian optimization algorithm, decision graphs, and Occam's razor. In: Genetic and Evolutionary Computation Conference (GECCO-2001), pp. 519–529. Morgan Kaufmann, San Francisco, California (2001)

11. Lima, C.F., Lobo, F.G., Pelikan, M., Goldberg, D.E.: Model accuracy in the Bayesian optimization algorithm. Soft. Comput. **15**, 1351–1371 (2011)

12. Henrion, M.: Propagation of uncertainty in Bayesian networks by logic sampling. In: Lemmer, J.F., Kanal, L.N. (eds.) Uncertainty in Artificial Intelligence, pp. 149–163. Elsevier, Amsterdam (1988)

13. Pelikan, M., Sastry, K., Goldberg, D.E.: Sporadic model building for efficiency enhancement of the hierarchical BOA. Genet. Program. Evolvable Mach. **9**(1), 53–84 (2008)

14. Hauschild, M., Pelikan, M., Sastry, K., Lima, C.F.: Analyzing probabilistic models in hierarchical BOA. IEEE Trans. Evol. Comput. **13**(6), 1199–1217 (2009)

15. Pappas, A., Gillies, D.F.: A new measure for the accuracy of a Bayesian network. In: Coello Coello, C.A., de Albornoz, A., Sucar, L.E., Battistutti, O.C. (eds.) MICAI 2002. LNCS (LNAI), vol. 2313, pp. 411–419. Springer, Heidelberg (2002)

Person Re-identification Based on Part Feature Specificity

Dengyi Zhang, Qian Wang[✉], Xiaoping Wu, and Yu Cao

School of Computer, Wuhan University, Wuhan 430072, China
dyzhangwhu@163.com, {hdwq,wuxp,caoyu}@whu.edu.cn

Abstract. Person re-identification has become one of the most important problems in video surveillance system. In a multi-camera video surveillance system with non-overlapped area, the appearances of one person are much difference according different cameras or in the same camera at different times. On the other hand, different person may appears similar in one camera, so which made person re-identification a challenging problem. This paper carried out a person re-identification algorithm based on part feature specificity. This algorithm extract color, texture and shape features of different part of body first, then gather statistic specificity weight of these features for each part. At last, doing feature weighting both part weight and feature specificity in distance calculating, which make features with strong specificity more important. This algorithm indicates some parts of body are more important than others in re-identification, and the same part from different people with different appearance, the features with strong specificity are more important than the others. We have done our experiments at public datasets VIPeR and iLIDS, and evaluate the result by CMC. Result indicates this algorithm has higher re-identification rate, and more robust to viewing condition changes, illumination variations, background clutter and occlusion.

Keywords: Person re-identification · Multi-camera · Video surveillance · Non-overlapped area · Part feature specificity

1 Introduction

Recently person re-identification in multi-camera video surveillance system with non-overlapped area has got even more attention. Since one person in different cameras may have different viewing condition, posture, illumination, background, occlusion and resolution, so appearance may variety according to different cameras or in the same camera at different times. On the other hand, different people may appear similar in the same camera. This make it difficult to distinguish people in different pictures. To solve this problem, two ways have been focused. One is feature describe method [1, 4, 5, 7, 8, 10, 11], finding more suitable features to coping with the variety of viewing condition, posture, illumination, background and resolution. The other is distance measure learning method [2, 3, 6, 9, 12, 14, 15], acquiring better distance measurement function by matching learning.

Feature describe method extract features from person first, then calculate distance of feature vectors using similarity function, which describe the similarity of people.

© Springer Science+Business Media Singapore 2016

K. Li et al. (Eds.): ISICA 2015, CCIS 575, pp. 219–229, 2016.

DOI: 10.1007/978-981-10-0356-1_22

The Symmetry-Driven Accumulation of Local Features (SDALF [1]), which is carried out by Farenzena et al., mix the following three features together: Weighted Color Histogram (WCH), Maximum Stabilize Color Region (MSCR) [13] and Recurrent High-Structured Patches (RHSP [1]). On the other hand, eSDC [8] method carried out by Zhao et al., using dense color histogram and dense SIFT operator to find most relevancy blocks between person's images, and doing person matching using these high relevancy blocks. Since overall features of person could not tolerate occlusion and background change, CPS [11] method had been proposed by Cheng et al. This method split body image into parts first, and extract HSV Color Histogram and MSCR feature of each part, then calculate similarity for each part using different weighting. Part split could weed out background from person's image effective, and matching based on part could tolerate part occlusion effective.

Machine learning method learns a similarity matrix by training samples of people, select more significant feature according to this matrix used for similarity calculation. The ELF [12] algorithm proposed by Gray et al., extract color and texture features from person's image, then using Adaboost algorithm to select best combo of features for similarity match. PRSVM method, proposed by Prosser, et al., trained an effective similarity function by sort-based SVM classifier, which combined weights of multi feature vectors. Compared to feature describe method, machine learning method has lower requirement for feature selection, could select better features automatically, archives better result. But, training similarity function could cost giant time and space, as well as over-fitting during training when samples isn't enough. The effect of feature describe method relays on the selection of features, as people appearance variety, the fusion of multi features could hardly adapt some appearance, influence result of re-identification.

For these reasons, we proposed a person re-identification algorithm based on part feature specificity. This algorithm uses PS [16] algorithm to split image into head, shoulder, torso, upper and lower arms, thighs and calf, totally 11 parts. Extract color, texture and shape features from each part, then calculate specificity weight of each feature, larger weight indicates this feature of the part is more different from others, so which could distinguish from other people better. After that, we calculate the similarity of each part between different people, using part feature specificity, sum the similarity of each part weighted to get final result. This algorithm combined multi features, which could resist viewing condition and illumination variations. Extract features from part separately reduced the influence of background, posture changes and occlusion. Using part feature specificity could increase the weight of specificity features, and increase re-identification rate.

2   Part Feature Specificity

This paper proposed a person re-identification algorithm based on part feature specificity, the flow diagram is shown as Fig. 1. This algorithm split image of person to 11 parts, and unify direction and scale, making it able to eliminate noisy environment of image efficiency, as well as reduce the influence of posture. Following we extract the color, texture and shape features from each part, fusing multi features refrain the instability of single feature caused by viewing angel, illumination or clothing. Then we calculate specificity weight of each feature and each part, which indicates the difference of the part between this person and others. Larger specificity weight means greater difference, so we could use it to distinguish this person from others better. Last we calculate the weighting similarity of the same part from different person. Since the influence of viewing angle, posture and illumination and scale is different, weighting each part could result better re-identification rate.

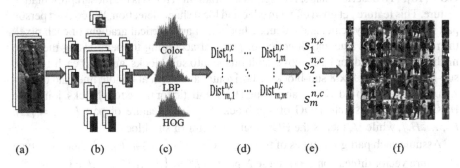

(a)   (b)   (c)   (d)   (e)   (f)

Fig. 1. Flowchart of person re-identification base on part feature specificity. (a) probe set, (b) body parts, (c) features of part, (d) similar matrix of parts, (e) part feature specificity weights, (f) distance ranking

2.1 Body Parts Extraction

Assume the body of person is combined by each parts, the upper part is depends on 7 different parts: head, shoulder, torso, upper arms, lower arms, and lower part is depends on 4 different parts: thighs and calf. So totally 11 parts. The structure could expressed as $L = \{l_0, l_1, \dots, l_N\}$, $l_n = (x_n, y_n, \theta_n, s_n)$ as the center coordinate is x_n, y_n, and the absolute direction is θ_n, the scale is s_n. Using PS algorithm to extract body parts. Rotate parts to same angle, and then zoom to same scale. The image of each part could describe as $P = \{p_0, p_1, \dots, p_N\}$.

2.2 Feature Extraction

Color Histograms. This paper mainly extracted three types of features: color, texture and shape, and the most representative feature of color is color Histograms. This paper used three color space of RGB, HSV and YCbCr, as HSV and YCbCr shares the brightness channel, so these color space consists 8 color channels, and calculate histograms for each channel. So, the color feature of part is $F_{Color} = \{H_R, H_G, H_B, H_S, H_V, H_Y, H_{Cb}, H_{Cr}\}$, while H indicates the color histogram.

LBP [17]. The most representative feature to describe texture is Local Binary Pattern (LBP), this feature is robust to locality rotation and illumination variations, and is widely used in person detect, face recognition, vehicle plate recognition and person tracks.

We used LBP$_{8,1}$ feature, which is 8 directions with radius 1. First convert person part image into gray scale, and then calculate LBP value of each pixel, split into 16 × 16 sub-blocks, while neighbor block is 4 pixels away in horizontal and vertical. Statistic LBP histogram for each sub-blocks, and it used as LBP feature vector of this sub-block. Thus, the LBP feature of part $F_{LBP} = \{H_1, H_2, \ldots, H_n\}$, while H_n means the LBP feature vector of sub-block n.

HOT [18]. To describe shape, Histogram of Gradient (HOG) is the most representative feature. This feature get good effect in edge and local shape detection. We convert person part image into gray scale, and calculate horizontal and vertical gradient of each pixel, then calculate the value and direction of gradient according to them. Using the same method as described for LBP above, spilt image into sub-blocks, and split each 16 × 16 sub-block into four 8 × 8 sub-block by horizontal and vertical middle line, and quantization direction into 9 division, and doing HOG statistic for these four blocks, join these HOG together constitute HOG of sub-block. So HOG feature of part $F_{HOG} = \{H_1, H_2, \ldots, H_n\}$, while H_n means the HOG feature vector of sub-block n.

Assume each part got C types of feature, as $C = \{C_{Color}, C_{LBP}, C_{HOG}\}$, then the c type of feature vector dimension set of these N parts is $D^c = \{d^{1,c}, d^{2,c}, \ldots, d^{N,c}\}$.

2.3 Part Feature Specificity Compute

This algorithm aims to find features those are more specificity according to different appearance of each part, which could distinguish person from others better. This distinguish is reflect by specificity weight of feature. For m unlabeled gallery image $\{I_i | i = 1, 2, \ldots, m\}$, after part detection each part of image i is $P_i = \{p_i^1, p_i^2, \ldots, p_i^N\}$. Assume feature extract function $F_c(I)$ is the feature extract function of type c, thus the type c feature vector of part n and image i is

$$X_i^{n,c} = F_c\left(p_i^n\right) = \left(x_i^1, x_i^2, \ldots, x_i^{d^{n,c}}\right)^T. \tag{1}$$

Using feature vector $X_i^{n,c}$ calculate the distance

$$Dist_{i,j}^{n,c} = \|X_i^{n,c}, X_j^{n,c}\| \tag{2}$$

between any two images in part n image set $P^n = \{p_1^n, p_2^n, \ldots, p_m^n\}$. As Dist is Barr's distance. Thus the similarity vector based on feature type c for part n image set P^n is:

$$M^{n,c} = \begin{pmatrix} Dist_{1,1}^{n,c} & \cdots & Dist_{1,m}^{n,c} \\ \vdots & \ddots & \vdots \\ Dist_{m,1}^{n,c} & \cdots & Dist_{m,m}^{n,c} \end{pmatrix} \tag{3}$$

Figure out specificity vector of type c feature of part n by sum the vector horizontal

$$S^{n,c} = \begin{pmatrix} \text{Dist}_{1,1}^{n,c} + \text{Dist}_{1,2}^{n,c} + \cdots + \text{Dist}_{1,m}^{n,c} \\ \text{Dist}_{2,1}^{n,c} + \text{Dist}_{2,2}^{n,c} + \cdots + \text{Dist}_{2,m}^{n,c} \\ \vdots \\ \text{Dist}_{m,1}^{n,c} + \text{Dist}_{m,2}^{n,c} + \cdots + \text{Dist}_{m,m}^{n,c} \end{pmatrix} \tag{4}$$

And normalization the vector $S^{n,c} = \left(s_1^{n,c}, s_2^{n,c}, \ldots, s_m^{n,c} \right)^T$, as the specificity weight of type c feature of part n in image i.

2.4 Feature Matching

Assume the vector of type c feature of part n in gallery person image I_i is $X_i^{n,c}$, and $X_j^{n,c}$ for probe person image I_j, so the distance of part n between person i and j is:

$$D\left(P_i^n, P_j^n \right) = \sum_c s_j^{n,c} \times \text{Dist}\left(X_i^{n,c}, X_j^{n,c} \right) \tag{5}$$

So the distance between I_i and I_j is:

$$D\left(I_i, I_j \right) = \sum_n w^n \times D\left(P_i^n, P_j^n \right) \tag{6}$$

As w^n means the weight of part n, the value determine by the result of person re-identification for each part, reference the Sect. 3.1 of this paper. The weight for each part in VIPeR dataset is shown as Fig. 2(b). Less the final distance between image I_i and I_j indicates more similar between image I_i and I_j.

3 Experimental Results

Using each person in probe set A as re-identification target to do similarity matching with gallery set B. Thus generated a similarity ranking table for each image in gallery set B, if the top ranked is exactly the person in B, the re-identification is absolutely right. The matching process is divided into three types [1]: (1) single-shot vs single-shot (S vs. S), Now there're only one frame of image exists in gallery set B for one person, but may be one (such as VIPeR dataset) or more (such as iLDIS dataset) images in gallery set A; (2) multiple-shot vs single-shot (M vs. S), There are more than one frame exists in gallery set B for one person under this situation, but only one frame exists in gallery set A; (3) multiple-shot vs multiple-shot (M vs. M), Here multiple-shot exists in both gallery set A and B for each person.

This section evaluate the method carried out in this paper by two public dataset for person re-identification VIPeR and iLIDS, where these datasets is most widely used and challenging nowadays. And the result will be shown by Cumulative Matching Characteristic (CMC). CMC indicated the expectation of correctly matched from first n matching objects. Area Under Curve (AUC) of CMC is another important index for evaluation the entire result of re-identification algorithm.

3.1   VIPeR Dataset

VIPeR dataset contains 632 pairs of images of people, from two different camera, all the images are 128 × 48 pixels. The viewing angle of most images is rotated by 90° from the two cameras for the same person, also the illumination variations. The images of the person from dataset is divided into group A and B randomly, with 632 images for each group, group B acts as gallery set, while A acts as probe set. We randomly select images of 316 people in A and B doing experiment, using every images from set A doing similarity match to B, using formula (6) to calculate the similarity between two images, and sorted, statistic CMC, repeat this for 10 times, and take the average as the final CMC for result. As there're only one image for each person in both sets, so this experiment is S vs. S.

PFS Algorithm Comparison Experiment on Each Part.   Using formula (5) to doing person re-identification experiment for each part of person independently, acquire weight set of body parts $W = \{w^1, w^2, ..., w^N\}$ according to experiment result, as w^n means the average area enclosed by the first 10 % of CMC and x axis of the part n, in other words, the average re-identification rate of first 10 %. The result is shown as Fig. 1(a). The re-identification of torso is best across all parts, the amount of images ranked before 20 is 61.65 % of all images, following with the upper arms and thighs. As lower arms is easily to be occlusion, the re-identification rate is lowest. The nAUC of every part constitute the weight set W for each part of formula (6).

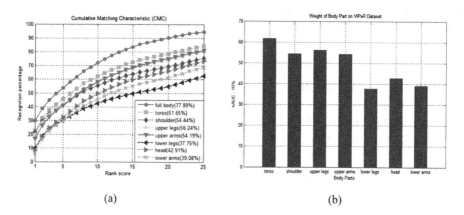

(a)   (b)

Fig. 2. Performance of PFS at All Parts on VIPeR dataset. (a) CMC curves of PFS at all parts on VIPeR dataset, (b) weights of body parts on VIPeR dataset

The Comparison Experiment of Part Feature Specificity Weighted.   The distance formula without part feature specificity weighted is:

$$D\left(I_i, I_j\right) = \sum_n w^n \times \text{Dist}\left(X_i^{n,c}, X_j^{n,c}\right) \tag{7}$$

And use formula (6) to calculate the similarity between two shots of person with part feature specificity weighted. The result is shown as Fig. 3(a). In the condition of part feature specificity weighted, the amount of shots which is ranked top 20 is 5.47 % higher than those do not. And the CMC area of first 10 % is 8.12 % higher.

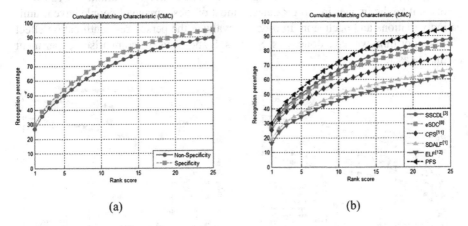

(a)   (b)

Fig. 3. Performances comparison using CMC curves on VIPeR dataset. (a) CMC curves of specificity and no specificity on VIPeR dataset, (b) the PFS compare with other methods on VIPeR dataset

The Comparison Experiment of Other Methods. This main objective of this experiment is compare PFS algorithm along with other 5 person re-identification algorithm, and effect is shown as Fig. 3(b). The person re-identification rate is significant higher than others in VIPeR dataset, indicated by CMC. The result is shown as Table 1. Using the PFS algorithm carried out by this paper, the amount of top ranked image is 11.07 % higher than SDALF. Although it is only 2.21 % higher than SSCDL algorithm which has better effect, using top ranked image to compare, but there're 6.95 % higher at top ranked 20, and nAUC is 5.59 % higher in first 10% CMC. Overall, the re-identification rate of PFS algorithm is significant higher than others, on VIPeR dataset.

Table 1. Matching rate (%) of different methods on VIPeR dataset

Method	rank = 1	rank = 5	rank = 10	rank = 20	nAUC-10 %
SSCDL [3]	27.88	49.93	66.59	83.19	72.30
eSDC [8]	26.93	47.97	63.54	79.07	69.10
CPS [11]	25.35	44.39	57.76	71.17	63.02
SDALF [1]	19.02	37.06	49.34	61.68	54.44
ELF [12]	16.37	33.97	45.51	57.16	49.03
PFS	**30.09**	53.64	71.95	**90.14**	**77.89**

3.2 iLIDS Dataset

iLIDS dataset is a public video dataset, from the video recorded by multi-cameras with non-overlapped area, in airport passageway at rush hours. Dataset contains 476 frames of image from 119 people. All the image in the dataset was normalizing to 128×48 pixels, unlike the images in VIPeR dataset, most of these image of people have illumination variations and occlusion. In order to evaluate the person re-identification effect of this algorithm, we have done S vs. S, M vs. S and M vs. M comparison experiment on this dataset.

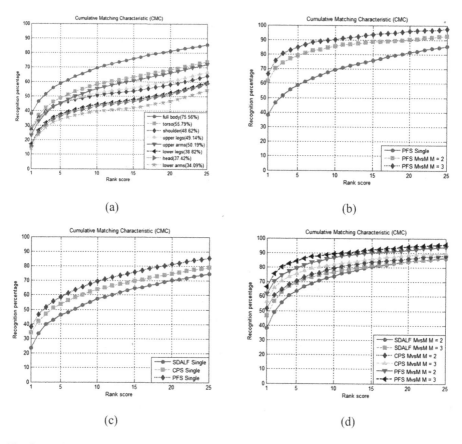

Fig. 4. Performances comparison using CMC curves on iLIDS dataset. (a) CMC curves of PFS at all parts, (b) PFS at single-shot and multiple-shot mode on iLIDS dataset, (c) PFS compare with other methods at single-shot mode on iLIDS dataset, (d) PFS compare with other methods at multiple-shot mode on iLIDS dataset

PFS Algorithm Comparison Experiment on Each Part. Doing person re-identification experiment on each part independently for every person in iLIDS dataset, using formula (5), the result is shown as Fig. 4(a), and the nAUC is the average area of top 30 % of CMC and x axis. The re-identification rate for the body is the best, as 68.5 % of total images is ranked top 20 %, following shoulder and upper legs, but the re-identification rate for the head is worst. The nAUS of every part consists the weight set W for each part in formula (6).

Comparison Experiment at Single-Shot and Multiple-Shot Mode. This experiment compare single-shot and multiple-shot mode in person re-identification rate for PFS algorithm. The result is shown as Fig. 4(b). The result of multiple-shot mode is signif-icant better than single-shot mode. In multiple-shot to multiple-shot mode, with $M = 3$, the correctly match percentage is 14.66 % higher than single-shot, and the nAUC of top 30 % is 17.2 % higher. Which indicates the re-identification rate of multiple-shot to multiple-shot is significant higher than single-shot. Taking different M for compare, the re-identification rate when $M = 3$ is 5.7 % higher than $M = 2$, which indicates the larger gallery set leads to higher re-identification rate.

The Comparison Experiment of Other Methods at Single-Shot Mode. This experi-ment compares re-identification rate of the PFS algorithm with the other two algorithm in single-shot mode, using iLIDS dataset. From Fig. 4(c) we could see, the PFS algorithm proposed by this paper is significant better than the other two. The result is shown in Table 2. The re-identification rate of the PFS algorithm is 14.6 % higher than the SDALF algorithm, and 4.21 % higher than the CPS algorithm in rank 1; and 11.06 % higher than the SDALF algorithm and 6.46 % higher than the CPS algorithm in rank 20. The nAUC of first 30 % of the PFS algorithm is significant higher than the other two. So, the re-identification rate for the PFS algorithm in single-shot mode is better than the SDALF algorithm and the CPS algorithm.

Table 2. Matching rate (%) of different methods at single-shot mode on iLIDS dataset

Method	rank = 1	rank = 5	rank = 10	rank = 20	nAUC-30 %
SDALF [1] single	23.53	46.22	57.48	70.25	64.3
CPS [11] single	33.92	53.75	63.84	74.85	69.82
PFS single	**38.13**	58.68	69.54	**81.31**	**75.56**

The Comparison Experiment of Other Methods at Multiple-Shot Mode. This experiment compares the re-identification rate of the PFS and the other two algo-rithms in multi-shot mode, still using iLIDS dataset. We could see from Fig. 4(d), the re-identification rate for the PFS algorithm is higher than the others. The result is shown as Table 3. When M = 3, the re-identification rate of the PFS algorithm is 66.47 % in rank 1, which is higher than 56.39 % of the CPS algorithm and 46.33 % of the SDALF algorithm. As well as taking M = 2 condition, the re-identification rate of the PFS algorithm is significant higher than the others. These indicates the person re-identification rate is better than the others in multi-shot mode.

4 Conclusion

This paper proposed a person re-identification algorithm based on part feature specificity. This algorithm split image of person to 11 parts, and extract color, texture and shape features of each parts, then calculate the similarity vector for each feature. Calculating the specificity absolute value of each feature according to the similarity vector, and finally the similarity of image by feature specificity weighting vector and part weighting. Experiment result indicates this algorithm has good re-identification rate under multi-camera non-overlapping view surveillance system, and robust to viewing condition changes, illumination variations, background clutter and occlusion. Propose better distance measurement function using similarity measure learning will be the point of our future research.

References

1. Farenzena, M., Bazzani, L., Perina, A., Murino, V., Cristani, M.: Person re-identification by symmetry-driven accumulation of local features. In: IEEE Conference on Computer Vision and Pattern Recognition, pp. 2360–2367. IEEE, San Francisco (2010)
2. Xiong, F., Gou, M., Camps, O., Sznaier, M.: Person re-identification using kernel-based metric learning methods. In: Fleet, D., Pajdla, T., Schiele, B., Tuytelaars, T. (eds.) ECCV 2014, Part VII. LNCS, vol. 8695, pp. 1–16. Springer, Heidelberg (2014)
3. Liu, X., Song, M., Tao, D., Zhou, X., Chen, C., Bu, J.: Semi-supervised coupled dictionary learning for person re-identification. In: Proceedings of the 2014 IEEE Conference on Computer Vision and Pattern Recognition, pp. 3550–3557. IEEE, Computer Society (2014)
4. Wang, X., Chen, F., Liu, Y.: Person re-identification by cascade-iterative ranking. In: Li, S., Liu, C., Wang, Y. (eds.) CCPR 2014, Part I. CCIS, vol. 483, pp. 335–344. Springer, Heidelberg (2014)
5. Yang, Y., Yang, J., Yan, J., Liao, S., Yi, D., Li, S.Z.: Salient color names for person re-identification. In: Fleet, D., Pajdla, T., Schiele, B., Tuytelaars, T. (eds.) ECCV 2014, Part I. LNCS, vol. 8689, pp. 536–551. Springer, Heidelberg (2014)
6. Li, W., Zhao, R., Xiao, T., Wang, X.: DeepReID: deep filter pairing neural network for person re-identification. In: IEEE Conference on Computer Vision and Pattern Recognition, pp. 152–159. IEEE, Computer Society (2014)
7. Li, W., Wang, X.: Locally aligned feature transforms across views. In: IEEE Conference on Computer Vision and Pattern Recognition, pp. 3594–3601. IEEE, Portland (2013)
8. Zhao, R., Ouyang, W., Wang, X.: Unsupervised salience learning for person re-identification. In: Proceedings of the IEEE International Conference on Computer Vision and Pattern Recognition (2013)
9. Kostinger, M., Hirzer, M., Wohlhart, P., Roth, P.M., Bischof, H.: Large scale metric learning from equivalence constraints. In: IEEE Conference on Computer Vision and Pattern Recognition, pp. 2288–2295. IEEE, Providence (2012)
10. Liu, C., Gong, S., Loy, C.C., Lin, X.: Person re-identification: what features are important? In: Fusiello, A., Murino, V., Cucchiara, R. (eds.) ECCV 2012 Ws/Demos, Part I. LNCS, vol. 7583, pp. 391–401. Springer, Heidelberg (2012)
11. Cheng, D.S., Cristani, M., Stoppa, M., Bazzani, L., Murino, V.: Custom pictorial structures for re-identification. In: British Machine Vision Conference, pp. 1–11. BMVC Press, Dundee (2011)

12. Gray, D., Tao, H.: Viewpoint invariant pedestrian recognition with an ensemble of localized features. In: Forsyth, D., Torr, P., Zisserman, A. (eds.) ECCV 2008, Part I. LNCS, vol. 5302, pp. 262–275. Springer, Heidelberg (2008)
13. Forssén, P.-E.: Maximally stable colour regions for recognition and matching. In: IEEE Conference on Computer Vision and Pattern Recognition, pp. 1–8. IEEE, Minneapolis (2007)
14. Dikmen, M., Akbas, E., Huang, T.S., Ahuja, N.: Pedestrian recognition with a learned metric. In: Kimmel, R., Klette, R., Sugimoto, A. (eds.) ACCV 2010, Part IV. LNCS, vol. 6495, pp. 501–512. Springer, Heidelberg (2011)
15. Prosser, B., Zheng, W.S., Gong, S., Xiang, T.: Person re-identification by support vector ranking. In: British Machine Vision Conference, pp. 1–11. BMVC (2010)
16. Andriluka, M., Roth, S., Schiele, B.: Pictorial structures revisited: people detection and articulated pose estimation. In: IEEE Computer Society Conference on Computer Vision and Pattern Recognition, pp. 1014–1021. IEEE, Computer Society (2009)
17. Mu, Y., Yan, S., Liu, Y., Huang, T., Zhou, B.: Discriminative local binary patterns for human detection in personal album. In: IEEE Computer Society Conference on Computer Vision and Pattern Recognition, pp. 1–8 (2008)
18. Dalal, N., Triggs, B.: Histograms of oriented gradients for human detection. In: IEEE Computer Society Conference on Computer Vision and Pattern Recognition, vol. 1, pp. 886–893. IEEE, Computer Society (2005)

A Gaussian Process Based Method for Antenna Design Optimization

Jincheng Zhang[1], Sanyou Zeng[1]([⊠]), Yuhong Jiang[1], and Xi Li[1,2]

[1] School of Computer Science, China University of Geosciences, Wuhan 430074,
Hubei, People's Republic of China
{zjccsg,sanyouzeng}@gmail.com, 503850769@qq.com
[2] School of Information Engineering, Shijiazhuang University of Economics,
Shijiazhuang 050031, Hebei, People's Republic of China

Abstract. In many expensive or time consuming engineering problems, like antenna design problems, it is unpractical to use the evolutionary algorithms directly. In recent years, Gaussian process has attracted more and more attention and had some successful applications. To further accelerate the speed of antenna design optimization process, a Gaussian process and fuzzy clustering assisted differential evolution algorithm(GPFCDEA) is presented in this paper. Four benchmark functions and two antenna design problems are selected as examples. Experimental results indicate that GPFCDEA performs much better than DE for low dimensions problems. However, for high dimensions problems, the performance of GPFCDEA still needs further research.

Keywords: Gaussian process · Antenna design optimization · Differential evolution · Expensive optimization

1 Introduction

In recent years, evolutionary algorithms(EAs) has been applied widely in many fields. Among them, DE is one of the most popular algorithms which has a very excellent global optimization efficiency for complex unimodal and multimodal problems.

However, many real world optimization problems are very difficult and expensive, especially in engineering optimization, such as antenna design problems. To solve these optimization problems, if we adopt the standard evolutionary algorithm, a great number of evaluations (often over 30000-200000 [1]) would be required. Such problems are usually along with time consuming simulations or expensive experiments. Obviously, using the standard evolutionary algorithm such as DE is unrealistic. By building models as a surrogate of the real fitness function has been proven to be a practical approach [2,3]. Generally speaking, the surrogate models are much more cheaper than simulations or experiments. Among those methods, the Gaussian stochastic process model(also called Kriging [5] or the design and analysis of computer experiments (DACE) stochastic

© Springer Science+Business Media Singapore 2016
K. Li et al. (Eds.): ISICA 2015, CCIS 575, pp. 230–240, 2016.
DOI: 10.1007/978-981-10-0356-1_23

process model [6]) is a newly developed machine learning technology based on strict theoretical fundamentals and Bayesian theory [4] to which more and more attention has been paid in recent years. And there are plenty of successful applications [7,8] of it. However, for most of them, the problems' dimensions are pretty low (In [8], high dimensions problems also included). Further research still be needed for the complex high dimensions problems.

In this paper, a Gaussian process and fuzzy clustering assisted differential evolution algorithm is presented. Four benchmark functions in different dimensions and two antenna design problems are selected to test the performance of the algorithm.

2   The Framework of GPFCDEA

The GPFCDEA optimization loop proceeds as follows: First of all, we need the initial number of individuals which are evaluated by the computationally expensive function as a training data set to build the GP model. These individuals are selected randomly in the solution. After this, we use the fuzzy clustering algorithm to cluster all the evaluated individuals into several small clusters and build GP model on each cluster. Then, DE algorithm is used to search for minima of the GP prediction. Finally, the best individual evaluated by the computationally expensive function will be added to the training data set.

The GPFCDEA includes three key components. In the following, brief introductions to these components will be provided.

2.1   GP Machine Learning

A Gaussian Process is a collection of random variables, any finite number of which have (consistent) joint Gaussian distributions [10]. GP machine learning assumes that the objective function is a sample of a Gaussian distribution.

Suppose that there is an expensive objective function $y = g(x), x \in R^n$, for any x, $g(x)$ is a sample of a Gaussian random variable

$$\mu + \epsilon(x) \tag{1}$$

where $\epsilon(x) \sim N(0, \sigma^2)$, μ and σ are two constants independent of x.

Using the GP model with the correlation function $c(x_i, x_i')$, the function value of a new point x^* can be predicted.

The correlation function shows as follows

$$c(x_i, x_i') = exp \left(-\sum_{i=1}^{n} \theta_i \mid x_i - x_i' \mid^{p_i} \right) \tag{2}$$

$\theta_i > 0, 1 \leq p_i \leq 2$ where n is the dimension of x, θ_i indicates the importance of x_i on $g(x)$ and p_i is related to the smoothness of $g(x)$ with respect to x_i [9].

The hyperparameters θ_i and p_i can be determined by maximizing the likelihood function:

$$\frac{1}{(2\pi\sigma^2)^{K/2} \mid C \mid^{1/2}} \times exp\left(-\frac{(y - \mu I)^T C^{-1}(y - \mu I)}{2\sigma^2}\right) \tag{3}$$

where K is the number of given points to build GP process model, I is a $K \times 1$ vector of ones, $y = (g(x_1), g(x_2), g(x_3), \ldots, g(x_K))^T$, C is a $K \times K$ matrix whose (i,j)-element is $c(x_i, x_j)$.

To maximize (3), the values of μ and σ^2 must be

$$\hat{\mu} = \frac{I^T C^{-1} y}{I^T C^{-1} I} \tag{4}$$

$$\sigma^2 = \frac{(y - I\hat{\mu})^T C^{-1}(y - I\hat{\mu})}{K} \tag{5}$$

By substituting (4) and (5) into (3), the likelihood function only depends on θ_i and $p_i, i = 1, 2, 3, \ldots, n$. We can choose an optimization method to maximize (3) (In this paper, DE algorithm is selected) to obtain the estimates of $\hat{\theta}_i$ and $\hat{p}_i$.

Once the hyperparameter estimates $\hat{\theta}_i, \hat{p}_i, \mu$, and $\hat{\sigma}^2$ are obtained, the function value of a new point can be predicted by the best linear unbiased predictor of $g(x)$

$$\hat{y}(x) = \hat{\mu} + r^T C^{-1}(y - I\hat{\mu}) \tag{6}$$

And the prediction uncertainty $s^2(x)$ is

$$s^2(x) = \hat{\sigma^2}\left(1 - r^T C^{-1} r + \frac{(1 - I^T C^{-1} r)^2}{I^T C^{-1} r}\right) \tag{7}$$

where $r = (c(x, x_1), c(x, x_2), c(x, x_3), \ldots, c(x, x_K))^T$.

2.2   Fuzzy Clustering Algorithm

As can be seen from the Eqs. (3), (4), (5) and (6), there are a great number of matrix operations in the calculation of the process. And the computing time mainly depends on K. When K is very large, the time which is spent on matrix calculations might be extremely long if all the points are directly used [9]. To deal with this problem, there are several solutions. One of them is to select a small part of evaluated points for estimation and prediction [11]. However, this will waste a lot of evaluated points. Another commonly used strategy is to cluster all the evaluated points into several small clusters and then will build models on all the small clusters.

There are two clustering methods can be selected. One is crisp clustering which makes each evaluated point belong to exactly one cluster. However, the prediction quality will be poor if the new point is in boundary areas among different clusters since it does not make full use of the evaluated points which close to it. Another is fuzzy clustering [12] in which a evaluated point can belong to several clusters. It will relieve the above drawback.

Fuzzy clustering needs two control parameters L_1 and $L_2, L_1 > L_2$. Where L_1 is the maximal number of points which a local model contains and L_2 is the number of points for adding one more local model.

When $K \leq L_1$, all the K evaluated points are directly used for building one model. When $K > L_1$, fuzzy clustering follows these steps.

(1) Calculate the number of cluster.

$$c_{size} = 1 + \lceil \frac{K - L_1}{L_2} \rceil \tag{8}$$

(2) Cluster the K evaluated points into c_{size} clusters. To do this, we should minimize the following function.

$$J = \sum_{i=1}^{K} \sum_{j=1}^{c_{size}} u_{ij}^{\alpha} \|x_i - v^j\| \tag{9}$$

where v^j is the center of cluster j, u_{ij} is the degree of membership of x_i in cluster j.

The algorithm works as follows.

Algorithm 1. The framework of fuzzy clustering algorithm

step 1 : Initialize $u_{ij}^0, i = 1, 2, \ldots, K, j = 1, 2, \ldots, c_{size}$. Set $t = 0$.

step 2 : Calculate the clustering centers v^j

$$v^j = \frac{\sum_{i=1}^{K} (u_{ij}^t)^\alpha x^i}{\sum_{i=1}^{K} (u_{ij}^t)^\alpha}$$

step 3 : Calculate u_{ij}^{t+1}

$$u_{ij}^{t+1} = \sum_{K=1}^{c_{size}} \left(\frac{\|x^i - v^j\|}{\|x^i - v^k\|} \right)^{-2/(m-1)}$$

step 4 : if $max_{1 \leq i \leq k, 1 \leq j \leq c_{size}} | u_{ij}^{t+1} - u_{ij}^t | < \mu$ where μ is the error of the precision, then output v^j and u_{ij}^t. Otherwise, set t=t+1 and go to Step 2.

In this paper, we set $\alpha = 2, \mu = 0.05$ and $\| * \|$ is Euclidean norm [13].

2.3   DE Algorithm

The DE algorithm has been widely used in many aspects of our life and it has been proved to be excellent optimization tools for complex problems since Storn and Price introduced the algorithm in 1997 [14]. The main steps of the DE algorithm are given below.

(1) Initialization
Suppose that there is a n-dimensional function $f(x)$, the size of the population is N_p. Then every individual of the generation can be expressed below:

$$x_{i,G}(i = 1, 2, \ldots, N_p)$$

Where G is the current generation and N_p is the population size.

We assume that the upper and lower boundaries of each dimension is $x_j^{(L)} < x_j < x_j^{(U)}$. After the initialization, the first generation is:

$$x_{ji,0} = rand[0, 1](x_j^{(U)} - x_j^{(L)}) + x_j^{(L)}$$

$$i = 1, 2, \ldots, N_p, j = 1, 2, \ldots, n$$

(2) Mutation
For each individual $x_{i,G}, i = 1, 2, \ldots, N_p$, the mutation operation is:

$$v_{i,G+1} = x_{r1,G} + F(x_{r2,G} - x_{r3,G})$$

Where F is the mutation probability from $[0, 2]$ and $x_{r1,G}, x_{r2,G}, x_{r3,G}$ is three different individuals.

(3) Crossover
In order to increase the diversity of the population, the crossover operation is:

$$u_{ji,G+1} = \begin{cases} v_{ji,G+1} & \text{if } (rand < CR) \text{ or } j = n - 1 \\ x_{ji,G+1} & \text{otherwise} \end{cases}$$

$$u_{i,G+1} = (u_{1i,G+1}, u_{2i,G+1}, \ldots, u_{ni,G+1})$$
$$i = 1, 2, \ldots, N_p, j = 1, 2, \ldots, n$$

Where CR is the crossover rate from $[0, 1]$.

(4) Selection
The DE algorithm uses a one-to-one competition scheme to select the next generation population. The operation is:

$$x_{ji,G+1} = \begin{cases} u_{ji,G+1} & \text{if } f(u_{i,G+1}) \leq f(x_{i,G}) \\ x_{ji,G} & \text{otherwise} \end{cases}$$

For more details about the DE algorithm, please see [15].

3   Experiments on Benchmark Functions

In this section, four famous benchmark functions are selected, they are sphere function, Griewank function, Rastrigins function and Rosenbrocks function respectively. All the functions' minimal value is $f = 0$ and the searching space is restricted to $[-2, 2]$.

3.1   Parameter Settings

DE Parameters. In this paper, the mutation probability(F) is randomly produced between 0.1 to 0.9 and the crossover rate(CR) is set to 0.9, population size of DE is set to 20.

The Initial Number of Individuals to Build Model. The initial number of individuals to build model has a great impact on the model quality. In this paper, it is set to $11n - 1$ [13], where n is the problem dimension.

3.2   Results Comparison and Discussion

Both the GPFCDEA and the DE algorithm solve the four benchmark function independently 20 runs and the number of exact fitness evaluation times in each run is 80 when n=3, 100 when n=5, 130 when n=8, 150 when n=10.

Table 1 shows the best, the worst and the average results of the two algorithms for the four benchmark functions. Experiment results in Table 1 showed the performance of GPFCDEA algorithm is better than that of DE algorithm when the function dimension is low(n=3,5,8). However, when the function dimension is high(n=10), the performance of GPFCDEA algorithm still needs further research.

Table 1. Statistics of the GPFCDEA and DE results for the four benchmark functions

F	n	GPFCDEA			DE		
		best	worst	average	best	worst	average
F_1	3	2.06e-6	9.93e-5	3.19e-5	0.013	0.180	0.086
	5	3.92e-4	8.15e-3	4.32e-3	0.219	0.805	0.399
	8	1.59e-4	7.67e-3	3.24e-3	0.115	1.633	0.775
	10	3.275	5.461	4.383	0.497	1.862	1.127
F_2	3	9.55e-7	1.51e-4	3.27e-5	1.70e-3	1.51e-2	7.06e-3
	5	0.019	0.099	0.044	0.139	0.403	0.265
	8	7.33e-4	1.56e-2	4.71e-3	0.077	0.171	0.115
	10	0.179	0.527	0.418	0.116	0.242	0.194
F_3	3	1.818	8.592	5.403	4.055	10.838	6.837
	5	6.197	15.821	12.622	10.815	23.626	15.843
	8	6.456	15.498	11.755	19.712	42.534	31.713
	10	44.983	69.586	56.687	20.781	48.232	41.766
F_4	3	4.38e-5	3.29e-4	1.95e-4	4.73e-2	0.336	0.169
	5	0.115	1.501	0.821	1.144	22.355	10.274
	8	2.58e-4	0.16	4.33e-2	29.629	78.148	52.988
	10	277.409	1035.962	551.955	51.289	150.332	94.156

4 Using the Algorithm to Design a Patch Antenna

In recent years, using evolutionary algorithms to design antennas has been widely applied [8, 16, 17]. To verify the algorithm's capacity in solving antenna design problems, we use it to design a patch antenna. The key requirements are summarized in Table 2. The structure is shown as Fig. 1, and Table 3 shows the range of the 5 design variables.

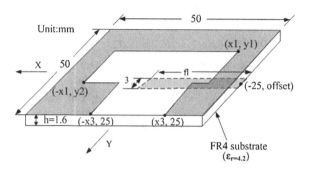

Fig. 1. The appearance of the patch antenna

Table 2. Requirement of the patch antenna

Property	Requirement
Frequency	2.3 GHz~2.5 GHz
VSWR	< 1.2 : 1
Input impedance	50 Ω

Table 3. Range of the 5 design (All size in mm)

Variables	x1	y1	y2	x3	Offerset
Lower bound	−23	−24	0	−23	−17
Upper bound	0	0	24	0	16

For GPFCDEA and standard DE algorithm, after 100 EM simulations, the variables of the best result are shown in Table 4 and the simulated performance is shown in Fig. 2. In fact, after 66 EM simulations, the objective was satisfied by GPFCDEA. The VSWR which obtained by GPFCDEA is smaller, which

Table 4. Values of the 5 design variables (All size in mm)

Variables	x1	y1	y2	x3	Offerset
Values(GPFCDEA)	−19.1	−22.6	14.7	−16.8	−12.2
Values(DE)	−19.1	−22.6	14.7	−16.8	−12.2

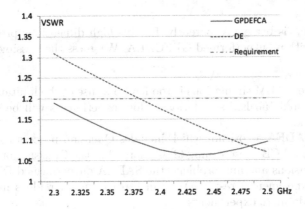

Fig. 2. VSWR of the antenna

means the performance is better. It can be observed that GPFCDEA clearly outperformed standard DE algorithm for this low dimensions problem.

Then the number of design parameters of the patch antenna is set to 10. The structure is shown as Fig. 3, and Table 5 shows the range of the 10 design variables.

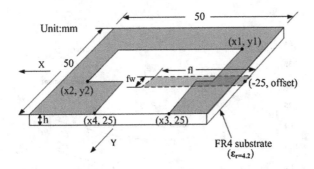

Fig. 3. The appearance of the patch antenna

For GPFCDEA and standard DE algorithm, after 150 EM simulations, the variables of the best result are shown in Table 6 and the simulated performance is shown in Fig. 4. The VSWR which obtained by DE is smaller, which means

Table 5. Range of the 10 design (All size in mm)

Variables	fl	fw	x1	y1	x2	y2	x3	x4	Offerset	h
Lower bound	2	2	−23	−24	0	0	−23	0	−25	1.5
Upper bound	50	6	0	0	24	24	0	24	25	2.5

the performance is better. Apparently, for this high dimensions problem, standard DE algorithm outperformed GPFCDEA. We guess the possible reasons as follows:

(1) The number of EM simulations is too few. Because of the limitation of time, we only tested 150 EM simulations. More experiments will be done in the future work.
(2) In [8], a SADEA algorithm which includes Gaussian process has been presented and a LCB prescreening [8] is adopted in the Gaussian process. For a high dimensions antenna problem, the SADEA outperformed DE according to their experimental results. We can learn from their ideas in the future work and do more experiments.

Table 6. Values of the 10 design variables (All size in mm)

Variables	fl	fw	x1	y1	x2	y2	x3	x4	Offerset	h
Values(GPFCDEA)	30.2	5.7	−10.9	−19.5	18.4	3.8	−21.1	24.8	9.9	2.2
Values(DE)	32.1	4.1	−6.4	−8.8	6.8	8.3	−21.0	13.5	15.7	1.8

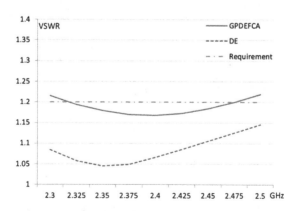

Fig. 4. VSWR of the antenna

5   Conclusions

In this paper, the Gaussian process and fuzzy clustering assisted differential evolution algorithm has been presented. Compared with DE algorithm by different dimensions of four benchmark functions and two antenna design problems, GPFCDEA clearly outperforms standard DE algorithm for low dimensions problems, however, for high dimensions problems, the performance of the GPFCDEA still needs further research. Future work will focus on doing more and more experiments, improving the performance of GPFCDEA on high dimensions problems and extending GPFCDEA to multi-objective problems.

Acknowledgment. This work was supported by the National Natural Science Foundation of China and other foundations(No.s: 61271140, 61203306, 2012001202, 61305086).

References

1. John, M., Ammann, M.: Antenna optimization with a computationally efficient multiobjective evolutionary algorithm. IEEE Trans. Antennas Propag. **57**(1), 1–24 (2009)
2. Jin, Y., Olhofer, M., Sendhoff, B.: A framework for evolutionary optimization with approximate fitness functions. IEEE Trans. Evol. Comput. **6**(5), 481–494 (2002)
3. Ong, Y.S., Nair, P.B., Keane, A.J.: Evolutionary optimization of computationally expensive problems via surrogate modeling. AIAA J. **41**(4), 687–696 (2003)
4. MacKay, D.J.C.: Introduction to Gaussian processes. NATO ASI Ser. F Comput. Syst. Sci. **168**, 133–166 (1998)
5. Stein, M.L.: Interpolation of Spatial Data: Some Theory for Kriging. Springer, New York (1999)
6. Sacks, J., Welch, W.J., Mitchell, T.J., et al.: Design and analysis of computer experiments. Stat. Sci. **4**(4), 409–423 (1989)
7. Su, G.: Gaussian process assisted differential evolution algorithm for computationally expensive optimization problems. In: Pacific-Asia Workshop on Computational Intelligence and Industrial Application, PACIIA 2008, pp. 1:272–1:276. IEEE (2008)
8. Liu, B., Aliakbarian, H., Ma, Z., et al.: An efficient method for antenna design optimization based on evolutionary computation and machine learning techniques. IEEE Trans. Antennas Propag. **62**(1), 7–18 (2014)
9. Rasmussen, C.E.: Gaussian Processes for Machine Learning. MIT Press, Cambridge (2006)
10. Rasmussen, C.E.: Gaussian processes in machine learning. In: Bousquet, O., von Luxburg, U., Rätsch, G. (eds.) Machine Learning 2003. LNCS (LNAI), vol. 3176, pp. 63–71. Springer, Heidelberg (2004)
11. Seo, S., Wallat, M., Graepel, T., et al.: Gaussian process regression: active data selectionand test point rejection. In: Mustererkennung 2000, pp. 27–34. Springer, Heidelberg (2000)
12. Bezdek, J.C.: Pattern recognition with fuzzy objective function algorithms. Kluwer Academic Publishers, Boston (1981)

13. Zhang, Q., Liu, W., Tsang, E., et al.: Expensive multiobjective optimization by MOEA/D with gaussian process model. IEEE Trans. Evol. Comput. **14**(3), 456–474 (2010)
14. Storn, R., Price, K.: Differential evolution: a simple and efficient heuristic for global optimization over continuous spaces. J. Global Optim. **11**(4), 341–359 (1997)
15. Price, K., Storn, R.M., Lampinen, J.A.: Differential Evolution: A Practical Approach To Global Optimization. Springer, Berlin (2006)
16. Zhang, L., Cui, Z., Jiao, Y.C., et al.: Broadband patch antenna design using differential evolution algorithm. Microw. Opt. Technol. Lett. **51**(7), 1692–1695 (2009)
17. Liu, Z., Zeng, S., Jiang, Y., et al.: Evolutionary design of a wide-band twisted dipole antenna for X-band application. In: 2013 IEEE International Conference on Evolvable Systems (ICES), pp. 9–12. IEEE (2013)

An Improved Algorithm of Watermark Preprocessing Based on Arnold Transformation and Chaotic Map

Dongbo Zhang[1(✉)] and Jingbo Zhang[2]

[1] Department of Computer Science, Guangdong University of Science
and Technology, Dongguan, China
neversurrender1314@163.com
[2] Department of Computer Science, Guangdong Innovative Technical College,
Dongguan, China
87229880@qq.com

Abstract. One of the main purposes of watermark preprocessing is to improve the robustness and security of the watermark. In this paper we proposed an improved image encryption algorithm, which combines position scrambling and gray scrambling. In order to achieve image location scrambling, first of all, the image is divided to even blocks, and the sub-blocks are scrambled according to Arnold transform. Then all of the pixels of each sub-block are scrambled by the proposed algorithm based on Logistic chaotic map. Finally, all of the Pixels are redistributed and scrambled totally. Basing on image location scrambling, it makes use of Logistic chaotic map and multi-dimensional Arnold transformation, image gray scrambling is attained. By histogram analysis, key sensitivity analysis and correlation analysis of adjacent pixels of the results of the simulation, indicating that the scrambling effect of the improved algorithm is good, and the key space is more larger.

Keywords: Arnold transformation · Logistic chaotic map · Image encryption · Histogram analysis · Correlation analysis

1 Introduction

With the development of networks and multimedia communications technology, the problems of digital multimedia information security, intellectual property protection and authentication issues become increasingly prominent. To solve those problems, digital watermarking technology becomes a focus of multimedia information security research. The early pretreatment of digital watermark plays an important part in the watermark technology. For example, it is good for the watermark's invisibility, robustness and security and so on. Watermark preprocessing has been researched widely. The generally method of watermark preprocessing is image scrambling technology. In recent years, many researches of watermark preprocessing only confined to the location scrambling, and didn't guarantee the security. This paper proposed an improved watermark encryption algorithm based on reference [1] combing position scrambling and gray scrambling. The algorithm integrates Arnold transformation [2] and Logistic chaotic map [3] well and attained a better effect. It can improve the robustness of digital watermark.

© Springer Science+Business Media Singapore 2016
K. Li et al. (Eds.): ISICA 2015, CCIS 575, pp. 241–247, 2016.
DOI: 10.1007/978-981-10-0356-1_24

2   Relevant Theory

Arnold Transformation. The digital images are expressed as a matrix form. Each element of the matrix corresponds image pixel gray value. Because of the good feature of periodicity of Arnold transformation [4, 5], when using it repeatedly, in a certain iteration, it will be able to restore original image, so it is used widely in image scrambling.

Arnold transformation [6] is extended to N dimension, expressed as following:

$$Y = AX(\mathrm{mod}\, N) \tag{1}$$

$A = \begin{bmatrix} 1 & 1 & 1 & \cdots & 1 \\ 1 & 2 & 2 & \cdots & 2 \\ 1 & 2 & 3 & \cdots & 3 \\ \vdots & \vdots & \vdots & \vdots & \vdots \\ 1 & 2 & 3 & \cdots & n \end{bmatrix}$ is the relevant transformation matrix. $Y = [x_1\, x_2\, x_3...x_n]^T$

expresses the original vector. $Y = [y_1\, y_2\, y_3...y_n]^T$ expresses the transformed vector. When it is exploited in the image scrambling, both location and gray scrambling can be achieved. In this case, N is the order of the image (or grayscale). X is the original image pixel coordinate (or the vector of pixel gray). Y is the pixel coordinates of the scrambled image (or the pixel gray vector of the scrambled image).

A Scrambling Algorithm Based on Logistic Chaotic Map. The reason that chaotic sequence is used to watermark preprocessing, mainly because of its sensitivity to initial values, and strong security features. *Logistic chaotic map* is a simple chaotic mapping. Logistic chaotic map [7] is defined as follows:

$$s_{i+1} = \mu s_i \left(1 - s_i\right) \qquad \mu \in [1,4] \qquad i = 0, 1, 2 \cdots \tag{2}$$

The initial choice is $0 < s_0 < 1$, then the resulting sequence s is unipolar, so $0 < s_i < 1$. Generally, μ is often chosen as the number close to 4.

According to the given initial value, it generates a chaotic sequence which length is M by formula (2), denoted as $F = \{f_1,f_2,f_3,...,f_n\}$. If $f_i > f_j (i,j \in \{1,2,...,M\}, i < j)$, then exchanging the two pixels(or rows and columns).

3   Image Encryption Algorithm

Location Scrambling. The location scrambled image has the same number of pixels as the original image, and its histogram isn't changed, but the order of its pixels has been disrupted. The more randomer the image is, the effect of scrambling is better. The location scrambled watermark image can not show certain characteristics and hide the original features. So it improves the security of the embedded watermark image. This paper adopts the method as follows:

(1) The image I (the order number is N) is divided into $b \times b$ sub-blocks. Each sub-block had N^2/b^2 pixels, denoted as

$$
\begin{bmatrix}
B_{11} & B_{12} & \cdots & B_{1b} \\
B_{21} & B_{22} & \cdots & B_{2b} \\
\vdots & \vdots & \vdots & \vdots \\
B_{b1} & B_{b2} & \cdots & B_{bb}
\end{bmatrix}
$$

which $B_{ij}(i = 1,2,...,b \; j = 1,2,...,b)$ is expressed as the sub-block. According to formula (2), it implements the scrambling of the sub-blocks, and then relieve their correlation-ship.

(2) According to the given initial value, it generates a chaotic sequence. In order to improve the effect of the encryption, the sequence isn't start from the initial segment. The chaotic sequence of the algorithm is started from the 101th number, and accesses N^2/b^2 real numbers composing the sequence $S = \{s_1, s_2, s_3, ..., s_n\}$ $(n = N^2/b^2)$. The two-dimensional matrix of the image produce one-dimensional sequence $T = \{t_1, t_2, t_3, ..., t_n\}(n = N^2/b^2)$. In line with the relationship of the bigness of each element in S, scrambling the position of each element in T. If $s_i > s_j(i < j)$, then exchanging t_i and t_j. After that it obtains the scrambled sequence $T' = \{t'_1, t'_2, t'_3, ..., t'_n\}(n = N^2/b^2)$. Finally, the sequence T' is produced to two dimensional image matrix.

(3) According to the method of step (2), each sub-block B_{ij} $(i = 1,2,...,b \; j = 1,2,..., b)$ of image I is scrambled, the resulting image is denoted as I'.

(4) Generate a new image I'', then divide it to $(N/b) \times (N/b)$ sub-blocks. Each sub-block has b^2 pixels. The first pixel of each sub-block B_{ij} $(i = 1,2,...,b \; j = 1,2,...,b)$ of image I' is placed in the first sub-block of image I'' in proper order. According this method, all of the pixels of sub-blocks in image I' are re-distribution. The final image is denoted as I''.

Gray Scrambling. The location scrambled image does not change the pixel gray value of the initial image, and can not resist a certain degree of statistical analysis attacks. Therefore, if it implement gray scrambled, the encrypted image has the capacity of resisting a certain degree of statistical analysis attack. So when it is used to watermark image, it improves the robustness and security. This paper adopts the method as follows:

(1) Take pixel grayscale values of each column(row) out of image I''. According to formula (1), they are transformated by N dimensional *Arnold transformation*, and then get the image H.

(2) According to the given initial value, it generates a chaotic sequence. Not consider the initial segment, take the 51th point as the initial point, and generate one chaotic sequence of length N, denoted as $S' = \{s'_1, s'_2, s'_3, ..., s'_N\}$. If $s'_i < s'_j(i < j)$, then the pixels of the i-line(row) and j-column(row) will be swaped. Finish the scrambling of all of lines (rows) by this method. The final image is denoted as H'.

The gray scrambling of rows will be achieved by the above method, but the gray value of pixels on row need to transfer-processing, and then according to formula (1), finish *Arnold transformation.*

4   Experimental Results and Analysis

Simulation Results. This paper used a gray image named cameraman.bmp which resolution was *256 × 256*. Through Matlab simulation. Firstly, the image was individed into *16 × 16* sub-blocks, and scrambled the sub-blocks by *Arnold transformation* for 20 iteration. The result image is the Fig. 1(b). Secondly, each sub-block was scrambled by Logistic chaotic algorithm. As Fig. 1(c) showing, each sub-block can be scrambled according to different initial value, and it expands the key space. When one or several sub-blocks are recovered, the initial image can not be recovered. Therefore it is difficult to decipher the encrypted image. Thirdly, each sub-block was redistribution according to the step (4) of the algorithm of location scrambling, and the resulting image is Fig. 1(d). Although each sub-block was scrambled in the last step, but there would be some blocking effect. So all of the pixels of each sub-block were redistributed and it brought a better effect. Finally, the grayscale values of pixels were scrambled by Arnold transformation, at the same time, the location of the pixels were scrambled according to Logistic chotic scrambling algorithm. The resulting images was Fig. 1(e).

(a)   (b)   (c)

(d)   (e)

Fig. 1. The process of image encryption algorithm (a) the original image, (b) the sub-blocks scrambled image, (c) the image of each sub-block scrambled, (d) the image of redistributed pixels, (e) the encrypted image

Histogram Analysis. Figure 2(a) and (b) show that the encrypted image has hidden the histogram of the original image and its histogram are uniform distribution and flat. This paper will analysis the histogram according to information entropy next.

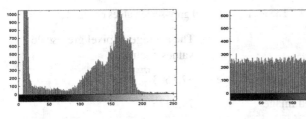

(a) the histogram of the original image   (b) the histogram of the encrypted image

Fig. 2. The histograms of the original and encrypted images

Image information entropy [7] can measure the distribution degree of grayscale values. The more uniformly the gray distributes, the bigger the information entropy of the image is. On the contrary, the information entropy is smaller. The biggest information entropy of 256 grayscale image is 8. Table 1 shows that the information entropy of the encrypted image is 7.9884, and almost 8. It certificates that the distribution of the gray is very uniform. Therefore it is more difficult for attacker to restore the original image through analysis the histogram of the encrypted image.

Table 1. Image information entropy

Image	Information entropy
The original image	7.0086
The location scrambled image	7.0086
The encrypted image	7.9884

Sensitivity Analysis of Key. In this paper, the sensitivity analysis of key uses the average of pixel grey-scale values changed [9] for quantitative analysis. The average of pixel grey-scale changed is used for measuring the degree of gray-scale changed well, and indirectly reflects the sensitively to the key. It is defined as:

$$gave(G, G') = \frac{\sum_{i=1}^{M} \sum_{j=1}^{N} |g_{ij} - g'_{ij}|}{MN} \tag{3}$$

Which $G = (g_{ij})_{M \times N}$ and $G' = (g'_{ij})_{M \times N}$ are the image which grayscale is L and size is $M \times N$. For an encrypted image, the value is big but may not show the higher security level. When Image encryption works best, it is $L/2$. As Table 2 shows, the average of pixel grey-scale values changed of the location scrambled image is worse than it of the final encrypted image. It indicates that the encryption algorithm which combines the position scrambling and gray scrambling achieves better encryption effect, and the encrypted image is sensitive to the key. If the key changes a small, the encrypted image in terms of point-to-gray pixel values will be changed dramatically.

Table 2. The average of pixel grey-scale values changed

Image	The average of pixel grey-scale values changed
The location scrambled image	67.1603
The encrypted image	79.5375

Correlation Analysis Between Neighboring Pixels. The effect of image in crambling have an inverse correlationship with the correlation coefficient [6] of adjacent pixels. The larger the correlation coefficient is, the worser the scrambling effect is. On the contrary, the effect is better. Extract a pair of adjacent pixels according to row (column) every two pixels, and calculate their correlation coefficient. As Table 3 shows, the correlation coefficient of adjacent pixels about the original image is close to 1, while the encrypted image's is close to 0, and the adjacent pixels is almost uncorrelated. So it indicates that the encrypted algorithm owns a high degree of scrambling.

Table 3. The comparison of correlation between adjacent pixels

Pixel relationship	Correlation coefficient of vertical pixels	Correlation coefficient of horizontal pixels
The original image	0.9638	0.9575
The location scrambled image	0.06402	−0.0340
The grayscale scrambled image	0.05862	−0.0030
The encrypted image	0.05113	0.00034932

5 Conclusion

This paper put forward a encryption algorithm about watermark image which combined Arnold transformation and Logistic chaotic map. It achieved not only the location scrambling, but also the grayscale scrambling. The encrypted image was analysed respectively from the histogram, the key sensitivity and the correlation of adjacent pixels. The result were good. Since the algorithm used the number of Arnold transformation and the initial value of Logistic chotic map as the key, the key space was big, and had a strong sensitivity. Although the encryption effect of watermark was good, when the original image was reset, the time complexity was relatively big. So this algorithm also needs to be improved in this side.

Acknowledgments. This project supported by Science and Technological Program for Dongguan's Higher Education, Science and Research, and Health Care Institutions (No.: 2012108102028) and scientific research project for Guangdong University of Science and Technology (No. GKY-2014KYYB-10).

References

1. Li, C., Li, S., Sim, M.A., Nunez, J., Alcarez, G., Chen, G.: On the security defects of an image encryption scheme. Image Vis. Comput. **24**, 926–934 (2006)
2. Chengmao, W., Xiaoping, T.: 3-dimensional non-equilateral Arnold transformation and its application in image scrambling. J. Comput.-Aided Des. Comput. Graph. **22**(10), 1831–1839 (2010)
3. Feng-Ying, W., Xiao-Li, H., Guo-wei, C.: An improved image encryption algorithm based on chaotic logistic map. J. Comput. Intell. Des. **2**, 105–107 (2012)
4. Kong, T., Zhang, D.: A new anti-Arnold transformation algorithm. J. Softw. **15**(10), 1558–1564 (2004)
5. Shao, L., Tan, Z., et al.: 2-dimension non equilateral image scrambling transformation. Acta Electronica Sinica **7**, 1290–1294 (2007)
6. Ren, H., Shang, Z., et al.: An algorithm of digital image encryption based on Arnold transformation. Opt. Technol. **35**(3), 384–390 (2009)
7. Li, Z., Hou, J.: DCT-domain fragile watermarking algorithm based on logistic maps. Acta Electronica Sinica **12**, 2134–2137 (2006)
8. Xue, F., Wang, C., et al.: Genetic and ant colony collaborative optimization algorithm based on information entropy and chaos theory. Control Decis. **26**(1), 44–48 (2011)
9. Fan, Y.J., Sun, X.H.: New algorithm to measure scrambling degree of scrambling images. Comput. Eng. Appl. **43**(29), 93–94 (2007)
10. Fang, W.S., Wu, L.L., Zhang, R.: A watermark preprocessing algorithm based on Arnold transformation and logistic chaotic map. Adv. Mater. Res. **341–342**, 720–724 (2011)

An Agent-Based Model for Intervention Planning Among Communities During Epidemic Outbreaks

Loganathan Ponnambalam[1(✉)], A.G. Rekha[2], and Yashasvi Laxminarayan[3]

[1] Institute of High Performance Computing, A*STAR, Singapore, Singapore
ponnaml@ihpc.a-star.edu.sg
[2] Indian Institute of Management Kozhikode, Kozhikode, India
[3] National Institute of Technology Surathkal, Surathkal, India

Abstract. We developed an agent-based model containing 50 communities, replicating the 50 states of USA. The age distribution, approximate household size and the socio-structural determinants of each community were modeled based on the US Census. The agent-based model was validated using in-silico seroprevalence data collection. Medical seeking behavior of individuals was parameterized based on the socio-structural determinants of the community. The interventions proposed in literature were tested and the optimal intervention strategy to counter an epidemic outbreak has been identified. In addition, we included novel interventions like coordination among the communities and increasing the awareness of individuals in the lower ranked communities based on information exchange between communities.

Keywords: Agent-based model · Epidemic outbreaks · Intervention planning · Coordination

1 Introduction

Epidemiology involves the understanding of factors affecting the health of populations and serves as the basis for interventions made in the interest of public health [1]. However, intervention studies using decision support models, available in the literature, often ignore the effect of socio-structural determinants on the spread of communicable diseases. The environment and its associated factors in which people reside are as important for communicable diseases as they are for non-communicable diseases [2]. Studies in the literature have shown the effect of socioeconomic status on the disease incidence during the 1918 influenza pandemic [3] and county-level socio-demographic and economic factors associated with the incidence of enteric diseases [4]. Hence, a socio-structural determinant perspective is essential. A decision support model that incorporates the socio-structural understanding is important for the development and precise evaluation of public health interventions and policies.

Modeling and simulation can help to systematically gauge the effectiveness of various pharmaceutical and non-pharmaceutical interventions [5]. Currently, the spread of an infectious disease outbreak is modeled [6] using two approaches: the equation based approach and the micro-simulation approach. The most commonly

© Springer Science+Business Media Singapore 2016
K. Li et al. (Eds.): ISICA 2015, CCIS 575, pp. 248–255, 2016.
DOI: 10.1007/978-981-10-0356-1_25

preferred equation based approach uses average parameter values for the entire population and does not address the interactions occurring at the micro level. The assumption is that the entire population moves and interacts in the same manner – this is not realistic. Agent based modeling [7–10] can handle the emergence property and our research exploits this advantage to develop realistic models and design optimal interventions to manage epidemic outbreaks. This work proposes an agent-based modeling approach to address the following questions: (i) How to use the community level surveillance data collected during the H1N1 2009 pandemic to understand the effect of socio-structural determinants on health outcomes?; (ii) How to use the proxy metrics (obtained from the data analysis) to rank the communities and quantify the heterogeneity in medical seeking behavior of the individuals?; (iii) How to integrate the heterogeneity in medical seeking behavior with agent based models and obtain realistic abstraction of the community/country of interest?; and (iv) How to use these models to explore novel targeted interventions and identify the optimal intervention policy to nullify the heterogeneity effect? The following section describes the agent-based model used in this study.

2   Agent Based Model

A stochastic agent based model containing a virtual environment of 50 communities, replicating 50 states of USA, was developed based on the integration of models proposed by Patel et al. (2005) [11] and Longini et al. (2004) [12]. The virtual environment contains numerous age-specific hubs of social interaction like schools, playgroups and workplaces. Each virtual state was populated with 1000 individuals. The age distribution, approximate household size and the socio-structural determinants of each community were modeled based on the US Census Bureau statistics [13]. Children in each state had access to high schools, middle schools, daycare centers and playgroups. Each state has a number of small playgroups and large day care centers for children. Four states share one high school and one middle school. Two states share one elementary school. The children were assigned to these schools based on their age. Ages 0–4 go to playgroups; 5–9 go to elementary schools; 10–14 go to middle school and 15–18 go to high school. Each state has 4–6 small playgroups and each small playgroup has 4 children in it. There are, on average, 14 children in each large day care. The total population of 50,000 is distributed among various families, each household consisting of 1–7 family members, with a mean of 2.3 family members per household. Uniform age distribution is assumed in the age groups 0–18 (children), 19–44 (young adults), 45–64 Middle adults) and 65–100 (aged). Adults in each state interact through workplaces present all over the virtual environment. The people in the virtual environment interact with each other, in accordance with their contacts per day [14]. The degree of interaction is highest in households, lower in schools, daycares and playgroups, still lower in states, and lowest across the country. Each person is geo-referenced to visit a doctor in his/her state, in case they are infected.

2.1   Disease Progression State Chart

The dynamics of the infection are defined by an extended version of the SEIR model, as shown in Fig. 1. A fixed number of people are randomly infected during initialization of the simulation.

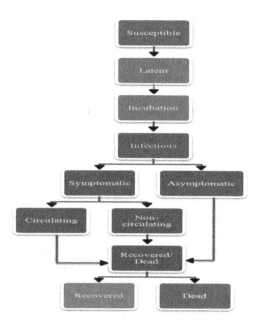

Fig. 1.   Disease progression state chart

Upon getting infected, the person passes through a latent and incubation stage, where they are not symptomatic. However, in the latent stage, they are not infectious whereas in the incubation stage, they are infectious. We assume that the latent and the incubation stages have the same lengths, ranging from 1 to 3 days with a mean of 1.9 days. This is followed by the infectious stage where the person may start showing symptoms. The probability that a person shows symptoms in this stage is 0.67. The length of the infectious stage varies from 2 days to 5 days with a mean length of 3.5 days. The infection transmissibility of an asymptomatic person is assumed to be half with respect to a symptomatic person. In the model, some of the symptomatic persons stop circulating and restrict themselves to their house, exposing only the other members of the family. If the person starts showing symptoms, the medical seeking behavior of the person dictates the spread. The person may go to the doctor after 1, 2 or 3 days, on showing symptoms. Medical seeking behavior is based on the socio-economic factors of the person, and was parameterized based on the income and population density of the state. If the person goes to the doctor, he/she receives antivirals, which reduces the infection transmissibility as well as the illness duration by one day. Also, it is assumed that everyone who goes to the doctor remains isolated from the society by restricting himself/herself to his/her house, thus exposing only his/her family members. The age dependency of vulnerability to the disease is portrayed by the usage of an appropriate multiplication factor [15]. Eventually, after the illness duration, the person either recovers immune to the disease, or dies due to it, depending on the mortality rate.

2.2 Medical Seeking Behavior

The 50 states of US were divided into 4 ranks, based on their population density and per capita income. Our prior studies have shown that these variables can be used as proxy metrics to quantify the socio-structural heterogeneity in health [16, 17]. To this end, we used H1N1 surveillance data collected from 102 US counties to model H1N1 case incidence, deaths using 28 socio-structural determinants and ranked the communities into 4 different classes. These 28 determinants were selected on the basis of the "the causes of the causes" model proposed by the Commission on Social Determinants of Health. These classification rules, extracted based on the analysis of socio-structural determinants on health inequity, were used to parameterize the medical seeking behavior of the individuals. The residents of well-ranked states were assumed to have high medical awareness; a majority of them approached the doctor quickly on falling ill. On the other hand, it was assumed that a majority of the residents of economically weaker ranked states either approached the doctor with a delay, or did not approach the doctor at all. The states with a population density higher than 88.8 per square mile and per capita income greater than $37000, were assigned a good rank, Rank 1. Whereas the states with low population density, lesser than 88.8 per square mile, and per capita income, lesser than $32500, were assigned Rank 4. The ranks of the individual states of US are listed in Table 1.

Table 1. Ranks of each state of US

Rank	Population density (per square mile)	Per capita income ($)	States
1	>88	>37000	California, Connecticut, Delaware, Hawaii, Illinois, Maryland, Massachusetts, New Hampshire, New Jersey, New York, Rhode Island, Virginia, Washington
2	>88	<37000	Florida, Georgia, Indiana, Kentucky, Louisiana, Michigan, North Carolina, Ohio, Pennsylvania, South Carolina, Tennessee, Wisconsin
3	<88	>32500	Wyoming, Colorado, Nevada, Minnesota, Alaska, Texas, Vermont, Kansas, Nebraska, Oregon, Iowa, Missouri, North Dakota
4	<88	<32500	Oklahoma, Maine, South Dakota, Arizona, Alabama, Montana, New Mexico, Idaho, Utah, Arkansas, West Virginia, Mississippi

2.3   Infection Transmission

Infection transmission probabilities are group-specific; highest in households, lower in day care centers and playgroups, still lower in schools, and lowest in the state and across the country [12]. Infection transmission also depends on the antiviral and vaccination status. Infection was introduced by randomly assigning 3 people in each state with infections. The mortality rates are assumed to be 0.0263 among kids, 0.0210 among children, 0.2942 among adults and 19.9797 among the aged.

2.4   Disease Spread Mechanism

An infected person will transmit the infection to a fixed number of people everyday depending on his/her contacts per day. Each person will have a fixed resistance towards the infection based on the infection strain. The resistance may be enhanced if the person is given antivirals or vaccines. If the infection is able to penetrate through the resistance offered by the individual, then the person turns infected. Mathematically, the infection probability is multiplied by an appropriate multiplication factor and (1-AVEs) and (1-VEs), where AVEs is the antiviral efficacy for susceptibility (AVEs = 0.3) and VEs is the vaccine efficacy for susceptibility (VEs = 0.7). The latter two terms are taken into account only if the person has been given antivirals or vaccinations. A random number between 0 and 1 is generated, and if this number is less than the calculated number, then the person is assigned infectious. Moreover if a person taking antivirals turns infectious, his illness duration will be 1 day lesser than normal. During transmission of the disease, the person will transmit the disease to another person if a random number generated between 0 and 1 is lesser than $(1 - AVEi) * (1 - VEi)$ where AVEi is the antiviral efficacy for infectiousness (AVEi = 0.80) and VEi is the vaccine efficacy for infectiousness (VEi = 0.80). This is done only if the person is given antivirals or vaccines.

2.5   Pharmaceutical and Non-pharmaceutical Intervention Schemes

The pharmaceutical and non-pharmaceutical interventions proposed in the literature were tested in the validated model and the optimal intervention strategy to counter an epidemic outbreak has been identified. In addition, we included novel interventions like coordination among the communities and increasing the awareness of individuals in the lower ranked communities based on information exchange between communities. If vaccination schemes are triggered in one state, similar vaccination schemes can also be triggered in other states. The same concept of coordination has been extended to awareness programmes across states with different ranks. Whenever a state registers high number of infections, all other states with different ranks initiate awareness programmes with an appropriate delay.

3 Results and Discussion

3.1 Scenarios Tested

The following non-pharmaceutical and pharmaceutical interventions were tested in the validated model, to identify the optimal intervention strategy to counter an epidemic outbreak.

3.1.1. Run 1: No intervention - The model was simulated for a reference case scenario, with no interventions. This scenario was used as a reference to quantify the effect of interventions for other trials.

3.1.2. Run 2: School closure - When 5 % of the population in a state was reported infected, school closure was activated in that particular state, for 14 days.

3.1.3. Run 3: Awareness - An awareness campaign was triggered after 14 days, which improved the medical seeking behavior of individuals of each state. 25 % of the population was targeted for the awareness campaigns.

3.1.4. Run 4: Coordinated awareness - When 1 % of the population in rank 1 states was reported infected, an awareness campaign was triggered in all other states, which targeted at improving the medical seeking behavior of 50 % of the state population. There was a delay of 2 days in implementing awareness programmes for rank 2 states, 4 days for rank 3 states and 7 days for rank 4 states.

3.1.5. Run 5: Vaccination – 50 % of the population of a state was independently vaccinated when 5 % of the population, of that particular state, was reported to be infected. There was a delay of 5 days for the implementation of the vaccination programmes, once the threshold was reached. This delay included delays due to data collection, policy making and vaccine delivery.

3.1.6. Run 6: Coordinated awareness with vaccination – 50 % of the population was vaccinated in a state when 5 % of the population of that particular state was reported infected. Also, when 1 % of the population in rank 1 states was reported infected, an awareness campaign was triggered in all other states, which targeted at improving the medical seeking behavior of 50 % of the state population. There was a delay of 2 days in implementing awareness programmes for rank 2 states, 4 days for rank 3 states and 7 days for rank 4 states.

3.1.7. Run 7: Coordinated school closure and vaccination - When 5 % of the population in a state was reported infected, 50 % of the population of that particular state was vaccinated and school closure was implemented in the state for 14 days. Also, when vaccinations and school closures are triggered in one state, they are initiated in all other states, with a coordination delay of 1 day.

3.2 Simulation Results

The simulation results (average results of 50 simulation runs) of total number of reported cases, number of vaccines/antivirals consumed, total number of deaths and actual cases, are shown in Table 2.

Table 2. Simulation results of each scenario

Run	Cases-reported	Vaccine count	AV count	Deaths	Cases-actual	% reduction in overall attack rate
1	6550	0	13231	50	24765	–
2	6235	0	12680	50	23866	3.63
3	6160	0	12871	41	23158	6.49
4	6326	0	12329	34	21204	14.38
5	4191	19682	8712	31	16476	33.47
6	3782	18176	7993	25	13843	44.10
7	3224	19618	6752	21	11659	52.92

Our results clearly indicate that the coordinated, targeted approach benefits not only the economically weaker communities, but also the overall virtual world. The significance of coordination is further illustrated by the reduction in deaths per 1000 and the consumption of antivirals. The combination of pharmaceutical and non-pharmaceutical interventions (coordinated action) reduces the attack rate in the lower ranked communities by 56 % (data not shown), as well as the overall attack rate in all communities by 53 %. In addition, such a strategy reduces the antivirals requirement by 49 % and deaths by 58 %, from 1 death per 1000 to 0.42 deaths per 1000. Coordination among the communities to trigger the non-pharmaceutical interventions alone resulted in 14.4 % reduction in the overall attack rate.

4 Conclusions

Through an agent-based model, we have clearly demonstrated as to how the proxy metrics obtained from the data collected during an epidemic outbreak can be used to rank the communities and quantify the heterogeneity in medical seeking behavior of the individual agents. In addition, the proposed approach also illustrates as to how the heterogeneity in medical seeking behavior can be integrated with the agent-based model so as to obtain realistic abstraction of the community/country of interest. From the results, it is evident that the shortage of resources and healthcare infrastructure in economically weaker communities can be combated by timely activation of awareness and school closure, aided by the coordinated communication between the communities. Our results for various intervention scenarios clearly emphasize the need for optimal, prioritized policies that addresses the heterogeneity of the communities and channelize the resources accordingly. H1N1 surveillance data have been documented at local, county, province, regional, and state level in many countries. This work provides a novel approach to use the collected data effectively to quantify the marginalization at each level, subgroup, the population to be targeted and optimize the distribution of public health resources efficiently.

References

1. Michael, M.: Social determinants of health inequalities. Lancet **365**, 1099–1104 (2005)
2. Sommerfeld, J., Kris, H.: Social dimensions of infectious diseases. In: Kris, H. (ed.) International Encyclopedia of Public Health, pp. 69–74. Academic Press, Oxford (2008)
3. Murray, C.J.L., et al.: Estimation of potential global pandemic influenza mortality on the basis of vital registry data from the 1918-20 pandemic: a quantitative analysis. Lancet **368**, 2211–2218 (2006)
4. Farmer, P.: Infections and Inequalities: The Modern Plagues. University of California Press, Berkley (1999)
5. Grassly, N.C., Fraser, C.: Mathematical models of infectious disease transmission. Nat. Rev. Microbiol. **6**, 477–487 (2008)
6. Arino, J., et al.: Simple models for containment of a pandemic. J. R. Soc. Interface **3**(8), 453–457 (2006)
7. Auchincloss, A.H., Roux, A.V.D.: A new tool for epidemiology: the usefulness of dynamic-agent models in understanding place effects on health. Am. J. Epidemiol. **168**(1), 1–8 (2008)
8. Marshall, B.D.L., Galea, S.: Formalizing the role of agent-based modeling in causal inference and epidemiology. Am. J. Epidemiol. **181**(2), 92–99 (2015). doi:10.1093/aje/kwu274.
9. Hernán, M.A.: Invited commentary: agent-based models for causal inference—reweighting data and theory in epidemiology. Am. J. Epidemiol. **181**(2), 103–105 (2015)
10. Nianogo, R.A., Arah, O.A.: Agent-based modeling of noncommunicable diseases: a systematic review. Am. J. Public Health **105**(3), e20–e31 (2015)
11. Patel, R., et al.: Finding optimal vaccination strategies for pandemic influenza using genetic algorithms. J. Theor. Biol. **234**, 201–212 (2005)
12. Longini, I.M., Halloran, M.E., Nizam, A., Yang, Y.: Containing pandemic influenza with antiviral agents. Am. J. Epidemiol. **159**, 623–633 (2004)
13. US-Census. http://www.census.gov/
14. Kretzschmar, M., Mikolajczyk, R.T.: Contact profiles in eight European Countries and implications for modelling the spread of airborne infectious diseases. PLoS ONE **4**(6), e5931 (2009). doi:10.1371/journal.pone.0005931
15. Cao, B., et al.: Clinical features of the initial cases of 2009 pandemic influenza A (H1N1) virus infection in China. N. Engl. J. Med. (2009). doi:10.1056/NEJMoa0906612
16. Loganathan, P., Lakshminarayanan, S, Lee, H.R., Ho, C.S.: Understanding the socioeconomic heterogeneity in healthcare in US counties: the effect of population density, education and poverty on H1N1 pandemic mortality. Epidemiol. Infect. (2011). doi:10.1017/S0950268811001464
17. Loganathan, P., Lakshminarayanan, S., Lee, H.R.: Evaluation of health systems performance and estimation of health-care inequality among US counties: a novel approach for priority setting by proxy metrics and decision trees. Published in the supplement to: GHME Conference Organizing Committee. Shared Innovations in Measurement and Evaluation. Lancet, 14 March 2011. doi:10.1016/S0140-6736(11)60169-4

The Comparisons Between the Improved Numerical Mode-Matching Method (NMM) and the Traditional NMM Using for Resistivity Logging

Dun Yueqin[1(✉)] and Kong Yu[2]

[1] School of Electrical Engineering, University of Jinan,
Jinan 250022, Shandong, China
dunyq828@163.com
[2] Information Center of Shandong Medical College,
Jinan 250002, Shandong, China
kongy@sdmc.net.cn

Abstract. Based on the analysis and comparison of existing numerical mode-matching method (NMM), an improved NMM is proposed in this paper. A new type of recursive formulas is derived by setting the proper positions for the unknown variables, and recursive formulas are unified to one form for the layers above and under source, avoiding the disadvantage of the traditional NMM to deduce the recursive formulas for the layers above and under source respectively. It is easy to understand and has concise physical meaning. But the incremental factor in the recursive formulas can't be eliminated thoroughly, which may affect the calculation accuracy and stability. Finally, the merits and demerits of the improved method and the traditional one are summarizes.

Keywords: Numerical mode-matching method · Resistivity logging · Recursive formulas

1 Introduction

To judge which layer of the earth is of petroleum or gas, a measurement sonde is put into a borehole and moved to measure the physical characteristic of the soil (such as, resistivity), which is called (resistivity) logging. In the logging simulation, the earth is simplified to multi-layer structures. Usually, they are axisymmetrical models. For inhomogeneous medium model with any number of horizontal layer and vertical layered, NMM is a fast numerical and analytical method [1–4]. The basic idea of NMM is decomposing a partial differential equation into two ordinary differential equations by using the variable separation method. Generally, numerical method (finite element method is chosen commonly) is used to solve the radial problem, and the eigen

This work was supported by the National Natural Science Foundation of China (Grant No. 51307090).

K. Li et al. (Eds.): ISICA 2015, CCIS 575, pp. 256–263, 2016.
DOI: 10.1007/978-981-10-0356-1_26

equation is formed in each layer, and then the corresponding eigen values and eigen vectors will be solved. In the axial direction, analytical method is used to deduce the recursive formulas. The recursive analytic expressions can be deduced by combining the interface conditions and the reflection and transmission characteristics between different layers. Therefore, the calculation speed of NMM is much higher than that of finite element method adopting numerical solution both in the radial and axial directions.

According to electromagnetic theory, the analytical formula of each layer can be expressed as superposition of up-going wave and down-going wave, and the analytical expression of each layer can be derived by matching the boundary conditions. Different scholars have taken different approaches to handle this process, so the different analytical expressions have been derived. And the analytical expressions of NMM can be divided into two types basing on the contrastive analysis of the existing literatures. The first type of analytical expression contains generalized reflection matrixes and transmission matrixes and a large number of matrix inversion [5–7]. The second type of analytical expression has been studied as a new NMM, in this new method, the generalized reflection matrixes and transmission matrixes are avoided, and the matrix inversion are also avoided [8]. The exponential parts in the first type of analytical expressions only have negative values which describes the attenuation characteristic of the wave (called attenuation factor), while the exponential parts in the second type have not only negative values but also positive values describing the incremental characteristic of the wave (called incremental factor).

The common point of the two types is that the recursive formulas should be deduced for the layers above and under source respectively, that is, there will be two sets of different analytical expressions. Now let's take the layers above source as example to analyze the characteristics of the two types respectively.

For the layers above source, the first type of analytic expressions take the field amplitude (called the expansion coefficient vectors to be determined in reference [5]) at interface in every layer as unknown variable (only one unknown variable in each layer), and the up-going wave and the down-going wave can be expressed by using the reflection and transmission characteristic, then matching the boundary conditions of each layer to get the recursive expressions. In this recursive process the relationship of the field between adjacent layers are reflected by the reflection and transmission matrixes, therefore, the calculation of the reflection and transmission matrixes cannot be avoided, and a large amounts of matrix inversion must be calculated. But the incremental factor can be changed into attenuation factor by some method, though the incremental factor appears in the analytic expressions.

For the layers above source, the second type of analytic expressions take the amplitude of down-going wave and the amplitude of up-going wave on the bottom interface as unknown variables (two unknown variables in each layer). The value of up-going wave at some field point in some layer can be obtained with the amplitude of up-going wave on the bottom interface of this layer multiplied by the attenuation factor. And the value of down-going wave at the field point can be obtained with the amplitude of down-going wave on the bottom interface of this layer multiplied by the incremental factor. Then the analytic expression of the potential field of the field point will be got by adding the value of up-going and down-going wave together. After that, the analytical expression will be derived by combining the interface conditions of each layer. In this process, the generalized reflection and transmission matrixes, and a large amounts of matrix inversion can be avoided, but the incremental factor can't be avoided.

For the layers under source, the analysis process of the two types are similar with the layers above source, except for the locations of the unknown variables are different. To simplify and unify the analytical expressions, we choose different locations of the unknown variables of each layer with the existing two types.

2 The Improved NMM

Figure 1 gives a multi-layer axisymmetric model with M-interfaces, each layer is composed of borehole, invasion and original layer. The conductivity σ of each layer has relation only with the radial radius ρ.

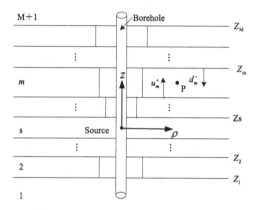

Fig. 1. The multi-layer axisymmetric model

If a point source is put on the axis of the borehole, the potential in the m-layer (without source) satisfy the Laplace Eq. (1).

$$\frac{1}{\sigma_m \rho} \frac{\partial}{\partial \rho} \left(\sigma_m \rho \frac{\partial \varphi_m}{\partial \rho} \right) + \frac{\partial^2 \varphi_m}{\partial z^2} = 0 \tag{1}$$

By using separated variable method [7, 8], the potential can be written as Eq. (2):

$$\varphi_m(\rho, z) = g_m^T C_m(\rho) [u_m^+(z) + d_m^-(z)] \tag{2}$$

Here, $g_m = (g_{m1}, g_{m2}, \ldots, g_{mN})$ is the basis function in the radial direction, C_m is a matrix composed of eigen vectors, and satisfy the eigen equation $A_m C_m = B_m C_m K_m^2$, and A_m, B_m are both $N \times N$ matrixes calculated as follows

$$A_m = \int_0^\infty \sigma_m \rho g_m' g_m'^T d\rho,$$

$$B_m = \int_0^\infty \sigma_m \rho g_m g_m^T d\rho.$$

$K_m^2 = diag(k_{m1}^2, \ldots, k_{mN}^2)$ is eigen value. $u_m^+(z)$ and $d_m^-(z)$ are the values of the up-going wave and the down-going wave.

Now, we define $u_m^+(z_b)$ and $d_m^-(z_t)$ as two unknown variables in each layer as shown in Fig. 1. $u_m^+(z_b)$ is the value of the up-going wave on the bottom interface of m-layer, and $d_m^-(z_t)$ is the value of the down-going wave on the top interface of m-layer. Then we can express the potential of an arbitrary point P in m-layer as Eq. (3).

$$\varphi_m(\rho_P, z_P) = g_m^T C_m(\rho)[e^{-K_m(z_P - z_{m-1})} \cdot u_m^+(z_b) + e^{-K_m(z_m - z_P)} \cdot d_m^-(z_t)] \qquad (3)$$

Because the locations of unknown variables of each layer are same, so we don't need to distinguish the layers above source and the layers under source to deduce the analytic expressions. Therefore, the analytic expressions can be unified to one form.

The expressions of the potential and the electric current density in m-layer are given in Eq. (4a and 4b),

$$\varphi_m(\rho, z) = g_m^T C_m[e^{-K_m(z - z_{m-1})} \cdot u_m^+(z_b) + e^{-K_m(z_m - z)} \cdot d_m^-(z_t)] \qquad (4a)$$

$$j_m(\rho, z) = -\sigma_m g_m^T C_m K_m[e^{-K_m(z - z_{m-1})} \cdot u_m^+(z_b) - e^{-K_m(z_m - z)} \cdot d_m^-(z_t)] \qquad (4b)$$

and Eq. (5a and 5b) is the expressions in (m + 1) layer.

$$\varphi_{m+1}(\rho, z) = g_{m+1}^T C_{m+1}[e^{-K_{m+1}(z - z_m)} \cdot u_{m+1}^+(z_b) + e^{-K_{m+1}(z_{m+1} - z)} \cdot d_{m+1}^-(z_t)] \qquad (5a)$$

$$j_{m+1}(\rho, z) = -\sigma_{m+1} g_{m+1}^T C_{m+1} K_{m+1}[e^{-K_{m+1}(z - z_m)} \cdot u_{m+1}^+(z_b) \\ - e^{-K_{m+1}(z_{m+1} - z)} \cdot d_{m+1}^-(z_t)] \qquad (5b)$$

Considering the interface condition $\varphi_m = \varphi_{m+1}$ and $j_m = j_{m+1}$ for the layers without source, we can get two equations, that is, Eq. (4a) is equal to Eq. (5a), and Eq. (4b) is equal to Eq. (5b). These two equations are left-multified by $C_{m+1}^T \sigma_{m+1} \rho g_{m+1}$ and $C_{m+1}^T \rho g_{m+1}$ on the both sides respectively, and integrated from 0 to ∞. And then the Eq. (6) can be derived by combing the orthogonality relation $C_m^T B_m C_m = I$.

$$\begin{bmatrix} u_{m+1}^+(z_b) \\ d_{m+1}^-(z_t) \end{bmatrix} = \begin{bmatrix} I & 0 \\ 0 & e^{K_{m+1} h_{m+1}} \end{bmatrix} \cdot RT_{m+1,m} \cdot \begin{bmatrix} e^{-K_m h_m} & 0 \\ 0 & I \end{bmatrix} \cdot \begin{bmatrix} u_m^+(z_b) \\ d_m^-(z_t) \end{bmatrix} \qquad (6)$$

Here, h_m and h_{m+1} are the thickness of the m layer and m + 1 layer respectively.

$$RT_{m+1,m} = \frac{1}{2} \begin{bmatrix} R_{m+1,m} + K_{m+1}^{-1} T_{m+1,m} K_m & R_{m+1,m} - K_{m+1}^{-1} T_{m+1,m} K_m \\ R_{m+1,m} - K_{m+1}^{-1} T_{m+1,m} K_m & R_{m+1,m} + K_{m+1}^{-1} T_{m+1,m} K_m \end{bmatrix}$$

And, $R_{m+1,m} = \int_0^\infty \sigma_{m+1} \rho g_{m+1} g_m^T d\rho$, $T_{m+1,m} = \int_0^\infty \sigma_m \rho g_{m+1} g_m^T d\rho$

Equation (6) gives the recursive relation between m + 1 layer and m layer.

The layer with source should be handled specially, that is, the interface zs should be inserted on the location of the point source as shown in Fig. 2.

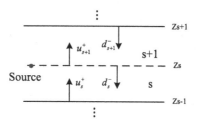

Fig. 2. The layer with source

Considering the interface condition $\varphi_s = \varphi_{s+1}$ and $j_{s+1} - j_s = -\frac{I\delta(\rho)}{2\pi\rho}$, the Eq. (7) will be derived by taking the similar processes of the layers without source.

$$\begin{bmatrix} u_{s+1}^+(z_b) \\ d_{s+1}^-(z_t) \end{bmatrix} = \begin{bmatrix} e^{-K_s h_s} & 0 \\ 0 & e^{K_{s+1} h_{s+1}} \end{bmatrix} \begin{bmatrix} u_s^+(z_b) \\ d_s^-(z_t) \end{bmatrix} + \begin{bmatrix} I & 0 \\ 0 & e^{K_{s+1} h_{s+1}} \end{bmatrix} \begin{bmatrix} b_s \\ -b_s \end{bmatrix} \quad (7)$$

Considering the natural boundary conditions, we know there has no down-going wave in the upmost layer, and no up-going wave in the bottommost layer, which can be described by Eq. (8).

$$\begin{cases} d_{M+1}^- = 0 \\ u_1^+ = 0 \end{cases} \quad (8)$$

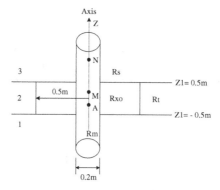

Fig. 3. A three layer model with borehole and invasion

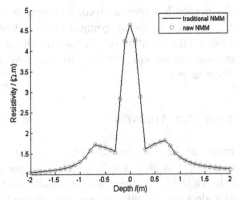

(a)   The result of the model without borehole and invasion

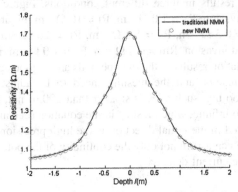

(b)   The result of the model with borehole

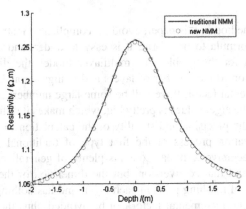

(c)   The result of the model with borehole and invasion

Fig. 4. The result comparison in different conditions

Hereto, all the formulas needed has been derived, an equation set will be formed according to the recursive formulas. So the up-going wave and down-going wave in each layer will be got by solving the equation set. Considering the analysis process of the new NMM, it is found that the dimension of equation set will increase with the increase of the number of the layers.

3 Results Comparison and Analysis

In order to compare the results of the new method with that of the traditional method, we take the model in reference [7] as an example, which is a three layer model with borehole and invasion as shown in Fig. 3. There are three point electrodes on the axis of the borehole, one emitter electrode A and two receiving electrodes M, N.

Figure 4 gives the results in three different conditions. Figure 4(a) is a model without borehole and invasion, $Rs = 1.0 \, \Omega \cdot m$, $Rt = 10 \, \Omega \cdot m$. Figure 4(b) is a model without invasion, $Rm = 0.5 \, \Omega \cdot m$, $Rs = 1.0 \, \Omega \cdot m$, $Rt = 2.0 \, \Omega \cdot m$. Figure 4(c) is a model with borehole and invasion, $Rm = 0.5 \, \Omega \cdot m$, $Rs = 1.0 \, \Omega \cdot m$, $Rxo = 1.0 \, \Omega \cdot m$, $Rt = 2 \, \Omega \cdot m$. The calculation results of the two methods are in good agreement, but it is not easy to get these results, and the meshing need to be adjusted to assure the eigenvalues K are not too big, such as K is smaller than 200, if not, there will appear very big number close to infinite, which results in the equation set can't be solved. The basis function we adopted in the radial direction is the high precision amplitude-slope base function [7], which can assure not only the continuity of the potential, but also the continuity of the electric current density.

4 Conclusions

The improved NMM method in this paper, avoiding complicated matrix inversion, has unified the recursive formula to one form, it is easy to understand and has concise physical meaning. But for the problem of multilayer model, the dimension of the equation set is proportion to the number of layers and is high, in addition, due to the influence of the incremental factor, there will be some large numbers in the coefficient of the equations when the eigenvalues is pretty big, which makes it difficult to solve the equations, and affects the precision and stability of the calculation to a large extent.

Although the derivation process of the first type of traditional NMM is pretty complicated, and the recursive formulas contains plenty of generalized reflection and transmission matrixes and matrix inversion, but the dimension of the matrix is only depends on the number of meshing grids in each layer, and has nothing to do with the number of layers, and the incremental factor can be avoided, thus the stability of the calculation can be ensured.

References

1. Chew, W.C.: Waves and Fields in Inhomogeneous Media. Van Nostrand Reinhold, New York (1990)
2. Tsang, L., Chan, A.K., Gianzero, S.: Solution of the fundamental problem in resistivity ogging with a hybrid method. Geophysics 49(24), 1596–1604 (1984)
3. Chew, W.C., Nie, Z.: An efficient solution for the response of electrical well logging tools in a complex environment. IEEE Trans. GRS 229(2), 308–313 (1991)
4. Dun, Y., Tang, Z., Zhang, F., et al.: Fast calculation and characteristic analysis of array lateral logging responses. Int. J. Appl. Electromagnet. Mech. 33, 145–151 (2010)
5. Fan, G., Liu, Q.: 3D numerical mode matching method (NMM) for resistivity well logging tools. IEEE Trans. Antennas Propag. 48(10), 1544–1553 (2000)
6. Dun, Y., Kong, Y., Zhang, L., et al.: Comments on "3-D numerical mode-matching (NMM) method for resistivity well-logging tools". IEEE Trans. Antennas Propag. 59(7), 2751–2752 (2011)
7. Zhang, G., Wang, H.: Solution of the normal resistivity logging with the numerical mode-matching method. J. Univ. Pet. China (Ed. Nat. Sci.) 20(2), 23–29 (1996). (in Chinese)
8. Tan, M., Zhang, G., Zhao, W.: An improved NMM method for electromagnetic field calculation in geophysics. J. Beijing Univ. Posts Telecommun. 29(5), 6–10 (2006). (in Chinese)

Negative Correlation Learning
with Difference Learning

Yong Liu[✉]

The University of Aizu, Aizu-Wakamatsu, Fukushima 965-8580, Japan
yliu@u-aizu.ac.jp

Abstract. In order to learn a given data set, a learning system often has to learn too much on some data points in the given data set in order to learn well the rest of the given data. Such unnecessary learning might lead to both the higher complexity and overfitting in the learning system. In order to control the complexity of neural network ensembles, difference learning is introduced into negative correlation learning. The idea of difference learning is to let each individual in an ensemble learn to be different to the ensemble on some selected data points when the outputs of the ensemble are too close to the target values of these data points. It has been found that such difference learning could control not only overfitting in an ensemble, but also weakness among the individuals in the ensemble. Experimental results were conducted to show how such difference learning could create rather weak learners in negative correlation learning.

1 Introduction

Supervised learning is often formulated as minimizing an error function defined by the absolute differences between the outputs of a learning model on a set of training points and their target values. By adjusting the weights of the learning model through the minimization, the outputs of the learned model would get near and near to the targets. From the point of minimizing the error function, the best solutions would be the models with weights that achieve the smallest absolute differences between the outputs of the models and the target values. When the number of training points is limited, there could be several such best models. Because these best models might give the different predictions on the unseen data, only those with the best predictions on the unseen data should be selected in application. Therefore, besides optimization, there is a model selection problem in learning, in which the basic principle of Occam's Razor is often adopted. It implies to use the simplest models among all the models that well fit the training points [1–8].

In the real applications, there often exist noises among the given training points. Therefore, the learned models should not fit the whole training points too well. This is another difference between learning and optimization. In order words, the differences between the outputs of the models and the target values

© Springer Science+Business Media Singapore 2016
K. Li et al. (Eds.): ISICA 2015, CCIS 575, pp. 264–274, 2016.
DOI: 10.1007/978-981-10-0356-1_27

should not be minimized to be the smallest values among the whole training points in learning. Otherwise, the noises would be learned as well.

These two problems in single model learning approaches also exist in multiple model learning approaches that use the combined outputs of a set of learned models to make the predictions. It also requires that the combined system should match the training points with a low cost in complexity. If the complexity of the combined system were higher than necessity, it could likely overfit the training points, and learn too much noises. Therefore, it is important to control the complexity of the combined system. When the complexity of the combined system is decided by the individual models in the ensemble, what kinds of the individual models should be created? As pointed by R. Schapire [9], combining a set of weak learners could create a strong learner. A weak learner can predict unseen data slightly better than random guessing while a strong learner can generate high-accuracy predictions. One way to generate such weak learners is through boosting. In the applications of boosting, it could either use a small number of little weak learners that are much better than random guessing, or a large number of truly weak learners in an ensemble. Since it might not be practical to use too many weak learners, small number of little weak learners is often set in applications. However, when the learners are not so weak, it could run a risk of overfitting as well. In order to prevent from such overfitting in boosting, it still has to depend on the methods used in the single model system, such as cross-validation.

Negative correlation learning is an ensemble learning method with correlation-based penalty [10–15], which is capable of creating weak learners in an ensemble. However, overfitting had still been observed in negative correlation learning. In order to learn a given data set, a learning system often has to learn too much on some data points in the given data set in order to learn well the rest of the given data. Such unnecessary learning might lead to both the higher complexity and overfitting in the learning system. In order to control the complexity of neural network ensembles, difference learning is introduced into negative correlation learning. The idea of difference learning is to let each individual in an ensemble learn to be different to the ensemble on some selected data points when the outputs of the ensemble are too close to the target values of these data points. It has been found that such difference learning could control not only overfitting in an ensemble, but also weakness among the individuals in the ensemble. Experimental results were conducted to show how such difference learning could create rather weak learners in negative correlation learning.

The rest of this paper is organized as follows: Sect. 2 describes ideas of negative correlation learning and difference learning for neural network ensembles. Section 3 compares the learning behaviors between negative correlation learning and negative correlation learning with difference learning on both the ensemble level and the individual level. Finally, Sect. 4 concludes with a summary of discussions.

2 Negative Correlation Learning and Difference Learning

Negative correlation learning [10] was developed to create negatively correlated neural networks for neural network ensemble. In negative correlation learning, the output y of a neural network ensemble is formed by a simple averaging of outputs F_i of a set of neural networks. Given the training data set $D = \{(\mathbf{x}(1), y(1)), \cdots, (\mathbf{x}(N), y(N))\}$, all the individual networks in the ensemble are trained on the same training data set D

$$F(n) = \frac{1}{M} \Sigma_{i=1}^{M} F_i(n) \tag{1}$$

where $F_i(n)$ is the output of individual network i on the nth training pattern $\mathbf{x}(n)$, $F(n)$ is the output of the neural network ensemble on the nth training pattern, and M is the number of individual networks in the neural network ensemble.

The error function E_i for individual i on the training data set D consists of two terms in negative correlation learning as follows.

$$\begin{aligned}
E_i &= \frac{1}{N} \Sigma_{n=1}^{N} E_i(n) \\
&= \frac{1}{N} \Sigma_{n=1}^{N} \frac{1}{2} \left[(F_i(n) - y(n))^2 - \lambda (F_i(n) - F(n))^2 \right]
\end{aligned} \tag{2}$$

where N is the number of patterns in D, $E_i(n)$ is the error function of network i at presentation of the nth pattern, and $y(n)$ is the desired output of the nth pattern. The first term in the right side of Eq. (2) is the mean-squared error of individual neural network i. The second term is a correlation penalty function. The purpose of minimizing the correlation penalty is to negatively correlate each individual's outputs with the outputs of the ensemble. The parameter λ is set to scale the strength of the penalty. With the correlation penalty in the error function, all the individual networks can be trained simultaneously and interactively in negative correlation learning.

The partial derivative of E_i with respect to the output of neural network i on the nth pattern is

$$\frac{\partial E_i(n)}{\partial F_i(n)} = (1 - \lambda)(F_i(n) - y(n)) + \lambda(F(n) - y(n)) \tag{3}$$

At $\lambda = 1$, the derivative of E_i becomes

$$\frac{\partial E_i(n)}{\partial F_i(n)} = F(n) - y(n) \tag{4}$$

where the derivative of E_i is only decided by the difference between $F(n)$ and $y(n)$.

In the negative correlation learning with $\lambda = 1$, the error signals for each individual neural network are only decided by the differences between the ensemble outputs and the target values. Changes on the weights of each individual in the ensemble would be continued as long as the differences between the ensemble outputs and the target values exist. For the classification problems, it might not be necessary to pursue the smallest difference between the ensemble outputs and the target values. For an example of a two-class problem, the target value y on a data point can be set up to 1.0 or 0.0 depending on which class the data point belongs to. As long as F is larger than 0.5 at $y = 1.0$ or smaller than 0.5 at $y = 1.0$, the data point will be correctly classified. Therefore, an ideal target should be a value between 1.0 and 0.5, or between 0.5 and 0.0. Unfortunately such ideal target values for the given data are usually unknown. If 1.0 and 0.0 would be set up as the target values in a two-class classification problem, some data would likely be learned too much without the constrains in optimization. If the data would be learned too much, the complexity of the learned system could grow too much as well than necessity.

In order to prevent negative correlation learning from overlearning, learning on the pattern k will be switched to difference learning from negative learning as soon as the absolute difference between $F(k)$ and $y(k)$ on the pattern k is not bigger than α. The parameter of α is a small value. In difference learning, the error function E_i for individual i is defined as

$$E_i = -\frac{1}{2}\beta(F_i(k) - F(k))^2 \tag{5}$$

The parameter of β can be adjusted in a certain range. Difference learning could make each individual be more different to the others in an ensemble. Meanwhile, different learning will also forbid the outputs of the ensemble to be too close to the learning targets so that the complexity of the learned ensemble could be well controlled.

3   Experimental Results

Three measurements have made in the learning process of both negative correlation learning (NCL) and negative correlation learning with difference learning (NCLDL), including the average error rates of both the ensembles and the individual neural networks in the ensemble, and the average overlapping rates of output between every two individuals in the ensembles. The first two measurements indicate how well the ensembles and the individuals perform. The lower the error rates are, the better performance they have. The third measurement shows the similarities among the individuals in an ensemble. If the overlapping rate between two individual neural networks were 1, these two individuals would have the same classification on all the measured data points. If the overlapping rate between two individual neural networks were 0, these two individuals would provide completely different classification on every measured data point.

3.1   Data Sets and Experimental Setup

The following four data sets from the UCI machine learning bench-mark repository were tested in the experiments. The first data set is the diabetes data set that is a two-class problem with 500 examples of class 1 and 268 of class 2. Each example has 8 attributes. The data set is quite hard to classify when there are high noises among the data points.

The second data set is the breast cancer data set whose purpose is to classify a tumour as either benign or malignant based on cell descriptions gathered by microscopic examination. The data set contains 9 attributes and 699 examples with 458 benign examples and 241 malignant examples.

The third data set is the heart disease data set on predicting the presence or absence of heart disease given the results of various medical tests carried out on a patient. This database includes 13 attributes and 270 examples.

The fourth data set is the Australian credit card assessment data set on assessing applications for credit cards based on a number of attributes. This database has 14 attributes and 690 examples.

5 runs of n-fold cross-validation were carried out to calculate the average values of the three measurements in which n is set to 12 for the diabetes data, and 10 for the rest three data sets. The ensemble architecture used in the simulations contains 10 individual neural networks. Each individual neural network has one hidden layer and 10 hidden nodes. $\lambda = 1$ was used in NCL. $\alpha = 0.1$ and the three values of β at 0.01, 0.1, 1 were tested in NCLDL(β) that represents NCLDL with the parameter β. The number of training epochs was set to 4000.

3.2   Results of NCLDL at the Ensemble Level

Table 1 described both the training and the testing error rates of the ensembles by NCL from 60 runs on the diabetes data set and 50 runs on the other three data sets at the different learning epochs. Although the training error rates showed different reductions on different data sets, all of them became smaller and smaller in the training process. Among the four data sets, the diabetes data is a rather harder to be learned, while the cancer data is quite easier to be learned. After 4000 learning epochs, the training error rates just touched 0.115 on the diabetes data. The training error rates dropped to as low as 0.025 just after the first 50 learning epochs on the cancer data. Although the training error rates on the heart data were just reduced to 0.079 in the first 50 learning epochs, they were further driven to 0.002 at learning epoch 4000. The training error rates on the card data fell gradually from 0.103 to 0.014 from learning epoch 50 to 4000. Different to the training error rates, the testing error rates increased moderately on the card data, and slightly on the cancer data while they were reduced significantly on the heart data, but had little changes on the diabetes data. Overfitting had actually been observed on the cancer data and the card data in NCL.

Table 2 described both the training and the testing error rates of the learned ensembles by NCLDL(0.01). Surprisingly, not only could NCLDL(0.01) converge faster, but also could learn better on the four data sets. Although no much differences were observed between NCL and NCLDL(0.01) in the first 50 learning epochs, NCLDL(0.01) reached the lower training error rates on all the four data sets than NCL. It should be noted that difference learning were just conducted by

Table 1. Average error rates of the ensembles by NCL.

No. of epochs	Diabetes		Cancer		Heart		Card	
	Training	Testing	Training	Testing	Training	Testing	Training	Testing
50	0.217	0.242	0.025	0.035	0.079	0.124	0.103	0.137
1000	0.137	0.242	0.016	0.039	0.019	0.061	0.037	0.142
2000	0.125	0.244	0.013	0.039	0.007	0.050	0.024	0.148
3000	0.120	0.241	0.011	0.039	0.004	0.046	0.018	0.145
4000	0.115	0.241	0.010	0.039	0.002	0.037	0.014	0.149

Table 2. Average error rates of the ensembles by NCLDL(0.01).

No. of epochs	Diabetes		Cancer		Heart		Card	
	Training	Testing	Training	Testing	Training	Testing	Training	Testing
50	0.217	0.242	0.025	0.035	0.0815	0.136	0.106	0.136
1000	0.112	0.241	0.015	0.039	0.0053	0.034	0.028	0.140
2000	0.089	0.239	0.009	0.039	0.0015	0.025	0.016	0.147
3000	0.080	0.243	0.007	0.040	0.0004	0.023	0.011	0.152
4000	0.074	0.242	0.005	0.038	0.0002	0.021	0.007	0.153

the individuals on those well-learned data by the ensemble rather than the whole training data. When learning on those well-learned data points was reversed by difference learning, learning on other not-well-learned data could actually gain more. Therefore, NCLDL(0.01) is able to learn the whole training data faster and better. On the testing data, NCLDL(0.01) had the similar results to those of NCL except for the better results on the heart data by NCLDL(0.01).

Table 3. Average error rates of the ensembles by NCLDL(0.1).

No. of epochs	Diabetes		Cancer		Heart		Card	
	Training	Testing	Training	Testing	Training	Testing	Training	Testing
50	0.218	0.238	0.023	0.031	0.096	0.147	0.112	0.139
1000	0.141	0.238	0.020	0.031	0.034	0.079	0.060	0.140
2000	0.130	0.239	0.019	0.032	0.030	0.068	0.053	0.141
3000	0.124	0.240	0.019	0.032	0.028	0.066	0.049	0.139
4000	0.121	0.240	0.019	0.033	0.027	0.063	0.047	0.139

Tables 3 and 4 described both the training and the testing error rates of the learned ensembles by NCLDL(0.1) and NCLDL(1). Compared to NCLDL(0.01), difference learning on the well-learned data was strengthened in NCLDL(0.1) and NCLDL(1). It was noticed that the strengthened difference learning could effectively stop learning too well on the whole training data. Even after 4000 learning epochs, NCLDL(0.1) and NCLDL(1) still learned less on the training data than what NCLDL(0.01) had learned after 1000 learning epochs. However, the strengthened NCLDL only showed the poorer performance on the heart data, but achieved the lower testing error rates on the other three data sets.

Table 4. Average error rates of the ensembles by NCLDL(1).

No. of epochs	Diabetes		Cancer		Heart		Card	
	Training	Testing	Training	Testing	Training	Testing	Training	Testing
50	0.224	0.238	0.022	0.029	0.103	0.147	0.119	0.143
1000	0.174	0.242	0.021	0.031	0.065	0.104	0.080	0.142
2000	0.166	0.244	0.020	0.029	0.062	0.099	0.078	0.139
3000	0.163	0.243	0.020	0.029	0.061	0.099	0.076	0.140
4000	0.161	0.244	0.020	0.029	0.060	0.098	0.075	0.141

3.3   Results of NCLDL at the Individual Level

Table 5 described the average results of both the training and the testing error rates among the learned individuals in the ensembles by NCL. Obviously, the performance of the learned individuals is worse than that of the learned ensembles on both the training and testing sets. On the training sets, the learned individuals became weak only on the diabetes data. The learned individuals were actually a little strong, whose training error rates were lower than 0.05 on the other three data sets at learning epoch 4000. It is interesting that the learned individuals by NCL showed a little weak on the testing sets of both the diabetes and card data. It suggests that the correlation penalty introduced in NCL is indeed helpful in creating weak learners especially when there are noises in the data.

Table 5. Average error rates of the individuals in the ensembles by NCL.

No. of epochs	Diabetes		Cancer		Heart		Card	
	Training	Testing	Training	Testing	Training	Testing	Training	Testing
50	0.274	0.296	0.032	0.042	0.127	0.173	0.148	0.176
1000	0.207	0.296	0.023	0.047	0.051	0.093	0.070	0.181
2000	0.191	0.294	0.020	0.049	0.038	0.078	0.055	0.189
3000	0.185	0.293	0.018	0.050	0.030	0.070	0.047	0.191
4000	0.181	0.295	0.017	0.051	0.026	0.065	0.043	0.195

As shown in Table 6, NCLDL(0.01) was able to make the individuals in the ensembles be weaker on both the training and the testing sets on all the

Table 6. Average error rates of the individuals in the ensembles by NCLDL(0.01).

No. of epochs	Diabetes		Cancer		Heart		Card	
	Training	Testing	Training	Testing	Training	Testing	Training	Testing
50	0.341	0.354	0.237	0.243	0.235	0.267	0.286	0.303
1000	0.290	0.360	0.223	0.240	0.197	0.218	0.211	0.291
2000	0.279	0.363	0.220	0.241	0.194	0.211	0.201	0.296
3000	0.272	0.364	0.217	0.240	0.192	0.207	0.196	0.296
4000	0.268	0.364	0.217	0.240	0.192	0.207	0.194	0.298

Table 7. Average error rates of the individuals in the ensembles by NCLDL(0.1).

No. of epochs	Diabetes		Cancer		Heart		Card	
	Training	Testing	Training	Testing	Training	Testing	Training	Testing
50	0.316	0.331	0.223	0.226	0.296	0.319	0.299	0.312
1000	0.277	0.330	0.221	0.228	0.248	0.269	0.266	0.308
2000	0.272	0.334	0.220	0.228	0.245	0.264	0.261	0.309
3000	0.271	0.336	0.220	0.228	0.242	0.262	0.258	0.308
4000	0.271	0.337	0.220	0.228	0.240	0.259	0.256	0.309

Table 8. Average error rates of the individuals in the ensembles by NCLDL(1).

No. of epochs	Diabetes		Cancer		Heart		Card	
	Training	Testing	Training	Testing	Training	Testing	Training	Testing
50	0.331	0.339	0.217	0.221	0.276	0.296	0.292	0.301
1000	0.291	0.328	0.215	0.220	0.253	0.276	0.274	0.297
2000	0.279	0.322	0.215	0.220	0.250	0.272	0.272	0.297
3000	0.273	0.317	0.215	0.220	0.249	0.271	0.270	0.296
4000	0.272	0.318	0.215	0.220	0.247	0.270	0.270	0.296

four data sets. When the amounts of the correlation penalty were increased in NCLDL(0.1) and NCLDL(1), the results presented in Tables 7 and 8 indicated that the learned individuals turned to be even weaker on the training sets, and kept their weakness on the testing sets.

3.4 Similarities Among the Ensembles by NCLDL

Table 9 provided the average results of overlapping rates between every pair among all the individuals in the ensembles by NCL at different training epochs. It suggests that NCL was good at generating different individuals in the ensembles when there are noises in the given data sets such as the diabetes data and the card data. The similarities among the individuals are near to 0.815 on the card data, while they decreased to around 0.726 on the diabetes data. In comparisons, the similarities among the individuals became over 0.9 after learning on both the

Table 9. Average cover rates between each pair of individuals in the ensembles by NCL.

No. of epochs	Diabetes		Cancer		Heart		Card	
	Training	Testing	Training	Testing	Training	Testing	Training	Testing
50	0.726	0.724	0.973	0.973	0.851	0.825	0.838	0.832
1000	0.742	0.719	0.974	0.962	0.922	0.885	0.902	0.835
2000	0.758	0.725	0.976	0.958	0.937	0.900	0.920	0.824
3000	0.764	0.726	0.977	0.957	0.948	0.910	0.928	0.817
4000	0.768	0.726	0.978	0.955	0.954	0.916	0.933	0.815

Table 10. Average cover rates between each pair of individuals in the ensembles by NCLDL(0.01).

No. of epochs	Diabetes		Cancer		Heart		Card	
	Training	Testing	Training	Testing	Training	Testing	Training	Testing
50	0.609	0.609	0.617	0.617	0.661	0.657	0.597	0.594
1000	0.599	0.586	0.627	0.623	0.655	0.647	0.650	0.622
2000	0.600	0.579	0.629	0.623	0.657	0.652	0.656	0.619
3000	0.604	0.581	0.629	0.623	0.659	0.655	0.660	0.621
4000	0.606	0.582	0.630	0.623	0.659	0.655	0.661	0.621

Table 11. Average cover rates between each pair of individuals in the ensembles by NCLDL(0.1).

No. of epochs	Diabetes		Cancer		Heart		Card	
	Training	Testing	Training	Testing	Training	Testing	Training	Testing
50	0.651	0.650	0.638	0.639	0.583	0.582	0.585	0.583
1000	0.663	0.655	0.637	0.637	0.607	0.603	0.598	0.590
2000	0.661	0.650	0.637	0.637	0.609	0.606	0.600	0.590
3000	0.656	0.645	0.637	0.638	0.611	0.607	0.602	0.590
4000	0.653	0.641	0.637	0.638	0.612	0.608	0.603	0.590

Table 12. Average cover rates between each pair of individuals in the ensembles by NCLDL(1).

No. of epochs	Diabetes		Cancer		Heart		Card	
	Training	Testing	Training	Testing	Training	Testing	Training	Testing
50	0.659	0.660	0.648	0.649	0.623	0.621	0.602	0.602
1000	0.679	0.679	0.649	0.650	0.628	0.626	0.608	0.609
2000	0.698	0.699	0.649	0.650	0.629	0.625	0.610	0.610
3000	0.709	0.709	0.649	0.650	0.629	0.626	0.612	0.612
4000	0.709	0.708	0.650	0.650	0.630	0.626	0.612	0.612

cancer data and the heart data. In comparisons, NCLDL(0.01) was capable of creating very different individuals in an ensemble. As shown in Table 10, the similarities among the individuals fell to around 0.65 or less on all the four data sets. The similarities among the individuals by NCLDL(0.1) and NCLDL(1) given in Tables 11 and 12 decreased further to around 0.6 on the heart data and the card data. However, it was surprised that neither NCLDL(0.1) nor NCLDL(1) had generated the more different individuals on the diabetes data and the cancer data.

4 Conclusions

With difference learning, NCLDL could not only learn well in classifying all the given data points, but also predict better the unknown data points. From the approximation point of view, the outputs of the ensembles by NCLDL were pushed away from the target values when they are too close each other. Therefore, there would be no risks of growing the complexity even in the longer learning such as 4000 learning epochs set in the experiments. After the difference learning in NCLDL, the individual learners in the ensembles became the weaker and more different each other than those in the ensembles by NCL.

Such complexity control is different to the early stopping approach. Early stopping approach prevents the learned models from having the higher complexity by observing the performance of the learned models on a separate validation set. When there are only a limited number of data points, using a separate validation set would lead to even less data points that could be used in learning. The learned models obtained from the less data might not perform well in the real applications when the number of the available data is rather small. Early stopping approach might also stop the learning too early to train the models well. Different to early stopping approach, NCLDL could use all the given data, and train the neural network ensembles safely in the long learning.

References

1. Fahlman, S.E., Lebiere, C.: The cascade-correlation learning architecture. In: Touretzky, D.S. (ed.) Advances in Neural Information Processing Systems 2, pp. 524–532. Morgan Kaufmann, San Mateo, CA (1990)
2. Śmieja, F.J.: Neural network constructive algorithms: trading generalization for learning efficiency? Circuits Syst. Sig. Process. 12(2), 331–374 (1993)
3. Kwok, T.-Y., Yeung, D.-Y.: Constructive algorithms for structure learning in feed-forward neural networks for regression problems. IEEE Trans. Neural Netw. 8(3), 630–645 (1997)
4. Mozer, M.C., Smolensky, P.: Skeletonization: a technique for trimming the fat from a network via relevance assessment. Connect. Sci. 1, 3–26 (1989)
5. Sietsma, J., Dow, R.J.F.: Creating artificial neural networks that generalize. Neural Netw. 4, 67–79 (1991)
6. LeCun, Y., Denker, J.S., Solla, S.A.: Optimal brain damage. In: Touretzky, D.S. (ed.) Advances in Neural Information Processing Systems 2, pp. 598–605. Morgan Kaufmann, San Mateo, CA (1990)

7. Hassibi, B., Stork, D.G.: Second derivatives for network pruning: optimal brain surgeon. In: Hanson, S.J., Cowan, J.D., Giles, C.L. (eds.) Advances in Neural Information Processing Systems 5, pp. 164–171. Morgan Kaufmann, San Mateo, CA (1993)
8. Tolstrup, N.: Pruning of a large network by optimal brain damage and surgeon: an example from biological sequence analysis. Int. J. Neural Syst. $6(1)$, 31–42 (1995)
9. Schapire, R.E.: The strength of weak learnability. Mach. Learn. 5, 197–227 (1990)
10. Liu, Y., Yao, X.: Simultaneous training of negatively correlated neural networks in an ensemble. IEEE Trans. Syst. Man Cybern. Part B Cybern. $29(6)$, 716–725 (1999)
11. Liu, Y.: A balanced ensemble learning with adaptive error functions. In: Kang, L., Cai, Z., Yan, X., Liu, Y. (eds.) ISICA 2008. LNCS, vol. 5370, pp. 1–8. Springer, Heidelberg (2008)
12. Liu, Y.: Balanced learning for ensembles with small neural networks. In: Cai, Z., Li, Z., Kang, Z., Liu, Y. (eds.) ISICA 2009. LNCS, vol. 5821, pp. 163–170. Springer, Heidelberg (2009)
13. Liu, Y.: Create weak learners with small neural networks by balanced ensemble learning. In: Proceedings of the 2011 IEEE International Conference on Signal Processing, Communications and Computing (2011)
14. Liu, Y.: Target shift awareness in balanced ensemble learning. In: Proceedings of the 3rd International Conference on Awareness Science and Technology
15. Liu, Y.: Balancing ensemble learning through error shift. In: Proceedings of the Fourth International Workshop on Advanced Computational Intelligence

SURF Feature Description of Color Image Based on Gaussian Model

Wen Sun, Qian Shen, and Chanjuan Liu[✉]

School of Information and Electrical Engineering,
Ludong University, Yantai 264025, China
luckycj80@sina.com

Abstract. To feature points description of color image, the fact that image color information has some effects on features of image is taken into consideration in this paper. And there is a novel method of SURF (speed up robust features) feature description of color image based on Gaussian color invariance model presented in this paper. During the stage of image feature description, the three kinds of color information in original color image are expressed by three components of Gaussian color invariance model respectively. Then, the matrix consisting of color invariants which are presented by Gaussian color invariance model represents original color image. Hereafter, the method of SURF feature description is used for describing the distribution of feature points. Finally, through our experiences, the correct matching ratio of feature point pairs of our method is higher than some typical algorithms represented in resent years when the image appears affine and blurring transformation.

Keywords: Color image · Feature description · Gaussian color invariance model · SURF algorithm

1 Introduction

Image visual feature as the low-level feature in image processing has already been applied in many fields widely, such as, target recognition, image retrieval, target tracking, image matching and image classification. The study of local image visual feature is mainly divided into three parts: feature extraction, feature description and image classification. Many research methods have been presented to the three parts. Li et al. [1] proposed a survey of recent progress and advances in visual feature detection. For blob feature detection, Lowe [2] presented SIFT (scale invariant feature transform) algorithm in 2004. Nevertheless, the dimension of feature description vector in SIFT algorithm is too high to computation in practice application. Hereafter, integral image and box filter were used by Bay in his proposed SURF (speed up robust features) algorithm [3] which greatly enhanced the accuracy of feature detection. Meanwhile, it also made the feature description dimension decline to 64 which obviously reduced the matching time. To increase the robustness of feature matching, Alcantarilla et al. [4] used gauge derivatives to build SURF descriptor and to calculate a single dominant orientation. Using SURF feature detection algorithm, Chen [5] proposed an image matching technique based on invariant methods. Bo [6] used SURF to achieve image

© Springer Science+Business Media Singapore 2016
K. Li et al. (Eds.): ISICA 2015, CCIS 575, pp. 275–283, 2016.
DOI: 10.1007/978-981-10-0356-1_28

registration. Among all the above approaches, image local feature is extracted from gray image, although which can optimize the performance of SURF algorithm to a certain extent. However, the corresponding approach of local feature description of color image is not given out.

To this problem, SIFT algorithm is expanded to color space domain in Refs. [7, 9]. They applied Gaussian color invariant model [8] to color image, and then describe image SIFT feature. In this paper, we will introduce the color invariance proposed in Gaussian color invariant model in Ref. [8], firstly, color information of three channels in RGB color space is expressed by three components of Gaussian color invariant model respectively; secondly, color invariance produced in Gaussian color invariant model as the input represents image pixel information; finally, the distribution of feature is described with the theory of SURF algorithm in the part of the construction of feature description vector. For color images with affine transformation and blurring change from standard Mikolajczik data set and universal two groups color images with affine transformation, compared with SURF algorithm and Ref. [9], the correct matching ratio of feature point pairs in our approach is higher.

In this paper, the introduction summarizes the researches of recent years briefly in Sect. 1. In Sect. 2, we will introduce the related theories of Gaussian color invariant model. In the third section, SURF feature description algorithm is introduced firstly. Then, the construction method of feature description vector in color image is recommended. In Sect. 4, the given experimental data verifies the effectiveness of our algorithm. We will conclude this paper in the last section.

2 Gaussian Color Invariant Model

Gaussian color invariant model [6] depends on the old Kubelka-Munk theory which is proposed by Geusebroek et al. in 2001. And the Kubelka-Munk theory models the photometric reflectance by:

$$E(\lambda, \vec{x}) = e(\lambda, \vec{x})\big(1 - \rho_f(\vec{x})\big)^2 R_\infty(\lambda, \vec{x}) + e(\lambda, \vec{x})\rho_f(\vec{x}) \tag{1}$$

where λ is the wavelength and $\vec{x}$ is a 2D vector of image position. $e(\lambda, \vec{x})$ denotes the illumination spectrum and $\rho_\lambda(\vec{x})$ is the Fresnel reflectance at $\vec{x}$. $R_\infty(\lambda, \vec{x})$ is the material reflectivity. $E(\lambda, \vec{x})$ represents the reflected spectrum in the view direction. This model is suitable for modeling non-transparent materials.

In most practical applications, the spectral components of source are constantly over wavelengths and variably over the position. Hence, a spatial component $i(\vec{x})$ which denotes intensity variations replaces $e(\lambda, \vec{x})$. And Eq. (1) will be:

$$E(\lambda, \vec{x}) = i(\vec{x})\Big[\rho_f(\vec{x}) + \big(1 - \rho_f(\vec{x})\big)^2 R_\infty(\lambda, \vec{x})\Big] \tag{2}$$

By differentiating Eq. (2) with respect to λ, we get:

$$E_\lambda = i(\vec{x})(1 - \rho_\lambda(\vec{x}))^2 \frac{\partial R_\infty(\lambda, \vec{x})}{\partial \lambda} \tag{3}$$

$$E_{\lambda\lambda} = i(\vec{x})(1 - \rho_f(\vec{x}))^2 \frac{\partial^2 R_\infty(\lambda, \vec{x})}{\partial \lambda^2} \tag{4}$$

Through Eqs. (3) and (4), we get:

$$H = \left(\frac{E_\lambda}{E_{\lambda\lambda}}\right) = \frac{\partial R_\infty(\lambda, \vec{x})}{\partial \lambda} \bigg/ \frac{\partial^2 R_\infty(\lambda, \vec{x})}{\partial \lambda^2}$$
$$= f(R_\infty(\lambda, \vec{x})) \tag{5}$$

Thus, $H = (E_\lambda/E_{\lambda\lambda})$ is the expression of color invariance, which is independent of viewpoint, illumination direction, illumination intensity and Fresnel reflectance coefficient.

The transformation relationship between Gaussian color model and RGB color model as follows:

$$\begin{bmatrix} \widehat{E} \\ \widehat{E}_\lambda \\ \widehat{E}_{\lambda\lambda} \end{bmatrix} = \begin{pmatrix} 0.06 & 0.63 & 0.27 \\ 0.3 & 0.04 & -0.35 \\ 0.34 & -0.6 & 0.17 \end{pmatrix} \begin{bmatrix} R \\ G \\ B \end{bmatrix} \tag{6}$$

The three components $\left(\widehat{E}, \widehat{E}_\lambda, \widehat{E}_{\lambda\lambda}\right)$ of Gaussian color model are substituted by the three components $(E, E_\lambda, E_{\lambda\lambda})$ of Eq. (5) at a given scale σ_x approximatively [6].

3 Feature Description of Color Image Based on SURF Algorithm

3.1 SURF Feature Descriptor

Orientation Assignment of Feature Point. In order to obtain the gradient information of each pixel in a circle region which is with feature point as a center and with 6 s as a radius (s is the scale of the feature point), the convolution operation is executed between image patch and haar wavelet operator. By using integral image, the calculation efficiency increases largely when calculating haar wavelet responses within image patch.

The calculated haar wavelet response of each feature point is weighted with a Gaussian ($\sigma=2s$). In current image region, designing a fan-shaped sliding window with centered feature point and interior angle $\pi/3$, the dominant orientation of this feature point is calculated.

Through rotating the sliding window and summing the haar wavelet response dx and dy at 0.2 radian step, the vector (m_w, θ_w) is calculated within current sliding window.

$$m_w = \sum_w dx + \sum_w dy \qquad (7)$$

$$\theta_w = \arctan\left(\sum_w dx \Big/ \sum_w dy\right) \qquad (8)$$

The orientation corresponding to longest vector is the dominant orientation of current feature point when accumulated value m_w of haar wavelet response is largest in current sliding window.

$$\theta = \theta_w | \max\{m_w\} \qquad (9)$$

Construction of Feature Description Vector for Feature Point. With the difference to the calculation of dominant orientation, it is a square region for obtaining description vector of feature point by calculating haar wavelet response of feature point in this image region, which is a $20s \times 20s$ region along the dominant orientation and around the feature point. To obtain the local spatial information, the square region is divided into 4×4 sub-blocks. The response value is calculated with haar wavelet template of size $2s$ in each sub-block. Figure 1 shows the division of image feature description region.

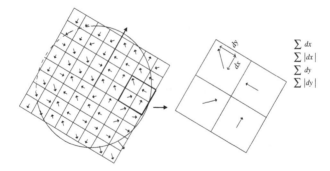

Fig. 1. The diagram of feature description vector

For $5s \times 5s$ pixel points in each divided sub-block, the responses dx along the dominant orientation and dy be perpendicular to the dominant orientation are calculated by using haar wavelet. And then, the vector of each sub-block is received by summing the weighted haar wavelet response ($\sigma = 3.3s$) dx and dy.

$$V_{Sub-block} = \left[\sum dx, \sum |dx|, \sum dy, \sum |dy|\right] \qquad (10)$$

For each feature point, the dimension of feature description is $4 \times 4 \times 4 = 64$ because the feature point region is divided into 4×4 sub-blocks.

3.2   SURF Feature Description Algorithm Based on Gaussian Color Model

When transplanting the theory of Ref. [9] into SURF algorithm directly, in image building (1) and image building (2), the number of detected feature points is 55 and 50 respectively. With the closest method, the number of correct matching feature point pairs is zero.

We only apply the Gaussian color invariance H in Ref. [9] to the part of image feature point description. The concrete procedure of our algorithm is presented in the following part:

1. According to Eq. (6), the RGB color model is transformed into Gaussian color invariant model, and the three color components R, G and B are represented by $\hat{E}$, $\hat{E}_\lambda$ and $\hat{E}_{\lambda\lambda}$ respectively;
2. For each pixel in color image, the color invariance in Eq. (5) expresses the corresponding pixel point. Then the color image changes to a new image matrix consisting of color invariance;
3. Using the new image matrix obtained in step (2) and the feature description in SURF algorithm, feature description vector is constructed;
4. Finally, feature points between two images are matched according to the theory in Ref. [10].

It not only keeps the number of detected feature points, but also describes and matches the feature points in greater range when we only apply the Gaussian color invariant model to the feature description part.

4   Results and Analysis

Compared with SURF algorithm, Ref. [9] and our algorithm, the number of corresponding feature point pairs and the ratio of correct matching feature point pairs in these approaches are contrasted for verifying the validity of our algorithm. The experimental data comes from part of the color image in standard Mikolajczik database which includes blurring and affine transformation color image, and two other universal groups of affine transformation color image. All of the experimental images are shown in the following part (Figs. 2, 3 and 4).

Using the feature point matching approach in Ref. [10], the experimental data of matching results among SURF algorithm, Ref. [9] and our algorithm is shown from Tables 1, 2, 3, 4, 5, 6, 7, 8, 9 and 10.

From Tables 1 and 5, the correct matching ratio of feature points in our algorithm is higher than SURF algorithm and Ref. [9] under blurring transformation of image. The constructed feature descriptor in this paper is not sensitive to image color variations because color image is expressed by color invariance. Meanwhile, this color invariance is independent of illumination direction and illumination intensity. Thus, compared to the grayscale of SURF algorithm, the performance of our algorithm is superior.

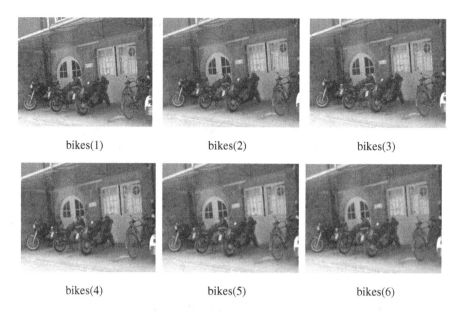

bikes(1)   bikes(2)   bikes(3)

bikes(4)   bikes(5)   bikes(6)

Fig. 2. Blurring transformation images from image bikes (1) to image bikes (6)

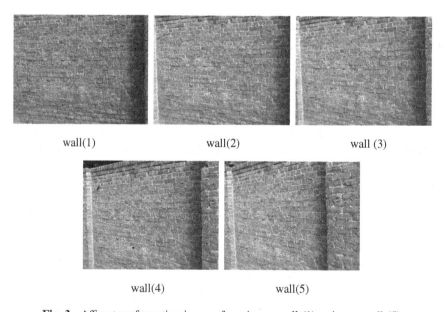

wall(1)   wall(2)   wall (3)

wall(4)   wall(5)

Fig. 3. Affine transformation images from image wall (1) to image wall (5)

From Tables 6, 7, 8, 9 and 10, meanwhile, for affine transformation of image, the correct matching of feature points in our approach is superior to the other two algorithms. Because of the independence to viewpoint of color invariance, the performance

building (1)   building (2)

Fig. 4. Image building (1) and image building (2) under affine transformation

Table 1. The comparing results among SURF algorithm, Ref. [9] and our algorithm between image bikes (1) and image bikes (2)

Algorithm	SURF algorithm	Ref. [9]	Our algorithm
The number of corresponding feature point pairs (right/all)	477/478	104/110	293/295
The correct matching ratio	99.79 %	94.54 %	99.32 %

Table 2. The comparing results among SURF algorithm, Ref. [9] and our algorithm between image bikes (1) and image bikes (3)

Algorithm	SURF algorithm	Ref. [9]	Our algorithm
The number of corresponding feature point pairs (right/all)	361/369	84/104	245/248
The correct matching ratio	97.83 %	80.77 %	98.79 %

Table 3. The comparing results among SURF algorithm, Ref. [9] and our algorithm between image bikes (1) and image bikes (4)

Algorithm	SURF algorithm	Ref. [9]	Our algorithm
The number of corresponding feature point pairs (right/all)	199/202	67/102	145/147
The correct matching ratio	98.51 %	65.69 %	98.64 %

Table 4. The comparing results among SURF algorithm, Ref. [9] and our algorithm between image bikes (1) and image bikes (5)

Algorithm	SURF algorithm	Ref. [9]	Our algorithm
The number of corresponding feature point pairs (right/all)	140/147	45/100	83/84
The correct matching ratio	95.24 %	45 %	98.81 %

Table 5. The comparing results among SURF algorithm, Ref. [9] and our algorithm between image bikes (1) and image bikes (6)

Algorithm	SURF algorithm	Ref. [9]	Our algorithm
The number of corresponding feature point pairs (right/all)	95/100	30/98	58/60
The correct matching ratio	95 %	30.61 %	96.67 %

Table 6. The comparing results among SURF algorithm, Ref. [9] and our algorithm between image wall (1) and image wall (2)

Algorithm	SURF algorithm	Ref. [9]	Our algorithm
The number of corresponding feature point pairs (right/all)	993/1002	162/163	572/573
The correct matching ratio	99.10 %	99.38 %	99.83 %

Table 7. The comparing results among SURF algorithm, Ref. [9] and our algorithm between image wall (1) and image wall (3)

Algorithm	SURF algorithm	Ref. [9]	Our algorithm
The number of corresponding feature point pairs (right/all)	686/698	82/94	392/394
The correct matching ratio	98.28 %	87.23 %	99.49 %

Table 8. The comparing results among SURF algorithm, Ref. [9] and our algorithm between image wall (1) and image wall (4)

Algorithm	SURF algorithm	Ref. [9]	Our algorithm
The number of corresponding feature point pairs (right/all)	332/353	51/67	204/214
The correct matching ratio	94.05 %	76.11 %	95.32 %

Table 9. The comparing results among SURF algorithm, Ref. [9] and our algorithm between image wall (1) and image wall (5)

Algorithm	SURF algorithm	Ref. [9]	Our algorithm
The number of corresponding feature point pairs (right/all)	88/118	24/42	61/68
The correct matching ratio	74.57 %	57.14 %	89.71 %

of feature descriptor in our algorithm is better than SURF algorithm, which keeps rotation invariance of descriptor through confirming the dominant orientation.

Table 10. The comparing results among SURF algorithm, Ref. [9] and our algorithm between image building (1) and image building (2)

Algorithm	SURF algorithm	Ref. [9]	Our algorithm
The number of corresponding feature point pairs (right/all)	71/83	0/0	32/36
The correct matching ratio	82.54 %	0 %	88.89 %

5 Conclusion

Focusing on color image, a novel feature description algorithm based on Gaussian color invariant model is proposed in this paper. After detecting SURF feature point in grayscale, the pixel information of color image is replaced by the color invariance presented in Gaussian model. And then, the feature description vector is constructed by using the theory of SURF feature description. Finally, through the experiments, the performance of our algorithm is superior to SURF algorithm and Ref. [9] when image under affine and blurring transformation.

Acknowledgements. This work is supported by National Science Foundation of China (No. 61170161, No. 61300155, No. 61503219), Outstanding Young Scientists Foundations Grant of Shandong Province (No. BS2014DX016), Ph.D. Programs Foundation of Ludong University (No. LY2014033, LY2015033). This work is partially supported by NSFC Grants (61503219).

References

1. Li, Y., Wang, S., Tian, Q., Ding, X.: A survey of recent advances in visual feature detection. Neurocomputing **149**, 736–751 (2015)
2. Lowe, D.G.: Distinctive image features from scale-Invariant keypoints. Int. J. Comput. Vis. **2**, 60 (2004)
3. Bay, H.: Speeded-up robust features (SURF). Comput. Vis. Image Underst. **3**, 110 (2008)
4. Alcantarilla, P.F., Bergasa, L.M., Davison, A.J.: Gauge-SURF descriptors. Image Vis. Comput. **1**, 31 (2013)
5. Chen, H.: Panoramic image mosaic based on SURF feature matching algorithm (2010)
6. Bo, K.: Research on image registration and mosaic based on SURF (2009)
7. Cui, Q.: Research on digital watermarking based on local features of color image (2013)
8. Geusebroek, J.M., Van, D.B.R., Smeulders, A.W.M.: Color invariance. IEEE Trans. Pattern Anal. Mach. Intell. **12**, 23 (2001)
9. Abdel-Hakim, A.E., Farag, A.: CSIFT: A SIFT descriptor with color invariant characteristics. In: 2006 IEEE Computer Society Conference on Computer Vision and Pattern Recognition, pp. 17–22. IEEE, New York (2006)
10. Mikolajczik, K., Schmid, C.: A performance evaluation of local descriptors. IEEE Trans. Pattern Anal. Mach. Intell. **10**, 27 (2005)

Data Mining and Cloud Computing

An Improved Adaptive Hexagon and Small Diamond Search

Fu Mo[✉] and Kangshun Li

Department of Computer Science, Guangdong University of Science and Technology,
Dongguan, China
zdb317710990@sohu.com

Abstract. Motion estimation have a significant impact in H.265 video coding systems because it occupies a large amount of time in encoding. So the quality of motion search algorithm affect the entire encoding efficiency directly. In this paper, a novel search algorithm which utilizes an adaptive hexagon and small diamond search is proposed to overcome the drawbacks of the traditional block matching algorithm implemented in the most current video coding standards. The adaptive search pattern is chosen according to the motion strength of the current block. When the block is in active motion, the hexagon search provides an efficient search means; when the block is inactive, the small diamond search is adopted. Simulation results showed that our approach can speed up the search process with little effect on distortion performance compared with other adaptive approaches.

Keywords: Adaptive hexagon search · Diamond search · Motion estimation · Motion strength

1 Introduction

Now, H.265 is a popular video codec standard, motion estimation is the key technology in video coding [1]. Motion estimation directly affects the coding efficiency and the quality of image restoration. For the video image sequence, as between adjacent frame macroblock significant time correlation, by reducing the time redundancy, can improve the video coding efficiency. Motion estimation based on block matching algorithm is an effective method, has been adopted by many video coding standards. Motion estimation occupies whole encoder 60 %–80 % [2] computation. Largely determines the encoder speed quality and compression efficiency encoding [3–5].

Many fast block matching methods have been proposed to speed up the search process [6–8]. The classical methods include new three-step search (NTSS), four-step search (FSS), hexagon based search, diamond based search. Ghanbari presented the cross-search algorithm for motion estimation. This method experiences a logarithmic step, and checks four points in each step. Cheung and Po proposed a new search algorithm combining cross and diamond search together [9], which uses a cross search pattern as the initial step; then a large or small diamond search pattern was chosen as the subsequent step. Banh and Tan designed an adaptive dual-cross search method composed of three steps. This algorithm searches for three motion vectors, and chooses

© Springer Science+Business Media Singapore 2016
K. Li et al. (Eds.): ISICA 2015, CCIS 575, pp. 287–293, 2016.
DOI: 10.1007/978-981-10-0356-1_29

the one with the least sum of absolute difference (SAD) as the initial search center. By comparing the SAD of the initial center with a pre-determined threshold, an early search termination strategy is used to reduce the computational cost. Finally, the global minimum is located through the dual-cross search scheme. The technique of block matching motion estimation has developed rapidly in recent years. The predictive motion vector field adaptive search technique (PMVFAST) was advanced by Tourapis et al. [11]; it uses several predictive motion vectors as the initial candidate vectors, and selects the best predictor to continue the diamond search. This method enhances the search speed greatly. Wong et al. [10] proposed an enhanced-PMVFAST (EPMVFAST) algorithm to further raise the speed. Unsymmetrical-cross multi-hexagon-grid search (UMHexagonS) [9] is a hierarchical approach consisting of four main steps, and most importantly, it has little loss in rate distortion performance compared to the full search method.

In this paper, we present a novel search algorithm, adaptive hexagon and small diamond search (AHSDS).

2 Early Termination Strategy

In H.265, the performance of the candidate search point is often evaluated by rate-distortion (RD) cost [10], which is composed of two parts:

$$J = SAD + \lambda_{motion}R \tag{1}$$

where SAD denotes the distortion measured as the sum of absolute differences between the original and referenced blocks, λ_{motion} is the Lagrange multiplier, and R represents the bits used to encode the motion information.

Assuming that the current block size is $M \times N$, SAD is the current block and reference block rate distortion difference, while the mean absolute error (Mean Absolute Difference) is the average difference of each pixel between the two blocks:

$$MAD\left(mv_x, mv_y\right) = \frac{1}{M \times N} \sum_{m=0}^{M-1} \sum_{n=0}^{N-1} \left|F_t\left(x + m, y + n\right)\right.$$
$$\left. -F_{t-1}\left(x + m + mv_x, y + n + mv_y\right)\right| \tag{2}$$

where F_t and F_{t-1} represent the current and the previous frames respectively, and (mv_x, mv_y) is the current motion vector. The motion search begins at the starting point. It comes from the predictive motion vector of the current block, and can be denoted as

$$mv_{st} = median\left(fmv_l, fmv_t, fmv_{tr}\right) \tag{3}$$

where fmv_1, fmv_t, fmv_{tr} are the final motion vectors of the adjacent left, top, respectively. As the search proceeds, the current search point progressively approaches the final optimal point. cmv and fmv denote the motion vectors at the current search point and the final search point, respectively. In most cases, the predictive motion vector mv_{st} is close to the final search point fmv. For the Foreman sequence, the mv_{st} of about 58 % of blocks is exactly equal to the *fmv*. To reduce the search burden, the early termination strategy is applied when the MAD at the starting point is less than threshold Th_1.

3   The Choice of Search Pattern

In our algorithm, the motion strength mos is defined as follows:

$$mos = MAD(cmv) \tag{4}$$

The motion strength, a dynamic variant, does not represent the absolute magnitude of the motion vector *cmv*, but reflects the deviation of *cmv* from *fmv*. When the block is at the starting point, i.e., $cmv = mv_{st}$, the value of mos is relatively high; when the current search point is close to the final optimal point, mos decreases gradually.

The choice of the search pattern is decided by the comparison between the motion strength and the threshold. A fixed threshold is not suitable for all sequences. Too large or too small a threshold will unnecessarily increase search costs, and may even lead to an error in the final optimal point. An adaptive threshold is adopted in our approach to meet the changes of sequences with various motion characters and different video quality. Tourapis et al. [11] present several forms of adaptive threshold. In our algorithm, the threshold Th_k is computed as follows:

$$\begin{cases} Th_k = a_k \times CMAD + b_k, \\ CMAD = mean(MMAD_l, MMAD_t, MMAD_{tr}) \end{cases} \tag{5}$$

where $MMAD_l, MMAD_t, MMAD_{tr}$ are the minimum *MAD* of the left, top, and top-right blocks respectively, and *CMAD* is the mean value of $MMAD_l, MMAD_t$, and $MMAD_{tr}$. a_k, b_k are both constant, K is the number of search. To prevent the threshold Th_k from being too large or too small, we use some limiting parameters to restrict the threshold:

$$Th_k = \min\left(\max\left(a_k \times CMAD + b_k, T_{jk}\right), T_{mk}\right) \tag{6}$$

where T_{mk} and T_{jk} are the upper and lower bounds for threshold Th_k, respectively. T_{mk} and T_{jk} are set usually according to the mean distortion of the previous frame.

4   The Design of AHSDS

The proposed algorithm AHSDS can be described as follows:

Step 1: Determine the activity of the current block.

In our algorithm, we use the *MAD* of the starting search point to describe the motion activity of the current block according to:

$$Th_k = \min\left(\max\left(a_k \times CMAD + b_k, T_{jk}\right), T_{mk}\right) \tag{7}$$

where Th_2 is a threshold which can be adaptively set according to the minimum *MAD* of the adjacent blocks, and mv_{st} is the motion vector of the starting point. If Equation a is satisfied, the current block is active; otherwise, it is inactive.

Step 2: The choice of the search pattern.

If the *MAD* at the starting point is smaller than Th_1, the search terminates immediately. If the block is active (which means there is a high probability that the block has a large motion vector), start with the hexagon search, and go to Step 3; otherwise, go to Step 4.

Step 3: Perform the hexagon search.

Repeat the hexagon search until the best point is at the center of the hexagon. Select the sub-optimal point among the seven points in Fig. 1.

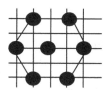

Fig. 1. Hexagon seven point distribution

If the center point A and the sub-optimal point B are in a horizontal direction, choose the newest and best point among point A and its three neighboring points as shown in Fig. 2(a). Similarly, Fig. 2(b) depicts the case where the points A and B are in a diagonal direction. Go to Step 5.

Step 4: Perform the small diamond search recursively until the best point with the minimum RD cost is at the center of the small diamond.

Step 5: Obtain the final optimal search point, and the whole search process then terminates.

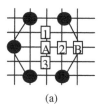 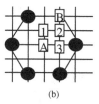

        (a)   (b)

Fig. 2. Different second optimal point of the three adjacent point position

In our algorithm, the motion vector at the starting search point comes from the predictive motion vector *pmv*, which is the median motion vector of three spatially adjacent blocks (the left, top, and top-right blocks). This simplifies the search procedure greatly. For a large majority of blocks, the final motion vector *fmv* is close to *pmv*, which means that small diamond is the main search pattern.

To analyze the similarity of motion vectors pmv and fmv, we calculate the probability P_n that the point (pmv_x, pmv_y) appeared in a square area, centered at (pmv_x, pmv_y), by

$$P_n = P\left\{ |fmv_x - pmv_x| \leq n, |fmv_y - pmv_y| \leq n \right\} \quad (8)$$

where pmv_x and pmv_y denote the x- and y-component of the motion vector fmv, respectively. P_n is worked out with a statistical method. For the Foreman sequence, when n equals 2, the corresponding probability is about 0.83. It does not mean that all these blocks have small final motion vectors. If the current block and the adjacent blocks all have similar large motion vectors, the small diamond search could still be used. Short-distance searches often occur whenever the final motion vector is close to the predictive motion vector, regardless of the absolute magnitude of the final motion vector. In contrast, long-distance searches usually appear in cases where the final motion vector of the current block is distinct from those of the adjacent blocks.

5   Implementation and Simulation

The proposed algorithm AHSDS was simulated on the platform H.265 reference software JM 14.2. We used different CIF video sequences, which contain various motion content, to test the performance of our algorithm. The search window was $[-16, 16]$. The number of reference frames was set to 5. The frame structure in our experiment was IPPPP..., and our proposed algorithm was compared with UMHexagonS (UMHS), simplified UMHexagonS (SUMH), and EPMVFAST in the same environment.

For different block matching algorithms, the number of search points (NSP) required for each block-matching was used to measure the computational efficiency. The average PSNR per frame is an important factor in evaluating the distortion performance. Different block matching approaches are compared in the following aspects.

A. *Search threshold selection*

Assume the threshold The can be denoted as

$$Th_1 = k_1 \times CMAD \quad (9)$$

In Fig. 3., when $|rmv|$ is equal to zero, the statistical ratio was near 1.3. We adopted the conservative threshold. The parameter k_1 was usually set to not greater than 1.0. The choice of search pattern is determined by Th_2. The small diamond search can achieve a smaller NSP, but it is not fit for long-distance searches because it might become trapped in a local optimal point and lead to the decrease of PSNR. In contrast, the hexagon search consumes more search points for short-distance searches.

A small diamond is composed of one center point and four edge points. When the minimum RD cost is located at one of the four edge points, the search will correctly detect the optimal point. When the distance between the real optimal point and the starting point is greater than 1, however, the search may not find the real optimal point successfully. Therefore, the hexagon search should be chosen when $|rmv|$ is greater than 1.0. The suggested range for k_2 is $[2.0, 2.6]$.

B. *Number of search points (NSP)*

With the increase of QP, the *PSNR* of the re-constructed frame dropped off progressively. Some details might be lost in the reference frame, affecting the number of search points. We tested each sequence on five different QP (24, 28, 32, 34, and 36), 100 frames for each QP, and then calculated the mean value of their NSPs and denoted it as an integer. The average NSPs for different video sequences and different methods are listed in Table 1.

Table 1. Three different QP of the average search points in different sequence and methods

Average search points / Sequence type	QP=24		QP=28		QP=36	
	UMHS	SUMH	UMHS	SUMH	EPMVFAST	AHSDS
Foreman	22	16	18	13	9	7
Coastguard	36	20	32	18	10	8
Flower	30	22	26	20	9	7
Mobile	34	25	30	19	8	6
Football	43	18	41	25	13	10
Silent	14	16	12	10	7	5

From Table 1, AHSDS generally had the lowest number of search points compared with other methods. For the different sequences, Silent had fewer average search points than Football, for example, because the latter had more blocks in active motion. Figure 4 plots the NSP curves with EMPMVFAST and AHSDS for the Foreman sequence when QP equals 28. We chose 100 frames for testing. The figure shows that the AHSDS method can speed up the process of block matching compared with EPMVAST. The main reason lies in the following two aspects:

- For EPMVFAST, several predictors are checked to obtain the optimal starting point, while in AHSDS, only the predictive motion vector is chosen as the starting point. On the other hand, in most cases, the final motion vector has a high level of similarity with the predictive one.

6   Conclusions

A new, fast approach AHSDS for block motion estimation is considered for the coding standard H.265. Using an initial classification on motion strength of the block according to its MAD, different search patterns were chosen to determine the optimal point with the minimum RD cost. When the block is in high activity, hexagon search is the preferred scheme. If the best point is located at the center of the hexagon, three points around the center are checked to obtain the optimal solution. Simulation was conducted with different methods and video sequences. Compared with EPMVFAST, AHSDS can enhance the search speed. Our approach is suitable for hardware realization due to its simplicity in algorithm complexity, and can be applied to real-time encoding systems owing to its small number of search points.

Acknowledgment. This project is supported by Science and Technological Program for Dongguan's Higher Education, Science and Research, and Health Care Institutions. Project number: 2012108102028.

References

1. Ugur, K., Alshin, A., Alshina, E., Bossen, F.: Motion compensated prediction and interpolation filter design in H.265/HEVC. IEEE J. Mag. 7(6), 946–956 (2013). Nokia Corp., Tampere, Finland
2. Zhu, S., Ma, K.K.: A new diamond search algorithm for fast block-matching motion estimation. IEEE Trans. Image Process. 9(2), 287–290 (2000)
3. Han, C.D., Liu, J.L., Xiang, Z.Y.: An adaptive fast search algorithm for block motion estimation in H.264. J. Zhejiang Univ. Sci. C 11(8), 637–644 (2010)
4. Sourabh, R., Neeta, T.: Hexagonal based search pattern for motion estimation in H.264/AVC. World Comput. Sci. Inf. Technol. J. 1(4), 162–166 (2011)
5. Yan, Y.X., Meng, S.L.: A new hybrid search scheme for video motion estimation. J. Convergence Inf. Technol. 6(3), 106–112 (2011)
6. Goel, S., Ismail, Y., Bayoumi, M.A.: Adaptive search window size algorithm for fast motion estimation in H.264/AVC standard. In: The 48th Midwest Symposium on Circuits and System, vol. 2, pp. 1557–1560 (2005)
7. Huang, Y.W., Hsieh, B.Y., Chien, S.Y., et al.: Analysis and complexity reduction of multiple reference frames motion estimation in H.264/AVC. IEEE Trans. Circuits Syst. Video Technol. 16(4), 507–522 (2006)
8. Kim, S.E., Han, J.K., Kim, J.G.: An efficient scheme for motion estimation using multireference frames in H.264/AVC. IEEE Trans. 8(3), 457–466 (2006)
9. Zhu, C., Lin, X., Chua, L.P.: Hexagon-eased search pattern for fast block motion estimation. IEEE Trans. Circ. Syst. Video Technol. 12(50), 349–355 (2002)
10. Wong, H.M., Au, O.C., Ho, C.W., et al.: Enhanced predictive motion vector field adaptive search technique (E-PMVFAST) -based on future mv prediction. In: Proceedings of IEEE International Conference on Multimedia and Expo, pp. 1–4 (2005)
11. Tourapis, A.M., Au, O.C., Liou, M.L.: Fast block-matching motion estimation using predictive motion vector field adaptive search technique (PMVFAST). In: ISO/IEC JTC1/SC29/WG11 MPEG99/m5866, Noordwijkerhout, The Netherland, March 2000
12. Cheung, C.H., Po, L.M.: Novel cross-diamond-hexagonal search algorithms for fast block motion estimation. IEEE Trans. Multimedia 7(1), 16–22 (2005)

A Research of Virtual Machine Resource Scheduling Strategy Based on Cloud Computing

Jun Nie[✉]

Department of Computer Science, Guangdong College of Science and Technology,
Dongguan 523083, China
13739149@qq.com

Abstract. For the load imbalance of resource scheduling, an algorithm based on improved genetic algorithm is proposed after the research of resource load scheduling model based on Cloud Computing. The algorithm designed the fitness function, which uses the spatial utilization rate, load changes and the weight, selected individual by the Roulette Wheel Method, and optimized the crossover and mutation operations. Experiment results demonstrate that the algorithm not only can accelerate convergence of load balance scheme, but also has less migration time. It provides a new solution for the research of load balance and virtual Machine Resource Scheduling Strategy.

Keywords: Cloud computing · Load balance · Scheduling strategy · Improved genetic algorithm

1 Introduction

As a new computing model, cloud computing is becoming more and more important. The virtual machine resource scheduling is one of the key technologies in cloud computing. The heterogeneity of the node resource processing capacity and the uncertainty of the application requirements will cause the load imbalance problem that some nodes have the low load and idle resources while some nodes do the opposite so that they cannot continue to accept the task. The load imbalance problem will greatly reduce the utilization and efficiency of whole resource. So, it becomes one of the key issues in the field of cloud computing to use a suitable resource load balance mechanism to coordinate the physical host load to improve the performance and utilization of the cloud system. Resource load balance mechanism is to allocate the resources such as disks and processes to obtain the best utilization rate. Load balance in cloud computing, which is based on the resource utilization and load balance to reduce the number of virtual machine migration, is to dispatch independent virtual machines to different physical hosts.

A lot of scholars have studied the problem of load balance in cloud computing. A hybrid genetic algorithm based on multi agent and genetic algorithm MAGA [1], which uses the binary encoding scheme, is proposed, but it doesn't clarify the corresponding relationship between virtual machines and hosts. A load balance algorithm based on genetic algorithm [2] is proposed, but the request of resource is coded with decimal

© Springer Science+Business Media Singapore 2016
K. Li et al. (Eds.): ISICA 2015, CCIS 575, pp. 294–304, 2016.
DOI: 10.1007/978-981-10-0356-1_30

which is complicated. The paper [3] proposes a load balance control strategy based on Berg model, and studies the equilibrium failure phenomenon and equilibrium recovery of model, but the literature does not implement the algorithm and only gives the idea of load regulation. On the basis of research of resource load scheduling model, this paper presents a virtual machine resource scheduling algorithm, which is based on genetic algorithm, for the problem of load imbalance of physical nodes. The algorithm designed the fitness function using the spatial utilization rate, load changes and the weight, selected individual by the Roulette Wheel Method, and optimized the crossover and mutation operations. Experimental results show that the proposed algorithm not only has the faster convergence speed, but also has the less time of virtual machine migration. It provides a new solution for the virtual machine resource scheduling strategy.

2 Scheduling Model and Related Definitions

The corresponding relationship between hosts and virtual machines in cloud computing is shown in Fig. 1.

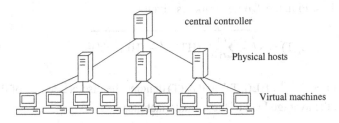

Fig. 1. The relationship between hosts and virtual machines

Several basic definitions are given below:

Definition 1: $H = \{H_1, H_2, ..., H_n\}$ is a collection of hosts in cloud data center. n represents amount of hosts. H_i represents the i'th host, where $1 \leq i \leq n$.

Definition 2: V_i is the set of the virtual machines on the host H_i, $V_i = \{V_i^1, V_i^2, ..., V_i^{mi}\}$. m_i represents amount of virtual machines on the i'th host. Let $M = m_1 + ... + m_i + ... + m_n$, which represents the number of available virtual machines. $V = \{V_1, V2, V3, ..., V_n\}$ represents the number of no deployed virtual machines.

Definition 3: Represents a set of mapping solutions for virtual machines represents a virtual machine mapping scheme on the i'th host.

The load of the i'th host equals to the sum of loads of all virtual machines which are running on the i'th host. If available resources of the host are D dimension vectors and each one dimension vector represents a kind of resource, then resources of each virtual machine are also D dimension vectors and its load equals to the sum of each one dimension value. Let T represents a load monitoring period, then it is divided into n time slots like $T = [(t_1 - t_0), (t_2 - t_1), ..., (t_n - t_{n-1})]$, where $t_k - t_{k-1}$ represents k'th time slot.

Definition 4: If the load of virtual machine is relatively stable in each time slot, let VT(i,k) denote the load of the i'th virtual machine in k'th time slot. The set, VT = {VT (1,k), VT (2,k), ... VT (i, k), ... , VT (M,k)}, represents the load of each virtual machine in the k'th time period. Then, the average load of the virtual machine Vi on the host Pi in the T period is:

$$\overline{VT_i(i, k)} = \frac{1}{T} \sum_{k=1}^{n} VT(i, k) \times (t_k - t_{k-1}) \tag{1}$$

The total load of the host Pi can be calculated by the average load of virtual machine, which is given by

$$HL(i, T) = \sum_{q=1}^{m_i} \overline{VT(q, T)} \tag{2}$$

When the virtual machine is dispatched to the host, the load changes. In this process, the load adjustment is required by virtual machine migration to achieve load balance. The standard deviation of load of the mapping scheme Si, which is that virtual machine set Vi to the host Pi in the T time period, is given by

$$\partial_i(T) = \sqrt{\frac{1}{n} \sum_{i=1}^{n} (HT(T)' - (HL(i,T)')^2} \tag{3}$$

where, $HL(T)' = \frac{1}{n} \sum_{i=1}^{n} HL(i, T)'$, and $\partial_i(T)$ denotes the discrete degree of the load of every host to the average load of the virtual machine.

Definition 5: If the current virtual machine provides a balanced solution Si to the mapping scheme Fi, then the mapping scheme S's collection corresponds to the equilibrium mapping scheme is S' = {S1', S2', S3', ..., Sn'}, F'(1 ≤ i ≤ n), which ensures that it is the optimal mapping scheme in the case that $\partial_i(S_i, T)$ satisfies the predefined load constraints of.

Definition 6: Let $\gamma(S_i)$ denote costs of load balance of virtual machine migration, which shows the number of the transitions from the mapping scheme Si to S'.

In order to make sure that the load index of the host with heterogeneous resources is comparable, the load index vector is normalized, and the host's memory, CPU and bandwidth are defined as follows:

Definition 7: Host Memory Load Indicator. Let the number of virtual machines on the host be denoted by n, the amount of memory occupied by the i'th virtual machine be denoted by VMemi, and the total memory of the host be denoted by Mem_{total}. The ratio between memory of all virtual machines and the total memory of host is given by

$$LI_{Men} = \frac{\sum_{i=1}^{n} VMem_i}{Mem_{total}} \times 100 \% \tag{4}$$

Definition 8: Host CPU Load Indicator. Let n denote the number of virtual machines on the host and C_i denote the utilization rate of the CPU of each virtual machine. Host CPU Load Indicator, LI_{cpu}, is given by

$$LI_{cpu} = \frac{\sum_{i=1}^{n} C_i}{CPU_{total}} \times 100\ \%$$

(5)

Definition 9: Bandwidth Load Indicator. Let n denote the number of virtual machines on the host, $VMbw_i$ denote the bandwidth used by virtual machine I and BW_{total} denote the total bandwidth of host. The ratio between the total bandwidth of each virtual machine on host and BW_{total} is given by

$$LI_{bw} = \frac{\sum_{i=1}^{n} VMbw_i}{BW_{total}} \times 100\ \%$$

(6)

Based on the above definition, the virtual machine load is

$$VT\,(i,k) = \varphi_1 LI_{CPU} + \varphi_2 LI_{Mem} + \varphi_3 LI_{bw}$$

(7)

where, φ_1, φ_2 and φ_3 are the load value of the resources, which satisfy $0 \leq \varphi_1, \varphi_2, \varphi_3 \leq 1$ and $\varphi_1 + \varphi_2 + \varphi_3 = 1$. The parameter φ varies from application to application.

The goal of this paper is to dispatch multiple virtual machines to physical hosts so that the host load is more balanced and the number of virtual machine migration is minimized. The form is given by

$$min\{f_1\,(x) = \sum_{i=1}^{m} \gamma\,(S_i), f_2\,(x) = \sum_{i=1}^{m} \partial_i(S_i, T)\}$$

(8)

Where, $f_1\,(x)$ represents the total number of virtual machine migration, and $f_2\,(x)$ represents the load variance.

When multiple virtual machines are dispatched to host K, the total amount of resources of the host K is not less than the total amount of resources used by the virtual machine. The constraints are followed by

$$\begin{cases} H_k^{Mem} \geq \sum_k r_i^{Mem} \cdot x_i^k \\ H_k^{CPU} \geq \sum_k r_i^{CPU} \cdot x_i^k \\ H_k^{bw} \geq \sum_k r_i^{bw} \cdot x_i^k \\ x_i^j \in \{0, 1\} \end{cases}$$

(9)

Here, H_k^{Mem}, H_k^{CPU} and H_k^{bw} represent the host K's memory, CPU and bandwidth, respectively. r_i^{Mem}, r_i^{CPU} and r_i^{bw} represent the virtual machine's memory, CPU and band width. If the virtual I is not dispatched to host J, then $x_i^j = 0$, otherwise it is 1.

3   Algorithm Implementation

In summary, the goal of this paper is to enable the system to achieve optimal load balance and the minimum number of virtual machine migration. Load balance of virtual machine scheduling belongs to the problem of the packing problem [4]. The traditional genetic algorithm is very poor to adaptability because of encoding problem [5]. In this paper, an improved genetic algorithm is proposed, which is based on the tree-shaped encoding and the bin packing decoding scheme. The fitness function counts on the space utilization, load variation and weight. The algorithm obtained by using roulette wheel to select individual and optimizing cross operation and mutation operation is effective and efficient.

3.1   Gene Encoding and Decoding

Taking into account that the relationship between the host and the virtual machine in the cloud computing is one to many mapping relations, Encoding for the virtual machine is based on the chromosome which works on the structure of tree. The chromosome is composed of gene point and gene value. The gene value, which is mapped to virtual machine, is the value of corresponding gene point, which is mapped to the host.

Here, each mapping scheme is represented as a three layer tree structure shown in Fig. 2, where the root node denotes a central controller, the 2nd layer nodes denote N physical hosts, and the 3rd layer nodes denote M virtual machines, and where a physical host node is corresponding to multiple virtual machine nodes.

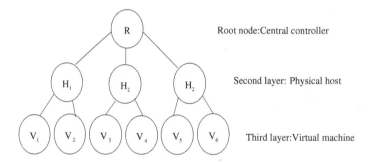

Fig. 2.   Encoding based on three layer tree

The chromosome is represented as a nested set of trees in Fig. 2, as follows:

$$H = \{R((H_1 (V_1, V_2), (H_2 (V_3, V_4), (H_3 (V_5, V_6)))\} \tag{10}$$

Where 6 virtual machines are divided into 3 sub trees, which are corresponding to 3 genes. Each gene contains a set of virtual machines. The number of virtual machines in different physical host is different. Therefore, the genetic operators need to deal with different lengths of chromosomes. Gene decoding is based on the multi dimensional space decomposition [6], which counts on the space layout of the virtual machines in the physical hosts. Because this algorithm is mainly concerned with three kinds of resources, such as CPU, memory and bandwidth, three-dimensional space packing method is adopted in paper.

As shown in Fig. 3, the 3 coordinates of the three-dimensional space represent the host's memory, CPU, and the bandwidth, respectively. When a virtual machine is dispatched to a physical host, the virtual machine will be placed at the right bottom of the current host space. It's mark is to delete from the chromosome nesting set. Then, the remaining host space is divided into A space, B space and C space. They are available for the current physical host. This process will be do until the resources meet the need of virtual machine or there is no space in physical host.

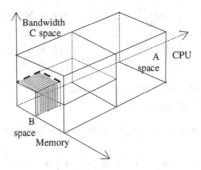

Fig. 3. Three-dimensional division decoding

3.2 Fitness Function

The selection of fitness function plays an important role to find the optimal solution and speed up the convergence speed in the genetic algorithm [7]. Taking into account the need of load balance and the limitation of multi dimension resources in the process of providing resource, Evaluation function is defined from three aspects: space utilization ratio, load change and weight in this paper and fitness function is realized by weighting method at last.

(1) Space utilization function

By the chromosome decoding, the relationship of space utilization between the host and the virtual machine is clear and the space utilization function is

$$f(x) = \frac{\sum_{i=1}^{n}(LI_{Mem}^{i} \times LI_{cpu}^{i} \times LI_{bw}^{i})}{HV} \times 100 \% \tag{11}$$

Where HV is the host volume, n is the number of virtual machines that are placed in the host space and $LI_{Mem}^{i} \times LI_{cpu}^{i} \times LI_{bw}^{i}$ is the volume of i'th virtual machine that has been palace in the host space.

(2) Load change function

$$f(y) = \varphi\left(\sigma(S,T) - \sigma_0\right) = \begin{cases} \theta & \sigma(S,T) - \sigma_0 > 0 \\ 1, & \sigma(S,T) - \sigma_0 \leq 0 \end{cases} \tag{12}$$

Where $\sigma(S,T)$ represents the actual load variation value of the host in the T time period. σ_0 represents load variation constraint for load balance. $\varphi(x)$ is a penalty function whose value is θ when the individual does not meet the constraints, otherwise it is 1.

(3) Weight function
The weight of the virtual machine is represented by its load VL and its weight function is

$$f(z) = \begin{cases} \frac{\sum_{i=1}^{n} VL_i}{ML} \times 100\%, & ML \geq \sum_{i=1}^{n} VL_i \\ 0 & ML < \sum_{i=1}^{n} VL_i \end{cases} \tag{13}$$

Where ML denotes the maximum load carrying capacity of the host and VL_i denotes the weight of the i'th virtual machine.

According to the previous analysis, the fitness function is defined as

$$F(S,T) = \alpha \times f(x) + \beta \times \frac{1}{f(y)} + \gamma \times f(z) \tag{14}$$

Where, α, β and γ are weighting factors, and the conditions $0 \leq \alpha, \beta, \gamma \leq 1$, $\alpha + \beta + \gamma = 1$ hold.

3.3   Select Operation

In this paper, the roulette wheel method is used to make Individuals with higher fitness values enter a generation population. The algorithm works in three stages. Firstly, according to the fitness function F(S, T) in Eq. 14, the sum of the fitness of the individual in the current population and the fitness of the population is obtained, where the individual's choice probability Q (Si) is proportioned to the sum. Then, the k'th individuals' cumulative selection probability $Q^*\left(S_k\right) = \sum_{i=1}^{k} Q(S_i)$ is calculated. Thirdly, a random number R between 0 to 1 is generated. If $Q^*\left(S_{k-1}\right) < r < Q^*\left(S_k\right)$, then k'th individual is selected, which makes the individual with the higher fitness value can well into the next generation of population.

3.4   Cross Operation

Cross operation promotes searching ability in the process of genetic evolution. In this paper, we introduce the selection probability P whose value is equal to the ratio of a single virtual machine load and the sum of all the loads. The cross operation process is

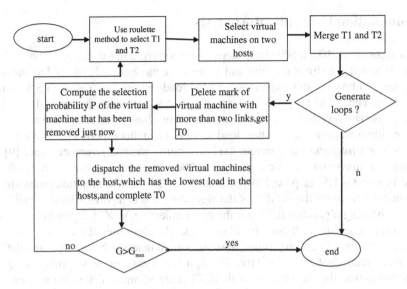

Fig. 4. Cross flow chart

shown in Fig. 4, where G is the current genetic algebra and G_{max} is the maximum genetic algebra.

3.5 Variation Operation

Variation is a very small probability of the individual changes so that the population is diversified. The variation operation of this paper is: (1) to set up a very small variation probability Tm, $T_m \in [0, 1]$ to produce the individual to carry out the variation. (2) to randomly select two host nodes from the resulting individual. (3) to generate a random number $\mu, 1 < \mu < q$, and exchange the u'th virtual machine on the one of two hosts for q'th virtual machine on the other hosts. In order to improve the spatial search ability and convergence performance, the adaptive mutation probability is introduced and the formula [8] is as follows:

$$T_m^{k'} = \exp\left[-\alpha \times \left(\frac{\left|F_{best} - F\left(S_k, T\right)\right|}{F_{best}}\right)\right] \times T_m^k \qquad (15)$$

Where T_m^k is the initial mutation probability of individual K, $F\left(S_k, T\right)$ is the fitness of the individual K and F_{best} is the current best individual fitness, $\alpha \in R^+$.

3.6 Termination Condition

The termination condition of the algorithm is to find a mapping scheme in the population, which makes the load change of the host meet the predetermined load conditions and have the minimum migration time in the load balance process for the virtual machine.

4   Simulation Experiment Analysis

In this experiment, 100 to 500 cloud services are deployed in a data center. Each cloud service is bound to a virtual machine and each virtual machine is dispatched to one host. There are 100 hosts in this experiment. Each cloud service has from 2 to 3.5 million instructions needed to be performed and the host's computing power is 10 million instructions. The way that host manage the virtual machine is the time share as the way that the virtual machine manage the cloud services is. In this experiment, web access is used as a test load and the concrete load generation method reference paper [9]. In addition, the maximum genetic algebra G is 200, the population size of N is 60, the fitness function is 1.5, the probability of mutation is 0.8, and the mutation probability is 0.15. In order to verify the validity of the algorithm, it is compared with the traditional heuristic packing algorithm BFH and the multi objective MOGA algorithm. Figure 5 shows the adaptability of these three algorithms. The algorithm of this paper uses tree structure encoding. In this algorithm, host and virtual machine has a mapping relationship and no redundant encoding. From Fig. 5, it is evident that the algorithm is significantly higher than the other two algorithms. BFH algorithm, which is due to the use of binary encoding method that is not suitable for the optimization of this kind of packing problem, has the lowest fitness. MOGA algorithm, which use the group encoding that is similar to the tree encoding and that adopts the selection algorithm with higher complexity, is lower than the algorithm of this paper too.

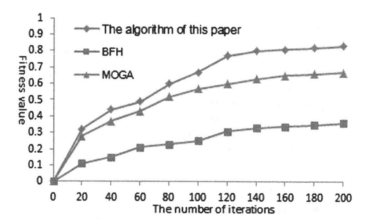

Fig. 5. Comparison of three algorithms

In some cases, due to the rapid increase in load at partial nodes, the frequent access to the system makes the whole system imbalance. Therefore, the load balance can be maintained by the migration of virtual machine. In the process of migration, the migration cost has to consider so that the cost can be minimized. In Fig. 6, the comparison of the migration times of the three algorithms in the optimization process is given. It is obvious that the number of virtual machine migration of this paper is significantly less than the other two algorithms. The BFH algorithm should be as far as possible to allocate virtual machines to less physical machines, which causes times of virtual machine migration is the most in the three algorithms. Considering the minimum number of physical hosts, the MOGA algorithm has more times of virtual machine migration.

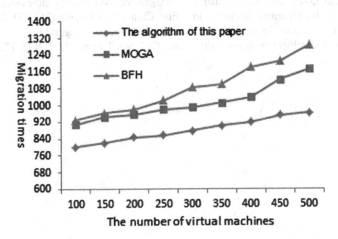

Fig. 6. Comparison of the migration times of the three algorithms

5   Conclusion

With the application of cloud computing is becoming more and more extensive, the Application presents the characteristics of multi tenancy and heterogeneity [10], which makes the distribution of different resources is not balanced. That seriously affects the efficiency of using system resources. The algorithm of this paper provides a new solution for the load balance strategy.

References

1. Cheng G.-J., Liu, L.-J., et al.: Application of a hybrid genetic algorithm in the load balance of cloud computing. J. Xi'an Shiyou Univ. (Nat. Sci. Ed.) **27**(2), 93–97, 122, 123 (2012)
2. Zhang, W., Zhang, H., Liu, N.: Design of server-end load-balance system based on genetic algorithm. Comput. Eng. **31**(20), 121–123 (2005)
3. Guo, P., Li, Q.: Load-balance scheduling algorithm base on classifying the servers by their load. J. Huazhong Univ. Sci. Technol. (Nat. Sci. Ed.) **Z1**, 62–65 (2012)
4. Chen, Z.: Resource allocation for cloud computing base on ant colony optimization algorithm. J. Qingdao Univ. Sci. Technol. (Nat. Sci. Ed.) **33**(6), 619–623 (2012)
5. Huu, T.T., Tham, C.K.: An auction-based resource allocation model for green cloud computing. In: Proceedings of the 2013 IEEE International Conference on Cloud Engineering, pp. 269–278. IEEE, Piscataway (2013)
6. Grossman, R.L.: The case for cloud computing. IT Prof. **11**(2), 23–27 (2009)
7. Liu, Z.-J.: A research into cloud-computing-based load balance technology. J. Guangxi Teach. Educ. Univ. Nat. Sci. Ed. **28**(2), 93–96 (2011)

8. Liu, Z.H., Wang, X.L.: Load balance algorithm with genetic algorithm in virtual machines of cloud computing. J. Fuzhou Univ. (Nat. Sci. Ed.) **40**(4), 453–458 (2012)
9. Li, Q., Hao, Q.-F., Xiao, L.-M.: Adaptive management and multi-objective optimization for virtual machine placement in cloud computing. Chin. J. Comput. **34**(12), 2253–2264 (2011)
10. Nie, J.: UAP cloud platform programmed control expansion algorithm based on swarm intelligence identification of linear difference. Bull. Sci. Technol. **31**(2), 125–127 (2015)

Multiple DAGs Workflow Scheduling Algorithm Based on Reinforcement Learning in Cloud Computing

Delong Cui, Wende Ke$^{(\boxtimes)}$, Zhiping Peng, and Jinglong Zuo

College of Computer and Electronic Information,
Guangdong University of Petrochemical Technology, Maoming, China
{delongcui,wendeke}@163.com, pengzp@foxmail.com,
oklong@gmail.com

Abstract. To the problem of scheduling multiple DAG workflow applications with multiple priorities submitted at different times in cloud computing environment, a novel workflow scheduling algorithm based on reinforcement learning is proposed in this paper. In the workflow scheduling scheme, the number of VMs in resources pool is defined as state space; the runtime of user task is defined as immediate reward, and then interactive with cloud computing environment to obtain the optimization policy. We use real cloud workflow to test the proposed scheme. Experiment results show the proposed scheme not only can solve the fairness of scheduling multiple DAGs with the same priority level submitted at different times, but also can ensure that the execution of the DAGs with higher priorities cannot be influenced by the DAGs with lower priorities. More importantly, the proposed scheme can reasonably schedule multiple DAGs with multiple priorities and improve utilization rate of resources better.

Keywords: Multiple DAGs · Reinforcement learning · Workflow scheduling · Cloud computing

1 Introduction

A system that defines, creates and manages the execution of workflows through the use of software, running on one or more workflow engines, which is able to interpret the process definition, interact with workflow participants and, where required, invoke the use of IT tools and applications [1]. A workflow can usually be described using formal or informal flow diagramming techniques, showing directed flows between processing steps. Single processing steps or components of a workflow can basically be defined by three parameters [2]:

Input description: the information, material and energy required to complete the step.

Transformation rules, algorithms, which may be carried out by associated human roles or machines, or a combination.

© Springer Science+Business Media Singapore 2016
K. Li et al. (Eds.): ISICA 2015, CCIS 575, pp. 305–311, 2016.
DOI: 10.1007/978-981-10-0356-1_31

Output description: the information, material and energy produced by the step and provided as input to downstream steps.

Recently years, Workflow Management System (WMS) is arranged in cloud computing environment became a new cloud computing application. Lots of system models and methods which aim at cloud workflow application or Directed Acyclic Graph (DAG) task scheduling are designed. Dornemann designed an on-demand resource provisioning for BPEL workflows using amazon's elastic compute cloud [3]. Byun proposed cost optimized provisioning of elastic resources for application workflows [4]. They all realized the resource provisioning schemes in real Amazon EC2 cloud computing platform.

But only few research works have been done in multiple DAGs workflow Scheduling problem [5–11]. In this paper, in order to solve the problem of scheduling multiple DAG workflow applications with multiple priorities submitted at different times in cloud computing environment, we proposed a multiple DAGs cloud workflow scheduling algorithm based on Reinforcement Learning (RL). The experiments results demonstrate the efficiency of our task scheduling scheme. The remaining of this paper is organized as follows. The concept of system model is briefly introduced in Sect. 2. In Sects. 3 and 4, the details of the proposed scheme and some experimental results are presented. A conclusion is drawn in Sect. 5.

2 System Model

2.1 DAG Workflow Scheduling Problem

A scheduling system model consists of an application, a target computing environment, and a performance criteria for scheduling. An application is represented by a directed acyclic graph, $G = (V, E)$, where V is the set of v tasks and E is the set of e edges between the tasks. Each edge $(i,j) \in E$ represents the precedence constraint such that task n_i should complete its execution before task n_j starts. Data is a $v \times v$ matrix of communication data, where $data_{i,k}$ is the amount of data required to be transmitted from task n_i to task n_k [7].

DAG scheduling object: node tasks are allocated object resources, which satisfied in chronological order constraint, could minimize the finished time.

2.2 Prototype Model of Workflow Scheduling System Based on Cloud Infrastructure

Aim at cloud computing workflow, some system models and DAG scheduling methods are proposed in recent years. Most of system models and scheduling methods common were constructed by virtualized, scaled cloud computing resources management and charging mode. Usually scheduling object is maximization cloud computing resources utilization or minimized user workflow implement cost. A prototype model of workflow scheduling system based on cloud infrastructure is shown in Fig. 1.

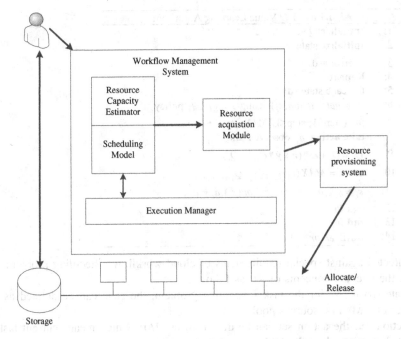

Fig. 1. A prototype model of workflow scheduling system based on cloud infrastructure

Workflow Management System mainly construct by resource capacity estimator, resource acquisition module, scheduling module and execution manager. Users can define and described workflow, deadline and resources constraint. Then users proposed executive request. WMS is responsible for resources obtained, scheduling and executive tasks.

3 Workflow Scheduling Scheme Based on RL

As we all know, the job scheduling in grid and cloud environment has proven to be an NP hard complete problem [12]. Reinforcement learning (RL) take actions in an environment so as to maximize some notion of cumulative reward, its online learning, model free etc. features make it became an important branch of machine learning. Especially suitable for the adaptive learning that the learner know and the dynamic complex environment.

We employ the Q learning algorithm, a popular reinforcement learning algorithm, to solve the workflow scheduling problem in cloud computing environment. The pseudo code of the basic Q value learning algorithm is illustrated in Algorithm 1 [13].

Algorithm 1 Q Value Learning Algorithm
1: **Initialize** Q value table
2: **Initialize** state s_t
3:     error = 0
4: **Repeat**
5: **for** each state s**do**
6:     $a_t = get\_action(s_t)$ using $\varepsilon - greedy$ policy
7: **for** (step=1; step<**LIMIT**; step++) **do**
8: **take action** a_t observe r and S_{t+1}
9:     $Q_t = Q_t + \alpha *(r + \gamma * Q_{t+1} - Q_t)$
10:    $error = MAX(error \mid Q_t - Q_{previous-t})$
11:    $s_t = s_{t+1}, a_{t+1} = get\_action(s_t), a_t = a_{t+1}$
12: **end for**
13: **end for**
14: **Until** $error < \theta$

Directed against special requirements of cloud workflow scheduling problem, we define the relevant concepts of RL as follows:

State space: As for the task scheduling problem, the state can be defined as the number of VMs in resources pool.

Action set: the action set can be described as (1/0), which means current task is accepted or rejected by the VM.

Immediate reward: immediate reward must correctly reflect the optimizations object and real runtime situation, the execution time of task is defined as immediate reward. We illustrate the workflow scheduling progress by employing real medical image analysis application. An example of medical image analysis application in cloud computing environment is shown in Fig. 2.

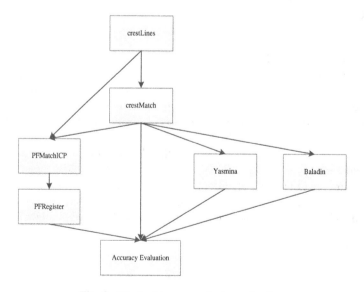

Fig. 2. Medical image analysis application

Algorithm 2. Single DAG cloud workflow scheduling algorithm

- Partition workflow by the task implementation progress. For example, the partition results of Fig. 2 can be described as follow:

crestLines	PFMatchICP	PFRegister		Accuracy evaluation
	creatMatch	PFMatchICP	PFRegister	
		\		
		Yasmina		
		Baladin		

- Estimate the execution time of each stage.
- Decompose cloud workflow and task queue.
- Set up priority according to the SLA constraint.
- Cloud workflow scheduling using algorithm 1.

Algorithm 3. Multiple DAG cloud workflow scheduling algorithm

- For each cloud workflow, decompose and estimate DAG by using algorithm 2.
- Set priority according to each DAG and SLA constraint.
- Task queue.
- Cloud workflow scheduling using algorithm 1.

4　Performance Evaluation

We employ CloudSim [14] as our workflow scheduling platform, and simulation the real cloud workflow in Fig. 2. The parameter settings of experiment environment is shown in Table 1. The experimental results are shown in Fig. 3.

Table 1. Parameter settings type 1

Parameter	Values
Total number of VMs	10
VM frequency	1,000–5,000 million instructions per second
VM memory (RAM)	1024–2048 mega byte
VM bandwidth	500–1,000 mega byte per second
Number of VM buffer	10–100
Number of PEs requirements	5
Number of datacenters	1
Number of hosts	2

To compare the performance of various schemes, we analyze the execution results of a randomly-selected batch jobs, under the deadline constraint of 60 min. From the experimental results show in Fig. 3, the proposed multiple DAG workflow scheduling scheme.

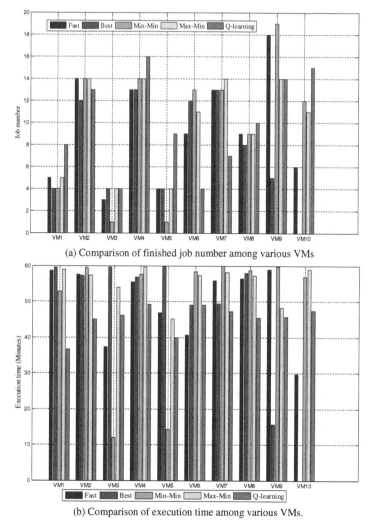

(a) Comparison of finished job number among various VMs

(b) Comparison of execution time among various VMs.

Fig. 3. Simulation comparison results

5   Conclusions and Future Work

In this study, we provide multiple DAGs workflow scheduling scheme based on rein-forcement learning for cloud computing platform. In light of the proposed system model, we theoretically analyze the execution process of jobs in cloud computing environment and design a novel task scheduling scheme using reinforcement learning to optimize the deadline under given cloud computing resources. We evaluate the proposed job scheduling optimization schemes in extensive simulations. Guided by the empirical observation, we find that the proposed job scheduling scheme can optimally utilize cloud resources and load balancing to obtain the minimal makespan with resource constraint.

Acknowledgments. The work presented in this paper was supported by National Natural Science Foundation of China (No. 61272382, 61402183). Key project of Guangdong Province in the research center of cloud robot (petrochemical) Engineering Technology (No. 650007). Guangdong Provincial Science and Technology Program (No. 2014A020208139); Distinguished Young Talents in Higher Education of Guangdong (No. 2013LYM-0057). 2014 Guangdong Provincial Technological Innovation Program (No. 650019). Wende Ke is corresponding author.

References

1. Workflow Management Coalition. http://www.wfmc.org/
2. Wikipedia. https://en.wikipedia.org
3. Tim, D., Ernst, J., Bernd, F.: On-demand resource provisioning for BPEL workflows using amazon's elastic compute cloud. In: 9th IEEE/ACM International Symposium on Cluster Computing and the Grid, pp. 140–147. IEEE Press, New York (2009)
4. Eun, K.B., Yang, S.K.: Cost optimized provisioning of elastic resources for application workflows. Future Gener. Comput. Syst. **27**(8), 1011–1026 (2011)
5. Guozhong, T., Yu, J., He, J.S.: Towards critical region reliability support for grid workflows. J. Parallel Distrib. Comput. **69**(12), 989–995 (2009)
6. Guozhong, T., Yu, J., He, J.S.: Scheduling and fair cost-optimizing methods for concurrent multiple DAGs with deadline sharing resources. Chin. J. Comput. **37**(7), 1607–1619 (2014)
7. Topcuoglu, H., Hariri, S., Min, W.: Performance-effective and low-complexity task scheduling for heterogeneous computing. IEEE Trans. Parallel Distrib. Syst. **13**(3), 260–274 (2002)
8. Ming, T., Shoubin, D., Liping, Z.: A multi-strategy collaborative prediction model for the runtime of online tasks in computing cluster/grid. Cluster Comput. **14**(2), 199–210 (2011)
9. Ming, T., Shoubin, D., Kejing, H.: A new replication scheduling strategy for grid workflow applications. In: 6th ChinaGrid Annual Conference (ChinaGrid 2011), pp. 74–80 (2011)
10. Jin, L., Qian, W., Cong, W., Ning, C., Kui, R., Wenjing, L.: Fuzzy keyword search over encrypted data in cloud computing. In: 29th IEEE International Conference on Computer Communications (INFOCOM 2010), pp. 441–445. IEEE Press (2010)
11. Li, J., Chen, X., Li, J., Jia, C., Ma, J., Lou, W.: Fine-grained access control system based on outsourced attribute-based encryption. In: Crampton, J., Jajodia, S., Mayes, K. (eds.) ESORICS 2013. LNCS, vol. 8134, pp. 592–609. Springer, Heidelberg (2013)
12. Ullman, J.D.: NP-complete scheduling problems. J. Comput. Syst. Sci. **10**(3), 384–393 (1975)
13. Sutton, R.S., Barto, A.G.: Reinforcement Learning: An Introduction. MIT Press, Cambridge (1998)
14. Calheiros, R.N., Ranjan, R., Beloglazov, A., De Rose, C.A.F., Buyya, R.: CloudSim: a toolkit for modeling and simulation of cloud computing environments and evaluation of resource provisioning algorithms. Softw. Pract. Experience **41**(1), 23–50 (2011)

An Improved Parallel K-Means Algorithm Based on Cloud Computing

Dongbo Zhang[1(⊠)] and Yanfang Shou[2]

[1] Department of Computer Science, Guangdong University of Science
and Technology, Dongguan, China
neversurrender1314@163.com
[2] Guangzhou Institute of Modern Industrial Technology, South China University
of Technology, Guangzhou, China

Abstract. In this paper we presented CK-means clustering algorithm based on improved K-means algorithm and the Canopy algorithm, which uses MapReduce programming model of Hadoop platform. The experimental results prove that the CK-means algorithm has a good advantage in the processing of large data sets, in the acceleration ratio, accuracy, expansion rate, and the effect of the algorithm after deploying on the Hadoop clusters.

Keywords: Cloud computing · MapReduce model · K-means clusters · Canopy algorithm

1 Introduction

Birds of a feather gather together, Clustering algorithm [1] is to research how to become a collection of physical or abstract objects to multiple classes or groups which Composed of similar objects. The objects in the same cluster are as similar as possible, while the objects in different clusters are as different [2] as possible. With the development of social information, the data on the network is growing exponentially, the global daily data generated by 2 EB [3]. How to retrieve valuable information from mass data has become our most urgent target. Clustering analysis, as an important part of data mining, has been widely used in various fields and has achieved some results, however, with the rapid growth of data scale, clustering algorithm has been transferred from serial to parallel, from single machine to clusters. At present, there are some literature about on the parallel clustering algorithms. The paper [4–6] proposed the idea of K-means clustering algorithm based on MapReduce. The paper [7] proposed a parallel clustering algorithm PK-means based on MapReduce, where the Map function computes the distance between the data object to the center point, and then re marks the new clustering category. The reduce function calculating the new clustering center based on the intermediate results. The paper [8–10] implemented the K-means clustering algorithm based on MapReduce.

At present, the classical K-means clustering algorithm [11, 12] has the advantages of simple structure, flexible change, easy hardware implementation, but the K-means algorithm still has the following disadvantages:

© Springer Science+Business Media Singapore 2016
K. Li et al. (Eds.): ISICA 2015, CCIS 575, pp. 312–320, 2016.
DOI: 10.1007/978-981-10-0356-1_32

(1) In the process of clustering, the local traps are prone to occur.
(2) In the clustering calculation, the number of iterations is increasing rapidly, and the time consuming is increasing. For better to solve these problems, this paper proposes a new algorithm of CK-means and Canopy based on the in-depth study of Hadoop technology.

2   Related Works

2.1   K-Means Algorithm

K-means is a classical clustering algorithm, Its main idea is we select k data points as the center of the initial K cluster randomly from the original target data sets, then calculates the distance between the other non central data points to the K cluster center. According to the distance from the center, select the nearest cluster, and then assign the data to the cluster, and then the data points are assigned to the cluster, and the process is repeated.

2.2   Canopy Algorithm

Canopy algorithm: For mass data, the data points are divided into some overlapping clusters by using the distance measurement method. And then, clustering data by using the method of calculating the accuracy of the points in Canopy.

3   CK-Means Clustering Parallel Algorithm

CK-means is a algorithm based on MapReduce programming model, which can be divided into two sub tasks. And then implemented based on MapReduce model in the natural order. The inputs of each subtask is the outputs of the previous subtask. The first subtask is to compute the similarity of the data set object based on the MapReduce programming model by using the Canopy idea. And then, put the data points with the similarity into a subset, which is Canopy. However, we no longer carry calculation of the similarity between objects of Canopy, which improved the efficiency of the clustering. Finally, we assign the number of Canopy in this subtask as the initial K value of the next subtask to avoid the blindness of the initial value setting. The second subtask is to make clustering analysis which are based on MapReduce programming model in the "Canopy" by Using the "internal error square and" (see Definition 3) "extreme point" principle (see Definition 2), and combined binary chop. The CK-means executing flow as Fig. 1(a) and (b).

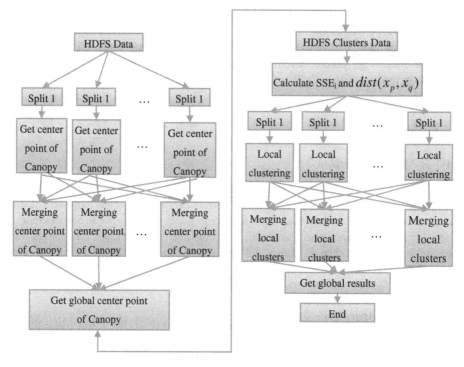

Fig. 1. (a). CK-means executing flow 1. (b). CK-means executing flow 2

3.1 Definitions and Concepts

Definition 1 (Canopy definition): Given data set $U = \{u_i | i = 1, 2, \ldots, n\}$, as to $\forall x_i \in U$ meet $\{c_j | \exists \| x_i - c_j \| \} \leq T_1, c_j \subseteq U, i \neq j\}$, The x_i collection called Canopy collection, c_j is the center point of Canopy, T_1 is the radius of Canopy collection.

Definition 2 (SSE_i): Sum of the squared errors of cluster C_i, also is the square of the distance between all points in cluster C_i and the center of the cluster. The calculation formula is as $SSE_i = \sum_{x \in c_i} dist(c_i, x)^2$.

Among them, c_i is the center of the cluster C_i.

Definition 3 (Limit point principle): Given cluster $C = \{u_i | i = 1, 2, \ldots, n\}$, $\exists x_p, x_q \in C$ make $dist(x_p - x_q) = \max\{dist(x_i - x_j) | \forall x_i, x_j \in C\}$,

Then x_p, x_q is the limit point of cluster C. And consider x_p, x_q as the initial center of cluster C is limit point principle. Among, $dist(x_p - x_q)$ is limit distance.

3.2 Canopy Algorithm Based on MapReduce

The Canopy algorithm starts with a collection that contains some data points and an empty Canopy list in the Map stage. Then, it iterating for classification according to the distance threshold, and generate Canopy collection in the iterative process. So we got the Canopy collection. In the Reduce stage, it merging the collection W_i with all node, and get the union set w, then making Global Canopy clustering in set w. Repeating the above process until the data set is empty. The algorithm output the number of clusters K finally. This value of K will be the input value for the next subtask.

In the Map phase, The data set of each node in the cluster is needed to clustering, and output Canopy collection W finally. The collection W in Canopy (V, N) will be used as the input to participate in the Reduce.

3.3   K-Means Algorithm Based on MapReduce

Algorithm process: First, the algorithm consider all data sets as cluster V and put it into cluster S in order to get the cluster k. Then the algorithm take out a cluster which must meet the limit point principle from cluster S by K-means clustering algorithm. It making two points clustering by the selected cluster, the sum of squared error (Definition 2) and the smallest 2 of the clusters through i times, then put these two clusters into cluster S. Repeating until generated K clusters.

In the two point, the algorithm find out the minimum sum of square errors in clustering results by using K-means clustering algorithm several times. Finally, Set the result as the initial center of CK-means clustering algorithm. The algorithm adopted the center which generated through two points search in optimization compared with the classical K-Means clustering algorithm. Therefore, it avoids the local optimization results obtained from random generation centers. At the same time, due to the "extreme point" principle, it can effectively reduce the number of clusters and improve the clustering efficiency.

1. The improvement of K-means based on the two - point search idea, and the combination of the "internal error square" and "the limit point principle"

Because of the K-means (V, K) algorithm has a large amount of computation in search of the target cluster and the determination of the target cluster, this paper divide the k-means algorithm into two steps, and implemented k-means by using the idea of the application of distributed computing and by a natural order based on Mapreduce programming framework.

The first process is to find the target cluster algorithm based on the MapReduce programming framework parallel implementation. The main principle is to calculate SSE_i of every node in the Map stage and find out the largest value of SSE_i and set it as target cluster.

The second step is to determine the limit point algorithm in the target cluster and implement the MapReduce programming framework. The main principle is that the cluster is divided into different nodes Mapper in Map phase, and then each node computes the distance from each point of the current node to each point of the cluster, and then distributes it to the Reduce; the Reducer stage get limit point by sorting these distances.

4 Experiment and Result Analysis

The experimental environment used in this paper is as follows: 5 sets of computers are used to build clusters. Configuration environment is as follows:

The above machines' hardware is consistent and configured as follows: I7 2.5 GHz 8 G memory, the operating system is: 14.04LTS Ubuntu (Table 1).

Table 1. Configuration list

IP	Name of nodes	Name of HDFS
192.168.1.101	Master	Namenode
192.168.1.102	Slave1	Datanode
192.168.1.103	Slave2	Datanode
192.168.1.104	Slave2	Datanode
192.168.1.105	Slave4	Datanode

4.1 CK-Means Algorithm Result Analysis

The experimental data sets are selected from UCI Machine Learning Repository, we also selected two data sets which are Breast Cancer sets and Synthetic Control Chart Time Series sets. The official description of the Cancer Breast data set has 286 samples, each of which has 9 attributes. Synthetic Control Chart Time Series data sets has 600 samples, however, how many attributes are not marked in each sample.

Table 2. The test results of the algorithm of the paper [13] and our algorithm in the cancer breast data sets

PNodes	CBK-means		Proposed algorithm	
	Accuracy/%	SSE_{min}	Accuracy/%	SSE_{min}
1	76	526	79	510
2	79	527	82	515
3	83	534	86	520
4	78	544	80	535
5	82	550	84	540

Tables 2 and 3 showed the experimental results of the accuracy of clustering results and average value of the minimum sum of square error respectively which based on Breast Cancel data sets and synthetic Control Chart Time Series data sets. The two comparison are corresponding to the algorithm of the paper [13] and this algorithm.

The CK-means algorithm, which is relative to the literature [13], determines the optimal number of clusters by improving the error square of the cluster and the principle of the limit point, and thus obtains higher accuracy of clustering and lower SSE_{imin}. Additional, it can be seen that, with the increase of the number of nodes, the computing advantages of the clusters are more obvious, and the Hadoop can be extended to ensure the high reliability of the program. Therefore, the clustering

Table 3. The test results of the algorithm of the paper [13] and our algorithm in synthetic data sets

Nodes	CBK-means		Proposed algorithm	
	Accuracy/%	SSE_{min}	Accuracy/%	SSE_{min}
1	80	711 762	81	689 862
2	81	716 832	82	691 962
3	83	722 698	84	697 700
4	81	726 926	81	706 826
5	82	730 982	84	710 740

accuracy of the CBK-means algorithm has no obvious changed compared CK-means algorithm, but the value of SSE_{imin} is increasing.

4.2 Analysis of Data Spreading Rate

From Figs. 2 and 3, it can be seen that the expansion rate of the algorithm is gradually reduced when the number of nodes and the size of the test data sets is increasing, which is because the number of nodes increases the communication cost between nodes. The more bigger the size of the data sets, the CBK-means parallel algorithm and CK-means parallel algorithm of paper [13] of expansion rate are better, and the expansion efficiency curve will fall more smoothly. The two parallel algorithms have good extendibility in large data sets.

Experiments show that the CK-Means parallel algorithm compared to the CBK-Means parallel algorithm of paper [13], which is suitable for the cloud computing platform running on large scale data sets and have different degrees of improvement in clustering accuracy, speed ratio, expansion rate, etc.

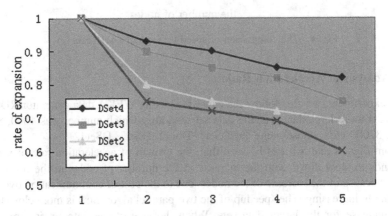

Fig. 2. The expression rate of CBK-means of 4 data sets in paper [13]

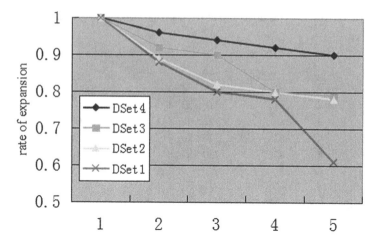

Fig. 3. The expression rate of CK-means of 4 data sets

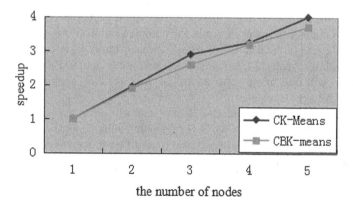

the number of nodes

Fig. 4. The comparison of speed-up ratio of Dset1 data set

4.3   Analysis of Acceleration Ratio

In this experiment, we also follow the paper [13] using WEKA data generator RDG1 to generate 4 more large data sets, the data set randomly which number was 4000, 8000, 16000, 32000 etc., whose name are DSet1, DSet2, DSet3, DSet4.

From Figs. 4 and 7 it can be seen that, the growth of acceleration rate of vertical axis tends to slow down with the increase in the number of nodes in the transverse, mainly because of the increase of the nodes lead to the cost of communication overhead is gradually increasing. The speedup of the two parallel algorithms is more close to the linear increase for the larger data sets. When the size of the data set is larger, the CK-means parallel algorithm is significantly higher than that of the CBK-means parallel algorithm (Figs. 5 and 6).

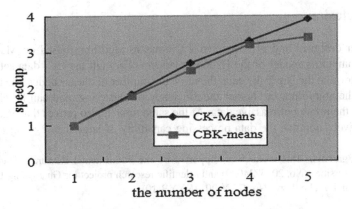

Fig. 5. The comparison of speed-up ratio of Dset2 data sets

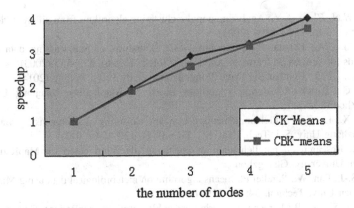

Fig. 6. The comparison of speed-up ratio of Dset3 data sets

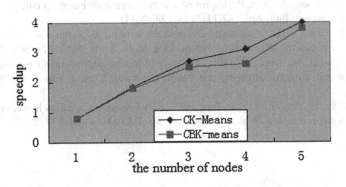

Fig. 7. The comparison of speed-up ratio of Dset3 data sets

5   Conclusion

This paper designed and implemented a CK-means algorithm based on MapReduce parallel computing model of the Hadoop platform. Through the UCI data set test, the results show that the larger the data, the more the number of cluster nodes, the better the effect of the algorithm, the higher the efficiency of clustering, and can be effectively applied to the mining of massive data. At the same time, it also proves that MapReduce can effectively improve the data processing capability of large data.

Acknowledgments. This research is supported by the Fundamental Research Funds for the Central Universities (No. 2015ZM039) and scientific research project for Guangdong University of Science and Technology (No. GKY-2014KYYB-10).

References

1. Qian, W.N., Zhou, A.Y.: Analyzing popular clustering algorithms from different viewpoints. J. Softw. **13**(8), 1382–1394 (2002)
2. Gustavo, E.A., Batista, P.A., Monard, M.C.: Annalysis of four missing data treatment methods for supervised learning. Appl. Artif. Intell. **13**(5/6), 519–533 (2003)
3. Bao, L., Li, Q.: Combat Big Data. Tsinghua University Press, Beijing (2014)
4. Wen, C.: Parallel Clustering Algorithm Based on MapReduce. Zhejiang University, HangZhou (2011)
5. Jiang, X., Li, C.: Parallel implementing k-means clustering algorithm using MapReduce. J. Huazhong Univ. Sci. Tech. (Nat. Sci. Ed.) **39**(1), 120–124 (2011)
6. Li, Y.: Research on parallelization of clustering algorithm based on MapReduce. Sun Yat-sen University, Guangzhou
7. Xue, S.-J., Pan, W.: Parallel Pk-means algorithm on meteorological data using MapReduce. J. Wuhan Univ. Technol. **34**(12), 139–142 (2012)
8. Ji, S.-Q., Shi, H.-B.: K-means clustering ensemble based on MapReduce. Comput. Eng. **39** (9), 84–87 (2013)
9. Xie, X., Li, L.: Reseach on parallel k-means algorithm based on cloud computing platform. Comput. Meas. Control **22**(5), 1510–1512 (2014)
10. Zhang, X., Zhang, G., Liu, P.: Improved k-means algorithm based on clustering criterion function. Comput. Eng. Appl. **47**(11), 123–128 (2011)
11. Su, M.C., Chou, C.H.: A modified version of the k-means algorithm with a distance based on cluster symmetry. IEEE Trans. Pattern Anal. Mach. Intell. **23**(6), 674–680 (2001)
12. Fu, N., Qiao, L.Y., Peng, X.Y.: Blind recovery of mixing matrix with sparse sources based on improved k-means clustering and hough transform. Chin. J. Electron. **37**(4), 92–96 (2009). (Ch)
13. Gao, R., Li, J., Xiao, Y., Zhu, S., Peng, W.: Parallel algorithm based on K-means clustering in cloud environment. J. Wuhan. Univ. (Nat. Sci. Ed.) **61**(4), 368–374 (2015)

Research on the Integration of Spatial Data Service Based on Geographic Service

Lei Shang[1(✉)], Shujing Xu[2], Wei Hou[3], and Lipeng Zhou[4]

[1] Key Laboratory of Evidence-Identifying in Universities of Shandong,
Shandong University of Political Science and Law, Jinan, China
leilishang@163.com
[2] Department of Science and Engineering,
Xinyang Normal University Huarui College, Xinyang, China
392825484@qq.com
[3] 66008 Unit, The Chinese People's Liberation Army, Tangshan, China
12969722@qq.com
[4] 95685 Unit, The Chinese People's Liberation Army, Kunming, China
zhouwuhu51@163.com

Abstract. Spatial data integration is the direction of the spatial data management and use. Because of the correlation between spatial data, the data redundancy and inconsistency appear, that is an important question in data integration. As a method based on user demand and data production, geographic service provides a method to solve the problem of data redundancy and inconsistency. The paper discusses the theory and method of the integration of spatial data based on geographic service and gives an example of MODIS vegetation index, showing the feasibility and superiority of this method.

Keywords: Geographic service · Spatial data · Data integration

1 Introduction

Nowadays the spatial data is widely used and has much production. The spatial data is complicated the application scenario is changing, which bring much difficulty to managers and users of the data. As a result, the integration of spatial data is the main method to solve this problem. Nowadays the research on the integration of spatial data is focused on the data form, including format conversion, projection transformation and so on. The problem is that it can produce large redundant data and potentially damage the data consistency. Actually producing derived new data through original data is the essence of the integration of spatial data. As a user-oriented method provided by geography information, geographic service include the re-organization and re-processing of the spatial data. Geographic service is suitable for the integration of spatial data, which can reduce data redundancy, keep data consistent and convenient to manage and use. Firstly, we introduce research status of spatial data and the integration. Secondly, we explain the content and characteristics of geographic service.

© Springer Science+Business Media Singapore 2016
K. Li et al. (Eds.): ISICA 2015, CCIS 575, pp. 321–328, 2016.
DOI: 10.1007/978-981-10-0356-1_33

Thirdly, we present the theory model that geographic service is applied to the integration of spatial data. Lastly, we build a prototype system of data integration service about MODIS vegetation index. The integration of spatial data service based on geographic service is superior and practicable.

2   Spatial Data and the Integration

Spatial data is the data relative to the location of the earth reference space and can express the property of entity and process state in geographical objective world [1]. Because the property of objective entity and process state is various, the spatial data is also various in expression form and content. With more production and using of the spatial data, the inter-operability of the spatial data is lower because of this variety. We must spend large human and material resources to preprocess in order to make the spatial data meet special application condition. Relative preprocess conclude the following two aspects.

1. Relative conversion which does not change the content of original data such as format conversion, coordinate conversion and the choice of spatial range.
2. Relative conversion which produce new data by processing original data such as buffer analysis and Vegetation index calculation.

To the manager and server of the data, relative preprocess above must be conducted if you want to provide direct data to users. So the new data produced by preprocessing derives from original data. To manage this new data will create data redundancy and the following inconformity of data. Data integration is needed to organize multiple spatial data, improve interoperability of data, eliminate redundancies and reduce inconformity.

Data integration is organic concentration of Geo-spatial data with different sources and different formats and different characteristics logically or physically. Organic means to consider the attributes of the data, time and space characteristics, the accuracy of data itself and the geographical features and process expressed [1,2]. The target of data integration is to improve management level of the managers, optimize service mode of data servers, lower difficult of using the data and make data have a strong interoperability. Finally it can provide massy data guarantee service for model integration next step.

It can be seen that there is a very broad application prospect after the data integration. But there are some difficulties now [3].

1. Data's isomerism.
2. Data redundancy and the following inconformity of data.

At present the researches of data integration mainly focus on the integration of heterogeneous spatial data and put forward some methods including the method based XML [4], the method based on GML [5,6], and the method based on WebGIS [7]. There is little attention on data redundancy and data inconformity problem in the data integration. Data redundancy increases the difficulty of integration and data inconformity would make the integration data become error directly.

3   Geographical Services

OGC (Open Geospatial Consortium) as the geospatial information standard definers has defined a series of definitions about spatial data standards which has played a positive role on spatial data specification and promoted the spatial data sharing strongly. Geographical services are a series of spatial data service and protocol based on network defined by OGC. These geographical service protocols include WFS (Web Feature Service) [8], WCS (Web Coverage Service) [9], WMS (Web Map Service) [10] and the newly proposed and being perfect WPS (Web Processing Service) [11].

WFS defined by OGC are about services of vector data on the Internet. It is used in the vector data transmission between the network.

WCS are about services of grid data on the internet as defined by OGC. Like WFS they are used in the grid data transmission between the network. They provide unified structure of grid data to make sure that any client supporting WCS protocol would transmit data with unambiguous understanding.

WMS are map show services defined by OGC. They are used for the output of web maps and the specific performance of spatial data while getting spatial data through the WFS or WCS and other means according to local data. They are visualization of spatial data and give users intuitive feelings.

WPS are network processing services defined by OGC. Their prime structure has come into being. And some parts are still amended. Relative to the mature WFS, WCS and WMS, researches and applications on WPS are in the preliminary and exploratory stage. And it has been applied in some situations [12].

WFS and WFS as data services integrate the existing heterogeneous data, provide a unified access structure, complete formal transformation of spatial data including the selection of data, filtering, format conversion and even projection transformation in a certain extent. This in a certain extent solves the problem of heterogeneous data in data integration and the first type in spatial data preprocessing as is above-mentioned. But for the second type in data integration as the transformations that reprocess raw data to generate new data, there isn't a corresponding method in either protocol. And this can't meet the needs of data integration. WPS provides GIS functions on the internet. In data integration perspective the way that raw data is reprocessed through web processing services to generate new data and the results are returned to the client could solve the second problem in data integration. But the WPS standards formed relatively late and related theory researches and technology applications are in the preliminary stage. There are only some researches on earthquake damage analysis [13] and Explosive bomb threats, etc. [14,15]. In fact WPS is a middleware which using XML language processes related spatial data, analyses functions and describes them, gives commands and parameters as customer demands to GIS software library in a standardized way and returns the results. In data integration perspective, through this kind of geographic service new data can be formed by raw data processing. The redundancy and inconformity brought by original data and new data deriving from it could be solved in spatial data integration. For spatial data integration new thoughts and methods are provided.

4   Data Integration Model of Geographic Service

Through the above analysis of the nature and characteristics of geographic service, we can see that geographic service not only can be used to release, share data, and also generate new data from spatial data processing. Synthesizing and improvement for these services can achieve smaller data redundancy, data integration keeping the data consistent, solve the redundancy and inconformity which are the most easy to appear in spatial data integration, optimize data management model and improve efficiency. The general model of this integration mode can be expressed as the following way, Fig. 1:

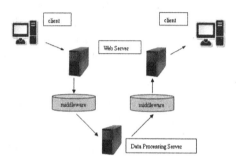

Fig. 1. Data integration model of geographic service

Here, the clients are users of data. They do not care about the organization form of the underlying data. What only need to do for them is providing a high-level view on the client side and data view in the perspective of user using data.

The Web Server mainly is used to provide friendly data application interface to users, collected user's data request message from the client, process the request message and then send formatted message to middleware. At the same time, the middleware transfer performance results of request message to Web Server. The Web Server does related encapsulation according to the results and then sends them to clients.

The middleware mainly transforms messages passed from the web server to geographic service requests for the corresponding data processing server according to the rules. These requests could be WFS, WCS or WPS requests in XML file format as a carrier that is defined by OGC, and could also be more simple requests set for a specific application scenario. Related messages are sent to the web server after getting the results from data processing server. The middleware is a bridge between the web server and the geographic data processing server.

The geographic data processing server is the core module of data integration. Except providing the users with WFS and WCS defined by OGC to meet the existing data integration and solving the redundancy and inconsistencies in the process of data integration according to the analysis of the above, it could also provide the function of generating new data based on raw data. After getting the required data, to the new data generated it could reduce repetitive operation and the data processing server's work burden to improve the efficiency of data services. Then data access method requested by client is provided to Web services by the middleware completing the procedure from the data request of the user to getting the data.

In this model data processing server produces data for users. This can effectively integrate existing spatial data, directly provide the user the required data and improve the interoperability of spatial data. In the following part a case in which this model was used to realize the integration of MODIS vegetation index is illustrated for applications of the model.

5   MODIS Vegetation Index Integration

Vegetation Indexes are simple, effective and experiential measures on the surface vegetation using satellite sensors of red light and near infrared channels. Now more than 40 Vegetation Indexes have been defined. Applications are very wide. The commonly used Vegetation Index: Ratio Vegetation Index (RVI), Difference Vegetation Index (DVI) and Normalized Difference Vegetation Index (NDVI) [16]. The MODIS sensor scans and obtains data around the world daily. The first and the second channel are red and near infrared wave bands. Two bands operation could get Vegetation Index of day by day and time series of Vegetation Index. These are applied to crop classification, vegetation classification [17], etc. Therefore, on the basis of the basic MODIS data services its derived service of vegetation index is more meaningful.

In the following part data integration model mentioned above based on geographic service was used. Data integration was done for MODIS day by day data of visible light and near infrared with spatial resolution of 250 m [18]. The users are provided with three Vegetation Indexes of NDVI, RVI and DVI through building geographic service. And it is easy to extend and provide Vegetation Index of other type.

The main technical route is: to build friendly UI on the client side using HTML 5 webpage markup language of a new generation, to provide parameters of accessing data, meanwhile using Javascript language and Openlayers [19] as data spatial extent customization and map display, to give users the best using feelings through the web front end technology.

Apache Server was chosen as the web server because of its wide use and mature technology. It was used to provide safe, stable and efficient web services. WSGI was chosen as the middleware. It is a kind of interface (WSGI) between Python applications or structure and Apache Web Server. It can connect Apache Server and data processing server to ensure unambiguous understanding between them.

The data processing server can be divided into data processing function and data providing, releasing function. Data processing and analysing function of GIS are done by GRASS software including Vegetation Index calculation, range extraction and output format control.

The final data providing and release function is using Mapserver to provide WFS, WCS and WMS access of data and file access provided by Web directly to release data after processing. This will effectively improve the interoperability of data. The whole server process was conducted organized and controlled by webpy web framework based on Python language (Fig. 2).

The web server converts requests of the client side through WSGI and sends them to the data processing server. The data processing server uses Python and uses GRASS as its GIS software library. Also it calls its internal algorithm to calculate the specified area and Vegetation Index specified. The relevant code is shown below;

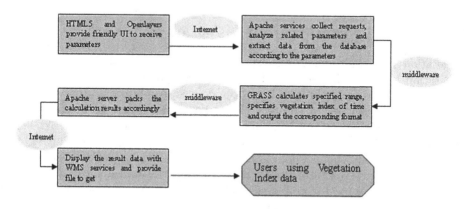

Fig. 2. Technical framework realized

```
if  vit=='ndvi ':
    grass.mapcalc('ndvi=1.0*(c3−c4)/(c3+c4)',overwrite=True)
    grass.run_command('r.out.gdal',overwrite=True,
        input='ndvi' ,output=outndvif)
elif  vit=='rvi ':
    grass.mapcalc('ndvi=1.0*c3/c4',overwrite=True)
    grass.run_command('r.out.gdal',overwrite=True,
        input='ndvi' ,output=outndvif)
else :
    grass.mapcalc('ndvi=c3−c4',overwrite=True)
    grass.run_command('r.out.gdal',overwrite=True,
        input='ndvi' ,output=outndvif)
```

UI (User Interface) using the new web front end technology provides friendly interface for users to choose the date and type of Vegetation Index and spatial extent they need to custom data.

Vegetation Index data that spatial extent and types are specified could be produced according to the needs of different users when this integration is used. Integrating the work of data service to the data server could improve farthest the convenient of the data users using data and make them focus on scientific questions pointed out by data instead of just processing data. This data integration system has served for ongoing research projects and can satisfy the actual needs of data integration.

6   Conclusion

The integration of spatial data is an important part of the data use management. The redundancy and inconsistency is a big question restricting the integration of spatial data. To solve this question, we used the geographic service generated from modern web technology and geographic information system to build a model, and verified the feasibility through the application of MODIS vegetation index integration. The practice has proved that the integration reduces the redundancy effectively and maintains the consistency of the data. It creates a prototype system and can easily spread to other types. It can also provide data preparation for the next model integration.

There are also some problems in the system, firstly, the data processing server has much burden. Secondly, a large number of computation is time-consuming for each request. Thirdly, it can not respond to the request of the client-side and is bad for user experience. The next work is to create new cached tables to manage the repeated data on the basis of this model. When the users ask for the data, if they can find them in the cached tables, return, if not, the data processing server does some conduct, then return the data to users and put them to the data processing server. It can reduce the burden of data processing server and promote response speed. It can also enhance user experience and the model of data integration.

Acknowledgment. This work was supported by Science & Technology Plan of Shandong High School (J14LN11).

References

1. Chen, J., Fei, C.Y.: Overview of study on Geo-spatial data integration. Prog. Geogr. **19**(3), 203–211 (2000)
2. Li, J., Zhang, D.F.: Theories and systems of Geo-spatial data integration. Prog. Geogr. **20**(2), 138–144 (2001)
3. Chen, Y.G., Wang, J.C.: A review of data integration. Comput. Sci. **31**(5), 48–51 (2004)
4. Cai, H.N., Xie, J., Wen, J.H., et al.: Research and application of geographic data integration based on XML. Comput. Eng. **34**(15), 77–79 (2008)
5. Chen, C.B., Wu, Q.Y., Chen, C.C.: Application of GIS in merchandise delivery information system of chainstores. Sci. Surv. Mapp. **30**(5), 53–55 (2005)
6. Ma, X.: Integration and share of spatial data based on web service. Shanghai Jiaotong University (2006)
7. Qi, Y., Chen, L., Zhang, R.X., et al.: A distributed web GIS data integration accessstrategy based on the dual-cache mechanism. Comput. Eng. Sci. **29**(5), 41–44 (2007)
8. Open Geospatial Consortium Inc. Web Feature Service Implementation Specification[EB/OL], 23 June 2013. http://www.opengeospatial.org/standards/wfs
9. Open Geospatial Consortium Inc. OGC® WCS 2.0 Interface Standard - Core[EB/OL], 23 June 2013. http://www.opengeospatial.org/standards/wcs

10. Open Geospatial Consortium Inc. OpenGIS® Web Map Server Implementation Specification[EB/OL], 23 June 2013. http://www.opengeospatial.org/standards/wms

11. Open Geospatial Consortium Inc. OpenGIS® Web Processing Service[EB/OL], 23 June 2013. http://www.opengeospatial.org/standards/wps

12. Foerster, T., Stoter, J.: Establishing an OGC web processing service for generalization processes. In: Workshop of the ICA Commission on Map Generalisation and Multiple Representation, vol. 6, p. 25 (2006)

13. Xiaoliang, M., Yichun, X., Fuling, B.: Distributed geospatial analysis through web processing service: a case study of earthquake disaster assessment. J. Softw. **6**, 671–679 (2010)

14. Gerlach, R., Schmullius, C., Nativi, S.: Establishing a web processing service for online analysis of earth observation time series data. Geophys. Res. Abstr. 10 (2008)

15. Stollberg, B., Zipf, A.: OGC web processing service interface for web service orchestration aggregating Geo-processing services in a bomb threat scenario

16. Guo, N.: Vegetation index and its advances. Arid Meteorol. **21**(4), 71–75 (2003)

17. Yang, H., Huang, W., Wang, J., et al.: Monitoring rice growth stages based on time series HJ-1A/1B CCD images. Trans. CSAE **27**(4), 219–224 (2011)

18. http://earthexplorer.usgs.gov//

19. Hazzard, E.: OpenLayers 2.10 Beginner's Guide, pp. 3–4. Packt Publishing, Birmingham (2011)

Predicting Maritime Groundings Using Support Vector Data Description Model

A.G. Rekha[1(✉)], Loganathan Ponnambalam[2], and Mohammed Shahid Abdulla[1]

[1] Indian Institute of Management Kozhikode, Kozhikode, India
agrekha64@gmail.com
[2] Institute of High Performance Computing, A*STAR, Singapore, Singapore

Abstract. This paper focuses on grounding prediction related to sea vessels. Grounding accidents are one of the most common causes for ship disasters. Hence, there is a growing need to assess and analyze probabilities as well as related consequences of ship running aground. Using a real world marine incident dataset obtained from the United States Coast Guard National Response Center, we have demonstrated that Support Vector Data Description based methods can be successfully used for grounding prediction. After preprocessing the raw data, a total of 15165 incidents were obtained out of which there were 291 cases of ship running aground and was used in our study. A prediction accuracy of 98.25 % was achieved using the Lightly Trained Support Vector Data Description.

Keywords: SVDD · LT_SVDD · Ship grounding · Maritime incident analysis

1 Introduction

Grounding or stranding refers to aground of ship into ashore, reef or wreck due to either under command or not under command [1]. It occurs when a vessel runs aground or otherwise makes contact with the sea bed (Fig. 1). This often results in a structural impact on the ship. The resultant damage to the ship may be catastrophic or minor. The depth of the sea differs from port to port. The larger the ship, higher the trouble in docking it to a port & hence, the chances of grounding is more. Grounding accident is one of the most frequent of marine accidents and is the major source of property loss. Ship groundings can create severe environmental and financial consequences and grounding accounts for about one-third of commercial ship accidents [2].

According to the annual overview report of Marine Casualties and Incidents of the European Maritime Safety Agency (EMSA) published in 2014, when the occurrence severity was serious, grounding/stranding was the event that represented the highest number (28 %), followed by loss of control (18 %) and collision (15 %). Moreover, almost each ship involved in grounding incidents damage to the hull or machinery heavily and more than half of grounding incidents result in pollution damage at sea [3]. Hence grounding prediction and prevention is one of the significant issues for improving maritime safety and also for the prevention of pollution.

© Springer Science+Business Media Singapore 2016
K. Li et al. (Eds.): ISICA 2015, CCIS 575, pp. 329–334, 2016.
DOI: 10.1007/978-981-10-0356-1_34

Fig. 1. Grounding, GELSO M, 10 March 2012, very serious casualty, no injuries, ship lost, no pollution (Source: [3])

Since grounding incidents are low probable events, one class classification methods such as SVDD, in which the data belonging to only one class, i.e. the normal class being available for training will be suitable for event prediction. Within this context, this paper presents the application of SVDD based method to predict whether the ship is aground or not. The approach is illustrated using a real world marine incident dataset obtained from the United States Coast Guard National Response Center [4].

2   Background

2.1   Marine Casualty and Grounding

Accident and incident reports have been previously studied by many researchers for risk modeling and mandatory incident reports are considered useful for understanding risk control measures [5]. Since the grounding incidents can cause serious damage to the vessel as well as can lead to environmental pollution, grounding prevention is considered as one of the significant issues for marine safety researchers. Various studies have been done in the past for accident and ship grounding analysis [6–9]. The hybrid approach incorporating Accident Analyze Mapping (AcciMap) and Analytical Network Process (ANP) methods to analyze causes of marine accidents analytically was proposed in [1]. A set of analytical expressions for the calculation of damage opening sizes in tanker groundings was discussed in [10]. The study in [11] presents an analysis on the factors contributing to groundings when ships transit in and out of ports. In [12], the authors have tried to establish a geometry model for the ship and hazard area and devised a GRI equation based on fuzzy theory to estimate the grounding danger probability along one ship's instantaneous track.

2.2   Review of Support Vector Data Description (SVDD)

Support vector data description (SVDD) method involves creation of a hyper-sphere around the target class data with the minimum radius to encompass almost all the target instances and exclude the non-target ones [13]. Given a set of data points, x_i; i = 1: N, the objective is to

$$\text{minimize:} F(R, a) = R^2 + C \sum_i \xi_i$$
$$\text{with the constraints } || x_i - a ||^2 \leq R^2 + \xi_i, \xi_i \geq 0 \, \forall i, \tag{1}$$

where R is the radius and a is the center of the sphere and the slack variables ξ_i allow the possibility of outliers in the training set. Kernel function maps the input space into a theoretical feature space where the pattern classes in the input space become linearly separable. $\phi : R^n \to F$, $x \to \phi(x)$. The objective function of SVDD can be translated into the following form by employing Lagrangian multipliers $\alpha_i \geq 0$ and $\gamma_i \geq 0$:

$$O(R, a, \alpha_i, \gamma_i, \xi_i) = R^2 + C \sum_i \xi_i - \sum_i \alpha_i \{ R^2 + \xi_i - (\|x_i\|^2 - 2a.x_i + \|a\|^2) \} - \sum_i \gamma_i \xi_i \tag{2}$$

O should be minimized w.r.t. R, a, ξ_i and maximized w.r.t α_i and γ_i. The parameter C controls the trade-off between the volume and the errors and ξ is the slack variable. Setting partial derivatives (w.r.t a and R) to zero gives the centre of the sphere as a linear combination of the objects as given by: $a = \sum_i \alpha_i x_i$; the distance from the center of the sphere, a, to (any of the support vectors on) the boundary, R^2 is given by: $R^2 = (x_k.x_k) - 2 \sum_i \alpha_i (x_i.x_k) + \sum_{i,j} \alpha_i \alpha_j (x_i.x_j)$. Also, $\sum_i \alpha_i = 1$ and $\alpha_i \geq 0$. The function becomes

$$L = \sum_i \alpha_i (x_i.x_i) - \sum_{i,j} \alpha_i \alpha_j (x_i.x_j) \tag{3}$$

Note that in all formulae, objects x_i appear only in the form of inner products with other objects $(x_i.x_j)$. These inner products that can be replaced by a kernel function to obtain more flexible methods. By placing kernel functions K(x, y) instead of the product (x, y) in the above equations we have: $L = 1 - \sum_{i,j} \alpha_i \alpha_j K(x_i, x_j)$. Only objects x_i with $\alpha_i > 0$ are needed in the description and these objects are therefore called the support vectors of the description (SV's). Support objects lie on the boundary in feature space that forms a sphere containing the transformed data points. Hence, an object in feature space, z, is accepted by the description (stated to be within the boundary of the sphere) when: $\|(z - a)\|^2 = (z - \sum_i \alpha_i x_i)(z - \sum_i \alpha_i x_i) \leq R^2$. Similarly, by applying the kernel function, the formula for checking an object z in the input space becomes:

$$1 - 2 \sum_i \alpha_i K(z, x_i) + \sum_{i,j} \alpha_i \alpha_j K(x_i, x_j) \leq R^2 \tag{4}$$

2.3   Review of Lightly Trained SVDD (LT-SVDD)

Lightly Trained SVDD (LT-SVDD) is a variant of SVDD which reduces the computational complexity of classical SVDD [14]. This method locates an approximate pre-image of the SVDD sphere's centre in the input space during the training phase (Fig. 2). In particular, what is obtained is an approximate solution to the SVDD problem since the exact pre-image of the feature space sphere's centre does not exist in input space. This method reduces the complexity from the initial O(N3) to O(d), where 'N' is the number of data points and 'd' is the dimension of the data set. In the training phase it uses the primal form of the kernelized SVDD problem as suggested in [15].

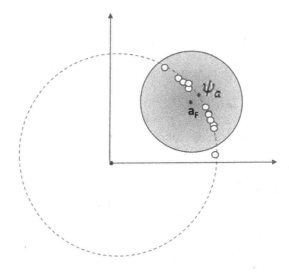

Fig. 2. Feature space SVDD Sphere Centre (a_F) and the approximate Centre (Ψ_a)

3   Experiments and Discussion

We have used a real world marine incident data obtained from the United States Coast Guard in our experiments. The data collected during marine casualty and pollution investigations are entered into the Marine Information for Safety and Law Enforcement (MISLE) database and is available via Homeport [4].

The data reflect information collected by U.S. Coast Guard personnel concerning vessel and waterfront facility accidents and marine pollution incidents throughout the United States and its territories. This database can be used to analyze marine accidents and pollution incidents by a variety of factors including vessel or facility type, injuries, fatalities, pollutant details, location, and date. The data collection period for marine casualties started from 1982. Table 1 describes the variables that are used in this study.

Table 1. Marine incident dataset-fields description

Field name	Field description
Incident date time	Date and time incident occurred, was discovered or planned
Vessel type	Type of vessel involved
Fire involved	Indicates if any fire was involved with the incident
Incident cause	Cause of the incident
Allision	Indicates if an Allision occurred
Weather conditions	Weather conditions at time of incident

After cleaning and removing the incidents having missing values, a total of 15165 incidents were obtained and was used in our analysis. Out of this there were 291 cases in which the vessel was aground. The predictor variables used in our study include time, fire involved, weather conditions, allusion, incident cause and vessel type. We have calculated the accuracies using classical SVDD method as well as LT_SVDD method and the results obtained are shown in Table 2.

Table 2. Prediction accuracy for CSVDD and LT_SVDD

Method	Accuracy of prediction
CSVDD	97.30 %
LT_SVDD	98.25 %

4 Conclusion

In this work, we have demonstrated that Support Vector Data Description based methods can be successfully applied for predicting whether the ship is aground or not. Also we have shown that LT_SVDD, the low complexity variant of SVDD can be used for making faster calculations in such scenarios. The prediction accuracy obtained using classical SVDD was 97.30 % while using LT_SVDD was 98.25 %. We have used a 5-fold cross validation in our experiments. The average time taken for the classical method was 60.67 s while for LT_SVDD was 0.82 s. Our future work will involve using more machine learning techniques for maritime incident analysis. Another possibility of extension of this work is to explore the application of the similar approach to the pollution database.

References

1. Akyuz, E.: A hybrid accident analysis method to assess potential navigational contingencies: the case of ship grounding. Saf. Sci. **79**, 268–276 (2015)
2. Kite-Powell, H.L., et al.: Investigation of potential risk factors for groundings of commercial vessels in US ports. International Journal of Offshore and Polar Engineering (1999)
3. EMSA. Annual Overview Report of Marine Casualties and Incidents (2014)
4. United States Coast Gaurd. Marine Casualty and Vessel Data for Researchers. https://homeport.uscg.mil/mycg/portal/ep/editorialSearch.do#
5. Mazaheri, A., Montewka, J., Nisula, J., Kujala, P.: Usability of accident and incident reports for evidence-based risk modeling – a case study on ship grounding reports. Saf. Sci. **76**, 202–214 (2015)
6. Pedersen, P.T.: Review and application of ship collision and grounding analysis procedures. Mar. Struct. **23**, 241–262 (2010)
7. Nguyen, T.-H., Amdahl, J., Leira, B.J., Garrè, L.: Understanding ship-grounding events. Mar. Struct. **24**, 551–569 (2011)
8. Wang, J., et al.: Use of advances in technology for maritime risk assessment. Risk Anal. **24**(4), 1041–1063 (2004)
9. Friis-Hansen, P., Simonsen, B.C.: GRACAT: software for grounding and collision risk analysis. Mar. Struct. **15**(4–5), 383–401 (2002)
10. Heinvee, M., Tabri, K.: A simplified method to predict grounding damage of double bottom tankers. Mar. Struct. **43**, 22–43 (2015)
11. Lin, S., Kite-Powell, H.L., Patrikalakis, N.M.: Physical risk analysis of ship grounding. Dissertation Massachusetts Institute of Technology, Department of Ocean Engineering (1998)
12. Yang, X., Liu, X., Xu, T.: Research of ship grounding prediction based on fuzzy theory. In: International Conference of Information Technology, Computer Engineering and Management Sciences (2011)
13. Tax, D.M.J., Duin, R.P.W.: Support vector data description. Mach. Learn. **54**, 45–66 (2004)
14. Rekha, A.G., Abdulla, M.S., Asharaf, S.: A fast support vector data description system for anomaly detection using big data. In: 30th Annual ACM Symposium on Applied Computing, Spain (2015)
15. Pauwels, E.J., Ambekar, O.: One class classification for anomaly detection: support vector data description revisited. In: Perner, P. (ed.) ICDM 2011. LNCS, vol. 6870, pp. 25–39. Springer, Heidelberg (2011)

Estimating Parameters of Van Genuchten Equation Based on Teaching-Learning-Based Optimization Algorithm

Fahui Gu[1,2(✉)], Kangshun Li[1], Lei Yang[1], and Wei Li[1]

[1] School of Mathematics and Informatics, South China Agricultural University,
Guangzhou 510006, Guangdong, China
gufahui@139.com
[2] School of Information Engineering, Jiangxi Applied Technology
Vocational College, Ganzhou 341000, Jiangxi, China

Abstract. The Van Genuchten Equation (VGE) is used to describe the characteristic of soil water movement, but it is super-set, nonlinear and containing many unknown parameters. Using the traditional method to estimate the parameters of VGE often results in a high margin of error because of complication. The teaching-learning-based optimization (TLBO) is a new swarm intelligent optimization method for solving complex nonlinear models. In this paper, the solution program of TLBO is compiled and used to estimate parameters of the VGE. The results show that the estimate method by TLBO is more efficient and accurate. Consequently, TLBO can be used as a new method to estimate parameters of VGE.

Keywords: Van Genuchten Equation (VGE) · Teaching-learning-based optimization algorithm · Estimating parameters · Soil water retention curve

1 Introduction

The Van Genuchten Equation (VGE) is used to describe the relationship between soil moisture and suction, so it is an important index to reflect the basic hydraulic characteristics of soil, and it is very important to study retention and migration of soil water [1]. As a question of soil water characteristic curve, the VGE is widely used because of its curve similarity with the measured data and its clear parameters meaning. The domestic and foreign scholars have put forward a number of mathematical models and calculation methods, such as using integral method to determine the parameter equation in [2]; Cui and Chi [3] proposed a simple approach to estimate parameters by the soil water content at saturation, soil water content and soil water suction at field capacity; a new artificial glowworm swarm optimization algorithm based biological parasitic behavior was proposed to solve these parameters in [4]; Du and Zhang [5] applied the particle swarm optimization to calculate the Parameters; an improved harmony search algorithm used to solve the parameters accurately was proposed by Xing et al. [6]; Wang et al. used pattern search algorithm to identify Van Genuchten equation parameter in [7]; Ma and Shen [8] estimated parameters of VGE using the damper least

© Springer Science+Business Media Singapore 2016
K. Li et al. (Eds.): ISICA 2015, CCIS 575, pp. 335–342, 2016.
DOI: 10.1007/978-981-10-0356-1_35

square method; Xu et al. [9] used differential evolution to calculate VGE's parameters; Liao et al. [10] Jun-zeng put forward three intelligent algorithms including genetic algorithm, simulated annealing algorithm and particle swarm optimization algorithm to estimate parameters of VGE.

However, the VGE is a super-set and nonlinear complex equation with many unknown parameters. Some of the above methods are complex calculation and weak general character, the accuracy of others needs to be improved.

The TLBO is a new swarm intelligent optimization method proposed by Rao et al. in 2010 [11]. It simulates the "teaching" process of the teachers and the "learning" process of the students. The purpose is to improve the students' learning achievement by the teacher's "teaching" and mutual "learning" between students. The TLBO is very simple and easy to understand. It has less parameter, high precision and strong ability of convergence, so it has attracted a lot of scholars' attention and has been a very good application [12].

Therefore, considering the simplicity and accuracy of the TLBO, it is proposed to estimate parameters of VGE for getting more efficient and accurate in this paper. The general implemented program of TLBO is made in MATLAB environment. Finally, a numerical example is analyzed.

2 Teaching-Learning-Based Optimization

The idea of the TLBO is derived from the knowledge of the students in the classroom by transmission of teachers and the knowledge of the students by exchanging among the students. Thus, the TLBO is divided into two parts: the teaching process and students' interaction, and measuring how much knowledge of the students acquired is in the form of score or grade.

2.1 Related Concepts

The optimization algorithm based on TLBO is to simulate the learning style of the class as a unit. The improved level of the students in the class needs to be guided by the teacher's teaching, and the students need to learn from each other to promote the absorption of knowledge. In TLBO, teachers and students are equivalent to the individuals in the evolutionary algorithm, and the teacher is one of the best individuals' adaptive values. A subject of each student learned is equivalent to a decision variable. Optimization problem is defined as [11, 12]:

$$z = \max_{x \in S} f(X) \tag{1}$$

Where S is search space with equal probability as $S = \{X | x_i^L \leq x_i \leq x_i^U, i = 1, 2, \ldots, d\}$, X is searching point in space with equal as $X = (x_1, x_2, \ldots, x_d)$, d is dimension number of dimension space, x_i^L and x_i^U are respectively the upper and lower bounds for each dimension, $f(X)$ is objective function.

If we suppose that $X^j = (x_1^j, x_2^j, \ldots, x_d^j)$ $(j = 1, 2, \ldots, NP)$ is a point in searching space, $x_i^j = (i = 1, 2, \ldots, d)$ is a decision variable of X^j, NP is the number of searching point in space. A class can be expressed as the following form:

$$
\begin{bmatrix} X^1 & f(x^1) \\ X^2 & f(x^2) \\ \vdots & \vdots \\ X^{NP} & f(x^{NP}) \end{bmatrix} = \begin{bmatrix} x_1^1 & x_1^1 & \cdots & x_d^1 & f(X^1) \\ x_1^2 & x_2^2 & \cdots & x_d^2 & f(X^2) \\ \vdots & \vdots & \vdots & \vdots \\ x_1^{NP} & x_2^{NP} & \cdots & x_d^{NP} & f(X^{NP}) \end{bmatrix} \tag{2}
$$

Where $X^j(j = 1, \ldots, NP)$ is the student of the class, NP is the number of students, d is the number of the subjects studied by the student. Therefore, the teacher can be defined as:

$$
X_{teacher} = \arg \max f(X^j)(j = 1, \ldots, NP) \tag{3}
$$

2.2 Algorithm Steps and Flow

The details of TLBO algorithm steps are given below.

Step 1: Initialize class X

Every student in the class $X^j = (x_1^j, x_2^j, \ldots, x_d^j)(j = 1, 2, \ldots, NP)$ is randomly generated in search space.

$$
x_i^j = x_i^L + rand(o, 1) \times (x_i^U - x_i^L) \quad j = 1, 2, \ldots, NP; i = 1, 2, \ldots d \tag{4}
$$

Step 2: Perform teaching

The following formula is used to realize the process of teaching.

$$
X_{new}^i = X_{old}^i + difference \tag{5}
$$

$$
difference = r_i \times (X_{teacher} - TF_i \times mean) \tag{6}
$$

Where X_{old}^i and X_{new}^i are distinct the before learning value and the after learning value, $mean = \frac{1}{NP}\sum_{i=1}^{NP} X^i$ is the average value of all students, $TF_i = round[1 + rand(0,1)]$ is teaching and learning factor, $r_i = rand(0,1)$ is learning step.

After completion of the teaching, it will update the score of every student according to score of the former and later of learning. The updating method is as following:

$$
\begin{aligned} &if f(X_{new}^i) > f(X_{old}^i) \\ &\quad X_{old}^i = X_{new}^i \end{aligned} \tag{8}
$$

Step 3: Perform learning

Choosing $X^j(j = 1, 2 \ldots, NP, j \neq i)$ randomly in the class as the learning object for every $X^i(i = 1, 2 \ldots, NP)$, X^i as a student will adjust learning through analysis of the differences between himself and the X^j student. The learning method is as following:

$$X^i_{new} = \begin{cases} X^i_{old} + r_i \times (X^i - X^j) & f(X^j) < f(X^i) \\ X^i_{old} + r_i \times (X^j - X^i) & f(X^i) < f(X^j) \end{cases} \qquad (8)$$

Where $r_i = U(0, 1)$ is learning step of the number i student.

After completion of the learning, it will perform the updating operator on the student. The updating method is as following:

$$\begin{aligned} if\, f(X^i_{new}) &> f(X^i_{old}) \\ X^i_{old} &= X^i_{new} \end{aligned} \qquad (9)$$

Step 4: If the ending condition is met, the optimization is finished. Otherwise, go to Step 2.

From the algorithm flow, we can see that the teaching stage of TLBO is similar as the social search section in the particle swarm algorithm. Each individual learns from the teacher, therefore, the population is easy to converge to the teacher and the search speed is very fast, but the diversity of the population is easy to lose and then fall into local search.

In the learning stage of TLBO, the students of the class complement each other through the exchange learning from between each other. Due to be carried out in the in a small range of students, the learning of each student is too early to global optimum direction of aggregation, so it can effectively maintain the diversity of students, thus ensuring the algorithm in the search space of the global exploration ability.

The TLBO algorithm flow is shown in Fig. 1.

3 Estimating Parameters of Van Genuchten Equation

3.1 Van Genuchten Equation

The VGE describes the relationship between the pressure head and soil water content, that is, the relationship between the amount of soil water energy and the amount of water, and the form of VGE is as following [13]:

$$\theta = \frac{\theta_s - \theta_r}{[1 + |\alpha h|^n]^m} + \theta_r \left(m = 1 - \frac{1}{n}, 0 < m < 1, n > 1\right) \qquad (10)$$

Where θ is volume content of soil, which unit is cm^3/cm^{-3}; h is soil water suction, which unit is cm; θ_s, θ_r, α, m and n are the parameters of soil water retention curve; θ_s is soil saturated water content, which unit is cm^3/cm^{-3}; θ_r is residual moisture content, which unit is cm^3/cm^{-3}.

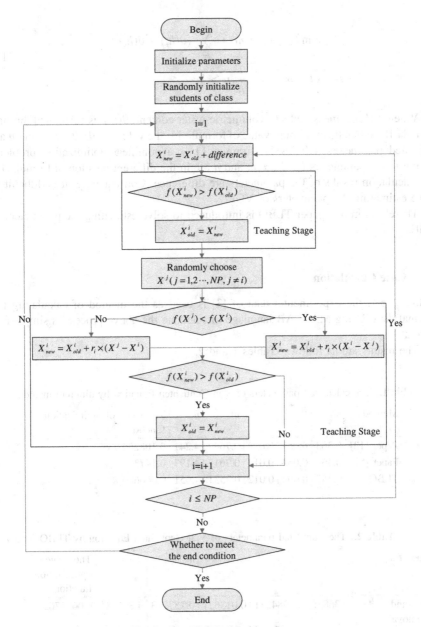

Fig. 1. The TLBO algorithm flow

3.2 Estimating Parameters of van Genuchten Equation Based on TLBO

In order to estimate the parameters of VGE, the following model can be established (the optimization criterion function):

$$\min f(\theta_s, \theta_r, \alpha, m) = \sum_{i=1}^{h} (\theta(h_i) - \theta(h_i)')^2$$

$$s.t. \quad m = 1 - \frac{1}{n}, n > 1 \tag{11}$$

Where $\theta(h_i)$ is measured soil volumetric water content, $\theta(h_i)'$ is calculated through formula 10. When the minimum value of formula 11, the values of θ_s, θ_r, α, m and n are estimated parameters. Obviously, formula 11 is a nonlinear optimization problem. Owing to the parameters θ_s, θ_r, α, m and n are in the different position of formula 11, the calculation results of the parameters are difference. Consequently, it is difficult to solve estimating the parameters of VGE.

Therefore, in this paper TLBO is introduced to solve estimating the parameters of VGE.

3.3   Case Calculation

In this paper, the experimental data of [3] is used as the method of calculating the original data. Using MATLAB language to calculate the above process is simple and general.

The results are as follows: Tables 1 and 2.

Table 1. The calculated parameters of Van Genuchten Equation by different methods.

Method	θ_s	θ_r	α	m	n	The value of optimization function
Paper [13]	0.363	0.053	0.013	0.7642	4.241	0.108234
Paper [8]	0.363	0.053	0.014	0.7617	4.197	0.147387
TLBO	0.357	0.064	0.012	0.7531	4.051	0.000689

Table 2. The calculated parameters of Van Genuchten Equation by TLBO.

Type of soil	θ_s	θ_r	α	m	n	The value of optimization function
Fine sand removal	0.368321	0.084231	0.013621	0.757213	4.397781	0.0003765
Fine sand moisture absorption	0.373268	0.079564	0.014843	0.789142	4.754387	0.0001621
Gravel dehydration	0.269456	0.058873	0.022987	0.669168	3.039765	0.0000608
Gravel moisture absorption	0.268786	0.055818	0.021951	0.698853	3.439871	0.0001089

4 Conclusions

From the results of the above analysis, we can draw the following conclusions:

(1) Using TLBO to optimize the parameters of VGE, it simplifies the difficulty of estimating parameters of VGE to some extent and has more efficient and accurate than some swarm intelligent optimization methods. Therefore, it is a new way for solving the parameters of VGE.

(2) The TLBO has the advantages of simple, fast computation speed, high accuracy, high automation degree and strong generality and is easy to understand. It can also be used to solve other complex nonlinear optimization problems.

Acknowledgments. This work is financially supported by the National Natural Science Foundation of China with the Grant No. 61573157, the Fund of Natural Science Foundation of Guangdong Province of China with the Grant No. 2014A030313454, the Education Department of Jiangxi province of china science and technology research projects with the Grant No. GJJ14807.

References

1. Van Genuchten, M.T.: A closed-form equation for predicting the hydraulic conductivity of unsaturated soils. Soil Sci. Soc. Am. J. **44**(5), 892–898 (1980)
2. Han, X.-W., Shao, M.-A., Horton, R.: Estimating van Genuchten model parameters of undisturbed soils using an integral method. Soil Sci. Soc. China **20**(1), 55–62 (2010). (in Chinese)
3. Cui, Y.-P., Chi, C.-M.: A simple method of estimating parameters in the Van Genuchten model of sand soil. Hubei Agric. Sci. **54**(7), 1696–1698 (2015). (in Chinese)
4. Mo, Y.-B., Liu, F.-Y., Ma, Y.-Z.: Improved artificial glowworm swarm optimization algorithm for solving parameters of Van Genuchten equation. Comput. Sci. **40**(11), 131–139 (2013). (in Chinese)
5. Du, G.-M., Zhang, Y.-L.: Calculate the van Genuchten equation parameters based on PSO. J. Irrig. Drainage **31**(6), 60–63 (2012). (in Chinese)
6. Xing, C.-M., Dai, Y., Yang, L.: Solving parameters of Van Genuchten equation by improved harmony search algorithm. J. Comput. Appl. **32**(8), 2159–2164 (2012). (in Chinese)
7. Wang, Y.-Q., Wang, Y.-H., Zhang, K.-D.: The parametric optimization of Van Genuchten equation based on pattern search algorithm. Rural Water Conservancy Hydropower China **9**, 8–14 (2010). (in Chinese)
8. Ma, Y.J., Shen, B.: Estimating parameters by solving Van Genuchten equation using the damper least square method. Trans. Chin. Soc. Agric. Eng. **21**(8), 179–180 (2005). (in Chinese)
9. Xu, X.-J., Tu, F.-F., Huang, X.-P.: Application of differential evolution in optimizing parameters of Van Genuchten equation. J. Hefei Univ. Technol. **31**(11), 1863–1866 (2008). (in Chinese)
10. Liao, L.-X., Shao, X.-H., Xu, J.-Z.: Estimating parameters of Van Genuchen equation based on intelligent algorithms. SHUILI XUEBAO **10**, 696–700 (2007). (in Chinese)

11. Rao, R.V., Savsani, V.J., Vakharia, D.P.: Teaching-learning-based optimization: a novel method for constrained mechanical design optimization problems. Comput.-Aided Des. **43**(3), 303–315 (2011)
12. Tuo, S.-H., Yong, L.-Q., Deng, F.-A.: Survey of teaching-learning-based optimization algorithms. Appl. Res. Comput. **30**(7), 1933–1938 (2013). (in Chinese)
13. Wang, J.-S., Yang, Z.-F., Chen, J.-J., Wang, Z.-M.: Study on water hysteresis in aerated soil. SHUILI XUEBAO **2**, 1–6 (2000). (in Chinese)

Analysis of Network Management and Monitoring Using Cloud Computing

George Suciu[1,2], Victor Suciu[2], Razvan Gheorghe[1], Ciprian Dobre[1], Florin Pop[1(✉)], and Aniello Castiglione[3]

[1] Faculty of Automatic Control and Computers Computer Science Department, University Politehnica of Bucharest, Bucharest, Romania
razvan.gheorghe@beia.ro, {ciprian.dobre,florin.pop}@cs.pub.ro
[2] R&D Department, BEIA Consult International, Bucharest, Romania
{george,victor.suciu}@beia.ro
[3] Department of Computer Science, University of Salerno, Salerno, Italy
castiglione@ieee.org

Abstract. In the near future the number of equipment connected to the Internet will greatly increase, so that further development of applications meant to verify their operations will be required. Monitoring represents an important factor in improving the quality of the services provided in cloud computing, given the fact that it allows scaling resource utilization in an adaptive manner. This paper aims to provide a solution for the monitoring of network devices and services, allowing administrator to verify connectivity of the equipment, their performances and network security. The main contribution of the paper consists in proposing an integrated solution that is deployed in the cloud for monitoring all the network components. Finally, the paper discusses the main findings and advantages for a reference implementation of the monitoring system using a simulated network.

Keywords: Network monitoring · Cloud computing · Nagios · Network management

1 Introduction

It is considered that in the near future the number of equipment which will be connected to the Internet will greatly increase, so that further development of applications meant to verify their operations will be required. Furthermore, it is expected that the number of 50 billion devices connected to the Internet will be reached in 2020, compared to 15 billion devices in 2015 [1].

Because networks of large operators are vast, troubleshooting the various problems that may arise can take a long time if a centralized solution is not used. A simple problem like the downtime of a web site can have multiple possible explanations: a faulty router, an out of service firewall or simply the server hosting the web site being down. All these are possible causes of a rather simple problem, but if the network is extensive, solving these problems can take even a few hours if the explanations presented above are taken one at a time.

© Springer Science+Business Media Singapore 2016
K. Li et al. (Eds.): ISICA 2015, CCIS 575, pp. 343–352, 2016.
DOI: 10.1007/978-981-10-0356-1_36

Cloud Computing represents a relatively new concept that refers to an integrated service offered as a whole application, which offers access to information and data storage without the user having to know the physical location and configuration of the systems providing these services. Cloud computing is a general term for anything that involves delivering services on the Internet. It can be divided into three categories namely, Software as a Service (SaaS), Platform as a Service (PaaS), Infrastructure as a Service (IaaS) in terms of classification by mode of delivery.

One of the main challenges associated to cloud computing is the resource monitoring, due to the lack of information and control regarding the customization of the parameters which describe the system. The current monitoring solutions are not entirely accurate in the cloud computing systems, given the fact that usually the resources are virtualized [2, 15].

We can thereby define the monitoring of a centralized network as being a system built to monitor the performances of a network at any time and also to notify the network administrator about arising problems.

The ideal monitoring system must meet lots of requirements, the most important being:

- The integration of all the network components into the monitoring system that is deployed in the cloud. Thus, essential information can be gathered from all the network entities, which leads to an accurate understanding of the network situation. Furthermore, problem solving is accelerated.
- Increasing productivity: a company will be more productive if the network problems are solved quickly. When the network is working poorly, a quality monitoring system will quickly detect the problem and alert the IT department of the company. Consequently, data loss in a network is prevented, which will determine the productivity to decrease.
- Efficiency: an ideal monitoring system produces a decrease in the network maintenance cost. Instead of multiple monitoring departments, each one supervising an area, it can all be reduced to the employment of a single system administrator who can monitor the network from a centralized location. Thus, the hours spent troubleshooting can become more productive.

The rest of the paper is organized as follows: Sect. 2 details some monitoring applications, Sect. 3 presents the components of the monitoring system and Sect. 4 presents the monitoring system implementation. Section 5 presents the results of simulations and the concluding section summarizes the contributions of this paper.

2 Related Work

In this section, we analyze related work and main monitoring applications used nowadays for networks. On today`s communications market, competition in monitoring applications is big, thus resulting in many companies dealing with their implementation. Among the most important applications are: Cacti, Zabbix, SCOM (System Center Operations Manager) 2012 and Wireshark.

We will further detail each approach, specifying the advantages and disadvantages.

2.1 Cacti

Cacti [3] is an open source monitoring application (free for users) based on a web server, being in fact a frontend for the standard monitoring technology RRDtool (Round Robin Database tool) [4]. Moreover, Cacti allows an easier use for inexperienced users of RRDTool.

Operation of Cacti can be reduced to three defined steps:

- Data processing: data is processed using a pooling system of the equipment connected to Cacti. More accurately, the network administrator can determine the status of a network equipment in real time, using the SNMP (Simple Network Management Protocol) protocol. Any equipment that has SNMP configured can be queried by Cacti.
- Data storage: data is stored using RRDtool. RRDtool is a data base which gathers information from the monitored network elements and then stores it efficiently and displays it as graphs.
- Presentation of data: at this point, data is processed in graphics and then presented to the network administrator.

2.2 Zabbix

Zabbix [5] is an open source monitoring application created in 2001 by Alexei Vladishev and it is used to monitor many network parameters, as well as the integrity of the servers. This uses a flexible notification mechanism which allows the administrator to receive an e-mail for every incident. The application supports both the pooling method outlined above and the trapping part (receiving data from various equipment without a prior request) [6].

In terms of software, the structure of the application is as follows:

- The Zabbix server, to which the agents report the information from various network equipment.
- The data base, where data from agents are stored.
- The web interface, where information about the network status can be accessed.
- The agent, which is installed on the equipment that is intended to be monitored. It sends the results to the main server.

2.3 SCOM 2012

System Center Operations Manager 2012 edition [7] represents the managing platform for the operating systems which features a single interface that can display data such as connectivity of the equipment, their performances, network security. It can also be used for computers that use Windows operating system and Linux. The application is used as client-server, an agent who collects and forwards data to the central SCOM server being installed on each equipment that is intended to be monitored. Later on, it can send notifications, depending on the event severity. For every type of monitored equipment (data base, Linux server, Apache web server) specific management packs are defined. For these management packs, filtering rules for monitored information are established, thus providing great flexibility.

2.4 Wireshark

Wireshark [8] is an application that analyses packets traversing the network, trying to display them to the user in a detailed and easy manner. It works as a measuring device, examining what happens at the protocol network level in order to be able to create graphs using the obtained results. The main advantage is the detailed manner in which it verifies the packets and, on the other hand, the main disadvantage is that it does not have a notification system like other monitoring applications. It is used especially for debugging.

3   Components of the Monitoring System

In this section we present the components and protocols that enable the cloud based operation for a network management and monitoring system. Furthermore, we present the components of a simulated network. Similar approach was presented in [16].

A. OpenStack

OpenStack [9] is a free open source cloud computing platform. The concept of open source describes how to produce or develop certain finished products, to allow users to engage in production or development. OpenStack is developed in the first stage model IaaS. The technology of this model consists in series of interrelated projects with, which activities are coordinated and monitored the processing, storage and network resources on activities. These activities are passed through a data center.

The main target of the open source project is to make OpenStack Cloud implementing a simple solution with a broad set of features. From this point of view, the IaaS model, OpenStack can provide many types of services such as basic services, warehousing services (Storage) sharing services (Shared Services) and service-level (High Level Services).

Furthermore, OpenStack provides an IaaS solution through a variety of complementary services, each service offering an API that helps the integration with other platforms and services.

On top of the OpenStack components there are deployed several services such as the Image one [10], which enables users to discover, register, and retrieve the virtual machine images.

B. Simulated Network

The main components of the simulated network are presented in Fig. 1 and described below.

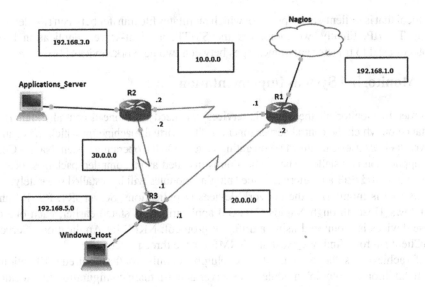

Fig. 1. The main components of the simulated network.

The main components of the simulated network are as follows:

- **Applications_Server:** A virtual machine having the Ubuntu 14.04 operating system, on which are installed HTTP (*Hypertext Transfer Protocol*), FTP (*File Transfer Protocol*), SMTP (*Simple Mail Transfer Protocol*) servers and also the monitoring agent used by Nagios Core, NRPE [11].
- **Windows_Host:** A virtual machine on which Windows 7 is installed as an operating system and NsClient++, used by Nagios Core, as a monitoring.
- **Nagios:** A virtual machine having the Ubuntu 14.04 operating system, on which the central server of Nagios Core is installed. Nagios Core is an open source version created by Nagios, representing now a standard for monitoring. Nagios monitors permanently the network equipment to verify the precision of their operation. The monitoring system used by Nagios serves two main components: hardware and software. The hardware component represents the physical equipment of the network: computers, printers, routers, servers, etc. The software component represents the processes run by the physical network equipment, such as: supported web sites, server applications, etc. By making this physical delimitation, Nagios manages to quickly identify network issues.
- **R2 and R3 routers** are responsible with traffic routing between equipment from different networks. The routing protocol used is RIP (*Routing Information Protocol*).

C. Used Protocols

In order to facilitate users' access to information on the Internet on World Wide Web (WWW) servers, HTTP (Hypertext Transfer Protocol) protocol is being used, where response time is the main parameter monitored [12]. Next, we monitored the FTP [13] protocol that is a client-server protocol which facilitates file transfer between two devices using TCP/IP. Finally, we monitored the SMTP, a client-server, application level protocol used to transfer email messages between two network devices [14].

4   Monitoring System Implementation

In order to monitor all the network services, we need to define a central monitoring instance on which the central server functions. The virtual machine on which the central server operates uses Ubuntu 14.04 operating system. On the operating system Nagios Core 3.0 application is installed. The application is installed as precompiled packages located at /etc/nagios3 and afterwards the "plugin" modules will be installed separately.

Its role is monitoring the network devices (Applications_Server, R1, R2, R3 and Windows_Host) through Nagios Core 3.0 application. As stated earlier, each one of these devices is monitored using a different protocol: NRPE for Applications_Server, NSClient++ for Windows_Host and SNMP for the three routers.

To achieve this, the server must have "plugin" modules for the required verifications.

In addition to "plugin" modules, the server needs a main configuration file, located at /etc/nagios3, named nagios.cfg. Within it are defined attributes, including the location where the application will write its logs, the location of the secondary configuration files for each new monitored equipment, the username used for GUI authentication and the interval between verifications.

After setting up all the aspects of connectivity between equipment and creating statements for each system, Nagios Core 3.0 will operate properly. Network administrators must be able to observe the behavior of services and equipment in real time and also to predict which services and equipment cause more problems.

All of this is achieved through a graphical interface that can be used by any existing browser. The graphical interface gives details about the status of services and equipment, their history, performances and many more besides.

For the graphical interface to function, setting up a web server using CGI is required to generate dynamic web pages and also a module using PHP and a programming language for faster functioning.

5   Measurement Results

For each of the routers is monitored the response time, the status of the port 1 and uptime of the router, as shown in Figs. 2, 3 and 4.

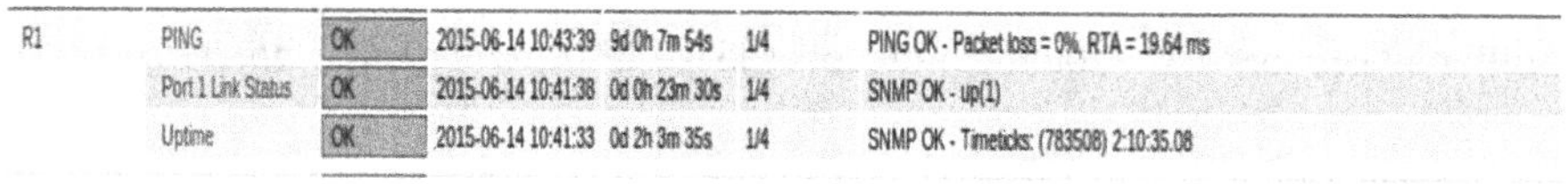

R1	PING	OK	2015-06-14 10:43:39	9d 0h 7m 54s	1/4	PING OK - Packet loss = 0%, RTA = 19.64 ms
	Port 1 Link Status	OK	2015-06-14 10:41:38	0d 0h 23m 30s	1/4	SNMP OK - up(1)
	Uptime	OK	2015-06-14 10:41:33	0d 2h 3m 35s	1/4	SNMP OK - Timeticks: (783508) 2:10:35.08

Fig. 2. Services monitored for the router R1.

From Fig. 2 it can be seen that for the router R1:

- PING command has 0 % packet loss and RTA of 19.63 ms;
- status 1 port is functional;
- the time that the system has been in continuous operation is 2 h 10 min and 35.8 s.

R2	PING	OK	2015-06-14 10:43:31	0d 1h 26m 37s	1/4	PING OK - Packet loss = 0%, RTA = 27.84 ms
	Port 1 Link Status	OK	2015-06-14 10:43:48	0d 0h 21m 20s	1/4	SNMP OK - up(1)
	Uptime	OK	2015-06-14 10:41:50	0d 2h 3m 18s	1/4	SNMP OK - Timeticks: (785076) 2:10:50.76

Fig. 3. Services monitored for the router R2.

From Fig. 3 it can be seen that for the router R2:

- PING command has 0 % packet loss and RTA of 27.84 ms;
- status 1 port is functional;
- the time that the system has been in continuous operation is 2 h 10 min and 50.76 s.

R3	PING	OK	2015-06-14 10:41:42	9d 0h 2m 5s	1/4	PING OK - Packet loss = 0%, RTA = 26.70 ms
	Port 1 Link Status	OK	2015-06-14 10:43:39	0d 0h 21m 29s	1/4	SNMP OK - up(1)
	Uptime	OK	2015-06-14 10:43:58	0d 2h 1m 10s	1/4	SNMP OK - Timeticks: (798750) 2:13:07.50

Fig. 4. Services monitored for the router R3.

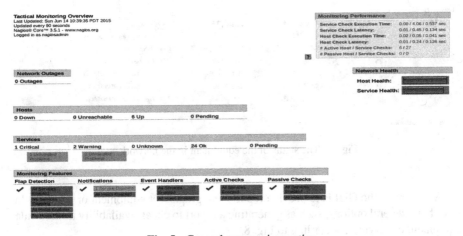

Fig. 5. General system view option.

From Fig. 4 it can be seen that for the router R3:

- PING command has 0 % packet loss and RTA of 26.70 ms;
- status 1 port is functional;
- the time that the system has been in continuous operation is 2 h 13 min and 7.50 s.

Through the network general view option (Fig. 5) we can see if there are common problems with equipment or network services how long it lasts the verifications or the duration between two consecutive checks.

There is also an option to view a map of the network, as shown in Fig. 6. Through its, the application knows the position of the equipment.

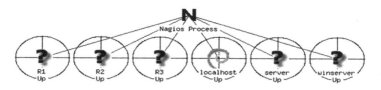

Fig. 6. Map of the monitored network.

Customers view option gives the opportunity to observe what equipment works or not, according to Fig. 7. We can also find out the date and time that occurred last check, the time when the equipment was placed into the application and effect of the command PING on equipment.

Host Status Details For All Host Groups

Limit Results: 100

Host ⬆⬇	Status ⬆⬇	Last Check ⬆⬇	Duration ⬆⬇	Status Information
R1	UP	2015-06-08 12:26:10	0d 1h 25m 51s	PING OK - Packet loss = 0%, RTA = 34.11 ms
R2	UP	2015-06-08 12:26:00	3d 1h 51m 43s	PING OK - Packet loss = 0%, RTA = 36.72 ms
R3	UP	2015-06-08 12:26:50	3d 1h 50m 53s	PING OK - Packet loss = 0%, RTA = 39.55 ms
localhost	UP	2015-06-08 12:27:40	211d 8h 50m 36s	PING OK - Packet loss = 0%, RTA = 0.03 ms
server	UP	2015-06-08 12:28:30	3d 1h 49m 13s	PING OK - Packet loss = 0%, RTA = 16.88 ms
winserver	UP	2015-06-08 12:29:20	2d 7h 11m 9s	PING OK - Packet loss = 0%, RTA = 36.34 ms

Results 1 - 6 of 6 Matching Hosts

Fig. 7. The status of the equipment in the network.

A feature of the GUI is to provide reports and graphs for equipment or services. In it we have several options, such as generating a report to check availability thrust while equipment or services, according to Fig. 8.

From the "problems" option can be seen warnings on the equipment or service for faster troubleshooting of network issues, as shown in Fig. 9.

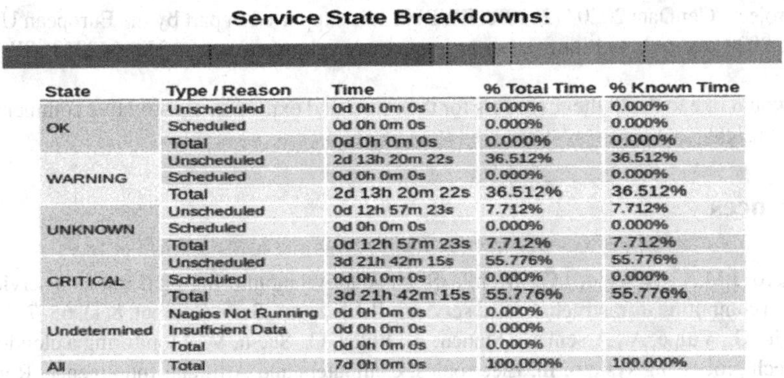

Service State Breakdowns:

State	Type / Reason	Time	% Total Time	% Known Time
	Unscheduled	0d 0h 0m 0s	0.000%	0.000%
OK	Scheduled	0d 0h 0m 0s	0.000%	0.000%
	Total	0d 0h 0m 0s	0.000%	0.000%
	Unscheduled	2d 13h 20m 22s	36.512%	36.512%
WARNING	Scheduled	0d 0h 0m 0s	0.000%	0.000%
	Total	2d 13h 20m 22s	36.512%	36.512%
	Unscheduled	0d 12h 57m 23s	7.712%	7.712%
UNKNOWN	Scheduled	0d 0h 0m 0s	0.000%	0.000%
	Total	0d 12h 57m 23s	7.712%	7.712%
	Unscheduled	3d 21h 42m 15s	55.776%	55.776%
CRITICAL	Scheduled	0d 0h 0m 0s	0.000%	0.000%
	Total	3d 21h 42m 15s	55.776%	55.776%
	Nagios Not Running	0d 0h 0m 0s	0.000%	
Undetermined	Insufficient Data	0d 0h 0m 0s	0.000%	
	Total	0d 0h 0m 0s	0.000%	
All	Total	7d 0h 0m 0s	100.000%	100.000%

Fig. 8. Changes of the system in the last 7 days.

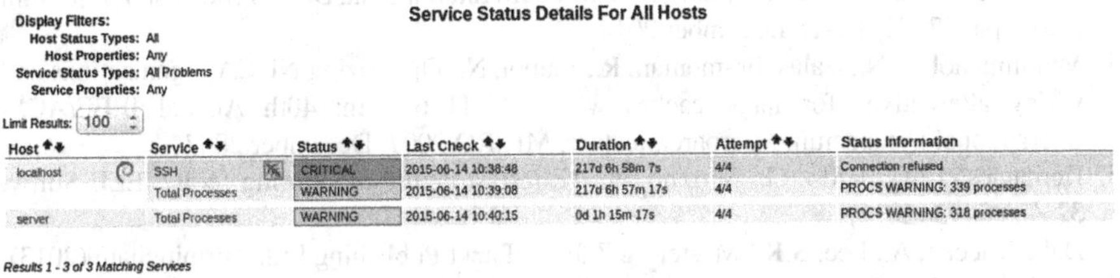

Display Filters:
  Host Status Types: All
  Host Properties: Any
  Service Status Types: All Problems
  Service Properties: Any

Service Status Details For All Hosts

Limit Results: 100

Host	Service	Status	Last Check	Duration	Attempt	Status Information
localhost	SSH	CRITICAL	2015-06-14 10:38:48	217d 6h 58m 7s	4/4	Connection refused
	Total Processes	WARNING	2015-06-14 10:39:08	217d 6h 57m 17s	4/4	PROCS WARNING: 339 processes
server	Total Processes	WARNING	2015-06-14 10:40:15	0d 1h 15m 17s	4/4	PROCS WARNING: 318 processes

Results 1 - 3 of 3 Matching Services

Fig. 9. The period during which the system will not issue notifications.

6 Conclusions

The results from simulations give us great flexibility on the multitude of devices and services that can be monitored and their presentation to the network administrator very easy to understand and very extensive as options.

For the average user, Nagios Core 3.0 shows its presence through a very good operating rate of networks and can monitor nearly any existing device, from a computer with a limited relevance to a router or server that is very important.

Thus, with this monitoring system we can perform the checking of very common contract clauses related to the total uptime of network components of 99.9 %.

As future work we envision to develop a portable container Docker-based solution.

Acknowledgments. The work has been funded by the Sectorial Operational Programme Human Resources Development 2007-2013 of the Ministry of European Funds through the Financial Agreement POSDRU/159/1.5/S/134398. The work is supported by in part by UEFISCDI Romania under grants: "Scalable Radio Transceiver for Instrumental Wireless Sensor Networks - SaRaT-IWSN" (20/2012), MobiWay (PN-II-PT-PCCA-2013-4), DataWay (PNII-RU-TE-2014-4-2731) and by European Commission by: grant no. 262EU/2013 - "eWALL" support project, grant no. 337E/2014 - "Accelerate" project and by FP7 IP project no. 610658/2013 "eWALL for Active Long Living". This work has been partially supported by the Italian Ministry of Research within

PRIN project "GenData 2020" (2010RTFWBH). It is supported in part by the European Union's Horizon 2020 research and innovation program under grant agreement No. 643963 (SWITCH project).

We would like to thank the reviewers for their time and expertise, constructive comments and valuable insight.

References

1. Calero, J.M.A., Aguado, J.G.: MonPaaS: an adaptive monitoring platform as a service for cloud computing infrastructures and services. IEEE Trans. Serv. Comput. **8**(1), 65–78 (2015)
2. Suciu, G., Vulpe, A., Arseni, S., Stancu, A., Butca, C., Suciu, V.: Monitoring a cloud-based speech processing system. In: Electronics, Computers and Artificial Intelligence, Romania (2015)
3. Arcaro, S., Di Carlo, S., Indaco, M., Pala, D., Prinetto, P., Vatajelu, E.I.: Integration of STT-MRAM model into CACTI simulator. In: 2014 9th International Design and Test Symposium (IDT), pp. 67–72, 16–18 December 2014
4. Muralimanohar, N., Balasubramonian, R., Jouppi, N.: Optimizing NUCA organizations and wiring alternatives for large caches with CACTI 6.0. In: 40th Annual IEEE/ACM International Symposium Microarchitecture, MICRO 2007, December 2007
5. Hernantes, J., Gallardo, G., Serrano, N.: IT infrastructure-monitoring tools. IEEE Softw. **32**(4), 88–93 (2015)
6. Dalle Vacche, A., Lee, S.K.: Mastering Zabbix. Packt Publishing Ltd., Birmingham (2013)
7. Marik, O., Zitta, S.: Comparative analysis of monitoring system for data networks. In: 2014 International Conference on Multimedia Computing and Systems (ICMCS). IEEE (2014)
8. Orebaugh, A., Ramirez, G., Beale, J.: Wireshark & Ethereal Network Protocol Analyzer Toolkit. Syngress, Rockland, MA (2006)
9. Lee, C.A., Desai, N.: Approaches for virtual organization support in OpenStack. In: 2014 IEEE International Conference on Cloud Engineering (IC2E), March 2014
10. Suciu, G., Halunga, S., Ochian, A., Suciu, V.: Network management and monitoring for cloud systems. In: Electronic, Computers and Artificial Intelligence International Conference, Romania (2014)
11. Magda, S.M., Rus, A.B., Dobrota, V.: Nagios-based network management for Android, Windows and Fedora Core terminals using Net-SNMP agents. In: 2013 11th Roedunet International Conference (RoEduNet) (2013)
12. Jestratjew, A., Kwiecien, A.: Performance of HTTP protocol in networked control systems. IEEE Trans. Industr. **9**(1), 271–276 (2013). doi:10.1109/TII.2012.2183138
13. Netto, J.E., Paulicena, E.H., Silva, R.A., Anzaloni, A.: Analysis of energy consumption using HTTP and FTP protocols over IEEE 802.11. In: Latin America Transactions. IEEE (2014)
14. Sureswaran, R., Al Bazar, H., Abouabdalla, O., Manasrah, A.M., El-Taj, H.: Active e-mail system SMTP protocol monitoring algorithm. In: 2nd IEEE International Conference on Broadband Network and Multimedia Technology, IC-BNMT 2009, October 2009
15. Ghit, B., Pop, F., Cristea, V.: Epidemic-style global load monitoring in large-scale overlay networks. In: 2010 International Conference on P2P, Parallel, Grid, Cloud and Internet Computing (3PGCIC), pp. 393–398. IEEE (2010)
16. Dobre, C., Pop, F., Cristea, V.: A simulation framework for dependable distributed systems. In: International Conference on Parallel Processing-Workshops, ICPP-W 2008, pp. 181–187. IEEE, September 2008

A Game-Theoretic Approach to Network Embedded FEC over Large-Scale Networks

Christian Esposito[1]([✉]), Aniello Castiglione[1], Francesco Palmieri[1],
and Massimo Ficco[2]

[1] Department of Computer Science, University of Salerno, Fisciano, Salerno, Italy
christian.esposito@dia.unisa.it, castiglione@ieee.org, fpalmieri@unisa.it
[2] Department of Industrial and Information Engineering,
Second University of Naples, Aversa, Caserta, Italy
massimo.ficco@unina2.it

Abstract. An efficient multicast communication is crucial for many parallel high-performance scientific (*e.g.*, genomic) applications involving a large number of computing machines, a considerable amount of data to be processed and a wide set of users providing inputs and/or interested in the results. Most of these applications are also characterized by strong requirements for the reliability and timely data sharing since involved in providing decision support for critical activities, such as genomic medicine. In the current literature, reliable multicast is always achieved at the expenses of violations of temporal constraints, since retransmissions are used to recover lost messages. In this paper, we present a solution to apply a proper coding scheme so as to jointly achieved reliability and timeliness when multicasting over the Internet. Such a solution employs game theory so as to select the best locations within the multicast tree where to perform coding operations. We prove the quality of this solution by using a series of simulations run on OMNET++.

Keywords: Application level multicast · Loss tolerance · Forward Error Correction · Game theory

1 Introduction

Current scientific (*e.g.*, genomic) applications are characterized by computation-intensive processes running on several distributed machines and a vast amount of data being processed in parallel so as to achieve high performance. Such applications witness a large number of users jointly collaborating by providing data to be processed and/or interested in accessing to the results of heavy parallel computations on such input data. To this aim, multicast is considered the pivotal technology which platforms for large-scale scientific applications are built upon, thanks to their peculiarity of offering decoupled and scalable communications. However, nowadays these QoS features are not sufficient alone so as to obtain an effective platforms for large-scale scientific applications. In fact, such applications are characterized by other non-functional requirements, such as reliability

© Springer Science+Business Media Singapore 2016
K. Li et al. (Eds.): ISICA 2015, CCIS 575, pp. 353–364, 2016.
DOI: 10.1007/978-981-10-0356-1_37

and timeliness. This is due to the use of the computed results in critical processing such as genomic medicine, and the monetizing access to scientific findings. Therefore, studying how to cope with faults in a timely manner when multicasting is a key issue to be resolved.

Despite nodes can be made as reliable as needed so as to face the most common and frequent errors in high performance computing applications, there is no control on the network dynamics, so message losses represent the main source of inefficiency and need to be particularly take care. Moreover, in the current solutions for reliable data dissemination, the gain of reliability is always achieved at the expenses of the achievable delivery time. Therefore, the goal of this paper is to fill the above mentioned gap in reliable publish/subscribe services by proposing an innovative reliability strategy that guarantees message multicasting without compromising the notification performances. Specifically, we propose a method based on Forward Error Correction [19], where the generation of redundant packets is not only performed at the root, but also in some interior nodes within the multicast tree. Our work aims at proposing a distributed algorithm to find the best nodes where performing FEC in order to support a good degree of reliability without incurring in a too high traffic overhead within the network.

The reminder of the paper is organized as follows. Section 2 provides an overview of the current techniques to provide loss-tolerance in multicast services deployable over Internet, and points out their main limitations to jointly provide timeliness and reliability. In Sect. 3, we describe in details our approach; while, in Sect. 4 we report experimental results of our simulation-based study to assess the quality of our approach. We conclude in Sect. 5 with conclusions and some remarks on future work.

2   Background and Open Issues

Reliable Multicast has been an active research topic in the last decades, and several protocols have been developed to support a reliable dissemination of messages both in TLM and ALM [10,12]. Such protocols are traditionally grouped in two main classes: the ones based on *Temporal Redundancy* use retransmissions to recover the lost messages, and the ones based on *Spatial Redundancy* forward additional information along to the application data so to recover dropped messages without using retransmission. The latter one represents the first solution investigated within the context of Computer Networks to tolerate message losses. In fact, such solution, known as Automatic Repeat reQuest (ARQ) [14], has met a considerable success for unicast communications, *e.g.*, TCP uses a variant of ARQ to ensure reliable transmission of data over IP. For this reason, it has been applied also in the most mature publish/subscribe services available in the market. For a concrete example, the products compliant to the OMG specification called Data Distribution Service (DDS) [8] adopt ARQ as the method for providing reliability. However, ARQ suffers from serious scalability and reliability limitations due to the centralization of the recovery duties at the publisher side.

To overcome such a drawback, retransmission-based approaches have evolved by performing recovery duties in a distributed manner. The widely-known practical example of such techniques is Gossiping [5], which distributes the recovery responsibility among the participants of a multicast group. There are different gossiping schemes, where the core logic is to have each node exchanging at a given time (e.g., when new data is received or at the expiration of a timeout) its state (e.g., the least received messages) with a set of randomly-chosen nodes among the ones constituting the system. Any possible data losses are detected by comparison and corrected through retransmissions.

The main criticism to all the retransmission-based schemes is that a high reliability degree is always obtained at the expenses of a worsening of the achievable performances. In fact, these approaches exhibit a reactive nature since recovery actions are taken only after a loss has been detected. These schemes are based on the certainty that it is only needed to retransmit a message once in order to correctly deliver it. However, there is no assurance to successfully receive the retransmitted message, so it may be needed to keep on retransmitting a given message several times before it can be correctly received by the interested destination. In fact, message deliveries over wide-area networks exhibit not-negligible bursty loss patterns [21], i.e., a message has a considerable probability P to be lost during the delivery and the succession of consecutive dropped messages has an average length ABL greater than two. In addition, [15] has shown that loss statistics are not constant during the day due to the higher amount of traffic, especially HTTP requests, corresponding to intense web surfing activities during specific hours of the day. So, the number of consecutive retransmissions that are needed to correctly deliver a message cannot be predicted beforehand, and the loss-affected delay is unpredictable and characterized by severe performance fluctuations. To have an empirical evidence of these issues we have considered the latency for the event notification between a publisher and three subscribers. The communication is realized by means of a publish/subscribe service complaint to the OMG DDS specification, which, as mentioned, adopts retransmissions on top of UDP for loss tolerance. We have plugged the four computers running the publisher and the subscribers by means of a network emulator called Shunra Virtual Enterprise (VE), which allows users to recreate a specific network behavior. We have considered three different scenarios with different network configuration starting with a network exhibiting scarce isolated losses to one with a high number of losses occurring in burst (more details on these scenarios are given in [3]). As it is possible to notice in Fig. 1(a), as soon as the network conditions worse we have an increase in the average of the delivery time and the occurrence of remarkable spikes in the measured latencies. In fact, in the figure we have a progressive augment of the interquartile rage in the delivery time, which is an evidence of severe performance fluctuations. The performance penalty can be found not only when ARQ is used, but also for more advanced retransmission-based schemes, such as gossiping, as seen in a previous work [4].

The most practicable solution to provide jointly reliability and timeliness is to proactively take reliability actions even if losses have not occurred by adopting a spatial redundancy rather than the temporal one. Specifically, multicast

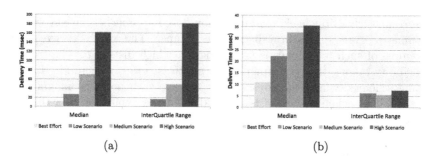

Fig. 1. Latency experienced in three different scenarios with (a) ARQ and (b) Used on top of a UDP-based publish/subscribe service.

services have to apply Forward Error Correction (FEC): the multicaster generates redundant data by encoding the one to be distributed, such redundancy can be used to reconstruct the original application data when some losses occurred. Unlike ARQ, the overhead imposed by FEC does not depend on the network conditions, making the delivery time more predictable. To have an empirical evidence of this statement, we have re-executed the previous experiments without using any fault-tolerant capability of the adopted publish/subscribe service and inserting a FEC approach in the publishing and subscribing applications. Figure 1(b) shows that the performance costs of the FEC-enabled loss recovery is moderate with compared with the ones illustrated in Fig. 1(a), and we do not find the same high values for the interquartile range, meaning that the performance fluctuations are minimal. In fact, we can see that the interquartile range is slightly larger than the one with only the best-effort transport, and the increasing of losses causes small changes in this range. Also, the increase in the median value is lower than the one in Fig. 1(a). The main criticism against FEC schemes is that the achievable reliability depends on the proper tuning of the applied redundancy. In the case such redundancy is lower than the loss rate applied by the network, then dropped packets cannot be reconstructed and a given event is irremediably lost. On the contrary, thanks to the closed control loop applied in retransmission.based approaches, we are always able to recover dropped packets.

Although FEC is strongly adopted in TLM solutions, in ALM ones it has not found the same enthusiastic use. Generally speaking, there are two available approaches [6] to embody FEC techniques in ALM solutions, depending on where the encoding of additional data is performed. On the one hand, End-to-End FEC consists in having the multicaster to encode the messages to be delivered, while all the destinations decode the received packets in order to obtain the original message even if the network was affected by losses. Such approach can be abstracted as follows: a FEC layer is placed between the ALM protocol and the business application, so messages generated by the application are encoded before been passed to the ALM, which will carry on the dissemination, while the packets received by the ALM will be decoded in order to be delivered to

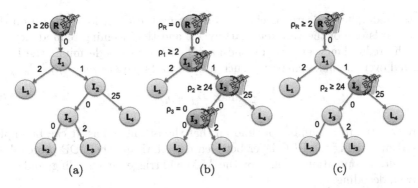

Fig. 2. The redundancy degree applied by the different introduced FEC approaches: (a) End-to-End FEC, (b) Link-by-Link FEC, and (c) Network-Embedded FEC. The average loss rate is indicated per each link in the multicast tree.

the application. This method is easy to implement, due to such layer separation, while it is strongly affected by an issue that we can refer as "the boat unbalanced by the heaviest" (BUH) problem. Specifically, FEC redundancy degree is usually decided by the encoder with respect to the worst loss pattern experienced along the path towards the i-th destination, namely λ_i:

$$\rho \geq \max_{i=0,\ldots,N} \lambda_i, \tag{1}$$

where N is the total number oh nodes within the tree. So, if only few paths exhibit heavy losses, the multicaster has to generate a large amount of redundant data, even if the other destinations do not need it. A concrete example is shown in Fig. 2(a), where coding applied at the root with a degree computed with Eq. 1. To evaluate the efficiency of such solution in terms of traffic load, we have to measure the link stress, *i.e.*, the number of redundant packets passing through a given link per each packet containing the application data to be delivered. In this example, such metric assumes a value equal to 20,5. This overwhelming redundancy degree, and the consequent traffic load, may overload the nodes that need less redundancy and/or cause serious congestion in certain portions of the network. Such a situation of highly-heterogeneous loss pattern is not rare and can be due to two main reasons:

1. The loss pattern of a node depends on the ones placed along the path to the root of the multicast tree. Therefore, the nodes closer to the root typically need a redundancy much lower than the ones that are further away.
2. Wide-area networks are made of the interconnection of different routing domains, each characterized by a certain network behaviour that varies over time; therefore, Internet-size ALM can have nodes in different domains experiencing heterogeneous loss patterns.

On the other hand, Link-by-Link FEC allows every node to perform encoding and decoding, in order to protect the delivery on each link from message losses, and the redundancy degree is chosen considering the worst loss pattern on the link towards one of the children, namely λ_j, and the incoming redundancy ρ_{in} (*i.e.*, the redundancy applied by the parent of the i-th node minus the losses occurred over the link connecting such a node to its parent: $\rho_{in} = \rho_j - \lambda_i$):

$$\rho_i \geq \max_{j=0,...,C} \lambda_j - \rho_{in}, \tag{2}$$

where C is the total number of children that the i-th node has. It can be implemented by putting the FEC layer between the ALM and the UDP protocol, so every delivery operation invoked by the ALM will trigger an encoding, and every receive a decoding.

This method is more flexible since the redundancy degree is chosen only with respect to the quality of an overlay link between two nodes. A concrete example is shown in Fig. 2(b), where coding applied by each interior node with a degree computed with Eq. 2. In this case, the link stress is almost halved to the value of 10.57 than the one in Fig. 2(a). Therefore, this resolves the BUH problem, but also causes strong degradations in performance due to the continuous execution of the two coding operations at every overlay node.

The severe drawbacks that affect these two approaches have limited so far the efficient usage of FEC techniques in the context of ALM solutions. This paper aims at proposing a novel FEC technique that can be effectively applied to ALM solutions in order to achieve both timeliness and reliability in a multicast communication.

3   An Optimized Distributed FEC Scheme

An appealing approach to mitigate the issues affecting the above mentioned ways to embody FEC-based spatial redundancy within an ALM solution is represented by Network-Embedded FEC [22]: allowing only a subset of interior nodes, called *codecs*, to perform encoding. In this way, the redundancy degree is chosen based on the worst loss pattern on the path towards one of the underlying codec in the tree, and C in Eq. 2 becomes the total number of nodes belonging to the subtree rooted at the i-th codec. This adapts the applied redundancy to better face the loss patterns exhibited by subtrees delimitated between two codecs to optimize the need of a flexible setting of the redundancy and to avoid excessive over-provisioning without worsening the achievable reliability degree in case of message losses. Such approach does not provide a coarse-grain control over the applied redundancy degree as seen for End-to-End FEC. On the other hand, since the number of codecs is kept lower than the total number of interior nodes, it does not incur the performance overhead experienced by the Link-by-Link FEC. Figure 2(c) shows a practical example where coding is applied only by an interior node, and link stress is equal to 10.85, close to the one of Link-by-Link FEC, but with less coding operations.

The problem of finding where to place codecs within a multicast tree can be formulated in the following way. Consider a set L of m possible locations in the multicast tree where a codec con be placed, a set U of n destinations of the multicast communication conveyed by the tree, and a n x m matrix D with the transportation costs $d_{i,j}$ for delivering a piece of information from the location j to the destination i, for all $j \in L$ and $i \in U$. In addition, let us define two sets of decision variables: (i) $y_j = 1$, if a codec is placed in $j \in L$, and 0, otherwise; and (ii) $x_{ij} = 1$, if the destination i receives redundancy from a codec located in $j \in L$, and 0, otherwise. The objective is to find codec locations so as to minimize the sum of these costs:

$$min \sum_{i \in U} \sum_{j \in L} d_{i,j} x_{i,j} + \sum_{j \in L} \alpha_j, \qquad (3)$$

subject to:

$$\sum_{j \in L} x_{i,j} = 1, \quad \forall i, \qquad (4)$$

$$x_{i,j} \leq y_j, \quad \forall i, j, \qquad (5)$$

$$\sum_{j \in L} y_j = p \leq n = |L|, \qquad (6)$$

$$x_{i,j}, y_j \in \{0, 1\}. \qquad (7)$$

where we mean as the transportation costs $d_{i,j}$ the redundancy that the codec in location j has to generate according to Eq. 2, and we indicate with α_j the costs of placing a codec in the j-th location. Such codec opening costs are needed to make unfeasible the trivial solution of placing a codec in all the interior nodes. The constraint in Eq. 4 indicates that each destination has to receive redundancy from only one codec; the one in Eq. 5 prevents each destination to receive redundancy from unactivated codec; and the total number of codecs in the multicast tree is set equal to p, lower than the total number of interior nodes within the multicast tree, by the constraint in Eq. 6. Such formulation belongs to a class of optimization problems known as the *P-Median Problem*, with the exception being the number of median not fixed a-priori, and can be solved directly as a Mixed-Integer Program (MIP), or it is possible to relax the integral constraints and solve the corresponding pure Linear Program (LP) [11,18]. The resolution of such an optimization problem can be centralized at a node of the tree, or distributed among all the nodes.

Although a centralized solution allows simplifying the problem resolution and taking optimal decisions, it is unfeasible in a large-scale distributed system since collecting global information is extremely difficult, if not impossible. In addition, even if assuming possible to acquire any global knowledge, it still exhibits serious scalability limits that prejudice its usage for systems composed by a large number of nodes. In fact, the time to collect loss statistics in a centralized point linearly increases with the number of nodes within the multicast tree. Moreover, the memory required to store all the collected data and/or the load to resolve the optimization may overwhelm the resources of the central decision node. A distributed approach has the strength to overcome the applicability issues of the previous solution for large-scale systems, since codec placement is performed using local computations at each node of the multicast tree avoiding turning to a central decision maker with global knowledge of the system. However, this advantage is paid at the expenses of obtaining a placement that is slightly far away from the optimal one.

The p-median problem, and more generally the facility location optimization problem, has been an active research topic within the context of game theory [17], leading to the so-called facility location games. Such a rich branch of the literature is rooted in the seminal work of Hotelling [9] and Downs [2]: we have a set of p players that simultaneously select a location, with the impossibility of two, or more, players to pick up the same location, to attract the maximum number of consumers and minimize the relative costs. More formally, let us consider an undirected graph (X, E), consisting in a set of nodes, namely X, and a set of edges connecting two nodes, namely E. Within the set of the available nodes, we define a subset of all the nodes, namely $Y \subseteq X$, containing the nodes where a player can be located. We consider a set of players $P := c_1, c_2, \cdots, c_p$ of finite size $p \geq 2$. Formally, the strategy set for each player $c \in P$ is defined as $S^c = Y$, such that a strategy of a player is the selection of a node s^c in Y. Combining the strategy sets of all the players, namely $S = S^{c_1} \times S^{c_2} \times \cdots \times S^{c_p}$, a strategy profile $s \in S$ implies a certain payoff to each player c, namely $\Phi^c(s)$, which are aggregated in the so-called profile of payoffs denoted as π. The payoff is the gain achievable by a player to serve a set of consumers considering the cost to open a facility and the transportation costs, as above mentioned. The scope of the game is to determine the best strategy profile that implies the maximum payoff for all the players. Despite the several possible formulations that came out within the literature, we can roughly describe facility location games in terms of a non-cooperative game [1] and a cooperative one [7].

In the first case, a game is defined non-cooperative since players are selfish, *i.e.*, there is no direct communication between the players, and each one only cares to maximize its own profit or to minimize its own costs without considering the state of the other players (with the eventuality of damaging them, even if it is not intentional). Then, the normal form for the noncooperative location game is given by $\Gamma = (P, S, \pi)$, with the objective of maximizing the payoff for all the players. One of the most studied aspects of such class of games is the existence of Nash equilibria, *i.e.*, given a certain strategy $s \in S$, it is not profitable for a player to select a different node than the one in the current strategy profile since moving to a neighbor node will not change or even reduce the achievable payoff, so a player has no incentive to change strategy:

$$\exists s \in S : \forall c \in P, \forall x \in Y, \pi^c(s^c, s^{-c}) \geq \pi^c(x, s^{-c}) \\ \rightarrow s \text{ is a Nash equilibrium.} \tag{8}$$

The demonstration of the existence of such equilibria is a known NP-hard problem and is resolved by means of theorems by making proper assumptions of the characteristics of certain elements of the game, such as the profit function or how customers prefers a player instead of another. For a concrete example, the authors in [20] demonstrate the existence of at leach one Nash equilibrium for such games and the conditions to induce such equilibrium are presented.

We can model our codec placement problem as a non-cooperative game with n players (where n is the number of interior nodes within the multicast tree), whose strategy is represented by a binary decision to play the role of codec or not. Rather than formalizing the payoff of each player, we consider its costs, according to Eq. 3. The global cost function specified in such an equation is decomposed in a set of local cost functions to be assigned to each player [13]. Specifically, let us indicate with S_i the binary value representing the strategy chosen by the i-th player (which is 1 if the i-th player decides to be a codec; otherwise, it is 0), and with U_i the set of nodes belonging to the subtree rooted at the node where the i-th player is located. The cost paid by the i-th player to follow its strategy can be formalized as follows:

$$C_i(S_i) = \alpha_i \cdot S_i + (\rho_i + \max_{j \in U_i} \lambda_j) \cdot (1 - S_i), \tag{9}$$

where ρ_i is the redundancy received by the i-th node, λ_j is the loss pattern experienced by the j-th node having the i-th node as parent within the multicast tree, and α_i is the cost to pay if the player has chosen to act as a codec. The game can start with a random strategy profile and evolve over the time where each player changes its strategy so as to minimize its costs formulated in Eq. 9. Such an evolution will bring to a stable solution represented by the Nash Equilibrium, where no player has an incentive to change its strategy. Based on the definition in Eq. 8, it is possible to see that a strategy profile s represents a Nash Equilibrium if and only if the two following conditions are guaranteed:

$$\exists i \in Y \text{ s.t. } \rho_i + \max_{j \in U_i} \lambda_j \leq \alpha_i \quad \nexists i \in Y \text{ s.t. } \rho_i < \alpha_i - \max_{j \in U_i} \lambda_j \tag{10}$$

The first condition indicates that the redundancy generated by a codec placed at the i-th node is never greater than the α factor of its children; so, none of its children has an incentive to act as a codec. While, the second condition states that, when the i-th node is the codec, it is not convenient to stop being a codec since the paid cost is already minimized. The second condition defines the control behavior of the agents to act as a codec or not, which can be formalized in Algorithm 1.

4 Experimental Evaluation

The scope of this section is to present experimental results that study the reliability and overhead (*i.e.*, the traffic generated by the approach in terms of number of packets per link) of the proposed strategic placement of FEC codecs. To achieve this aim, we implemented our solution by using the OMNET++[1] simulator, and decided to not use any real wide-area networks, such as PlanetLab, due to the uncontrollable loss patterns that make the obtained results

---

[1] www.omnetpp.org.

Algorithm 1. Agent behaviour with a non-cooperative game

Require: τ, α
 1: Function run()
 2: find_Noncoop_Solution(∞);
 3: EndFunction
 4: Function find_Noncoop_Solution(TermCond)
 5: isCodec ← false
 6: t ← 0;
 7: **while** $t < TermCond$ **do**
 8:    wait(τ);
 9:    demands ← collectDemands();
10:    **if** isCodec == true **then**
11:       ρ ← demands.max();
12:    **else**
13:       ρ ← getRedundancy(Parent);
14:       **if** ρ + demands.max() > α **then**
15:          isCodec ← true;
16:          ρ ← demands.max();
17:       **else**
18:          sendMessage(Parent, ρ + demands.max());
19:       **end if**
20:    **end if**
21:    t ← t +1;
22: **end while**
23: return isCodec;
24: EndFunction

---

non reproducible [3]. The used workload is characterized by exchanged messages with a size of 23 KB, the publication rate of 1 message per second and the total number of nodes equal to 256. The network behavior has $50\,ms$ as link delay, and 0.02 as PLR, based on a measurement campaign described in [3], with message losses not independent as proved by [16]. We have assumed that the coding and decoding time are respectively equal to $5\,ms$ and $10\,ms$. We have also considered the block size equal to 1472 bytes, that is the size of MTU in Ethernet, so that a single event is fragmented in 16 blocks. Finally, without loss of generality, we considered a system with 1 publisher and 39 subscribers, all subscribed to the same topic, *i.e.*, members of the same multicast tree. We have published 1000 events for each experiment, executed each experiment 3 times and reported the average. We have compared our approach with the delivery quality achievable with Scribe (with no reliability means) and with Scribe equipped with the Gossip strategy. As it is possible to notice in Fig. 3(c), our approach and Gossip are able to increase the reliability of the delivery protocol offered by Scribe, where Gossip is more reliable due to its retransmission behavior. From the rest of Fig. 3(c), we can notice that the higher reliability of the Gossip is obtained at the cost of worse performances, as it has also been observed in [4], while our approach have performances comparable to the one of Scribe with no reliability.

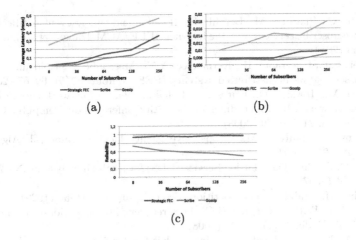

(a)   (b)

(c)

Fig. 3. Experiments results.

5 Conclusions

We have presented an innovative approach to guarantee a resilient and timely multicast communication in publish/subscribe services built on top of an application-level multicast infrastructure. Such approach adopts a Forward Error Correct approach that adds redundant packets to recover lost packets and does not get into the same timeliness limitations of approaches that use retransmission. On the other hand, we avoid the typical problems with coding approaches, *i.e.*, either the drawback of a coarse-grain control on the redundancy degree or a strong performance overhead. Specifically, we propose an optimal approach that places encoding functionality only in a selected subset of the interior nodes and the root of the tree. The problem of finding where to place the coding nodes as an optimization problem, which has been resolved by means of game theory. However, future work is to approach such a problem with other possible ways so as to find the best ones in terms of achievable reliability and costs.

Acknowledgment. This work has been partially supported by the Italian Ministry of Research within PRIN project "GenData 2020" (2010RTFWBH).

References

1. Cardinal, J., Hoefer, M.: Non-cooperative facility location and covering games. Theor. Comput. Sci. **411**(16–18), 1855–1876 (2010)
2. Downs, A.: An economic theory of political action in a democracy. J. Polit. Econ. **65**(2), 135–150 (1957)
3. Esposito, C.: Data Distribution Service (DDS) Limitations for Data Dissemination w.r.t. Large-scale Complex Critical Infrastructures (LCCI). Mobilab Technical report, March 2011. www.mobilab.unina.it

4. Esposito, C., Platania, M., Beraldi, R.: Reliable and timely event notification for publish/subscribe services over the internet. IEEE/ACM Trans. Networking **22**(1), 230–243 (2014)
5. Eugster, P.T., Guerraoui, R., Kermarrec, A.M., Massoulié, L.: Epidemic information dissemination in distributed systems. IEEE Comput. **37**(5), 60–67 (2004)
6. Ghaderi, M., Towsley, D., Kurose, J.: Reliability gain of network coding in lossy wireless networks. In: Proceedings of the 27th Conference on Computer Communications, pp. 2171–2179 (2008)
7. Goemans, M., Skutella, M.: Cooperative facility location games. J. Algorithms **50**(2), 194–214 (2004)
8. Group, O.M.: Data Distribution Service (DDS) for Real-Time Systems, v1.2. OMG Document (2007)
9. Hotelling, H.: Stability in competition. Econ. J. **39**(153), 41–57 (1929)
10. Jin, X., Yiu, W.P.K., Chan, S.H.G.: Loss recovery in application-layer multicast. IEEE MultiMedia **15**(1), 18–27 (2008)
11. Klose, A., Drexl, A.: Facility location models for distribution system design. Eur. J. Oper. Res. **162**(1), 4–29 (2005)
12. Levine, B.N., Garcia-Luna-Aceves, J.J.: A comparison of reliable multicast protocols. Multimedia Syst. **6**(5), 334–348 (1998)
13. Li, N., Marden, J.: Designing games for distributed optimization. IEEE J. Sel. Top. Signal Proc. **7**(2), 230–242 (2013)
14. Lin, S., Costello, D., Miller, M.: Automatic-repeat-request error-control schemes. IEEE Commun. Mag. **22**(12), 5–17 (1984)
15. Loguinov, D., Radha, H.: Measurement study of low-bitrate internet video streaming. In: Proceedings of the 1st ACM SIGCOMM Workshop on Internet Measurement, pp. 281–293 (2001)
16. Markopoulou, A., Iannaccone, G., Bhattacharyya, S., Chuah, C.N., Ganjali, Y., Diot, C.: Characterization of failures in an operational IP backbone network. IEEE/ACM Trans. Networking (TON) **16**(4), 749–762 (2008)
17. Osborne, M., Rubinstein, A.: A Course in Game Theory. The MIT Press, Cambridge (1994)
18. Reese, J.: Solution methods for the p-median problem: an annotated bibliography. Networks **48**(3), 125–142 (2006)
19. Rizzo, L., Vicisano, L.: RMDP: an FEC-based reliable multicast protocol for wireless environments. ACM SIGMOBILE Mobile Computing and Communications Review (MC2R 98) **2**(2), 23–31 (1998)
20. Vetta, A.: Nash equilibria in competitive societies, with applications to facility location, traffic routing and auctions. In: Proceedings of the 43rd Annual IEEE Symposium on Foundations of Computer Science, pp. 416–425 (2002)
21. Wang, F., Mao, Z., Wang, J., Gao, L., Bush, R.: A measurement study on the impact of routing events on end-to-end internet path performance. Comput. Commun. **36**(4), 375–386 (2006)
22. Wu, M., Karande, S.S., Radha, H.: Network-embedded FEC for optimum throughput of multicast packet video. J. Signal Process. Image Commun. **20**(8), 728–742 (2005)

The Research on Large Scale Data Set Clustering Algorithm Based on Tag Set

Qiang Chen[✉]

Department of Computer Science, Guangdong University of Science and Technology,
Dongguan 523083, Guangdong Province, China
cqjxnc@qq.com

Abstract. This paper proposes a set of SSLOKmeans algorithm that helps to guide the clustering before using tag memory resident, this algorithm can further improve the large-scale data sets clustering efficiency and clustering results of quality.

Keywords: Clustering · Tag set · SSLOKmeans algorithm · Data set

1 Introduction

Clustering has become the focus of research in the field of data mining, statistical analysis and compression algorithm. The application of clustering technology has data analysis and modeling, image segmentation, marketing, and so on. The classical clustering algorithms include Kmeans, PAM, DIANA and so on [1].

With the popularity of database technology and increasing requirement of modern commercial application, the clustering data set of this stage becomes very large. The data set that storage in large capacity hard disk and disk array cannot read into main memory for clustering. In order to achieve high efficiency and high quality clustering mining in limited memory space, many clustering algorithms for large scale data sets are optimized and improved. Some clustering algorithms have their own limitations. The problem is more prominent when the large scale data set is clustered, such as Kmeans algorithm. The algorithm itself is greatly affected by the initialization parameters.

In order to deal with large-scale data sets, large scale data sets clustering algorithm generally uses the method of sampling. Obviously, the sample data can only be used to represent the original data no matter what kind of sampling method. It cannot be used to replace the original data completely. Therefore, the sampling method based on sampling method and sampling data is very large.

In the practical work of clustering, there is little processing and analysis. It is directly on the original data set clustering approach. If combined application background in practice, more or less have the relevant background knowledge. For example, the expert gives a small amount of labeled data, industry background knowledge, etc. The idea of semi supervised learning is absorbed by scientific research personnel, and a small amount of labeled data and constrained information are used to guide the clustering process. So as to improve the quality of clustering results and the efficiency of clustering

K. Li et al. (Eds.): ISICA 2015, CCIS 575, pp. 365–372, 2016.
DOI: 10.1007/978-981-10-0356-1_38

algorithm. The algorithm uses a small number of labeled data to guide the clustering. It is based on a small number of labeled data to learn the distance measure, and the target function is modified according to the constraint information.

2   Related Work

In this paper, we choose the clustering algorithm for large scale data mining based on single computer environment. In clustering mining research, the idea of semi supervised learning is introduced into the research of large scale data set clustering algorithm. The whole process of large scale data set is guided by a small amount of labeled data and some constraints.

We studied the Scalable K-mean sand BIRCH algorithm and obtained the difference between Scalable K-means and BIRCH algorithm [2, 3].

(1) BIRCH algorithm is in the process of clustering is likely to occupy the memory cannot be estimated, only through the parameters of the T gradually adjust. The Scalable K-means algorithm has the upper limit value of the memory footprint of the cluster, it can be artificially set in advance and the large scale data set cannot exceed the memory occupancy.
(2) Although the BIRCH algorithm scans the original data set and the results can be obtained, but the quality of the clustering results is not high. In the actual use, the general will use the upgrade stage 4, once again scanned the original data set to improve the quality of clustering results. Therefore, the BIRCH algorithm generally needs to scan the original data set two times. The Scalable K-means algorithm can obtain high quality clustering results only by scanning the original data set.
(3) CS, DS and RS can be considered as a more general form of data compression in Scalable K-means algorithm.
(4) BIRCH algorithm in the process of clustering need to maintain the CF- tree. The process is relatively more consumption of resources, the need for more computing. The Scalable K-means algorithm is relatively simple and the clustering process is not required to maintain the complex data structure.
(5) Scalable K-means algorithm in the data compression stage need to sub clustering, in order to identify the compressed region and the BIRCH algorithm without the process.

3   Maths Fuzzy Theory—Fuzzy Assessment Method

3.1   SSLOKmeans Algorithm

SSLOKmeans Algorithm full name is semi-supervised labels one scan Kmeans. First, the data point set is Kmeans clustering when clustering is performed in memory. Clustering results based on MC (Main Cluster) [4]. Then the main cluster MC data set are discarded and compression and clearing the main memory space, to once again to be read into the data clustering to cluster until the completion of all the work of the cluster.

SSLOKmeans Algorithm by resident memory marker set to guide the clustering process and makes the clustering efficiency and clustering results of quality has been further improved [5]. Details of the process are as follows:

(1) Read (or otherwise read) data points from the data set to the finite memory, called the pipeline until the pipeline space is filled.
(2) Semi supervised clustering is limited in memory of the pipeline data points, SS, OUTS, labels, until convergence.
(3) The data points in the pipeline are compressed and discarded. Meet the conditions of the compressed data set replaced by SS, the corresponding data point is removed from memory. Meet the conditions of the discarded data points are removed from the main memory, enter the OUTS collection. In order to clear space in memory, read in the data set.
(4) The data points are still there if there is a data point. The process jump 1 if there is. Otherwise, the clustering process is over.

SSLOKmeans use the labels set can greatly improve the quality of clustering, at each stage of the limited memory of semi supervised clustering can as a kind of special E-M (Expectation-Maximization) algorithm.

At each stage of the discarded and compression processing, clearing out the memory of the point set by triples (sum, sumsq, Num) preserves the data information and clustering and algorithm SSLOKmeans has been retained labels sets in memory. So labels set can always guide for each stage of the clustering process. Therefore, SSLOKmeans uses the labels set to improve the quality of the whole data set.

3.2 Data Point Set Compression and Discard

1. Discarding process
   The discarding process is divided into two steps, which are respectively called as the main discard and the discarded. The main idea of this process is to discard two does not change the ownership point out cluster memory that stored in the discarded data set OUTS. The point of the Labels tag is not involved in the process, so as to guide the clustering process.
   (1) Main discarding process
   ① Set up a data point set in a MC cluster for x (see formula 1). Calculate the center point of the cluster for μ (see formula 2) and the co-variance matrix for Σ (see formula 3). The obtained data points to the cluster center of the Mahalanobis distance for D (see formula 4).

$$x_i = \{x_{i1}, x_{i2}, \ldots x_{in}\}^T, i = 1, 2, \ldots, N \tag{1}$$

$$\mu = \frac{1}{N} \sum_{i=1}^{N} x_i \tag{2}$$

$$\Sigma = \text{cov}(x_i^T) \tag{3}$$

$$D_M(x_i, \mu) = \sqrt{(x_i - \mu)^T \sum^{-1} (x_i - \mu)}$$ (4)

② Each data point to the cluster center Mahalanobis distance is D. Set a discard rate p %. According to the D distance to the ascending order. A P % data point in the front is removed from the main memory and save the statistical information into OUTS. Because they are closest to the clustering center, it is not easy to change the clustering properties. The dropping rate of P is given when the algorithm is initialized.

(2) Secondary discarding process

Set the jitter confidence interval parameters α, the center of each dimension ν for MC to meet P (see formula 5). According to Bonferroni inequality can be obtained (see formula 6). So the use of α/ν for each confidence interval for the dimension of confidence interval.

$$P((v_1 \in [L_1, U_1] \wedge ...\wedge v_n \in [L_n, U_n])) \geq 1 - \alpha$$ (5)

$$P((v_1 \in [L_1, U_1] \wedge ...\wedge v_n \in [L_n, U_n])) \geq 1 - \sum_{j=1}^{n} P(v_1 \notin [L_j, U_j]$$ (6)

For each dimension of each MC center point u calculated jitter interval for each data point in MC, center mu of each MC is in accordance with the corresponding jitter interval change, the center mu of the MC data points belong to deviate from the data points, the other center MC Mu bias the data, if the data points of the cluster relationship does not change is considered this point in the process of clustering is not changed, removed from the main memory into outs.

2. Compression process

The main idea is the high density compression process in the clustering process area is replaced by SS, in order to clear the main memory space. The data points within the region have the same category of ownership and the overall occurrence of the category changes. Therefore, it can be considered as a whole, and it can be used to store the clustering information with sufficient statistics.

SS can be used to replace what data point to determine the memory set, parameter setting. Kmeans clustering of the memory set, clustering number is n times that of the target number of clusters that do is formed than the main cluster of cluster area in order to perform compression process. The clustering results are satisfied if the specifications are met it (see formula 7).

$$\max_{j=1...n} \left\{ \sqrt{\frac{sumsq_j}{num} - \frac{sum_j^2}{num}} \right\} \leq \beta$$ (7)

It is considered that the clustering density is large enough, it can be replaced by SS, and if the SS is merged, it should be merged with SS. N value is too large, clustering efficiency drops quickly. The value is too small. The compression process is not obvious.

3.3 Semi Supervised Clustering Is Limited in Memory

SSLOKmeans use labels set clustering initialization and guide the limited memory and overcome the Kmeans is sensitive to the initial cluster center, cluster quality is not high, are described in detail such as algorithm.

(1) Initialization

Using labels set to initialize clustering center(see formulas 8 and 9). If the lack of a label, the class type were randomly generated.

$$labels = U_{i=1}^{k} label_i \tag{8}$$

$$\mu_h \leftarrow \sum \frac{1}{|label_h|} x, (x \in label_h, h = 1, \dots, k), t \leftarrow 0 \tag{9}$$

(2) Repeat until convergence or exceed the specified number of cycles
① Clustering distribution
Assign each data point D_i, SS, OUTS, labels to clustering (see formula 10). Formation set x_h^{i+1} Use three tuple instead of SS and OUTS as the center of the weight of the center point to participate in clustering.

$$h^* = \arg_h \min \left\| x_i - \mu_h^i \right\|^2 \tag{10}$$

② Calculating cluster center (see formula 11).

$$\mu_h^{i+1} \leftarrow \frac{1}{|x_h^{i+1}|} \sum_{x \in x_h} (i+1)^x \tag{11}$$

③ Calculating t (see formula 12).

$$t = t + 1 \tag{12}$$

For main memory data points, labels, SS and outs (in accordance with the afore-mentioned three tuple in clustering) for semi supervised clustering, contained in the labels of data points does not participate in the SS compression and discarded, so that the labels set can have guided clustering in main memory. In this process, the data points of the algorithm design labels can change the clustering attribution, and also allows the original labels to focus not on the new clustering; to prevent the initial labels concentration has the wrong tag, so that the algorithm has a certain error tolerance.

4 Experimental Results and Analysis

In order to verify the validity of the algorithm SSLO Kmeans, the basic idea of Random-Kmeans algorithm and the basic idea of Scalable-Kmeans algorithm are also proved.

Random-Kmeans, Scalable-Kmeans, SSLOKmeans algorithm is implemented using C++ language. Discovery and Data Mining (KDD98) Knowledge Conference was used to test the algorithm. KDD98 data set records the user's response to the request for charitable donations [8]. There are 95413 records in the data set, each of which has 481 dimensions, and 56 dimensions are selected from the data set.

The weights of each dimension of the data points are equivalent in this experiment. So the standard of each dimension is standardized (see formula 13).

$$normal_i = \frac{data_i - mean_i}{\sqrt{(sumsq_i/num) + (sum_i/num)^2}}, i = 1, 2, \ldots, n \tag{13}$$

Because it is not the data set generated by the program, the real clustering center of KDD98 is not known, in order to obtain the labels set, it can use the expert's domain knowledge to carry on the manual annotation. This experiment adopts the way that the Kmeans algorithm is used to cluster 100 times, and the clustering quality is the best when the distance clustering center is the closest of the 10 data points labels set. The parameters of the experiment are shown in Table 1, and the test results are shown in Figs. 1 and 2.

Table 1. Relative parameters setting

Parameter	Value
Confidence level for cluster means	95 %
Max std. Dev for β	1.5
Number of secondary cluster (N = 4)	20
Fraction of points discarded P %	20 %

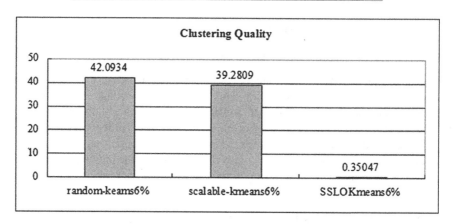

Fig. 1. Comparing analysis of clustering quality

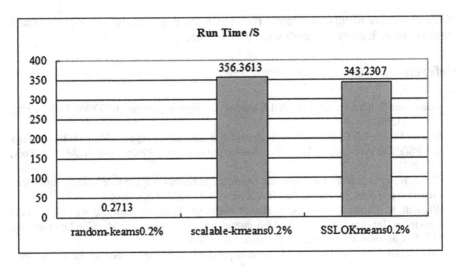

Fig. 2. Comparing analysis of run time

As can be seen from the results of clustering test (see Fig. 1), random-Kmeans 6 % can be seen that the quality of the cluster is the worst. This is a random sampling of the original data set, and any method of sampling cannot completely represent the original data. So the clustering results are affected by the original data set and the initial conditions. Scalable-Kmeans has achieved better clustering quality, but the quality of the clustering is still not high because of the sensitivity to the initialization parameters. The SSLOKmeans algorithm uses the labels set to guide the clustering, which greatly improves the clustering quality.

Finally, the effectiveness of the SSLOKmeans algorithm in the large scale data set is tested on the 32 bit platform. Single file can be processed under the Linux platform cannot exceed 2 GB, so the generated data set contains 4608159 data points. Each data point of the dimension is 100, a total of 1.7 GB. Set the number of clusters 5 and throughout the experiment randomly generated three different artificially generated data sets. The pipeline memory limit set for data set (10000 data points) of 0.2 %, for each data set, respectively, to perform Random-Kmeans, Scalable-Kmeans, SSLOKmeans five times each, running a total of 15 times, and finally to the average value as the evaluation of the quality of clustering algorithms. It can be seen from Fig. 2 that the SSLOKmeans algorithm for large scale data set is effective, which improves the quality of clustering results and accelerates the convergence and improves the efficiency of the algorithm.

5 Conclusion

Algorithm SSLOKmeans use labels set for large-scale data clustering, absorbing the idea of semi supervised learning in the limited memory iterative semi supervised clustering, and clustering to complete the work of the entire data set is proposed in the paper. The results show that the SSLOKmeans algorithm can improve the quality of the clustering results, accelerate the convergence of the clustering process and improve the efficiency of the clustering algorithm.

References

1. Kantabutra, S., Couch, A.L.: Parallel K-means clustering algorithm on NOWs. Tech. J. **1**(6), 243–248 (2000)
2. Alsabti, K., Ranka, S., Singh, V.: An efficient K-means clustering algorithm. In: Proceedings of IPPS/SPDP Workshop on High Performance Data Mining, 1998, pp. 1–6. ACM, New York, NY (2011)
3. Chang, H., Yeung, D.Y.: Locally linear metric adaptation with application to semi-supervised clustering and image retrieval. Pattern Recogn. **39**(7), 1253–1264 (2006)
4. Wagstaff, K., Cardie, C., Rogers, S., et al.: Constrained K-means clustering with background knowledge. In: Proceedings of the 18th International Conference on Machine Learning, pp. 577–584. Morgan Kaufmann, San Francisco, CA (2001)
5. Basu, S., Banerjee, A., Mooney. R.: Semi-supervised clustering by seeding. In: Proceedings of the 19th International Conference on Machine Learning, pp. 27–34. Morgan Kaufmann, San Francisco, CA (2002)
6. Basu, S., Bilenko, M., Mooney R.J.: A probabilistic framework for semi-supervised clustering. In: Proceedings of the Tenth ACM SIGKDD International Conference on Knowledge Discovery and Data Mining, pp. 59–68. ACM, New York, NY (2004)
7. Lu, Z., Leen, T.K.: Semi-supervised clustering with pair wise constraints: a discriminative approach. In: Proceedings of the 11th International Conference on Artificial Intelligence and Statistics, AISTATS 2007, pp. 299–306. Microtome Publishing, USA (2007)
8. Bradley, P.S., Fayyad, U., Reina, C.: Scaling clustering algorithms to large databases. In: Proceedings of the Fourth International Conference on Knowledge Discovery and Data Mining (KDD 1998), pp. 9–15. AAAI Press, New York, USA (1998)

Partitioned Parallelization of MOEA/D for Bi-objective Optimization on Clusters

Yuehong Xie[1], Weiqin Ying[1（✉）], Yu Wu[2], Bingshen Wu[1], Shiyun Chen[1], and Weipeng He[1]

[1] School of Software Engineering, South China University of Technology, Guangzhou 510006, China
yingweiqin@scut.edu.cn
[2] School of Computer Science and Educational Software, Guangzhou University, Guangzhou 510006, China

Abstract. The multi-objective evolutionary algorithm based on decomposition (MOEA/D) has a remarkable overall performance for multi-objective optimization problems, but still consumes much time when solving complicated problems. A parallel MOEA/D (pMOEA/D) is proposed to solve bi-objective optimization problems on message-passing clusters more efficiently in this paper. The population is partitioned evenly over processors on a cluster by a partitioned island model. Besides, the sub-populations cooperate among separate processors on the cluster by the hybrid migration of both elitist individuals and utopian points. Experimental results on five bi-objective benchmark problems demonstrate that pMOEA/D achieves the satisfactory overall performance in terms of both speedup and quality of solutions on message-passing clusters.

Keywords: Evolutionary algorithm · Bi-objective optimization · Decomposition · Parallelization · Message-passing clusters

1 Introduction

Many real-world optimization problems involve simultaneous optimization of multiple, possibly conflicting objectives [1]. Each of such multi-objective optimization problems (MOPs) requires to reconcile several conflicting objectives. So it's difficult to solve MOPs with deterministic methods. Evolutionary algorithms (EAs) are suitable and widely used for solving MOPs, since they can achieve a set of trade-off solutions simultaneously in one single run. Many multi-objective optimization evolutionary algorithms (MOEAs) have been developed during the last twenty years. Most of those are based on Pareto dominance. In recent years, some decomposition-based MOEAs, such as the multi-objective evolutionary algorithm based on decomposition (MOEA/D) [2] and the conical area evolutionary algorithm (CAEA) [3], have become popular and widely used for MOPs. MOEA/D explicitly decomposes a MOP into several scalar optimization subproblems and the offspring for each subproblem only needs to update

© Springer Science+Business Media Singapore 2016
K. Li et al. (Eds.): ISICA 2015, CCIS 575, pp. 373–381, 2016.
DOI: 10.1007/978-981-10-0356-1_39

the individuals of its neighboring subproblems. As a result, MOEA/D has a remarkable overall performance againts the dominance-based MOEAs.

But one possible drawback of MOEAs is that they may consume much running time before figuring out good results. That is particularly true for MOPs with complicated evaluation procedure, like protein structure prediction problems [4], truss structure design problems [5], wireless sensor network layout problems [6] and so on. To improve the efficiency of MOEA/D, several thread-based parallel versions of MOEA/D have been proposed by Nebro et al. [7,8]. These parallel versions are designed for shared-memory multiprocessors, but not yet available for distributed-memory, message-passing clusters.

To ulteriorly improve the efficiency of MOEA/D, a partitioned parallelization of MOEA/D, referred to as pMOEA/D, is presented in this paper in order to achieve remarkable time reduction on message-passing clusters for bi-objective optimization problems (BOPs). A partitioned island model is introduced in this paper for evenly partitioning the population over processors on a cluster. Furthermore, a hybrid migration policy is proposed for the cooperation between sub-populations among separate processors on the cluster in pMOEA/D. Finally, experiments on five bi-objective benchmark problems are presented to validate the effectiveness of pMOEA/D in this paper.

2   Partitioned Parallelization of MOEA/D

In this section, the partitioned parallelization of MOEA/D is designed for distributed-memory clusters using the message passing interface (MPI). A partitioned island model is at first introduced to partition a population evenly over processors on a cluster while a hybrid migration policy is adopted to share critical evolutionary information including both elitist individuals and utopian points among processors.

2.1   Partitioned Island Model

In MOEA/D, a BOP is decomposed into a series of subproblems, and each of them is assigned a weight vector which aggregates both individual objectives to one single scalar objective. In this paper, a partitioned island model based on weight vectors is proposed to evenly split the entire population for a BOP into several partitions, among which each evolves separately on an individual processor. Let N denote the size of the entire population P and q be the number of participating processors, then the r-th partition $P^{(r)}$, $r \in [0..q-1]$, has the size $N^{(r)} = N/q + 1$ if $r < N \bmod q$, and otherwise $N^{(r)} = N/q$ where mod indicates the modulus operation.

Besides, a certain amount of overlapping individuals from the adjacent partitions are also appended to each side of a partition in the partitioned island model so that the entire operation of updating neighboring individuals could be completed independently within its own processor. A partition $P^{(r)}$ and its overlapping portions constitute a separate sub-population or island $P^{\{r\}}$. Figure 1

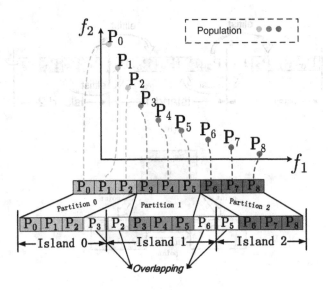

Fig. 1. Partitioned island model

presents the partitioned island model in a simple case where the population size $N=9$, the number of participating processors $q=3$, the size of overlapping portion $k=1$, and the sizes of resulting partitions are $N^{(0)}=N^{(1)}=N^{(2)}=3$. It's worth noting that the first partition $P^{(0)}$ only has k overlapping individuals appended to its rightmost side, the last partition $P^{(q-1)}$ only has k overlapping individuals appended to its leftmost side, and any other partition $P^{(r)}$, $r\in[1..q-2]$, has not only k ones to its leftmost side but also k ones to its rightmost side as shown in the case of Fig. 1. As a result, every partition is responsible for converging towards a segment of pareto front.

2.2 Hybrid Migration Policy

The sub-populations on separate processors exchange evolutionary information by a hybrid migration policy in the partitioned parallelization of MOEA/D. In this hybrid policy, the elitist individuals and the updated utopian points migrate between processors respectively conforming to an elitist migration policy and a utopian migration policy. On one hand, the elitist migration policy can share the fruits of evolution between adjacent sub-populations and accelerate the evolutionary convergence by regularly exchanging the updated overlapping individuals at each side of each sub-population, referred to as elitist individuals. According to the elitist migration policy, the elitist individuals at the leftmost side of any sub-population $P^{\{r\}}$, $r\in[1..q-1]$, emigrate to its previous one $P^{\{r-1\}}$ while those at the rightmost side of $P^{\{r\}}$, $r\in[0..q-2]$, will emigrate to its next one $P^{\{r+1\}}$, as shown in Fig. 2(a). The elitist individuals immigrating to any sub-population $P^{\{r\}}$, $r\in[0..q-1]$, would be utilized to update its neighbors in $P^{\{r\}}$ in the same manner as the update operator of neighbors in MOEA/D.

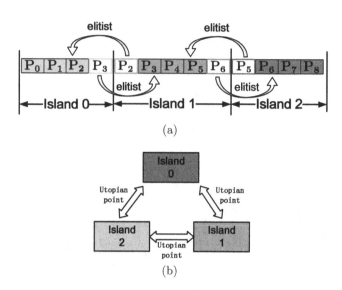

Fig. 2. Elitist migration policy (a) and utopian migration policy (b).

On the other hand, the utopian migration policy can propagate the best utopian points throughout sub-populations, which widen the search scope for the pareto front segment in the charge of each sub-population. A utopian point, often also called an ideal point, is composed of the best attainable value out of the given solutions set for each objective. The better utopian point can promote the population of MOEA/D to approximate the wider pareto front. Since a utopian point has to be maintained separately for any sub-population in pMOEA/D, the updated utopian point for any sub-population $P^{\{r\}}$ emigrates into both its previous one $P^{\{prev\}}$ and its next one $P^{\{next\}}$ where $prev=(i-1+q)\ mod\ q$ and $next=(i+1-q)\ mod\ q$ according to the utopian migration policy, as shown in Fig. 2(b). The utopian point immigrating to any sub-population $P^{\{r\}}$ helps improve the local utopian point for $P^{\{r\}}$ according to the update operator of utopian point in MOEA/D. In this paper, the hybrid migration policy is finally adopted so that the advantages of both pure policies can be taken.

3   Procedure of pMOEA/D

In this section, the procedure of pMOEA/D using the above partitioned island model with the hybrid migration policy is presented in Algorithm 1. The key idea behind pMOEA/D is that it distributes the evolution of population in Line 1–21 among a number of concurrent processes running on separate processors according to the partitioned parallelization. A sub-population is initialized on each process in Line 2–6. The function INITIALIZESUBPOPSIZE(N, q, r) in Line 2 generates the size $N^{\{r\}}$ of the subpopulation $P^{\{r\}}$ on the process r, the indexes I_f and I_l of the first and last individuals of the partition $P^{(r)}$ in $P^{\{r\}}$. Before

---

Algorithm 1. The main procedure of pMOEA/D

```
1: for each r∈[0..q−1] in parallel do //q : the number of processors
2:     N^{r}, I_f, I_l ← INITIALIZESUBPOPSIZE(N, q, r)
3:     λ ← INITIALIZESUBPOPWEIGHT(r)
4:     B ← INITIALIZESUBPOPNEIGHBORHOOD(λ)
5:     P^{r} ← INITIALIZESUBPOP()
6:     z ← INITIALIZELOCALUTOPIANPOINT(P^{r})
7:     z ← SYNCHRONIZEUTOPIANPOINT(z)
8:     gen ← 0; prev ← (r−1+q)%q; next ← (r+1−q)%q
9:     while gen≤MaxGen do //evolutionary loop
10:        for each i∈[I_f..I_l] do
11:            parents ← SELECTION(P^{r}, i)
12:            child ← CROSSOVER(parents)
13:            child ← MUTATION(child)
14:            EVALUATEFITNESS(child)
15:            INTERNALUTOPIANPOINTUPDATE(child, z)
16:            INTERNALSOLUTIONSUPDATE(child, i, P^{r})
17:        end for
18:        HYBRIDMIGRATION() //Migration Policy
19:     end while
20:     P ← SYNCHRONIZEPOP(P^{(r)})
21: end for
```

Algorithm 2. Hybrid migration

```
1: function HYBRIDMIGRATION()
2:     ELITISTMIGRATION()
3:     UTOPIANMIGRATION()
4: end function
5: function ELITISTMIGRATION()
6:     if gen%10 == 0 then
7:         for each i∈[0..I_f−1]∪[I_l+1..N^{r}−1] do //Elitist Migration
8:             if isIndivUpdated[i] then
9:                 nearest ← (i∈[0..I_f−1])?prev : next
10:                SENDELITISTINDIV(P_i^{r}, i, nearest)
11:                isIndivUpdated[i] ← false
12:            end if
13:        end for
14:        EXTERNALSOLUTIONSUPDATE(P^{r}, prev, next)
15:     end if
16: end function
17: function UTOPIANMIGRATION()
18:     if isUtopianUpdated then //Utopian Migration
19:         SENDUTOPIANPOINT(z, prev); SENDUTOPIANPOINT(z, next)
20:         isUtopianUpdated ← false
21:     end if
22:     z ← EXTERNALUTOPIANPOINTUPDATE(z, prev, next)
23: end function
```

the evolution, the root process 0 gathers local utopian points from all processes to update the global utopian point and then broadcasts it to all processes in SYNCHRONIZEUTOPIANPOINT(z) in Line 7. During the evolution, the major difference against the sequential MOEA/D is that those separate processes of pMOEA/D have to cooperate through the hybrid migration policy in Line 18, whose pseudocode is given in Algorithm 2. After the evolution, the root process 0 gathers local partitions from all processes to compose the final population in SYNCHRONIZEPOP($P^{(r)}$) in Line 20.

In Algorithm 2, the elitist migration policy in Line 6–15 is responsible for emigrating each elitist individual to its nearest adjacent process only once every 10 generations since the migration traffic of elitists is fair heavy. Besides, EXTERNALSOLUTIONSUPDATE($P^{\{r\}}, prev, next$) in Line 15 receives the elitist individuals from two adjacent processes and updates all their neighboring solutions in the local sub-population. Similarly, the utopian point migration policy in Line 18–22 is responsible for emigrating the updated utopian point to both adjacent processes and updating the local utopian point by receiving the utopian points from both adjacent processes.

4   Experimental Results and Discussion

In this section, several numerical experiments on five benchmark BOPs are carried out to validate the effectiveness of the partitioned parallelization with hybrid migration in pMOEA/D. The simulations are carried out on a cluster, which has a total of 20 computing nodes using Intel Core I5 3.20 GHz CPU and 8GB RAM and connected by Gigabit Ethernet. All the sequential and parallel MOEA/Ds are implemented in C++ using MPI. Each sequential or parallel MOEA/D adopts the aggregation function of Tchebycheff [2], the population size N=200, and the neighborhood size T=20 of subproblems, and terminates after the evolution of 400 generations in our experiments. The size of overlapping portion at each side of each subpopulation is set as $k=\frac{1}{2}T$ in each parallel MOEA/D. For the sake of fairness, a total of 30 statistically independent runs of each algorithm have been conducted for each problem.

In our experiments, pMOEA/D is firstly tested and compared with the sequential MOEA/D as well as the two variants of pMOEA/D, respectively with the elitist migration and the utopian migration, referred to as pMOEA/D^e and pMOEA/D^u, for five widely used benchmark BOPs: ZDT_1 (convex), ZDT_2 (non-convex), ZDT_3 (non-convex, disconnected), ZDT_4 (convex, multimodal) and ZDT_6(non-convex, nonuniformly spaced) [9]. 30 decision variables are used for ZDT_{1-3} while 10 decision variables are used for $ZDT_{4,6}$. The operators of crossover and mutation used in each sequential or parallel MOEA/D in this paper are the same as used in [3,10] for these benchmark problems. Besides, the inverted generational distance (IGD) metric used in [3] is also adopted to assess the quality of solutions discovered by each algorithm for each benchmark problem in our experiments.

The number of participating processors is firstly set as q=8. Figure 3 presents the convergence curve of average IGD score (in log scalar) in 30 runs of each

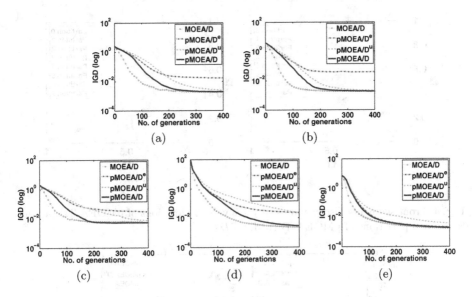

Fig. 3. Convergence curve of average IGD score in 30 runs of each sequential or parallel MOEA/D for ZDT_1 (a), ZDT_2 (b), ZDT_3 (c), ZDT_4 (d) and ZDT_6 (e).

sequential or parallel MOEA/D for each benchmark problem on the cluster. Figure 4 plots the front segments discovered, respectively, by 8 partitions of pMOEA/D^e in the run with the lowest IGD value and those by 8 partitions of pMOEA/D for ZDT_1. It can be inferred from Fig. 3 that pMOEA/D^e converges faster in the early stage of evolution than pMOEA/D^u for all benchmark problems, which indicates that the migration of elitist individuals can accelerate the evolutionary convergence effectively. However, it is also evident from Fig. 3 that pMOEA/D^u achieves the obviously better fronts for most of problems in the terminal stage of evolution than pMOEA/D^e. It's attributed to the fact that the migration of utopian points helps guide the evolutionary search towards the wider front by propagating the best utopian point throughout separate sub-populations.

Furthermore, as demonstrated in the case of ZDT_1 in Fig. 4, pMOEA/D^e without the migration of utopian points often converges towards only several narrow front segments and obtains an incomplete front that usually breaks between adjacent partitions due to the worse utopian points on some separate processors. Meanwhile, it can be inferred from Figs. 3 and 4 that pMOEA/D can converge to nearly the same complete front as the sequential MOEA/D by propagating utopian points. Figure 3 also indicates that pMOEA/D not only converges faster than pMOEA/D^u does in the early stage, but also achieves more complete fronts than pMOEA/D^e does in the terminal stage.

Figure 5 shows the curves of average CPU time in second spent by each parallel MOEA/D with the increase of participating processors (or partitions) from 1 to 8 for five problems on the cluster. Figure 5 indicates that all three parallel

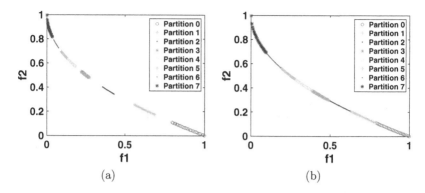

(a)   (b)

Fig. 4. Front segments discovered, respectively, by 8 partitions of pMOEA/D^e (a) and those by 8 partitions of pMOEA/D (b) for ZDT_1.

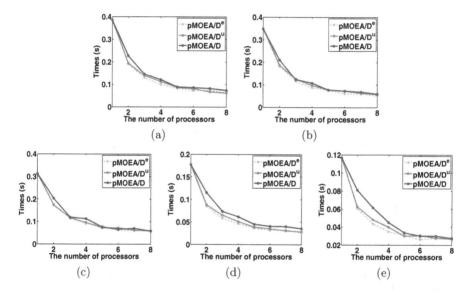

Fig. 5. Curves of average CPU time (in second) spent by each parallel MOEA/D with the increase of processors for ZDT_1 (a), ZDT_2 (b), ZDT_3 (c), ZDT_4 (d) and ZDT_6 (e).

MOEA/Ds can reduce the running time obviously for all five problems and pMOEA/D achieves nearly the same speedups as pMOEA/D^e and pMOEA/D^u though it uses the hybrid policy to migrate both elitist individuals and utopian points. Overall, it can be claimed that pMOEA/D performs obviously better in terms of quality of solutions than pMOEA/D^e and pMOEA/D converges faster than pMOEA/D^u while they obtains the similar speedups for benchmark problems.

5 Conclusions

In this paper, a partitioned parallelization of MOEA/D is presented to solve bi-objective optimization problems on message-passing clusters efficiently. In the partitioned island model of MOEA/D, the entire population of MOEA/D is partitioned into a certain number of sub-populations over separate processors on a cluster. Besides, the sub-populations cooperate among separate processors to guide the search towards the complete front quickly by the hybrid migration of elitist individuals and utopian points. Experimental results on bi-objective benchmark problems verify the satisfactory overall performance of pMOEA/D in terms of both speedup and quality of solutions on message-passing clusters. Our future research will focus on the parallelization of the conical area evolutionary algorithm (CAEA) on clusters.

Acknowledgments. This work is supported by the National Natural Science Foundation of China (No.61203310, 61503087), the Fundamental Research Funds for the Central Universities, SCUT (No.2013ZZ0048), the Pearl River S&T Nova Program of Guang-zhou (No.2014J2200052), the Natural Science Foundation of Guangdong Pro-vince, China (No.2015A030313204) and the China Scholarship Council (CSC) (No.201406155076, 201408440193).

References

1. Zhou, A., Qu, B., Li, H., Zhao, S., Suganthan, P.N., Zhang, Q.: Multiobjective evolutionary algorithms: a survey of the state of the art. Swarm Evol. Comput. **1**(1), 32–49 (2011)
2. Zhang, Q., Li, H.: MOEA/D: a multiobjective evolutionary algorithm based on decomposition. IEEE Trans. Evol. Comput. **11**(6), 712–731 (2007)
3. Ying, W., Xu, X., Feng, Y., Wu, Y.: An efficient conical area evolutionary algorithm for bi-objective optimization. IEICE Trans. Fundam. Electr. Commun. Comput. Sci. **E95-A**(8), 1420–1425 (2012)
4. Cutello, V., Narzisi, G., Nicosia, G.: A multi-objective evolutionary approach to the protein structure prediction problem. J. Royal Soc. Interface **3**(6), 139–151 (2006)
5. Coello, C.A., Christiansen, A.D.: Multiobjective optimization of trusses using genetic algorithms. Comput. Struct. **75**, 647–660 (2000)
6. Molina, G., Alba, E., Talbi, E.G.: Optimal sensor network layout using multi-objective metaheuristics. J. Univ. Comput. Sci. **14**(15), 2549–2565 (2008)
7. Nebro, A.J., Durillo, J.J.: A study of the parallelization of the multi-objective metaheuristic MOEA/D. In: Blum, C., Battiti, R. (eds.) LION 4. LNCS, vol. 6073, pp. 303–317. Springer, Heidelberg (2010)
8. Durillo, J.J., Zhang, Q., Nebro, A.J., Alba, E.: Distribution of computational effort in parallel MOEA/D. In: Coello, C.A.C. (ed.) LION 2011. LNCS, vol. 6683, pp. 488–502. Springer, Heidelberg (2011)
9. Zitzler, E., Deb, K., Thiele, L.: Comparison of multiobjective evolutionary algorithms: empirical results. Evol. Comput. **8**(2), 173–195 (2000)
10. Ying, W., Xu, X.: Bi-objective optimal design of truss structure using normalized conical-area evolutionary algorithm. Int. J. Adv. Comput. Technol. **4**(15), 162–171 (2012)

A Double Weighted Naive Bayes
for Multi-label Classification

Xuesong Yan[1(✉)], Wei Li[1], Qinghua Wu[2], and Victor S. Sheng[3]

[1] School of Computer Science, Chin University of Geoscience,
Hubei 430074, Wuhan, China
yanxuesongcug@gmail.com
[2] Faculty of Computer Science and Engineering,
WuHan Institute of Technology, Hubei 430074, Wuhan, China
wuqinghua@sina.com
[3] Department of Computer Science, University of Central Arkansas,
Conway, AR 72035, USA
ssheng@uca.edu

Abstract. Multi-label classification is to assign an instance to multiple classes. Naïve Bayes (NB) is one of the most popular algorithms for pattern recognition and classification. It has a high performance in single label classification. It is naturally extended for multi-label classification under the assumption of label independence. As we know, NB is based on a simple but unrealistic assumption that attributes are conditionally independent given the class. Therefore, a double weighted NB (DWNB) is proposed to demonstrate the influences of predicting different labels based on different attributes. Our DWNB utilizes the niching cultural algorithm to determine the weight configuration automatically. Our experimental results show that our proposed DWNB outperforms NB and its extensions significantly in multi-label classification.

Keywords: Multi-label classification · Naive Bayes · Cultural algorithm · Double weighted Naive Bayes

1 Introduction

Bayesian network is a popular data mining technique used in classification [1]. A highly practical Bayesian network model is Naive Bayes [2]. Naive Bayes is based on the Bayesian theorem and the simple but unrealistic assumption that attributes are conditionally independent given the class. Despite its simplicity, Naive Bayes can often outperform more sophisticated classification methods. Since it assumes that all the attributes are independent, the posterior probabilities of an instance belonging to a class are proportional to the product of the conditional probabilities of each attribute value. Figure 1 presents the framework of the structure for a simple Bayesian network [3].

This paper is to extend NB for multi-label classification. Before we discuss our extensions, we provide some notations which will be used later. We assume that a multi-label classification data set D with n attributes. And each instance can be described by the tuple of attribute values $X_i = \langle x_1, x_2, \cdots, x_n \rangle$, where x_n denotes the

© Springer Science+Business Media Singapore 2016
K. Li et al. (Eds.): ISICA 2015, CCIS 575, pp. 382–389, 2016.
DOI: 10.1007/978-981-10-0356-1_40

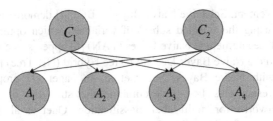

Fig. 1. The framework of a simple Bayes

value of the *n-th* attribute. The multi-label data set contains N instances, i.e., $D = \{X_1, X_2, \cdots, X_N\}$. We also assume the multi-label data set has m class labels. That is, each instance in D can be described by the tuple of its attribute values and its class labels $X_i = \langle x_1, x_2, \cdots, x_n | c_1, c_2, \cdots, c_m \rangle$.

Following Bayesian theorem and the label independence assumption in multi-label classification, NB is naturally extended to perform multi-label classification. It can determine whether an unseen instance belongs to every m class labels. Typically, we know that each training in stance has the information whether it belongs to the m class labels, presenting in a 0 and 1 vector with m items. According to Bayesian theorem and the label independence assumption, for each class label we can calculate $P(c_i = 1 | x_1, x_2, \cdots, x_n)$, which is the probability of an unseen instance $X = \langle x_1, x_2, \cdots, x_n \rangle$ belonging to the class label c_i. We can calculate $P(c_i = 0 | x_1, x_2, \cdots, x_n) = 1 - P(c_i = 1 | x_1, x_2, \cdots, x_n)$, which denotes the probability of the unseen instance not belonging to the class label c_i. When $P(c_i = 1 | x_1, x_2, \cdots, x_n) > P(c_i = 0 | x_1, x_2, \cdots, x_n)$, the unseen instance can be determine to belong to the class label c_i. That is, the extended NB makes predictions for each class independently. Its prediction for each class can be defined as:

$$P(c_i | X) = \arg \max_{c_i \in C} P(c_i) P(x_1, x_2, \ldots, x_n | c_i) \tag{1}$$

It is easy to estimate the prior probability $P(c_i)$ for each label from a multi-label data set. Based on the attribute independence assumption and the label independence assumption, we can calculate $P(x_1, x_2, \ldots, x_n | c_i)$ by directly following the way NB does. Thus, our extended NB can be defined as follows.

$$P(c_i = 1 | X) = \arg_{c_i \in C} P(c_i = 1) \prod_{j=1}^{n} P(x_j | c_i = 1)$$

$$P(c_i = 0 | X) = \arg_{c_i \in C} P(c_i = 0) \prod_{j=1}^{n} P(x_j | c_i = 0)$$

$$\tag{2}$$

As we know, the attribute independence assumption made by Naive Bayes harms its classification performance when it is violated in reality. In order to weak the attribute independence assumption of Naive Bayes while at the same time retaining its simplicity and efficiency, researchers have proposed many effective methods to further improve the performance of Naive Bayes. Three methods have demonstrated remarkable improvement. Selective Naive Bayes (SBC) [4] demonstrates a remarkable improvement by using the selected subset of variables, which optimizes its classification accuracy. Tree augmented Naive Bayes (TAN) [5] appears as a natural extension to Naive Bayes. And a Naive Bayes Decision-Tree Hybrid (NB-Tree) [6] has combined a decision tree with Naive Bayes. Sucar et al. In paper [7] proposed a method for chaining Bayesian classifiers that combine the strengths of classifier chains and Bayesian networks for multi-label classification. Quet et al. [8] proposed to tackle multi-label classification using Bayesian Theorem. They proposed two approaches, coined as Pair-Dependency Multi-label Bayesian Classifier (PDMLBC) and Complete-Dependency Multi-label Bayesian Classifier (CDMLBC).

Besides, another major way to help mitigate its primary weakness—attributes independence assumption by assigning weights to important attributes in classification. Since attributes do not play the same role in many real world applications, some of them are more important than others. A natural way to extend NB is to assign each attribute different weight to relax the conditional independence assumption. This is the main idea of weighted Naïve Bayes (i.e., WNB). Figure 2 presents the framework of the structure for a simple weighted Bayesian network.

$$P(c_i|X) = \arg_{c_i \in C} P(c_i) \prod_{j=1}^{n} P(x_j|c_i)^{w_j} \tag{3}$$

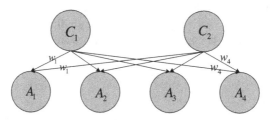

Fig. 2. The framework of a simple weighted Bayes

In recent years, how to learn the weights of weighted Naive Bayes to improve its performance has attracted many attentions. Researchers have proposed many useful methods to evaluate the importance of attributes for single-label classification, like Gain Ratio [9], CFS (Correlation-based Feature Selection) attribute selection algorithm [10], Mutual Information [11], Relief F attribute ranking algorithm [12] and so on. Hall [13] proposed an attribute weighted methods based on the degree to which they depend on the values of other attributes. Wu and Cai [14] uses a differential evolution algorithm to determine the weights of attributes, and then use these weights in the previously developed weighted Naïve Bayes (DE-WNB). Later on, they further proposed a dual weighted Naïve Bayes [15]. They firstly employed an instance similarity based method to weight each training instance. After that, they built an attribute weighted model based on weighted training data.

As we said before, this paper focuses on multi-label classification. We focus on reducing the effects of the attribute independence assumption by assigning weights to important attributes for different class labels in multi-label classification. Different from single-label classification, an instance belongs to more than one label from the class label set in multi-label classification. What's more, according to the influences of predicting different labels which depends on different attributes, we will propose a double weighted Naïve Bayes for multi-label classification (denoted as DWNB). The framework in Fig. 3 illustrates the structure of our simple double weighted Bayesian network.

$$P(c_i|X) = \underset{c_i \in C}{\arg} P(c_i) \prod_{j=1}^{n} P(x_j|c_i)^{w_{ji}} \tag{4}$$

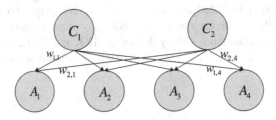

Fig. 3. The framework of a simple double weighted Bayes

How can we obtain the optimal weight configuration is the main task of our DWNB? In this paper, we will utilize the niching cultural algorithm (i.e., NCA) to determine the double weights for our DWNB. We simply call this specific DWNB as DWNB-NCA since then. Our experimental results show that DWNB-NCA produces considerably satisfying average accuracy for multi-label classification.

2 Reviewing Niching Cultural Algorithm

Cultural Algorithms are a class of computational models derived from observing the cultural evolution process in nature [16–18]. In this algorithm, individuals are first evaluated using a performance function. The performance information represents the problem-solving experience of an individual. An acceptance function determines which individuals in the current population are able to impact, or to be voted to contribute, to the current beliefs. The experience of these selected individual is used to adjust the current group beliefs. These group beliefs are then used to guide and influence the evolution of the population at the next step, where parameters for self-adaptation can be determined from the belief space. Information that is stored in the belief space can pertain to any of the lower levels, e.g. population, individual, or component. As a result, the belief space can be used to control self-adaptation at any or all of these levels. The cultural algorithm is a dual inheritance system with evolution taking place at the population level and at the belief level. The two components interact through a communications protocol. The protocol determines the set of acceptable individuals that are able to update the belief space. Likewise the protocol determines how the updated beliefs are able to impact and influence the adaptation of the population component.

According to the geographical isolation technology of nature [19], the basic concept of the isolation niching cultural algorithm (NCA) technique and evolution strategy is produced, which divides the initial population space of the cultural algorithm into several sub populations according to every class label in a multi-label data set, and these sub populations perform evolutions independently. In NCA, the evolution speed and scale of each sub population space depends on the average fitness value of each sub population(denoted as SPOP). We use two kinds of knowledge in the belief space, situation and normative knowledge, denoted as SK and NK respectively. According to different class labels, we divide the SK and NK into m parts (note that we have m classes), each for one class label, simply defined as SKK and NKK, where k means the kth class label. The framework diagram of the niching cultural algorithm is shown in Fig. 4.

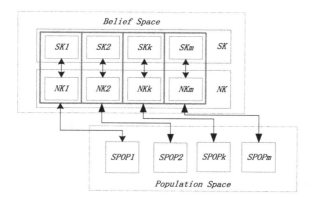

Fig. 4. The framework diagram of niching cultural algorithm

3   Experiment and Results

DWNB-NCA were tested on four benchmark multi-label data sets; each of them with different dimensions ranging from 6 to 174 labels, and from about 500 examples to more than 2400. All class variables of the data sets are binary. However, in some of the data sets the feature variables are numeric. The details of the data sets are summarized in Table 1. N is the size of a data set, m is the number of binary classes or labels, and d is the number of features.

Table 1. Charateristics the four multi-label data sets

No	Data set	N	d	m	Domain
1	CAL500	502	68	174	Music
2	Emotions	593	72	6	Music
3	Scene	2407	294	6	Vision
4	Yeast	2417	103	14	Biology

In our DWNB-NCA during our experiment, there are five artificial parameters: the maximum iterations, the sub-population initial size, the initial accept proportion of the acceptance function, the maximum allowable size, and the minimum allowable size of sub-population space. Their values are set as follows (Table 2).

Table 2. Parameters setting of NCA

Parameters name	Sub-population initial size	Maximum iterations	Initial accept proportion	Maximum size	Minimum size
value	40	200	0.2	50	30

We conduct extensive empirical comparisons for NB, WNB-CA (using the weighting base on cultural algorithm), and DWNB-NCA. The classification accuracy of each algorithm on each data set was obtained via 10 runs. And our algorithm is carried out on the training data set and evaluated on the test data set. We select the training set through a hierarchical sampling method. 70 % of the instances are used as a training set, and the remaining part as a testing set. Taking the importance of data fitting way in Naive Bayes in consideration, we use three common fitting ways of continuous attributes (Gaussian probability distribution, Cauchy probability distribution, discretization of continuous attributes), for the same data set. When using discretization of continuous attributes as the data fitting way, we choose ten (P1) and twenty (P2) discrete intervals respectively. Here, we use the classification accuracy as the criterion to assess the performance of algorithms. The higher the classification accuracy, the better the corresponding classification effect is.

Table 3 show the detailed maximum, minimum and average results to best of the experiments respectively for Gaussian and Cauchy probability distribution about the

Table 3. Experimental results to best of Gaussian and Cauchy probability distribution

Data set	Algorithm	Gaussian			Cauchy		
		MAX	MIN	AVE	MAX	MIN	AVE
CAL_500	NBMLC	0.8732	0.8574	0.8622	0.8721	0.8554	0.8635
	CA-WNB-P1	0.8893	0.8679	0.8813	0.8871	0.8737	0.8800
	CA-WNB-P2	0.8897	0.8716	**0.8825**	0.8890	0.8744	**0.8811**
	NCA-DWNB-P1	0.8757	0.8535	0.8636	0.8835	0.8534	0.8672
	NCA-DWNB-P2	0.8757	0.8535	0.8636	0.8835	0.8534	0.8672

(*Continued*)

Table 3. (*Continued*)

Data set	Algorithm	Gaussian			Cauchy		
		MAX	MIN	AVE	MAX	MIN	AVE
Emotions	NBMLC	0.6976	0.6798	0.6884	0.6976	0.6787	0.6892
	CA-WNB-P1	0.8059	0.7853	0.7938	0.8215	0.7850	0.8040
	CA-WNB-P2	0.8115	0.7900	0.7993	0.8215	0.7869	**0.8044**
	NCA-DWNB-P1	0.8604	0.8257	0.8377	0.8571	0.8075	0.8355
	NCA-DWNB-P2	0.8604	0.8257	**0.8379**	0.8590	0.8075	0.8358
Scene	NBMLC	0.8239	0.8195	0.8212	0.8195	0.8098	0.8151
	CA-WNB-P1	0.8693	0.8564	0.8630	0.8841	0.8652	0.8732
	CA-WNB-P2	0.8714	0.8592	0.8654	0.8848	0.8615	**0.8744**
	NCA-DWNB-P1	0.8773	0.8591	**0.8673**	0.8901	0.8421	0.8725
	NCA-DWNB-P2	0.8773	0.8591	**0.8673**	0.8901	0.8421	0.8725
Yeast	NBMLC	0.7749	0.7636	0.7688	0.7739	0.7688	0.7673
	CA-WNB-P1	0.8045	0.7787	0.7901	0.8129	0.7873	0.7948
	CA-WNB-P2	0.8051	0.7851	0.7933	0.8126	0.7831	0.7952
	NCA-DWNB-P1	0.8191	0.8029	**0.8101**	0.8139	0.7987	**0.8058**
	NCA-DWNB-P2	0.8191	0.8029	**0.8101**	0.8139	0.7987	**0.8058**

multi-label classification and standard deviation of the DWNB-NCA, WNB-CA, NB. The optimal predicted results were shown in boldface. DWNB-NCA achieves a significant improvement over NB. No matter what method of fitting, or what prediction technique is selected, compared with NB, DWNB-NCA wins in all data sets.

4 Conclusion

In this paper, we firstly reviewed typical attribute weighted methods for WNB, and then presented a DWNB in multi-label classification. After that, we developed an improved DWNB algorithm based on the niching cultural algorithm DWNB-NCA. Firstly, DWNB-NCA learns the best double weights using the niching cultural algorithm. And then it classifies the test instances using the DWNB multi-label classifier based on the double weights learned. We experimentally tested our new algorithm using the four benchmark multi-label data sets, and compared it to NB and WNB-CA. The experimental results showed that the accuracy of DWNB-NCA is much higher than NB, and a little higher than WNB-CA.

Acknowledgement. This paper is supported by Natural Science Foundation of China (No. 61402425, 61272470, 61305087, 61440060, 41404076), the Provincial Natural Science Foundation of Hubei (No. 2013CFB320, 2015CFA065).

References

1. Pearl, J.: Probabilistic Reasoning in Intelligent Systems: Networks of Plausible Reasoning. Morgan Kaufmann Publishers, Los Altos (1988)
2. Langley, P., Iba, W., Thompson, K.: An analysis of Bayesian classifiers. AAAI **90**, 223–228 (1992)
3. Xie, Z., Hsu, W., Liu, Z., Li Lee, M.: SNNB: a selective neighborhood based Naïve Bayes for lazy learning. In: Chen, M.-S., Yu, P.S., Liu, B. (eds.) PAKDD 2002. LNCS (LNAI), vol. 2336, pp. 104–114. Springer, Heidelberg (2002)
4. Langley, P., Sage, S.: Induction of selective Bayesian classifiers. In: Proceedings of the 10th International Conference on Uncertainty in Artificial Intelligence, pp. 399–406. Morgan Kaufmann Publishers Inc (1994)
5. Friedman, N., Geiger, D., Goldszmidt, M.: Bayesian network classifiers. Mach. Learn. **29**(2–3), 131–163 (1997)
6. Kohavi, R.: Scaling up the accuracy of Naive-Bayes classifiers: a decision-tree hybrid. In: KDD, pp. 202–207 (1996)
7. Sucar, L.E., Bielza, C., Morales, E.F., et al.: Multi-label classification with Bayesian network-based chain classifiers. Pattern Recogn. Lett. **41**, 14–22 (2014)
8. Qu, G., Zhang, H., Hartrick, C.T.: Multi-label classification with Bayesian theorem. In: 4th International Conference on Biomedical Engineering and Informatics (BMEI), vol. 4, pp. 2281–2285. IEEE (2011)
9. Zhang, H., Sheng, S.: Learning weighted Naive Bayes with accurate ranking. In: 4th IEEE International Conference on Data Mining, pp. 567–570. IEEE (2004)
10. Hall, M.: Correlation-based feature selection for discrete and numeric class machine learning. In: Proceedings of the 7th Intentional Conference on Machine Learning, Stanford University (2000)
11. Jiang, L., Zhang, H.: Weightily averaged one-dependence estimators. In: Yang, Q., Webb, G. (eds.) PRICAI 2006. LNCS (LNAI), vol. 4099, pp. 970–974. Springer, Heidelberg (2006)
12. Robnik-Šikonja, M., Kononenko, I.: Theoretical and empirical analysis of ReliefF and RReliefF. Mach. Learn. **53**(1–2), 23–69 (2003)
13. Hall, M.: A decision tree-based attribute weighting filter for Naive Bayes. Knowl.-Based Syst. **20**(2), 120–126 (2007)
14. Wu, J., Cai, Z.: Attribute weighting via differential evolution algorithm for attribute weighted Naive Bayes (wnb). J. Comput. Inf. Syst. **7**(5), 1672–1679 (2011)
15. Wu, J., Pan, S., Cai, Z., et al.: Dual instance and attribute weighting for Naive Bayes classification. In: 2014 International Joint Conference on Neural Networks (IJCNN), pp. 1675–1679. IEEE (2014)
16. Reynoids, R.: An introduction to cultural algorithms. In: Sebald, A.X., Fogel, L.J. (eds.) Proceedings of the 3rd Annual Conference on Evolutionary Programming, pp. 13 1–139. World Scientific Publishing, River Edge, NJ (1994)
17. Chung, C.: Knowledge-based approaches to self-adaptation in cultural algorithms. Ph.D thesis, Wayne State University, Detroit, Michigan, USA (1997)
18. Zhang, Y.: Cultural algorithm and its application in the portfolio. Master thesis, Harbin University of Science and Technology, Harbin, China (2008)
19. Horn, J., Nafpliotis, N., Goldberg, D.E.: A niched Pareto genetic algorithm for multi-objective optimization. In: Proceedings of the IEEE World Congress on Computational Intelligence, pp. 82–87 (1994)

An Improved Keyword Search on Big Data Graph with Graphics Processors

Xiu He[1(✉)] and Bo Yang[2]

[1] Department of Computer Science, Guangdong University of Science and Technology, Dongguan, China
hexiushow@163.com
[2] Information Engineering Department, Huashang College, Guangdong University of Finance and Economics, Zengcheng, Guangdong Province, China
375442583@qq.com

Abstract. With the development of database research, keyword search on big data graph have attracted many attentions and becoming a hot topic. However, most of existing works are studied on CPU. An important problem is efficiently generating answers for keyword search. In this paper, we research an method of keyword search under graphical processing unit. An improved algorithm based on interval coding is proposed. It includes two main tasks, which are finding root nodes and getting shortest paths from root to keyword nodes. To find root nodes quickly, we judge the reachability between any two nodes based on interval assigned to every node. To speed up finding root nodes and getting shortest paths from root to keyword nodes, we provide data parallel processing for compute-intensive tasks based on intervals assigned to every node and Floyd-Warshall algorithm. Experiment results show the high performance of the proposed solution both on CPU and graphical processing unit.

Keywords: Keyword search · Data graph · Graphic processing unit

1 Introduction

Due to its flexibility and no requirement on background knowledge of searched data, keyword search has been applied to find not only useful web documents, but also relational and graph data. Especially recently, quite a lot efforts have been put for keyword search over graphs [1–3]. However, in these works, all Graphs in the database are assumed to be certain or accurate, and in real-life applications, this assumption is often invalid. For example, Resource Description Framework (RDF) is a standard language for describing web resources, and RDF data are modeled as graphs. In practice, RDF data can be highly unreliably [4] due to errors in the web data or data expiration. In the application of the data integration, we need to incorporate such RDF data from various data sources into an integrated database. In this case, uncertainties/inconsistencies often exist. In social networks, each link between any two persons is often associated with a probability that represents the uncertainty of the link [5] or the strength of influence a person has over another person in viral marketing [6]. In XML data (a tree or graph

K. Li et al. (Eds.): ISICA 2015, CCIS 575, pp. 390–397, 2016.
DOI: 10.1007/978-981-10-0356-1_41

structure), uncertainties are incorporated in XML documents known as probabilistic XML document (p-document) [7, 8]. There are quite some works having been proposed to manage p-documents [7–9]. Keyword search over RDF data, social networks and XML data have many important applications [10]. Therefore, in this work, we want to relax the strict assumption of deterministic graphs and study keyword search over big and uncertain graphs.

Concrete objects and relations between them can be modeled as a directed, labeled graph. Graph is a nature tool with the characters of expressing abundant semantic and good understanding. It is used widely in different fields, such as protein interactions net (proteins as nodes and edges as interactions between proteins in graph), World Wide Web, chemical compounds topology graph and social networks [11].

More existing query languages are difficult for applying to general graph. Moreover, nonexpert users are hard to learn and master these languages. Keyword search on graph-structured data has attracted many attentions in recent years [12]. Users can get the required information without knowing any query languages (such as SQL, Xpath, Xquery) and underlying schema by keyword search technique.

2   Preliminaries

A graph db-model is that the structure of schema or instance can be modeled as a directed and possibly labeling graph. Similar to [13–15], we give the following definitions.

Definition 1 (Data Graph). A data directed graph G can be defined as two-tuples $\{V, E\}$, namely $G = \{V, E\}$, where V is the set of nodes, E is the set of edges, $E = V \times V$.

Definition 2 (Keyword Search). Given a keyword query $q = \{t_1, \cdots t_n\}$ where t_i denotes search term, $1 \leq i \leq n$ and $n \geq 1$. Let M_i are the sets of nodes and each node $v \in M_i$ is matched with t_i. Then, the answer to a graph G is a minimal rooted directed tree with the following properties:

(1) The tree is a subgraph of G.
(2) Every keyword is contained in some nodes of the tree. That is to say, the tree contains at least one node from each M_i. Let r is the root of the tree and r can reach to every keyword node in a shortest path.

Definition 3 (Keyword Node). Keyword node is a node of graph G that contains some search terms in query q.

A graph G example is shown in Fig. 1. Each vertex which $v \in V$ contains a list of keywords. For example in Fig. 1, vertex $v8$ contain two keywords $\{d, e\}$. In Fig. 2, given a query $q = \{a, d\}$ and a graph G, two answers $T1$ ($v4 \leftarrow v2 \rightarrow v5 \rightarrow v8$) and $T2$ ($v9 \leftarrow v6 \rightarrow v10$) are shown.

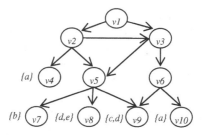

Fig. 1. Graph

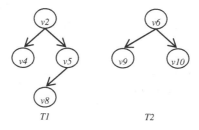

Fig. 2. Presentation graphs

To users, ranking and sorted search results are helpful to quickly find the most interesting answers. Many scoring functions for sorting have been studied. The basic idea is computing weights of nodes and edges with different formula. It is not our focus in this paper. In here, we adopt scoring function which has been used in [20]. An answer is measured by the size of a corresponding tree. The lower the size, the better the answer.

Definition 4 (Uncertain Graph). Let $g^c = (V, E)$ be a directed deterministic graph. Each node $v \in V$ is associated with a set of keywords and a weight. Each edge $e \in E$ is directed and weighted. An uncertain graph is defined as $g = (g^c, \text{Pr})$, where $\text{Pr} : E \to (0, 1]$ is a function that determines the probability of existence of edge e in the graph.

Definition 5 (Possible World Graph). A possible world graph $g' = (V', E')$ is a deterministic graph which is a possible instance of the random variables representing the edges of the uncertain graph $g = (V, E, \text{Pr})$, where $V' \subseteq V, E' \subseteq E$. We denote the relationship between g' and g as $g \Rightarrow g'$.

3   Keyword Search on Data Graph

Generally speaking, the processing of keyword search over data graph involves two main steps. They are identifying keyword nodes and finding a minimal rooted directed tree.

In the first step, keyword nodes can be easily retrieved from disk resident inverted keyword index. The index is created originally and stores a keyword t_i and a list of nodes that contain t_i.

In the second step, generating a result tree should output the root node and shortest paths from the root to all keyword nodes. We know that finding the root node of the answer to a query faces two problems. One is how to determine which node can reach to all the keyword nodes and get rid of its ancestor nodes. The other is to generate a tree that connects two nodes with a shortest path. There is a way that to find ancestor node of multi-keyword nodes using backward search. In here, we integrate a labeling scheme in [17] finding the root node of an answer.

The labeling scheme is an extended version of coding scheme in [21] for general digraph. The reachability between any two nodes in graph G can be judged efficiently by interval coding. It is also applicable in the context of a cyclic graph and a graph with a large number of nodes. Interval coding is a technique of building index. It works as follows:

1. Find an optimal tree-cover T of a graph G in a depth-first manner. During spanning tree, assign an interval $[x, y]$ to every node n, where x is the postorder number of n, y is the smallest postorder number in the set of all n's descendants.

2. Next, for a non-tree arc (A non-tree arc in G is an arc that is not in T), a node b will inherit $[x2, y2]$ associated with node d through the non-tree arc $b \rightarrow d$. Then, do some operations for each node, such as merge or discard intervals.

The detail about how to code can be seen in [17]. Let the postorder number of node u be i, and interval $[x, y]$ associated with node v. There exists a path from u to v if and only if $x \le i \le y$. Interval coding for G in Fig. 1 is shown in Fig. 3. It can be immediately concluded that $v2$ can reach $v8$ since $1 \le 3 \le 6$ (3 is the postorder number of node $v8$, an interval $[1, 6]$ associated with node $v2$). For the same reason, we know that $v2$ can reach $v4$. Let $v4$ and $v8$ be keyword nodes, we quickly find that $v2$ is the root node.

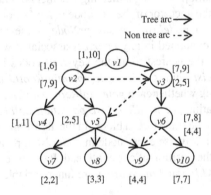

Fig. 3. Interval coding for G

Based on this, we propose a new technique for keyword search, which we call Keyword Search based on Interval Coding and GPU (KSICG, for brevity). KSICG includes two main compute-intensive tasks: finding root nodes and generating answers. After finishing the previous tasks, we can generate an answer by constructing tree. The tree is composed of root node, keyword nodes and shortest paths between root forwarding to keyword nodes.

A. *Finding root node*

To quickly find root nodes for answers, we design an algorithm using GPU with its advantage of massive parallel computation.

Algorithm 1 GPU-Implementation of finding root nodes
Input: *keyList, nodeList*   Output: *rootNodeList*
begin
state[i]<-0; potentialRootList $\leftarrow \phi$;

tid_in_grid<-blockDim.x*blockIdx.x+threadIdx.x;
prNode<- nodeList [tid_in_grid]//get node data controlled by thread
for each *keynode* in *keyList*
for each *interval* in *prNode.intervals*
if *keynode.postNumber* is contained in *interval*
state[i]=1;break;
if all the value in the array *state[]* is 1 then
update(potentialRootList, prNode)
__syncthreads
*rootNodeList<- potentialRootList.*end

Algorithm 1 depicts how to find the nearest common ancestor of all nodes in *keyList* from *nodeList* in GPU-implementations. For each search term t_i in q, we find the sets of nodes M_i. Let $M = \{M_1, \cdots M_n\}$. *keyList* is a list of keyword nodes matching different terms. It can be precomputed by performing cartesian product of $M_1, M_2, ..., M_n$ in host end. *nodeList* is a structure array with the data type of *NodeInterval*. *NodeInterval* is a predefined structure, which includes *vID* and *intervals* associated with node v. *vID* is the identifier of node v. *state[i]* is a flag indicating whether *nodeList [tid_in_grid]* (node controlled by thread) is the ancestor of the i-th node in *keyList* or not. For each *keynode* in *keyList* and *prNode* in *nodeList*, we say *prNode* is the ancestor of *keynode* if *keynode.postNumber* is contained in any *intervals* associated with *prNode*. To ensure the common ancestor is the nearest to all the keyword nodes in *keyList*, we define a variable *potentialRootList* and call *update(potentialRootList, prNode)*. Function *Update()* supports judging whether each *potentialRoot in potentialRootList* is *prNode*'s ancestor or not and updating *potentialRootList*. *potentialRootList* is shared and can be accessed by all threads in the same thread block. In the algorithm, with using *tid_in_grid* (threadId in grid), we can traverse every node in *nodeList* in parallelism.

tid_in_grid denotes the position of current thread. It can be computed by a formula, where *blockDim.x, blockIdx.x* and *threadIdx.x* are inner variables to positioning thread in device end.

Algorithm 1 reduces the searching loop of node interval sets through the thread in parallelism. For example in Fig. 1, let a query $q1 = \{b, c\}$, the matching nodes are $v7$ and $v9$. $S1 = \{v7\}$, $S2 = \{v9\}$. For *keyList* = $\{v7, v9\}$, it is feasible to compute that *potentialRootList* = $\{v5\}$, which can both reach $v7$ and $v9$ according to the algorithm based on interval coding in Fig. 3. That is mean to say $v5$ is the nearest common ancestor.

B. Getting shortest paths

Once find root node, we may get shortest paths from root to keyword nodes. It is crucial for generating an answer. As a classical algorithm problem in graph theory, the common algorithms are Dijkstra algorithm, Floyd-Warshall algorithm, Bellman-Ford algorithm. In [3, 4], each keyword node, run Dijkstra's single source shortest path algorithm. In here, we employ Warshall.

Warshall based on GPU-implementation is not difficult to realize, so we don't discuss more detail here. To be convenient for thread control, we translate adjacent matrix in host end to one-dimension array in device end. The distance data in adjacent matrix is controlled by thread. The time complexity can reduce from $O(n^3)$ to $O(n)$ through traversing adjacent matrix in parallelism. For the same example in Fig. 2, the shortest path between $v8$ and $v8$ is $v2 \rightarrow v5 \rightarrow v8$, not $v2 \rightarrow v3 \rightarrow v5 \rightarrow v8$.

4 Experiments

We have implemented the above proposed algorithm with the *NVIDIA CUDA* SDK and VS2010. Experiments are working on NVIDIA GeForce 9600 GT, Intel i7CPU and 4 GB memory. For comparison, we have also implemented KSICG on CPU, an approach based on backward search on CPU. The information about graph's structure is storing in database. CUDA is NVIDIA's parallel computing architecture that enables dramatically increasing in computing performance by harnessing the power of the GPU [12]. In heterogeneous computing model, compute-intensive tasks can be performed on using GPU. The dataset we use is DBLP. The original data of DBLP is actually a XML tree. XML data can be seen as a digraph with label, where node is corresponding to element or attribute, edge is corresponding to the nested relationship or reference relationship of element to sub-element and element to attribute [10]. To get a graph, some measures of adding two types of non-tree edges in [5] are also used in here, such as randomly making citations between papers and connecting these nodes.

Table 1 shows some queries for test. The third column is the number of keyword nodes which contain corresponding search term. In Fig. 4, for KQ1 to KQ6, each square

Table 1. Query examples

ID	Query	#(keyword nodes)
KQ1	{Dany, David}	(519, 597)
KQ2	{Steven, David}	(2420, 588)
KQ3	{Steven, Dany}	(2410, 528)
KQ4	{David, Rifiul}	(582, 3268)
KQ5	{Hector, David, Dany}	(90, 588, 507)
KQ6	{Betty, Hector, Dany}	(225, 92, 517)

bar shows the execution time by using KSICG on CPU and GPU, KSICG on CPU, Backward Search on CPU. KSICG on CPU and GPU outperforms others in most cases.

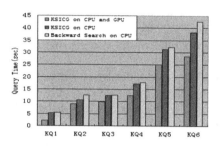

Fig. 4. Execution time

5   Conclusion

Big Data graphs become more and more important in modeling complex things with the rapid development of bioinformatics, web analysis and other application domains. As a topic of data graphs research, the study emphasis of keyword search is still improving performance of query processing.

To quickly find the root of an answer, we introduce the technique of judge reachability based on interval coding for every node in a graph instead of backward search. Based on the technique, we design an efficient algorithm to get shortest paths from root to keyword nodes and generate answers using GPU. Experiments show the high performance using GPU. Our next work is to improve our algorithm and study graph partitioning both on GPU and CPU.

Acknowledgment. This project is supported by Guangdong Province's Quality engineering construction project in 2015.

References

1. Bhalotia, G., Hulgeri, A., Nakhe, C., Chakrabarti, S., Sudarshan, S.: Keyword searching and browsing in databases using banks. In: Proceedings of the 18th International Conference Data Engineering (ICDE), pp. 431–440 (2002)
2. Kacholia, V., Pandit, S., Chakrabarti, S., Desai, R., Karambelkar, H.: Bidirectional expansion for keyword search on graph databases. In: Proceedings of the 31st International Conference Very Large Data Bases (VLDB), pp. 505–516 (2005)
3. Kimelfeld, B., Sagiv, Y.: Finding and approximating top-k answers in keyword proximity search. In: Proceedings of the 25th ACMSIGMOD-SIGACT-SIGART Symposium Principles Database Systems (PODS) (2006)
4. Huang, H., Liu, C.: Query evaluation on probabilistic RDF databases. In: Vossen, G., Long, D.D., Yu, J.X. (eds.) WISE 2009. LNCS, vol. 5802, pp. 307–320. Springer, Heidelberg (2009)

5. Liben-Nowell, D., Kleinberg, J.: The link prediction problem for social networks. In: Proceedings of the 12th International Conference Information Knowledge Management (CIKM), pp. 556–569 (2003)
6. Adar, E., Re, C.: Managing uncertainty in social networks. IEEE Data Eng. Bull. **30**(2), 15–22 (2007)
7. Nierman, A., Jagadish, H.V.: ProTDB: probabilistic data in XML. In: Proceedings of the International Conference Very Large Data Bases (VLDB) (2002)
8. Senellart, P., Abiteboul, S.: On the complexity of managing probabilistic XML data. In: Proceedings of the 26th ACM SIGMOD-SIGACTSIGART Symposium Principles Database Systems (PODS) (2007)
9. Kimelfeld, B., Kosharovsky, Y., Sagiv, Y.: Query efficiency in probabilistic XML models. In: Proceedings of the ACM SIGMOD International Conference Management of Data (2008)
10. Golenberg, B.K.K., Sagiv, Y.: Keyword proximity search in complex data graphs. In: Proceedings of the ACM SIGMOD International Conference Management of Data (2008)
11. Ke, Y., Cheng, J., Yu, J.X.: Querying large graph databases. In: 15th International Conference on Database Systems for Advanced Applications (2010)
12. Dalvi, B.B., Kshirsagar, M., Sudarshan, S.: Keyword search on external memory data graphs. In: VLDB, pp. 1189–1204 (2008)
13. Bhalotia, G., Nakhe, C., Hulgeri, A., Chakrabarti, S., Sudarshan, S.: Keyword searching and browsing in databases using BANKS. In: International Conference on Data Engineering (ICDE), pp. 431–440 (2002)
14. Kacholia, V., Pandit, S., Chakrabarti, S., et al.: Bidirectional expansion for keyword search on graph databases. In: Proceedings of 31st International Conference on Very Large Data Bases, pp. 505–516 (2005)
15. Hao, H., Wang, H., Yang, J., Yu, P.S.: BLINKS: Ranked keyword searches on graphs. In: SIGMOD, pp. 305–316 (2007)
16. Zhou, G., Feng, H., He, G., Chen, H.: Survey of data management on graphics processor units. J. Front. Comput. Sci. Technol. **4**, 289–303 (2010). (in Chinese)
17. Wang, H., Wang, W., Lin, X., Li, J.: Labeling scheme and structural joins for graph-structured XML data. In: Zhang, Y., Tanaka, K., Yu, J.X., Wang, S., Li, M. (eds.) APWeb 2005. LNCS, vol. 3399, pp. 277–289. Springer, Heidelberg (2005)
18. DBLP XML Repository. http://dblp.uni-trier.de/xml/. Accessed September 2010
19. Angles, R., Gutierrez, C.: Survey of graph database models. ACM Comput. Surv. **40**, 1–39 (2008)
20. Hristidis, V., Papakonstantinou, Y., Balmin, A.: Keyword proximity search on XML graphs. In: Conference on Data Engineering, pp. 367–378. IEEE Press, Bangalore (2003)
21. Jagadish, H.V., Agrawal, R., Borgida, A.: Efficient management of transitive relationships in large data and knowledge bases. In: Proceedings of the 1989 ACM SIGMOD International Conference on Management of Data (SIGMOD 1989), Portland, Oregon, pp. 253–262 (1989)
22. NVIDIA: The cuda toolkit. http://www.nvidia.com/object/what_is_cuda_new.html. Accessed September 2010

Applications and Security

Rural Micro-credit Decision Model Based on Principle of Risk Control

Jiali Lin[1], Dazhi Jiang[2(✉)], and KangShun Li[3]

[1] Business School, Shantou University,
Shantou 515063, China
jllin@stu.edu.cn
[2] Department of Computer Science, Shantou University,
Shantou 515063, China
jiangdazhi111007@sina.com
[3] College of Mathematics and Informatics,
South China Agricultural University, Guangzhou 510642, China
likangshun@sina.com

Abstract. Scientific and effective risk control is a core part of the implementation of agriculture-related loans business for micro-credit companies. In this paper, a rural micro-credit decision model is presented based on maximizing the expected rate of return while reducing the investment convergence. Use a unified multi-parent combination algorithm to solve the model and results show that the proposed method and model is scientific and easy operation which can provide a referential solving idea for decision management in micro-credit companies.

Keywords: Risk control · Micro-credit · Decision making model · Unified multi-parent combination algorithm

1 Introduction

In May 4, 2008, China Banking Regulatory Commission (CBRC) and the People's Bank of China jointly issued "Guidance on small loan company pilot", in order to promote the development of pilot micro-credit company all over the nation. According to statistics released by the People's Bank of China, there were 5267 micro-credit companies with 58441 employees up to the end of June, 2012. Meanwhile, rural small loan increased 9.77 billion Yuan in the first half of 2012. (http://finance.chinanews.com/fortune/2012/07-30/4069057.shtml). The establishment of microfinance company system provides viable methods for solving financial dilemma in rural [1]. However, throughout the rural development and current situation of micro-credit companies we can find that rural small loan is faced with two following problems:

Problem I: the quantity of rural small loan business is large but small single amount. Rural small loans are faced with the rural population, accounting for 70 % of total population, which requires micro-credit companies to invest a lot of manpower, material and time in household loan data collection, collation, evaluation and review. However, the fact is that the majority of their clerks are not only in low professional

© Springer Science+Business Media Singapore 2016
K. Li et al. (Eds.): ISICA 2015, CCIS 575, pp. 401–408, 2016.
DOI: 10.1007/978-981-10-0356-1_42

levels but a shortage. According to statistics released by the People's Bank of China, there are about 11 clerks per loan company in average. In general shortage of manpower, the huge number of business brings both the increase of implementation cost and the risk of rural small loan.

Problem II: It is easy to lead to the "herd behavior" resulting from the implementation of rural small loan. The aim to issue rural small loan is to solve the difficulty of loan and rural development, the biggest beneficiaries should be farmers. But most of them live in rural area with relatively backward economy and culture, which means that they are not familiar with financial knowledge, strange to the change of policy and legal concepts, fail to grasp the correct market demand information, and even advocate demonstration. Therefore, "herd behavior" appears in population center, causing convergence of invest project, irrational breach from borrowers as well [2]. Once somebody shows default behavior, no matter what reason leads to this, there also likely occur chain reaction, increasing the risk of small loan [3].

The research paper on small loan risk control mainly measures risk of borrowers by setting credit risk indicators [4, 5], but, due to the uncertainty of micro-credit companies operation, rural natural risk and the credit risk on subjective intention and external interference factors [6], it affects the utility of credit risk indicators. In addition, it still has not aroused the attention of scholars on loan risk due to herd behavior of rural small loan, while herd behavior has already been researched widely in the field of investment in securities market [7–10]. Generally, a simple and effective method to prevent herd behavior is that investors own their independent spirit, which means that they do not believe and follow others unceremoniously. However, the premise of this method is that investors are rational. Considering the characteristic of rural small loan borrowers, it is not applicable to choose such method.

Business objective of commercial bank is to maximize the profit and minimize the risk. According to this principle, we can establish commercial bank loan portfolio optimization model, transforming into model to deal with loan problem [11–13]. This thought provides a creative and practical way for micro-credit companies to make a decision, which is defined as selecting the most suitable loan borrowers to form a group of object with taking into account the benefits and risks of the loan. After that, we can determine the loan objectives following with the aim of maximizing the profit and minimizing the risk. Therefore, this paper sets the profit of micro-credit companies as a starting point, reducing the loan risk as a way; propose a solution for micro-credit companies' loan process for effective decision support. This solution consists of three main elements, to estimate the expected rate of return (cover the shortage of borrowers statistics and information, reduce the loan risk of incorrect operations done by salesmen), to prevent "herd behavior" and to establish the small loan decision model to solve it by applying a unified multi-parent hybrid algorithm. It helps micro-credit companies control small loan risk scientifically by establishing and operating model, providing valuable reference solution ideas and operations management solutions through its decision-making.

2 The Estimation of Expected Yield

In terms of loan portfolio optimization and portfolio problems, expected yield is an important variable in decision-making model. The expected rate of return used to calculate model has capital asset pricing model, arbitrage pricing model. But the problems that can't be ignored are, firstly, small loans, loan portfolio and portfolio optimization problems are essentially different, you can't simply copy the loan portfolio and the portfolio optimization method for estimating the expected rate of return; secondly, compared with stock and medium-sized enterprises, small loan business doesn't own complete and comprehensive information and data for reference; moreover, the expected yield calculation model is needed to consider more factors, which requires to fully collect these factors. Once clerks fail to collect fully qualified professional data, it will set off great credit risks. At present, the method of risk evaluation that most micro-credit companies being used is based on clerks for the field evaluation, which is subjective and low maneuverability [4]. To compensate for the lack of borrower information and data, but also to reduce credit risk because of improper operations caused by the clerks, this paper proposes a combination of sales and field evaluation of the history of micro-credit companies yield data to estimate the expected yield algorithm, whose role is to assess the results of the clerk of the field correction. Described as follows: assuming the total lending history data is m, represented as, $X_i = \{x_{i,1}, x_{i,2}, x_{i,n}\}$, $i = 1, \cdots, m$, n is the number of attributes borrowers, $x_{i,j}$ is represented as the j-th attribute of the i lenders, $1 \leq j \leq n$. $x_{i,a}$ is the attribute of use of proceeds by loan borrowers, $x_{i,n}$ is expected yield of lenders, $a \in \{1, \cdots, n\}$. For a new small loan borrower X_{new}, $x'_{new,n}$ is the estimation for small loan borrower X_{new} about loan yield measured by operators, then the final estimate of the expected yield of the algorithm shown in Fig. 1.

1.	**Begin**
2.	$val1=0.0$, *Set*, $t=0$;
3.	**For** $i=1$ to m **Do**
4.	**If** ($x_{i,a} = x_{new,a}$) **Then**
5.	$val1 = val1 + x_{i,n}$
6.	$Set \leftarrow X_i$
7.	$t++$
8.	**End If**
9.	**End For**
10.	$\overline{x}$ =average yield= $val1/t$
11.	Through formula $\dfrac{<X_\alpha, X_\beta>}{\|X_\alpha\|\|X_\beta\|}$ select vector X_l which is the most similar to X_{new} from a group of Set
12.	$x_{new,n} = a\overline{x} + \beta x_{l,n} + \gamma x'_{new,n}$
13.	**End**

Fig. 1. The final estimate of the expected yield of the algorithm

In the algorithm, α, β and γ are the linear combination of parameters (for a linear combination of the three vectors ratio control) where $\alpha, \beta, \gamma \in [0, 1]$, $\alpha + \beta + \gamma = 1$.

In the estimate of expected yield method, what we should do first is to get the loan record yields mean (average yield) of new loan from the historical data of micro-credit companies. After that, use the similarity function to find a new loan with the most similar loans recorded in the historical data, by doing a linear combination between yield calculated in site assessment, average yield and yield which is most similar to the loan records, whose aims are to determine excepted yield of the ultimate new loans, to make up the shortage of objective reality and manual operation in micro-credit companies.

3 The Prevention of Herb Behavior

A necessary condition of "herb behavior" is that the decision-making process appears to others can be observed by imitators; otherwise it is impossible to be imitated. Therefore, in terms of small loan, it will be an effective method to prevent "herb behavior" by reducing the convergence between borrowers and the probability of which can be observed. According to our research and analysis about rural small loan business, we believe that convergence of borrowers can be measured by two parameters, namely "loan purpose" and "borrower locality". "Loan Purpose" refers to the investment project loans used for the loan. The lender can be positioned in an industry based on investment projects by expanding the difference of the "use of the loans", in order to reduce the interaction between a certain extent and the imitable the chance. When it comes to the "borrower locality", even though the information technology develops rapidly, due to means on Internet communication and survival habits of farmers, it reduces the interaction between farmers in different regions. At the meantime, it reduces the affect from opinion surrounding by expanding the "local" differences.

Research has found that the causes of herd behavior are triggered by "loan concentration" or called "convergence" [7–10]. Affected by subjective and objective factors, if the investors follow others to invest the same item, the probability of the herd behavior will be higher while if the investments are diversified, the probability of the herd behavior will be lower, reducing convergence can be achieved to a certain extent diversification. We use the entropy to describe the degree of convergence here. Entropy was first used to describe the system in a state of disorder, now has been considered as the best diversity and uncertainty measure, which is applied to natural, social, and economic management and many other disciplines.

Based on the above analysis, the process can be set in rural micro-credit corresponding herd behavior prevention strategies, described as follows:

For the same group of borrowers, $X_i = \{x_{i,1}, x_{i,2}, \cdots, x_{i,n}\}$, $i = 1, \cdots, M$, M is represented as the whole number of this groups, while $x_{i,c}$ and $x_{i,d}$ are represented "loan purpose" attribute and "local" properties (properties are represented by two consecutive integers) respectively. Take "local" properties for an example, Assume that X' is a subset of X, X' in the sub Q-dimensional subsets have $(x_{i,c})$, denoted $S'_1, \cdots, S'_Q$, the number of individuals included in each subset are referred to as $\left|S'_p\right|$, $p \in \{1, \cdots, Q\}$, and for any $p, q \in \{1, \cdots, Q\}$, $S'_p \cap S'_q = \emptyset$. This attribute defines the object on the loan entropy as follows:

$$E_1 = -\sum_{i=1}^{Q} p_i \lg(p_i)$$

Where $p_i = \left|S'_i\right| \bigg/ \sum_{j=1}^{Q} \left|S'_j\right|$. Similarly, the entropy can be obtained on the sub-dimension of a and b, denoted by E2 and E3. Loans object is defined as the total convergence of E = E1 + E2 + E3. E greater the diversity of lenders, the smaller the convergence trend, the more diversified investment. Similarly, the entropy can be obtained on the sub-dimension of $x_{i,d}$ and $(x_{i,c}, x_{i,d})$, denoted by E2 and E3. Loans object is defined as the total convergence of E = E1 + E2 + E3. E is greater the diversity of lenders, the smaller the convergence trend, and the more diversified investment.

4 The Model of Small Loan Decision-Making

Assumed that lending positions are available to L, the number of borrowers to apply for loans is M, $x_{i,n}$ means the expected yield of the i-th borrowers, $x_{i,n} = a\bar{x}_i + \beta x_{i,l,n} + \gamma x'_{i,n}$. b is 0-1 variable, where $b = 1$ means that the i-th object is selected borrowers to obtain loans, $b = 0$ means that the i-th object is not selected borrowers to obtain loans. Combined with the expected yield and the "herd behavior" prevention strategies, we can get small loans on the consolidated decision optimization model, which is expressed as follows:

$$\max f = \alpha \frac{\sum_{i=1}^{M} x_{i,n} Y_i}{|X|} + \beta E \;\; s.t. \quad L_{low} \le \sum_{i=1}^{M} x_{i,0} Y_i \le L_{up}$$

$$E = E_1 + E_2 + E_3 \qquad\qquad T_{low} \le |X| \le T_{up}$$

$$E_i = \sum_{j=1}^{Q_i} p_{i,j} \lg(p_{i,j}) \qquad\qquad Y_i \in \{0, 1\}(i = 1, 2 \cdots, M)$$

$$|X| = \sum_{i=1}^{M} Y_i$$

Where L_{low} and L_{up} represent the upper and lower limits of total loans issued, $0 \le L_{low} < L_{up} \le L$. T_{low} and T_{up} represent the upper and lower limits access to credit number, $0 \le T_{low} < T_{up} \le M$. $x_{i,0}$ is one of the properties of the i-th borrower, indicating that the lender the loan amount applied for.

5 The Analysis of Case

We collected microcredit businesses in a microcredit company in Shantou from March 1, 2014 to March 31, 2014. During this period, we acquired 25 valid samples, in which we conduct empirical analysis.

5.1 Sample Data

All borrowers are represented as G (G1, G2,......, G25), and the company's clerks collate statistics about loans loan amount, location, loan purpose, according to calculation need, "local" is divided by "rural" jurisdiction, represented in real numbers (1, 2,......) in system; loan purpose means the business category with loans classified as farming, aquaculture, processing industry, handicrafts and services five, represented in real number (1,2,3,4,5) similarly. Estimation algorithm is used to calculate the expected yield, where $\alpha = 0.25$, $\beta = 0.4$, $\gamma = 0.35$, then we can get the basic properties of each borrower as shown in Table 1.

Table 1. The basic properties of borrowers

Borrowers	The amount of loan (10 thousand yuan)	Local	Loan purpose	Loan term (month)	The expected yield
G_1	2	1	1	6	0.52
G_2	10	1	1	7	0.55
G_3	3	2	1	7	0.53
G_4	12	2	1	6	0.61
G_5	9	3	1	6	0.61
G_6	7	4	1	6	0.57
G_7	12	1	2	8	0.97
G_8	8	2	2	8	0.81
G_9	8	4	2	9	0.80
G_{10}	4	3	2	8	0.81
G_{11}	7	4	2	9	0.86
G_{12}	3	4	2	8	0.72
G_{13}	2	1	3	8	0.70
G_{14}	2	1	3	6	0.72
G_{15}	4	1	3	6	0.71
G_{16}	5	3	3	8	0.75
G_{17}	8	3	3	8	0.70
G_{18}	4	3	3	8	0.74
G_{19}	6	1	4	5	0.63
G_{20}	5	2	4	5	0.78
G_{21}	5	3	4	6	0.62
G_{22}	6	4	4	6	0.84
G_{23}	4	2	5	12	0.85
G_{24}	12	1	5	14	0.78
G_{25}	5	4	5	12	0.87

We analyzed this case through small loan decision-making model. In the constraints of the decision model, we set: $L_{low} = 70$, $L_{up} = 130$, $T_{low} = 12$, $T_{up} = 20$, $a = \beta = 10$.

5.2   Algorithm

Evolutionary algorithm is an efficient algorithm for solving complex non-linear programming problems. This paper uses a unified multi-parent hybrid algorithm [14, 15] to solve the small loan decision-making model, which is a binary constrained optimization problem. In order to adjust the problem needs, we decode the rules for individual algorithms modified as follows:

For $X = \{x_1, x_2, \cdots, x_D\}$ real vector of each component of the formula which is executed $b_j = round((x_j - x_j^L)/(x_j^U - x_j^L))$, $j \in \{1, 2, \cdots, D\}$, D is the dimension of the individual, $[x_j^L, x_j^U]$ is the range of x_j, $round(x)$ represents the x rounded off. Thus the real vector is mapped to a binary vector $B = \{b_1, b_2, \cdots, b_D\}$.

5.3   Model Solution and Analysis

Parameter of UMCACT follows: $N = 100$, $M = 3$, $\alpha_{low} = -1.0$, $\alpha_{up} = 1.5$, $S_{low} = -1$, $S_{up} = 2$, $C_r = 0.9$. The total number of loans is 25, so setting $D = 25$. We assumed MAX_NFC = 10000 (max number of function calls) as the algorithm termination condition. $[x_j^L, x_j^U] = [0, 1]$, $j \in \{1, 2, \cdots, D\}$. The calculated results show that the best loan program is the G2, G4, G5, G7, G10, G11, G14, G16, G20, G22, G23 and G25, with the total loans of 81 million yuan and the total number of loans is 12, both of which satisfy the constraints. Through the results we can find that the optimal loan program does not use the same area and the same loans lenders repeated loans. In addition, in the region of 1 to 4, the number of loans are 3, 3, 3 and 3 respectively, while in the use of the loans of 1 to 5, the number of loans are 3, 3, 2, 2 and 2 respectively, showing the diversity of large borrowers, the small convergence trend. Therefore, in order to verify the effectiveness of the model, we worked with the real repayment loan companies for comparison (as of September 31, 2014), found that G4, G5, G14, G20 and G22 are successful in repayment, although it cannot tell the whole story problem, the validity of this model may be embodied in a certain extent.

6   Conclusion

It is an important issue that providing financial services to farmers and rural small or medium enterprises to be addressed in rural economic development in China. In recent years, the occurrence of micro-credit companies is an innovative achievement of developing small loan in China, which is also an essential carrier in the field of rural finance incremental reform for Chinese government. This paper builds a decision-making model for micro-credit companies, which providing a new way to solve the credit business risk management issues. Although this paper is subjected to sampling conditions which leads to small number of samples, the fitness and practical operation of the model have been verified effective in practice as the advantage of this model will gradually highlight with the micro loan portfolio increasing.

Acknowledgements. The authors would like to thank anonymous reviewers for their very detailed and helpful review. This work was supported by National Natural Science Foundation of China (No.: 61502291), in part by the Project for Outstanding Young Teachers in Higher Education Institutions of Guangdong Province, in part by the Characteristic Innovation Project in Higher Education Institutions of Guangdong Province, and partly by the Fund of Natural Science Foundation of Guangdong Province of China (No.: 2014A030313454).

References

1. Li, Y.: Micro credit corporations: mechanism and efficiency. PhD thesis, Zhejiang Gongshang University (2010)
2. Ninglan, W.: Study on the micro-credit risk management based on the covariant theory. J. Anhui Agric. Sci. **38**(35), 20363–20364 (2010)
3. Gao, S.: Evaluation of rural consumption credit risk. PhD thesis, China University of Mining and Technology (2010)
4. Chen, Y.: Credit risk control study of rural small loans. Master thesis, Southwestern University of Finance and Economics (2006)
5. Hu, Y., Xu, H., Wang, X.: Probe into the credit degree appraisal of the micro-credit for farmer. Theor. Pract. Finance Econ. **28**(145), 30–33 (2007)
6. Zhu, X.: Risk management of rural micro-credit loans. Econ. Rev. **12**, 80–83 (2010)
7. Xiangnong, Y., Sihui, L.: Study of the convergence of investment projects based on the theory of herding behavior. J. Hunan Univ. Sci. Technol. (Soc. Sci. Ed.) **15**(5), 58–61 (2012)
8. Chen, H., Zhang, L., Sun, Y.: Empirical research on commercial bank credit concentration—based on perspective of herding. Technol. Manag. Res. **1**, 86–89 (2012)
9. Li, W., Liu, C., Guo, M.: Structural surveys, loan concentration and value investing: an empirical study of China's commercial bank credit to invest in policies. Manag. World **10**, 174–175 (2010)
10. Zou, X., Zhen, W., Zhao, Y.: Credit transmission mechanism of commercial banks under the influence of herd behaviour. Finan. Forum **10**, 17–24 (2011)
11. Chi, G., Zhu, Z.: Decision-making model of loan portfolio optimization based on comprehensive risk restriction. J. Dalian Univ. Technol. **40**(2), 245–248 (2000)
12. Yang, Z.: Optimization model of loan portfolio based on return per unit of risk. Sci.-Technol. Manag. **12**(2), 104–107 (2010)
13. Chi, G., Haowen, W., Yan, D.: Optimal model of loan portfolio based on the higher central-moment constraints. Syst. Eng.-Theor. Pract. **32**(2), 257–267 (2012)
14. Jang, D., Lin, J.: A unified multi-parent combination algorithm. J. Huazhong Univ. Sci. Technol. (Nat. Sci.) **38**(12), 98–101 (2010)
15. Lin, J., Jang, D.: Development of enterprise downsizing model and its optimization solution. East China Econ. Manag. **26**(4), 144–148 (2012)

A Video Deduplication Scheme with Privacy Preservation in IoT

Xuan Li[1,2(✉)], Jie Lin[1,2], Jin Li[1,2], and Biao Jin[1,2]

[1] School of Software, Fujian Normal University, Fuzhou, China
{jessieli24,jinbiao}@fjnu.edu.cn, linjie891@163.com, jinli71@gmail.com
[2] School of Computer Science, Guangzhou University, Guangzhou, China

Abstract. In recent years, the Internet of Things (IoT) has received considerable attentions and the overall number of connected devices in IoT is growing at an alarming rate. The end-terminals in IoT usually collect data and transmit them to the data-processing center. However, when the data involve user privacy, the data owners may prefer to encrypt their data for security consideration before transmitting them. This paper proposes a video deduplication scheme with privacy preservation by combining the techniques of data deduplication and cryptography. In the scheme, the data owner divides every frame of video into blocks of the same size. Then the blocks are encrypted and uploaded to the cloud platform. On the server side, identical blocks which have been already stored in the database are eliminated for saving the storage space.

Keywords: Internet of thing · Cloud storage environment · Data deduplication · Privacy preservation · Cryptosystem

1 Introduction

In the last few years, the concept of connecting things, which is called The Internet of Things, has drawn a lot of attention from many researchers [1,14]. The idea behind the IoT concept is to include objects in the IT data flows and enable them to interact with other objects. The connected and interacted smart objects in IoT include mobile devices, wearable sensor devices and environmental sensors. By integrating wireless sensors in common scenarios, the object can project a digital snapshot of its state, activity and the surroundings. Then the digital information will be collected and communicated among the interconnected objects or provided to the users locally and remotely.

As IoT has received considerable attention in recent years, the overall number of connected devices such as wearable sensor and mobile devices is growing at a very fast rate. However, the capabilities of storage and computation in most of the IoT terminals are extremely limited. Consider that the cloud platform is resource-abundant and capable of finishing designated tasks [8,11,15], the terminals can exploit the storage and computational power of the cloud to relieve from storing and processing the data.

© Springer Science+Business Media Singapore 2016
K. Li et al. (Eds.): ISICA 2015, CCIS 575, pp. 409–417, 2016.
DOI: 10.1007/978-981-10-0356-1_43

Nevertheless, the technologies of IoT which introduce a new kind of automation and remote interaction have to face many security risks [12,18]. For example, during the transmission of sensor data from the peripheral part to the central structure, the data may be intercepted by third parties. In fact, the protection, of privacy has already become one of the most important concerns when people consider exploring the technology of Internet [2,16]. A good solution for this problem should allow the users in control of their own privacy data. Therefore, the creation of a secure channel between the sensor node and the Internet host is necessary in the applications of IoT. In this context, data encryption is a practical secure method for privacy preserving [6,9,11]. By transmitting the encrypted sensor data from the peripheral device to the central structure, even if other unauthorized devices including the potentially malicious devices on the sharing wireless physical medium can intercept the information, they are not able to decrypt the data. Such encryption treatment in the terminal side can efficiently reduce the risk of data leakage. However, on the cloud side, the processing and computing operations have to be conducted in the encrypted form. This brings new challenges to the applications of IoT.

Video is a typical multimedia type and there are a lot of redundancy between the frames [7]. Actually, when a smart node in IoT tries to record its state, activity or the surrounding, video-recording must be one of the most effective ways. After smart devices take a video, the video can be transmitted to the cloud platform for storage and further processes. For the purpose of privacy preservation, the video will be encrypted before being transmitted and then all the processes on the cloud side must be conducted in the encrypted form. In this paper, we target the problem of privacy-preserving video deduplication on the servers of cloud center.

The remainder of this paper is organized as follow. In Sect. 2, we describe some necessary preliminaries. In Sect. 3, we give the implementation of our deduplication schemes. Section 4 presents experimental results. Finally, conclusion and future work are given in Sect. 5.

2 Preliminaries

In this section, we formally introduce the primitives that are relevant to the proposed scheme in this paper.

2.1 Data Deduplication

With the fast growth of data volume on the internet, data deduplication becomes an impactful technique to be widely adopted in storage systems, which offers significant savings of storage and bandwidth for both the service providers and the users [4,20]. The term deduplication refers to the methods for eliminating redundant copies of duplicate data and replacing them with a pointer to the unique copy [9]. The deduplication methods can be divided to two categories according to the granularities they handle: file-level deduplication and block-level deduplication [13]. In the file-level method, the files are identified as duplicated copies if each bit of them exactly matches [3,5]. Then only a unique file is reserved while other copies are all deleted. In the block-level method, each file is first divided to several blocks and then the deduplication is conducted on the blocks. In consequence, a single copy of each block is stored and other identical blocks in a file or different files are totally deleted [5].

2.2 Symmetric Encryption

In cryptosystem, the original message that is commonly called plaintext is encoded to be unintelligible context for the people not having access to the correct decryption key [10,19]. The encryption method can be categorized into symmetric and asymmetric schemes which generally differ in what keys are used by the encryption and decryption sides.

In the symmetric encryption, the user encrypts the plaintext message P with the secret key K, yielding the ciphertext by the following equation:

$$C = SymEnc(P, K) \tag{1}$$

If the recipient receives the ciphertext, he can decrypt it with the same key K to recover the plaintext:

$$P = SymDec(C, K) \tag{2}$$

The key K should be shared securely between communication parties before encryption process starts. And it must be kept secret from everyone except the sender and receiver.

The detailed mechanism of symmetric encryption is generally based on permutation and diffusion operations on the input data. The symmetric encryption can be further classified to stream cipher and block cipher based on the encryption granularity. The stream cipher encrypts the plaintext as an entirety whereas the block cipher divides the plaintext to several blocks and encrypts it block by block.

2.3 Convergent Encryption

The convergent encryption is a specific encryption method which guarantees that the same ciphertexts are generated from the same plaintexts even if by different users. The crucial point of convergent encryption is the generation of the key. In convergent encryption, the cipher key is derived from the plaintext, which is usually called convergent key.

$$K = CovGen(P) \tag{3}$$

Then the key can be deployed to encrypt the plaintext and output the ciphertext.

$$C = CovEnc(P, K) \tag{4}$$

From the above formulas, the convergent key is entirely dependent on the plaintext. Thus, the ciphertexts encrypted from identical plaintext are certainly the same.

In this paper, convergent encryption is adopted to realize the private-preserving multimedia deduplication model.

3   System Implementation

This section will describe the detailed implementation of privacy-preserving video duplicate removal storage model in IoT. Multimedia data usually have intrinsic features such as high redundancy and high correlation. Moreover, the collections of multimedia context by a particular terminal in IoT are often carried out in some fixed environments such as the users' working site and home. This means that the collected multimedia context must contain lots of duplicate context. If the redundancy can be eliminated, the space required for storing videos must be significantly reduced.

The basic deduplication methods generally falls into two categorical types based on different granularities: file-level deduplication and block-level deduplication. The file-level deduplication system with privacy preservation can be realized based on the convergent encryption as illustrated in Fig. 1. We can see that the file-level scheme is only able to duplicate the identical copies. Considering that the frames of video are usually not completely identical but strong correlated and locally similar, the reduction of storage by the file-level scheme is very limited. So the file-level deduplication may not be appropriate in this case. Therefore, we consider to adopt the block-level deduplication for a greater saving of storage space comparing to the file-level deduplication.

For this purpose, we consider using the block-level deduplication which is performed on the granularity of blocks system. Although the block-level deduplication has been studied in recent years, a private-preserving realization is still largely ignored, which should take both storage and security performance into consideration. In this paper, we propose a privacy-preserving block-level deduplication systems which can be applied to eliminate the redundancy within video frames. The proposed framework adopts the convergent encryption and hash indexes for protecting privacy.

In this system, the server is able to perform the deduplication in the encrypted form by making use of convergent encryption. The main steps of this system on the terminal side can be described as following.

Step 1: Each frame of the video are all divided to blocks of size 16×16, and denoted as

$$P = \{b_1, b_2, ..., b_n\} \tag{5}$$

Here, n is the total number of blocks. If the size of the frame is not an integer multiple of 16×16, it can be extended by adding zeros.

Step 2: Each block b_i is used to calculate a convergent key k_i . Thus, the convergent key is entirely dependent on the block.

$$k_i = CovGen(b_i), i = 1, 2, ..., n \tag{6}$$

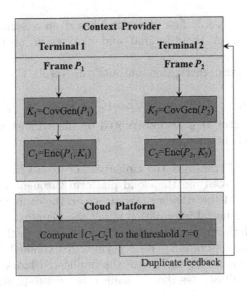

Fig. 1. The private-preserving file-level deduplication scheme

Step 3: The terminal encrypts every block b_i with the corresponding key k_i.

$$c_i = CovEnc(b_i, k_i), i = 1, 2, ..., n \tag{7}$$

Here, $C = \{c_1, c_2, ..., c_n\}$ is obtained as the result of convergent encryption. Note that the ciphertext is the same size as the plaintext P.

Step 4: To improve the speed of deduplication in the encrypted form, $C = \{c_1, c_2, ..., c_n\}$ is first mapped to hash values $H = \{h_1, h_2, ..., h_n\}$.

$$h_i = hash(c_i), i = 1, 2, ..., n \tag{8}$$

However, when convergent encryption is directly applied in each block of frames, it will lead to the leakage of statistic characteristics of plaintext. The cause of this problem is that the same result is obtained from the same block by the convergent encryption. Therefore, the eavesdropper over the public channel may learn some statistical information from the ciphertext.

In order to avoid such information leakage, we consider about exploiting a second round of encryption on the result of convergent encryption, which is able to disturb the statistic characteristics. Here, we adopt the stream encryption scheme, which is a typical class of symmetric cryptosystem, to disturb the statistic characteristics by permutation and diffusion processes. The property of stream cryptosystem guarantees that the statistic characteristics existing in the encrypted blocks after the first round of encryption can be disturbed completely. Therefore, the output of convergent encryption C should be further encrypted by the symmetric encryption algorithm in our system. The procedure takes as follows:

Step 1: A keystream KS is first generated based on a secret key k' , which is shared only between the terminal and the server at the beginning of communication.

Step 2: The symmetric encryption outputs the result C' from C with the keystream KS .

$$C' = SymEnc(C, KS) \tag{9}$$

Step 3: The hash vector H is also encrypted by the symmetric encryption and $H' = h'_1, h'_2, ..., h'_n$ is obtained.

Then the terminal transmits the ciphertext C' to S-CSP and the ciphered hash function H' to I-CSP on the cloud platform through the public channel. By this scheme, even if the attacker eavesdrops on the communication channel, he cannot obtain any useful information about the plaintext.

On the cloud platform, the block-level deduplication in the encrypted form can be conducted. The operation of decryption is first conducted on H' to obtain H. Then, the server can find and delete the duplicated blocks by checking the value of h_i. Figure 2 illustrates the mechanism of private-preserving block-level image deduplication system based on the convergent encryption. For simplicity, we suppose the size of image is 256×256, which means that the number of block is 256. For block-level deduplication, once identical elements are found, the redundant hash elements in I-CSP and the corresponding ciphered blocks in S-CSP will be eliminated and replaced with pointers pointing to the unique copy. Note that the comparison can be conducted among different frames as well as in the single frame, which follows the same procedures.

4   System Evaluation

We randomly extracted frames from a traffic video sequence recorded by a stationary camera [17]. Each frame is resized to 256×256 and divided into 256 blocks of size 16×16. The process of encryption can be performed on the frame as described in last section. Each block of the frame b_i is first encrypted using convergent encryption to obtain c_i. Then c_i is ciphered by the symmetric encryption. Finally, the ciphertext can be transferred to the cloud through public channel.

To perform block-level deduplication in the cloud, the system should check and eliminate duplicated blocks. This task can be accomplished based on the indexes that have already been stored in the I-CSP. Because if blocks are identical, the hash value of encrypted blocks must be the same. This can be inferred from the property of scheme described in last section. Figure 3 shows the results of two neighboring frames which are divided to 256 blocks of 16×16 pixels.

Finally, the block-level deduplication of frames is performed in the encrypted form. We apply this block-level scheme on the video and report the result of from 100 to 2000 frames in Table 1. The ratio of deduplication increases to 29.51 % when the number of frames achieves 2000, which means that the space savings is 29.51 %. So the requirement of storage space for these 2000 frames is only about 70.49 % of the required storage space before deduplication.

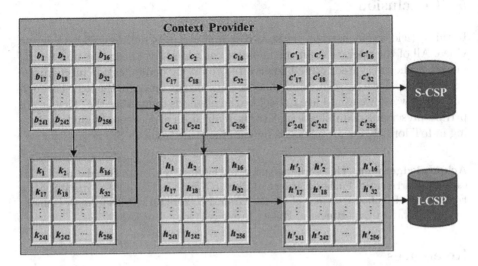

Fig. 2. The private-preserving block-level image deduplication scheme

Fig. 3. The frames which are divided to 256 blocks of 16 × 16 pixels

Table 1. Deduplication ratio using the secure block-level scheme

Frames	Divided blocks	Duplicated blocks	Deduplication ratio
100	25600	5866	22.91 %
500	128000	33778	26.38 %
1000	256000	70680	27.61 %
2000	512000	151069	29.51 %

5   Conclusion

In this article, we focused on the secure deduplication of multimedia data as the video. All of the operations in the file-level and block-level deduplication can be moved to the encrypted form. The private-preserving deduplication mechanisms are of practical use for secure and efficient centralized storage in the IoT. Through our future work, we would like to explore the methods to improve the efficiency of private-preserving deduplication. Also the realization of encrypted signal processing in IoT for privacy protection is by an interesting research problem.

Acknowledgement. This work was supported by Research Youth Foundation of Education Department of FuJian Province of China(Grant No. JA15116), Natural Science Foundation of FuJian Province of China (Grant No. 2014J01220) and Youth Foundation of Education Department of FuJian Province (Grant No. JA14091).

References

1. Atzori, L., Iera, A., Morabito, G.: The internet of things: a survey. Comput. Netw. **54**(15), 2787–2805 (2010)
2. Bellare, M., Keelveedhi, S., Ristenpart, T.: Message-locked encryption and secure deduplication. In: Johansson, T., Nguyen, P.Q. (eds.) EUROCRYPT 2013. LNCS, vol. 7881, pp. 296–312. Springer, Heidelberg (2013)
3. Chen, M., Wang, S., Tian, L.: A high-precision duplicate image deduplication approach. J. Comput. **8**(11), 2768–2775 (2013)
4. Douceur, J.R., Adya, A., Bolosky, W.J., Simon, P., Theimer, M.: Reclaiming space from duplicate files in a serverless distributed file system. In: Proceedings of the 22nd International Conference on Distributed Computing Systems, pp. 617–624 (2002)
5. Harnik, D., Pinkas, B., Shulman-Peleg, A.: Side channels in cloud services: deduplication in cloud storage. IEEE Secur. Priv. **8**(6), 40–47 (2010)
6. Kaaniche, N., Laurent, M.: A secure client side deduplication scheme in cloud storage environments. In: Proceedings of the 6th International Conference on New Technologies, Mobility and Security, pp. 1–7 (2014)
7. Katiyar, A., Weissman, J.: Videdup: an application-aware framework for video de-duplication. In: Proceedings of the 3rd USENIX Conference on Hot Topics in Storage and File Systems, p. 7. USENIX Association (2011)
8. Kiani, S.L., Anjum, A., Antonopoulos, N., Knappmeyer, M.: Context-aware service utilisation in the clouds and energy conservation. J. Ambient Intell. Humanized Comput. **5**(1), 111–131 (2014)
9. Li, J., Chen, X., Li, M., Lee, P., Lou, W.: Secure deduplication with efficient and reliable convergent key management. IEEE Trans. Parallel Distrib. Syst. **25**(6), 1615–1625 (2014)
10. Li, X., Zhang, G., Zhang, X.: Image encryption algorithm with compound chaotic maps. J. Ambient Intell. Humanized Comput. **6**, 563 (2014)
11. Litwin, W., Jajodia, S., Schwarz, T.: Privacy of data outsourced to a cloud for selected readers through client-side encryption. In: WPES Proceedings of Annual Acm Workshop on Privacy in the Electronic Society, pp. 171–176 (2012)

12. Medaglia, C.M., Serbanati, A.: An overview of privacy and security issues in the internet of things. In: Proceedings of TIWDC, pp. 389–395 (2009)
13. Meyer, D.T., Bolosky, W.J.: A study of practical deduplication. ACM Trans. Storage **7**(4), 14 (2012)
14. Roman, R., Alcaraz, C., Lopez, J., Sklavos, N.: Key management systems for sensor networks in the context of the internet of things. Comput. Electr. Eng. **37**(2), 147–159 (2011)
15. Wang, C., Cao, N., Li, J., Ren, K., Lou, W.: Secure ranked keyword search over encrypted cloud data. In: 2010 International Conference on Distributed Computing Systems, pp. 253–262 (2010)
16. Wang, C., Zhang, B., Ren, K., Roveda, J.M.: Privacy-assured outsourcing of image reconstruction service in cloud. IEEE Trans. Emerg. Top. Comput. **1**(1), 166–177 (2013)
17. Wang, M., Li, W., Wang, X.: Transferring a generic pedestrian detector towards specific scenes. In: 2012 IEEE Conference on Computer Vision and Pattern Recognition, pp. 3274–3281 (2012)
18. Weber, R.H.: Internet of things-new security and privacy challenges. Comput. Law Secur. Rev. **26**(1), 23–30 (2010)
19. Zhang, G., Liu, Q.: A novel image encryption method based on total shuffling scheme. Opt. Commun. **284**(12), 2775–2780 (2011)
20. Zhu, B., Li, K., Patterson, R.H.: Avoiding the disk bottleneck in the data domain deduplication file system. In: Proceedings of the 6th USENIX Conference on File and Storage Technologies, pp. 269–282 (2008)

Accurate 3D Reconstruction of Face Image Based on Photometric Stereo

Yongqing Lei[1], Yujuan Sun[1(✉)], Zeju Wu[2], and Zengfeng Wang[1]

[1] Department of Information and Electrical Engineering,
Ludong University, Yantai 264025, China
syj_anne@163.com
[2] College of Communication and Electronic Engineering,
Qingdao Technological University, Shandong, China

Abstract. The average face model has been used to correct the reconstructed results of the classical photometric stereo. By fusing the low frequency of the average face model and the results of photometric stereo, the accuracy of the face three dimensional shape has been largely improved. Moreover the results of the proposed method are more robust under different lighting conditions than those of the classical photometric stereo.

Keywords: 3D reconstruction · Human face · Photometric stereo

1 Introduction

Three dimensional face shape can perfectly show the personal detail information at different vision angles and help face recognition and pedestrian tracking to achieve more accurate results. This paper mainly focuses on 3D reconstruction of photometric stereo (abbreviated PMS), which has good high-frequency characteristics, but poor low-frequency characteristics. An effective low-frequency information of average face model has been fused with the reconstructed result of PMS. The main works of this paper is as follows:

Firstly, with the known lighting conditions, the face surface normal is estimated using PMS. By computing the mean of the random ten face surface normal in YaleB database, the average human face model can be built.

Secondly, by using the low frequency of the average face model, the low-frequency characteristics of the reconstructed results of PMS has been refined.

2 Related Work

Parker et al. established the first 3D face model [9], which caused the concern of the 3D reconstruction technologies. Blanz and Vetter built a statistical model, which was called the 3D morphable model [2,3]. The model can be divided into two matrices, one is the texture matrix, and the other is the shape matrix. The

© Springer Science+Business Media Singapore 2016
K. Li et al. (Eds.): ISICA 2015, CCIS 575, pp. 418–423, 2016.
DOI: 10.1007/978-981-10-0356-1_44

3D morphable model of the target face can be obtained by a linear combination of the two vector matrices.

In order to facilitate human face research, many institutions already had their own 3D face databases. For example: the general used method is the principal component analysis (abbreviated PCA)[4,7,10], which was used to transform the human face shapes in the training database to a shape matrix, which contains only the main components of the information of human face.

The above methods all required the training database. However the image-based method can reconstruct the 3D shape only from the input images. There were two common used methods. The first was the method of restoring the 3D structure from motion (abbreviated SFM)[5]. The second is the method of color-forming, which includes the method of shape from shading (abbreviated SFS) [1,8,11] and the method of PMS [6]. Liao et al. estimated the depth of the human face using SFS algorithms [8], and then computed the partial facial deformation to fit the realistic 3D model of the human face. Dong and Chantler [6] estimated the surface normal of the object with the images in different light environments.

This paper presents an algorithm to compensate the low-frequency component of the reconstruct results based on PMS, and can improve the accuracy of the reconstructed 3d face shape.

3   PMS and the Proposed Method

3.1   The Classical PMS

The classical photometric stereo [6] estimates the normal of the object by computing the light matrix of multiple input images, the intensity of the object surface can be expressed by Eq. (1):

$$I_j = \rho_j(n_j \cdot l^T) \tag{1}$$

where, n_j represents the surface normal at jth pixel of the object; l represents the lighting direction. By adjusting the direction of the light source, the n images of the object can be captured under different lighting directions $(l_1, l_2, ..., l_n)$. The each captured image can be scanned as one column and be built as the image matrix, as shown in Eq. (2).

$$I = [I_1, I_2, ..., I_n] = \rho(n \cdot L) = \rho(n \cdot [l_1, l_2, ..., l_n]^T) \tag{2}$$

where $I_1, I_2, ..., I_n$ represent the scanned image columns; $L=[l_1, L_2, ..., L_n]$ represents the light matrix. If the inverse matrix of exists, the surface normal and albedo of the object can be estimated according to Eqs. (3) and (4).

$$\rho n = IL^{-T} \tag{3}$$

$$\rho = \sqrt{(\rho n_x)^2 + (\rho n_y)^2 + (\rho n_z)^2}$$
$$n = \frac{\rho n}{\rho} \tag{4}$$

However the estimated surface normal of the object exists low frequency deviation, and then large error will be caused in the reconstructed 3D shape of the object.

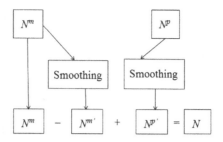

Fig. 1. The framework of the proposed method

3.2 The Proposed Method for Improving the Accuracy of PMS

There are good high-frequency characteristics of the surface normal computed using PMS, but many low frequency deviation. we use the low-frequency information of the average face model to improve the reconstructed results of PMS.

Build the Average Face Model. Classical PMS can used to build the average face model [12]. The average face model can be built by averaging the different human faces:

$$
\begin{aligned}
\rho_{ref} &= \frac{1}{n} \sum_{i=1}^{n} \sqrt{(\rho_i n_{ix})^2 + (\rho_i n_{iy})^2 + (\rho_i n_{iz})^2} \\
n_{ref} &= \frac{1}{n} \sum_{i=1}^{n} \rho_i n_i / \rho_i
\end{aligned}
\tag{5}
$$

where ρ_{ref} represents the albedo of the average face model; n_ref represents the surface normal of the average face model. In experiments, the random ten human faces have been used to compute the average face model.

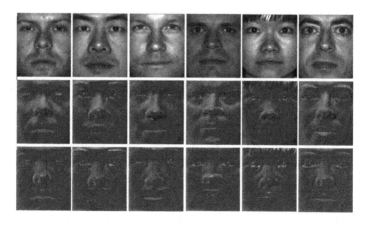

Fig. 2. The face surface normals of PMS and the proposed method

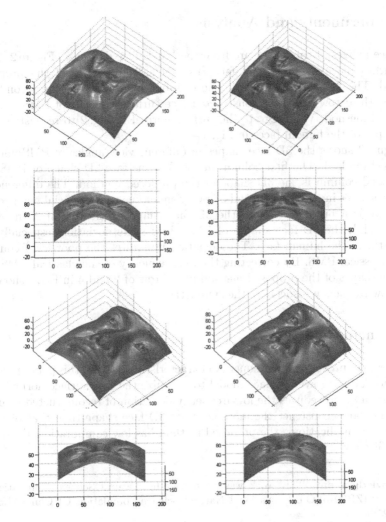

Fig. 3. The face 3D shapes in different view angles of PMS and the proposed method

Improve the Accuracy of PMS. Our basic framework of the proposed algorithm is shown in Fig. 1:

Suppose N^p represents the surface normal of the average face model; N^m represents the reconstructed the surface normal of the proposed method. In this paper, the low-frequency component of N^p has been used to replace the low-frequency component of N^m. The smoothing method such as the mean filter has been used to separate the low-frequency and the high-frequency components of N^m and N^p.

4   Experiments and Analysis

The face images of this paper are from the YaleB face database. Figure 2 shows the surface normals of human faces using classical PMS and the proposed method. The first row in Fig. 2 is one of input human faces. The second row and the third row are the reconstructed face surface normals of PMS and the proposed method. We integrate the surface normal in Fig. 2 to reconstruct the 3D shapes of the human faces in Fig. 3.

Figure 3 shows the 3D face shapes in different view angles of PMS and the proposed method. The first column in Fig. 3 is the face 3D shape of PMS and the second column is the 3D shape of the proposed method. The difference of 3D face shapes in the row of 1 and 3 is very small due to the special perspective, but there is a big difference in the 3D face shape in the row of 2 and 4. In row 2 and 4, the 3D face shapes of PMS show some deviation, especially the symmetry of the human face. The deviation illustrates the poor low-frequency of the classical PMS. By correcting the low-frequency of the classical PMS, the 3D face shapes of the proposed method in the row of 2 and 4 in Fig. 3 show the good low-frequency (especially the symmetry).

5   Conclusions

In order to improve the low-frequency characteristics of PMS, this paper presents a method of using average face model to optimize the face surface normal. The proposed method reduces the low-frequency deviation of reconstructed result of PMS and can reconstruct the more accurate 3D face shape than that of PMS. In practical applications, the proposed method has low computational cost and high efficiency.

Acknowledgement. This work was supported by Natural Science Foundation of Shandong (ZR2015FQ013); National Natural Science Foundation Of China (NSFC) (61501278).

References

1. Atick, J.J., Griffin, P.A., Redlich, A.N.: Statistical approach to shape from shading: reconstruction of three-dimensional face surfaces from single two-dimensional images. Neural comput. **8**(6), 1321–1340 (1996)
2. Blanz, V., Vetter, T.: A morphable model for the synthesis of 3d faces. In: Proceedings of the 26th Annual Conference on Computer Graphics and Interactive Techniques, pp. 187–194. ACM Press/Addison-Wesley Publishing Co. (1999)
3. Blanz, V., Vetter, T.: Face recognition based on fitting a 3d morphable model. IEEE Trans. Pattern Anal. Mach. Intell. **25**(9), 1063–1074 (2003)
4. Castelán, M., Hancoc, E.R.: A simple coupled statistical model for 3d face shape recovery. In: 18th International Conference on Pattern Recognition, ICPR 2006, vol. 1, pp. 231–234. IEEE (2006)

5. Chellappa, R., Krishnamurthy, S., Vo, T., et al.: 3d face reconstruction from video using a generic model. In: Proceedings of the 2002 IEEE International Conference on Multimedia and Expo, ICME 2002, vol. 1, pp. 449–452. IEEE (2002)
6. Dong, J., Chantler, M.: Capture and synthesis of 3d surface texture. Int. J. Comput. Vis. **62**(1–2), 177–194 (2005)
7. Kemelmacher-Shlizerman, I., Basri, R.: 3d face reconstruction from a single image using a single reference face shape. IEEE Trans. Pattern Anal. Mach. Intell. **33**(2), 394–405 (2011)
8. Liao, H.B., Chen, Q.H., Zhou, Q.J., Guo, L.: Rapid 3d face reconstruction by fusion of sfs and local morphable model. J. Vis. Commun. Image Represent. **23**(6), 924–931 (2012)
9. Parke, F.I.: Computer generated animation of faces. In: Proceedings of the ACM annual conference, vol. 1, pp. 451–457. ACM (1972)
10. Russ, T., Boehnen, C., Peters, T.: 3d face recognition using 3d alignment for pca. In: 2006 IEEE Computer Society Conference on Computer Vision and Pattern Recognition, vol. 2, pp. 1391–1398. IEEE (2006)
11. Samaras, D., Metaxas, D., Fua, P., Leclerc, Y.G.: Variable albedo surface reconstruction from stereo and shape from shading. In: Proceedings of the 2000 IEEE Conference on Computer Vision and Pattern Recognition, vol. 1, pp. 480–487. IEEE (2000)
12. Sun, Y., Dong, J., Jian, M., Qi, L.: Fast 3d face reconstruction based on uncalibrated photometric stereo. Multimedia Tools Appl. **74**(11), 3635–3650 (2015)

Business Process Merging Based on Topic Cluster and Process Structure Matching

Ying Huang[1(✉)] and Ilsun You[2]

[1] Department of Mathematics and Computer,
Gannan Normal University, Ganzhou, JiangXi, China
nhwshy@whu.edu.cn
[2] Department of Information Security Engineering,
Soonchunhyang University, Asan-si, South Korea
ilsunu@gmail.com

Abstract. This article presents an approach for automating business process consolidation by applying process topic clustering based on business process libraries, using graph mining algorithm to extract process patterns, find out frequent sub-graphs under the same process topic, then filling sub-graph information into the table of process frequent sub-graph, finally merging these frequent sub-graphs to get merged business processes on the basis of process merge algorithm. We use compression ratio to judging the capability of our merge methods, the compression ratios of integrated processes in same topic cluster are much lower than the different topic processes, and our method achieves similar compression ratio compare with previous work.

Keywords: Correlated Topic Model · Topic distillation · Business process merge · Gspan · Process sub-graph

1 Introduction

In the context of company merge and restructuring, several processes maybe reduce to single process for improve efficiency. This model comparison and merging task is intricate, time- consuming and error-prone. In one instance reported in this article, it took a team of three analysts 130 man-hours to merge 25 % of two variants of an end-to-end process model, so (semi-) auto business process merging has very high practice value [1].

Most of process merging method only uses similarities of activities to judge whether tow processes can be consolidated, but they ignored the isomorphism of process graphs from the view of process structure. In most condition processes to be merged almost under the same topic, such as medical process can hardly merge with student manage processes, so topic cluster has great help to improve the efficiency of business process merge.

We use CTM (Correlated Topic Modeling) to cluster similar processes in the process library. Then, we adopt high efficient sub graph searching algorithm in light of the graphic structure of process, we use a graph-mining technique to extract the patterns from the business process repository. The minimum depth-first search (DFS) codes [2]

© Springer Science+Business Media Singapore 2016
K. Li et al. (Eds.): ISICA 2015, CCIS 575, pp. 424–434, 2016.
DOI: 10.1007/978-981-10-0356-1_45

are served as the tags for business process models or fragments. Based on these codes, we employ similarity metric based on string edit distance (SED) [3], which transformed the graph-matching problem into string matching problem. Generally speaking, our process consolidation system include following steps: process topic cluster, processing and analyzing a large number of business processes, finding out frequent process subgraphs, then consolidate these frequent sub-graphs.

2 Business Process Topic Cluster

We use CTM (Correlated Topic Model) [4] to construct a topic process clustering model (TPCM), after obtaining all labels of activities in process repository, texts extracted from a process are regarded as a document, all of labels are written by short texts. TPCM regard processes as documents, extract topics contained by these documents, and then aggregate processes to different topic clusters according to the probability that processes belonging to topics.

2.1 Overview of TPCM

Definition 1: Topic Process Cluster Model TPCM $= (P, T, F, M)$, $P = \{p_1, p_2, \cdots p_n\}$ is set of process cluster; $T = \{t_1, t_2, \cdots t_m\}$ is set of topics contained by certain process cluster; $F = \{f_1, f_2, \cdots f_k\}$ is set of features contained by certain topics; $M = \{merge\}$ is merger strategies, e.g. maximum common regions consolidation strategy [5], dynamic distance consolidation strategy [6], and reconcile messaging technology consolidation strategy [7], we use based on frequent sub-graph consolidation strategy.

Definition 2: Process $P \leftrightarrow F$, namely a process can be represented by multiple feature words in feature set F. These feature words are extracted from process labels, feature words can reflect the characteristics of the process. The types of procesess are unlimited, its include: EPC, BPMN, Petri net. These words are ordered, due to the nature of business process itself.

Definition 3: Topic $T \leftrightarrow F$, namely one topic can be represented by a collection of multiple features in F. Different topics maybe involve same feature words, however probability of the same feature words belong to different topic may be different.

We use for clustering process topic, which explicitly models the correlation between the latent topics in the collection, and enables the construction of topic graphs and document browsers that allow a user to navigate the collection in a topic-guided manner. The correlated topic model builds on the earlier latent Dirichlet allocation (LDA) model of Blei, Ng and Jordan [7], which is an instance of a general family of mixed membership models for decomposing data into multiple latent components.

CTM use more flexible lognormal distribution instead of Dirichlet distributions, compared with LDA, it provide a more practices latent topic structure, reflected a latent topic is associated with the presence of the other relevance topics. CTM is a hierarchical model of document collections, it models the words of each document from a mixture model. The mixture components are shared by all documents in the collection; the mixture proportions are document-specific random variables. The CTM allows each document to exhibit multiple topics with different proportions. It can thus capture the heterogeneity in grouped data that exhibit multiple latent patterns. Here we treats each process as a document, each topic is constituted by a set of feature distribution, the procedure is shown in Fig. 1.

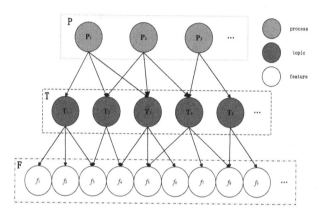

Fig. 1. Three layer CTM descriptive model of business process

2.2   Procedure of Modeling TPCM

The procedure of using CTM to achieve the relationships between processes and topics, topic and feature in certain business process collection $P = \{p_1, p_2, \cdots p_n\}$ as follows:

M is total number of processes in the set of process, p denote the pth process, N_p is the number of features in the pth process. Every process is comprised of k latent topics, CTM applies polynomial $Multi(\theta_d)$ to defied topic distribution, $\theta_p = (\theta_{p1}, \cdots \theta_{pk})$, $\sum_k \theta_{pk} = 1. \theta_{pk}$ is the probability of a word in the topic k pth process.

For each topic $k \in \{1, \cdots, K\}$, draw parameters $\phi_k = (\phi_{k1}, \cdots \phi_{kw})$ of a multinomial distribution $Multi(\phi_k)$ defined over words $\{1, \cdots, W\}$ from the corpus-wide Dirichlet prior distribution $Dir(\beta)$, $\sum_w \phi_{kw} = 1. \beta = (\beta_1, \cdots, \beta_w)$ is a set of its hyperparameters. A topic β is a distribution over the vocabulary, a point on the W-1 simplex. The model will contain K topics $\beta_{1:K}$.

Let $w_{d,n}$ denote the nth word in the dth document, which is an element in a W -term vocabulary. The topic assignment $z_{d,n}$ is associated with the nth word and dth document. Each process is association with a group of topic proportions θ_d. Thus θ_d is a distribution over topic indices, and reflects the probabilities with which words are drawn from each topic in the collection. We will typically consider a natural parameterization of this multinomial $\eta = \log(\theta_i / \theta_K)$

CTM assumes that an N-word process arises from the following generative process. Given topics $\beta_{1:K}$, a K-vector μ and a $K \times K$ covariance matrix Σ.

1. For every process, getting K-dimensional η_d from Gaussian distribution $\mathcal{N}(\mu, \Sigma)$
2. For $n \in \{1, \cdots, N_d\}$:
   (a) Draw topic assignment $Z_{d,n}|\eta_d$ from $\text{Mult}(f(\eta_d))$
   (b) Draw word $W_{d,n}|\{z_{d,n}, \beta_{1:K}\}$ from $\text{Mult}(\beta_{zd,n})$
   where $f(\eta)$ maps a natural parameterization of the topic proportions to the mean parameterization,

$$\theta = f(\eta_i) = \frac{\exp(\eta)}{\sum_i \exp(\eta_i)}$$

$\exp(\eta_i) = \exp(\lambda_i + \frac{v_i^2}{2})$, variational parameters λ_i, v_i^2 are mean and covariance of logistic normal distribution.

The key issue is how to calculate the posterior probability of latent variations to certain processes, we use variational expectation–maximization (EM).

Given a document w and a model $\beta_{1:K}$, the posterior distribution of the per-document latent variables is

$$p(\eta, z|w, \beta_{1:K}, \mu, \Sigma) = \frac{p(\eta|\mu, \Sigma \prod_{n=1}^{N} p(z_n|\eta)p(w_n|z_n, \beta_{1:K}))}{\int p(\eta|\mu, \Sigma) \prod_{n=1}^{N} \sum_{z_n=1}^{K} p(z_n|\eta)p(w_n|z_n, \beta_{1:K})d\eta}$$

It is intractable to compute due to the integral in the denominator, First, the sum over the K values of z_n occurs inside the product over words, inducing a combinatorial number of terms. Second, the distribution of topic proportions $p(\eta|\mu, \Sigma)$ is not conjugate to the distribution of topic assignments $p(z_n|\eta)$.

We used Jensen's inequality to bind the log probability of a process,

$$\log p(w_{1:N}|\mu, \Sigma, \beta) \geq E_q[\log p(\eta|\mu, \Sigma)] + \sum_{n=1}^{N} E_q[\log p(p(z_n|\eta))]$$

$$+ \sum_{n=1}^{N} E_q[\log p(w_n|z_n, \beta)] + H(q)$$

Where the expectation is taken with respect to q, a variational distribution of the latent variables, and $H(q)$ denotes the entropy of that distribution.

$$q(\eta, z|\lambda, v^2, \emptyset) = \prod_{i=1}^{K} q(\eta_i|\lambda_i, v_i^2) \prod_{n=1}^{N} q(z_n|\emptyset_n)$$

The objective function of variational EM is the likelihood bound given by summing Eq. (3) over the document collection $\{w_1, \cdots, w_D\}$

$$\mathcal{L}(\mu, \Sigma, \beta_{1:K}; w_{1:D}) \geq \sum_{d=1}^{D} E_{q_d}[\log p(\eta_d, z_d, w_d | \mu, \Sigma, \beta_{1:K})] + H(q_d)$$

The variational EM algorithm is coordinate ascent in this objective function. In the E-step, we maximize the bound with respect to the variational parameters by performing variational inference for each document. In the M-step, we maximize the bound with respect to the model parameters. This amounts to maximum likelihood estimation of the topics and multivariate Gaussian using expected sufficient statistics, where the expectation is taken with respect to the variational distributions computed in the E-step, where n_d is the vector of word counts for document d.

$$\hat{\beta} \propto \sum_d \phi_{d,i} n_d \quad \hat{\mu} = \frac{1}{D} \sum_d \lambda_d \quad \hat{\Sigma} = \frac{1}{D} \sum_d I v_d^2 + (\lambda_d - \hat{\mu})(\lambda_d - \hat{\mu})^T$$

3   Frequent Sub-Graph Mining

In order to reflect the universality of the method, we defined the Process Structure Graph to denote different types of processes.

Definition 4: Business process graph: It is a triple set BPG $= (V, \tau, \ell)$

V : is set of node types;
$\tau : V \rightarrow T$ determine the type of nodes;
$\ell : V \rightarrow \Omega$ determine the label of nodes;

Let type set T, process node text set Ω. Let $P = (V, \tau, \ell)$ is business process graph, P is directed graph. $E : V \times V$ is set of direct edges, x, y are nodes in P, $x \in V$, $y \in V$, $(x, y) \in E$, that is x is predecessor of y, y is successor of x.

Definition 5: Process Structure Graph (PSG): PSG $= (T, L, f)$, T is set of types, L is set of node labels, subjective function $f : \Omega \leftrightarrow L$, makeing the text of the node in the process has the unique corresponding label..

Definition 6: Business Process Isomorphism. Let $P = (V, \tau, \ell)$ and $P' = (V', \tau', \ell')$ are two business process graphs. A business process isomorphism between P and P' (denoted by $P \cong P'$)) is a mapping $f : V \leftrightarrow V'$ such that:

$$\forall x \in V, \tau(x) = \tau'(x),$$
$$\forall x \in V, \ell(x) = \ell'(x),$$
$$\forall x, y \in V, (x, y) \in E \Rightarrow (f(x), f(y)) \in E'.$$

We use high-efficiency graph mining algorithm (gSpan [9]) to carry out business process subgraph mining, then finding frequent sub-graphs (frequent process fragments). The canonical label for a graph (denoted as Cl(G)) is a unique code which is a sequence of bytes, characters, or numbers.

DFS $= (i, j, f, l_i, l_{(i,j)}, l_j)$, where l_i and l_j are the labels of v_i and v_j, and $l_{(i,j)}$ is the label of the edge connecting them. $l_{(i,j)}$ represents the direction of $l_{(i,j)}$: $f = 1$ represents $v_i \rightarrow v_j$, and $f = -1$ represents $v_i \leftarrow v_j$ [8]. Let the edge order take the first priority, the vertex label l_i take the second priority, the edge label $l_{(i,j)}$ take the third and the vertex label l_j take the fourth to determine the order of two edges. The ordering based on above rules is called DFS lexicographic order.

Definition 7: Frequent process table. The process table is a quintuple FPT $=$ (F, P, A, D, m)

F: is frequent process fragment, $F = (N, \mathcal{L}(n_i, n_j))$;
P: is original process;
A: is predecessor node of frequent process fragments in original process;
D: is successor node of frequent process fragments in original process;
m: $F \rightarrow P$ is surjection from process fragment to correspond process.
N is set of nodes in frequent sub-graph, $n_i \in N, n_j \in N, \mathcal{L}(n_i, n_j)$ is the label on the edge between n_i, n_j different edges maybe have different labels.

Given a collection of business processes $P = \{p_1, p_2, \cdots p_n\}$, a collection of process fragments $F = \{f_1, f_2, \cdots f_n\}$. frequency$(f_i)$ is the total number of process fragment f_i occurs in the process collection P,frequency(f_i, p_i) is the number of process fragment f_i occur in a process p_i.

Frequency is the total number of a process fragment occurs in the process collection. confidence equals to frequency(f_i, p_i) divide frequency(f_i), the precondition is frequency$(f_i) \geq$ threshold. If confidence less than 1, indicates this process fragment appear at different processes; If confidence equals to 1 indicates this process fragment appear at same processes.

$$\text{frequency}(f_i) \geq \text{threshold}$$
$$\text{confidence} = \frac{\text{frequency}(f_i, p_i)}{\text{frequency}(f_i)}, \quad 0 < confidence < 1$$

4 Business Process Consolidation Algorithm

Given two business process structure graphs G_1, G_2, the merging algorithm (Algorithm 1) starts by initial version of the merged graph MG by calculating the union of the edges of G_1 and G_2, excluding the edges of G_2 that are substituted (2–3). Then setting the annotation of each edge in CG that originates from a substituted edge, with the union of the annotations of the two substituted edges in G_1 and G_2 (4–5). We connect frequent process subgraph by adding xor-split connector to the predecessor node of frequent process subgraph in G_1 and G_2, adding edges to link them at the same time (6–13). We connect frequent process subgraph by adding xor-joint connector to the successor node of frequent process subgraph in G_1 and G_2, adding edges to link them at the same time (14–20). Finally, we merge connectors in the process frequent sub-graph (Algorithm 1).

	Algorithm 1:Merge
1	Function merge (PSG G_1, G_2)
2	$MG = G_1 \cup G_2 \setminus (G_2 \cap comme)$
3	PSG MG, Frequent pattern fp,
4	foreach $(x, y) \in$ (MG $\cap$ comme)
5	$\lambda_{MG} = \mathcal{L}(x, y)$
6	foreach fp do
7	foreach $ap_1 \in fp.A$ and $ap_2 \in fp.A$, $ap_1 \in G_1, ap_2 \in G_2$
8	xs=new connector("xor",true)
9	$MG = MG \setminus (\{(ap_1, sp_1), (ap_2, sp_2)\})) \cup \{(ap_1, xs), (ap_2, xs), (xs, sp_1)\}$
10	$\lambda_{MG}(ap_1, xs) = \lambda_{G_1}(ap_1, sp_1)$
11	$\lambda_{MG}(ap_2, xs) = \lambda_{G_2}(ap_1, sp_1)$
12	$\lambda_{MG}(xs, sp_1) = \lambda_{G_1}(ap_1, sp_1) \cup \lambda_{G_2}(ap_2, sp_2)$
13	end
14	foreach $dp_1 \in fp.A$ and $dp_2 \in fp.A$, $dp_1 \in G_1, dp_2 \in G_2$
15	xe=new connector("xor",true)
16	$MG = MG \setminus (\{(ep_1, dp_1), (ep_2, dp_2)\})) \cup \{(xe, dp_1), (xe, dp_2), (dp_1, xe)\}$
17	$\lambda_{MG}(xe, dp_1) = \lambda_{G_1}(ep_1, dp_1)$
18	$\lambda_{MG}(xe, dp_2) = \lambda_{G_2}(ep_2, dp_2)$
19	$\lambda_{MG}(nsc, sp_1) = \lambda_{G_1}(ep_1, dp_1) \cup \lambda_{G_2}(ep_2, dp_2)$
20	end
21	end
22	$MG=MergeConnectors(MG)$

comme is the common edges in G_1, G_2, we must mark which processes they come from; ap_1, ap_2 are corresponding predecessor nodes of frequent process subgraphs in G_1, G_2; dp_1, dp_2 are corresponding successor nodes of frequent process subgraph in G_1, G_2; sp_1, sp_2 are corresponding start nodes of frequent process subgraph in G_1, G_2; ep_1, ep_2 are corresponding end nodes of frequent process subgraphs in G_1, G_2.

	Algorithm 2: MergeConnectors
1	$MergeConnectors(MG)$
2	$foreach\ c \in fp$
3	$if\ \tau(c) == "c"$
4	$S_c = \{x \in fp(c) \cdot \| x \notin fp(c)\}$
5	$P_c = \{x \in \cdot fp(c) \| x \notin fp(c)\}$
6	$MG = (MG \backslash \bigcup_{x \in S_c}\{(fp(c),x)\} \cup \bigcup_{x \in P_c}\{(x,fp(c))\})$ $\bigcup_{x \in S_c}\{(c,x)\} \cup \bigcup_{x \in P_c}\{(x,c)\}$
7	$foreach\ x \in S_c$
8	$\lambda_{MG}(x,c) = \lambda_{G_2}(fp(c),x)$
9	$foreach\ x \in P_c$
10	$\lambda_{MG}(x,c) = \lambda_{G_2}(x,fp(c))$
11	$if\ (\lambda_{G_1}(c) \neq \lambda_{G_2}(c))\ then$
12	$\lambda_{MG}(c) = "or"$
13	$\tau(m) = \tau(c) = "c"$
14	$If(\|m \cdot\| > 1 \&\& \|n \cdot\| > 1)$
15	$MG = (MG \backslash \{(m,n)\} \cup \bigcup_{x \in n \cdot}\{(n,x)\}) \cup \bigcup_{x \in n \cdot}\{(m,x)\}$
16	$foreach\ x \in n \cdot$
17	$\lambda(m,x) = \lambda(m,x) \cup \lambda(n,x)$
18	$if\ \lambda(m) \neq \lambda(n)\ then\ \lambda(m) = "or"$
19	$If(\|\cdot m\| > 1 \&\& \|\cdot n\| > 1)$
20	$MG = (MG \backslash \{(m,n)\} \cup \bigcup_{x \in \cdot m}\{(x,m)\}) \cup \bigcup_{x \in \cdot m}\{(x,n)\}$
21	$foreach\ x \in n \cdot$
22	$\lambda(x,n) = \lambda(x,m) \cup \lambda(x,m)$
23	$if\ \lambda(m) \neq \lambda(n)\ then\ \lambda(n) = "or"$
24	$if\ \tau(m) = "c"\ \&\ \|\cdot m\| = 1, \|m \cdot\| = 1$
25	$MG = (MG \backslash \{(\cdot m, m),(m, m \cdot)\}) \cup (\cdot m, m \cdot)$
26	$\lambda(\cdot m, m \cdot) = \lambda(\cdot m, m)$

G_1, G_2 except connectors are aligned by function MergeConnectors (Algorithm 2). For connectors in the frequent process subgraphs to be merged, if the connectors are start connector or end connector of frequent process subgraphs, if these two connector are different, reconnect their predecessors and successors, transfer type of connectors to "or" (3–12); then merge consecutive connectors (13–23); Finally, remove single-in-single-out connectors (24–25).

5 Experiment and Evaluation

5.1 Overview of Framework

Our work contains three main blocks: preprocessing module, process pattern discovery module, and process consolidate module.

Preprocessing module handles the text information in process set, the domain of process is determined by text information of business process, such as logistic process, medical process etc.. We cluster process collections according to their topics first, in order to improve the efficiency of the process matching in the procedure of process merge. Texts need to be clean before cluster, include remove stop words, stemming words and word frequency counting, producing feature words set which describes the core functions of processes. In this paper, the CTM is used to cluster process topics, because CTM reflects the relationship between a latent topic and other topics.

Process pattern discovery module is applied to find frequent process subgraph. Processes in the same cluster have higher similarity. There are many kinds of process modeling method, so we unify these process models to process structure graph, then using gSpan algorithm to mining frequent process subgraph, these subgraph frequency(f_i) be equal or greater than 2 and confidence less than 1 greater than 0. The outputs of this module are collection of frequent process subgraphs and their corresponding frequent process sub-graph tables.

5.2 Experiment Preparation

All experiments were conducted using an Intel(R) Core(TM) i5, 2.6 GHz PC with 8 GB main memory, running OSX 64 bit. Next, we focus on studying the efficiency and effectiveness of our method. In addition, the confidences for pattern discovery in both synthetic and real world datasets were fixed to 0.15 and pattern locations were restricted to a maximum of 5.

We conducted tests to compare the sizes of the models produced by the merging operator relative to the sizes of the input models. For these tests, we took the SAP reference model, consisting of 604 EPCs, and constructed every pair of EPCs from among them. We used a similarity algorithm from [3] to search for the similarity threshold over 20 % between input models,

We conduct topic cluster for the 382 processes, then merging the processes within the same cluster. Table 1 is clustering results of CTM to 382 SAP reference model.

Table 1. Clustering results of CTM to SAP reference model

T1	T2	T3	T4	T5	T6
47	64	68	50	79	74

5.3 Size of Merged Processes

Size of merged processes is defined as the number of edges, which is a key factor affecting the understandability of process models [9]. It is thus desirable that merged models are as compact as possible. We expect that the size of the merged model will be almost equal to the sum of the sizes of the two input models, when mereging very different models. We expect to obtain a model of size close to that of the larger of the two models, when mergeing very similar models.

Next, we merged each of these model pairs and calculated the compression ratio [10], which in our context is the ratio between the size of the merged model and the size of the input models, i.e. $CR(G1, G2) = |CG|/(|G1| + |G2|)$, where $CG = Merge(G1, G2)$. A compression ratio equals one means that the input models are totally different.

Table 2 is the average compression ration of 382 SAP business processes, which within the same topic cluster after process topic cluster. Table show that the average compression ration of topic 2 is the highest, demonstrate there are exist small the process frequent subgraphs in topic 2, so fewer parts of fragments can be merged. The average compression ration of topic 6 is the lowest demonstrate there are exist many the process frequent subgraphs in topic 6, so more parts of fragments can be merged. The average compression ratio of SAP reference process model is 0.76, the optimized average compression ratio is 0.69 in reference [3]. The highest compression ratio obtains by our method to is 0.726, slightly lower than the average compression ratio in the literature [3], the minimum compression ratio is 0.122, far below the optimized compression ratio in the literature [3].

Table 2. Average compression ration in same topics

Topic1	Topic2	Topic3	Topic4	Topic5	Topic6
0.628	0.726	0.482	0.659	0.657	0.122

6 Conclusion

In this paper, we proposes an process consolidate approach, it utilizing CTM to cluster business processes to different topic clusters, then using frequent sub-graph tables to store frequent sub-graphs which are achieved by gSpan algorithm, finally applies merging algorithm to achieving merged business processes. The operator ensures that the merged model subsumes the original model and that the original models can be derived back by individualizing the merged model. We do not consider the entanglement problem in the merged models because we deal with isomorphism phenomenon before processes consolidate, and the frequent process tables showing the recurrent fragments in the input models.

Much work still needs to be conducted in the future. For example, further work is required in order to support the coevolution of process variants based on a merged model. We outlined a set of merging principles in Algorithm 1. The proposed algorithm is designed to ensure a syntactic correctness criterion on the merged models and on the variants. More sophisticated consolidate operator could be defined to guarantee semantic correctness in addition to syntactic correctness.

Acknowledgment. This work is supported by the National Basic Research Program of China undergrant No. 2014CB340404, the National Natural Science Foundation of China under grant Nos. 61373037, 61562073, the Natural Science Foundation of Jiangxi Province the grant No. 20142BAB217028, the Basic Science Research Program through the National Research Foundation of Korea funded by the Ministry of Science, ICT and Future Planning under Grant 2014R1A1A1005915.

References

1. Gottschalk, F., van der Aalst, W.M., Jansen-Vullers, M.H.: Merging event-driven process chains. In: Meersman, R., Tari, Z. (eds.) OTM 2008, Part I. LNCS, vol. 5331, pp. 418–426. Springer, Heidelberg (2008)
2. Yan, X., Han, J.: gSpan: graph-based substructure pattern mining. In: Proceedings of the ICDM, Maebashi City, Japan, pp. 721–724 (2002)
3. Levenshtein, V.I.: Binary codes capable of correcting deletions, inser- tions and reversals. Sov. Phys. Dokl. **10**(8), 707–710 (1966)
4. Blei, D., Lafferty, J.: Correlated topic models. Adv. Neural Inf. Process. Syst. **18**, 147 (2006)
5. La Rosa, M., Dumas, M., Uba, R., et al.: Business process model merging: an approach to business process consolidation. ACM Trans. Softw. Eng. Method. (TOSEM) **22**(2), 11 (2013)
6. Ma, D.C., Lin, J.Y.-C., Orlowska, M.E.: Automatic merging of work items in business process management systems. In: Abramowicz, W. (ed.) BIS 2007. LNCS, vol. 4439, pp. 14–28. Springer, Heidelberg (2007)
7. Li, C., Reichert, M., Wombacher, A.: The minadept clustering approach for discovering reference process models out of process variants. Int. J. Coop. Inf. Syst. **19**(3–4), 159–203 (2010)
8. Li, Y., Cao, B., Xu, L., et al.: An efficient recommendation method for improving business process modeling **10**(1), 502–513 (2014)
9. Mendling, J., Reijers, H., van der Aalst, W.: Seven process modeling guidelines (SPMG). Inf. Softw. Technol. **52**(2), 127–136 (2010)
10. Salomon, D.: Data Compression: The Complete Reference, 4th edn. Springer, Heidelberg (2006)

The BPSO Based Complex Splitting of Context-Aware Recommendation

Shuxin Yang[✉], Qiuying Peng, and Le Chen

School of Information Engineering,
Jiangxi University of Science and Technology, Ganzhou 341000, China
yimuyunlang@sina.com

Abstract. Item Splitting splits an item into two items rated under two alternative contextual conditions respectively for improving the prediction accuracy of contextual recommendations. To get more specialized rating data, Complex Splitting is proposed to further improve the accuracy of recommendations. The key of the approach is to select multiple contextual conditions for splitting user or item. We translate it into a contextual conditions combinatorial optimization problem based on discrete binary particle swarm optimization (BPSO) algorithm. The item or user is split into two different items or users according to those contextual conditions in optimal combination. We evaluate our algorithm through a real world dataset and the experimental results demonstrate its validity and reliability.

Keywords: Context-aware recommendation · Complex splitting · Particle swarm optimization · Collaborative filtering

1 Introduction

The key of Context-Aware Recommender System [1] is how to incorporate the contextual information into recommender system. The existing approaches can be divided into three categories [2]: pre-filtering, post-filtering and contextual modeling. Item Splitting [3] is a very effective contextual pre-filtering approach, it holds the idea that one item can be regarded as two different items under different contextual conditions. For instance, as to the two groups of movie rating about a horror movie: one was given by users when they watched the movie in day time, and the other one was given when users watched the film at night. If there exits significant difference which is called impurity criteria [3] between the two groups of rating, this movie should be regarded as two different movies. Item Splitting has gained wide attention in recent years, and been expanded into the User Splitting [4] and the User-Item (UI) Splitting [5]. Study [4] inferred two contextual user profiles for each user automatically, and only one of the profiles instead of the full user profile was used for recommendation. The User Splitting put forward in research [5] shares the same principle with Item Splitting, and it splits one user into two different users based on single contextual condition. Meanwhile, this research conducted User Splitting on the basis of Item Splitting, and put forward the UI Splitting, which can achieve higher accuracy of recommendations according to the research achievements. However, the existing context splitting

K. Li et al. (Eds.): ISICA 2015, CCIS 575, pp. 435–444, 2016.
DOI: 10.1007/978-981-10-0356-1_46

approaches merely conduct simple binomial splitting using single contextual condition, which is called the Simple Splitting approach. In order to get more specialized data, Complex Splitting was proposed to further improve the accuracy of recommendations in this paper. Multiple contextual conditions is taken into consideration for splitting user or item in Complex Splitting approach. This research focused on the complex binomial splitting and mainly used Complex Splitting for splitting the user. Then, the key point of Complex Splitting is how to select contextual condition combination, and determine the number of contextual conditions to be selected, as well as determine whether or not the contextual condition number is uniform or self-adaptive for all users. In order to solve these problems, the discrete binary particle swarm optimization (BPSO) was utilized in the research to dynamically select the optimal contextual condition combination, perform self-adaption for the number of contextual conditions, and split the user. The data set of movie rating containing contextual information was employed in the experiment to verify validity of the proposed approach. Based on the calculated root mean squared error (RMSE) of the predicated rating, the advantages and disadvantages of the proposed approach were evaluated. The experimental results demonstrate that the proposed approach in the paper is able to further improve the accuracy of recommendations.

2 The Principle of Complex Splitting

In order to deepen the understanding, part of the movie rating data under different contextual conditions was provided, as illustrated in Table 1.

Table 1. Movie ratings in contexts

User	Item	Rating	Time	Social	Day type
U1	T1	5	Night	Alone	Weekend
U1	T1	2	Morning	Friends	Weekday
U1	T1	?	Afternoon	Colleagues	Weekday
U1	T2	4	Night	Alone	Holiday

As shown in Table 1, there are one user U1, two items T1 and T2, three known ratings and one unknown rating. Taking Item Splitting for example, as a context variable, time has three contextual conditions including the morning, afternoon and night. Accordingly, there are three groups of alternative choices: morning or not morning, afternoon or not afternoon, night or not night. Investigation [3] suggested calculating the difference degree using multiple methods, so as to judge whether or not there exists statistical difference in the split two parts of each selection. Item splitting iterates over all contextual conditions of each context factors and evaluates the splits based on the impurity criteria. It finds the optimal split for each item in the rating matrix and then split the item into two new ones, then the original multi-dimensional rating matrix is transformed to a two-dimensional matrix as a result by eliminating contexts.

Table 2. Item splitting

User	Item	Rating
U1	T11	5
U1	T12	2
U1	T12	?
U1	T2	4

Table 3. User splitting

User	Item	Rating
U12	T1	5
U11	T1	2
U11	T1	?
U12	T2	4

Table 4. UI splitting

User	Item	Rating
U12	T11	5
U11	T12	2
U11	T12	?
U12	T2	4

Table 5. Complex splitting

User	Item	Rating
U11	T1	5
U12	T1	2
U12	T1	?
U11	T2	4

Assume that the item T2 in Table 1 does not satisfy the splitting conditions, and the optimal contextual condition for splitting item T1 is night, then T2 is kept unchanged while T1 is split into T11 and T12. That is to say, the items of the rating data at night and other time are T11 and T12, respectively. In this way, the multidimensional rating matrix in Table 1 can be converted into the two-dimensional rating matrix in Table 2. Similarly, suppose that the optimal contextual condition for splitting U1 is weekday, and then U1 is split into U11 representing the rating data at weekday, and U12 representing the rating data in holiday. Finally, the rating matrix in Table 1 is converted into the matrix shown in Table 3. Based on the Item Splitting, UI Splitting splits the user in a further way, and the data finally obtained are shown in Table 4.

In order to further improve the accuracy of recommendations, this research puts forward conducting binomial splitting on the user by taking multiple contextual conditions into consideration. For example, if the user U1 is split by considering two factors–night and alone simultaneously, that is to say, splitting the rating data when U1 watches the film alone at night as group one, U1 with data meeting the two conditions changed to U11 accordingly. Similarly, the rating data that fail to satisfy the two factors belong to the other group, where U1 is turned into U12. If the difference degree of the two groups of rating data is greater than that obtained by any possible contextual condition combination for splitting U1, the combination night and alone is the optimal integrated contextual condition for splitting U1. Then the rating matrix in Table 1 is finally converted into the matrix shown in Table 5.

The key of Complex Splitting lies in how to find the optimal combination of contextual conditions for splitting the user. With respect to the rating data containing more context factors, it is unpractical to traverse all the possible combinations using the exhaustive search method adopted by the Simple Splitting. This is because the time complexity is expected to grow exponentially along with the increase of the number of context factors. Therefore, the discrete BPSO is put forward in the paper to optimize the contextual condition combination, so as to converge the optimal combination more quickly.

3 The BPSO Based Complex Splitting

BPSO [6] was put forward to solve the optimization problem of discrete or binary space. This algorithm maps the target optimization problem to a D-dimensional search space, and signifies the particle location through binary encoding. If the value of x_{id} (dimension d) of particle x_i in the search space is 1, it indicates that dimension d is selected, otherwise if the dimension is 0, it indicates that it is not selected. The velocity v_{id} is employed to express the possibility that the value of dimension d converting into 1. The velocity value range of this possibility is [0, 1], and realized by the sigmoid() function. Therefore, the updates of the velocity and particle location in BPSO are respectively shown in formulas (1) and (2):

$$v_{id}^{n+1} = wv_{id}^n + c_1 r_1^n (p_{id}^n - x_{id}^n) + c_2 r_2^n (p_{gd}^n - x_{id}^n) \tag{1}$$

$$\begin{aligned} if \ (rand() < s(v_{id}^{n+1})) \ then \ x_{id}^{n+1} = 1; \\ else \ x_{id}^{n+1} = 0 \end{aligned} \tag{2}$$

Where $d = 1, 2, \ldots, D; i = 1, 2, \ldots, m$, m represents the scale of particle swarm; w is the non-negative number, which is called the inertial weight; c_1, c_2 are normal numbers, which are called learning factors and used to adjust the weights of the individual and global optimum, generally $c_1 = c_2 = 2$ [7]; p_{id}^n denotes the value of the optimal location at the d th dimension after iterating for n times for the particle x_i; p_{gd}^n indicates the value of the global optimal location at the dth dimension after iterating for n times for the whole particle swarm. $r_1, r_2, rand()$ are random numbers in the range of [0, 1]. The $s(v_{id}^{n+1}) = \frac{1}{1+e^{-v_{id}^{n+1}}}$ in formula (2) represents the possibility that the value of a location is 1. v_{id}^{n+1} is within the range of $[-v_{max}, v_{max}]$, and v_{max} limits the velocity of particles within an appropriate range to avoid too fast and too slow velocity.

Since the original BPSO is lack of the ability of local search and has little possibility to converge to the global optimal location. Therefore, the modified BPSO [8] is utilized in the research. The velocity update formula of the particle in the modified algorithm keeps unchanged while location update formula is improved, as shown in formulas (3) and (4):

$$\begin{aligned} when \ v_{id}^{n+1} \prec 0, x_{id} = \begin{cases} 0 & if \ rand() \prec s(v_{id}^{n+1}) \\ x_{id} & otherwise \end{cases} \\ when \ v_{id}^{n+1} \succ 0, x_{id} = \begin{cases} 1 & if \ rand() \prec s(v_{id}^{n+1}) \\ x_{id} & otherwise \end{cases} \end{aligned} \tag{3}$$

$$s(v_{id}^{n+1}) = \begin{cases} 1 - \frac{2}{1+\exp(-v_{id}^{n+1})} & when \ v_{id}^{n+1} \leq 0 \\ \frac{2}{1+\exp(-v_{id}^{n+1})} - 1 & when \ v_{id}^{n+1} \succ 0 \end{cases} \tag{4}$$

By comparing the particle velocity with 0, the modified algorithm deduces the location relationships between the current particle and the individual optimum and the

Algorithm 1. The BPSO based Complex Splitting
Input: multi-dimensional contextual rating matrix R, impurity threshold d, and iteration number N
Output: the two-dimensional rating matrix R'
Method:
for each user u in R do
1. Initiating m particles $x_1, x_2, \cdots, x_m$;
2. Calculating the initial values of $p_1, p_2, \cdots, p_m$;
3. The global optimum $p_g = \max\{p_1, p_2, \cdots, p_m\}$;
4. while $n \prec N$ do
    for each particle x_i do
       1) if $n \prec \text{rand}() \times N$
           updating particles $x_i^n \leftarrow \text{BPSO1}(x_i^{n-1})$ using formulas (1) and (2);
         else
           updating particles $x_i^n \leftarrow \text{BPSO2}(x_i^{n-1})$ using formulas (1) and (3);
         endif
       2) calculating all the possible contextual condition combinations $C \leftarrow \text{Compose}(x_i^n)$
       3) for each $c \in C$ do
           splitting all the rating data of the user u into the rating r_c that satisfies the contextual condition and $r_{\bar{c}}$ that does not.
           calculating the impurity criteria $t(x_i, c)$;
         endfor
       4) the adaptive value of particle x_i in the nth iteration $fit_{x_i, n} \leftarrow \arg \max \{t(x_i, c) | c \in C\}$;
       5) The corresponding contextual condition combination of the adaptive value of particle x_i in the nth iteration $c_{x_i, n} \leftarrow \arg \max_c \{t(x_i, c) | c \in C\}$;
       6) if $fit_{x_i, n} \succ p_i$ then
           $p_i = fit_{x_i, n}$; $c_i = c_{x_i, n}$; // c_i represents the corresponding contextual condition combination of the optimal individual value for particle x_i
         endif
       7) if $p_i \succ p_g$ then
           $p_g = p_i$; $c_{max} = c_i$; // c_{max} denotes the corresponding contextual condition combination of the global optimal value
         endif
    endfor
    end while
5. if $p_g \succ d$ then splitting the u in R into two different users $u_{c_{max}}$ and $u_{\bar{c}_{max}}$ and a new rating matrix R' is then generated;
Endfor;

global optimum at the dth dimension. The modified particle location update formula ensures that when velocity of the particle is 0, the possibility of the location change tends to be 0, which enables the particle swarm to readily get close to the global optimal particle.

In this research, BPSO is not directly used to select contextual conditions, but to select the contextual factors. The permutation and combination of all the contextual conditions of the contextual factors selected in each iteration constitute the solution set of the adaptive values for solving the objective function. In order to meet the requirements that the algorithm is endowed with global searching capacity in early period and stronger local searching capacity in later period, the original BPSO was combined with the modified BPSO which was proposed in [8]. The specialized process is shown in algorithm 1:

The BPSO1() and BPSO2() in algorithm 1 are functional functions for realizing the discrete BPSO. Compose() is the solution set of the adaptive values for particle in this iteration constituted by the contextual condition combinations, which are obtained using all the selected contextual factors of each particle. The time complexity of this algorithm is O($UNmC$), where U, N, m and C are respectively the numbers of the user, iteration, particle and contextual condition combination. Therefore, it can be seen that compared with the exhaustive search method, the proposed algorithm can significantly improve the operating efficiency, which reflects exactly PSO suffers little from the dimension problems.

4   Experiment and Results Analysis

4.1   The Setting of Experimental Parameters

The actual movie rating dataset LDOS-CoMoDa containing contextual information were adopted in the experiment of Complex Splitting [9]. After filtering out the rating records with incomplete contextual information, the final dataset containing 113 users, 1190 items, 2099 ratings and 12 contextual variables was obtained. The specialized description on the 12 context variables can be seen in investigation [9]. After conducting complex splitting on these rating data, some new and specialized data were obtained, among which 80 % were regarded as the training set, and 20 % were the test set. Afterwards, 5-fold cross validation was carried out to test the algorithm.

In the experiment, the BPSO was performed on each user to optimize the optimal contextual condition combination for splitting. Owing to the smaller v_{max} value readily increases the variation possibility of particle location in the BPSO algorithm, the velocity of particle was therefore limited in the range of $[-2, 2]$, where, v_{max} was 2. As to the inertial weight w in the velocity update formula (1), when it descends linearly from the value close to 1 to 0.4 in the iteration process, its optimization effect is much better than that when it is set as a fixed value [10]. Therefore, the value of w in the nth iteration can be updated according to formula (5), where w_{start} and w_{end} are respectively 0.9 and 0.4, and N represents the total number of iteration.

$$w_n = w_{end} + \frac{(N - n)}{N} \times (w_{start} - w_{end}) \tag{5}$$

Five methods for calculating the impurity criteria were introduced in study [3], and they were respectively t_{mean}, t_{prop}, t_{chi}, t_{IG} and t_{random}. As demonstrated in the experimental results [5], t_{chi}, t_{prop} and t_{mean} perform preferably in calculation. Therefore these three methods were employed in this experiment to calculate the impurity criteria. To make the splitting meaningful, a threshold is required to be assigned to t_{chi}, t_{prop} and t_{mean} respectively. In this experiment, the significance level of difference was set as 0.05, and this value is regarded as threshold of $p - value$ in the three test methods respectively. When one user is split, all the possible contextual condition combinations are traversed. If the calculated $p - value$ is less than 0.05, then the impurity criteria satisfies the splitting criteria, and the corresponding contextual condition combination of the maximum impurity criteria is considered as the optimal contextual condition combination to split the user. Alternatively, the user that does not satisfy the splitting criteria is expected to keep unchanged.

This research verifies validity of the proposed Complex Splitting by using the greatly modified User Based Collaborative Filtering (UBCF) [11], Item Based Collaborative Filtering (IBCF) [12] and SVD++ [13] respectively. The number 5, 10, 20, and 40 are selected for K respectively as the neighborhood number in collaborative filtering algorithms (UBCF and IBCF), and the experimental results demonstrate that when K is 10, the splitting effect is the optimal. In the SVD++ algorithm, iteration is conducted for 100 times at the learning rate of 0.01 with 10 latent classes.

4.2 Analysis of Experimental Results

The 5-fold cross validation is performed in the research to calculate the RMSE of the predicated rating, so as to measure the accuracy of recommendations. Compared with simple splitting, the new rating dataset obtained through Complex Splitting are more specialized, and the recommendations are therefore well-directed to the personalization of target user and the contextual environment, which can further improve the accuracy of recommendations. As is shown in Figs. 1, 2 and 3, combined with IBCF, UBCF and SVD++ respectively, Complex Splitting all achieves lower RMSE than that of Item Splitting and UI Splitting. This result fully demonstrates the validity of the proposed algorithm.

Among those methods for calculating the impurity criteria, t_{chi} achieves the best performance, followed by the t_{prop}, and t_{mean} ranks the last. In contrast, by adopting the chi-square test of fourfold table, t_{chi} calculates the impurity criteria by taking the proportions of high rating and low rating in two parts of data into account simultaneously. While t_{mean} only focuses on the mean value difference of the rating data in two parts, however, the overall mean value difference sometimes fails to reflect the local interest of the user. Therefore, t_{chi} can further distinguish the user preference. Owing to t_{prop} only calculates one-sided proportion, such as the proportion of high rating, the overall effect is thus imbalanced. Therefore, the t_{chi} calculation method is slightly superior to others according to the experimental results.

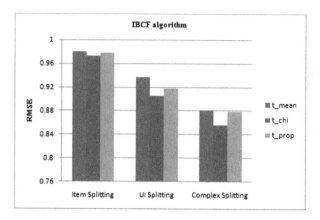

Fig. 1. RMSE comparison of splitting approaches using IBCF

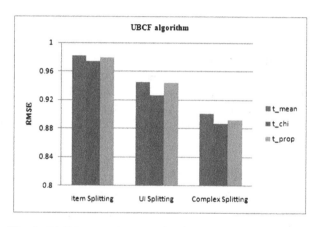

Fig. 2. RMSE comparison of splitting approaches using UBCF

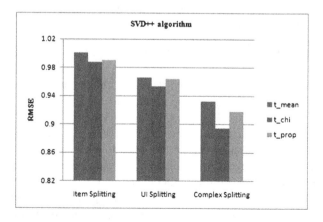

Fig. 3. RMSE comparison of splitting approaches using SVD++

5   Summary and Future Work

Based on the existing studies on the Simple Splitting, the discrete BPSO based Complex Splitting is put forward in the paper to further improve the accuracy of recommendations. The selection of multiple contextual conditions is to optimize the contextual condition combination using the discrete BPSO, and the optimal adaptive value of objective function eventually decides whether or not to split the user. As observed in the experiment, the proposed Complex Splitting is capable of improving the accuracy of recommendations. However, merely binomial splitting is conducted on the user in the paper using multiple contextual conditions, it needs to develop a method that can split one user into multiple users which satisfy different contextual conditions in the future and dispense with the over-fitting caused by the restriction of data sparsity. The another following research is to detect the influence of various contextual conditions on the user's interests, and delete the invalid context conditions to avoid interference.

Acknowledgement. This research is supported by the National Natural Science Fund (Grant no. 41362015), a Science and Technology Project of Education Department of Jiangxi Province (Grant nos. GJJ14431, GJJ14432, and GJJ14458), and the Youth Science Foundation Project of the Science and Technology Department of Jiangxi Province (Grant no. 20122BAB211035).

References

1. Wang, L., Meng, X., Zhang, Y.: Context aware recommender systems. J. Softw. **23**(1), 1–20 (2012)
2. Adomavicius, G., Tuzhilin, A.: Context-aware recommender systems. In: Ricci, F., Rokach, L., Shapira, B., Kantor, P.B. (eds.) Recommender Systems Handbook, pp. 217–253. Springer, New York (2011)
3. Baltrunas, L., Ricci, F.: Experimental evaluation of context-dependent collaborative filtering using item splitting. User Model. User-Adap. Inter. **24**(1–2), 7–34 (2014)
4. Said, A., De Luca, E.W., Albayrak, S.: Inferring contextual user profiles-improving recommender performance. In: Proceedings of the 3rd RecSys Workshop on Context-Aware Recommender Systems (2011)
5. Zheng, Y., Mobasher, B., Burke, R.D.: The role of emotions in context-aware recommendation. In: Decisions@RecSys, pp. 21–28 (2013)
6. Kennedy, J., Eberhart, R.C.: A discrete binary version of the particle swarm algorithm. In: Computational Cybernetics and Simulation, pp. 4104–4108 (1997)
7. Shi, Y., Eberhart, R.C.: Empirical study of particle swarm optimization. In: Proceedings of the 1999 Congress on Evolutionary Computation-CEC99, pp. 1945–1950. IEEE (1999)
8. Liu, J.: The Basic Theory of PSO and its Improved Research. Central South University, Changsha, China (2009)
9. Odic, A., Tkalcic, M., Tasic, J.F., et al.: Relevant context in a movie recommender system: users' opinion vs. statistical detection. In: Proceedings of the 4th International Workshop on Context-Aware Recommender Systems (2012)
10. Shi, Y., Eberhart, R.C.: Empirical study of particle swarm optimization. In: Proceedings of the 1999 Congress on Evolutionary Computation-CEC99, pp. 1945–1950. IEEE (1999)

11. Choi, K., Suh, Y.: A new similarity function for selecting neighbors for each target item in collaborative filtering. Knowl.-Based Syst. **2013**(37), 146–153 (2013)
12. Wang, C., Wang, C., Xu, L.: User-adaptive Item-based collaborative filtering recommendation algorithm. Appl. Res. Comput. **30**(12), 3606–3609 (2013)
13. Koren, Y.: Factorization meets the neighborhood: a multifaceted collaborative filtering model. In: Proceedings of the 14th ACM SIGKDD International Conference on Knowledge Discovery and Data Mining, pp. 426–434. ACM, Las Vegas (2008)

A Method for Calculating the Similarity of Web Pages Based on Financial Ontology

Lu Xiong[1,2(✉)], Kangshun Li[2], and Suping Liu[1]

[1] Department of Computer Science,
Guangdong University of Science and Technology, Dongguan, China
[2] College of Mathematics and Informatics, South China Agricultural University,
Guangzhou, Guangdong Province, People's Republic of China
317771184@qq.com

Abstract. The search results of the traditional concept similarity algorithm in search engine is not accurate, and can not support search results query for search results. A method of calculating the similarity of web pages based on financial ontology is proposed to solving the above problem. First of all, the concept of the financial ontology based on WordNet is constructed, and the corresponding concepts are obtained; Secondly, a new improved strategy of extracting financial key words is proposed in Key words mining according to the characteristics of the web pages (give different weights to different parts of the web pages), which is based on the traditional TF*IDF algorithm, and this strategy can better represent the subject of a web page; Then calculate the semantic distance between keywords, the depth of the levels and the degree of semantic overlap. Finally, the optimal computation of the similarity is realized by the comprehensive weighted processing of multiple similarity. Experimental results showed that the method compared with SSRM in recall ratio and precision ratio has greatly improved. At the same time, the method improved the algorithm performance.

Keywords: Data mining · Financial ontology · Similarity

1 Introduction

The concept of semantic similarity is used to measure the semantic content or meaning of a document or term [1]. At present, similarity computation has been widely used in ontology learning and merging, semantic annotation, information extraction in knowledge management and natural language understanding and other related fields. Compared with the retrieval on keywords, the retrieval based on semantic can greatly improve the precision ratio [2] and recall ratio in information retrieval.

The similarity calculation of the concept determines the accuracy of the semantic matching, which is the basis of the semantic retrieval. Therefore, it is the key to improve the accuracy of the calculation of the concept similarity.

At present, domestic and foreign scholars have carried out extensive research on the concept similarity calculation, and proposed a lot of methods to calculate the similarity. The similarity algorithm can be roughly divided into based on HowNet method, based on RDFS method and based on ontology method according to the data source we used.

© Springer Science+Business Media Singapore 2016
K. Li et al. (Eds.): ISICA 2015, CCIS 575, pp. 445–455, 2016.
DOI: 10.1007/978-981-10-0356-1_47

The method based on HowNet consist of three similarity of sememe semantic, concept semantic and words semantic; All of the concepts of HowNet are expressed by the sememe, so the calculation of sememe similarity is the basis of concept similarity. Each concept of words is represented by multiple sememe.

We calculate the similarity through the simple semantic distance according to sememe hierarchy tree which is constructed based on relationship between the upper and lower sememe [3]. Zhu [4] proposed a calculation method of similarity based on the domain ontology concept. The concept similarity is calculated by the semantic distance in domain ontology, which more reflects the domain experts on the concept of the category of the division.

Varelas [5] proposed a retrieval method SSRM based on ontology. In this paper, we proposed a new semantic similarity algorithm based on financial ontology after considering the financial concept in the classification of the sub node information, depth information, public father node information according to the structure of WordNet dictionary.

The solution method based on WordNet structure is not required other corpora to participate and easy implemented. At the same time, the paper constructs the ontology of the financial domain by using the WordNet dictionary, and proves that the algorithm can effectively improve the accuracy of the semantic similarity computation between concepts.

2 Retrieval Method Based on Notology

A retrieval method SSRM based on ontology by using Ontology semantic expansion of query terms. It's the concrete operation is the first to using TF*IDF [9, 16] algorithm for document, the purpose is put the keywords into one dimension vectors in the extraction process, then making concept extension of the vector of the query sentence by synonymous. The concept extension is based on the WordNet ontology, at the same time, we recalculate the TF*IDF weights after the extension of the vector and finally use the formula of calculating the document similarity:

$$Sim(q,d) = \frac{\sum_i \sum_j q_i d_j Sim(i,j)}{\sum_i \sum_j q_i d_j} \tag{1}$$

Where, q refers to the query statement, also can refer to the document; d refers to a document that is compared with q. i refers to the key words in document q; q_i refers to the weight value of TF*IDF of keywords in q; j refers to the key words in document d; d_j refers to the weight value of TF*IDF of keywords j in d. $sim\,(i,j)$ refers to the similar values of keywords j and i in WordNet.

The traditional TF*IDF algorithm formula is as follows

$$tf\left(w_k, d_j\right) = \frac{n(w_k)}{\sum\limits_{w \in d_j} n(w)} \tag{2}$$

$$idf = \log\left[\frac{|D|}{d_j \supset w_k}\right] \tag{3}$$

Where, $n(w_k)$ refers to the number of word w_k appeared, $\sum_{w \in d_j} n(w)$ refers to the sum of the number of all words that appeared in the sentence d_j, $tf(w_k, d_j)$ refers to the probability of word w_k appeared in a sentence d_j, D refers to the number of all sentences in the document, $|d_j \supset w_k|$ refers to the number of sentences in which words w_k appeared.

Formula (2) * formula (3) refers to weights value of the TF*IDF in the document.

According to the TF*IDF algorithm, the order of the TF*IDF weights can be concluded as the most important key words in a document. However, TF*IDF only highlights the relatively high frequency of words in the document, eliminating the influence of some common words in the document. It is aimed at the general web pages, which has better effect, but it is used to deal with the professional field of financial and economic web pages, when considering the impact of the characteristics of the financial web pages, the effect of the content feature extraction should be weaker than the keyword extraction. Therefore, combining the traditional TF*IDF algorithm and the characteristics of the financial web pages to give a new algorithm for extracting financial keywords.

3   Improved Ontology Similarity Calculation Algorithm

3.1   The Framework of Financial Ontology Similarity Computation Model

The main steps of similarity calculation based on the financial ontology is shown in Fig. 1.

The main step of the similarity computation is to construct the concept tree by means of tools from the ontology of the financial domain. Calculating the weight of each concept according to the premise of the custom. and the concept and the weights are saved to the library. Additional, extracting a web page from the web sites, and analysis the web page by financial keyword extraction algorithm. Finally, calculating the similarity value between the financial and financial words in the ontology library by using the method of similarity between keywords, compared with the setting of the threshold, to judge whether it is a financial keyword.

3.2   Building the Concept Tree of Financial Ontology

In order to calculate the similarity between common web pages and financial web pages, all the concepts in the financial ontology should be quantified, and the concept of weighted value is formed. Because there are a lot of shortcomings of human operation, this paper presents a method for calculating the weight of the concept. First, according to the relationship between the concept of the financial ontology to build a

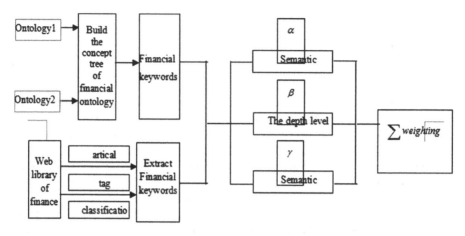

Fig. 1. The similarity calculation algorithm flow

financial ontology concept tree; secondly, according to the level of each concept in the tree to calculate its weights; Finally, form a {concept, weight} list, which is stored in the ontology library.

The following premises should be complied with:

There are two kinds of relations between the concept of the financial domain ontology: inheritance and attribute; Inheritance represents a concept of the relation between the parent and the child, and attribute represents a description of a concept to another concept [10, 11].

If the concept belongs to inheritance or attribute, it can accurately express the concept of concrete and abstract relations, in general, the concept of the child's weight than the parent class, because the concept of the child is more specific than the parent class [12–14].

Considering the depth of the tree and hierarchical relations for financial ontology concept tree hierarchy to define the concept of weight, the definition of the concept of weight values, as shown in formula 4

$$W = I + \frac{L}{2H} \qquad (4)$$

Where, W refers to the weight of concept, I refers to the initial value of the root node in the concept tree, L refers to the hierarchy of concepts, H refers to the height of concept tree [15].

We construct the financial domain ontology concept tree as shown in Fig. 2, where, the existing inheritance relationship is a "stock" and "Securities circle", "futures" and "securities circle", "business celebrities" and "financial circle", "economic celebrities" and "financial circle"; Additional, the "A shares" and "stock", "U.S. stocks" and "stock", etc. belongs to attribute relationship.

Thus, determining the weight of each concept by the financial ontology concept tree, assign 0.5 to I generally, specific examples are shown in Table 1.

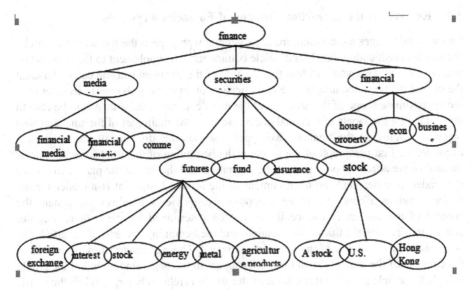

Fig. 2. The hierarchy diagram of financial concept class

Table 1. Calculation examples of the concept of weight

Concept	Relation	hie.	Weights
Root	No	0	0.5
Securities circle	Inheritance	1	0.5 + 1/2*4
Futures	Inheritance	2	0.5 + 2/2*4
Foreign exchange	Attribute	3	0.5 + 3/2*4
Interest	Attribute	3	0.5 + 3/2*4
Stock index	Attribute	3	0.5 + 3/2*4
Energy	Attribute	3	0.5 + 3/2*4
Metal	Attribute	3	0.5 + 3/2*4
Agriculture products	Attribute	3	0.5 + 3/2*4
Fund	Inheritance	2	0.5 + 2/2*4
Insurance	Inheritance	2	0.5 + 2/2*4
Stock	Inheritance	2	0.5 + 2/2*4
A stock	Attribute	3	0.5 + 3/2*4
U.S. stocks	Attribute	3	0.5 + 3/2*4
Hong Kong stocks	Attribute	3	0.5 + 3/2*4
Financial circle	Inheritance	1	0.5 + 1/2*4
Business celebrity	Inheritance	2	0.5 + 2/2*4
Economic celebrity	Inheritance	2	0.5 + 2/2*4
Media circle	Inheritance	1	0.5 + 1/2*4
Financial media people	Inheritance	2	0.5 + 2/2*4
Financial media	Inheritance	2	0.5 + 2/2*4
Commentator	Inheritance	2	0.5 + 2/2*4

3.3   Key Words the Extraction Strategy of Financial Keywords

Financial web pages are different from the general web page is the former is the article, the article classification, labels and article comments as a supplement to the web. When you browse the web, you will find that the title of the financial website can understand the central theme of the article. The classification of financial web pages can determine the approximate scope of the article, and the main content of the article is to be able to understand by the labels of financial pages. Because the main part of the financial web pages is articles, so the financial web page also has the structure of the article: for example, the first paragraph of the article is the beginning of the article, the last part is the end of the article, if only look at the beginning and the end of the part, it can make the reader to understand the main content of the article. Comment is a reader for the financial web page content to agree or oppose or other views, it helps to understand the content of the article. But since it can not be determined by the author, it is not considered. The labels, titles, classification, and the beginning and end of the article are determined by the author of the financial web page. These characteristics are related to the contents of the article. When the words appear in the beginning or end of the article is high, the article shows that the value of the article is relatively large. When the words appear in the number of tags is higher, it shows that the value of the label is large. In the same way, when the classification of words is large, the probability of the key words is large.

From the above we can know, the value of the characteristics of the word, the value of the label and the value of the classification of the value determines the word is or not the keyword, so the definition the sum of the three to the weight of the word. The greater the weight of the word, the higher the probability the word is the keyword of the financial web page. The following financial keyword extraction algorithm is discussed based on this idea.

Each financial web page is obtained is the collection of all the vocabulary through the Chinese word segmentation. The algorithm needs to enter the vocabulary of the WordSet is obtained by the financial web after word segmentation, and the output is the financial keywords set KeywordSet.

The keywords extraction algorithm is completed by the following steps:

(1) calculate the Eigenvalues of the article of word w_k

$$SW(w_k) = \log \left[\frac{|\mathbf{P}|}{\sum\limits_{s_i \in (p - p_i)} tf(w_k, p_i)} \right] \tag{5}$$

Where, $V = \xi_1 \cdot tf(w_k, t_s) + \xi_2 \cdot tf(w_k, para_1) + \xi_3 \cdot tf(w_k, para_2)$

(2) Calculate the Eigenvalues of the tags of word w_k

$$TW(w_k) = \log \left[\frac{|P|}{\sum\limits_{S_i \in P} tf(w_k, p_i)} \right] \bullet tf(w_k, T) \tag{6}$$

(3) Calculate classification Eigenvalues of word w_k

$$CW(w_k) = \log \left[\frac{|P|}{\sum_{S_i \in P} tf(w_k, p_i)} \right] \bullet tf(w_k, C) \tag{7}$$

(4) Calculate weight of word w_k

$$weight(w_k) = SW(w_k) + TW(w_k) + CW(w_k) \tag{8}$$

(5) Repeat steps 1 to 4 until the weight of each word in the WordSet is calculated.
(6) Compared the weight of each word, selecting the most n representatives keywords of the financial web content as set $KeyList = \{K_1, K_2, \ldots K_n\}$ according to the size of the weights.

Therefore, the financial keywords of financial web pages extracted from financial web sites.

3.4 The Similarity Calculation Between Keywords

Actually, the similarity calculation of financial web pages is the similarity calculation between keywords form Fig. 1. The similarity calculation between the financial keywords which is extracted through the financial Web database and the financial keywords in the ontology library. The similarity value is the way to judge whether the financial web is a financial web page. Using ontology to calculate the similarity between words is actually the similarity between concepts in the ontology. Therefore, the similarity calculation between concepts mainly consider the following factors:

(1) Semantic distance [6]. Calculating the shortest path of two concepts in the tree by using the ontology hierarchy tree. The shortest path can also be expressed as the semantic distance between them. Generally speaking, the semantic distance between the two concepts is more and more different, the less the semantic distance between the two concepts, and the more the similarity between them. Let a and b be any two nodes (that is the concept of ontology) in the ontology hierarchy tree. $d(a, b)$ indicates the path length from a to b.
(2) The depth level [7]. The difference between the two concepts of the level of depth is usually used to measure the similarity. The ontology level tree follows the top-down principle, and the concept is also followed from the large to small classification method. The concept similarity of large classes is generally smaller than that of small class. In the hierarchical tree, the similarity between the two concepts is higher than that of the root, and the difference between the two concepts is smaller under the same circumstances. Let a and b be any two nodes in the ontology hierarchy tree, $l(a)$ indicates the level of node a, $l(b)$ indicates the level of node b, $|l(a)-l(b)|$ represents the level difference between node a and node b.

Semantic overlap [8]. The similarity between two concepts is the proportion of the total number of nodes in the two concepts of ontology, which is the common ancestor node. Let the root of the tree is a, a and b are two arbitrary nodes in the ontology hierarchy tree, $n(a)$ starts with a, up until set of the nodes which through by root. n (a) represents the number of set of nodes, $n(a) \cap n(b)$ represents the intersection of nodes set from a and b to A. $n(a) \cup n(b)$ represents the union of nodes set from a and b to A. $\frac{|n(a) \cap n(b)|}{|n(a) \cup n(b)|}$ represents the Semantic overlap between concepts a and b. On the basis of the above three factors, the similarity calculation formula (9) of any two concepts in the ontology is proposed as follows:

$$sim(a,b) = \begin{cases} 1 & a = b \\ \frac{\alpha \times \beta \times |n(a) \cap n(b)|}{(d(a,b) + \alpha) \times |n(a) \cup n(b)| \times (\gamma \times |l(a) - l(b)| + 1)} & a \neq b \end{cases} \qquad (9)$$

Where, α, β, γ are adjustable parameters, used to reflect the semantic distance and semantic similarity $\alpha > 0$. β used to adjust the effect of semantic overlap on similarity, $1 \leq \beta \leq \frac{D(O)}{D(O)-1}$, $D(O)$ represents the hierarchical tree depth of Ontology. γ used to adjust effect of hierarchy difference on similarity, $0 < \gamma < 1$.

We can see the following conditions must applied with from the formula:

(1) if $d(a, b) = 0$, $sim(a, b) = 1$;
   else if $d(a, b) = \infty$, $sim(a, b) = 0$;
(2) $0 \leq sim(a, b) \leq 1$;
(3) Two concept hierarchy is inversely proportional to the similarity, the higher the hierarchy of the two concepts, the smaller the similarity.
(4) Two concept semantic overlap and similarity is inversely proportional. If the higher the two concepts of semantic overlap, the smaller the similarity.

Preparation: Table 1 is used to store the ontology concept information generated after Jena parsing, including (node number, node name, parent node number, node level). Table 2 is used to store the similarity of concepts and the comprehensive value, fields including (concept 1, concept 2, similarity, and comprehensive value).

Input: OWL file.

Output: List the similarity between concepts in the ontology, and write the data to Table 1.

   Step 1: Initialize and generate the concept hierarchy.
   Generating the three element combination on the concept of the relationship between the upper and lower in the memory by using the tool API Jena to parse the ontology structure. In the combination of three elements, from the root node, the depth first strategy is used to traverse each node of the hierarchical tree, and the node's information is stored in the data Table 2.
   Step 2: Combination of all possible concepts and choose the pair of concept.
   Read concepts from the data Table 2 and combination of all possible concepts pair, and write data into table a according to the concept of information, in which the last three values are taken as 0. Finally, the concept is selected from the data Table 1.

Step 3: Computing concept semantic distance.

Generating a node path through concept of data Table 2 to the root, sequential storage by array. In the concept of the formation of the two path array, to find the first public nodes of the two array, add the value of the two array, you can calculate the semantic distance.

Step 4: Computing semantic overlap.

Calculating the intersection and union of paths in the two path array we got in step 3. We can calculate the semantic overlap: the number of nodes of intersection/the number of nodes of union.

Step 5: Computing concept hierarchy.

Get the concept hierarchy from the concept hierarchy Table 2, the absolute value of the difference between the two concepts is also known as the concept hierarchy.

Step 6: Calculate concept similarity.

According to concept semantic distance, semantic overlap and concept hierarchy, the concept similarity is calculated by using the formula (9), and modify the value of the "similarity" field corresponding to the record in Table 1.

Step 7: To determine whether there is concept pair of non computing.

Check if there is an pair for no calculation, if you have, go to Step 2, else go to Step 8.

Step 8 The end of algorithm.

Output data Table 1, the end.

4 Experimental Analysis

4.1 Evaluation Criteria of Performance

In the experimental evaluation, this paper adopts the traditional precision ratio, recall ratio as the acceptance standard.

$$precisionratio = \frac{\text{The number of correct matches found}}{\text{The number of matches found}} \times 100\% \qquad (10)$$

$$recallratio = \frac{\text{The total number of matches found}}{\text{The total number of matches in existence}} \times 100\% \qquad (11)$$

4.2 Experiment Setting

Experiments using API JENA to resolve the ontology in Java language in the financial ontology similarity, the use of WordNet Java to provide the API operation Word-Net2.0, access to the ontology of the similarity value. In the similarity calculation of the financial and financial ontology, the API Java provided by WordNet WordNet2.0 is used to obtain the similarity value of the ontology.

OAEI 2015 recommend 3 of the test data set, respectively is benchmark, conference and Anatonmy. In this paper, benchmark is used as the test data set.

4.3   Experimental Results and Analysis

The algorithm name is CSFO; Experiments were carried out by using CSFO and SSRM to test the ontology set benchmark, and executing computational testing by using the TempLoadRunner tool.

The experimental results are shown in Figs. 3 and 4 (Note: "r1 = stock", "r2 = fund", "r3 = futures", "r4 = real estate", "r5 = insurance").

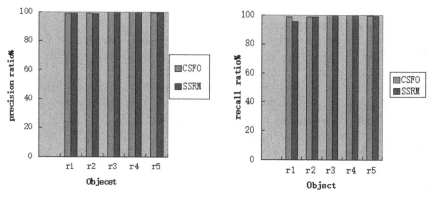

Fig. 3. Comparison of precision ratio of CSFO method and SSRM method

Fig. 4. Comparison of recall ratio CSFO method and SSRM method

The experimental results showed that the accuracy of CSFO method is slightly higher than that of the SSRM algorithm. But the CSFO method is better than the SSRM algorithm in the case of a large set of instances when mapping benchmark.

This is because the CSFO method, the financial keyword extraction strategy improved the accuracy of keyword extraction and the effective offset error according to the characteristics of web pages, the paper introduces the characteristics of the article, labels and classification. In addition, the CSFO method introduced the semantic distance, the depth of the level, the degree of semantic overlap to calculate the similarity, narrowing the scope of keywords, reduce the time spent in the calculation process, more efficient.

5   Conclusion

In this paper, we proposed a algorithm for calculating the similarity of web pages based on financial ontology after comprehensively considering the data mining theory and domain ontology. The similarity calculation of the financial ontology is to calculate the similarity between the keywords in the semantic distance, the depth of the level and the degree of semantic overlap and obtained a comprehensive similarity value of this paper. The experimental results show that the algorithm is satisfactory in the matching effect, but it is not particularly ideal because of the need of the artificial adjustment parameter setting. At the same time, with the development of ontology, the ontology evolution [15, 16] is becoming more and more serious, which is also the next step to take into account and make the appropriate improvements.

Acknowledgment. This work is supported by the National Natural Science Foundation of China with the Grant No. 61573157, the Fund of Natural Science Foundation of Guangdong Province of China with the Grant No. 2014A030313454.

References

1. Li, H.S.: Knowledge Management Technology and Application. Beijing University of Posts and Telecommunications Press, Beijing (2012)
2. Li, W., Yang, N.-N.: The research of similarity based on semantic of ontology. Sci. Technol. Eng. **12**(21), 5328–5331 (2012)
3. Liu, Q., Li, S.-J.: The calculation of semantic similarity based on vocabulary of Hownet. Comput. Linguist. **7**(2), 59–76 (2002)
4. Zhu, L.-J., Tao, L., Liu, H.: The calculation of concept similarity based on ontology in the domain. J. S. China Univ. Technol. (Nat. Sci. Ed.). **32**(2), 147–150 (2004)
5. Varelas, G., Voutsakis, E.P., Raftopoulou, P.: Semantic similarity methods in WordNet and their application to information retrieval on the web. In: Proceedings of WIDM 2005, pp. 151–155 (2005)
6. Xu, D.-Z., Wang, H.-M.: The research of semantic similarity calculation method based on concept of ontology. Comput. Eng. Appl. **43**(8), 154–156 (2007)
7. Andreasen, T., Bulskov, H.: From Ontology Over Similarity to Query Evaluation. Elsevier Science, Philadelphia (2003)
8. Knappe, R., Bulskov, H., Andreasen, T.: Similarity graphs. In: Zhong, N., Raś, Z.W., Tsumoto, S., Suzuki, E. (eds.) ISMIS 2003. LNCS (LNAI), vol. 2871, pp. 668–672. Springer, Heidelberg (2003)
9. Uschold, M., Gruninger, M.: Ontologies: principles, methods and applications. Knowl. Eng. Rev. **11**(2), 93–136 (1996)
10. Uschold, M.: Building ontologies: towards a unified methodology. AIAI Technical reports, United Kingdom (1997)
11. Chen, J.: The Creation and Application of Domain Ontology. University of International Business and Economics, Beijing (2006)
12. Li, J., Su, X.-L., Qian, P.: The method of constructing domain ontology. Comput. Agric. **7**(1), 7–10 (2003)
13. Fernandez, M,. Gomez-Perez, A., Juristo, N.: Methontology: from ontological art towards ontological engineering. In: AAAI 1997 Spring Symposium on Ontological Engineering, Standford University (1997)
14. Liu, R.-N., Li, Y.-S.: The method of constructing domain ontology. J. Wuhan Polytech. Univ. **27**(1), 73–77 (2008)
15. Liu, J.-H., Zhang, Y.-H., Li, S.-W.: Research progress of ontology evolution. Appl. Comput. Syst. **20**(7), 239–243 (2011)
16. Xiong, L.: The Design and Implementation of Chinese Financial and Economic Blog Search Engine Based on Domain Ontology. Jiangxi University of Science and Technology, Ganzhou (2012)

An Expert System for Tractor Fault Diagnosis Based on Ontology and Web

Chunyin Wu[1(✉)], Qing Ouyang[1], Shouhua Yu[1], Chengjian Deng[1],
Xiaojuan Mao[1], and Tiansheng Hong[2]

[1] College of Mathematics and Informatics,
South China Agricultural University, Guangzhou, China
{wuchunyin,ouyangqing,segrad,
dcj,yiyang}@scau.edu.cn
[2] College of Engineering,
South China Agricultural University, Guangzhou, China
tshong@scau.edu.cn

Abstract. This paper proposed an Expert System for Tractor Fault Diagnosis (ESTFD) based on ontology and web technologies. The ESTFD consists of several components such as diagnosis interface, OWL reasoner, explanation module, ontology base and database etc. The diagnosis interface was designed as web interface, which could support users to access the ESTFD by internet anytime and anywhere. A domain ontology, tractor fault diagnosis ontology, was constructed to build ontology base. The OWL API was called to manipulate the ontology base. The OWL reasoner, Pellet, was used to make logical reasoning and generate explanations for the process of logical reasoning. The ESTFD could provide tractor fault diagnosis service via internet to tractor maintenance personnel and drivers who located in the wide rural areas in China. Since the ESTFD has explanation module to explain how the diagnostic results was obtained, it also could be used as a training tool.

Keywords: Expert system · Tractor · Fault diagnosis · Ontology · Web

1 Introduction

China is a large agricultural country which has a rural population of about 900 million. An important way to improve the living conditions of the rural population is to develop the rural economy. In order to speed up the development of rural economy, the Chinese government promotes the application of agricultural machinery in agriculture production in recent years. Thus, the number of agricultural machinery has increased rapidly in rural areas. However, the maintenance industry of agricultural machinery in China has been suffering from some critical problems, which limits the application of agricultural machinery [1]. One of the critical problems is the lack of maintenance technicians, another is that many maintenance personnel have not sufficient skills and experience. A solution to these problems is to develop an expert system for agricultural machinery fault diagnosis.

© Springer Science+Business Media Singapore 2016
K. Li et al. (Eds.): ISICA 2015, CCIS 575, pp. 456–463, 2016.
DOI: 10.1007/978-981-10-0356-1_48

Edward Feigenbaum, professor of Stanford University, defined the conception of expert system in year 1982 as "an intelligent computer program that uses knowledge and inference procedure to solve problems that are difficult enough to require significant human expertise for their solutions". That is said, an expert system is a computer system that emulates the decision-making ability of a human expert [2]. An expert system for fault diagnosis could be used not only to aid the maintenance personnel to diagnose faults, but also to train the maintenance personnel to improve their skills. The research and development of expert system for agricultural machinery fault diagnosis could ease the maintenance personnel problems in some degree.

Tractor is one of the most important and most widely used agricultural machinery, the research and development of an expert system for tractor fault diagnosis will promote the development of maintenance industry of agricultural machinery in China.

Ontology technologies, which have outstanding advantages in knowledge expression, are gradually adopted to develop expert systems in recent years. This paper proposed an Expert System for Tractor Fault Diagnosis (ESTFD) based on ontology and web technologies.

2 Tractor Fault Diagnosis Ontology

The word ontology stems from the philosophy domain. In philosophy domain, ontology is a branch of metaphysics, the nature of existence, that is what is the most general sense of real existence, and how to describe them. In computer science area, the most cited definition of ontology is: ontology is an explicit, formal specification of a shared conceptual system, which proposed by Gruber [3] and refined by Studer [4]. Ontology is very suitable for organization and representation of domain knowledge because firstly ontology could accurately and formally describe the relationships between different concepts, secondly the implicit relationships between different concepts described by ontology also could be obtained by logical reasoning.

This paper designed the Tractor Fault Diagnosis Ontology (TFDO) to construct knowledge base for ESTFD. TFDO was developed by using Ontology Web Language version 2 (OWL 2) which was proposed by World Wide Web consortium (W3C) in year 2009. There are three semantic building blocks in OWL 2: class, individual and property. Class is a collection of resources. Individual is a kind of resources, and it should be a member of at least one class. Property is used to describe resources. There are two main kinds of properties: object property and data property. Object property could associate a group of individuals with another group of individuals. The implicit relationships between classes and individuals could be deduced by OWL reasoner [5].

The TFDO ontology was separated into two parts: the domain knowledge of the structure of tractor and the operational knowledge of tractor fault diagnosis. The former was constructed based on the well defined hierarchical structure of tractor parts. The latter was constructed based on fault tree analysis. The TFDO ontology could support bi-directions reasoning, which could deduce both from reasons to results and from results to reasons.

In TFDO, tractor parts were modeled as a class named TractorPart. The class Tractor-Part has two object properties isPartOf and hasPart, the former represents a tractor part could be a part of other bigger tractor part, the later represents a tractor part could be constituted by other smaller tractor parts. The property isPartOf is an inverse property to the property hasPart, vice versa. In order to trace the composite relationships between different parts, the object properties isPartOf and hasPart are modeled as transitive properties. The TractorPart class was used to indicate the location of the fault events occurred.

In TFDO, fault events in fault tree are modeled as FaultEvent class, FaultEvent class has two object properties isReasonOf and isCausedBy, which's domains and ranges are both set to FaultEvent class. The isReasonOf object property expresses the meaning that one fault event could be the reason of another fault event. The isCausedBy object property expresses the meaning that one fault event could be the result of another fault event. The isReasonOf object property is the inverse property of the isCausedBy object property, vice versa. In order to trace the causal relationships between fault events, the isReasonOf property and isCausedBy property are modeled as transitive properties. Thus, the TFDO could support bi-directions reasoning, which could deduce both from reasons to results and from results to reasons. FaultEvent class also has other properties such as hasPhenomenon, judgementMethod, hasDiagnosticPriority, repairMethod, and preventiveMeasure etc., which express the phenomenon of the fault events, the methods to judge whether the fault events occurred, the diagnostic priority of the fault events, the repair methods and preventive measures for the fault events etc. accordingly.

3   Structure of ESTFD

Traditional rule based expert system consists of several components such as user interface, explanation facility, working memory, inference engine, agenda, knowledge base and knowledge acquisition facility [6]. Ontology based expert systems are similar as traditional rule based expert systems in many aspects; the main difference between them is the representation technologies for knowledge base. Ontology based expert systems use ontology technologies to represent knowledge. This paper proposed the structure of the ontology based expert system ESTFD referencing to the structure of traditional rule based expert systems. The structure of ESTFD, shown in Fig. 1, consists of several components such as user interface, explanation module, OWL reasoner, ontology base and database etc.

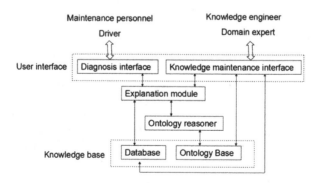

Fig. 1. The structure of ESTFD

3.1 User Interface

The user interface is the shell by which the user and the system communicate. There are two kinds of user interfaces of ESTFD, one is the diagnosis interface which used by tractor maintenance personnel and drivers, another is the knowledge maintenance interface used by knowledge engineers and domain experts. Tractor maintenance personnel and drivers input fault phenomenon in the diagnosis interface to get diagnostic results and explanations for the results from the system. Knowledge engineers and domain experts use the knowledge maintenance interface to create, read, update, and delete knowledge in the knowledge base. The knowledge maintenance interface also should provide functions to find and delete the duplicate knowledge in knowledge base.

3.2 Explanation Module

The explanation module provides facilities to trace the process of reasoning, records the axioms of ontology which used to deduce the diagnostic results and gives plain explanation about how the diagnostic results were obtained. This module would use ontology reasoner and knowledge base to generate plain, easy understood explanations.

3.3 Ontology Reasoner

The ontology reasoner is the core module of ontology based expert systems, which used to deduce implicit relationships between different concepts. Ontology reasoner should provide standard reasoning services such as consistency checking, concept satisfiability, classification, and realization. It also should provide query engines to support ABox queries and SPARQL [7] queries.

3.4 Ontology Base

The ontology base contains the domain knowledge of tractor and operational knowledge of tractor fault diagnosis needed to solve problems. These knowledge was coded in the form of classes, individuals, properties, axioms and SWRL [8] rules.

3.5 Database

The database contains detailed information about tractor and the tractor fault diagnosis, such as tractor composition principle, component structure diagrams of tractor, technology and performance indicators of tractor's parts etc.

4 Process of Tractor Fault Diagnosis

The process of tractor fault diagnosis is shown in Fig. 2.

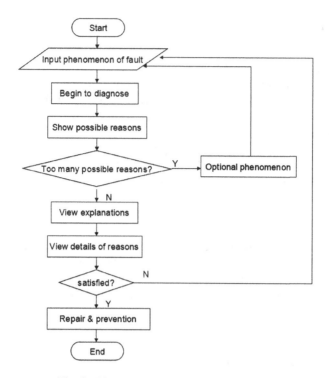

Fig. 2. The process of tractor fault diagnosis

The user (tractor maintenance personnel or drivers) should input the phenomenon of the tractor fault through diagnosis interface at first; the system would begin to diagnose to get the reasons by using knowledge in the knowledge base. Then the system would show the possible reasons, if there are too many possible reasons, the user could input optional phenomenon of the fault to refine the diagnostic results. The user could make the system either to show the explanations about how the diagnostic results were obtained, or to show the detailed information about the reasons, if the user is satisfied with the diagnostic results, he (she) could repair the tractor according to the diagnostic results and take preventive measures provided by the system to prevent the fault from occurring again in the future.

5   Software Architecture of ESTFD

This paper adopted the layers pattern to design software architecture of ESTFD, shown in Fig. 3.

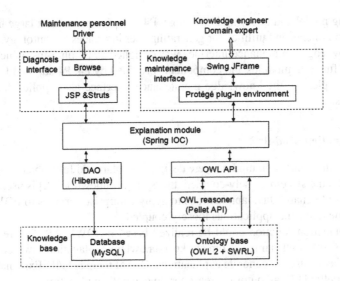

Fig. 3. The software architecture of ESTFD

5.1 Diagnosis Interface

The Diagnosis interface provides the interface for the tractor maintenance personnel and the drivers to diagnose tractor faults. In general, the tractor maintenance personnel and drivers are distributed at different physical locations, in order to facilitate their access to ESTFD system, a web interface was developed for tractor fault diagnosis. JSP and Struts [9] technology were used to develop the diagnosis interface.

Struts, an open source software, is one of the main stream web presentation layer technologies. Struts adopt MVC pattern to separate model, viewer and controller, thus make web applications have better maintainability, robustness and scalability.

Through web browser, the tractor maintenance personnel and drivers could access ESTFD anytime and anywhere.

5.2 Knowledge Maintenance Interface

The knowledge maintenance interface should provide powerful tools to manage the knowledge base. In order to reduce the development workload, it's a good idea to utilize the existing tools.

The software Protégé [10] is a powerful ontology editor which was developed by the Stanford University School of Medicine. Protégé could be used to create, read, update and delete ontology elements. The newest stable version of Protégé is Protégé 4 (P4). P4 uses the OSGi framework as a plug-in infrastructure, developers could write plug-ins in Java IDEs such as Eclipse to extend the functions of P4.

P4 provides API for developers to access the model and various utilities of P4. P4 API is built on top of the OWL API [11], which provides full functionality for manipulating model and querying knowledge. P4 API also provides a large number of reusable components and utilities for generating user interfaces for ontology.

The knowledge maintenance interface was designed as P4 plug-ins, which utilized the existing function modules and components provided by P4. P4 adopted Java Swing GUI technologies, so the knowledge maintenance interface also applied Java Swing technologies.

5.3   Explanation Module

The services in explanation module were designed as plain old Java objects (POJO). In order to make the system loosely-coupled, the Spring framework [12] is used.

Spring could enable Java applications to apply enterprise services to POJO, and at the same time make the applications loosely-coupled.

The explanation function could be realized based on OWL API. There is a Java package in OWL API named "com.clarkparsia.owlapi.explanation", which includes some Java classes used to generate explanations. One of them is PelletExplanation class which uses Pellet [13], an ontology reasoner, to generate explanation.

5.4   DAO and Database

The POJO is object oriented, in order to facilitate to store and retrieve objects from rational database, the DAO (Data Access Objects) layer is provided. The DAO layer uses Hibernate [14], a popular ORM (Object-Relational Mapping) tool, to access the back-end database.

The back-end database uses MySQL [15] database, which is a very popular open source database software at present.

5.5   OWL Reasoner

The system uses Pellet as OWL reasoner. Pellet is an open-source Java based OWL DL reasoner. It can be used to co-work with OWL API. Pellet provides functionalities to check the consistency of ontology, classify the taxonomy, check entailments, give explanations, and answer SPARQL-DL queries.

5.6   Ontology Base

The ontology base was built by using OWL 2 ontology language and SWRL.

6   Conclusion

In order to ease the maintenance personnel problems of the maintenance industry of agricultural machinery in China, such as the lack of maintenance technicians and many maintenance personnel have not sufficient skills and experience, this paper proposed an ontology based expert system ESTFD.

ESTFD includes several components such as diagnosis interface, OWL reasoner, explanation module, ontology base and database etc. The software architecture of ESTFD was designed based on layers pattern. The SSH frameworks were used in the software architecture, which make the system has better maintainability, robustness and scalability. The diagnosis interface was designed as web interface, which could support users access the system by internet anytime and anywhere. The knowledge maintenance interface was designed as plug-ins based on P4, which could utilize the existing powerful functions of P4. In order to build ontology base, the TFDO ontology was constructed. The OWL API was called to manipulate the ontology base. The OWL reasoner, Pellet, was used to make logical reasoning and generate explanations for the process of logical reasoning. MySQL database was applied as back-end database to store detail information about tractor and tractor faults.

ESTFD could provide tractor fault diagnosis service through internet for the tractor maintenance personnel and drivers who distributed in wide rural areas in China. As ESTFD has explanation module to explain how the diagnostic results was obtained, it also could be used as a training tool to improve tractor maintenance personnel and drivers' skills.

Acknowledgements. This project was granted financial support from Agriculture Department of Guangdong Province, China (5600-H09278). This work was conducted using the Protégé resource, which is supported by grant LM007885 from the United States National Library of Medicine.

References

1. Yu, Z.: The problems and countermeasures of the construction of basic maintenance service system for agricultural machinery in China. China Agric. Inf. **5**, 125–126 (2015)
2. Cai, Z., John, D., Gong, T.: Advanced Expert Systems: Principles, Design and Applications. Science Publication, Beijing (2005)
3. Gruber, T.R.: A translation approach to portable ontology specifications. Knowl. Acquisition **5**(2), 199–220 (1993)
4. Studer, R., Benjamins, V.R., Fensel, D.: Knowledge engineering: principles and methods. Data Knowl. Eng. **25**(1), 161–197 (1998)
5. Hebeler, J., et al.: Semantic Web Programming. Wiley Publishing Inc., Indianapolis (2009)
6. Giarratano, J.C., Riley, G.D.: Expert Systems Principles and Programming, 4th edn. China Machine Press, Beijing (2005)
7. SPARQL Query Language for RDF. http://www.w3.org/TR/rdf-sparql-query/
8. SWRL: A Semantic Web Rule Language Combining OWL and RuleML. http://www.w3.org/Submission/SWRL/
9. The Apache Struts Project. http://struts.apache.org/#Project
10. Protege 4 User Documentation. http://protegewiki.stanford.edu/wiki/Protege4UserDocs#Protege-OWL_Editor
11. The OWL API. http://owlapi.sourceforge.net/
12. Spring Framework. http://static.springsource.org/spring/docs/3.1.0.M1/
13. Pellet: OWL 2 Reasoner for Java. http://clarkparsia.com/pellet/
14. Relational Persistence for Java and .NET. http://www.hibernate.org/
15. MySQL Database 5.5. http://www.mysql.com/products/enterprise/database/

A Kuramoto Model Based Approach to Extract and Assess Influence Relations

Marcello Trovati[1]([✉]), Aniello Castiglione[2], Nik Bessis[1,3], and Richard Hill[1]

[1] Department of Computing and Mathematics, University of Derby, Derby, UK
`M.Trovati@derby.ac.uk`
[2] Dipartimento di Informatica, Universita' di Salerno, Salerno, Italy
`castiglione@acm.org`
[3] Department of Computing, Edgehill University, Ormskirk, UK
`Nik.Bessis@edgehill.ac.uk`

Abstract. In this paper, we introduce a novel method to extract and assess influence relations between concepts, based on a variation of the Kuramoto Model. The initial evaluation focusing on an unstructured dataset provided by the abstracts and articles freely available from PubMed [7], shows the potential of our approach, as well as suggesting its applicability to a wide selection of multidisciplinary topics.

Keywords: Knowledge discovery · Information extraction · Data analytics

1 Introduction

Discovering and assessing relationships among concepts is a crucial step in the investigation of the topology and dynamics of semantic networks [1]. In particular, due to their efficacy in modelling a variety of contextualised scenarios, semantic networks provide a valuable tool in knowledge discovery [2]. Furthermore, the mutual relationships among concepts can also be used in the definition of suitable graphical models, such as Bayesian Networks and decision trees, where the assessment of the corresponding relationships is an important aspect of their investigation [3]. Causal relationships provide an excellent example of the relevance of a relationship between concepts [4]. In fact, consider the sentence *"smoking causes fatal diseases"*. The associated causal relationship links *smoking* with *fatal diseases* with a well-defined, non-commutative direction, namely

$$smoking \; \rightarrow \; fatal \; diseases$$

In other words, "*A causes B*", is not the same as "*B causes A*".

Typically, the extraction of concepts and their relationships is a complex task due to the specific constraints which govern the type and properties of the different relationships between the corresponding concepts. As a simple example, the temporal connotation of a causal relation must be consistent. In fact,

© Springer Science+Business Media Singapore 2016
K. Li et al. (Eds.): ISICA 2015, CCIS 575, pp. 464–473, 2016.
DOI: 10.1007/978-981-10-0356-1_49

in *"smoking causes fatal diseases"*, the act of *"smoking"* needs to happen *before* the occurrence of *"fatal diseases"*, if the former does indeed cause the latter. In this paper, we will not restrict our investigation to causal relations as we will consider *influence* relations. These, loosely speaking, capture any semantic connection between two concepts, which are, by definition, semantically more general, and perhaps vaguer types of relations [3,4]. However, such vagueness does not require to address all the constraints of other types of relationships, such as causal ones. In particular, this aspect can facilitate the definition and analysis of such relations especially when addressing large datasets. In fact, the lack of strict constraints enables a more efficient computational approach.

The main motivation of this paper is based on the intuition that the theory of coupled oscillators, modelled by the Kuramoto model, can be suitably modified to investigate the properties of influence relationships in semantic networks. The Kuramoto model was introduced to investigate the behaviour of systems of chemical and biological oscillators, with widespread applications such as in neuroscience [5]. In particular it models the behaviour of a large set of coupled oscillators.

Suppose we have a set of interconnected oscillators. Their mutual connections can be measured by investigating their dynamics, or in other words, is oscillator i (eventually) *synchronous* with oscillator j? If so, this might suggest that they are connected via a measurable and clear relationship.

In other words, if we "perturb" concept A and one of its neighbours, say concept B, reacts to this event, then we could say that B *depends on* A, or $A \to B$.

2   Theoretical Background

Network theory has increasingly attracted much interest from a variety of inter-disciplinary research fields, including mathematics, computer science, biology, and the social sciences. More specifically, a semantic network $G = G(V, E)$ consists of a *vertex-set* $V = \{v_i\}_{i=1}^n$ and an *edge-set* $E = \{e_{ij}\}_{i \neq j=1}^n$, which captures the relationships between the corresponding nodes.

In this paper, we assume that nodes are connected via relations extracted from both structured and unstructured datasets, as well as from pre-populated databases. More specifically, each node is a semantics concept and as a consequence, it has semantic attributes interconnected with other ones belonging to other nodes, as described in [6] (Fig. 1).

Definition 1. *Let $A \in V$ be a node. We define the set of attributes of A as $\mathcal{A}$. These can be quantified via a map $g : \mathcal{A} \to [0,1]^n$, so that the $n-$dimensional vector $\mathbf{a} \in [0,1]^n$ defines the attribute vector of A, and we write $A(\mathbf{a})$.*

In general, such attributes might include

- Location
- Size

- Semantic Components
- Temporal information
- Type
- Action

For example, the concept *sofa* could include the following attributes:

- Location: *lounge*
- Size: $1.5\,m^3$
- Semantic Components:
  - *legs*
  - *arm support*
  - *seat*
- Temporal information: *none*
- Type: *furniture*
- Action: *none*

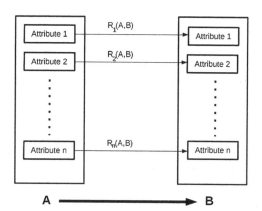

Fig. 1. A depiction of the individual relationships between the attributes of A and B, as in [6].

2.1   Kuramoto Model

As mentioned above, synchronisation can be found almost everywhere in nature, for example the synchronisation rising from the coupling of the front and back wheels of a bicycle, as well as the circadian rhythms observed in animals [5].

One specific mathematical model, which captures the synchronisation occurring between coupled oscillators is the *Kuramoto model* [5]. It consists of a set of n coupled phase oscillators, $\theta_i(t)$, with natural frequencies ω_i distributed with a given probability density $g(\omega)$, whose dynamics is described by

$$\dot{\theta}_i = \omega_i + \sum_{i=1}^{n} k_{ij} \sin\left(\theta_j - \theta_i\right), \quad i = 1, \dots, n \tag{1}$$

Each oscillator attempts to run independently at its own frequency, while the coupling tends to synchronise it to all the others [5].

An important parameter is the *order parameter* $r \in [0, 1]$ [5] defined as

$$re^{i\psi} = \frac{1}{n} \sum_{i}^{n} e^{i\theta_i}, \qquad (2)$$

where ψ is the average phase of all the n oscillators. The parameter r indicates how synchronised two oscillators are. Therefore, when investigating complex systems characterised by coupled oscillators, such parameter can provide an insight into their overall dynamics.

In this paper, we will introduce a novel method based on the intuition that semantic concepts can be modelled as coupled oscillators. Crucially, their synchronisation properties can be analysed to understand and assess any influence relation between them.

3   Description of Method

In [6], a method to assess the direction of influence extracted from textual sources is discussed, which is based on the properties of dynamical attractors. More specifically, suppose we have a set of physical objects with "attracting" properties. Their mutual connections can be measured by investigating their dynamics, or in other words, how strongly they attract other objects. In other words, if we "peturb" a concept A and one of its neighbours, say concept B, feels an attracting force towards A, then we could say that B *depends on* A, or $A \to B$.

In this paper, as in [6], we assume that the attracting properties determine the direction of dependence relations. If A is affected by B, it means B is not, and so we say that A is influenced by B, or $B \to A$. Figure 2 depicts a simple network corresponding to the sentence

"The grass is wet if either the water sprinkler is on or it has been raining."

In particular, the directed edges describe the directions of the influences specified by the above sentence. Each node has its attributes updated throughout the extraction of their relevant information [6], and the overall direction of the interaction between two corresponding attributes of A_i and A_j is defined as

$$d(a_k)_{t_l} = \frac{1}{l} \sum_{h=1}^{l} R_{a_k}(A_i, A_j)_{t_h}, \qquad (3)$$

where $R_{a_k}(A_i, A_j)_{t_h}$ refers the individual relationships between the attributes of A_i and A_j at time t_h.

Furthermore, the "gravity field" $G(A_i, A_j)$ with respect to time t_m is

$$G(A_i, A_j)|_{t=t_l} = \frac{1}{n} \sum_{k=1}^{n} d(a_k)_{t_k}$$

$$= \frac{1}{n} \sum_{k=1}^{n} \sum_{h=1}^{l} R_{a_k}(A_i, A_j)_{t_h}. \qquad (4)$$

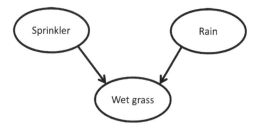

Fig. 2. The graphical model of "*The grass is wet if either the water sprinkler is on or it has been raining.*"

For a full explanation of Eqs. 3 and 4, please refer to [6].

However, this approach might not be suitable to every context, due to the specific characterisation of its concepts and attributes. In fact, the precise assessment of the parameters of Eqs. 3 and 4 might not be always possible. Therefore, rather than an *accurate* assessment of the parameters of an influence relation, we will consider the synchronisation properties which are based on the general behaviour, rather than specific parameters. In fact, although such equations still play an important role in the characterisation of the proposed model, their overall trend is the main focus of this investigation.

In the next section, a different approach based on the Kuramoto model, describing the synchronisation properties of two oscillators is introduced.

3.1   An Interpretation of Kuramoto Model

As discussed above, the synchronisation of the properties of two concepts can contribute to the understanding of the direction of the mutual relationships.

Equation 1 can be re-written in a more general form as

$$\dot{\theta}_i = \omega_i + \sum_{i=1}^{n} k_{ij} f(\theta_j - \theta_i), \quad i = 1, \ldots, n \tag{5}$$

where f is the *interaction* between oscillators and is periodic. In this paper, the idea of oscillators will be exploited, where these correspond to nodes in a (semantic) network, which are governed by the parameters as discussed in Definition 1. In particular, Eq. 5 is modified as follows

$$\dot{A}_i(\mathbf{a}) = A_i(\tilde{\mathbf{a}}) + \sum_{|p(A_i, A_j)|} \sum_{l=1}^{n-1} f(A_{i_l}(\mathbf{a}), A_{i_{l+1}}(\mathbf{a})), \tag{6}$$

where

- f measures the attributes of the node A_j with respect to those of A_j. In other words, f assess the level of synchronisation of the attributes of the corresponding two nodes,

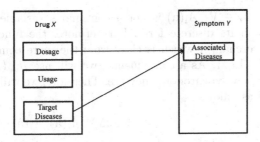

Fig. 3. The example described in Sect. 3.1.

- $A_i(\tilde{\mathbf{a}})$ is the initial value of A_i with respect to its attributes,
- $A_i = A_{i_1}, \ldots A_{i_n} = A_j$ is a path connecting A_i with A_j,
- $p(A_i, A_j)$ is the set of paths between A_i and A_j.

In other words, Eq. 6 measures the level of synchronisation of two connected nodes $A_i(\mathbf{a})$ and $A_j(\mathbf{a})$. Note that the paths connecting two nodes can be identified via the topological properties of the network. However, as discussed in [3,4] when such network is large, it is a computationally demanding task. As a consequence, we suggest a simplification of (6), as follows

$$\dot{A}_i(\mathbf{a}) = A_i(\tilde{\mathbf{a}}) + \max_{p(A_{i_l}(\mathbf{a}), A_{i_{l+1}}(\mathbf{a})) \in \tilde{P}(A_i, A_j)} \left(f(p(A_{i_l}(\mathbf{a}), A_{i_{l+1}}(\mathbf{a}))) \right), \quad (7)$$

where $\tilde{P}(A_i, A_j)$ is the set of shortest paths between A_i and A_j, and f is defined as above.

Consider, for example, two concepts *Drug X* and *Symptom Y*. Such two concepts might or might not be related depending on their specific attributes. Furthermore, the context into which they are embedded would provide much of the information necessary to assess their mutual dependence. Assume now that these two concepts have the following attributes, as depicted in Fig. 3.

- *Drug X*
  - Dosage
  - Usage
  - Diseases targeted
  - ...
- *Symptom Y*
  - Associated diseases
  - ...

If *Drug X*, by varying its dosage, was effective for a specific disease, which would affect *Symptom Y* accordingly, then we could see a certain level of synchronisation. This would contribute to identify and assess their mutual relationship, suggesting *Drug X* → *Symptom Y*. However, the other direction of such relationship might not apply as by perturbing *Symptom Y*, *Drug X* might not be affected.

In general, for $A_i \in V$, $A_i(\mathbf{a})$ is not a continuous variable. Therefore, (7) will be re-written in its discrete form. Furthermore, the function f in Eq. 5 compares the phase change, which is clear from the more common Kuramoto model described by Eq. 1. As a consequence, we will use $R_{a_k}(A_i, A_j)$ for each attribute to assess how synchronous they are. Therefore, we will have that Eq. 6 can be formulated as follows

$$\Delta_{t_m}(A_i(\mathbf{a}), A_j(\mathbf{a})) =$$
$$\frac{1}{n} \max_{|\tilde{P}(A_i, A_j)|} \left(\sum_{l=1}^{n-1} (G_{t_m}(A_i, A_j)_{p_l} - G_{t_{m-1}}(A_i, A_j))_{p_l} \right), \tag{8}$$

where the subscript p_l refers to the shortest path $p_l \in \tilde{P}(A_i, A_j)$. We then define the *overall trend of synchronisation* $S(A_i(\mathbf{a}))$ as

$$S((A_i(\mathbf{a}), A_j(\mathbf{a})) = \frac{1}{M} \sum_{i=1}^{M} \Delta_{t_i}(A_i(\mathbf{a}), A_j(\mathbf{a})). \tag{9}$$

Note that $0 \leq S((A_i(\mathbf{a}), A_j(\mathbf{a})) \leq 1$.

Intuitively, $S((A_i(\mathbf{a}), A_j(\mathbf{a}))$, or $S(A_i, A_j)$ for short, measures how much the relationship between $A_i(\mathbf{a})$ and $A_j(\mathbf{a})$ varies dynamically. In this paper we use $S(A_i, A_j)$ to assess the corresponding relationship, so that when $S(A_i, A_j) = 0$, such relationship is absent, whereas $S(A_i, A_j) = 1$ indicates a *certain* relationship.

4   Evaluation

The structure of the evaluation carried out is similar to [4]. In particular, we considered PubMed [7], which includes over 24 million citations within the biomedical field, with numerous life science journals, and online books.

Approximately 245,000 abstract freely available from PubMed were extracted by considering the following keywords: "breast cancer" and their variations (such as cancer of the breast). Table 1 shows the trend of their occurrences in these abstracts over a 15 year period. In particular, 268 freely available articles were available, of which we considered 20 randomly selected. These were pre-processed and parsed using the Python NLTK toolkit [8] to determine their syntactic structure.

In particular, all the abstracts were POS-tagged to identify the syntactic role of each of the different tokens in every sentence, to populate a semantic network. For example the sentence *"smoking causes fatal diseases"* generates the following POS-tagging

```
(ROOT
  (S
    (NP (NN smoking))
    (VP (VBZ causes)
      (NP (JJ fatal) (NNS diseases)))
    (. .)))
```

Table 1. The trend of the occurrence of "breast cancer" in the abstracts available from PubMed over a period of 15 years

Year	Breast Cancer keywords count
2016	5
2015	14806
2014	16889
2013	15967
2012	15001
2011	13392
2010	12407
2009	11171
2008	10789
2007	9841
2006	9531
2005	9016
2004	8531
2003	7897
2002	7301
2001	7000

Subsequently, the output of the text analysis was merged with achieved by merging existing semantic knowledge from biomedical databases containing ontological terms and WordNet [9]. Note that two terms within the same sentence linked by a verb were identified as semantically connected. Even though this is rather simplistic, the integration of the results of the text extraction with existing semantic knowledge was assumed to be sufficiently accurate for the scope of this validation.

Subsequently, we considered the following concepts:

– Age
– Gender
– Menopause
– Hormones
– Smoking
– Diet

which were manually assessed when determining $R_{a_k}(A_i, A_j)$. The dynamical component was simulated by considering sequentially text fragments, and each iteration corresponding to a time snapshot.

The manual evaluation was carried out by an expert who suggested an interval regarding the specific influence relations. Table 2 shows the results of our evaluation.

Table 2. Evaluation results

A_i	A_j	$S((A_i, A_j)$	Manual assessment
Breast cancer	Age	0.68	0.6–0.7
Breast cancer	Gender	0.82	0.8–0.9
Breast cancer	Menopause	0.71	0.7–0.8
Breast cancer	Hormones	0.29	0.3–0.5
Breast cancer	Smoking	0.54	0.5–0.7
Breast cancer	Diet	0.43	0.5–0.7

We note that among all the pairs, *Breast Cancer* and *Diet* are the only two concepts which do not match the manual extraction. In fact, even though the influence relation between *Breast Cancer* and *Hormones* is evaluated as 0.29, it is very close to the suggested range, which was likely caused by the assumptions in the text analysis stage. In future research, we are aiming to provide a deeper semantic analysis to address a variety of text analysis aspect, such as the identification of verbs and cue phrases capturing influence, as well as specific challenges including anaphora resolution and text entailment [3].

5   Conclusion

In this paper, we have discussed a novel method to assess influence relationships between concepts in semantic networks,by considering some dynamical properties of synchronisation. In future research, we are aiming to expand this method to fully integrate the methods in [3,4,6] to provide a reliable approach to influence relation extraction from unstructured datasets.

In particular, it will be part of a wider line of research, whose aim is to automatically build Bayesian networks (BNs) from both structured and unstructured data sets. Since our knowledge of the world focuses on unknown parameters based on scenarios which typically lack of certainty, BNs provide a very powerful graphical tool with a wide range of applications to facilitate the decision making process.

References

1. Trovati, M., Bessis, N., Huber, A., Zelenkauskaite, A., Asimakopoulou, E.: Extraction, Identification and Ranking of Network Structures from Data Sets. In: Proceedings of CISIS, pp. 331–337 (2014)
2. Trovati, M.: Reduced topologically real-world networks: a big-data approach. Int. J. Distrib. Syst. Technol. **6**(2), 13–27 (2015)
3. Trovati, M.: An influence assessment method based on co-occurrence for topologically reduced big datasets. Soft Comput. 1–10 (2015). Springer, Berlin, Heidelberg

4. Trovati, M., Bessis, N., Palmieri, F., Hill, R.: Dynamical extraction and assessment of probabilistic information between concepts from unstructured large data sets. Submitted to IEEE transactions (2015)
5. Kuramoto, Y.: International Symposium on Mathematical Problems in Theoretical Physics. Lecture Notes in Physics, vol. 39. Springer, New York (1975)
6. Trovati, J., Trovati, P., Larcombe, L., Liu, A.: Semi-automated assessment of the direction of influence relations from semantic networks: a case study in maths anxiety. In: The Proceedings of IBDS (2015)
7. PubMed Website. http://www.ncbi.nlm.nih.gov/pubmed. Accessed September 2015
8. Natural Language Toolkit Website. http://www.nltk.org/. Accessed September 2015
9. Miller, G.: WordNet: a lexical database for English. Commun. ACM **38**(11), 39–41 (1995)

PEMM: A Privacy-Aware Data Aggregation Solution for Mobile Sensing Networks

Zhenzhen Xie[1], Liang Hu[1], Feng Wang[1], Jin Li[2], and Kuo Zhao[1(✉)]

[1] Jilin University, Changchun, People's Republic of China
zhaokuo@jlu.edu.cn
[2] School of Computer Science, Guangzhou University,
Guangzhou, Guangdong, People's Republic of China

Abstract. By more and more, privacy preservation problem is widely discussed among users and researchers. For mobile sensing network, an imperfect privacy preservation scheme will directly put participants into a dangerous situation. The better privacy protection applied, the better sensing data quality will be achieved. In this paper, we present a privacy-aware data aggregation scheme for mobile sensing networks. We considered both the smart nodes like smart-phone and dumb nodes like wearable device or GPS device. We take the location information and the sensing content into consideration separately. And this thought will make sure the sensing content will be k-anonymous and the accurate location will be protected well either. We use erasure coding technology to slice the sensing data record according to the k-anonymity rules. For the sake of efficiency and stability, we compare two coding technology in two sensing data types and give the experiment results and explanations in detail. After that, we give a social model to describe the social relation and a security data sharing protocol among the participants. The introduction of the participants' social relation may give a new way to the reputation and data trustworthy evaluation mechanism.

Keywords: Privacy preservation · Sensors · Mobile sensing · Participatory sensing

1 Introduction

As a new type of participatory sensing systems, mobile sensing network is attracting interests among users and researchers at present. The mobile nodes can be smart phones, vehicle sensors, or other devices with all kinds of sensors to gathering information of our daily life and physical environments. In this paper, we classified the mobile nodes into two types: first is the intelligent mobile terminals such as smart phones and tablet computers which has higher processing capability, we call these as 'smart nodes' for short; the second is the dumb terminals such as vehicle sensors and environment noise sensors which have more resource constraint and energy consumption requirement, we call these as 'dumb nodes' for short. Through those mobile sensing nodes, mobile sensing systems can gather a variety of data for different applications. In the past, a large and complicated sensing task cannot be

© Springer Science+Business Media Singapore 2016
K. Li et al. (Eds.): ISICA 2015, CCIS 575, pp. 474–482, 2016.
DOI: 10.1007/978-981-10-0356-1_50

accomplished with single device or fixed wireless sensor networks. But these goals all can be achieved in the mobile sensing systems by a participating way [1].

However, there are some challenges at the back such as the reputation evaluation, incentive mechanism construction and energy saving problem. Among those challenges, the privacy and security problems are always the most important issue. If the sensing record is in the risk, it may cause privacy data abuse or leakage. Since the sensing records are always having time and location feature, some data about the address, trace, identities may be revealed directly. And with the development of data mining and other data analysis technique, some personal habits will be revealed indirectly by inference. These problems not only pose highly risk to the system participants but also limited the development of mobile sensing. Every participant want a secure and privacy-preserving solution to keep the private information safely and it definitely affect their data sharing motivation. In mobile sensing systems, there are some major problems and challenges that we listed below:

The mobile sensing systems are in the risk of external attack and internal attack. The privacy information can be revealed in both the two ways.

1. As the same with WSNs networks, the data aggregation process brings new challenges for privacy preservation [2].
2. In a mobile sensing system, the mobile computing devices usually have more than one sensor such as GPS, camera, audio device and so on. So the multimedia data should be considered together which make the research more complicated.

And there are some important point should be noticed when starting to design a privacy-preserving plan:

1. The privacy preserving solution for mobile sensing systems should be energy efficient since both the smart nodes and the dumb nodes have limited computing resource.
2. Maybe more noise adds to data, more privacy guarantee, but that means the data value may lost completely. Because of that, data quality should take into consideration seriously.
3. Different kind nodes should have different ways for privacy and security requirement. The systems have mixed type of nodes needs a better architecture which is privacy-aware.

In this paper, we present a privacy-aware data aggregation scheme for the mobile sensing system. The mobile nodes in our scheme are classified into two groups: smart nodes and dumb nodes. At first, the accurate location data will be replaced by a anonymous location; Then, the sensing record with anonymous location will be sliced by the erasure code for the content privacy preservation. In this step, we evaluated two RS code algorithm and the performance and validly is verified by experiment results. At last, the data slice will be transferred by a social information based strategy.

2 Related Work

In the past few years, there is a lot of work on privacy preserving since both users and researchers realized it is a serious problem. At first, most study concentrate on the location privacy since it is one of the most sensitive information of private data.

Nowadays, users and researchers realized that the content privacy is important either. The traditional ways are k-anonymization, differential privacy or other solutions based on identity authentication. And the emergence of new type of networks, such as Internet of things, mobile sensing networks and mobile crowd sensing networks give researchers new application scenarios and new requirements. The privacy problems in these three networks are similar problems and some technique can be used in common. Ling Hu et al. [3] used a Hot-potato-privacy-protection algorithm to protect users privacy in a participatory sensing networks and urban sensing; Xinlei Wang et al. [4] proposed a frame work ARTSense for 'trust without identity' problem. The framework provide a data trust assessment and a anonymous protocol without using trusted third party support which simplified the architecture. [5] using temporal cloaking technique to protect participant's location information which is based on k-anonymity. In spite of the work focused on mobile sensing and participatory sensing area, some work for wireless sensor networks are also give us some summary and thoughts. [6] proposed a privacy-preserving data aggregation function based on SUM and two aggregation protocols: PDAAS and PDACAS for a sensor networks with mobile nodes.

And there are also some good surveys to summarize the current work in different respective. [7] reviewed some works which focused on data-oriented and context-oriented privacy and discussed the future challenges. [2] gave a survey that concentrates the privacy-preserving data aggregation, they classified current schemes and compare the algorithms by performance, communication consumption, data accuracy and other targets.

And there is some new development in location privacy area either: [8] present a new attempt for location privacy problem in Internet of things environment. They used a cross-layer, distributed beamforming approach to boost BS (Base Station) anonymity and it minimized the communication energy overhead. [9]'s work aimed at the source location privacy and sink location privacy, they provided a framework called FAC (Fortified Anonymous Communication) protocol with end-to-end anonymity thought for WSN (Wireless Sensor Network). Except of solutions with anonymity thoughts, there are some work based on fake source to solve the source location privacy problem, for instance, in [10]'work, they give the fake source selection problem a new formalization and appropriate parameterizations and the experiment results are convincingly.

Different from aforementioned work, our study aims at both location privacy and content privacy. Our novel framework also take social relationship among participants into privacy preserving problem's consideration since the social based mobile sensing network will be a new trend.

3   System Model

The mobile sensing system models are different due to the applications, service type and data storage platform. In this paper, we consider the system model as a commonly used one. The nodes type can be smart nodes such as smart-phone or dumb nodes as smart band or other wearable devices. The model is cloud-based participatory sensing for better processing capability. And different from others work, we take the social relationship of participants into consideration for better data gathering and data evaluation in the future. The detailed definitions are presented in Sect. 3.1.

3.1   Related Definition

The system model discussed in our paper is facing two classes of users: one is the service providers who aggregating and using sensing data, and the other one is the mobile device users who carrying mobile phone, wearable devices or other sensors. A basic architecture of a mobile sensing system is shown in Fig. 1.

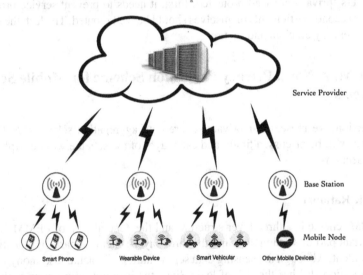

Fig. 1. System model

The mobile sensing system is usually participatory-sensing oriented due to the better data gathering ability. So in our model, the mobile devices users are participant who gathering all kinds of data for service providers. Some definitions are list below for better understanding:

- *Mobile node:* We considered each mobile device of a single user is a single mobile node. It can be phones, pads or wearable device. According to the storage ability, processing ability and the battery life, they are divided into two groups: the smart ones and the dumb ones.
- *Mobile device users/participants:* One mobile device users could have serval kinds of devices. The privacy problem is one of the most important things which will

directly cause the reduction participants. Users' privacy information can be the vehicle trace, heart beat, personal address and so on.

- *Smart Agency:* Each smart node has a smart agency to accomplish some simple location anonymity work for sensing data and it also manage the friend user list. The smart agency not only deals with the sensing data of the smart phones selves but also the data from dumb nodes such as wearable devices or GPS devices.
- *Service Provider:* The sensing data will pass through the smart agency for elementary anonymity need and then go to the data centre for advanced data processing work. The service provider stores the sensing data and provide the requests from service provider.

3.2   Objectives

The privacy-aware mobile sensing data aggregation scheme should guarantee the sensing data quality and meet the energy consumption requirement under the promise that the users 'privacy are well protected. First, it needs to prevent service providers' data abuse; second, participants themselves should be well reputed. The last, the scheme needs to be energy efficient and secure.

4   PEMM: A Novel Privacy Protection Scheme for Mobile Sensing Systems

In this section, we present our privacy-aware data aggregation scheme-*PEMM*. The general idea will be discussed firstly and each layer of the scheme will be explained in the latter sections.

4.1   Basic Rational

The *PEMM* scheme is a three layer structure and the general idea of PEMM is taking two important part of sensing data into the three-layer-scheme separately: location and sensing content. Although the sensing data segmentation will achieve k-anonymity, the accurate location still has the risk of leak. Since the location information is the users mainly focused on, we take it seriously in our scheme: before the slicing task, we make the location anonymous firstly and then the sensing content will no longer have the accurate location data. In the slicing step, we use slicing method to segment the location-anonymous sensing data into k parts so that the service providers identity the data origin at least $1/k$. The data slicing task should take the nodes' ability into consideration and the data transferring strategy is energy-efficient and privacy-aware. To obtain these goals, we classified the node into two kinds: smart node and dumb node. The smart agency on the smart node can do the location anonymity work and the dumb node in charge of generating the sensing data only. For the participants, we used a social model to explain the social relationship and data sharing protocol among them. The social model introduced gives a new solution to motivate new participants and give more information for the reputation evaluation and trust mechanism construction. The basic work flow of this scheme is shown in Fig. 2.

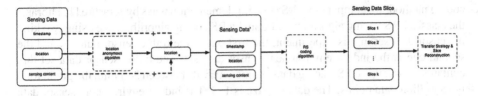

Fig. 2. Basic work flow of *PEMM* scheme

4.2 Sensing Data Slicing Based on K-Anonymous Rule

K-anonymous rule simplified the work of sensing data privacy protection in a large extent. However, we cannot just slice data into several pieces or simply transfer the data for several times. To guarantee the data quality and not add too much redundant data, the anonymous method needs to be well considered and the related transferring strategy needs to be well designed. The erasure code is an adaptive way for the slicing task since the replication policy often needs more storage space. There are many kinds of erasure code such as RS, Tornado and Ceph, we use RS code in our work because it is general and stable.

4.3 Sensing Data Slicing Solution

In [11]'s work, they using erasure code technology to slice the sensing data in their SLICER framework to meet the k-anonymity requirement. They take the sensing data as entirety and then using the erasure coding technology to code the sensing data which is inspired us. In our work, we use RS code for the sensing data slicing work. Different from [11]'s work, the location data will be replaced by location anonymous algorithm like before the slicing step. That because the location is one of the most sensitive data, if the accurate location can be anonymous after data sensing step, it means that if the data slice are obtained by illegible users, they cannot acquired the true location data. And it will also prevent the information deduction risk: some adversary using the location data to generate the users' daily life trace. Depend on this thought, we proposed a new slicing strategy: using location anonymous algorithm to make sure the location can be well protected at the beginning and then start to code the sensing data with the new location information.

In *PEMM* scheme, we divided the nodes into two kinds: smart nodes and dumb nodes. The smart nodes can gather sensing data from their neighboring dumb nodes, and the dumb nodes' exact location can be replaced by new location information by anonymity algorithm. After the neighbor dumb nodes searching and sensing data gathering step, the whole sensing data will be sliced by erasure code and the data slice transferring will only occur among the smart nodes.

In this paper, we use RS coding technique to complete the slicing task. RS coding algorithm has many ways to be implemented, to choose an energy efficient way, we used two wireless sensor data set to carry out this experiment. To find out a suitable erasure coding method, we compare two RS code implementation ways by Jerasure [12] library: the classical RS coding algorithm and Cauchy RS coding algorithm. We using two kind of sensor data set in this experiment, the first one is the traffic sensor data set and the second one is the indoor environment sensor data set. The traffic sensor data set that contains 7,648 taxis GPS tracing data at time intervals between 3–5 min, the grand total is 18 million data points. The data set size is 1 G. The indoor environment sensor data set we gathered from ten sensors in three rooms. And this data set has two parts, each part of data are gathered from different time period and their data size are 30 M and 50 M. We using Jerasure Library 1.2 edition to compare the coding time between classic RS code and Cauchy RS code, the result shows in Fig. 3. The results indicated that Cauchy RS coding algorithm has better performance than RS coding algorithm on the two data sets.

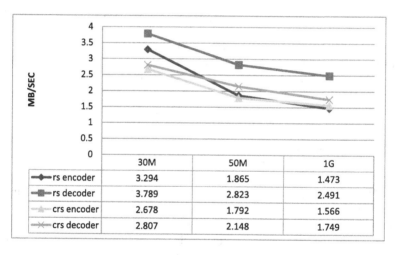

Fig. 3. Comparison result

4.4   The Social Interaction Model for Mobile Participants

In most mobile sensing networks, researchers mainly aimed at sensing data processing, data security or network architectures. Few research to combine social interaction with mobile sensing network. However, we think it maybe a new way to provide exuberant data sources and keep users' stickiness. In [13]'s work, the authors build a s-WBANs network which combine wireless body area networks (WBANs) and social networks feathers together. The new attempt inspired us and that is a proof of availability at one side. If we take users' social interaction into consideration, the privacy preservation solution, data trust evaluation and reputation computing mechanism will have richer information to be developed.

There are two important modules in our model: social relation building and privacy settings. And the preconditions of our model are listed below:

1. Users' social relationship is the foundation of our model. A user can build his relationship with others. But this relationship not mean that all your friends' sensor nodes' sensing data will pass your devices for the consideration of privacy.
2. Each user can have multiple smart sensor and dumb sensors. Each user can have multiple friends and can use different privacy settings on different friend.
3. As the smart node, mobile phone, pad or laptop are the most important entities in our social model. The smart nodes have the social interaction information and they can realize which a friend node is or not. In this work, we assume that each user has one smart node only.
4. A users dumb nodes' sensing data will be pre-process at users' own smart nodes. There may be some users has dumb nodes only but it is out of our scope of this work.
5. If user A gets his friend B's permission, the friend nodes' sensing data slices transferring routine will take B as a transmit node.

In the social relation maintain module, users can enable the 'friend discovering' function to find friend node, if they get the permission; it means the social relationship is built. In the privacy setting module, each user can specified which sensors can be opened and open to which friend. To explain the whole process in detail, we give an example below:

User Alice and User Bob enable the friend node discovering function, and Bob received the invitation from Alice. Bob agree with the invitation and he begins to choose open which sensors of his own to Alice. Only the open sensors can transfer and sharing data with Alice. So that each user maintains a friend list to record the friend and the corresponding open sensors. For better understanding, the social interaction model is shown in Fig. 4.

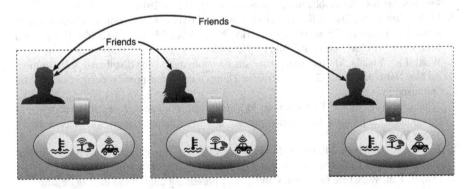

Fig. 4. Social interaction model

5   Conclusion and Future Work

In this paper, we discussed the privacy preservation problems in mobile sensing networks. A number of important related works give the problem a basic foundation.

The privacy scheme called *PEMM* is presented. It takes the nodes features into consideration and gives a cooperating data slicing method to comply with the k-anonymous rule. In the future, the real-environment evaluation will be required in order to better prospect of applications. And we will give more attention to the content privacy protection and the privacy-ware inference scheme.

Acknowledgements. This work is funded by: European Framework Program (FP7) under Grant No. FP7-PEOPLE-2011-IRSES, and by National Natural Science Foundation of China under Grant No. 61073009, and by National Sci-Tech Support Plan of China under Grant No. 2014BAH02F03.

References

1. Jose, J., Princy, M., Jose, J.: PEPPDA: power efficient privacy preserving data aggregation for wireless sensor networks. In: 2013 International Conference on Emerging Trends in Computing, Communication and Nanotechnology (ICE-CCN), pp. 330–336, March 2013
2. Jian, X., Geng, Y., Zhengyu, C., Qianqian, W.: A survey on the privacy-preserving data aggregation in wireless sensor networks. Commun. Chin. **12**(5), 162–180 (2015)
3. Hu, L., Shahabi, C.: Privacy assurance in mobile sensing networks: go beyond trusted servers. In: 2010 8th IEEE International Conference on Pervasive Computing and Communications Workshops (PERCOMWorkshops), pp. 613–619, March 2010
4. Wang, X., Cheng, W., Mohapatra, P., Abdelzaher, T.: Enabling reputation and trust in privacy-preserving mobile sensing. IEEE Trans. Mob. Comput. **13**(12), 2777–2790 (2014)
5. Kalnis, P., Ghinita, G., Mouratidis, K., Papadias, D.: Preventing location-based identity inference in anonymous spatial queries. IEEE Trans. Knowl. Data Eng. **19**(12), 1719–1733 (2007)
6. Yao, Y., Liu, J., Xiong, N.N.: Privacy-preserving data aggregation in two-tiered wireless sensor networks with mobile nodes. Sensors **14**(11), 21174 (2014)
7. Li, N., Zhang, N., Das, S.K., Thuraisingham, B.: Privacy preservation in wireless sensor networks: a state-of-the-art survey. Ad Hoc Netw. **7**(8), 1501–1514 (2009). Privacy and Security in Wireless Sensor and Ad Hoc Networks
8. Ward, J.R., Younis, M.: Increasing base station anonymity using distributed beamforming. Ad Hoc Netw. **32**, 53–80 (2015). Internet of Things security and privacy: design methods and optimization
9. Abuzneid, A.-S., Sobh, T., Faezipour, M., Mahmood, A., James, J.: Fortified anonymous communication protocol for location privacy in wsn: a modular approach. Sensors **15**(3), 5820 (2015)
10. Jhumka, A., Bradbury, M., Leeke, M.: Fake source-based source location privacy in wireless sensor networks. Concurrency Comput. Pract. Exp. **27**(12), 2999–3020 (2015)
11. Qiu, F., Fan, W., Chen, G.: Privacy and quality preserving multimedia data aggregation for participatory sensing systems. IEEE Trans. Mob. Comput. **14**(6), 1287–1300 (2015)
12. Plank, J.S., Simmerman, S., Schuman, C.D.: Jerasure: a library in C/C++ facilitating erasure coding for storage applications - Version 1.2. Technical report CS-08-627, University of Tennessee, August 2008
13. Wang, H., Zhang, Z., Lin, X., Fang, H.: Socialized WBANs in mobile sensing environments. IEEE Netw. **28**(5), 91–95 (2014)

SmartNV: Smart Network Virtualization Based on SDN

Xiaodi Yu[1], Hu Liang[1], Fu Tao[1], Li Jin[2], and Zhao Kuo[1(✉)]

[1] College of Computer Science and Technology, Jilin University,
Changchun, People's Republic of China
zhaokuo@jlu.edu.cn
[2] School of Computer Science, Guangzhou University, Guangzhou,
Guangdong, People's Republic of China

Abstract. Nowadays virtual networks (VN) need an easier way to configure and manage. There are already plenty works about VN mapping but there are still no a proper approach can meet these requirements no single approach currently can meet these requirements simultaneously. Software-defined networking (SDN) is an emerging networking pattern that gives hope to change the limits of current network. Some researcher use SDN to deal with VN problem such as Flowvisor and Openvirtex, but these works only have VN platform's core virtualization feature. In this paper, we proposed a SmartNV (Smart Network Virtualization) platform using our enhance Openvirtex. Based on our SDN-NV architecture, SmartNV can collect information from physical infrastructure and choose appropriate mapping algorithm to calculate the virtualization scheme. The evaluation experiments on our prototype system can satisfy the requirement of the NV based SDN.

Keywords: SDN · NV · Resource slice

1 Introduction

Cloud computing has proven its worth by providing bigger storage, faster calculation and more elastic network. Different from traditional networks, cloud computing use network virtualization technology to make the network flexible, easy to configure and manage, and it can be used in the same underlay physical infrastructure [1]. The existing network virtualization research pays much attention to allocate the resources based on the existing physical resources and how to divide the virtual network reasonably, but have no good solutions on the complex management and control problems generated in the deployment of virtual network. SDN is a new kind of network architecture, which is a kind of realization method of network virtualization [2]. The main theoretic of SDN is to separate the network device control surface and data surface, so as to realize the flexible control of network traffic. Using SDN technology to achieve VN can solve these problems. However, existing VN schemes based on SDN depends on the existing OpenFlow protocol and they provide only the basic framework for mapping the underlay network to the virtual network. Finding a more intelligent approach for a more scalable NV based on SDN can provide a more flexible management and resource allocation method.

© Springer Science+Business Media Singapore 2016
K. Li et al. (Eds.): ISICA 2015, CCIS 575, pp. 483–488, 2016.
DOI: 10.1007/978-981-10-0356-1_51

This paper introduces a SDN-NV architecture that constructs VN on top of a SDN network. Based on the SDN-NV architecture, it ensures that we can create and manage virtual Software Defined Networks intelligently according to the underlay resources and virtual network resource requirements. We present a SmartNV platform based on this architecture, which is a flexible SDN network virtualization platform, we can use it in a rich Openflow network in which all the core switches use Openflow protocol. The NV-strategy engine can choose the proper algorithm according to the virtualization requirement and support intelligent network resource allocation.

This paper has three major works as follows:

(1) We propose a SDN-NV architecture which can support flexible network resource allocation.
(2) We design SmartNV, a smart network virtualization platform based on SDN which use different NV algorithm (such as D-DIVE, R-DIVE [3] and vnmFlib [4]) in different situation.
(3) We describe the implementation of SmartNV.

The rest of this paper is organized as follows. We introduce the background and previous work in Sect. 2, and then propose SDN-NV architecture and SmartNV platform in Sect. 3. We discuss the implementation of SmartNV in Sect. 4 and conclusions are written in Sect. 5.

2   Background and Previous Work

Virtual network is a feasible and resilient way to allow different network architectures coexist. The concept and technology of virtual network make the network structure dynamic and diversified possible, which is considered to be a solution to the ossification in existing network system, and to build the ideal scheme of the next generation network.

SDN is an innovative networking architecture. It separates control planes and data planes, provide a logically centralized control function, and the underlay network infrastructure is abstracted from relevant applications. Owning to its advanced nature SDN has developed rapidly since its birth. SDN enables operators to construct highly scalable, flexible networks that are readily adapted to fluctuating network requirements. Obviously, SDN can play an important role in the management and configuration of NV. There are three main points of contact between SDN and network virtualization. SDN can work as a realization technology of NV. NV can serve as a evaluation and simulation method of SDN, such as mininet. SDN can also do virtual network slicing, it works easier than traditional overlay slicing scheme. There are already some related works about NV based on SDN.

FlowVisor [5] is a network virtualization platform based on OpenFlow. It can be divided into multiple logical networks, so as to realize open SDN. It provides manager with a broad definition of rules to manage the network, rather than adjusting the routers and switches to manage the network. Currently FlowVisor is deployed in the production environment of the United States, especially in large campuses (e.g. Stanford). In addition, the two major networks that focus on research - the global network of innovative environment (GENI) and Internet2 participants are also using FlowVisor.

Similarly to FlowVisor, Openvirtex (OVX) works between NOS and SDN switches. OVX is a network virtualization platform proposed by On.lab [6], OVX establishes a view of the physical network infrastructure by means of topology discovery, and provides the view representation of the virtual network through the configuration information provided by API OVX.

FlowN [7] present by J. Rexford is a lightweight virtualization scheme. Different with the Flowvisor and OVX, FlowN doesn't provide whole virtualization technology, it can be seen as a container-based virtualization by provide a database storage system to maintain a mapping between the physical and virtual layer.

The existing virtualization scheme based on SDN are mainly provide network virtualization layer that functions as a proxy between OpenFlow switches and controllers. They only have the core virtualization features, which can not choose the suitable virtualization strategy according to the situation.

3   The SmartNV Platform

It has been identified that the current resource slicing network virtualization approach lacks a native network information collection mechanism and causes complex management issue in large-scaled networks virtualization with complex network topologies. However, not all the flows in a virtual network need the same NV strategy. Simply using NV algorithm to facilitate all virtual network virtualization requirements is improper. Implement every NV algorithm to manage different requirement of different network virtualization in classical way in inefficient. Our approach decouples network virtualization and resource slicing, we use the NV strategy layer for network virtualization and the OVX support resource mapping.

As shown in Fig. 1, our SDN-NV architecture constructs enhance OVX on top of a enhance Openflow, The information of network and computing resource from physical infrastructure such as bandwidth and CPU will be collected by enhance Openflow, and the NV-strategy management will choose the appropriate strategy according to the underlay information and the virtualization requirement.

As shown in Fig. 2, we design SmartNV network virtualization platform. SmartNV has three layers, physical layer, virtualization layer and virtual network layer. In physical layer, the SDN physical network contain enhanced SDN switches which support enhanced Openflow protocol, we gather physical network information and use local database to manage these information. In virtualization layer, physical network information will be upload to the global network database, enhance OVX and NV strategy engine will use those information to make mapping strategy to fit the NV request of virtual network layer, global network database stores network configuration and information. In order to keep the data synchronization, there will have a channel between the local database and global network database. NV strategy engine can choose different NV algorithm according to the physical network information and the NV request. SmartNV use enhance OVX to realize the mapping calculate by NV strategy. In this way, our proposed SmartNV platform maintains an intelligent and flexible NV configuration and management.

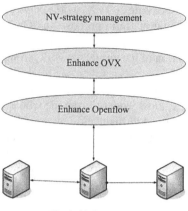

Physical infrastructure

Fig. 1. SDN-NV architecture

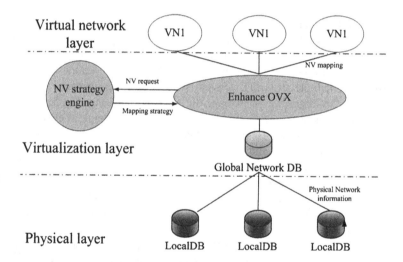

Fig. 2. The SmartNV network virtualization platform architecture

4   Implementation and Evaluation

In order to evaluate the performance of our SmartNV platform, a prototype system is implemented. The work is mainly focus on NV strategy module and the data synchronization between global network database and local database. We improve the OVX, use the existing API of OVX to achieve resource mapping.

NV Strategy Module. The NV strategy module is implemented by Java, according to different circumstances choose different NV algorithm. In the prototype system, NV strategy module makes a choice between MCF and D-ViNE [8]. According to the algorithm characteristic, if the node mapping is already known in the NV request content, MCF is choose, otherwise we chose D-ViNE. OVX will use the NV scheme calculate by the NV algorithm to build the virtual network automatically.

Two-Level Database. The SmartNV has a centralized mechanism and designed for a large-scale network virtualization, it has a distributed synchronized two-level database. In order to ensure data synchronization between the global database and local database, we use Redis work as both local database and global network database. It stores the local network information, virtual request and mapping scheme. Redis supports master-slave synchronization, Data can be synchronized from the master server to any number of servers. Therefore, the local database can aware the changed in the global network database and make the appropriate changes to keep the data synchronization.

Evaluation. SmartNV is designed for large scale network virtualization, but it is hardly practical to do build a large scale testbed to fit the need. We do a small prototype use four computer as physical network and one computer as the server of the enhance OVX. Due to the limitations of the evaluation environment, the CPU and bandwidth is random generated, and the NV requests are read from file. The experimental results show that the NV strategy engine can NV canselect the appropriate virtualization algorithm.

5   Conclusion and Future Works

This paper propose a SmartNV virtualization platform based on SDN network. Current network virtualization techniques based on SDN major focus on provide the mapping function, lack intelligent to provide flexible management without human intervention. We make improvements to this by introducing a SDN-NV architecture and a new network virtualization platform called SmartNV based on enhance OVX and NV strategy engine. The proposed SmartNV network virtualization platform will provide a nice way to improve the scalability and flexibility. We have introduced the SDN-NV architecture, and explained how the problem can be solved by using SDN-NV and the SmartNV platform. The implementation of SmartNV is also discussed.

Because of the diversity of virtualization algorithms, we should improve the NV strategy engine to accept more NV algorithm. According to the different NV request and different physical network architecture, SmartNV need to find a better evaluation method to choose proper algorithm. Besides, the two-level database system should be improved to collect the heterogeneous physical network information.

Acknowledgments. This work is funded by: European Framework Program (FP7) under Grant No. FP7-PEOPLE-2011-IRSES, and by National Natural Science Foundation of China under Grant No. 61073009, and by National Sci-Tech Support Plan of China under Grant No. 2014BAH02F03.

References

1. Anderson, T., Peterson, L., Shenker, S., Turner, J.: Overcoming the Internet impasse through virtualization. Computer **38**(4), 34–41 (2005)
2. McKeown, N., Anderson, T., Balakrishnan, H., Parulkar, G., Peterson, L., Rexford, J., Shenker, S., Turner, J.: Openflow: enabling innovation in campus networks. ACM SIGCOMM Comput. Commun. Rev. **32**(2), 69–74 (2008)
3. Chowdhury, M., Rahman, M.R., Boutaba, R.: ViNEYard: virtual network embedding algorithms with coordinated node and link mapping. IEEE/ACM Trans. Netw. **20**(1), 206–219 (2012)
4. Lischka, J., Karl, H.A.: A virtual network mapping algorithm based on subgraph isomorphism detection. In: Proceedings of the 1st ACM Workshop on Virtualized Infrastructure Systems and Architectures, New York, pp. 81–88 (2009)
5. Sherwood, R., Gibb, G., Yap, K., Appenzeller, G. Casado, M., McKeown, N., Parulkar, G.: Can the production network be the testbed? In: Operating Systems Design and Implementation, October 2010
6. Drutskoy, D., Keller, E., Rexford, J.: Scalable network virtualization in software-defined networks. IEEE Internet Comput. **17**, 20–27 (2013)
7. Al-Shabibi, A., De Leenheer, M., et al.: OpenVirteX: make your virtual SDNs programmable. In: Proceedings of Workshop of SIGCOMM on Topic of HotSDN 2014, Chicago, IL, USA, pp. 25–30, 22 August 2014
8. Chowdhury, N.M., Rahman, M.R., Boutaba, R.: Virtual network embedding with coordinated node and link mapping. In: Proceedings of the IEEE INFOCOM, April 2009

Image Feature Extract and Performance Analysis Based on Slant Transform

Jinglong Zuo[1], Delong Cui[1(⊠)], Hui Yu[2], and Qirui Li[1]

[1] College of Computer and Electronic Information, Guangdong University
of Petrochemical Technology, Maoming, China
oklong@gmail.com, delongcui@163.com,
liqirui@foxmail.com
[2] Information Centre, Dongguan Power Supply Bureau,
Guangdong Power Grid Co, Dongguan, China
hy0769@21cn.com

Abstract. In order to improve the efficient and simple the steps of generation an image hashing, a security and robustness image hashing algorithm based on Slant transform (ST) is proposed in this paper. By employing coefficients of Slant transform, a robust hashing sequence is obtained by preprocessing, feature extracting and post processing. The security of proposed algorithm is totally depended on the user-key which are saved as secret keys. For illustration, several benchmark images are utilized to show the feasibility of the image hashing algorithm. Experimental results show that the proposed scheme is robust against perceptually acceptable modifications to the image such as JPEG compression, mid-filtering, and rotation. Therefore, the scheme proposed in this paper is suitable for image authentication, content-based image retrieval and digital watermarking, etc.

Keywords: Image hashing · Slant transform · Normalized Hamming distance · Encryption keys

1 Introduction

With the rapid development of information-communication and personal computers, copyright protection of digital media as image, video and audio becomes a more and more important issue. Image hash as one of content-based image authentication techniques has become an important research topic recently. An image hash function maps an image to a short binary string based on the image's appearance to the human eye, so it can be used in authentication, content-based image retrieval and digital watermarking.

Security and robustness are two important requirements for image hash functions. By security, it means that one image should have different function values according to the different applications. By robustness, it means that the hash function should keep invariable by common image processing operations such as additive noise, filtering, compression, etc. The underlying techniques for constructing image hashes can roughly be classified into property-based, such as statistics [1]; interaction relation of transform field decomposition coefficients [2]; vision-based feature points [3], etc.; and content-based, such as transform field significantly coefficients [3–5]. Typical case of

© Springer Science+Business Media Singapore 2016
K. Li et al. (Eds.): ISICA 2015, CCIS 575, pp. 489–494, 2016.
DOI: 10.1007/978-981-10-0356-1_52

property-based image hash is proposed by Venkatesan et al. [1], which gets statistics by using decomposition coefficients of discrete wavelet transform (DWT). A relation-based technique generates a hash sequence by employing invariant relation of image blocks, which is expect robust to JEPG compression. For sensitive to geometric attacks, the vision-based image hash is inefficacy under common image processing. On the contrary, the content-based algorithm for its super robustness performance became more and more popular, but an expensive search is needed to handle manipulations. Recently, a new kind of robust image hashing algorithms based on non-negative matrix factorization (NMF) [6] is proposed. It is robust against most image processing, minimizes collision probability and identified the tampered place.

In this paper, an image hashing algorithm based on Slant transform (ST) is proposed for certain application. The scheme extracts a robust feature vector to generate a content-based hash sequence, which includes three-step preprocessing, feature generation and post processing. To improve the security of the proposed scheme, the user-keys are used as the encryption keys. Experimental results show that the proposed scheme is robust against common image processing. The remaining of this paper is organized as follows. The concept of Slant transform is briefly introduced in Sect. 2. The framework and the details of the proposed scheme are described in Sect. 3. In Sect. 4, some experimental results and comparison are presented. Security analysis is presented in Sect. 5. A conclusion is drawn in Sect. 5.

2 Slant Transform

The concept of an orthogonal transformation containing Slant basis vector was introduced by Enomoto and Shibata [7, 8]. The Slant vector is a discrete saw tooth waveform decreasing in uniform steps over its length. It has been seen that Slant vectors are suitable for efficiently representing gradual brightness change in a face image line.

Slant Matrix Construction: If $S(n)$ denotes the $N \times N$ Slant matrix ($N = 2^n$), then

$$S(1) = \frac{1}{\sqrt{2}} \begin{bmatrix} 1 & 1 \\ 1 & -1 \end{bmatrix} \tag{1}$$

The Slant matrix for $N = 4$ can be written as

$$S(2) = \frac{1}{\sqrt{4}} \begin{bmatrix} 1 & 1 & 1 & 1 \\ a+b & a-b & -a+b & -a-b \\ 1 & -1 & -1 & 1 \\ a-b & -a-b & a+b & -a+b \end{bmatrix} \tag{2}$$

where a and b are real constants to be determined subject to the following conditions:

(1) step size must be uniform;
(2) S(2) must be orthogonal.

The properties of Slant transform are as follows:

(1) The Slant transform is real and orthonormal

$$S = S^* \tag{3}$$

$$S^{-1} = S^t \tag{4}$$

(2) It is a fast algorithm reducing the complexity to $O(N \log_2 N)$ for $N \times 1$ vector;
(3) It is very good in energy compaction for facial images. Very few coefficients are sufficient for recognizing the stored image of face in database.

The above Slant matrices [9] are used to define the ST as:

$$D_x(n) = S(n)X(n) \tag{5}$$

where $D_x(n)' = [D_x(0)D_x(1)\cdots D_x(N-1)]$, $X(n)' = [X(0)X(1)\cdots X(N-1)]$ and S (n) is $N \times N$ Slant Matrix.

The 2D forward and inverse Slant transform filter bank are shown in Fig. 1.

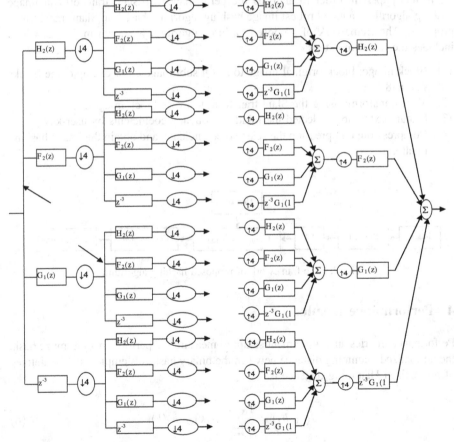

(a) Filter bank of 2D Slant Transform   (b) Filter bank of 2D Slant Inverse Transform

Fig. 1. 2D forward and inverse Slant transform filter bank

3 Proposed Algorithms

For analysis the performance of Slant transform in image hashing algorithm. An image hashing algorithm based on Slant transform is proposed in this paper. The image hash scheme can be constructed by preprocessing, extracting and post processing appropriate image features. In order to improve the property of feature extracting [10–12], the preprocessing of image is always used. The common image preprocessing includes applying a low-pass filter, rescaling, or adjusting the components of image, and so on. But in this paper, for analysis the performance of Slant transform, all the processing of image seemed as attacks to original image. To achieve robustness, security, and compactness, the feature extraction is the most important stage of constructing an image hash. A robust image feature extraction scheme should withstand various image processing that does not alter the semantic content. Various image hashing schemes mainly differ in the way randomized features and extracted. For post-processing, the aim is compression the length of hash sequence and without less the magnitude feature.

In this paper, in order to analysis the performance of Slant transform in image hashing algorithm, a novel robust image hashing algorithm based on Slant transform is proposed. The framework of proposed hashing algorithm is shown in Fig. 2, which includes the following steps:

(1) Block image: block original image to $n \times n$ sub-images. For example, the blocks size is 8×8.
(2) Slant transform: using the Slant transform to every block images.
(3) Feature extraction: selected the middle-frequency coefficients by user-key.
(4) Compression: compression the selected coefficients, and obtain the image hashing finally.

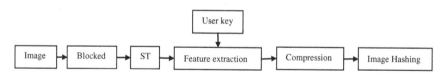

Fig. 2. The framework of proposed hashing algorithm

4 Performance Results

Performance metrics and experiment setup: to measure the performance of image hash, the normalized Hamming distance between the binary hashes is employed. The defined of normalized Hamming distance is:

$$d(h_1, h_2) = \frac{1}{L} \sum_{k=1}^{L} |h_1(k) - h_2(k)| \qquad (6)$$

where $h_1(k)$, $h2(k)$ are different image hash sequence values; L is the length of image hash. The normalized Hamming distance d has the property that for dissimilar images, the expected of d is closed to 0.5, else the expected is closed to 0.

Several benchmark images (such as Lena, Baboon, Peppers, F16, Cameraman, etc.) are used to test the performance of the proposed scheme. Images used in this paper are shown in Fig. 3 and Table 1.

| baboon | cameraman | canal | elaine |
| frog | lena | peppers | sailboat |

Fig. 3. Several benchmark images

Table 1. Normalized Hamming distances of proposed algorithm between benchmark image

Attacks		Factor	Normalization hamming distance							
Jpeg comp.	2	0.0918	0.0811	0.1182	0.1289	0.1191	0.1182	0.0645	0.0713
	3	0.0869	0.0859	0.1348	0.1338	0.1084	0.1318	0.0586	0.0605
	4	0.0947	0.0977	0.1387	0.1387	0.1104	0.1533	0.0791	0.0742
	5	0.0928	0.0986	0.1523	0.1494	0.1162	0.1543	0.0801	0.0771
	6	0.0928	0.1484	0.1553	0.1611	0.1201	0.2031	0.1396	0.1348
	7	0.0928	0.5010	0.4326	0.4502	0.1094	0.5010	0.4883	0.4775
	8	0.4521	0.4990	0.2021	0.2021	0.1094	0.1611	0.1289	0.1260
	9	0.0928	0.4990	0.5674	0.5498	0.1094	0.4990	0.5117	0.5225
median-filter	3	0.0576	0.0459	0.0615	0.0654	0.0713	0.0703	0.0313	0.0410
	5	0.0723	0.0674	0.0908	0.0908	0.0859	0.0928	0.0479	0.0479
	7	0.0850	0.0684	0.1074	0.1006	0.0977	0.1055	0.0615	0.0527
	9	0.0967	0.0791	0.1162	0.1182	0.1025	0.1094	0.0664	0.0635
Gaussian low-pass	/	0	0	0	0	0	0	0	0
Gaussian noise	0.05	0.0474	0.0495	0.0481	0.0479	0.0479	0.0520	0.0501	0.0499
	0.01	0.0474	0.0495	0.0481	0.0479	0.0479	0.0520	0.0501	0.0499
Peppers and slat noise	/	0.0151	0.0174	0.0396	0.0424	0.0220	0.0277	0.0103	0.0121
Cut	1/8	0	0.0156	0.0156	0.0156	0	0.0117	0.0156	0.0156
Rescaling	0.5	0.0459	0.0898	0.0684	0.0947	0.0811	0.1299	0.0059	0.0459
	2	0	0	0	0	0	0	0	0
Rotation	10^0	0.0818	0.0501	0.0356	0.0364	0.0742	0.0567	0.0384	0.0388
	15^0	0.0765	0.0501	0.0354	0.0368	0.0698	0.0555	0.0394	0.0383

5 Conclusion

In this work, a novel robust image hash scheme for certain application is proposed. The Slant transform is used for constructing robust image hashes, and the user-key are used as the encryption key. The image is first blocked to sub-images, then after the Slant transform of every block image, feature vector is extracted from the transform field coefficients, finally the resulting statistics vector is quantized and the binary hashes sequence is obtained. Experimental results show that the proposed scheme is robust against common image processing such as JPEG compression, mid-filtering, and rotation. Therefore, the scheme proposed in this paper is suitable for image authentication, content-based image retrieval and digital watermarking.

Acknowledgments. The work presented in this paper was supported by Guangdong Provincial Science and Technology Program (No. 2014A020208139); Distinguished Young Talents in Higher Education of Guangdong (No. 2013LYM-0057). Delong Cui is corresponding author.

References

1. Soo, C.P., Jian, J.D.: Closed-form discrete fraction and affine Fourier transforms. J. IEEE Trans. Signal Process. **48**(5), 1338–1353 (2000)
2. Ozaktas, H.M., Arikan, O.: Digital computation of the fractional Fourier transform. J. IEEE Trans. Signal Process. **9**, 2141–2149 (1996)
3. Pei, S.C., Yeh, M.H.: Two Dimensional discrete fractional Fourier transform. J. Signal Process. **67**, 99–108 (1998)
4. Venkatesan, R., Koon, S.M., Jakubowski, M.H., Moulin, P.: Robust image hashing. In: IEEE Conference on Image Processing, pp. 664–666 (2000)
5. Fridrich, J., Goljan, M.: Robust hash functions for digital watermarking. In: IEEE International Conference on Information Technology: Coding and Computing, pp. 178–183 (2000)
6. Mihcak, K., Venkatesan, R.: New iterative geometric techniques for robust image hashing. In: ACM Workshop on Security and Privacy in Digital Rights Management Workshop, pp. 13–21 (2001)
7. Enomoto, H., Shibata, K.: Orthogonal transform coding system for television signals. IEEE Trans. Electromagn. Compat. **13**(3), 11–17 (1971)
8. Vaid, S., Mishra, D.: Comparative analysis of palm-vein recognition system using basic transforms. In: IEEE International Advance Computing Conference, pp. 1105–1110 (2015)
9. Gupta, J., Chanda, B.: An efficient slope and slant correction technique for off-line handwritten text word. In: 4th International Conference of Emerging Applications of Information Technology, pp. 204–208 (2014)
10. Jin, L., Qian, W., Cong, W., Ning, C., Kui R., Wenjing, L.: Fuzzy keyword search over encrypted data in cloud computing. In: the 29th IEEE International Conference on Computer Communications (INFOCOM 2010), pp. 441–445. IEEE Press (2010)
11. Jin, L., Xiaofeng, C., Mingqiang, L., Jingwei, L., Patrick, L., Wenjing, L.: Secure deduplication with efficient and reliable convergent key management. IEEE Trans. Parallel and Distrib. Syst. **25**(6), 1615–1625 (2014)
12. Jin, L., Man, H.A., Willy, S., Dongqing, X., Kui, R.: Attribute-based signature and its applications. In: 5th ACM Symposium on Information, Computer and Communications Security (ASIACCS 2010), pp. 60–69. ACM (2010)

Offline Video Object Retrieval Method Based on Color Features

Zhaoquan Cai[1], Yihui Liang[2(✉)], Hui Hu[1], and Wei Luo[1]

[1] Huizhou University, Huizhou, China
{cai, luowei}@hzu.edu.cn, 2296434036@qq.com
[2] South China University of Technology, Guangzhou, China
395265585@qq.com

Abstract. At present, video retrieval has been applied to many fields, for example, security monitoring. With the development of the technique of content-based video retrieval, video retrieval will be applied to more areas. The article mainly do research on offline video retrieval based on color features and realize offline video color features retrieval. The research realized Algorithm for Video Objective Tracking based on Adaptive Hybrid Difference and was focused on designing color features range calculation scheme with the combination of RGB and HSL color model. And extract and judge the color feature of the blob in the video then analyze and process the retrieval result. According to the result of this test, the success rate of detection of the system have reached ninety percentage upon. The realization of offline video object retrieval system based on the color features can decrease the time of Manual Retrieval to the color features object in the video, help people filter information and have benefits on the realization of intelligent and automatic video retrieval.

Keywords: Content-based video retrieval · Color features extraction · Computer vision

1 Introduction

With the development of social economy, more and more multimedia data are obtained by camera and stored with the form of video files. Manual type can't satisfy the demand of real-time multi-window monitoring for a long time and find the desired objectives from a huge amount of video data effectively. So the automatic detection, track and recognition of moving video targets is a key research subject of computer vision, and it has been used widely in many fields such as intelligent traffic system and video monitoring [1].

The original video object preprocess way is based on the thought of moving object and video segmentation. There are two kinds of methods in common use, one is based on statistic, and the other is difference method. The way of background modeling based on statistic is seen in Temporal Average Model [2] and the single gauss background model [3] at the earliest. The calculated amount of the two ways is small, but the effect is instability. It's not fit for this situation where video background Pixel luminance is fluctuating. Gaussian Mixture Model [4] is a relatively classical method. The calculated

© Springer Science+Business Media Singapore 2016
K. Li et al. (Eds.): ISICA 2015, CCIS 575, pp. 495–505, 2016.
DOI: 10.1007/978-981-10-0356-1_53

amount of this way is large, and it is easy to lack time to response when applied, the accuracy rate of Gaussian Mixture Model estimated will decline when meet with fast moving rigid object. Nonparametric background model [5] can achieve better effect than GMM by means of online estimate method. But the problem of large calculated amount hasn't been improved. New methods such as Kalman filtering [6] and particle filter [7] has been brought in the study for background molding late, but the calculated amount is also large.

Many objectives have complicated shapes, and these complicated shapes can't be described with simple geometry such as hand, head and shoulder of man. The track method based on outline has come up in order to solve the track problem for this target. And this method can be divided into two categories: one is shape matching, the other is appearance tracking. Zhao [8] describe the target with the combination of color histogram and border. Pan [9] make use of the border information obtained through target outline to describe the appearance of target and use K-L distance to measure match degree. Cha [10] use Hough Transformation for the outline of target to track target. The method for target appearance tracking first initializes the target with a shape. We should update the target shape according to changes of state of target in the later each frame. The method has two solver status forms: one is to apply state model to describe the state and shape of target, the other is to obtain the optimal shape for target through minimizing energy equation of describing target shape. Li [11] define target state as dynamic control point, then adjust these control points to describe target state. Tao [12] improve contour tracing method based on particle filter to track multi-objective.

In conclusion, the current video target identification research situation mainly exists following problems: (1) We need to perfect mathematical model to reflect the characteristics of video object further [13]; (2) the current detection algorithm can only identify a motion object, but can't judge whether the object is target [14]; (3) the existing tracking algorithms is hard to classify and deal with moving target; (4) if there are multiple moving targets in the video, the existing method cannot extract target effectively. Research trend tend to automatically identify and analyze the special events of video data, and provide real-time alarm and decision aid.

2   Adaptive Hybrid Difference Algorithm

The realization of Adaptive Hybrid Difference Algorithm first is to read video input by users, then initialize these videos according to the parameter N which has been set before, use Hybrid Difference Algorithm to obtain the static background model of the video (that's no motion object in these background images) from these video images in the first N frames. We can get video background model through the initial estimate and the scope of the threshold value of T of binaryzation image frame at the same time. In order to execute video retrieval, We should go on binary image processing to get foreground (that's moving object in the video) contour according to the scope of adaptive threshold T obtained before, reach the basic effect for detecting moving object in the video. The flowchart of Adaptive Hybrid Difference Algorithm is drawn below (Fig. 1).

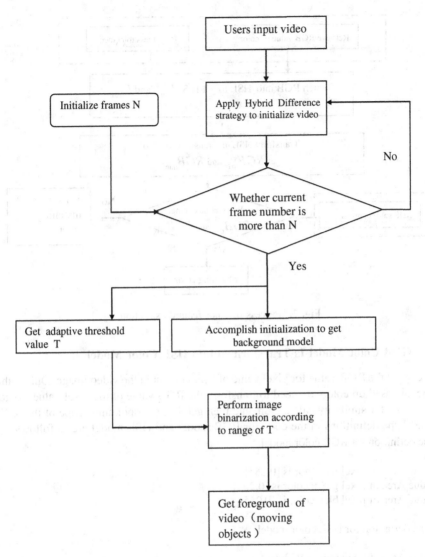

Fig. 1. Process of Adaptive Hybrid Difference Algorithm

3 Color Feature Extraction and Retrieval

The retrieval based on color feature, first research color composition and how it performs in the video image, then the computer realize retrieval for moving object in the video according to the color composition. The process of color feature extraction is shown in Fig. 2.

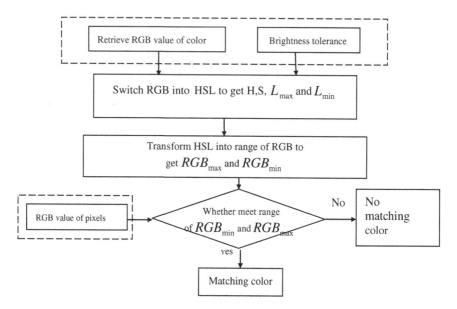

Fig. 2. Process of color feature extraction.

3.1   GRB Color Model Is Transformed into HSL Color Model

We can get the RGB value for pixel value of a pixel point in the video image. During the retrieval based on color, we need to transform the RGB value of the pixel values to get HSL value for similarity calculation, and then get the corresponding scope of the RGB value. Type definitions in the code for RGB model and HSL model are as follows:
Type definition for RGB color model:
{
Value Area of pixel red color is [0,255]
Value Area of pixel green color is [0,255]
Value Area of pixel blue color is [0,255]
}
Type definition for HSL color model:
{
Value Area of pixel tone is [0,360]
Value Area of pixel saturation is [0,100]
Value Area of pixel brightness is [0,100]
}
We assume there is a pixel $C(i,j)$, (r, g, b) is color information of the pixel: red, green and blue coordinates, the value is the real number between 0 to 1. Set max is equivalent to the maximum in r, g and b. We assume that min is equivalent to the minimum in these values. To find value (h, s, l) in the HSL space, $h \in [0, 360)$ is Hue angle, $s, l \in [0, 1]$ refers to saturation and brightness respectively.

Calculation Formula:

$$h = \begin{cases} 0 & \textbf{\textit{When max}} = \textbf{\textit{min}} \\ 60 \times \frac{g-b}{max-min} & \textbf{\textit{When max}} = \textbf{\textit{r}} \textbf{ and } \textbf{\textit{g}} \geqslant \textbf{\textit{b}} \\ 60 \times \frac{g-b}{max-min} + 360 & \textbf{\textit{When max}} = \textbf{\textit{r}} \textbf{ and } \textbf{\textit{g}} < \textbf{\textit{b}} \\ 60 \times \frac{b-r}{max-min} + 120 & \textbf{\textit{When max}} = \textbf{\textit{g}} \\ 60 \times \frac{r-g}{max-min} + 240 & \textbf{\textit{When max}} = \textbf{\textit{b}} \end{cases}$$

$$1 = 1/2 \, (\mathbf{max} + \mathbf{min})$$

$$s = \begin{cases} 0 & \textbf{\textit{When l}} = 0 \textbf{ or } \textbf{\textit{max}} = \textbf{\textit{min}} \\ \frac{max-min}{max+min} & \textbf{When } 0 < l \leqslant 1/2 \\ \frac{max-min}{2-(max+min)} & \textbf{When } l > 1/2 \end{cases}$$

3.2 HSL Color Model Is Transformed into RGB Color Model

A color in the given HSL color model is defined in value (h, s, l), the range of parameter h which refers to Hue angle is $h \in [0, 360)$, the range of parameters s and l which refers to saturation and brightness respectively is $s, l \in [0, 1]$, correspondingly Three Primary Colors (r, g, b) in the RGB space, R corresponds to red, g corresponds to green, b corresponds to blue, their value area is $[0, 1]$, their computing method is below: First, if s equals to zero, then resultant color is achromatic or gray. Under the special circumstances, r, g and b all equal to one. When s doesn't equal to zero, we can use the following formula to switch HSL color model into RGB color model:

$$q = \begin{cases} i \times (l+s) & l < \frac{1}{2} \\ l+s-(l \times s) & l \geq \frac{1}{2} \end{cases}, \quad p = 2 \times l - q, \quad h_k = \frac{h}{360}$$

$$t_R = h_k + \frac{1}{3} \text{ is used to get the value of R,}$$

$$t_G = h_k \text{ is used to get the value of G,}$$

$$t_B = h_k - \frac{1}{3} \text{ is used to get the value of B.}$$

When t is less than zero, there exists $t = t + 1$, for every color value belonging to the RGB including the values of r, g and b. When t is more than one, there exists $t = t - 1$ for every color value belonging to the RGB including the values of r, g and b. There is a $ColorRGB(R, G, B)$ Vector (r, g, b) for every color.

$$ColorRGB(R, G, B) = \begin{cases} p + ((q-p) \times 6 \times t) & \textbf{When } t < 1/6 \\ q & \textbf{When } 1/6 \leqslant t < 1/2 \\ P + ((q-p) \times 6 \times (\frac{2}{3} - t)) & \textbf{When } 1/2 \leqslant t < 2/3 \\ p & \textbf{others} \end{cases}$$

3.3   HSL Color Model Is Transformed into RGB Color Model

During the video retrieval based on color, color detection is aimed at moving object in the video, after applying Adaptive Hybrid Difference Algorithm, we can get foreground in the video(that's video moving object), the information of foreground are saved in the form of blob. Blob refers to relatively homogeneous nonlinear area which is different from around region. Blob is defined as follows in OPENCV:

Structure: Blob
{
   Blob position: coordinates of x and y,
   High and wide of blob
   Blob ID number
}

Pixels in the video, which are generally considered to be the smallest completely sampled pixels in the image, contain color information of these pixels. Generally, the pixel in the video is saved and processed in RGB color model. RGB value of a pixel in the image can be used to represent coordinate in the image as RGB color value of pixel. So for region which is a size of $m \times n$, RGB color value set is $S(m,n)$, $C(i,j) \in S_{(m,n)}$ $i \in (0,m)$, $j \in (0,n)$. Figure 3 is the process and schematic plot of getting color attribute of blob.

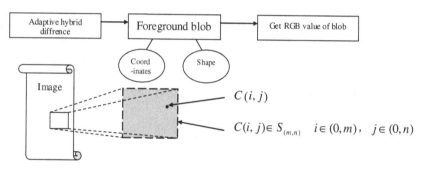

Fig. 3. Process of getting color attribute of blob

3.4   The Description of Realization of Retrieval Based on Color Features

After the user choose the color terms that need to retrieval, then transform the color from RGB to HSL and calculate to get. Adjust the brightness L in the allowable tolerance range of the brightness, then get and two values. And convert Color HSL into RGB value respectively to get, that's the similar scope value of color that needs to be retrieved. When moving object (namely the information of the foreground blob) is detected in the video, compare RGB value of each pixel with and, go on statistics for those that meet the limits, use statistics methods to calculate matching rate of blob to color retrieved. Then gather information of plaque accord with certain matching rate into retrieval results lists. The flowchart of color retrieval in the offline video retrieval based on color is drawn below (Fig. 4).

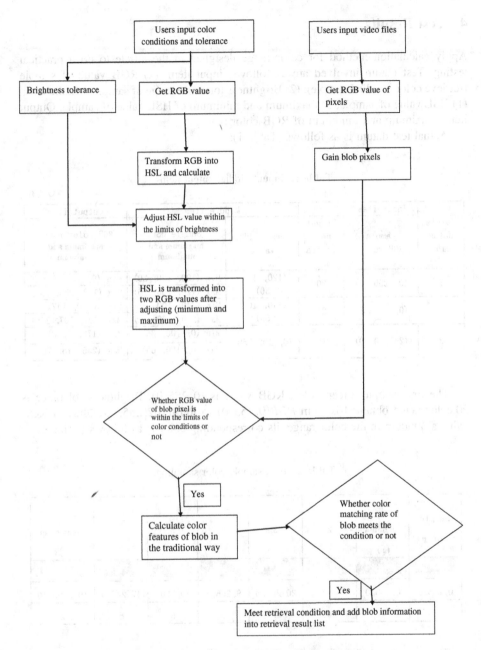

Fig. 4. Process of offline retrieval based on color features

4 Test Results

Apply calculation method for color range designed in the article to do a practical testing. Test datum involved are as follows, input item: (1) RGB value of sample (retrieve color is called sample); (2) Brightness tolerance range of sample. Middle term. (1) HSL value of sample;(2) maximum and minimum of HSL value of sample. Output item: maximum and minimum of RGB color range.

Actual test datum is as follows (Table 1):

Table 1. Input, middle, output items

| Sample datum number | Input item | | Middle item | | Output item |
	sample RGB value	Bright ness tolera nce	Sample HSL value	Sample HSL maximum and minimum	RGB color range maximum and minimum
1	(0, 255, 0)	20	(120, 100, 50)	Min (120, 100, 30)	Min (0, 153, 0)
				Max (120, 100, 70)	Max (102, 255, 102)
2	(0, 0, 255)	21	(240, 100, 50)	Min (240, 100, 29)	Min (0, 0, 147)
				Max (240, 100, 71)	Max (107, 107, 255)
3	(255, 0, 0)	15	(0, 100, 50)	Min (0, 100, 35)	Min (178, 0, 0)
				Max (0, 100, 65)	Max (255, 76, 76)

The first sample: green color, RGB value is $(0, 255, 0)$. Brightness tolerance is 20.color range obtained is from $RGB(0, 153, 0)$ to $RGB(102, 255, 102)$.draw a RGB value at random in the color range, its corresponding color is as follows (Table 2):

Table 2. First sample color sampling

Sample color	1 (minimu m)	2	3	4	5	6	7 (maximum)
0, 255, 0	0, 153, 0	60, 160, 5 4	20, 200, 5 4	9, 206, 2 9	80, 230, 5	45, 247, 9 6	102, 255, 10 2

The second sample: blue color, RGB value is $(0, 0, 255)$, Brightness tolerance is 21.color range obtained is from $RGB(0, 0, 147)$ to $RGB(107, 107, 255)$.draw a RGB value at random in the color range, its corresponding color is as follows (Table 3).

The third sample: red color, RGB value is $(255, 0, 0)$, Brightness tolerance is 15.color range obtained is from $RGB(178, 0, 0)$ to $RGB(255, 76, 76)$.draw a RGB value at random in the color range, its corresponding color is as follows (Table 4).

Table 3. Second sample color sampling

Sample color	1 (minimum)	2	3	4	5	6	7 (maximum)
0, 0, 255	0, 0, 147	14, 49, 169	20, 79, 219	11, 100, 181	51, 39, 156	98, 2, 233	107, 107, 255

Table 4. Third sample color sampling

Sample color	1 (minimum)	2	3	4	5	6	7 (maximum)
255, 0, 0	178, 0, 0	180, 50, 32	184, 40, 76	215, 64, 26	255, 76, 7	240, 0, 60	255, 76, 76

On the basis of above datum, We can conclude that first, color scheme of color range obtained fit sample color; second, close brightness tolerance value(for instance, first sample and second sample), although sample color is different, color change is approximately same; third, brightness tolerance is more small, range of color change is more small (Fig. 5).

Fig. 5. Color detection screenshot

Fourteen results are detected, As screenshot shows, The success rate for detection of blue vehicles appeared in the video have reached ninety percentage, but there also exist some error, for instance, in the above images, a white tour bus sprayed blue pattern advertisement is also detected, but the color matching rate is lower, it declares that the sensitivity of the system is quite high.

5 Conclusions

The article mainly expounds the algorithm of the video retrieval based on color features, first we begin to summary various color models and select the way of the combination of the RGB and HSL color model according to the visual feature of man. And design the structure of the RGB and HSL color model and its corresponding applicable range detailedly and present two color model structures. Come up with how to get color attributes of pixel of the blob during the video retrieval process. Tell the definition of color similarity and judge process as well as conversion between two color model objects in the system detailedly. Finally provide the process of realization of the module of detection for color features offline video retrieval method based on color features designed in the article is easy to realize and implement stably and realized the video object retrieval functions based on color features accurately.

References

1. Friedman, N., Russell, S.: Image segmentation in video sequences: a probabilistic approach. In: Proceeding of Thirteenth Conference on Uncertainty in Artificial Intelligence, pp. 175–181 Morgan Kaufmann Publishers, Providence (1997)
2. Wren, C.R., Azarbayejani, A., Darrell, T., Penyland, A.: Pfinder: real-time tracking of the human body. IEEE Trans. Pattern Anal. Mach. Intell. **19**(7), 780–785 (1997)
3. Stauffer, C., Grimson, W.: Adaptive background mixture models for real-time tracking. In: Proceedings of IEEE Conference on Computer Vision and Pattern Recognition, pp. 245–252. IEEE, Fort Collins (1999)
4. Youfu, W., Shen, J., Dai, M.: Traffic object detections and its action analysis. Pattern Recogn. Lett. **26**(13), 1963–1984 (2005)
5. Colombari, A., Fusiello, A.: Segmentation and tracking of multiple video objects. Pattern Recogn. **40**(4), 1307–1317 (2007)
6. Ming, X., Ellis, T.: Augmented tracking within complete observation and probabilistic reasoning. Image Vis. Comput. **24**(11), 1202–1217 (2006)
7. Hung, M.-H., Hsieh, C.-H.: Event detection of broadcast baseball videos. IEEE Trans. Circ. Syst. Video Technol. **18**(12), 1713–1726 (2008)
8. Zhao, Y., Liu, Y.: Video synthesis from still images using 3-D flow models. Sig. Process. Lett. **15**, 509–512 (2008)
9. Pan, P., Schonfeld, D.: Dynamic proposal variance and optimal particle allocation in particle filtering for video tracking. IEEE Trans. Circ. Syst. Video Technol. **18**(9), 1268–1279 (2008)

10. Zhang, C., Yin, P., Rui, Y., Cutler, R., Viola, P., Sun, X., Pinto, N., Zhang, Z.: Boosting-based multimodal speaker detection for distributed meeting videos. IEEE Trans. Multimedia **10**(8), 1541–1552 (2008)
11. Li, H., Ngan, K.N., Liu, Q.: FaceSeg: automatic face segmentation for real-time video. IEEE Trans. Multimedia **11**(1), 77–88 (2009)
12. Xiang, T., Gong, S.: Video behavior profiling for anomaly detection. IEEE Trans. Pattern Anal. Mach. Intell. **30**(5), 893–908 (2008)
13. Kasturi, R., Goldgof, D., Soundararajan, P., Manohar, V., Garofolo, J., Bowers, R., Boonstra, M., Korzhova, V., Zhang, J.: Framework for performance evaluation of face, text, and vehicle detection and tracking in video: data, metrics, and protocol. IEEE Trans. Pattern Anal. Mach. Intell. **31**(2), 319–336 (2009)
14. Kim, W., Kim, C.: A new approach for overlay text detection and extraction from complex video scene. IEEE Trans. Image Process. **18**(2), 401–411 (2009)

A Traffic-Congestion Detection Method for Bad Weather Based on Traffic Video

Jieren Cheng[1,2], Boyi Liu[1(✉)], and Xiangyan Tang[1]

[1] Hainan University,
Haikou 571101, China
1165995505@qq.com
[2] Xiangnan University,
Chenzhou 423000, China

Abstract. In order to solve the problem that the result of traffic congestion detection in bad weather is inaccurate, we analyzed current vehicle identification algorithms and image processing algorithms. After that, we proposed a detection method of traffic congestion based on histogram equalization and discrete-frame difference. Firstly, this method uses discrete-frame difference algorithm to extract the images that have vehicle information. Secondly, this method uses the histogram equalization algorithm to eliminate the noise of the images. Finally, this method recognizes the vehicle from the video and computes the traffic congestion index by the calculation method based on discrete-frame difference. It has proved by experiments and theoretical analysis that this method decreases false-negative rate and increases the accuracy rate of automatic traffic congestion detection in bad weather.

Keywords: Traffic congestion · Histogram equalization · Video processing · Inter-frame difference

1 Introduction

Traffic congestion detection is of great significance in city vehicle, road designing, traffic lights setting, preventing traffic congestion and other application fields. Therefore, the traffic congestion detection plays an important role in transportation field. Unfortunately, the results of an automatic vehicle identification system for detecting traffic is prone to appear high false-negative rate in bad weather such as the bad weather such as fog, mist, and rain. It may lead cause of errors in statistics of traffic condition index and increases the risk or economic loss in relevant departments and industry [1]. Therefore, we designed and implemented a traffic-congestion detection method for bad weather to improve the accuracy of traffic congestion detection in bad weather.

This work was supported by the National Natural Science Foundation of China (Project no. 61363071, 61379145), The National Natural Science Foundation of Haian (Project no. 614220), Hunan Province Education Science Planning Funds (Project no. XJK011BXJ004), Hainan University Youth Fund Project (Project no. qnjj1444).

K. Li et al. (Eds.): ISICA 2015, CCIS 575, pp. 506–518, 2016.
DOI: 10.1007/978-981-10-0356-1_54

2 Vehicle Identification and Image De-noising Algorithms

Traffic information acquisition-devices include contact traffic information acquisition-device and non-contact traffic information acquisition-device. Contact devices put the sensing device below the pavement. When the vehicle passes, sensors collect related data to realize traffic information acquisition. Those methods have high accuracy, simple programs and easy to implement. On the other hand, their high economic costs, short life, complicated installation and maintenance restrict their developments. Non-contact detection devices do not require installation and maintenance and they are becoming more and more popular. Non-contact detection methods include wave frequency detection and video detection. The former method includes ultrasonic wave frequency detectors or infrared detection monitors. Video detection method processes the video recorded by the camera to obtain traffic information. Its installation and maintenance are simple. Especially, with the rapid developments of video equipment and computer hardware, video-based traffic detection systems will be more widely used. Therefore, the traffic congestion detection method proposed by this paper uses video detection devices.

The automatic detection systems of traffic congestion based on video [2–4] are mainly use the methods of virtual detection line [5, 7], vehicle tracking [6, 7] or background subtraction nowadays. In the method of virtual detection line, a video measuring line is set on the top of every image. It can collect vehicles passing information and then recognize the vehicle. The matching method of the vehicle features get the similarity between the characteristic of moving object and the characteristic of sampling vehicles, and compares the similarity with the threshold value to identify vehicle in the images. Background subtraction method recognizes the vehicle by the subtraction between the current image and the background image obtained through Gaussion Mixture Models (GMM). However, only using these methods cannot recognize the vehicles accurately in bad weather. It is necessary to add programs to increase the contrast of the images. Fortunately, there have been some methods, such as the dark channel prior method that was present by He [8]. It takes the minimum in the color field and some fields firstly. After that, the images will serve as the dark channel of original images. It will increase the contrast by combining the light intensity and priori conditions finally. The neighborhood average method is replacing the pixels with the average of points gray value in its neighborhood. Some other effective methods [9, 10] also base on it. There are also many methods such as the conversion based on retinex [11, 12], media fuzzy filter [13] or by readjusting brightness of images [14]. O'gorman gives the optimum definition of histogram equalization [15, 16]. After analyzing these methods above, we proposed a new method to realize accurate detection of traffic congestion index in bad weather. Analysis of advantages and disadvantages of some popular image de-noising algorithms present in Table 1.

Table 1. Advantage and disadvantage of methods to image defogging

Algorithm	Advantages	Disadvantages
Histogram equalization	The shortest time	Loss of gray levels
Conversion based on retinex	Better image processing effect on images with low brightness	Influenced by luminance
Media fuzzy filter	Big gray difference	Not suitable for real-time processing
Readjust brightness method	Good at treatment on light	Taking a long time
Dark channel prior	Stable and reliable	Taking a long time
Neighborhood Average	Suitable for real-time processing	May video delay

3   Traffic-Congestion Detection Method for Bad Weather

This paper uses the method of histogram equalization to increase the contrast of images. Then the method alleviates the fuzzy caused by bad weather. Using improved inter-frame difference (discrete-frame method) to extract the video part that has vehicle information. It not only can save system energy but also make the whole system be fit for real time processing.

3.1   Process of the Traffic Congestion Detection Method

The traffic-congestion detection method based on histogram equalization and discrete-frame algorithms receives every image frame with a short memory. The next step is using discrete-frame method to recognize whether there are vehicles or not. To the images that have vehicle information, we use the histogram equalization method to increase its contrast. Lastly, the method recognizes vehicles and calculates traffic congestion index. Process of this method illustrated in Fig. 1.

As shown in the Fig. 1, the system divided into video reading-in procedure, control-load procedure, image-processing procedure, vehicle detection procedure, traffic congestion index computing procedure. Among them, video reading-in procedure makes the video load in image frames and store shortly. Controlling-load procedure loads the vehicle images into cache by using the discrete-frame difference method. Increasing contrast procedure makes the images more distinctly by using the method of histogram equalization. Vehicle detection procedure distinguishes the car form the image sequences. Traffic congestion index computing procedure is to statistic the number of the cars and then compute the traffic congestion index. We will discuss the key technology of this method in the following.

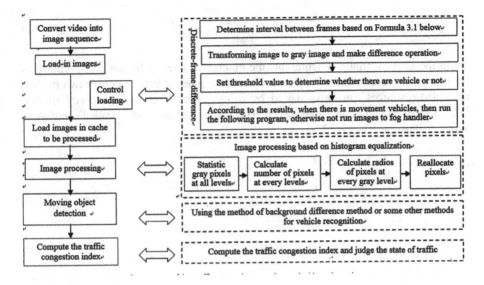

Fig. 1. Process of the traffic congestion detection method in bad weather

3.2   Image Extracting Algorithm Based on Discrete-Frame Difference

If we handle every frame of the video like inter-frame difference method [17–19] in the bad weather to extract the vehicle images, system resources will be large occupied and it can't be fit for the real time processing. There is also no significance that we handle the video in which doesn't have any car image. So this paper improves the inter-frame difference method and present discrete-frame difference algorithm to save system energy and realize the extraction of vehicle images. Inter-frame difference method is making difference calculation between the interfacing images. That is to say, this method analysis the motion feature by comparing the gray difference between inter-facing frames. It can be determined preliminary when the gray difference is beyond the threshold value. At this time, loading the image frame into cache to increase its contrast by using histogram equalization. In bad weather, vehicles speed is slow, and the time interval between every two frames is short. So vehicle travelling distance is small between every two frames. Then, we can easily infer that we don't need to compute gray difference between every two frames. So the resources and time can be saved by intervaling several frames. In order to guarantee every car can be recognized, we give the following formula to compute the number of interval frames in discrete-frame method.

$$Z < \frac{S_{space}}{V_{max} \times t} \tag{1}$$

In the above formula, Z represents the number of interval frames. V_{max} represents the highest speed of vehicle. S_{space} represents time interval between every two frames. Occupied resources can be increased by using ways of intervaling several frames. What's more, the time can also be saved. This paper used the toolbox called simulink to simulate illustrated in Fig. 2.

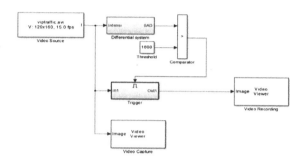

Fig. 2. Simulation implementation for discrete-frame algorithm

In the figure above, computing the grayscale difference between background image and current image after inputting the image sequences. Then comparing with the threshold value in a comparator to judge whether there is vehicle information in the current image. If the grayscale difference is greater than the threshold value, We record this image frame. Otherwise, not to record it at all. If you want to achieve more frame difference apart, add a delay timer.

We can also calculate the sum of the pixels that are different from its former or next frame. After computing the interval number of frames, we can judge whether the current image contains vehicle information by the calculation method of discrete-difference method. As shown in the formula 2, $P(x, y)$ represents the pixel gray level difference. When the pixel points of the gray level difference is less than the pre-configured threshold value of T, c will be valued 0 in the processing. Otherwise b will be valued 1. In the end, in accordance with the formula 3, compute the sum of pixel points of current image that are different from the background image roughly.

$$P(x, y) = \begin{cases} 1 & c \geq T \\ 0 & c < T \end{cases} \tag{2}$$

$$S = \sum_{i=1}^{a} \sum_{j=1}^{b} P_{ij}(x, y) \tag{3}$$

When S is greater than the threshold value, we can deem that there is vehicle information in the current image. If S is less than the threshold value, we can deem that there is no vehicle information the image and do not require image processing programs for this image.

3.3 Image Processing Algorithm Based On Histogram Equalization

Histogram equalization changes the condition that the gray level histogram of the original image are concentrated in some certain gray level internal. It redistributes image elements and improves the contrast of the image by putting some certain pixels of gray level in a similar state. Table 1 indicates that the method of histogram equalization has advantages of low cost of time, less time complexity and space complexity. Therefore, it is more suitable for real-time processing compared with other methods so this paper proposes a method of using histogram equalization to improve the contrast of the vehicle images. Algorithm as shown on Table 2.

Table 2. Histogram equalization program

Histogram equalization program(Realized in matlab 2014)
Input: vehicle images a.jpg
Output: after histogram equalization images grey value of the corresponding matrix C
A=imread("a.jpg")   // Read in the image a and deposited in the array
For k=0 to 255
B(k+1)=length(find(A==k))/(m*n);   /*The function called find can output where elements of the element to meet the requirements*/
End
//Original histogram each gray-scale frequency calculation, and deposited in the array B
For k=1 to 256
For j=1:k
S(k)=B(j)+S(k)
End
End
/*Using the formula $f_i = (m-1)\sum_{k=0}^{i}\frac{q_k}{Q}$ [20,21] to equalize the histogram about the gray frequency of images. Among them, r_j represents gray level, B_r represents frequency of gray level, k represents there are k gray levels in the image，S_k represents the frequency of gray level after histogram equalization.*/
For k=1 to 256
S(k)=round((S(k)*256)+0.5)
C(find(A==k))=S(k+1)
End for   //update images pixels

3.4   Vehicle Identification and Calculation Method of Traffic Congestion Index

We can adopt visual detection line method, vehicle tracking method or background subtraction method. Virtual test line method sets a virtual detection line above the detection area described in the literature [5, 7]. Then, we can judge and track the vehicle according to mutative intensity of test line. Tracking vehicle method described in the literature [6, 7], which cuts the pixels consistent with vehicle characteristics in order to achieve identification and track of vehicle by matching vehicle characteristics. Background subtraction method that this paper adopted recognizes the moving vehicles by making gray-scale difference between the current image and the background image.

Traffic congestion index is a conceptual index value that shows the degree of traffic congestion. There is no unified calculation method for traffic congestion index in current international. The domestic or foreign experts and scholars present many effective evaluation algorithms or protocols, such as the COC protocol [22] present by Fukumoto. This protocol of traffic congestion assessment is to use the operation data from the vehicle information real-time transmission to calculate the area of vehicle density. Then vehicle density will reflect the degree of the current traffic congestion. There is a voting protocol present by Padron who regarded that every car driving condition was likely to affect the transportation and operation of the whole area [23]. Therefore, every car data based on the speed of the vehicle driving must transfer to other vehicles. The congestion and vehicle running status can be determined through the contrasts between different vehicles data. What's more, many researches such as the ECODE protocol that shows the situation of traffic running based on traffic flow characteristics and present by Younes [24], etc. There is an evaluation method to calculate Beijing traffic congestion index proposed by Beijing traffic committee [25]. It evaluates traffic congestion states through data from real-time transmission of taxis, mainly considering the speed and the vehicle density. It realizes the real-time monitoring of the city traffic condition.

These methods are not suitable for traffic congestion detection method based on discrete-frame difference, so this paper presents a method (traffic congestion index calculation method for discrete-frame difference) to calculate the traffic congestion index in bad weather. It supports breaking down the image process when there is no vehicle information, and the traffic-congestion index can calculated by the formula 7 raised in this paper. Derivation process is as follows:

$$W = \sum_{i=1}^{k_1} C_i \tag{4}$$

$$W_{max} = \frac{ST_{max}}{Zmt} \tag{5}$$

$$\rho_y = \frac{W}{W_{max}} \tag{6}$$

In above formulas, W represents the total number of identified vehicles. k_1 represents the total number of frames of video section containing information of vehicles. C_i represents the number of identified vehicles in i frames in the video that has vehicle information. W_{max} represents the maximum number of identified vehicles. S represents length of the video detection region. Z represents the number of interval frames in discrete-frame algorithm. T_{max} represents the total time of the video. t represents the difference of two frames. m represents the minimum length of a vehicle. ρ_y represents the traffic congestion index without unit conversion [20].

Calculating Traffic congestion index for inter-frame difference is setting a limit value. It means that the minimum length of vehicles replaces real length of vehicles in the most congested state. Meanwhile, we regard no distance between vehicles as the most terrible state. That is to say, when the vehicle speed is high, the number of recognized cars will decrease and the traffic congestion index will decrease. Therefore, the number of vehicles recognized in all frames can reflect the degree of traffic congestion accurately, so the paper regards the quotient of the vehicle number detected and the number of total vehicles in worst traffic congestion condition. It can reflect the degree of traffic congestion. After analysis of formulas 4 and 5 and transformed the units, we can obtain the formula 7.

$$\rho = 10m \times \sum_{i=1}^{k_1} \frac{C_i t Z}{S T_{max}} \qquad (7)$$

After calculating the traffic congestion index, we can judge the degree of traffic congestion by its traffic congestion index. Table 3 shows the relation between the traffic congestion index and the degree of traffic congestion condition.

Table 3. Relation between degree of traffic congestion and traffic congestion index

Traffic congestion index	Degree of traffic congestion situation	Running condition
0~2	Smooth	Fine
2~4	Barely smooth	Little fine
4~6	Mild congestion	Little worse
6~8	Moderate congestion	Worse
8~10	Severe congestion	Very worse

4 Experiments and Results Analysis

4.1 Experiment of Images Processing

We conducted the experiment test and data collection by using the traffic video data of reference 7 on MATLAB 2014a software under windows 7 system. The program runs in the processing speed of 2.5 GHz dual-core CPU host. The running time about each algorithm presented in Table 4, and their effects are in Table 5.

Based on the effects of images processing and running time data collection and distribution of the above algorithms, we can obtain the conclusion that the running time of

Table 4. Time comparison of image increasing algorithm

Method	Running time
Histogram equalization	7.7252 ms
Conversion based on retinex	40.7518 ms
Media fuzzy filter	25.3611 ms
Readjust brightness method	10.8834 ms
Dark channel prior	13.5112 ms
Neighborhood average	20.9842 ms

Table 5. Effects comparison of de-noising algorithms

Methods	Dark channel prior	Media fuzzy filter	Neighborhood Average
Effects			

Methods	Conversion based on retinex	Histogram equalization	Readjust brightness
Effects			

histogram equalization is minimum and image-processing effects conform to the requirements of the standard. Under the comprehensive consideration, histogram equalization method is more suitable for images processing in automatic traffic-congestion detection system for bad weather.

4.2 Experiment of Discrete-Frame Method to Extract Traffic Images

The paper designs the experimental to validate the efficiency of discrete-frame method. Using formula 1, we valued the interval frame number 3. Some result of screen on image sequences illustrated in Table 6.

Time comparison between before discrete-frame difference and after discrete-frame difference illustrated in Table 7.

Table 6. Sifting condition for video images by discrete-frame difference algorithm

Saved frame	1th frame	8th frame	11th frame	14th frame	17th frame
Image					
Saved frame	20th frame	23th frame	26th frame	32th frame	......
Image					......

Table 7. Time comparison of discrete-frame difference algorithm

Index	Running time of discrete-frame method	Average time of histogram equalization method on an image	Number of the whole video	Number of frames after running discrete-frame method	Saved time
Times	2.055 s	0.234 s	120	31	18.771 s

It can be saw from above Table 7 that the running time have been saved roughly after using discrete-frame algorithm and achieved a good result.

4.3 Experiment of Vehicle Identification

We adopt the experimental data in the reference 7. In order to simulate bad weather, we carry out the fuzzy processing to the video. As shown in Fig. 3, it is the operating condition of traffic-congestion detection methods based on background subtraction in bad weather. Figure 4 is the operating condition of traffic-congestion detection methods that present by this paper.

Figure 3 is simulation of the real-time detection condition in bad weather for the traffic-congestion detection method based on background subtraction method. We can see that the detection results appear greater errors. Figure 4 shows the effect of the method proposed in the paper for the traffic congestion detection in bad weather. We can see that the latter is more suitable for bad weather. Table 8 is statistical comparison of the two methods.

The calculation method for accuracy rate in Table 8 is shown in formula 8:

Fig. 3. Operation of the method based on background subtraction

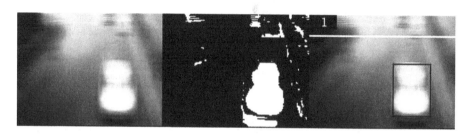

Fig. 4. Operation of the method based on discrete-frame difference

Table 8. Comparison of experimental data in two methods

	Running time	Accuracy rate	Resources occupation	Algorithm complexity
Traffic congestion detection method based on background subtraction	4.298 s	0.14	1	O(n)
Traffic congestion detection method based on discrete-frame difference	6.852 s	0.88	0.89	between 2 O(n) and O(n)

$$Z = 1 - \frac{\sum_{i=1}^{n} |C_i - E_i|}{E}, \left(n = \frac{T}{T_0}\right) \tag{8}$$

In above formula 8, Z represents the accuracy of each method, T represents the total time of video, T_0 represents the time interval between two adjacent video images, C_i represents the number of cars that is in the ith video frame when recognizing the car in video that has vehicle information. E_i represents the number of vehicles in the ith frame of the video that have vehicle information actually. E represents the total number of vehicles that recognized in the traffic video. The ratio of the resources occupation in the Table 8 refers to the radio of processed occupation of video resources of cache in the two different methods.

The result of experiments shows that the threshold value is general to take 0.4 to 0.6, and if the number takes higher, the accuracy of our detection method will be higher. Whether there is fog at that time, the greater the number, the smaller the possibility of miss checking. In bad environment, even if the current methods do not consider the causes mistakenly identified and make the value very small, ordinary method can appear complete miss recognition for a or more cars. In bad weather, our methods will be lower error detection rate and wider applicable weather space by improving the decision threshold value.

Comparing the results of the detection methods, we can conclude that the advantages of our traffic-congestion detection methods are as follows:

(1) Capturing images of vehicle information video by discontinuous frame difference is faster than before. Shorten the time, less resource consumption, and higher operation efficiency.
(2) We use de-noising processing procedure based on histogram equalization to eliminate the interference of bad weather causes. Image processing procedure not only can enhance the contrast of the image containing vehicle information but also decrease false-negative rate. What's more, it increases accuracy rate of traffic congestion detection in bad weather.

5 Conclusion

In bad weather, the false-negative detection rate of existing traffic detection methods is high for low visibility. To solve this problem we research the current traffic-congestion detection methods in bad weather, and compared and analyzed the advantages and disadvantages of mainstream image de-noising algorithms. We proposed a method for traffic congestion detection in bad weather. Firstly, we present the definition of discrete-frame difference algorithm, which can decrease the resource consumption of detection methods through extracting the video that has vehicle information. Then, we use the histogram equalization algorithm to process vehicle images and decrease the interference of bad weather causes. Finally, we can recognize the vehicles in images and calculate the traffic congestion index. In a word, experiments and theoretical analyses show that this method decreases the false-negative rate effectively and improves the accuracy of traffic congestion detection in bad weather.

References

1. Mu, Q., Zhang, S.: An evaluation of the economic loss due to the heavy haze during January 2013 in China. China Environ. Sci. **33**(11), 2087–2094 (2013). (in Chinese)
2. Xia, C.: A number of key technology research of highway network operation monitoring. South China University of Technology, Guangzhou (2013). (in Chinese)
3. He, Y.: Study on traffic video surveillance vehicle detection and tracking methods. Xi'an: Chang'an University (2009). (in Chinese)

4. Yang, G., Du, Q.: MATLAB Digital Image Processing. Tsinghua University Press, Beijing (2009)
5. Zhang, H.J., et al.: An integrated system for content based on video retrieval and browsing. Pattern Recogn. **30**(4), 643–657 (1997)
6. Stark, K., Ihle, T.: Visual tracking of solid objects based on an active contour model. In: Proceedings of 8th British Machine Vision Conference, pp. 640–649 (1997)
7. Zhao, X., He, H., Miu, Y., et al.: MATLAB Image Processing Real, p. 6. Mechanical Industry Press, Beijing (2013). (in Chinese)
8. He, K., Jian, S., Tang, X.: Guided image filtering. In: The 11th European Conference on Computer Vision, pp. 1–14 (2010)
9. Ataman, E., Aatre, V.K., Wong, K.M.: Some statistical properties of median filters. IEEE Trans. Acoust. Speech Sig. Process. **28**(4), 415–421 (1980)
10. Xie, Q.: Combined with image noise filtering large bilateral filtering algorithm and multi frame mean. Comput. Eng. Appl. **45**(27), 154–156 (2009). (in Chinese)
11. Wang, Y., Wang, X., Peng, Y.: Adaptive center weighted modified trimmed mean filter. Tsinghua Sci. Technol. **9**, 76–78 (1999). (in Chinese)
12. Rahman, Z., Jobson, D.J., Woodell, G.A.: Multi-scale retinex for color image enhancement. Image Process. **3**, 1003–1006 (1996)
13. Cover, S.A.: Another look at land's retinex algorithm. In: IEEE Proceedings of Southeastcon 1991, pp. 351–355 (April 1991)
14. Tarel, J.P., Hautiere, N.: Fast visiblity restoration from a single color or gray level image. In: IEEE International Conference on Computer Vision (ICCV 2009), pp. 2201–2208 (2009)
15. O'gorman, L.: Computer vision. Graph. Image Process. **41**(2), 229–282 (1988)
16. Wu, C.: Studies on mathematical model of histogram equalization. Chin. J. Electron. **41**(3), 598–602 (2013). (in Chinese)
17. Takaba, S., Sakauchi, M., Kaneko, T., et al.: Measurement of traffic flow using real time processing of moving pictures. In: 32nd IEEE Vehicular Technology Conference, vol. 32, pp. 488–494 (1982)
18. Chen, X.: Research of moving object-tracking method. Chin. J. Sci. Instr. **S1**, 49–53 (2014). (in Chinese)
19. Lin, J., Yu, Z., Zhang, J., et al.: Video motion detection based on background subtraction and frame difference method. Chin. J. Instr. **29**(4), 111–114 (2008). (in Chinese)
20. Zhu, Y., Huang, C.: An adaptive histogram equalization algorithm on the image gray level mapping. Phys. Procedia **25**, 601–608 (2012). ISSN 1875-3892
21. Gonzalez, R.C., Woods, R.E.: Digital Image Processing, 3rd edn. Pearson Education, London (2008)
22. Fukumoto, J., Sirokane, N., Ishikawa, Y., Wada, T., Ohtsuki, K., Okada, H.: Analytic method for real-time traffic problems by using contents oriented communications in VANET. In: 7th International Conference on ITS Telecommunications, pp. 1–6 (2007)
23. Padron, F.M.: Traffic-Congestion Detection Using VANET. Florida Atlantic University, Florida (2009)
24. Younes, M.B., Boukerche, A.: A performance evaluation of an efficient traffic-congestion detection protocol (ECODE) for intelligent transportation systems. Ad Hoc Netw. **24**, 317–336 (2015)
25. Beijing Municipal Commission of Transport. Interpretation of traffic performance index [EB/OL] (1 October 2014). http://www.bjjtw.gov.cn. (in Chinese)

Uncertainty-Based Sample Optimization Strategies for Large Forest Samples Set

Yan Guo[1]($\boxtimes$), Wenyi Liu[2], and Fujiang Liu[3]

[1] College of Computer Science, China University of Geosciences, Wuhan, Hubei, China
323110966@qq.com
[2] School of Remote Sensing and Information Engineering,
Wuhan University, Wuhan, Hubei, China
Lwy9306@163.com
[3] Faculty of Information Engineering, China University of Geosciences, Wuhan, Hubei, China
felixwuhan@163.com

Abstract. Our study was focused on the optimization of large training samples set selected from the global forest cover change detection system. Automatically delineating training samples procedure labeled tens of millions of samples representing forests and non-forests. To improve the precision, reduce the computational complexity and avoid over-fitting, we need to select samples from the large set of tens of millions of samples that are helpful for training a classifier. In this paper, two methods were used to optimize a large sample set from the Landsat-7 ETM+ data and obtain samples for training the classifier. The first method was the traditional stratified system sampling strategy. The second was uncertainty-based sample set optimization that selects training samples based on uncertainty by examining the uncertainty measure of samples and the distribution of their feature space, and involving the subtractive clustering, KNN and support vector machine. Through precision evaluation, our experiments validated that the uncertainty-based sampling strategy can achieve better results than the stratified system sampling strategy.

Keywords: Sample optimization · Uncertainty · Clustering · KNN · Remote-sensing image classification

1 Introduction

In the supervised classification of remote sensing images, in order to obtain a classification model, sample data have to be collected as training data for this model [1]. As training data are an important contributor to the precision of a supervised classification system for remote-sensing images, how to obtain a high-quality training samples set is a concern for the remote sensing community.

Traditional classification systems for remote sensing image obtain normally train data by manual marking. This method is both time and effort-consuming and expensive. Besides, manual interpretation is also highly subjective and easily leads to misjudgment. As such, the processing of global large remote sensing images has to involve prior

© Springer Science+Business Media Singapore 2016
K. Li et al. (Eds.): ISICA 2015, CCIS 575, pp. 519–530, 2016.
DOI: 10.1007/978-981-10-0356-1_55

knowledge automation to build a training data sample library. An example is the global forest cover change detection project by the team of Professor John Townshend and Chengquan Huang of the University of Maryland, which introduced the spectral knowledge of features into the computer interpretation algorithm, and worked out an automatic acquisition algorithm for training samples [2]. Using this algorithm, nearly "ten million" forest and non-forest samples can be automatically generated from one Landsat ETM+ image.

As the computational complexity of many classification models increases rapidly with the number of samples in a training set, if a large sample set is directly used to train a classification model, the computation will face problems such as the demand of a large memory, low execution speed, and even non-executability. Besides, concentration of large data could involve a lot of high-similarity data or what we call redundant data, which do not contribute much to the processing result of the data set, but increase the processing time [3]. The essential goal of sample selection is to reduce the storage demand and time consumption for the classification model while retaining its generalization performance.

The traditional way of sampling is the stratified system sampling strategy. However, experiments have discovered that the image classification effect of this strategy is yet to be further improved [3]. In 2010, Angiulli managed to reduce large data with FCNN and improved the classification speed of the vector machine. However, while the idea of FCNN is highly referential, the condensation ratio is very small.

Recently experts consider introducing active learning strategy to optimize samples. Optimization strategies for samples sets based on uncertainty performs well, which select the samples close to boundary to determine more precise boundary and increase generalization performance of the model.

2 Optimization Strategies for Sample Sets

2.1 Stratified-System-Sampling Strategy

Principle of Stratified-Systematic-Sampling. Simple random sampling, stratified random sampling and stratified systematic sampling are commonly used recently. Because of the random setting of the training samples in the feature space, simple random sampling would intensive distribute in some zones but sparse in another zones, result in samples that are not representative. Stratified-random-sampling and stratified-systematic-sampling can effectively avoid intensive distributing and reduce the loss of useful information, acquiring higher precision than simple random sampling generally when using the same number of the training samples. Particularly, if stratification variable can be used to stratify, stratified-random-sampling will surely acquire higher classification precision than simple-random-sampling and systematic sampling [4].

In stratified systematic sampling the population are divided into smaller subgroups called strata. Systematic sampling is then applied within each subgroup or stratum [5]. Due to the large number of labeled pixel had been acquired, among which cloud, water and shadow pixel were extracted entirely and should be masked, the remaining large labeled training set can be stratified into forest and non-forest based on the label of each pixel.

SVM Classifier. The training sample set used to training classification model has been ready, and then a classification model should be chosen. The SVM method outputs a hyperplane that has the largest distance to the nearest training data, whose optimization involves the minimization of a quadratic programming (QP) problem [6]. Many researches show SVM can get higher precision than decision tree algorithm and neural network algorithm, especially when used to binary classification. Considering that the Landsat ETM+ image masked cloud, water and shadow is required to classified into only two classes, forest and non-forest, the research use SVM as classification model.

SVM hyperplane described by WX + b = 0 lies midway between the bounding. Hyperplanes given by WX + b = −1 and WX + b = 1. The margin of separation between the two classes is given by 2/|W|. Support vectors are those training examples lying on the above two hyperplanes. The mathematical model of the theory is getting minimum of (2), when (1) can be satisfied.

$$y_i(W^T xi + b) \geq 1 - \xi \quad \xi_i \geq 0 \tag{1}$$

$$\frac{1}{2}W^T W + C \sum_{i=1}^{n} \xi_i \tag{2}$$

Where c is a penalty parameter and ξ_i are the slack variables.

By the use of the kernel trick, SVMs can learn a nonlinear decision function which is linear in a potentially high-dimensional feature space [6]. Kernel trick can prevent SVM from curse of dimensionality. Different kernel function lead to different SVM, the following four kernel functions, linear kernel function, polynomial kernel function, RBF kernel function, S kernel function, perform well.

The research uses LIBSVM toolbox. Parameter should be sat when using SNM. Firstly, we should determine using RBF kernel function, which performs well with only one parameter γ should be set. Another parameter should be set is c. c is a regularized parameter balancing the tradeoff between the margin and the error [7]. c and γ should be traversed from low to high regularly and each group of c and γ should make cross-verification to assess the precision. Finally, the group of c and γ with highest precision are the final parameter.

2.2 Uncertainty-Based Optimization Strategy for Training Samples

While random sampling is easy to perform and reduces the sampling time, the blindness nature of this method can lead to poor and unstable performance of the classification result.

In a large set of training samples, the contribution made by near-boundary samples to the training of the classifier is different from that made by their non-boundary counterparts at the class center. Boundary samples typically work on the classification precision and contribute greatly to the classifier, while the contribution of non-boundary samples at the class center to the classifier is mainly displayed in the early stage of the classifier [8]. As such, sample training has to involve an adequate number of training samples with typicality and uncertainty, before a classifier with good generalization performance can be trained.

To select some meaningful samples from a large set of samples with which to train the classifier, first, some samples with typicality were selected using subtractive clustering. Second, samples with uncertainty were selected according to the KNN rule. Finally, SVM was selected as the classifier and trained with all the samples selected to obtain an initial SVM classification plane, which was then refined with samples near the plane selected from a random small sample pool we built. The flowchart is given below (Fig. 1):

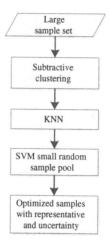

Fig. 1. The flowchart of uncertainty-based optimization strategy

Subtractive Clustering. Subtractive Clustering is a clustering algorithm based on density. We consider each point is a potential cluster center and define the density of the point to serve as a cluster center. The circle should be carried out until present density divided by first density is smaller than a setting value ε.

The steps of Subtractive Clustering are described as follows:

1. Computing the density of each point, getting density index:

$$\rho(v_i) = \sum_{k=1}^{n} e^{-\alpha * d(v_k, v_i)} \quad \alpha = \frac{2}{\tau_1} \tag{3}$$

2. The first cluster center v_1 is chosen as the point having the highest density. The density of v_1 denoted as ρ_1^*. Then the density of each point is revised as follow:

$$\rho_k(v_i) = \rho_{k-1}(v_i) - \rho_{k-1}^* e^{-\beta * d(v_{k-1}^*, v_i)} \quad \beta = \frac{1}{\tau_2} \tag{4}$$

3. The one with highest density is selected as the second cluster center circling until:

$$\frac{\rho_{k-1}^*}{\rho_1^*} < \varepsilon \tag{5}$$

τ_1 is a positive constant defining a neighborhood radius [9]. The higher τ_1 is, the larger range can be influenced, and the lower τ_1 is, the more classes will be gotten. The circle will end until present density divided by first density is smaller than a setting value ε. If ε is low, the algorithm will find the center in corner. On the contrary, if ε is high, it will remains large area cannot be clustered. ε is an important factor to compression ratio.

KNN. The idea underlying the KNN method is to assign new unclassified examples to the class to which the majority of its K nearest neighbors belongs [10]. This method determines which class samples belong to only depending on which class the nearest k samples most belong to, instead of distinguishing which class zones it in. So, the method is more adapt when it comes to the sample set with heavy overlap area.

The algorithm describes as follow. Firstly, setting parameter k represents the number of neighbors. Secondly, dividing sample set into Training phase in which samples have class label and Classification phase in which samples remains to be classified. Thirdly, we get the k neighbors through computing the distance. Finally, determining which class the sample belongs based on majority of its K nearest neighbors belongs [11]. The research introduces the concept of KNN and put forward the method to acquire the consistency subset form large training set.

For convenience, some concepts should be introduced in advance. T represent large sample set, S represent the consistency subset of it, p represent a sample point in S, q represent a sample point in T.

- $nn_k(p, T)$: The k samples nearest to p in T.
- $nns_k(p, T)$: The set containing all nearest samples of p in T.
- $NN_k(p, S)$: KNN rule which determines the sample belongs to the class that most k samples belong to.
- $l(p)$: The class p belonging to.
- $Vor(p, S, T)$: The set $\{q \in T | p = nn(q, S)\}$.
- $Voren_k(p, S, T)$: The set $\{q \in (Vor(p, S, T) - \{p\}) | l(p) \neq NN_k(q, S)\}$.

A subset S of T is said to be a k-training-set-consistent subset of T if for each $r \in (T\text{-}S)$, $l(r) = NN_k(r, S)$. Thus, this definition guarantees that the object T-S is correctly classified by S through the KNN rule [12].

The algorithm above can get the consistent subset S form large sample set T. The set $Voren_k(p, S, T)$ is composed of the point lying in the $Vor(p, S, T)$, which are misclassified by S through the KNN rule [13]. In other word, these samples have uncertainty, and we add them to consistency subset S to train the classifier.

Constructing Small Random Sample Pool. The sample set gotten through above to steps can be used to train the classifier, getting two parameters c and γ to construct the classifier, whose performance can satisfy the basic requirements. To increase the performance of classifier, a number of samples close to boundary, which can be judged by (6), should be added to train classifier.

$$\arg\min(|h_a(x_i) - h_b(x_i)|) \tag{6}$$

$h_a(x_i)$ and $h_b(x_i)$ mean probability of class A and class B [13].

It is a problem when selecting samples close to boundary that if the plane go through a group of intensive points, which are all close to plane and uncertain, selection of all these point would lead to redundancy and increase of computing complexity because of similarity to these point. To solve this problem, Sanjoy used stratified sampling method to select samples based on density [14]. The method can reduce intensive samples around boundary, but the computing complexity will be increased if taking density into consideration. The research introduces random small sample pool to save this problem. Using the random small sample pool, we find boundary samples among the small sample pool, saving the time and avoiding selecting a group of point with similar characteristic.

The research assumes that L samples are selected randomly in sample set $X_N = (x_1, x_2, x_3,\ldots,x_n)$ to construct X_L in each circle. The probability of xi in the small sample pool, p% nearest to the boundary, is $1-\eta$. There at least exist a sample x_i that p% nearest to the boundary whose probability is $1-(1-p\%)^L$ in X_L. So, Eq. (7) is established.

$$1 - (1 - p\%)^L = 1 - \eta \tag{7}$$

The number of random small sample pool XL can be reasoned from (7)

$$L = \log\eta / \log(1 - p\%) \tag{8}$$

The efficiency will be increased 4 to 6 times according to (8) [15].

3   Data Sets and Experimental Design

3.1   Data Sets

In this study, we used the multi-spectral remote sensing data acquired by the Landsat Enhanced Thematic Mapper Plus (ETM+) sensor. The study area is a part of Myanmar. A Landsat ETM+ image of 56815101 pixels covering the study area was selected (as shown in Fig. 2). The automatic sample marking system marked all the water bodies, clouds, shadows and backgrounds on the image, as well as 26200166 forest samples and 9508632 non-forest samples. Though the automatic sample marking system had marked approximately 80 % of the elements of the image, as there were pixels of unknown classes in the image, classification was still necessary [13]. The classification scheme included two types: (1) forest; and (2) non-forest. For efficiency concerns, we needed to select some meaningful training samples from the 30 million forest and non-forest samples marked. In our experiment, these two schemes were used to select valuable samples from the large set of samples.

Fig. 2. A Landsat ETM+ image of Myanmar in 2010 year

3.2 Stratified System Sampling Strategy

The classes of the image marked by the automatic marking system included backgrounds, water bodies, shadows, clouds, forests and non-forests. First, we integrally collected 8000 samples. According to the labels assigned by the automatic sample marking system, the samples collected were divided into a forest layer, a non-forest layer and a miscellaneous layer. 3726 training samples were obtained from the forest layer as illustrated by Fig. 3 (the green pixels represent forests; the red pixels are forest layer samples obtained by stratified system sampling; the black pixels represent the miscellaneous layer). 1360 samples were obtained from the non-forest layer as illustrated by

Fig. 3. Forest layer samples obtained by stratified system sampling (Color figure online)

Fig. 4 (the brown pixels represent non-forest; the green pixels are non-forest samples obtained by stratified system sampling; the black pixels represent the miscellaneous layer).

Fig. 4. Non-forest layer samples obtained by stratified system sampling (Color figure online)

SVM classifier was selected as our classifier model. 5086 forest and non-forest samples were inputted into the SVM classifier to train it. When c = 2048, γ = 0.5, the highest precision, 98.64 %, was obtained by cross validation. Hence, this group of parameters was used as the preset parameters for training SVM. After the SVM training model was yielded, the SVM model was inputted into the automatic classification system. The resultant classification of forest cover is illustrated by Fig. 5.

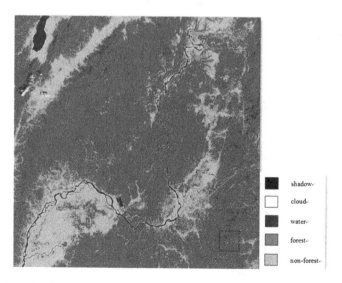

Fig. 5. The classification result of Landsat ETM+ image by stratified system sampling

3.3 Sampling Strategy Based on Uncertainty

We took four steps to perform uncertainty-based sampling.

First, we used subtractive clustering to select samples with typicality. We used the function subclust of the MATLAB, of which two parameters, τ_1 and ε, had to be set. As a rule of statistics, the statistical histogram of an image is in normal distribution, and the gray scale of 99.7 % of its pixels is within $[\mu - 3\sigma, \mu + 3\sigma]$. Now we assume that all the pixels are distributed uniformly in the characteristic space. As the class identifier is ignorable since it is infinitely small relative to the wavebands, the class identifier makes up a seven-dimensional feature space with the six wavebands. If each of these dimensions is equally divided into four parts, the characteristic space of the seven dimensions will be divided into 47 parts. Then the τ_1 of each waveband is 1.5σ. Consequently, parameter τ_1 was set to be [1000, 277, 347, 428, 806, 761, 742], and parameter ε was set to be [2, 0.15, 0.8]. Finally, the characteristic vector of the 13963 subtractive cluster centers was derived.

Second, we applied the KNN rule to select samples with uncertainty. We used the function knnclassify of MATLAB to mark the classes of the pixels first, and then selected the pixels whose classes marked by the function knnclassify were different from those marked by the automatic sample marking system. There were a total of 235069 pixels assumed to be highly uncertain. To reduce the number of samples and improve the work efficiency, we selected 35000 samples randomly from those high-uncertainty samples with inconsistent marks as the final samples with uncertainty.

Third, we built a random small sample pool to select boundary samples. We used SVM classifier as the classifier model, and inputted into the classifier the 13960 samples with typicality and 35000 samples with uncertainty resulting from the two steps above to train the SVM and obtain its classification plane. To work out a classifier with good generalization performance, we needed to add some samples near the SVM classification plane to refine the plane. We used the svm-predict of the libsvm toolkit to predict the classes, and selected samples to perform probability estimation prediction, i.e. the probability of this vector belonging to a certain class. The closer the probability is, the more uncertain the vector is and the closer it is to the class boundary. Samples with the probability of belonging to a class between 0.5 and 0.6 were assumed to be near-boundary ones and used to build a random small sample pool. The capacity of this pool was set to be 59, and one sample was selected from each 59 samples as the boundary sample.

After completing all steps of uncertainty-based sampling, the resultant final samples included samples with typicality, samples with uncertainty and boundary samples. These samples were used to train SVM as the training samples for SVM. The grid tool of LIBSVM was used to search parameters c and γ regularly in a descending manner. When $c = 512, \gamma = 0.5$, the highest precision, 98.82 %, was obtained by cross validation. Hence, this group of parameters was used as the preset parameters for training SVM. After the SVM training model was yielded, the SVM model was inputted into the automatic classification system. The resultant classification of forest cover is illustrated by Fig. 6.

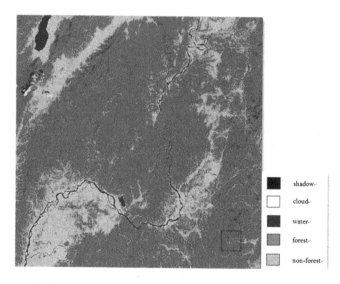

shadow·

cloud·

water·

forest·

non-forest·

Fig. 6. The result of remote-sensing image classification by sampling strategies based on uncertainty

4   Precision Evaluation

In our study, the quality of our optimization strategy for sample selection was evaluated from two aspects: the precision of the training samples, and the precision of the classifier. The former was validated with the function svm-predict of libsvm, which has the same essentials as the traditional precision evaluation. The latter was evaluated by 10-fold cross validation with the function svm-train of libsvm. The resultant precisions are given in Table 1.

Table 1. Precision of training sample and precision of classifier

	Precision of training sample	Precision of classifier
Stratified system sampling	96.82 %	79.53 %
Sampling strategy based on uncertainty	98.97 %	87.74 %

As can be observed from the result, when the classifier model was trained with optimized samples selected by uncertainty-based sampling strategy, the precision of the classifier was much higher.

Stratified system sampling strategy features simple process, high condensation ratio and fast computation. However, while this strategy ensures that adequate samples are obtained for each class, and the sample capacity is manually controllable, as the optimized samples obtained from each layer are highly random, the classified model trained with them does not provide high precision, making it difficult to distinguish elements on the classification boundary well. As the training of a classifier is more concerned about the element values in the characteristic space, while the characteristic space is not considered by stratified system sampling, there can be redundant information or blank areas in the characteristic space. All these could directly affect the training of the classifier model.

Uncertainty-based sample optimization strategy, on the other side, selects samples with typicality in the characteristic space as well as samples with high uncertainty, i.e. samples within the isolation zone in the vicinity of the classification plane, by taking into account both the sample distribution in the characteristic space and the uncertainty of samples, which effectively avoids information redundancy or "overlearning", refines the classification boundary, and increases the generalization capacity of the classifier so that each class has the largest coverage and no overlap. However, as this method is quite complex with high condensation ratio and it is hardly possible to control the sample capacity manually.

5 Conclusions

Using the large set of remote sensing training samples obtained by the automatic sample marking module of the global forest cover change detection system from the Landsat ETM+ remote sensing data, a sample selection strategy based on sample uncertainty and distribution was investigated. The merits of the two methods were summarized by examining the precision of the training samples and the classifier with a training sample set obtained by stratified system sampling and one by uncertainty-based sample optimization.

As we did not increase the number of samples when building the random small sample pool, in the next step, we will add boundary samples for further precision evaluation so as to examine the relationship between the number of samples and the precision when more samples are involved and eventually identify the optimum number of samples.

References

1. Donmez, P., Carbonell, J.G., Bennett, P.N.: Dual strategy active learning. In: Kok, J.N., Koronacki, J., Lopez de Mantaras, R., Matwin, S., Mladenič, D., Skowron, A. (eds.) ECML 2007. LNCS (LNAI), vol. 4701, pp. 116–127. Springer, Heidelberg (2007)
2. Huang, C., Song, K., Kim, S., et al.: Use of a dark object concept and support vector machines to automate forest cover change analysis. Remote Sens. Environ. **112**, 970–985 (2008)
3. Li, Y., Wang, Q., Li, Z.: The research for the influence of the classification precision caused by sampling method. J. Huhai Insts. Technol. (Nat. Sci. Ed.) **S1**, 67–69 (2011)
4. Zhang, H.: Study on reliable classification method based on remotely sensed image. China University of Mining and Technology (2012)
5. Ye, Y., Wu, Q.: Stratified sampling for feature subspace selection in random forests for high dimensional data. Pattern Recogn. **46**(3), 769–787 (2013)

6. Xie, X., Sun, S.: Multitask centroid twin support vector machines. Neurocomputing **149**, 1085–1091 (2014)
7. Yang, X., Tan, L., He, L.: A robust least squares support vector machine for regression and classification with noise. Neurocomputing **140**, 41–52 (2014)
8. Dasgupta, S., Hsu, D.: Hierarchical sampling for active learning. In: International Conference on Machine Learning, pp. 208–215 (2008)
9. Chen, J.Y., Qin, Z., Jia, J.: A weighted subtractive clustering algorithm. Inf. Technol. J. **7**, 356–360 (2008)
10. Saini, I., Singh, D., Khosla, A.: QRS detection using k-nearest neighbor algorithm (KNN) and evaluation on standard ECG databases. J. Adv. Res. **7**(4), 331–344 (2013)
11. Anggiuli, F.: Fast nearest neighbor condensation for large data sets classification. IEEE Trans. Knowl. Data Eng. **19**(11), 1450–1464 (2007)
12. Bay, S.: Nearest neighbor classification from multiple feature sets. Intell. Data Anal. **3**, 191–209 (1999)
13. Hu, Z., Gao, W.: The algorithm of sampling optimizing based on FCNN and SVM nearest to boundary rule. J. Yanshan Univ. **5**, 421–425 (2010)
14. Huang, C., Thomas, N., Gaward, S.N.: Automated masking of cloud and cloud shadow for forest change analysis using Landsat images. Int. J. Remote Sens. **31**(20), 5449–5464 (2010)
15. Zhang, L., Guo, J.: A method for the selection of training samples based on boundary samples. J. Beijing Univ. Post Telecommun. **4**, 77–80 (2006)

A PCB Short Circuit Locating Scheme Based on Near Field Magnet Specific Point Detecting

Shuqiang Huang[1](✉), Jielin Zeng[2], Hongchun Zhou[1], Zhusong Liu[3], and Yuyu Zhou[1]

[1] Network and Education Technology Center of Jinan University,
Guangzhou 510632, China
hsq@jnu.edu.cn
[2] Guangdong Planning and Designing Institute of Telecommunications Co.,
Ltd., Guangzhou 510632, China
[3] Faculty of Computer Science and Technology of Guangdong University
of Technology, Guangzhou 510006, China

Abstract. The existing print circuit board (PCB) short circuit fault detection schemes can only detect the presence of a short circuit network, but not the specific location of the short circuit. To solve this problem, we have determined a scheme based on near field magnet specific point detecting. First, we test the existence of a short circuit network using the short circuit detection algorithm. Second, a high frequency voltage is applied to excite the detected short circuit network, and the arranged near-field magnetic sensors above the PCB detect the peak value of the magnetic field. This technique can be utilized to precisely position the short circuit point on the PCB. The simulation results show that the scheme is operational and can effectively locate the short circuit point position.

Keywords: PCB · Short circuit · Near field magnet detecting · Impendence mismatch · Hall sensor

1 Introduction

PCB fault testing and verification technology faces enormous challenges. Bateson pointed out that the possibility of the presence of a fault in a circuit that is shorted maybe greater than 50 % [1], therefore, to detect errors and locate the short circuit point quickly on PCB is particularly important. The most mature industrial non-destructive testing method to test PCB standard is known as the ICT (In circuit testing), atypical application of this method is a flying-probe short circuit tester with utilizing relatively mature detection algorithms to test the short circuit fault [2–5], by inserting the external excitation signal to the test points and find out the electric connections of wires. PCB circuit point positioning can be theoretically divided into two steps: detecting the existence of a short circuited network and determining the precise location of the short circuit point(s). In this paper, we propose a PCB short circuit location scheme based on near field magnet specific point detection, which compared to existing methods of short circuit detection, is both operational and effective.

© Springer Science+Business Media Singapore 2016
K. Li et al. (Eds.): ISICA 2015, CCIS 575, pp. 531–540, 2016.
DOI: 10.1007/978-981-10-0356-1_56

2 The Overall Scheme and Short Circuit Detection Algorithm

Currently, there are few short circuit fault detection schemes based a magnetic field to focus the accurate location of a short point on a PCB. Socheatra and Soeung [6, 7] and his team proposed to use the eddy current principle to position a short circuit point. However, this scheme does not specifically consider the complex interconnected structures that may also lead to an impedance mismatch on other non-short points. These structures may lead to distrusted detection results. In order to improve the detection accuracy, this paper presents a complete PCB circuit point positioning solution. The scheme can be theoretically divided into two steps: detecting the existence of a short circuited network and determining the precise location of the short circuit point(s). The flowchart of this scheme is shown as Fig. 1:

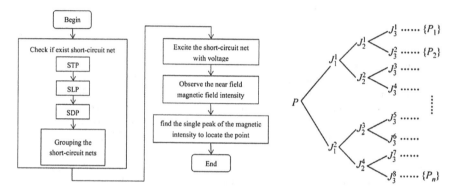

Fig. 1. Flowchart of the overall scheme   **Fig. 2.** Schematic of set partitioning

Figure 1 shows a short circuit network which exists on a PCB can be determined by using a flying-probe short circuit tester, consider a PCB with N-networks includes unknown short circuit network(s), beforehand, set it as k short networks. A short circuit detection algorithm proposed by Leung [5] is as follows: the main steps include short testing procedure (STP), short localization procedure (SLP) and short diagnosis procedure (SDP). If P is the networks set on a PCB, this algorithm first makes an average separation of P by dividing it into two disjoint subsets J1 1 and J2 1. Next, further separate J1 1 into J1 2 and J2 2, J2 1into J4 2 and J4 2 using the same method. The subdivisions are ended when a subset only contains one network or an empty element. As shown in Fig. 2, for each division, the number of networks is reduced by half of the original set. After $\log_2 N$ times of division, only one circuit network may contained in every subset.

Consider giving a voltage impulse into a split network, while the other networks remain inactive. If there is a short among the active network and at least one of the others, an impulse signal can be detected by a short circuit detector as response in these shorted networks. We define a function as Fun (X,Y) and the input parameters X, Y are sets of the networks, and $X \cup Y = P$, $X \cap Y = \varnothing$. If the network X receives an impulse, Y remains inactive, and Fun (X,Y) has no response, then either X and Y are considered as independent from each other or they are shorted. If there is no response, then Fun (X,Y) is defined as output "0", otherwise the output is defined as "1". The short circuit detecting algorithm can be indicated as follows:

STP:

Step 1. Define p = 1;

Step 2. If Fun $\left(H_L^p = \bigcup_{k=1}^{2^{p-1}} J_{2k-1}^p, \ H_R^p = \bigcup_{k=1}^{2^{p-1}} J_{2k}^p \right) = 0$, and p < log$_2$N, then p = p+1;

Otherwise there exists a short between networks, stop.

SLP:

Step 1. Define $L^0 = H_L^p, \ R^0 = H_R^p$, k = 1;
Step 2. If Fun (L^k = the first half of L^{k-1}, R^k = the first half of R^{k-1}) = 0, execute step 3, otherwise, stop;
Step 3. If Fun (L^k = the second half of L^{k-1}, the second half of $R^k = R^{k-1}$) = 0, then k = k+1, otherwise stop, repeat until k = p.

SDP:

Step 1. Define $L^1 = J_{2q-1}^p, \ R^1 = G_{2q}^p$, k = 1;
Step 2. If Fun (L^{k+1} = the first half of L^k, R^{k+1} = the first half of R^k) = 0, execute step 3, otherwise, stop;
Step 3. If Fun (L^{k+1} = the first of, R^{k+1} = the second half of R^k) = 0, then k = k+1, otherwise stop, repeat until k = p.

Through the above three stages, the short circuit networks can be detected.

3   A PCB Short Circuit Locating Scheme Based on Near Field Magnet Specific Point Detecting

To determine the specific location of the short point on the PCB, this paper proposes a scheme based on magnetic field detection. If a high-frequency excitation is added to a differential wire on the PCB, it can be analyzed using the transmission line model. Consider a transmission line using the load-end model shown in Fig. 3:

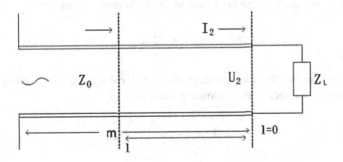

Fig. 3. Transmission line with load-end model

The reflection coefficient is defined as the ratio of reflected voltage to the incident voltage ratio that is 1 from the load, we hold:

$$\Gamma(l) = \frac{U_{re}(l)}{U_{in}(l)} = \frac{U_2 - I_2 Z_0}{U_2 + I_2 Z_0} e^{-j2\beta z} = \Gamma_2 e^{-j2\beta z} \tag{1}$$

Where

$$\Gamma_2 = \frac{U_2 - I_2 Z_0}{U_2 + I_2 Z_0} = \frac{Z_L I_2 - I_2 Z_0}{Z_L I_2 + I_2 Z_0} = \frac{Z_L - Z_0}{Z_L + Z_0} = |\Gamma_2| e^{-j\phi_2} \tag{2}$$

$Z_0 = \sqrt{\frac{R_0 + j\omega L_0}{G_0 + j\omega C_0}}$ is the characteristic impedance and ϕ_2 is the phase difference between the incident signal and the reflected signal (Fig. 4).

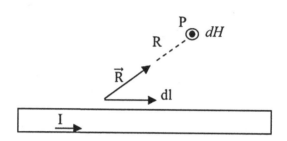

Fig. 4. Single-wire model of the magnetic field intensity

With a constant current in the wire, the magnetic field intensity at a point P outside the wire is in accordance with the following formula:

$$H = \frac{I}{4\pi} \int_l \frac{dl \times \vec{R}}{R^2} \tag{3}$$

We hold that if the wires on the PCB were given a fixed voltage excitation and this was added to the basis of a high-frequency voltage, the voltage on the wire incident certain micro transmission line conforms to the following formula:

$$U_{in} = R_0 i + L_0 \frac{\partial i}{\partial t} + U_{stat} \tag{4}$$

The U_{stat} describes the fixed voltage. Reflected voltage (only reflecting the high frequency part) is subject to the following formula:

$$U_{re} = \Gamma(R_0 i + L_0 \frac{\partial i}{\partial t}) + U_{stat} \tag{5}$$

The magnetic field intensity over the two voltage lead wires in a space outside of a point P can be described as follows:

$$H_{in} = \frac{1}{4\pi} \int_l \frac{U_{in} * dl \times \vec{R}}{Z_0 R^2} = \frac{1}{4\pi} \int_l \frac{(R_0 i + L_0 \frac{\partial i}{\partial t} + U_{stat}) * dl \times \vec{R}}{Z_0 R^2} \qquad (6)$$

$$H_{re} = \frac{1}{4\pi} \int_l \frac{U_{re} * dl \times \vec{R}}{Z_0 R^2} = \frac{1}{4\pi} \int_l \frac{\Gamma(R_0 i + L_0 \frac{\partial i}{\partial t} + U_{stat}) * dl \times \vec{R}}{Z_0 R^2} \qquad (7)$$

The magnetic field intensity is generated at this point:

$$H = H_{in} + H_{re} = \frac{(1 + \Gamma)}{4\pi} \int_l \frac{(R_0 i + L_0 \frac{\partial i}{\partial t} + U_{stat}) * dl \times \vec{R}}{Z_0 R^2} \qquad (8)$$

It can be seen that, under the same conditions, a transmission line with a larger reflection coefficient Γ will produce a stronger radiation in space. Short-circuit point on the PCB obviously holds mismatch impedance, compared to the non-short circuit position on the circuit. After the above analysis, this point has a greater reflection coefficient and magnetic field intensity; it can be used to determine the specific location of the short circuit on the PCB.

4 Eddy Current Testing Method and Near-Field Magnetic Field Detection Method Simulation Results

Before setup a simulation model, first we give an actual PCB model as Fig. 5 shown:

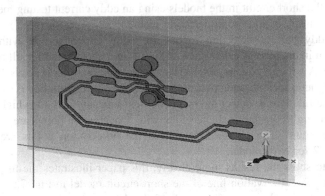

Fig. 5. Actual PCB model

According to the Fig. 5, we can conclude these basic elements of a PCB: parallel wires, width-mutation wires and direction changing wires (corner). Then, we can setup simplified models of a PCB. In order to compare the superiority between the near-field magnetic field detection method and the eddy current effect detecting method which has been proposed in the literature [7], a comparative test model was established in Fig. 8 using the CST simulation environment:

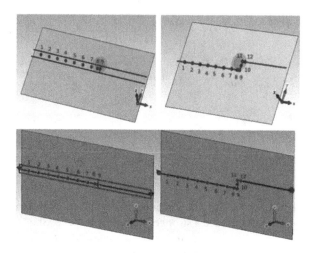

Fig. 6. Comparative test model

In Fig. 6, two test models were established. The two models on the left side simulate a shorted situation in a 50 mm*30 mm area, illustrating the magnetic field intensity when these two parallel wires short for each other. The length of these two wires is 50 mm, with a width and thickness of 300 μm and 32 um, respectively. The short circuit point is at X = 30 mm, and the short circuit point disposed in this model is a 300 μm wide and 32 μm thick small piece of wire. The two models on the right side simulate in a same area but with a right corner angle on the wire at X = 30 mm. This basis detects the short circuit in the models using an eddy current testing method and a magnetic field near singular point detection method.

In the eddy current detecting model, an eddy current probe with shape in 1.5 mm*5 mm in diameter is established at X = 30 mm above the angle. It contains the current of 0.1 A with a frequency of 1 kHz, and the number of coil ties is 1000. In the near-field magnetic specific point detecting model, the observation line is established at 1.5 mm above the wire to detect the magnetic field intensity and adds high-frequency signals continuing to the terminal end of the wire. Observation points are set as shown in the above figure. The obtained data from the observation points are shown in Tables 1 and 2 for each model:

In order to show the results more clearly, this paper illustrates the magnetic field intensity along the observation line of the short circuit model in Fig. 7.:

According to the above simulation, the following results can be observed: in the short circuit model, with the use of an eddy current detection method, the peak value (the observation point 9) compared to the average value of the normal points (observation point 1- observation point 7), demonstrates only a 0.1 % difference; however, with the use of the near-field magnetic specific point detection method, its peak value (observation point 8) compared to the average value of the normal points (observation

Table 1. Short circuit model simulation data

Observing point number	Eddy current testing method (Vrms)	Near field magnet specific point detecting method (A/m)
1	1.9680	6.0721
2	1.9680	5.8444
3	1.9680	5.8222
4	1.9680	5.8193
5	1.9680	5.8222
6	1.9680	5.8431
7	1.9693	6.0431
8	1.9705	7.6296
9	1.9714	2.2264
Average of point 1–7	1.9682	5.8952

Table 2. Corner model simulation data

Observing point number	Eddy current testing method (Vrms)	Near field magnet specific point detecting method (A/m)
1	1.9703	4.3050
2	1.9702	3.8844
3	1.9702	3.8094
4	1.9702	3.7863
5	1.9702	3.7778
6	1.9702	3.7765
7	1.9702	3.8335
8	1.9708	5.3161
9	1.9714	3.1679
10	1.9714	2.9003
11	1.9713	3.7232
12	1.9712	4.8430
Average of point 1–7	1.9702	3.8818

point 1 - observation point 7) demonstrates a 29.4 % enhancement. Similarly, in the corner model, with the use of the eddy current detection method, its peak value (observation point 9) compared to the average value of the normal points (observation point 1 - observation point 7), demonstrates an increase of only 1.6 %, whereas if the near-field magnetic field detection method is applied, then compared to its peak value (observation point 8 and observation point 12) and the average value of the normal points (observation point 1 - observation point 7), shows a 36.9 % and a 24.8 % increase, respectively.

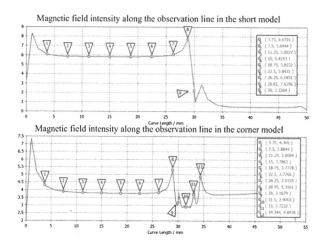

Fig. 7. Magnetic field intensity along the observation line in two models

It can also be noted that the difference of the peak value is small between the short model and the corner model with an eddy current detection method; it is hard to distinguish between these two cases, and may lead to a mistaking judge because of the slight thinness of wires on PCB. In contrast, using a magnetic near-field detection method, a short circuit and other impendence mismatch points can be identified by observing the difference between the values of the near-field magnetic field intensity, thereby improving the success rate of the detection.

5   Near-Field Magnetic Specific Point Detection Method Simulation Results

In practical applications, a short circuit network on the PCB generally appears in the adjacent networks. We established a short-circuit model with a right angle corner in a wire on PCB, as is shown in Fig. 8:

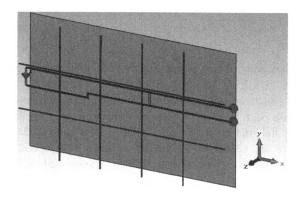

Fig. 8. Parallel wires shorted model with right angle corner

As shows in Fig. 8, this model simulates the magnetic field generated by two parallel wires that occurs a short-circuit in a 50 mm*30 mm area. These two wires have a length of 50 mm, width of 300 μm, and thickness of 32 μm. The short circuit point is at the position X = 30 mm, and the short circuit point disposed in this model is 300 μm wide with the small pieces of wire having a thickness of 32 μm; in order to simulate the corner of mutations, one of the wires offset 1000 μm in Y direction at 15 mm (X = 15 mm) from the beginning.

Then we set two horizontal magnetic fields' observing lines 3 mm above the wires, spacing 10 mm parallel to the X axis, and four vertical observing lines 3 mm above the wires, spacing 10 mm parallel to the Y axis either. The horizontal lines were named as horizontal line1 - horizontal line2 from the bottom upward and the vertical lines are named vertical line1-line4 from the left to right direction. The simulation results obtained from this model are shown in Fig. 9:

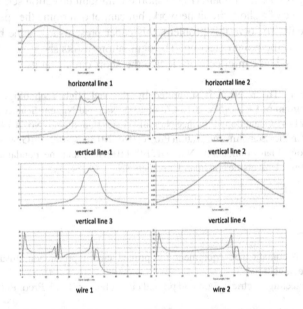

Fig. 9. Magnetic field intensity of short model with right angle corner

Figure 9 shows the simulation results for the horizontal and vertical magnetic intensities determined from the observation lines. Because of the offset at 15 mm in front of another wire, the corresponding position of wire 1 and wire 2 in Fig. 9 should be postponed 1000 μm from the abscissa. As shown in figure, after the addition of the corner point mutation, the peak value appears in the vicinity of a right angle corner, and the values from the other observation position are consistent with the previous line. The reason is that at a right angle corner, the wire is cut off suddenly in the direction of travel, resulting in a short circuit similar to that leading to an impedance mismatch. The right angle corner generates standing waves that are relatively large, greater in magnitude than the actual standing wave ratio at the short circuit point.

6 Summary

From the above discussion and simulation results, we obtained the following conclusions: When detecting the magnetic field intensity along a wire, rapid decay will appear after the short circuit points with a maximum value occurs above the short-circuit point. The impedance mismatch caused by a right corner leads to a greater impact on the results than a short circuit point; the magnetic field intensity along the wire will meet a greater value, compare to a short one, and return to the steady value after a corner point rather a meet a fast decay.

Based on the conclusions drawn above, we figure out a practical scheme to locate a short-circuit point: after judging the existence of a short-circuit network, a magnetic field intensity detection device that detects a specific point (such as a Hall sensor) can be used to obtain the position of a short point.

The existing print circuit board (PCB) short circuit fault detection schemes can only detect the presence of a short circuit network, but cannot determine the specific location of the short circuit. To solve this problem, this paper presents a scheme based on near field magnetic specific point detecting. The simulation results show that the scheme of operation is strong and can effectively locate the short circuit faults.

Acknowledgement. The research is supported by National High Technology Research and Development Program of China (No. 2013AA040404), National Natural Science Foundation of China (No. 61572144), Guangdong Province Natural Science Foundation (No. 2014A030313386), Guangdong Universities Scientific Innovation Project (No. 2013KJCX0018), Guangdong Province Science Technology Plan Foundation (No. 2012B091100161), and the Fundamental Research Funds for the Central Universities of Jinan University (No. 21615439).

References

1. Batenson, J.: Automatic test equipment for printed circuit production. Insulation/Circuits pp. 40–76 (November 1982)
2. Fang, S.C.: Detecting electrical shorts on printed circuit boards. Int. J. Prod. Res. **28**(6), 1031–1037 (1990)
3. Leung, Y.-W.: A parallel algorithm for locating short circuits on printed circuit boards. IEEE Trans. Instrum. Meas. **42**(3), 746–751 (1993)
4. Leung, Y.-W.: An algorithm for locating short circuits among N signal paths using K-port parallel short detector. IEEE Trans. Instrum. Meas. **43**(6), 918–921 (1994)
5. Leung, Y.-W.: A signal path grouping algorithm for fast detection of short circuits on printed circuit boards. IEEE Trans. Instrum. Meas. **43**(1), 80–85 (1994)
6. Socheatra, S., Ali, N.B.Z., Khir, M.H.M.: CST simulation of magnetic field for PCB fault investigation. In: 2012 4th International Conference on Intelligent and Advanced Systems, pp. 407–411 (2012)
7. Soeung, S., Ali, N.B.Z., Khir, M.H.M.: 3D electromagnetic simulation of interconnect fault inspection based on magnetic field behavior. In: RSM 2013 Proceedings of the 2013, Langkawi, Malaysia, pp. 387–390 (2013)

Prosodic Features Based Text-dependent Speaker Recognition with Short Utterance

Jianwu Zhang[1], Jianchao He[1(✉)], Zhendong Wu[1], and Ping Li[2]

[1] School of Communication Engineering,
Hangzhou Dianzi University, Hangzhou, China
gaarahe@163.com, wzd@hdu.edu.cn
[2] School of Mathematics and Computational Science,
Sun Yat-sen University, Guangzhou, China
liping26@mail.sysu.edu.cn

Abstract. Over the past several years, Gaussian mixtures models have been the dominant approach for modeling in text-independent speaker recognition field. But the recognition accuracy for these models declines when utterances' length becomes short. Presently Mel-frequency cepstral coefficients are generally used to characterize the properties of the vocal tract and widely applied in speech recognition. In addition, prosodic features, such as pitch and formant, are generally considered to describe the glottal characteristics. However, the efficiency of those approaches remain unsatisfactory. In text-dependent short utterances speaker verification systems, prosodic features can assist to improve the recognition result theoretically. In order to optimize the performance of speaker verification systems under the framework of adapted GMM-UBM, we adopt a variant speaker verification system based on prosodic features, in which a dual-judgment-mechanism is used in order to integrate vocal tract features with prosodic features. Experimental results showed that the new speech recognition system led a better consequence.

Keywords: Speaker verification · Text dependent · Prosodic features · Dual judgment mechanism

1 Introduction

As one of the most natural biometric identification methods, speaker recognition has great potential in the field of convergent key [1, 2], ordinary digital signatures, biometric key [3], and so on. Speaker recognition technology [4], aiming to recognize the speaker identities automatically, is becoming more and more attractive. In the meantime, Short utterance speaker recognition (SUSR) has been hotspot.

GMM-UBM and GMM-SVM [5, 6], based on clustering and subspace, are two popular speaker recognition methods. In systems based on such structures, [7] illustrates the performance change with different valid test utterance lengths on the NIST SRE 2005 database, where it can be seen that the Equal Error Rate increases sharply when the test utterances become shorter.

© Springer Science+Business Media Singapore 2016
K. Li et al. (Eds.): ISICA 2015, CCIS 575, pp. 541–552, 2016.
DOI: 10.1007/978-981-10-0356-1_57

In order to solve the problem of large data requirements, research has lead to Joint Factor Analysis (JFA), Support Vector Machine (SVM) and i-vector based technologies. The factor analysis subspace estimation and the i-vector method introduced in [8, 9] decrease the number of redundant model parameters to develop more accurate speaker models. Some methods try to improve the performance by selecting segments with higher discriminability on speaker characteristics. In other works performing short utterance speaker recognition, such as [10], dimension decoupled GMM is applied. Training and testing with 10 s of speech on variations of GMM and SVM have been researched. Since SUSR is a complicated task, it has also been used in combination with video aid for better results. The study suggests that phoneme level speaker recognition can be improved when using larger amount of training data [11]. A study of formant contours at consonant-vowel boundary shows that a speaker can be identified by using the consonant to vowel transition information. The speaker information at the transition was named as the "speaker-style". Other than [12, 13], most of the researches make use of utterances no less than 2 s.

Speaker verification systems for security purposes typically employ a text-dependent approach where enrolled users utter either a specific pass-phrase or a string of prompted digits. The phonetic constraint is imposed by requiring the tester to pronounce certain pass-phrases, generated during the authentication. Text-dependent speaker verification offers several advantages. First by constraining the phonetic content of the test utterances, text-dependent speaker verification reaches higher accuracy than its text-independent counterpart; especially in dealing with short-duration utterances [14]. Secondly, in cases where by the users are assigned or allowed to choose their personalized pass phrases, security is reinforced as both the pass phrase and the voice of the user have to match in order to be authenticate. Recently, JFA [15, 16] and i-vectors have been used for text-dependent verification and showed in [17, 18] that can take advantage of the phonetic constraint imposed on short duration utterances.

For the past few years, the researches about feature extraction based on the vocal tract characteristics, such as MFCC and LPCC, stepped into a bottleneck. The investigators had turned their attention to immature prosodic features. There are many kinds of prosodic features, such as pitch, frame energy, formant etc. Prosodic features perform better in both noise and time robustness than short-time vocal tract features generally. In spite of that nevertheless, prosodic features have their own weakness in complex extraction algorithm, maldistribution, untoward tracking etc. Pitch, which reflects periodical vibrate between the vocal cords with air, is an important characteristic parameter of sound source in phonetic pronunciation model. A formant is a concentration of acoustic energy around a particular frequency in the speech wave. In our paper, formant refers to the broad peak in the spectral envelope. Therefore, how to fuse short-time channel characteristic parameters and prosodic feature parameters is a topic worth discussing.

In light of these researches, we strive to find the unconspicuous usefulness of prosodic features in speaker recognition using short utterance under the framework of baseline GMM-UBM. A dual-judgment-mechanism is adopted to handle the relationship between short-time channel features and prosodic features. In this mechanism, short utterances will undergo a second sentence while the initial score from adapted GMM-UBM system located in dual-judgment-area. Prosodic features will be used as new parameters instead of MFCCs. How to confirm dual-judgment-area will be discussed in later paper. In second time, gender verification will first be executed recur to mean pitch. Then, we will used DTW algorithm to compute the distance of formant trajectories line-by-line between the tested utterance and the hypothesized template utterance. The results of pitch and detection have a contribution to generate a new score. The threshold will conformed through DET curves by amended scores. The experiment results showed that our text-dependent speaker verification based on prosodic features helped to improve performance under dual-judgment-mechanism.

This paper is organized as follows. The baseline adapted GMM-UBM system and two kinds of prosodic features are described in Sects. 2 and 3 respectively. The new speaker verification based on prosodic features under dual-judgment-area engine is given in Sect. 4 while the results and analysis are presented in Sect. 5. Perspective of this work are discussed in Sect. 6.

2   Preliminary

Speaker Recognition process is composed of three steps namely acoustic feature extraction, statistical modeling and score calculation. We have experimented with GMM-MAP-UBM frame using feature vectors comprised of 24-dimension MFCCs. After training the UBM, the special speaker model parameters are usually estimated through Expectation Maximization algorithm (EM) [19]. This algorithm iteratively refines the GMM parameters to monotonically increase the likelihood of the estimated model for the observed feature vectors, i.e., for iterations k and $k + 1$, $p(X \mid \lambda^{(k+1)}) > p(X \mid \lambda^k)$. The whole process of parameters estimation is updating latest model parameters $\lambda*$ iteratively until the convergence. Define $Q(\lambda, \lambda*)$, according to *Jensen* inequality, the parameters estimation is equal to maximization $Q(\lambda, \lambda*)$.

$$Q(\lambda, \lambda*) = \sum_{k=1}^{m} \sum_{i=1}^{M} \frac{\omega_i p(x^k \mid \lambda, i)}{p(x^k \mid \lambda)} \cdot \left[\log \omega_i^* + \log p(x^k \mid \lambda*, i) \right] \tag{1}$$

Where $\omega_i\, p(x^k \mid \lambda, i) = p(x^k, i \mid \lambda)$. i.e. let $\partial Q(\lambda, \lambda*)/\partial \omega_i^* = 0$, then we will get the best mean estimate. Similarly, take its partial respect to weight and covariance respectively and we will get the updating formulas.

$$b_i(x^k) = \frac{\omega_i}{(2\pi)^{D/2} |\Sigma_i|^{1/2}} \exp(-\frac{1}{2}(x^k - \mu_i)' \Sigma_i^{-1}(x^k - \mu_i)) \tag{2}$$

$$\frac{\partial Q(\lambda, \lambda*)}{\partial \theta} = \nabla_\theta \sum_{k=1}^{m} \sum_{i=1}^{D} \phi_i^k \cdot \log \frac{b_i(x^k)}{\phi_i^k} \tag{3}$$

Where $\theta \in \{\omega_i, \mu_i, \Sigma_i\}$ is inherent parameter of the model λ, $i = 1, 2, \dots, D$ is the number of Gaussian component, $k = 1, 2, \dots, m$ is the number of features. In E-step, speaker model λ is random initialization. And ϕ_i is a posteriori probability computed by sample data primordially and updated iteratively in M-step. The process of using maximum a posteriori (MAP) criterion to fit prediction model is optimization problem, is defined as follows:

$$\eta^* = \arg\max_{\eta} p(\eta|X,\lambda_u) = \arg\max_{\eta} p(X|\eta,\lambda_u)p(\eta) \tag{4}$$

Like the EM algorithm, the adaption is a two step estimation process. The first step is identical to the expectation step, where estimates of the sufficient statistics, such as n_i, $E_i(x)$, $E_i(xx')$ and a posteriori probability $p(i \mid x^k)$, of the training data are computed for each component in the UBM. Unlike the second step of EM algorithm, for adaptation these new sufficient statistic estimates are then combined with the old sufficient statistics from the UBM mixture parameters using a data-dependent mixing coefficient. The adaption equations are as follows:

$$\begin{aligned}
\omega_i^* &= [\alpha_i^\omega n_i/m + (1 - \alpha_i^\omega)\omega_i] \cdot \gamma \\
\mu_i^* &= \alpha_i^m E_i(x) + (1 - \alpha_i^m) \cdot \mu_i \\
(\sigma_i^*)^2 &= \alpha_i^v E_i(xx') + (1 - \alpha_i^v)(\sigma_i^2 + \mu_i\mu_i') - (\mu_i^*)^2
\end{aligned} \tag{5}$$

We derive the hypothesized speaker model by adapting the parameters of UBM using target speaker's speeches and a form of Bayesian adaptation. The adaptation coefficients controlling the balance between old and new estimates are $\{\alpha_i^\omega, \alpha_i^m, \alpha_i^v\}$ for the weights, means and variances, respectively. The best overall performance is from adapting only the mean vectors. In adapted GMM-UBM system, we use a single adaptation coefficient for all parameters ($\alpha_i^\omega = \alpha_i^m = \alpha_i^v = n_i/(n_i + r)$) with a relevance factor of $r = 14\sim16$.

In first stage, we chose 24-dim Mel Frequency Cepstrum Coefficients as acoustics features used baseline UBM-MAP-GMM. The complete MFCCs are composed of primary 12-dim MFCCs and their 12-dim first-differences. Because of mute frames against prosodic feature detection, we will filter them after framing. The frames whose energy less than one percent average energy of the utterance were regard as mute frames. Considering the limitation of training utterances, speaker special model can but only model the phoneme information appeared in training utterances that failed to describe the phonemes firstly appeared in testing utterances. In allusion to solve the problem above, adapted GMM-UBM was proposed. The time-normalization outputs are as follows:

$$\Lambda_0(X) = \frac{1}{m} \cdot \sum_{k=1}^{m} [\log p(x^k|\lambda) - \log p(x^k|\lambda_{ubm})] \tag{6}$$

3   Prosodic Features

It has been proven that the fusion of different features would be useful to improve the recognition performance in many research fields. The simplest way of feature combination is to concatenate all the feature vectors into a bigger one for each frame followed by dimensionality reduction. But these different features are not orthogonal, so practically redundant information should be removed. There are two aspects: one is to de-correlate the concatenated feature vectors into individual ones from multiple feature streams; the other one is to eliminate the coefficients with redundant and unimportant information.

3.1 Pitch Detection

Pitch [20] is one of the most important parameters of speech signal, which describes an important characteristic of voice source. The ups and downs of the pitch trajectories are loaded with discriminability information. The pitch distributions of male and female are quite different. Pitch can be extracted from the voiced sound with strong noise resistance. We will use modified autocorrelation method (AUTOC) to compute the pitch. The detection consequences are shown in Fig. 1.

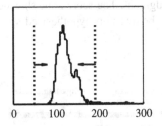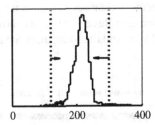

| 0 | 100 | 200 | 300 | 0 | 200 | 400 |

Fig. 1. The mean pitch histogram of male (left) and female (right) of testing corpora

3.2 Formant Detection

To some voice researchers, the formant refers to a peak in the spectral envelope (a property of the sound of the voice), to others it refers to a resonance of the vocal tract (a physical property of the tract), while to a third group it refers to the pole in a mathematical filter model (a property of a model). We extracted four trajectories showed in Fig. 2 from utterance by LPC algorithm according to first definition. Dynamic time warping (DTW) is an classical algorithm that based on dynamic programming for measuring similarity between two sequences, which may vary in time or speed during the course of an observation. The minimum cumulative distance and the optimal path are calculated through back tracking algorithm. The constraint that the ratio of compared lengths must be between 0.5 and 2.0 to make sense.

The formant comparison results are follows:

$$\varphi(S, S^h) = \frac{1}{4n} \sum_{i=1}^{4} f(s_i, s_i^h) \tag{7}$$

Where $n = \min(n_1, n_2)$ is the length of formant, $f(s_i, \mathbf{s}_i^h)$ is DTW calculations distance for i th trajectory (not normalized), the DTW algorithm is extended to deal with four trajectories in the same time. The output can be used to describe the similarity of the formant trajectories. $\varphi(S, S^h)$ is a real number will be adopted to produce a new score.

4   Dual Judgment Mechanism

Repeated observation and comparison have been done to experimental results. We found that the short utterances from hypothesized speaker have much higher scores than threshold in most times, while fewer lower than threshold but close to it. We also found that the utterances from imposters speaker have much lower scores than threshold for the most part, while fewer higher than threshold but close to it. In other words, the errors occurred easily when the scores tight distributed around the threshold. Unter dieser Bedingung, we consider that the MFCCs as acoustic features decreased their ability to distinguish in short utterances when the score located near the threshold. So we adopted a dual-judgment-mechanism to reduce misjudgment. The score zone showed in Fig. 2 will be divided into three sections as shown below: non-hypothesized speaker zone, hypothesized zone and dual-judgment-area.

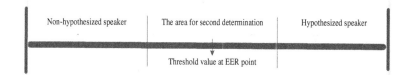

Fig. 2. The regional division for dual-judgment-mechanism

Though prosodic features contain the useful discriminability information as well, many experiments confirmed that they are unsuitable as unique or independent features. According to our observation, the amount of utterance fall into dual-judgment-area is limited and controllable. So it is feasible theoretically to extract prosodic information as our features to distinguish different people. While the score falls into the dual-judgment-area, the MFCCs are failed to describe discriminability information in special speaker. Therefore, we considered that new feature parameters to describe short utterance in different angles. According to the above analysis, the prosodic features describe the information of glottis and perform better in both noise and time robustness than short-time vocal tract features. From this perspective, we think prosodic features will help to improve recognition results that showed in Fig. 3.

In consideration of the fact that pitches exist in voiced sounds, and failed to simply concatenate pitch to MFCCs. It's a statistical fact that female's pitch mean is higher than male's obviously. The pitch will be extracted and used to gender verification on this account in our algorithm. If the gender verification failed, the score will turn to a much lower place. If the gender verification succeed or in fuzzy region, the score will turned slightly. The formant trajectories extracted from short utterances will be compared with corresponding template through extended DTW algorithm. The model scores based on prosodic features under dual-judgment-area engine are modified as follows:

$$\Lambda(X) = \Lambda_0(X) + \gamma_1 \cdot \phi(p) + \gamma_2 \cdot \varphi(S, S^h) \tag{8}$$

Where $\Lambda_0(X)$ is an old score of baseline UBM-MAP-GMM, and we choose MFCCs as our acoustic features. γ_1 and γ_2 represent the credibility and given empirically. And p is the mean pitch extracted from the test utterance. We compared the old score and the system threshold in the first decision stage. We would compute the new score while the output in first-stage failed. In seconded stage, the old score $\Lambda_0(X)$ is renewed by the new score $\Lambda(X)$. The likelihood ratio used in either case.

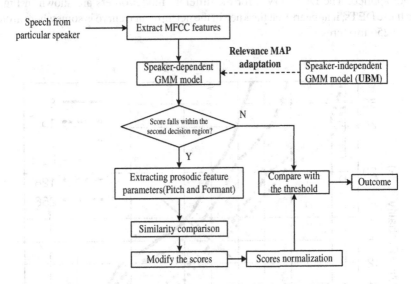

Fig. 3. Logical procedure diagram of dual-judgment-mechanism

5 Experiments

5.1 The DET Curves of Classic Adapted GMM-UBM System

In this section, we present experiments and results using our classic adapted GMM-UBM. Experiments are conducted on a specific corpus including 50 people (34 males and 16 females). Each tester read 25 different short Chinese words at a normal speed. Each word would be recorded ten times. Each short utterance lasts 2~3 s. This corpus is designed for text-depend speaker verification task. For each speaker, the training data consist of two minutes long utterances. 50 independent trials by 50 different hypothesized speakers are conducted independently in each test file. The sample rate is 16 kHz, mono, 16 bit quantization. The recording environment are mostly classrooms and laboratories. The recording tools contain mobile phone and PC recorder. Training data for UBM has half an hour's males' utterances and half an hour's females' utterances. In our case, a single high order UBM was trained by pooling all the training data together roughly. And the GMM comes from UBM by adapting its parameters. Therefore, both GMM and UBM have the same mixture number. We used system order to describe the mixtures.

A single, speaker independent, threshold is swept over the two sets of scores and the probability of miss (FR) and probability of false alarm (FA) are computed for each threshold. The error probabilities are then plotted as detection error tradeoff (DET) curves to show system performance. The next set of experiments examined the effect of model size on performance. Using the same training data as in the previous experiment and pooling the data, UBMs of sizes 8–256 were trained and evaluated. Again, HNORM was not applied. The DET curves for the different model orders are shown in Fig. 4. From these DETs, it appears that the knee in the performance curve is somewhere around 128 and 256 mixtures.

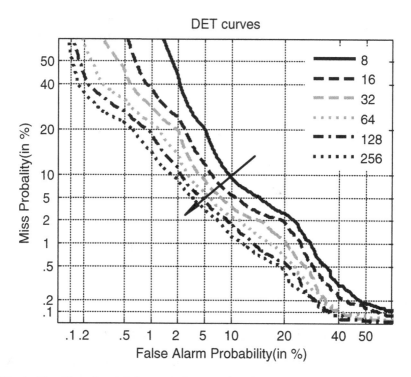

Fig. 4. The effect of model size on performance based on baseline adapted GMM-UBM

5.2 The Determination of Dual-Judgment-Area

In first set of experiments examined the effect of model size on performance. In this section, the prosodic features and dual-judgment-mechanism were unused. Figure 5 shows DET curves in detail. In the next set of experiments demonstrated the chosen of the dual-judgment-area. The whole possible outcomes of judgment in adapted GMM-UBM system were displayed in Table 1.

Table 1. All possible situations to test outcomes

Determination results / Voice from	The hypothesized speaker	The non-hypothesized speaker
Is hypothesized speaker	Right to Accept (Arr)	Wrong to Accept (Awr)
Is non-hypothesized speaker	Wrong to Refuse (Arw)	Right to Refuse (Aww)

For example, when the system order is 128, the dual-judgment-area we choose is [−2.360 4.130], and the threshold is 0.40 (normalized). The consequences are showed in Fig. 5. Actually, we can obtain it through DET curves.

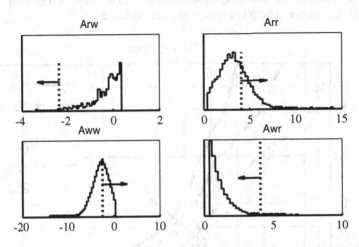

Fig. 5. The choose of dual-judgment-area (e.g. system order is 128, threshold is 0.40, dual-judgment-area is [−2.360 4.130])

5.3 The Extraction of Prosodic Features

The third part of experiments examined the mean pitch distribution for both male and female used training utterances. The mean pitch will be used for gender verification. We used modified autocorrelation method (AUTOC) to extract the pitch. We concluded that the female's overall mean pitch higher than man's in Fig. 1.

Table 2. The results of pitch detection

Style	Pitch/Hz
Threshold	162.2
Uncertainty interval	144.7~166.8

The mean pitch can distinguish men and women obviously while sometimes failed in uncertainty interval. In other words, cases where the mean pitch located in uncertainty interval are less significant. Then we extracted the formants from testing utterances and the hypothesized speakers' utterances (as templates) one by one. And count the distance (non-Euclidean distance) through DTW algorithm. The extensional DTW was taken to compute the four formant trajectories simultaneously which is similar to contour comparison (Table 2).

5.4   The DET Curves of New System Under Dual-Judgment-Mechanism

In the last set of experiments examined the effect of model size on performance based on prosodic features under dual-judgment-mechanism. The results in Fig. 6 showed that the revised algorithm can help to improve the performance.

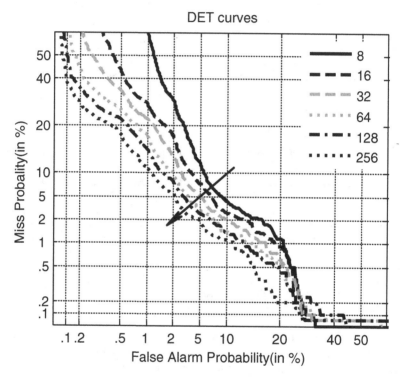

Fig. 6. The effect of model size on performance based on prosodic features under dual-judgment-mechanism

In the first set of experiments displayed the effect of model order on system performance. It appears that the best equal error rate occurred when the mixtures are 128 and 256. The EER are 4.80 and 4.23 respectively. In the next set of experiments we determined the dual-judgment-area through DET curves when the system order various from 8 to 256. In the third part, we obtained the mean pitch distribution through large amount short utterances supplied by different gender and age. And determine the distribution range and decision threshold for both male and female. We also extracted the formant trajectories from test utterances. Compute the cumulative distance between the tester's and hypothesized speaker's short utterances through DTW algorithm. The distance divided by shorter length will be normalized. Both mean pitch and distance after the normalization will change into similarity and used to compute the new score empirically. In the last set of experiments examined the effect of model size on performance based on prosodic features under dual-judgment-mechanism. DET appears that the knee in the performance curve is somewhere around 128 and 256 mixtures. The EER are 3.87 and 3.11 respectively which are promoted than before. The results are showed in Tables 3 and 4.

Table 3. The threshold and dual-judgment-area

System order	8	16	32
Threshold	0.37	0.41	0.49
Dual-judgment-area	−2.43~5.15	−2.30~4.74	−2.70~4.99
New threshold	0.40	0.42	0.48
System order	64	128	256
Threshold	0.46	0.40	0.26
Dual-judgment-area	−2.60~4.40	−2.36~4.13	−2.29~3.95
New threshold	0.50	0.39	0.31

Table 4. The EER (%) before and after improvement

System order	8	16	32
MFCCs	9.7611	7.7370	6.5036
Prosodic features	7.0814	6.0718	5.0391
System order	64	128	256
MFCCs	5.4654	4.7965	4.2266
Prosodic features	4.0917	3.8679	3.1071

6 Acknowledgements and Futures Directions

This research was supported by Zhejiang Science Fund (NO. LY16F020016) and Zhejiang Province Science and Technology Innovation Program under Grant Number (2013TD03). For future work, we will find a more suitable way to fusion the MFCCs and the prosodic features rather than a rough linear combination [21]. And the potential prosodic features needs to be further tapped. And we can experiment on the text-independent short utterances.

References

1. Jin, L., Xiaofeng, C., Mingqiang, L., et al.: Secure deduplication with efficient and reliable convergent key management. IEEE Trans. Parallel Distrib. Syst. **25**(6), 1615–1625 (2014)
2. Jin, L., Yatkit, L., Xiaofeng, C., et al.: A hybrid cloud approach for secure authorized deduplication. IEEE Trans. Parallel Distrib. Syst. **26**(5), 1206–1216 (2015)
3. Zhendong, W., Bin, L., et al.: High dimension space projection-based biometric encryption for fingerprint with fuzzy minutia. Soft Comput. (2015, in Press). doi:10.1007/s00500-015-1778-2
4. Campbell, J.P.: Speaker recognition: a tutorial. Proc. IEEE **85**, 1437–1462 (1997)
5. Reynolds, D.A., Quatieri, T., Dunn, R.: Speaker verification using adapted gaussian mixture models. Digital Signal Process. **10**, 19–41 (2000)
6. Reynolds, D.A.: Channel robust speaker verification via feature mapping. In: ICASSP, pp. 53–56 (2003)
7. Vogt, R., Sridharan, S., Michael, M.: Making confident speaker verification decisions with minimal speech. IEEE Trans. ASLP **18**(6), 1182–1192 (2010)
8. Kenny, P., Boulianne, G., Dumouchel, P.: Eigenvoice modeling with sparse training data. IEEE Trans. Speech Audio Process. **13**(3), 345–354 (2005)
9. Dehak, N., Dehak, R., Glass, J., Reynolds, D., Kenny, P.: Cosine similarity scoring without score normalization techniques. In: Proceedings of Odyssey 2010 - The Speaker and Language Recognition Workshop (2010)
10. Nosratighods, M., Ambikairajah, E., Epps, J., Carey, M.J.: A segment selection technique for speaker verification. Speech Commun. **52**(9), 753–761 (2010)
11. Fattah, M.A.: Phoneme based speaker modeling to improve speaker recognition. Information **9**(1), 135–147 (2010)
12. Davis, S.B., Mermelstein, P.: Comparison of parametric representation for monosyllabic word recognition in continuously spoken sentences. IEEE Trans. ASLP **28**(4), 357–366 (1980)
13. Chow, D., Abdulla, W.H.: Robust speaker identification based perceptual log area ratio and Gaussian mixture models. In: INTERSPEECH (2004)
14. Matthieu, H.: Text-Dependent Speaker Recognition. Springer, Heidelberg (2008)
15. Vogt, R.J., Lustri, C.J., Sridharan, S.: Factor analysis modelling for speaker verification with short utterances. In: Odyssey Speaker and Language Recognition Workshop. IEEE (2008)
16. Vogt, R., Baker, B., Sridharan, S.: Factor analysis subspace estimation for speaker verification with short utterances. In: INTERSPEECH 2008, pp. 853–856 (2008)
17. Kanagasundaram, A., Vogt, R., Dean, D., Sridharan, S., Mason, M.: I-vector based speaker recognition on short utterances. In: Annual Conference of the International Speech Communication Association (2011)
18. Larcher, A., Bousquet, P.M., Lee, K.A., Matrouf, D., et al.: I-vectors in the context of phonetically-constrained short utterances for speaker verification. In: IEEE International Conference on Acoustics, Speech, and Signal Processing, ICASSP (2012)
19. Bilmes, J.A.: A gentle tutorial of the EM algorithm and its application to parameter estimation for Gaussian mixture and hidden Markov models. Int. Comput. Sci. Inst. **4**, 126 (1998)
20. Rabiner, L., Cheng, M., Rosenberg, A.E., McGonegal, C.: A comparative performance study of several pitch detection algorithms. IEEE Trans. Acoust. Speech Signal Process. **24**(5), 399–418 (1976)
21. Zhendong, W., Jie, Y., Jianwu, Z., Huaxin, H.: A hierarchical face recognition algorithm based on humanoid nonlinear least-squares computation. J. Ambient Intell. Humanized Comput. (2015, in Press). doi:10.1007/s12652-015-0321-8

User Oriented Semi-automatic Method of Constructing Domain Ontology

Chao Qu[1,2], Fagui Liu[2(✉)], Hui Yu[3], Ruifen Yuan[1],
and Anxiong Wang[1]

[1] School of Computer, Dongguan University of Technology, Dongguan, China
chaos7@126.com
[2] School of Computer Science and Engineering, South China University
of Technology, Guangzhou, China
fgliu@scut.edu.cn
[3] Information Centre, Dongguan Power Supply Bureau,
Guangdong Power Grid Co., Dongguan, China
hy0769@21cn.com

Abstract. Based on the analysis of the existing main ontology construction methods and tools, we proposed a user-oriented method of semi-automatic domain ontology construction. The method needs to build a descriptor set according to the user's requirements, and establish a hierarchy structure based on it. Users can construct ontologies by using this method in the case without the participation of experts. And the method still has advantages for expansion, collaborative development of ontologies.

Keywords: Domain ontology · Ontology construction · Semi-automatic · User oriented

1 Introduction

In the information field, ontology is defined as a formal explicit specification of a shared conceptualization [1]. As a highly abstract concept model, ontology has been widely applied to various information systems [2]. In practices, consider a specific problem or a project, there are variety methods of constructing domain ontology. There is not a standard ontology construction method so far, but during the process of building ontology, researchers reached a consensus that the ontology construction is an engineering problem that requires a set of scientific criteria to be guided. Drawing on the basis of previous experience, researchers summed up some useful guidelines for guiding ontology build. The most influential criteria were proposed by Gruber in 1995. Clarity, coherence, extendibility, minimal encoding bias and minimal ontological commitment [3]. In fact, it is difficult to satisfy all criteria and often requires trade-offs in the course. For example, clarity and coherence criteria demand limit the definition of the term has multiple interpretations, while the minimal ontological commitment meant the ontology requires the ability to accommodate a variety of possible models. In cloud computing environment many algorithms have been used [4] and the ontology will be widely used also.

© Springer Science+Business Media Singapore 2016
K. Li et al. (Eds.): ISICA 2015, CCIS 575, pp. 553–561, 2016.
DOI: 10.1007/978-981-10-0356-1_58

2   Related Work

Ontology construction methods can be divided into three categories: artificial, semi-automatic and automatic method. The artificial method needs the participation of experts, concepts and their relationships are established manually. The common methods include: IDEF5 (Integrated DEFinition Methods) proposed by KBSI (Knowledge Based System, Inc) [5], Skeletal Methodology proposed by Mike Uschold and Michael Gruninger [6], TOVE from enterprise integration laboratory of University of Toronto [7] and METHONTOLOGY which combined Skeletal Methodology and TOVE [8]. The semi-automatic ontology construction methods usually implement simple subtasks automatically but the remaining work still need experts to complete. The representative methods are Cyclic Acquisition Process proposed by Alexander Maedche [9], Five steps cycle [10], Theme Extraction and Text Mining [11]. The automatic method has gradually been in a hot research for the reason of effectiveness. Besides Text2onto and ACORN [12], there are still many methods such as text pattern matching [13], clustering [14] and machine learning [15].

Various organizations have developed a variety of ontology building tools to edit, view and maintain ontologies, such as Ontosaurus, Ontolingua, WebOnto, WebODE, OntoEdit, OLEd and Protégé etc.

3   User Oriented Method of Constructing Domain Ontology

It is difficult to customize the ontology to meet the user's needs with the existing ontology construction methods, so we proposed the concept of description set and designed an User Oriented Semi-automatic Method of Constructing Domain Ontology (UOSM) based on it. Users can use it to build ontology according their needs or to expand the existing ontology.

3.1   Description Set

Maedche et al. defined ontology structure as a 5-tuple $O := \{C, R, H^c, rel, A^0\}$, where C is a concept set, each element in C is a concept; R is a relationship set and each element in R represents a kind of relationship. H^c is a concept structure or classification structure $H^c \subseteq C \times C$ is a kind of directed relationship. $H^c(C_1, C_2)$ means C_1 is a sub-concept of C_2. $rel: R \rightarrow C \times C$ is a function to represent non-classified relations between concepts. A^0 is an axioms described by some kind of logic language. Based on the above definition, we expand the ontology structure and made a new definition as follow:

- **Definition 1.** Ontology Structure is a 5-tuple $O := \{C, R, H^c, rel, A^0\}$, where C is a concept set, each element in C is a concept; R is a relationship set and each element in R represents a kind of relationship. H^c is a concept structure or classification structure, $H^c \subseteq C \times C$ is a kind of partial order. $H^c(C_1, C_2)$ means C_1 is a sub-concept of C_2. Concepts in C constitute different levels $\{C^0, C^1, C^2, \ldots, C^n\}$ by category relationship. It means the concepts in level C^{i+1} at least have one relationship $H^c(C^{i+1}, C^i)$ with concepts in the level C^i and have no relationship with higher level. *rel: R → C × C* is a function to represent non-classified relations between concepts. A^0 is an axioms described by some kind of logic language.

Introduction of ontology classification hierarchy between the concepts makes the tree structure constructed by partial order more obvious, the performances of its advantages are as follow. First, it converts the complex graph-like structure of relationship classification into a hierarchical tree structure and conductive to the use of more efficient retrieval algorithms. Secondly, the hierarchy in favor of a measure of the distance between concepts and facilitate the realization of ontology mapping and ontology reasoning. Third, the hierarchy directly reflects the degree of refinement of concepts, as well as for the expansion and evaluation of ontologies.

- **Definition 2.** Top concepts are concepts in layer C^0 which are the smallest unit of concepts and cannot be described by other concepts. Top concepts can convert to non-top concepts during the ontology construction.

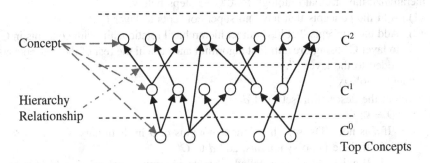

Fig. 1. Ontology structure

Based on the definitions above, the ontology structure is shown in Fig. 1. Besides, in order to achieve semi-automatic ontology construction, we need some knowledge base for the vocabulary retrieval and identification.

- **Definition 3.** *Dictionary* is a knowledge base for vocabulary retrieval. It can identify vocabularies for providing their part of speech and their synonyms and antonyms. For multiple POS vocabularies, it can also discriminate them.
- **Definition 4.** *Description set* is a word-pair <*r,w*> set that the word-pairs are used to describe a word *d*. The pair mean there is a relationship *r* between the word *d* and *w*. The relationship *r* is called Description Prefix (Dp) and the word *w* is called Description Suffix (Ds). A description set can be represented as *d* = {<*r₁,w₁*>, <*r₂, w₂*>, <*r₃,w₃*>,...,<*rₙ,wₙ*>}. The description set can be extracted from ontology.
- **Definition 5.** Specially, we defined a pair of reciprocal relationship {*sof,pof*} to construct the hierarchy. If the description set of *d* contains a word-pair <*sof,w*>, it means *d* is a sub-concept of *w*. If the description set of *d* contains a word-pair <*pof, w*>, it means *d* is a super-concept of *w*.

By Definition 4, a description set can be broken down into multiple sub-described sets. Users can define their own description set to meet their requirements by determining the concepts and the relationships between concepts. This process is entirely oriented to user needs and it can be achieved without the participation of experts.

3.2   Ontology Construction Algorithm

The main idea of construct ontology is to establish the concept hierarchy (classification relationship) and non-classified relations. Therefore, this section introduces a kind of ontology construction algorithm based on equivalence relation evaluation by dictionary. The main idea of the algorithm is to use the dictionary as a means of matching concepts and vocabulary. Extend the ontology according to the match degree of word and its description set with the concepts already in the ontology. The construction algorithm is list below:

1. Initialize ontology
   If there is no original ontology, the 5-tuple $\{R, Hc, Rel, A^0\}$ are all empty. Only create C^0 and C^1 in concept set C and set them empty. Otherwise we should hierarchize the original ontology follow the steps below:
   (1) Add the concepts that have no super-concepts to layer C^0.
   (2) Add the concepts that have direct hierarchical relationship with concepts in C^0 to layer C^1 using breadth-first algorithm and so will the rest of the concepts are added to the layer C^i.
2. Expand ontology
   (1) Input the description set D of d
   (2) If $D = \emptyset$:
       (a) If d is not a Ds, search for the synonyms of d in dictionary
           i. If there is no synonyms, add d to C^0
           ii. If exist a synonyms called c, create an equivalence relationship between d and c
       (b) If d is a Ds of word d'(d' is a word not a concept, because of has not been added to the ontology) and r is the Dp of pair $<r,d>$, search for the synonyms of d in dictionary:
           i. If there is no synonyms

              If $r \neq \{pof, sof\}$, add d to C^0 as a top concept and create a r relationship between d' and d.
              If $r = pof$, create a super-concept relationship between d' and d, $level(d) = level(d') + 1$.
              If $r = sof$, add d to C^0 as a top concept and create a sub-concept relationship between d' and d,

              $$level(d') = max\{level(d'), level(d) + 1\}$$

ii. If exist a synonyms called c,create an equivalence relationship between d and c and:

If $r \neq \{pof, sof\}$, create a r relationship between d' and c.
If $r = sof$, create a *sub-concept* relationship between d' and c.

$$level(d') = max\{level(d'), level(c) + 1\}$$

If $r = pof$, If c has no super-concept then add d' to C^o, else:Suppose that the super-concept set of c is $Fc = \{c_1, c_2, c_3, \ldots c_n\}$, (for c_i is sub-concept of c_{i+1}, and the description set of non-hierarchical relationships of sub class contains the parent's), compare the non-hierarchical relationship sets of c_i and d' ($i = n..1$):
If $D_{d'} \supseteq D_{c_i}$ and $D_{d'} \subset D_{c_{i-1}}$, then add d' as sub-concept of c_{i-1}, else add d' to C^o as a new concept.

iii. If there is a sub-concept or super-concept relations between d and d' detecting whether the relationship transferred from another path. If exists, delete the direct hierarchy relationship of d and d'.

(3) If $D \neq \emptyset$

  (a) Relationship set $R = R \cup \{r_1, r_2, r_3, \ldots r_n\}$.

  (b) Search for the synonyms of d in concept set C by dictionary and the synonyms of w_i (Ds of d):

    i. If there is no synonym of d, add d to C^o.

    ii. If exists, create an equivalence relationship between d and c

(4) Take each Ds of d (w_i) as a new input and go to step (1).

3.3 Comparison and Analysis

There is no mature theory of guiding for building ontology for now. The current method of constructing ontology is proposed for specific projects, which led to the emergence of various ontology construction methods. Compared with the main existing methods of constructing ontology, the features of the semi-automatic method of constructing domain ontology we proposed are user oriented and the description set. The comparisons of our method with others are shown in Table 1.

It can be seen from the table that the artificial construction methods usually need the participation of experts and a long development cycle, but the quality of ontology is high and the ontology expansion still needs experts' help. Semi-automatic construction methods need experts also and the productions have a general quality, but the problems of long development cycle and hard to expand still exist. The automatic methods extract ontology from date by algorithm model, so it is not need the participation of experts but the quality of ontology commonly unsatisfactory. In contract, the proposed method of constructing domain ontology is not involved experts, shorter the development cycle, the quality meet the customer needs and expand easily.

Table 1. Comparison of methods of constructing ontology

Methods	Build mode	Experts	Scalability	Build cycle	Flexibility	Ontology quality	Specific method
IDEF5	Artificial	Yes	Hard	Long	Low	High	NO
Skeletal methodology	Artificial	Yes	Hard	Long	Low	High	NO
TOVE	Artificial	Yes	Hard	Long	Low	High	NO
METHONTOLOGY	Artificial	Yes	Hard	Long	Low	General	NO
Cyclic acquisition	Semi-auto	Yes	Hard	General	Low	General	NO
Five steps cycle	Semi-auto	Yes	Hard	General	Low	General	NO
Text2onto	Auto	No	Hard	Data depend	Low	Low	Yes
ACORN	Auto	No	Hard	Data depend	Low	Low	Yes
UOSM	Semi-auto	No	Easy	short	High	user oriented	Yes

4 Experiments

In this section, we use the ontology of "Pizza" one demo of Protégé as test object and take some concepts and relationships as input of our algorithm. We initialized the ontology with the first input:

```
Pizza={<hasBase,PizzaBase>,<hasTopping,PizzaTopping>}
```

The reault is shown in Fig. 2. It can be seen that according to the algorithm, "Pizza", "PizzaBase" and "PizzaTopping" are added to C^0 as basic concepts. Relationships such as "hasBase" and "hasTopping" are added to the ontology, too.

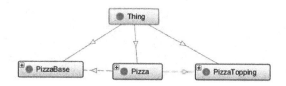

Fig. 2. The testing of concepts and relationships

Next, we expanded the ontology with the inputs as follow (Fig. 3):

```
PizzaBase={<pof, DeepPanBase>,<pof ThinAndCrispyBase>}
PizzaTopping={<pof,cheeseTopping>,<pof,VegetableTopping>}
HamTopping={<sof,PizzaTopping>}
PepperoinTopping={<sof,PizzaTopping>}
AmericannaPizza={<sof,Pizza>}
```

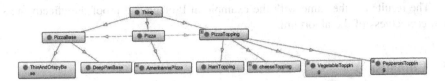

Fig. 3. The testing of extended concepts and relationships

It can be seen that the relationships "pof" and "sof" work well in constructing ontology hierarchy. The test result of changing the existing hierarchy of ontology with the follow inputs is shown in Fig. 4. The result shows the effectiveness of changing the ontology structure.

```
NamedPizza={<sof,Pizza>,<pof,AmericannaPizza>}
MeetTopping={<sof,PizzaTopping>,<pof,PepperoniTopping>
,<pof,HamTopping>}
```

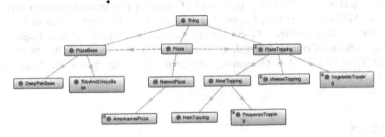

Fig. 4. The testing ontology of modified hierarchy

With the same function, we added "MozzarellaTopping" and "ParmezanTopping" as sub-classes of "cheeseTopping", as well as added "TomatoTopping" and "OnionTopping" as the sub-classes of "VegetableTopping". According to the example of "Pizza" we added an attribute of "AmericannaPizza" with the follow input and the final result is shown in Fig. 5.

```
AmericannaPizza={< hasTopping, MozzarellaTopping >,
< hasTopping, TomatoTopping >,< hasTopping,
PepperoinTopping >}
```

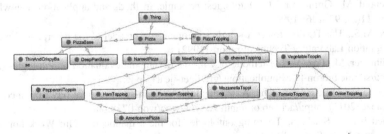

Fig. 5. The final testing ontology

The results are the same with the example in Protégé and proof the effectiveness and correctness of the algorithm.

5 Conclusion

The user oriented semi-automatic method of constructing domain ontology is based on the distinction of vocabularies by retrieving the dictionary. It establishes the hierarchy and no-hierarchical relationships between concepts with description sets. The proposed algorithm constructs the structure of ontology to meet the user's needs without involving experts. The method also applies to domain ontology integration and extension. However, the method does not provide a way for setting attributes of relationships.

Acknowledgements. This paper is supported by the 3th strategic rising industry program of Guangdong Province, (Project No. 2012556003); International Cooperation Special Program for platform, (Project No. 2012J510018); the key lab of cloud computing and big data in Guangzhou (Project No. SITGZ[2013]268-6); Engineering & Technology Research Center of Guangdong Province for Big Data Intelligent Processing (Project No. GDDST[2013]1589); Engineering & Technology Research Center of Guangdong Province for Big Data Intelligent Processing (Project No. GDDST[2013]1513-1-11). Guangdong Universities Scientific Innovation Project (Project No. 2013KJCX0177).

References

1. Borst, P., Akkermans, H.: An ontology approach to product disassembly. In: Plaza, E., Benjamins, R. (eds.) Knowledge Acquisition, Modeling and Management. LNCS, vol. 1319, pp. 33–48. Springer, Heidelberg (1997)
2. Chao, Q., Fagui, L., Ming T.: Ontologies for the transactions on IoT. Int. J. Distrib. Sens. Netw. **2015**, 1–12 (2015)
3. Gruber, T.R.: Towards principles for the design of ontologies used for knowledge sharing. Int. J. Hum. Comput. Stud. **43**(5–6), 907–928 (1995)
4. Jin, L., Qian, W., Cong, W., Ning, C., Kui, R., Wenjing, L.: Fuzzy keyword search over encrypted data in cloud computing. In: Proceedings of the 29th IEEE International Conference on Computer Communications (INFOCOM 2010), pp. 441–445. IEEE (2010)
5. Benjamin, P.C., Menzel, C.P., Mayer, R.J., et al.: IDEF5 method report[R/OL] (2010). http://www.idef.com/pdf/Idef5.pdf
6. Uschold, M., Gruninger, M.: Ontologies: principle, methods and application. Knowl. Eng. Rev. **11**(2), 93–136 (1996)
7. Fox, M.S.: The TOVE project: towards a common-sensemodel of the enterprise. Enterprise Integration Laboratory Technical report (1992)
8. Grüninger, M., Fox, M.S.: Methodology for the design and evaluation of ontologies (2011). http://stl.mie.utoron-to.ca/publications/gruniger-ijcai95.pdf
9. Kietz, J.U., Volz, R., Maedche, A.: Extracting a domain-specific ontol-ogy from a corporate intranet (2011). http://aclweb.org/anthology/w/woo/woo-0738.pdf
10. Maedcbe, A., Staab, S.: Learning ontologies for the semantic web. In: Workshop on the Semantic Web (SemWeb) (2001)

11. Guangqing, T.: Research on Knowledge Organization Based on Concept Lattice of Digital Library. Jilin University, Changchun (2012)
12. Ming, X.: Linked Data and Knowledge Representation Automatic Semantic Annotation. Wuhan University, Wuhan (2012)
13. Degeratu, M., Hatzivassiloglou, V.: Building automatically a business registration ontology. In: Proceedings of the 2nd National Conference on Digital Government Research (2002)
14. Khan, L., Wang, L.: Automatic ontology derivation using clustering for image classi canon. In: Proceedings of the 8th International Workshop on Multimedia Information Systems, IWMIS (2002)
15. Dong, W.: The Research of Chinese Ontology Learning Based on Web Mining. Taiyuan University of Technology, Taiyuan (2007)

Research and Implementation for Rural Medical Information Extraction Method

Yutong Gao[1], Feifan Song[2,3(✉)], Xiaqing Xie[4], Shengnan Geng[3], and Wenling Tang[3,5]

[1] Department of Computer Science and Information Engineering,
National Taipei University of Technology, Taibei, China
2293243834@qq.com

[2] School of Economics and Management, Beijing University of Posts and Telecommunications,
Beijing, China
songfeifanhkl@outlook.com

[3] Key Laboratory of Trustworthy Distributed Computing and Service (BUPT),
Ministry of Education, Beijing, China
15210528096@126.com, tangwl@bupt.edu.cn

[4] Beijing University of Posts and Telecommunications Library, Beijing, China
xiexiaqing@bupt.edu.cn

[5] School of Computer Science, Beijing University of Posts and Telecommunications,
Beijing, China

Abstract. Currently, the input of rural clinic system in china is the symptom from patients' descriptions. However, the existing problems in patients' descriptions are irregular and colloquial description, too much irrelevant information, etc. We need to use information extraction technology to extract more useful information as the input information of the following procedure for better matching between symptom and illness and enhancing the accuracy of the clinic. We designed an information extraction method for rural clinic using open source tools. Based on the machine learning methods, we extract time and degree information was extracted from the patients' descriptions. We designed and implement parallelization of the algorithms to speed up the response.

Keywords: Rural medical · Information extraction · Naïve bayes parallelization

1 Introduction

Information extraction technology helps us to find what we need among massive data. Based on the extracting and reorganization the disordered information, we obtain the structured information for further process. In a medical clinic system, patients' descriptions are regarded as the input information. But patients' descriptions contain irregular and colloquial description along with too much irrelevant information, so that it is unable to treat them as the input information for further clinic. Using the information extraction technology, we get the key information such as the degree and time of the disease from the patients' descriptions, and enhance the matching between symptom and illness as well as improve the accuracy of the clinic. To improve the level of rural medical, we designed and implemented a system including text pretreatment, extracting key words

© Springer Science+Business Media Singapore 2016
K. Li et al. (Eds.): ISICA 2015, CCIS 575, pp. 562–571, 2016.
DOI: 10.1007/978-981-10-0356-1_59

of symptom, time, degree and corresponding relation as well as the function of giving weight of certain degree. It improve current extracting technology and in a certain extent, our method can improve the accuracy and response speed of information extraction. In this way, we try to provide a basis for the further diagnosis and treatment.

2   The Present Application of Information Extraction Technology in the Medical Field

With the development of the society, more and more institutions regard electronic medical records as an important modern tool of managing medical information. As there are many different kinds of electronic medical records with complex content and format used by various agencies, moreover, the narrative of the electronic medical records is colloquial and of no unified discipline, it is hard to extract complete and accurate records information from the medical records [2].

Several information extraction research work has been carried out in the field of medicine both in China and abroad [3, 4]. The MEDDLEE system is one of the successful medical information extraction system [11]. Fung Kin Wah use MetaMap and other publicly available resources from the structured drug label indications to extract the drug information, and describe the information in standard medical term coding [6].

The patients' descriptions are mostly short text, with regard to long text, short text has its own characteristics. It makes some methods such as computing word frequency are not available. Song et al. put forward a kind of information classification method based on the spatial concept which is superior to the normal vector space model [7]. Zelikovitz et al. put forward a method that can classify short text by using latent semantic index [8]. Chang Peng and Ma Hui raised a method of feature selection which can overall consider the word frequency, document frequency the location weight of the word and word co-occurrence to achieve better effects of information extraction [9].

At present, there is few activities of nutrition and health care in domestic rural areas. The research of systematic nutrition and health service related to key crowd national wide has not yet been thoroughly conducted. In addition, there is no relatively application in China about information extraction methods facing to rural medical care, and the related research is scant abroad.

3   The Design of Information Extraction and Algorithm Improvement

3.1   The Framework of Information Extraction

Aiming at the actual demand of rural medical information extraction, we design a functional framework of information extraction including text pretreatment, information extraction, dependency analysis and algorithm parallelization improvement. It is shown in Fig. 1.

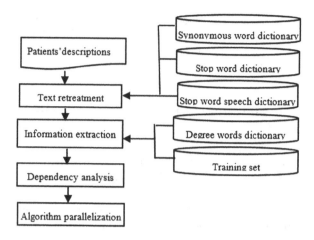

Fig. 1. Functional framework of information extraction

Text pretreatment: process the Chinese word segmentation and delete stop word.

Information extraction: extract the information of patients' condition, time and degree from free text and assign appropriate weight to the degree word.

Dependency analysis: obtain the dependency between words through syntax analysis and find the corresponding relationship between illness condition and time and degree information.

Algorithm parallelization: parallelized algorithm to achieve the distributed processing of information extraction.

Stop word dictionary: the Chinese words in the text which automatically filtered out after processing deposit here.

Stop word speech dictionary: a collection of speech of word which cannot be illness deposit here.

Degree words dictionary: common degree words which are classified according to different degrees deposit here.

Training set: manual annotation text deposit here.

3.2 The Design of Information Extraction Process

3.2.1 Extraction Process of Disease Conditions, Time and Degree

We extract the information of disease conditions, time and degree based on the results of word pretreatment. The design of information extraction process is shown in Fig. 2.

Every word in the word set after filtering out by word pretreatment can be directly judged into disease condition if the word completely match the disease name or the text in training set, otherwise, we have to judge if it belongs to a disease based on the naïve Bayesian classification method. The probability that the word belongs to all classes can be computed through classification method. However, a word may not related to a disease or belongs to any class, in this case, there will always be a maximum value of all probabilities that the word belongs to all classes. It is not accurate if we directly select the corresponding category that the largest probability value point to. So we need to set

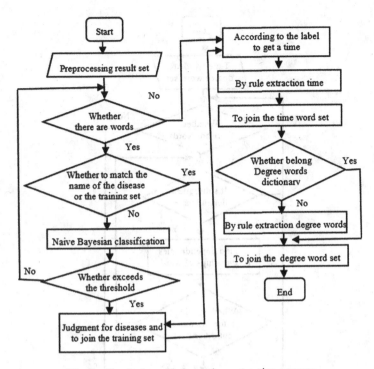

Fig. 2. The design of information extraction process

a threshold which should be determined through experiments. There is reference value only the largest probability is greater than the threshold.

In the patients' descriptions there will be some degree modifier of the disease, such as "have a little headache" or "have a terrible headache", "fever 38°" of "fever 39°" etc. Although these descriptions remain the same, but the diagnosis and treatment scheme will not remain the same because of different disease levels. So it is one of essential function to extract degree modifier of the disease condition and to give proper weight. It can help system to analyze the disease conditions better and to judge the stage of disease conditions.

3.2.2   The Process Design of Obtaining Dependency

We judge the corresponding relationship between disease conditions and time and degree after getting the word set including information of disease conditions, time and degree. We turn to help doctors to judge disease conditions better by using time and degree information to add to the disease condition information. The process design of obtaining dependency is shown in Fig. 3.

We obtain the dependence relationship between words through natural language processing tool, in order to determine the corresponding weights of disease time and degree for storage and output. For instance, in the description "I have a headache today, and I had a fever yesterday", the disease "headache" corresponds to the time "today", not "yesterday". The corresponding relationship is obtained through dependencies.

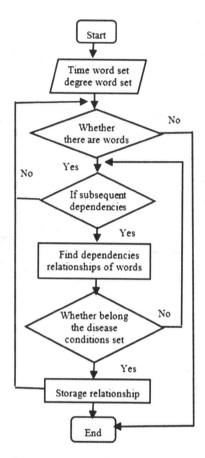

Fig. 3. Process design of obtaining dependency

For instance, as the description of time is almost an adverbial modifier, we can find the corresponding disease description by seeking out the verb-object relationship and ADV relationship.

3.3 The Algorithm Distributed Design

In rural medical care, the health data which contain patients' case over the years, patients' descriptions, all kinds of disease diagnosis, key technology related and some relevant documents is very huge and these data will grow huger over the time. It is not appropriate when dealing with big data if using circulation and judgment method to extract the information of time, degree and relationship between them. So, we can improve the processing speed and meet the requirement of real-time by designing and implementing a parallel processing of extraction algorithm.

We parallelized the machine learning algorithm and improve the extraction algorithm into distributed algorithm, training set, test set and dictionary which are deployed on HDFS [10] using Hadoop cluster and MapReduce programming framework.

The process of disease conditions extraction is designed to three MapReduce process.

(1) To calculate number of text in every category

The Map process: organize text according to the name of the class and the text style.
The Reduce process: accumulate to get the number of every text in every category.

(2) To calculate the quantity of key words in the text in every category

The Map process: when a key word appears, the system forms a <keyword, 1> value pair.
The Reduce process: organize the input value pair according to the key words, calculate them separately to get the quantity of key words in every category.

(3) To calculate the probability, to get the word of disease conditions through the classification method.

The Map process: input the result of the first two processes, calculate the probability of each keyword belongs to each category, then return key value set to the system.
The Reduce process: accumulate and sort to get the classification result which conform to the requirement.

4 The Technology Realization of Information Extraction and Algorithm Improvement

4.1 The Extraction of Disease Conditions Based on the Naive Bayesian Classification

We obtain disease conditions candidate word set in the process of text pretreatment via a call to a segmentation tool NLPIR [11] from the Chinese Academy of Science. We calculate the probability of all the candidates in the word set with the exiting class, then comparing all probability to get the maximum. If the maximum is more than the threshold, we judge it belongs to this class, otherwise the word will be abandoned.

We prepare for the probability calculation by using the getTrainingClissifications method gotten from the TrainingDateManager class, including getting the name of the class, getting the number of files in each class, getting path of all the training text in the class, getting the number of training text in which contain key words in the given category and judging if there are files that completely match the given string. We calculate the probability that the candidate belongs to each text, and the system returns the maximum value. By calling the method such as calculatePc(String c) in the class ClassConditionalProbability, the probability that the given text belongs to each class will be calculated and the maximum probability will be returned.

4.2   The Extraction of Relationship

The word refering to disease conditions, time and degree can be gotten through the degree(D) method calling from the Degree class based on the method of combining the dictionary and the rules. Then, we confirm the related time and degree word to each disease condition by using LTP-Cloud through which can get the dependencies between words. For instance, the dependency between degree word and disease condition is always ADV,as the degree word is always an adverb.

In LTP-cloud class the match of disease condition and time and degree word is reached by calling the methods of the open source tool LTP-Cloud. The final results are represented as the Result class, which is used to store and process the triple group of disease-time-degree. There are several ways constructing the Result class which reloaded the Result() method, dealt with the structure with different input values separately.

4.3   The Process of Algorithm Parallelization

The procedure is as follows. First use the Map method to divided the algorithm into small pieces, then use the Reduce method to integrate the step intermediate results, finally use the Job method to schedule. The value pair as the input in map process will be received in reduce process. Meanwhile, the maximum value which is the category that the disease belongs to will be selected and output through the sorting operation by the internal method.

5   Test and Conclusion

We select the data of high blood pressure disease as the experimental sample data [12] and divide the data into test set and training set after obtaining the patients' descriptions. The date of the training set is labeled according to ICD-10 disease classification which includes 530 text in 65 categories. The test set include 50 text.

5.1   Selection of the Threshold in Naïve Bayesian Classification

First we adjust the output of the system and output the probability value and the judgement of classification of all the disease candidate words. The system is test by the date in the test set. The output results is under analyzing and comparing. Then, we selected a roughly range of the threshold. Finally, the result of the output example is shown in Fig. 4.

We adjust the output that only the disease candidate word which has been classified successfully will output from the system. The result is shown in Fig. 5.

```
In process.
胃灼热: 0.0
In process.
胃肠气胀及有关情况: 0.0
In process.
腹水: 0.0
In process.
腹部和盆腔痛: 0.0
In process.
血压读数升高: 0.10256411
In process.
血尿: 0.0
In process.
言语障碍: 0.002913753
In process.
语音障碍: 0.0
In process.
诵读困难和其他象征性机能障碍: 0.0
In process.
遗尿: 0.0
In process.
非特异性低血压读数: 0.008484163
高血压
此项属于[血压读数升高]
```

Fig. 4. The sample of the classification result analysis before threshold selecting

```
第1篇文档
出血
此项属于[出血]
血
此项属于[泌及循环和呼吸系统的其他症状和体征]
头晕
此项属于[头晕和眩晕]
高血压
此项属于[血压读数升高]
第2篇文档
糖尿病
此项属于[泌及循环和呼吸系统的其他症状和体征]
高血压
此项属于[血压读数升高]
```

Fig. 5. The sample of the classification result after threshold selecting

5.2   The Testing of the Accuracy and the Recall Rate of the Extraction Method

After adjusting the system output, the output is composed of time word, degree word and a triple form of disease-time-degree. Then, we test the system using the test set and analyze the accuracy and the recall rate of the output result which is shown in Fig. 6.

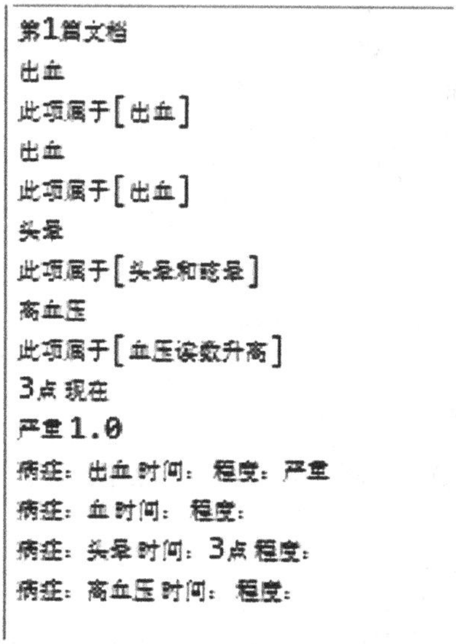

Fig. 6. The sample of the output result

5.3   Experimental Result

According to the result of classification statistic, we calculate the accuracy and the recall rate of the system under different thresholds for comparison and select the optimal result. The experiment shows that the threshold should be 0.05. In the same way, we test the extraction of time and degree word and the obtaining of corresponding relationship. The experiment shows the extraction accuracy of disease conditions is 0.89, the recall rate is 0.86; the extraction accuracy of time is 0.95, the recall rate is 0.88; the extraction accuracy of degree is 0.75, the recall rate is 0.86; the extraction accuracy of corresponding relations is 0.88, the recall rate is 0.74. As is shown in the result, the extraction effect of time word is best.

6   Summary and Outlook

In this article, we design and implement an information extraction method for rural medical treatment. We adopt the open source software NLPIR as the segmentation tool. We use the naïve Bayes method to extract disease conditions from patients' descriptions and make the improvement for rural medical treatment. We extract the information of time and degree through the method that combine the dictionary and the rules and obtain the corresponding relationship between them using the open source tool LTP-Cloud. We implement the algorithm parallelization based on the Hadoop platform and the MapReduce programming model and increase the response speed of the method. From now on, scholars should further enrich dictionary and rules and to do further analysis aimed at patients' descriptions. We should formulate more relevant rules in order to enhance the accuracy of the extraction.

Acknowledgment. This paper is supported by Internet Science Popularization construction technical scheme for China Association of Science and Technology and National High-tech R&D Program (863 Program) No. 2012AA01A404.

References

1. Stefanov, T.: System for information extraction from news sites. Math. Softw. Eng. **1**(1), 25–30 (2015)
2. Kreuzthaler, M., Schulz, S., Berghold, A.: Secondary use of electronic health records for building cohort studies through top-down information extraction. J. Biomed. Inform. **53**, 188–195 (2015)
3. Opitz, T., Azé, J., Bringay, S., et al.: Breast cancer and quality of life: medical information extraction from health forums. In: Medical Informatics Europe, pp. 1070–1074 (2014)
4. Deleger, L., Molnar, K., Savova, G., et al.: Large-scale evaluation of automated clinical note de-identification and its impact on information extraction. J. Am. Med. Inform. Assoc. **20**(1), 84–94 (2013)
5. Bakken, S., Hyun, S., Friedman, C., et al.: A comparison of semantic categories of the ISO reference terminology models for nursing and the MedLEE natural language processing system. Medinfo **11**(Pt 1), 472–476 (2004)
6. Fung, K.W., Jao, C.S., Demner-Fushman, D.: Extracting drug indication information from structured product labels using natural language processing. J. Am. Med. Inform. Assoc. **20**(3), 482–488 (2013)
7. Song, P., Bruza, D.: Based on Information Inference.In: Proceedings of the 14th International Symposium on Methodologies for Intelligent Systems, pp. 297–306 (2003)
8. Zelikovitz, S.: Transductive LSI for short text classification problems. In: Proceedings of the 17th International FLAIRS Conference (2004)
9. Chang, P., Ma, H.: Efficient short texts keyword extraction method analysis. Comput. Eng. Appl. **47**(20), 126–128 (2011)
10. Dittrich, J., Quiané-Ruiz, J.A.: Efficient big data processing in Hadoop MapReduce. Proc. VLDB Endowment **5**(12), 2014–2015 (2012)
11. Zhang, H.P.: NLPIR Chinese Segmentation System [EB/OL]. http://ictclas.nlpir.org/. Accessed 16 August 2015
12. High Blood Presure [EB/OL]. http://club.xywy.com/list_284_all_1.htm#h. Accessed 26 September 2014
13. Baidu Baike ICD-10[EB/OL]. http://baike.baidu.com/link?url=vI5gBTTm812sbDpWf5zojq1wiL_1K386raqgWey6CodhPtnw1lJ-GwBp0UHEi0vXix_T0xxplOnF_yJiObhtga. Accessed 26 September 2014

A Finger Vein Recognition Algorithm Using Feature Block Fusion and Depth Neural Network

Cheng Chen[1], Zhendong Wu[1(✉)], Ping Li[2], Jianwu Zhang[1], Yani Wang[1], and Hailong Li[1]

[1] School of Communication Engineering, Hangzhou Dianzi University, Hangzhou 310018, China
chencheng1992@outlook.com, wzd@hdu.edu.cn,
{13750894036,18767137938,18268883436}@163.com
[2] School of Mathematics and Computational Science, Sun Yat-sen University, Guangzhou, People's Republic of China
liping26@mail2.sysu.edu.cn

Abstract. Along with the development of biometric recognition, the technology of finger vein recognition possesses better anti forgery performance and identification stability in collecting and certificating information of human bodies. The available finger vein recognition method is mainly based on template matching or whole feature recognition, suffering from light instability of the acquisition equipment which leads to low robustness. In order to strengthen the robustness, we adopt a finger vein recognition algorithm using Feature Block Fusion and Deep Belief Network (FBF-DBN), in which the nonlinear learning ability of deep neural network is used to recognize the features of finger veins. Meanwhile, we improve deep network input by using feature points set in vein images, sharply reducing the time in learning and detection, meeting the practical needs of biometric recognition specifically applied to embedded equipment. Experimental results showed that FBF-DBN algorithm presented better recognition performance and faster speed.

Keywords: Feature block fusion · Depth neural network · Finger vein recognition

1 Introduction

As a rapid development of new technology, biometric authentication is becoming more and more important in social life. Vein recognition technology is a highly reliable biometric identification method, which has a good specificity, uniqueness and is difficult to falsify. The finger vein recognition has become a high security of the second generation of biometric authentication technology because of its outstanding advantages, earn growing popularity and acceptance among the public and scholars. As a kind of living creature feature recognition technology, finger vein recognition has great potential in the field of biometric key [4], Hybrid Cloud Secure [7], Attribute-based Encryption [5, 6], and so on. Yu et al. [17] proposed a simple template matching method that directly

© Springer Science+Business Media Singapore 2016
K. Li et al. (Eds.): ISICA 2015, CCIS 575, pp. 572–583, 2016.
DOI: 10.1007/978-981-10-0356-1_60

calculate the similarity between the matching and the registered finger vein image features, Setting the appropriate threshold by experiment, if the similarity is greater than the threshold, it is considered that the matching authentication is successful, and vice versa. Because of the exhaustive coordinate matching operation, the process required to identify longer. Li et al. [2] used the invariant matrix to extract the feature of the finger vein, Hausdorff distance (HD) to measure the similarity between the finger vein images and the registered finger vein images, while there still has the problem of large amount of computation. Wang et al. [3], proposed a method of principal component and analysis (PCA) and Linearity Distinction Analysis (LDA), which overcome the low accuracy rate on the single feature recognition, but large computational quantity also exists.

To solve the problems above, we proposed a finger vein recognition algorithm using Feature Block Fusion and Deep Belief Network (FBF-DBN) according to the characteristics of vein image. First, we extracted one of the finger vein images and normalized its size and gray scale; then filtered the finger vein profile by cutting the rest; finally the feature points were extracted. Then DBN is used for training feature points. Neural network has the ability to extract the important information from the image, generate the model of the finger vein image, and achieve the matching and recognition of finger vein image. The number of nodes of the visible layer in the neural network can affect the complexity and accuracy of the experiment. In this paper, we proposed Feature Block Fusion (FBF) algorithm, which is no need to input a whole image for training neural network, therefore, the amount of computation is greatly reduced.

The organization of this paper is as follows: Sect. 2 provides the process of finger vein image. In Sect. 3, we describe the main contents of FBF-DBN and display the whole process of FBF-DBN algorithm. In Sect. 4, we introduce the experimental result. In Sect. 5, we summarize the whole paper and put forward to a direction of further research.

2 Finger Vein Image Process

To improve the recognition efficiency of this system, we need to cut out the area with the most abundant information from those collection images for subsequent processing. Image smoothing and enhancement is to reduce noise in finger vein images, and improve the contrast of the venous area and background area, meanwhile providing venous information, extracting feature points set accurately.

2.1 Extraction of the Interest Region

Through the transmission light, we can collect finger vein image samples which are showed in Fig. 1a, the gloomy area in the picture is vein texture. But in addition to the area of the finger we can still find a wide background area on both sides from this figure. Therefore, it is necessary to extract the finger vein image. Extraction can be divided into the following three steps:

1. Finger location. The different positions of finger produce different locations and areas of finger vein image, which means finger positioning is very important for the next step of the interception. Comparison of vein image before and after finger positioning is illustrated in Fig. 2b and c.

2. The extraction of the figure contour. Image after finger positioning shows that there exist obvious edges between the finger region and the background region, moreover, the finger region is in the axial direction in the image. Thus, the contour of the finger can be detected only by using the directional operator, detected image is illustrated in Fig. 2c. If we change the detected edge contour bones into a single pixel, the result is shown in Fig. 2a.

3. The region of interest extraction from the contour. Specific steps: search from the center line of the image to the left and right sides of the finger until the first non-zero pixel point is encountered. From the position we plot two straight line parallel to the center line, inscribed to the fingerprint profile, which shows in Fig. 2b and c. Similarly showed in Fig. 2d, we put the following border as the bottom line, then draw two parallel lines in the position of 0.125 and 0.875, so as to cut out the finger vein image. Figure 2e is the region of interest.

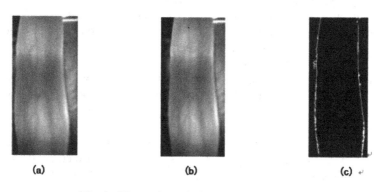

(a)   (b)   (c)

Fig. 1. Finger picture before and after position

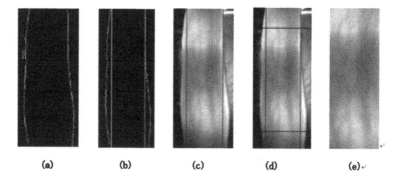

(a)   (b)   (c)   (d)   (e)

Fig. 2. Extraction process of interest region

2.2 Image Segmentation

Image segmentation is one of the important parts of image processing, of which the purpose is to extract the finger vein lines from the background area. Therefore, the selection of the segmentation threshold directly determines the segmentation effect. The segmentation algorithm are shown in this section, which shows a good effect on the extraction of low quality finger vein images.

Through the finger vein image we can find the image on the darker regions are mostly vein texture features, while the background area is brighter, which makes the vein feature in the valley region of the image, moreover the "Valley" is deeper and the possibility of belonging vein is greater. Thus, intravenous lines feature extraction is equivalent to the detection of the valley-area image, and then dividing up the valley-shaped region. The overall process steps are as follows:

1. Take any one of the pixels, expressed by *, and decompose it on 8 ridge line direction of which the ridge direction is 45 degrees respectively. As shown in Fig. 3, it indicates the position of the 8 directions.

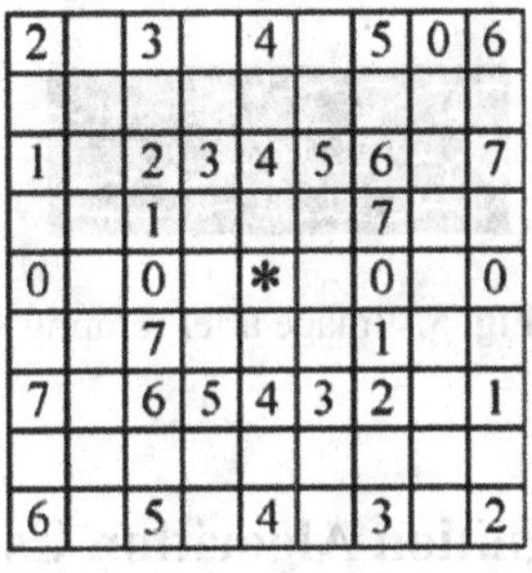

2	3	4		5	0	6
1	2	3	4	5	6	7
	1				7	
0	0	*		0		0
	7			1		
7	6	5	4	3	2	1
6	5	4		3		2

Fig. 3. 8 directions of a pixel

2. The convolution operation is performed by using the template operator of the 8 directions and the pixels in the neighborhood of the pixel, then regards the maximum value after the operation as the gray value of the pixel, that is:

$$Q(x, y) = Max(F\,gray(i)) \tag{1}$$

Finally, we will obtain a vein feature image containing the pseudo vein and noise after calculating for each pixel gray value by this method.

3. The obtained vein feature images contain pseudo veins and noise, so it is required to be divided into 3 regions: the most likely region which contains only venous characteristics (target area), the least likely region which contains only venous characteristics (background area), region with noise and vein characteristics (fuzzy area),the purpose of this step is to process each area separately, reserve target area, remove background area, finely cut apart fuzzy region. Finally, the image only with the characteristics of the vein is obtained, and the effect is shown in Fig. 4.

Fig. 4. Effect of vein segmentation

2.3 Image Thinning

Vein image lines after binarization still have a certain width, refining for further compression data. In this paper, a simple look-up table method is used to refine the image, which computes the 8 neighborhood of each pixel first, and then check table to determine whether the pixel can be deleted. There exist various "burr" in the vein skeleton image after thinning, so we need to cut and remove them to avoid the bad impact of the feature points extracted and identified, shown in Fig. 5.

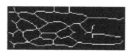

Fig. 5. Image after refining

3 A Finger Vein Recognition Algorithm Using Feature Block Fusion and Deep Belief Network (FBF-DBN)

Deep belief network (DBN) has been widely used in features recognition and the accuracy rate is really satisfying such as handwritten numbers. However, in other databases such as images with large size, as shown in Fig. 5. Recognition rata of DBN algorithm is just passable and it takes a long time for training. In this paper, we use DBN for finger vein recognition, due to large size of image and a large number of training samples cannot be provided, DBN needs to adjust on input layer to deal with the performance on recognition and efficiency, and we further suggest that, features in finger vein may be accurately identified by DBN. So we designed a new finger vein recognition algorithm named FBF-DBN.

FBF-DBN has three steps:

Step 1: Feature extraction. The feature points are extracted by using the detail feature extraction method based on 8 neighborhoods, that is, endpoint and intersection, which will be used for the next operation.

Step 2: Feature block fusion. After handling the step 1, we extract feature points further by using a new proposed algorithm by FBF, which can reduce the number of feature points and image sizes .

Step 3: Judge by DBN. We use DBN algorithm for training feature points by FBF, which promotes the recognition rate and speed.

The flow diagram of FBF-DBN is shown in Fig. 6.

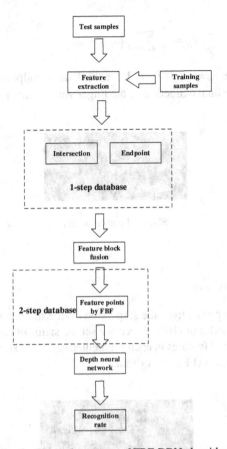

Fig. 6. The main process of FBF-DBN algorithm

a_1	a_2	a_3
a_4	a_0	a_5
a_6	a_7	a_8

Fig. 7. Feature point template

3.1 Feature Extraction

Define a region of 3×3, as is shown in Fig. 7. a_0 means current detection point, clockwise order around 8 pixels, successively set to $a_1, a_2, \ldots, a_8$, and the values are 0 or 1.

From a_1 to a_8 the number of alternating transformation between 0 and 1, we define it as follow:

$$T_N = \frac{1}{2} \sum_{i=1}^{8} |a_{i+1} - a_i| \qquad (2)$$

When i = 8, we define $a_9 = a_1$. If $T_N = 1$, we think a_0 as endpoint; if $T_N \geq 3$, we think a_0 as intersection. The final extraction of the finger vein feature points shown in Fig. 8.

Fig. 8. Feature points

3.2   Feature Block Fusion

We extract the feature points from the step 1 and divide the image into n same size, n also is the number of nodes in visible layer, we set the status of each segmentation area by judging the number of features in each segmented region as shown in Fig. 9. We call it as Feature Block Fusion (FBF) algorithm.

Fig. 9. FBF process

3.3   Judge by DBN

In this paper, we study deep belief network (DBN) for finger vein recognition. The DBN is a multilayer neural network that can gradually learn complex structures of data and work as probabilistic generative model [1]. A typical deep network as shown in Fig. 10. It is composed of several layers of hidden variables and one layer of visible units. There is a connection between the layers, but there is no connection between the units in the layer. The basic of a DBN is restricted Boltzmann machine, we will explain it next.

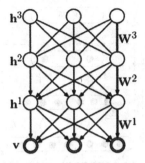

Fig. 10. DBN composed with three RBMs

Restricted Boltzmann Machine. As shown in Fig. 11. The RBM is a two-layer neural network with one layer of visible units and one layer of hidden units. There is no effects between nodes in a single layer of RBM (Fig. 13).

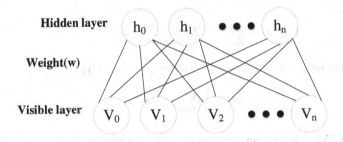

Fig. 11. An example of restricted Boltzmann machine

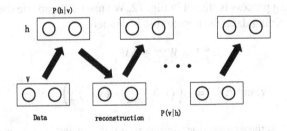

Fig. 12. Training RBM with CD

If a RBM has n units of visible layer and m units of hidden layer, the energy function of the RBM is defined as:

$$E(v, h; \theta) = -\sum_{i=1}^{n} a_i v_i - \sum_{j=1}^{m} b_j h_j - \sum_{i=1}^{n} \sum_{j=1}^{m} v_i w_{ij} h_j \tag{3}$$

Where $\theta = (W, a, b)$ and w_{ij} represents the weights between visible layer and hidden layer, a_i and b_j are their bias.

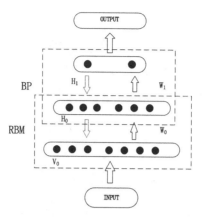

Fig. 13. Training process of DBN

Probability Distribution:

$$p\,(v, h) = \frac{1}{Z}e^{-E(v,h)}Z\,(\theta) = \sum\nolimits_{v,h} e^{-E(v,h)} \tag{4}$$

Through summation we can get the probability of visible layer v:

$$p\,(v) = \frac{1}{Z}\sum\nolimits_{h} e^{-E(v,h)} \tag{5}$$

Training RBM. The learning process of a RBM is unsupervised. We regard p(v) as likelihood function, since maximizing the likelihood needs a long time, Contrastive Divergence(CD) algorithm is put forward by Hinton [1], it is a fast learning way for RBM. The training process is shown in Fig. 12. We use CD to update the weights of the RBM to fit the training data, the update rule is defined as:

$$W_{ij}^{new} = W_{ij}^{old} + \Delta W_{ij} \tag{6}$$

$$\text{Where } \Delta W_{ij} = \varepsilon\left(\left\langle v_i h_j \right\rangle_{data} - \left\langle v_i h_j \right\rangle_{re}\right) \tag{7}$$

Here $\left\langle v_i h_j \right\rangle_{data}$ is the expectation through training data and $\left\langle v_i h_j \right\rangle_{re}$ is the expectation through reconstructed data, ε is the learning rata.

Training DBN. We learn the weights of the DBN though a greedy layer-wise unsupervised algorithm, when the weights of an RBM have been learned, we have an output of hidden layer and it can be used as visible data for training next RBM, in this way, we can train a whole DBN, this is called pertaining. This will make the system have a better accuracy. Training process is as follows:

1. Training process of an RBM is an unsupervised learning process. First initialize a new DBN with 1 layer of RBM.

2. Input the original data by FBF and train the first RBM, then we use the output of first RBM as the input of the second BP network.
3. Use gradient-descent algorithm to adjust the weight of the whole network.

4 Experiments and Performance Analysis

In order to evaluate the effectiveness of method proposed in this paper. We use the finger vein image of 64 people, including 15 pieces per person, a total of 960 images, the image size is 70 × 150, and we used a class of 12 pictures for training, the remain 3 pictures as a test sample. The architecture of DBN was n-100-64. It is consisted of 1 RBM. The DBN was trained with algorithm described in Sect. 3. We set different size of Division region to get n for training is given in Table 1. The result reveals that size of 6 × 13 has a better performance on classification accuracy.

Table 1. Performacne of different size

Size	Nodes in visible layer	Err rate	Identification accuracy
4 × 4	450	9.9 %	85.9 %
5 × 5	312	5.6 %	93.7 %
6 × 13	100	1.5 %	96.9 %

In order to see the Efficiency of using Feature Block Fusion, we also trained another DBN using the whole image as input. As shown in Table 2. It can be clearly observed that the average training time for using feature points is less than using whole image.

Table 2. Performance of different approaches

Method	Training time(s)
Training for using FBF	5.62
Without using FBF	183.4

Table 3. Performance comparisons with other scheme

Method	Training time(s)	Identification accuracy	Err rate
FBF-DBN	0.0049	96.9 %	1.5 %
Hausdorff Distance(HD)	0.2235	92.4 %	5.2 %
PCA	0.3536	95 %	4.3 %

We compare the recognition performance and efficiency performance of FBF-DBN with other methods, Hausdorff Distance (HD) schemes proposed by Li et al. [2] and Principal Component and Analysis (PCA) proposed by kejun Wang [3]. The comparison is shown in Table 3. The result shows that FBF-DBN approach can achieve higher efficiency and better accuracy.

5   Conclusion

This research was supported by Zhejiang Science Fund (NO. LY16F020016) and Zhejiang Province Science and Technology Innovation Program under Grant Number (2013TD03). In this paper, we investigated a feature learning based on FBF-DBN for finger vein recognition. We also used traditional learning approach to show the effectiveness of deep learning. The main focus of this paper is to demonstrate the power of Feature Block Fusion. In this approach, there is unnecessary to input a whole training image, it can reduce training time effectively, and recognition result is satisfactory. Our finger vein dataset we used is not sufficient, it was possible for training by using larger dataset and performance would be better. When we add another hidden layer in network, the recognition result is not satisfactory. For future work, we will find a more suitable approach to training neural network with several hidden layers and achieve a better performance. And the potential features needs to be further tapped.

References

1. Hinton, G.E., Osindero, S., Teh, Y.-W.: A fast learning algorithm for deep belief nets. Neural Comput. **18**(7), 1527–1554 (2006)
2. Li, X.-Y., Guo, S.-X., Gao, F.-L. et al.: Vein pattern recognition by moment invariants. In: 1st International Conference on Bioinformatics and Biomedical Engineering, pp. 612–615 (2007)
3. Kejun, W., Zhi, Y.: Finger vein recognition based on Wavelet moment fused with PCA transform. Pattern Recogn. Artif. Intell. **20**(5), 692–697 (2007)
4. Wu, Z., Liang, B., You, L., Jian, Z., Li, J.: High dimension space projection-based biometric encryption for fingerprint with fuzzy minutia. Soft Comput. (2015). doi:10.1007/s00500-015-1778-2
5. Li, J., Wang, Q., Wang, C., Cao, N., Ren, K., Lou, W.: Fuzzy keyword search over encrypted data in cloud computing. In: Proceeding of the 29th IEEE International Conference on Computer Communications (INFOCOM 2010), pp. 441–445. IEEE (2010)
6. Li, J., Wang, Q., Wang, C., Ren, K.: Enhancing attribute-based encryption with attribute hierarchy. Mob. Netw. Appl. (MONET) **16**(5), 553–561 (2011)
7. Li, J., Li, Y., Chen, X., Lee, P., Lou, W.: A hybrid cloud approach for secure authorized deduplication. IEEE Trans. Parallel Distrib. Syst. **26**(5), 1206–1216 (2015)
8. LeCun, Y., Kavukcuoglu, K., Farabet, C.: Convolutional networks and applications in vision. In: Proceedings of IEEE International Symposium on Circuits and Systems, pp. 253–256 (2010)
9. Rajput, G.G., Anita, H.B.: Handwritten script recognition using DCT, Gabor filter, and wavelet features at word level. In: Proceedings of International Conference on VLSI, Communication, Advanced Devices, Signals & Systems and Networking, pp. 363–372 (2013)
10. Jinfeng, Y., Minfu, Y.: An improved method for forger-vein image enhancement. In: International Conference on Signal Processing Proceeding, Beijing, China, pp. 1706–1709 (2010)

11. Pi, W., Shin, J., Park, D.: An effective quality improvement approach for low quality finger vein image. In: 2010 International Conference on Electronics and Information Engineering, ICEIE 2010, Kyoto, Japan, (1)V 1–424-V, pp. 1-427 (2010)
12. Song, W., Kim, T., Kim, H.C., et al.: A finger-vein verification system using mean curvature. Pattern Recogn. Lett. **32**(Compendex), 1541–1547 (2011)
13. Tang, D., Huang, B.-N., Li, W.-X., et al.: A method of evolving finger vein template. In: 2012 International Symposium on Biometrics and Security Technologies, ISBAST 2012, Taipei, Taiwan, pp. 96–101 (2012)
14. Yuan, A., Bai, G., Yang, P., Guo, Y., Zhao, X.: Handwritten English word recognition based on convolutional neural networks. In: Proceedings of IEEE International Conference on Frontiers in Handwriting Recognition, pp. 207–212 (2012)
15. Hinton, G.: A practical guide to training restricted Boltzmann machines. Momentum **9**(1), 926 (2010)
16. Hinton, G.E.: Training products of experts by minimizing contrastive divergence. Neural Comput. **14**(8), 1771–1800 (2002)
17. Liu, Z., Yin, Y.-L., Wang, H.-J., et al.: Finger vein recognition with manifold learning. J. Netw. Comput. Appl. **33**(3), 275–282 (2010)
18. Chan, L.-H., Salleh, S., Ting, C.-M., Ariff, A.K.: PCA and LDA based face verification using back-propagation neural network. In: 2010 10th International Conference on Information Sciences Signal Processing and their Applications (ISSPA), pp. 728–732 (2010)
19. Ranzato, M., Susskind, J., Mnih, V., Hinton, G.: On deep generative models with applications to recognition, Neural Computing and Applications (2011)
20. Yu, C.-B., Zhang, D.-M., Li, H.-B., et al.: Finger-vein image enhancement based on mutithreshold fuzzy algorithm. In: Proceedings of the 2009 2nd International Congress on Image and Signal Processing, CISP 2009, Tianjin, China, pp. 1–3 (2009)

A New Process Meta-model
for Convenient Process Reconfiguration

Xin Li[(✉)] and Chao Fang

Department of Computer Science, University of Shantou,
Shantou City 515063, Guangdong Province, China
lixin@stu.edu.cn

Abstract. With the analysis of WFMC (Workflow Management Coalition) process meta-model, it's indicated that the limitations of that model cause the difficulty for convenient reconfiguration of business process. So, a new process meta-model, ESR (Event-State-Rule) meta-model, is presented in this paper as the substitution to WFMC's. Some elements are added in the new model, such as the event, the state and the rule with which dynamic relevance between the process and the business can be normalized expressed in process definition. Also, the boundary between process logic and business logic becomes much more explicit with the benefit from that model, which means that process logic can be separated from business logic quite effectively. Now, it becomes possible that rigid process and flexible process are modeled in a unified process framework. When process logic varies, the reconfiguration of the process may be fulfilled only with the corresponding variation of process definition so that convenient process reconfiguration is implemented.

Keywords: Worflow · Process meta-model · Process logic · Dynamic relevance between the process and the business · Convenient reconfiguration

1 Introduction

The uncertainty problem in business process which is caused by the relevance between the process and the business has been studied from different angles for a long time [1]. In the domain of the workflow, the concept of flexible workflow was presented. Here, the term of flexible means the ability of being adapted to the dynamic varieties from the environment and that of dealing with the exception. The researches of flexible workflow mainly include two aspects: flexible strategy and flexible model. In flexible strategy, the process model is usually existent. Some techniques are adopted to solve the problems of routing selection, structural modification and exception handling under certain model [2–4]. Relatively, the method of flexible model is more radical, new process model is tried to be put forward to support flexible workflow architecture [5–7]. The two flexible methods seem quite different apparently. However, the achievements of those researches often assist each other in reality. Flexible strategy can provide referenced technical means for flexible model and flexible model can establish the infrastructure for the realization of flexible strategy.

© Springer Science+Business Media Singapore 2016
K. Li et al. (Eds.): ISICA 2015, CCIS 575, pp. 584–595, 2016.
DOI: 10.1007/978-981-10-0356-1_61

There're a wide variety of theories and techniques in the realm of flexible workflow presently, each of which is aimed at some troubles in the business process, and new theories and techniques are still proposed constantly with the appearance of new problems. Although the development of flexible workflow has gone through a not short time, the realm is quite immature now. The following problems are very remarkable at least.

(1) The tactics of the combination of the rigidity and the flexibility are adopted by a lot of workflow systems. Namely, traditional process model such as WFMC model is used for rigid process and some other flexible techniques are taken for uncertain process as the supplement. That scheme makes that two or more different modeling methods coexist in one system which increases the complexity of process modeling much more. Especially, when the application expands, the promotion of the system will become highly difficult. In comparison, it may be much easier if there's a unified modeling method available for both rigid process and flexible one.

(2) In existing workflow systems, the ability to express process logic is deficiency. Even for a respectable number of logics which are frequently used, the expression is exertive. Workflow pattern is the foundation of the technique about process logic. There're a great variety of workflow patterns. Usually, a workflow system only supports a subset of the pattern. Yet even if a system which can support all the patterns appears, it still can't satisfy the requirement of expressing process logic because workflow pattern itself doesn't cover the scope of process logic. So, a thinking different from workflow pattern may be needed which not only can improve the ability to express process logic substantially but also is simply constructed to be used easily.

(3) The relevance between the process and the business haven't been taken into account sufficiently. Process logic is affected by business elements on a number of aspects. However, it hasn't been well induced how business elements affect process logic up to now. Formalized and abstract expression for business elements is absent in most workflow systems. Thus, the form of program code is widely used in the occasion where the process is related with the business so that the boundary between them becomes vague.

The above problems are the reflections of the uncertainty in business process in different sides. The root of them is process logic hasn't been separated from business logic well, which leads to process reconfiguration is hard to implemented usually. A suit of new techniques aimed at those problems in this paper. The foundation is a process meta-model. Then, process modeling method and expression method for process logic are set up on the base of the model. Also, the corresponding operating mechanism of workflow system is discussed. As a result, convenient process reconfiguration can be realized with the techniques.

2 An Example of Process Reconfiguration

In order to explain the concept of dynamic relevance between the process and the business, and the difficulty of process reconfiguration, an example is given firstly. There's a segment of a business process in Fig. 1.

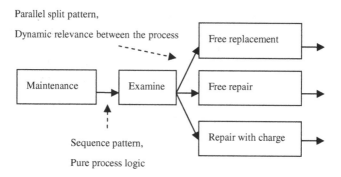

Fig. 1. A business process application

The description of Fig. 1 is as follows. There's something wrong in the product which the customer has bought. So, he submits a maintenance application to the company. Then, the application is examined, and the sale date and the product type are abstracted. If the date is within a month by now, the product will be replaced by a new one free of charge. If the product type is computer mainframe and the date is between one month to three years, the product will be repaired free of charge. If the product type is peripheral device and the date is between one month to one year, the product will be repaired free of charge too. Otherwise, the product will be repaired with charge.

It's easy to see that there're two process logics in above process segment. One is between the activity of the maintenance application and that of the examine, the other is between the activity of the examine and the latter three activities (replacement and repair). The former is of sequence pattern and the latter is of parallel split pattern. Sequence pattern and parallel split pattern both belong to workflow patterns. Here, another question is worthy of attention. The first process logic is a pure process logic, which means that no business elements in the application activity will affect the process logic. But the second process logic has something to do with the product type and the bought date, which are both business elements. Namely, business elements will affect the second process logic. The phenomenon is called dynamic relevance between the process and the business in this paper. Now, precise expression can be given to some concepts.

Definition 1. Business process. Business process is the whole that includes a group of business activities and the logic relations among those activities.

A triad can be used to describe a business process. Let BP denotes business process, A denotes the set of activities, and P denotes the set of process logics. AF denotes the affections which show how business elements act on process logics, namely AF is the dynamic relevance between the process and the business. Then, BP = <A, P, AF>

Definition 2. Dynamic relevance between the process and the business. Process logic is related with the value or the structure of business elements and only when the process instance is running, the value of the elements can be obtained. Such relation is called dynamic relevance between the process and the business. Namely, let BS denotes the set of business elements, BSP denotes the power set of BS, P denotes the set of process logics. if $\exists pl : P \exists bs : BSP \bullet relate(pl, bs)$, then, it's called that pl is dynamically related with bs. Let AF denotes the set of all dynamic relevance between the process and the business. Then, AF is a relation on BSPXP, namely, $AF \in BSP \leftarrow \rightarrow P$.

Definition 3. Pure process logic. It means that any set of business elements can't be found to be related with any process logic. Namely, let pl denotes a process logic, $\neg \exists bs : BSP \bullet relate(pl, bs)$, then, pl is called a pure process logic.

Because workflow pattern can't support dynamic relevance between the process and the business directly, program code must be used to express the process logic. For example, in Fig. 1, the second process logic usually expressed by such codes shown in Fig. 2, which is in examine activity.

```
duration = presentdate() - Actapplication.getsaledate();
if ( duration <= 1month ) then start(ActFreeReplace);
...... ( Other business code )
componenttype = Actapplication.gettype();
if ( componenttype is mainframe )   and ( duration > 1month ) and ( duration <= 3year )
    then start(ActFreerepair);
...... (Other business code )
if ( componenttype is peripheral )   and ( duration > 1month ) and ( duration <= 1year )
    then start(ActFreerep);
...... (Other business code )
start(ActChargerep);
```

Fig. 2. The sample of the code for dynamic relevance between the process and the business

In the figure, ActXXX denotes an activity, start(ActXX) denotes that ActXX can be started. Other elements in the code aren't difficult to understand according to the literal meaning and the context. They aren't explained because of limited space.

Now, suppose that the maintenance regulations varied.

(1) Two examine activity is set after application activity and following activity can be continued only ater both examines passed.
(2) The period of free replacement is extended to three month, of course, the period of the repair is changed correspondingly.

According to the method of workflow pattern and WFMC model, such process reconfiguration as the follows is needed. Firstly, some activities should be added. New structure diagram for the process is shown in Fig. 3.

For the reason of the limited space, shorthand is used in Fig. 3. It's noticed that two activities are added in the process. One is an exam activity, the other is a converge

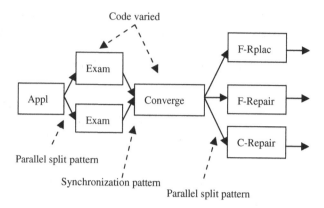

Fig. 3. Process reconfiguration on workflow pattern and WFMC model

activity. Subsequently, the patterns related to the activities are also changed. But the variation requirement is much more than the above. Program code should be varied too. There're two aspects about the code variation.

(1) The code in Fig. 2 should be varied according to the business requirement obviously. The contents of varied code is omitted not to write here.
(2) The program code should be moved from exam activity to converge activity. It means that the program will be redeployed.

Evidently, recoding and redeployment means compiling, linking … etc. once more. That's why process reconfiguration usually is a big burden in the system of business process.

Compared to traditional workflow system, most flexible systems just make the code in Fig. 2 more regular, in addition to which, little improvement can be acquired in the circumstances like above example. So, a suit of new techniques ought to be proposed to solve the problem which is quite important to the platform of business process.

3   The Techniques for Business Process Reconfiguration

Convenient process reconfiguration means that the implementation of process reconfiguration is carried out fully or mainly through the variation of process definition. The prerequisite of convenient reconfiguration is that process logic is well separated from business logic. However, process logic is often coupled with business logic in fact, which is the characteristic of business process. In the previous chapter, it's very clear to see that. So, a contradiction is in front of the researchers. The defect of existent workflow technique just lies in the unsatisfied performance on the coupling.

The barrier of process logic being separated from business logic is considered mainly on two aspects in this paper. One is workflow pattern and the other is WFMC process meta-model, which are both foundations of most existent workflow system. Those two questions will be expounded respectively.

3.1   Workflow Pattern and Logic Rule

Workflow pattern has been regarded as the constitutional unit for process logic all along. There're tens of patterns already and more ones may be proposed in the future. However, no matter how many patterns there are, it seems that there always exist process logics which can't be covered by the patterns. So, a different perspective is tried on the question of process logic here.

A workflow pattern can be divided into two more fundamental units, which is shown in Fig. 4.

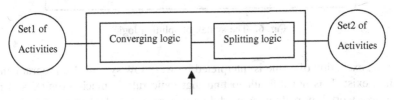

Logic rule substituted for workflow pattern

Fig. 4. Logic rule to express process logic

Now, a new entity called logic rule is proposed to be substituted for workflow pattern. A rule can be expressed as the form of P->Q. P is the condition of a rule, which is called converging logic and Q is the result of a rule, which is called splitting logic. Two kinds of logics have similar mathematical form and mathematical properties, but the contents in the two logics are different. Converging logic mainly includes the states and splitting logic mainly includes the actions.

The syntax of converging logic is depicted in Fig. 5.

$$
\begin{aligned}
&P::=A\theta A \\
&A::=Activity.State|Constant|F|Count(P(,P)^*)\ |others \\
&P::=\sim P|P\Phi P|(P) \\
&\theta::=\text{'>'}|\text{'<'}|\text{'='}|\text{'}\geq\text{'}|\text{'}\leq\text{'}|\text{'}\neq\text{'} \\
&\Phi::=\text{'}\wedge\text{'}|\text{'}V\text{'}|\text{'}\oplus\text{'} \\
&F::=Founc()|Founc(A(,A)^*)
\end{aligned}
$$

Fig. 5. The syntax of converging logic

Correspondingly, The syntax of splitting logic is depicted in Fig. 6.

An example of logic rule is as follows,

Rule: ActExam.finished ∧ Actapply.sDuration≤1month ->ActF-Rplac.start()

The function of the rule is the same as the following code,

if (duration <= 1month) then start(ActFreeReplace)

$$Value::=Activity.State|Constant|F|Count(Q(,Q)^*)$$
$$Q::=Activity.F|SetState(Activity.State, Value)$$
$$\quad|Value\theta Value$$
$$F::=Founc()|Founc(Value (,Value)^*)$$
$$Q::=Q\Phi Q|(Q)$$
$$\theta::='>'|'<'|'='|'\geq'|'\leq'|'\neq'$$
$$\Phi::='\wedge'|'\vee'|'\oplus'$$

Fig. 6. The syntax of splitting logic

But the execution of a rule is interpreted by workflow system. No program and no compiling exist. It is not difficult to find that logic rule is much stronger and more flexible than workflow pattern. A workflow pattern can be looked upon as the combination of two logic. Supposing there're ten kinds of logics, we'll get one hundred of workflow patterns. It explains well why workflow patterns always aren't enough for process logic.

A rule can express any combination of converging logic and splitting logic. The form in Fig. 7 is unable to express with workflow pattern directly, which is like the example in Fig. 3 where an individual converge activity has to be imported into the process.

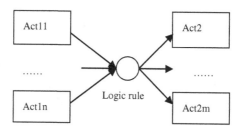

Fig. 7. Any combination of converging logic and splitting logic

Besides, the state in converging logic includes business elements. But in workflow pattern, only abstract informations such as the finish of an activity and the start of an activity ... etc. are contained. If process logic is concerned about business elements, program code has to be cooperated with the pattern.

3.2 WFMC Meta-model and ESR(Event-State-Rule) Meta-model

WFMC has released a workflow meta-model which nearly acts as a criterion in workflow realm [8]. The model is shown in Fig. 8. A new model called ESR process meta-model is presented in this paper, which is shown in Fig. 9.

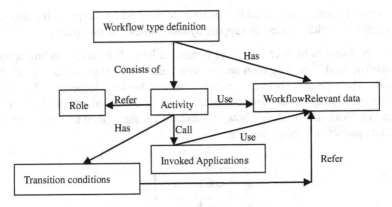

Fig. 8. WFMC Meta-model

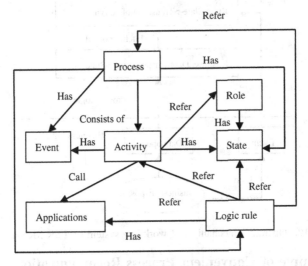

Fig. 9. ESR Meta-model

Because of the limitation of the space, the difference between two above models is only briefly narrated here.

(1) Workflow relevant data in WFMC Meta-model is removed, whose meaning and usage are both vague, so that it becomes the central issue where the coupling between process logic and business logic happens. State is used to replace workflow relevant data, which has more general meaning and is strictly limited to appear in logic rule.

(2) Transition conditions in WFMC Meta-model is removed, which is only an affiliation of the activity and whose usage is vague too. More general logic rule is used to replace transition conditions, which isn't affiliated with the activity but lies in the same layer as the activity. Logic rule is strictly limited to be written in the form of process definition, not that of program code.

(3) A new element, event is added in ESR Meta-model to express the happen of something, which is used to support dynamic nature of the system.

The formalization of state and event is omitted here. The corresponding operating mechanism of workflow system on new process meta-model is shown in Fig. 10. It can be seen that the separation between process logic and business logic is well carried out now. The description to the process is much more enhanced on the base of ESR Meta-model. Next, it'll be shown how new techniques are applied to process definition to facilitate process reconfiguration greatly.

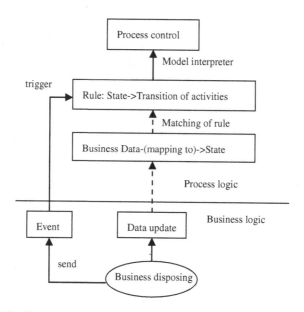

Fig. 10. Operating mechanism of workflow system on ESR Meta-model

4   An Example of Convenient Process Reconfiguration

Still, the example in Chapter 2 is used for the comparison. New process structure diagram on ESR Meta-model is shown in Fig. 11.

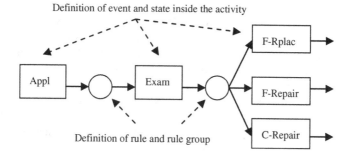

Fig. 11. Process structure diagram on ESR Meta-model

Firstly, the definition of each activity is given.

Activity Appl

Event: ActAppl.eActcommit   ActAppl.eActcommit -trigger-> RGAppl-Exam

(RGappl-Exam is a rule group which will be illustrated afterwards)

State: ActAppl.sDuration, ActAppl.sType

presentdate()-saledate -MapTo-> ActAppl.sDuration;   componenttype -MapTo-> ActAppl.sType

Activity Exam

Event: ActExam.eActcommit   ActExam.eActcommit   -trigger-> RGExam-Service

Then, the definition of rule and rule group is given.

RuleGruop RGAppl-Exam

Rule: ActAppl.finished -> ActExam.start()

RuleGruop RG Exam-Servive

Rule1: ActExam.finished $\wedge$ ActAppl.sDuration$\leq$1month -> ActF-Rplac.start()

Rule2:   ActExam.finished $\wedge$ ActAppl.sType=mainframe $\wedge$ ActAppl.sDuration>1month

$\wedge$ Actapply.sDuration$\leq$3year -> ActF-Repair.start()

Rule3:   ActExam.finished $\wedge$ ActAppl.sType=   peripheral $\wedge$ Actapply.sDuration>1month

$\wedge$ Actapply.sDuration$\leq$1year -> ActF-Repair.start()

Rule4: ActExam.finished $\wedge$ others -> ActC-Repair.start()

When process reconfiguration happens, for this example, only process definition needs to be changed. Varied process structure diagram is shown in Fig. 12.

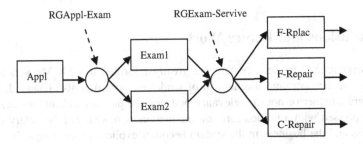

Fig. 12. Process structure diagram after reconfiguration

Varied process definitions is as follows,

Activity Exam1

ActExam1.eActcommit -trigger-> RGExam-Service

Activity Exam2

ActExam2.eActcommit -trigger-> RGExam-Service

RuleGruop RGAppl-Exam

Rule: Actapply.finished -> ActExam1.start() $\wedge$ ActExam2.start()

RuleGruop RG Exam-Servive

Rule1: ActExam1.finished $\wedge$ ActExam2.finished $\wedge$ ActAppl.sDuration$\leq$3months
          -> ActF-Rplac.start()

Rule2: ActExam1.finished $\wedge$ ActExam2.finished $\wedge$ ActAppl.sType=mainframe
          $\wedge$ ActAppl.sDuration>3months $\wedge$ Actapply.sDuration$\leq$3years ->
ActF-Repair.start()

Rule3: ActExam1.finished $\wedge$ ActExam2.finished $\wedge$ ActAppl.sType= peripheral
          $\wedge$ Actapply.sDuration>3months $\wedge$ Actapply.sDuration$\leq$1year ->
ActF-Repair.start()

Rule4: ActExam1.finished $\wedge$ ActExam2.finished $\wedge$ others -> ActC-Repair.start()

It can be seen that process reconfiguration has becomed much more convenient on the base of ESR Meta-model. Concrete improvements include,

(1) Only process denitions need to be varied. Program code isn't involved here.
(2) It no need to add individual convering activity now because logic rule has the function of the convergence.
(3) A unified model framework is provided for both riggid process and flexible process.

Because of the limitation of the space, no more analysis about process reconfiguration is deployed.

5 Conclusions and Future Work

Model elements and model structure are greatly reformed in ESR Meta-mode. Correspondingly, the operating mechanism of workflow system is also changed. Process logic where whether or not the relevance between the process and the business exists can be expressed with a unified form on the new meta-model. The boundary between the process and the business in the system becomes explicit. Process logic is executed through the interpreter, which doesn't depend on program code any more. So, when process logic varies, process reconfiguration can be carried out mainly by the means of the modification of process definitions, which is far more convenient than the past.

The expression of process logic is the core of process modeling. The syntax of logic rule is just briefly discussed in this paper. There're still lots of questions worth being probed deeply for the syntax, which affects the implementation of WFMS (Workflow management system) directly.

Acknowledgment. This work is supported by Campus Foundation of Shantou University.

References

1. Raik, H., Bucchiarone, A., Khurshid, N.: Dynamic context-aware adaption for service-based systems. In: Proceedings of IEEE the 8th World Congress on Service, pp. 385–392 (2012)
2. Ferreira, J., Wu, Q., Malkowski, S.: Towards flexible event-handling in workflows through data states. In: Proceedings of 6th World Congress on Services, pp. 344–351 (2010)
3. Knuplesch, D., Reichert, M., Pryss, R.: Ensuring compliance of distributed and collaborative workflows. In: Proceedings of the 9th IEEE International Conference on Collaborative Computing, pp. 133–142 (2013)
4. Paul, D., Henskens, F., Hannaford, M.: Dynamic transactional workflows in service-oriented environments. In: Proceedings of the 9th International Conference on Web Information Systems and Technologies, pp. 26–36 (2013)
5. Zhang, J.-L., Yang, Y., Zeng, M.: Method for flexible workflow modeling supporting task change and performance analysis. In: Proceedings of 2009 WRI World Congress on Software Engineering, pp. 51–55 (2009)
6. Poppe, O., Giessl, S., Rundensteiner, E.A., Bry, F.: The HIT model: workflow-aware event stream monitoring. In: Hameurlain, A., Küng, J., Wagner, R., Amann, B., Lamarre, P. (eds.) TLDKS XI. LNCS, vol. 8290, pp. 26–50. Springer, Heidelberg (2013)
7. Van der Aalst, W., Adriansyah, A., Van Dongen, B.: Replaying history on process models for conformance checking and performance analysis. Data Min. Knowl. Discov. 2(2), 182–192 (2012)
8. Workflow Management Coalition: The workflow reference model. Technical report, WFMC-TC00-1003, Hampshire: Workflow Management Coalition (1995)

Efficient ORAM Based on Binary Tree
without Data Overflow and Evictions

Shufeng Li, Minghao Zhao, Han Jiang$^{(\boxtimes)}$, Qiuliang Xu,
and Xiaochao Wei

School of Computer Science and Technology,
Shandong University, Jinan, China
{lishuhuafeng,weixiaochao2008}@163.com,
zhaominghao@hrbeu.edu.cn, {jianghan,xql}@sdu.edu.cn

Abstract. ORAM is a useful primitive that allows a client to hide its data access pattern and ORAM technique as a wide range of applications nowadays. In this paper, we propose a verified version of binary-tree-based ORAM with less data access overhead. We provide a new method to reselect the leaf node and write data back to the tree, and accordingly, avoid complicated evict operation. Besides, the bucket capacity is reduced to a constant level. Overall, our scheme improves the efficiency meanwhile maintains security requirement of ORAM.

Keywords: ORAM · Storage security · Binary-tree · Access patterns protection

1 Introduction

Oblivious Random-access Memory (ORAM) is a cryptographic scheme that can be used to completely hide the data access pattern for IO operations. It has been extensively used for secure Multi-party computation [1–6] (SMPC), secure processors [7–10], and secure cloud storage system [11–13], especially for Searchable Encryption (SE) [14].

In research field of SE, which allow a client to store a collection of encrypted document on a remote server and latterly perform search and retrieve the document it needed, the widely accepted security require is that nothing should be leaked to the server, except for the search pattern and access pattern [15, 16]. However, this security definition is not perfect, since recent researches show that access pattern disclosure may leads leakage of sensitive data exposure. Specifically, Islam [17] shows that under the prerequisite that the attacker have known some underground knowledge, it can perform inference attack to get the user's sensitive data. Latterly, Islam [18] proposes two concrete attacks. Having gotten the access pattern, the attacker can perform this attack

---

S. Li—This work is supported by the National Natural Foundation of China (No. 61173139 and 61572294), Foundation of Science and Technology on Information Assurance Laboratory (No. KJ-14-002), and Doctoral Foundation of Ministry of Education of China (No. 20110131110027).

© Springer Science+Business Media Singapore 2016
K. Li et al. (Eds.): ISICA 2015, CCIS 575, pp. 596–607, 2016.
DOI: 10.1007/978-981-10-0356-1_62

to know which keywords the user have been searched. All of these indicate that access pattern should be protected.

In addition, ORAM is a significant material for SMPC. A SMPC protocol allows distributed parties to jointly compute $f(x, y)$, without revealing any information other than the output of $f(x, y)$ and information can be inferred from it. Normally, generic SMPC protocols require an oblivious presentation of $f(x, y)$ (i.e. represent f with Garbled Circuits as in [19–21]). Also, f can be given as a RAM program. Gordon [22] proposes a method to compile Oblivious RAM to SMPC protocols, which provide a new approach for design secure multi-party protocol.

Besides ORAM, especially Path-ORAM can be used for designing secure processor, which could resistant side-channel attack [23] (esp. timing attack [24, 25]) and provide integrity verification.

Overall research on ORAM protocols is of great significance. ORAM is first proposed by Ostrovsky in [26], in which Square root ORAM and Hierarchical ORAM is proposed. Afterwards, inspired by the work of Ostrovsky, tree-Based ORAM [27] and Partition ORAM [28] are constructed. Path ORAM [29] is a simper version of Tree-Based ORAM. In this work, we make a modification of tree based ORAM.

1.1 Our Contributions

This paper provides a modified Tree-based ORAM scheme that achieves $O(1)$ client-side storage, $O(N)$ server-side storage, and $O(\log N)^2$ cost of data access. We mainly make the following contributions.

(1) **Free of data overflow.** We modify the method of reselecting the leaf node in the algorithm ReadAndRemove(u). Specifically, every selection we get an unoccupied lead node and fill it with new data.
(2) **No eviction operation.** We put forward a new principle to write data back to the tree in the algorithm Add(u,data), and thus avoid complicated evict operation.
(3) **Less data access overhead.** We propose a modified way to reselect *l** and write data back to the tree, which reduces the bucket capacity to a constant level, thus reducing server storage and overhead of data access.

2 Preliminary

2.1 System Architecture

We assume that the client has N data to be stored at a remote untrusted server. We organize the O-RAM storage into a binary tree over data buckets at the server, while moving data blocks obliviously along tree edge. The binary tree has N leaf nodes and every node is a bucket ORAM. Let Z denote the bucket capacity, which means a bucket has Z blocks. The client allocates a random leaf node l for every data block, and then the data block can be stored randomly on the path from the leaf bucket l to the root bucket. To ensure the security of the ORAM, in our construction, we often rely on dummy blocks and dummy operations to hide certain information from the untrusted server, such as whether a bucket is loaded, and where in the tree a block is headed. In addition to the real data, we will use dummy data to pad to the other blocks. We will think of the dummy block as a regular but useless data block.

2.2   Notations

Let N denote the O-RAM capacity, i.e., the maximum number of data blocks that an O-RAM can store. We assume that data is stored in atomic units called blocks. Let B denote the block size in terms of the number of bits. We assume that the block size $B \geq c\log N$, for some $c > 1$. Notice that this is true in almost all practical scenarios. We assume that each block has a global identifier $u \in \{0,1,2...N\}$. We dedicate a certain block identifier (e.g. $u = 0$ to serve as the dummy block). Other specifically symbols used in our algorithm are listed in Table 1.

Table 1. Notations in our algorithm

$L = \log N$	The height of the binary tree
$u \in \{0,1,...,N\}$	The identifier of a block
$l \in \{0,1,...,N\text{-}1\}$	The address(path)of the data stored
$l^* \in \{0,1,...,N\text{-}1\}$	The new address(path)of the data stored
index	The index of the client
index[u] $\in \{0,1\}^{L}$	The leaf node associated with block u
state	Global variant
p(l)	The path from root to leaf node l
Z	The capacity of a bucket
Bucketb, b $\in \{0,1\}$	The b-th child node of the current bucket
UniformRandom(S)	Randomly sample function on set S
$\perp$	Dummy data
count(l^*)	The number of real data blocks on the path p(l^*).

2.3   Definition of ORAM and Security of ORAM

We adopt the definition of ORAM and secure ORAM form the Shi's scheme [27].

Definition 1 [27]: An Oblivious RAM is a suite of interactive protocols between a client and a server, comprising the following two basic operations which are aimed at a bucket.

Read(u): Given a searched block u, the client will sequentially scan a bucket. If the current block matches identifier u, the client remembers its content and overwrites it with dummy data; otherwise, the client writes back the original block read.

Write(u,data): the client will sequentially scan a bucket. When the first time the client see a dummy block, the client overwrites it with (u,data), and if u $\neq$ 0 update the client's address table, then replace (u,data) with (0,$\perp$); Otherwise, the client writes back the original block read.

As mentioned earlier, whenever blocks are written back to the server, they are re-encrypted using a randomized encryption scheme in order to hide its contents from the server.

Definition 2 [27]: Let $y=((op_1,arg_1),(op_2,arg_2)........(op_M,arg_M))$ denote a data request sequence of length M. Each opi denotes a ReadAndRemove(u) or Add (u,data) (in the Sect. 3). Moreover, if op_i is a ReadAndRemove(u) operation, then $arg_i=u_i$ else if op_i is an Add(u,data) operation, then $arg_i=(u_i,argi)$, where u_i denotes the identifier of the block being read or added, and $data_i$ denotes the data content being written in the second case. Recall that before each Add(u,data) operation there is a ReadAndRemove(u) operation.

We use the notation ops(y) to denote the sequence of operations associated with y, i.e., $ops(y) :=(op_1,op_2,........op_M)$. Let A(y) denote the (possibly randomized) sequence of accesses to the remote storage given the sequence of data requests y. An ORAM construction is said to be secure if for any two data request sequences which have the same length y and z, and ops(y) = ops(z), their access patterns A(y) and A(z) are computationally indistinguishable by anyone but the client.

3 Basic Construction

3.1   Client-Side Storage and Server-Side Storage Organization

The client stores an address index table and a table that counting the number of real data block on the path p(l*). The client can get the address index l* associated with block u, and its concrete position stored.

The data blocks are stored in a binary-tree structure in the server (as described in the Sect. 2.1).

3.2   Implementation of the Basic Algorithm

In this paper, we use two algorithm ReadAndRemove(u) (as shown in Fig. 1) and Add(u,data) (as shown in Fig. 2) to implement standard read and write operation. When a data block is accessed, its memory address will be revealed to the server. We will protect the client's privacy by writing the data back to a fresh address. Now we describe details of our algorithm.

We use the following Figs. 3 and 4 to demonstrate the procedure of algorithm ReadAndRemove(u) and Add(u,data).

When fetching the block u, the client first looks up the leaf node associated with block u from the client's address index. To look up the block u, it checks every bucket on the path l from the root to the leaf (denoted by the shaded buckets in the Fig. 3). After a data block is accessed; it will be logically assigned to a fresh random leaf node. When the block u is found, the client does not stop until it access to the last block of the leaf node.

Every time a block is accessed, it will be logically assigned to a fresh random leaf node l* by the client. After exerting the algorithm ReadAndRemove(u), the client

ReadAndRemove(u) and Add(u,data).

```
ReadAndRemove(u)
1. l*←UniformRandom({0,1,...,N-1})
2. i=0
3. While(count(l*)≥Z(L+1))
4. l*←UniformRandom({0,1,...,N-1}-the chosen l*)
5. l←index[u],index[u]←l*
6. state ←l*
7. data ←⊥
8. for each bucket on p(l) do
9. if ((data0//10) ← bucket.Read(u))  ≠⊥then
10. data ← data0
11. end if
12. end for
13. return data
```

Fig. 1. ReadAndRemove(u)

```
Add(u,data)
1. l←state
2. for(l=0,1,2.......L)
3. if(u≠0)
4. b=1-st bit of l
5. Bucketb.write(u,data)
6. end if
6. else
7. b∈{0,1}
8. Bucketb.write(u,data)
9. end else
10. end for
```

Fig. 2. Add(u,data)

also needs to write the data back to the binary-tree. It finds a bucket which has empty position along the path $p(l^*)$, then it writes the data back. If the client finds a bucket with empty position and has written the data back to it, it will stop access the bucket along the path $p(l^*)$; alternatively, it randomly access along the current bucket's left or right child node until reaching a leaf node. The server does not know that, among $p(l^*)$ and $p(l^{*'})$ (as shown in Fig. 4), which path has the client accessed. Therefore the server does not know the data's fresh address.

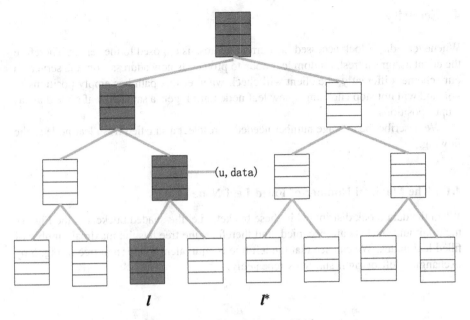

Fig. 3. Searching for a data block

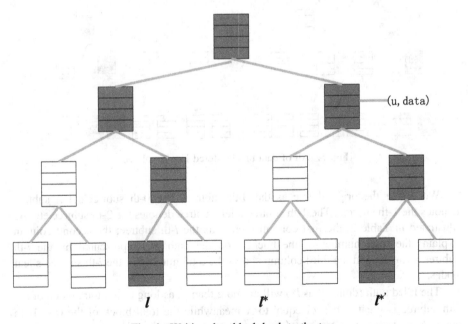

Fig. 4. Writing data block back to the tree

4 Security

Whenever a data block accessed, its current address is exposed to the server. Therefore the client assigns a fresh random leaf node to protect its new address from the server. In our scheme, differently, the client will check whether this path has empty positions or not, and will not stop choosing a new leaf node until it gets a satisfied leaf node that has empty positions.

We describe the average number needed to reselect a satisfied fresh leaf node as the flowing.

4.1   The Maximal Number of Filled Leaf Node

When the data block distributed in these buckets (i.e. the shaded buckets in the Fig. 5), the minimum buckets are occupied, and therefore, the tree has the maximal number of filled leaf nodes. We can see that the left tree is equivalent to the right tree in Fig. 5 by exchanging left or right sub-trees repeatedly.

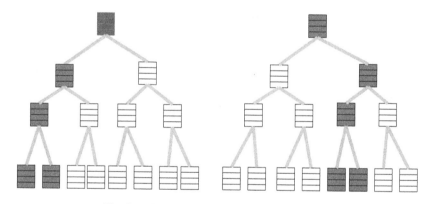

Fig. 5. All of data blocks stored in one subtree

We define the original tree as the 0-th subtree. The 0-th subtree's left subtree denotes the 1-th subtree. The 1-th subtree's left subtree denotes the 2-th subtree etc. As illustrated in Table 2, the first column represents the i-th subtree; the second column explains the total number of the buckets on the path p(l) represented by the i-th subtree's leaves; and the third column shows the total number of the i-th subtree's leaf nodes.

The filled path (donated as l^*) will no more than $\frac{N}{2^t}$, as long as the total block of the t-th subtree is greater than or equal to N meanwhile the total block of the $(t+1)$-th sutree is no larger than N (i.e. $Z\left(\frac{N}{2^{t-1}} + t - 1\right) \geq N$ and $Z\left(\frac{N}{2^t} + t\right) \leq N$ are satisfied) So we have

Table 2. The number of leaf node in the i-th subtree

i-th subtree	Number of bucket	Number of leaf node
0	$2N-1$	N
1	N	$\frac{N}{2}$
2	$\frac{N}{2}+1$	$\frac{N}{2^2}$
...	...	...
t	$\frac{N}{2^{t-2}}+t-1$	$\frac{N}{2^t}$
$t+1$	$\frac{N}{2^t}+t$	$\frac{N}{2^{t+2}}$
...	...	...

$$\frac{N}{\frac{N}{2^{t-1}}+t-1} \le Z \le \frac{N}{\frac{N}{2^t}}$$

$$\Pr[\text{count}(l^*) \ge Z(L+1)] \le \frac{\frac{N}{2^t}}{\frac{N}{2}}$$

$$\Pr[\text{count}(l^*) \ge Z(L+1)] \le \frac{1}{2^t}$$

$$\Pr[\text{count}(l^*) \ge Z(L+1)] \le \frac{1}{2^t} \approx \frac{1}{Z}$$

The maximal number of filled leaf node is $\frac{N}{2^t}$, and we mark it as s. Meanwhile we mark the average number needed to reselect a satisfied fresh leaf node as s'. We demonstrate the calculating formula of s' as the following.

$$s' = 1 \cdot \frac{N-s}{N} + 2 \cdot \frac{s}{N} \cdot \frac{N-s}{N} + 3 \cdot \frac{s}{N} \cdot \frac{s-1}{N-1} \cdot + \cdots$$
$$+ s \cdot \frac{s}{N} \cdot \frac{s-1}{N-1} \cdots \frac{2}{N-(s-2)} \cdot \frac{N-s}{N-(s-1)} + (s+1) \cdot \frac{s}{N} \cdot \frac{s-1}{N-1} \cdots \frac{2}{N-(s-2)} \cdot \frac{1}{N-(s-1)} \cdot 1$$

The expression above indicates that s' approach to a constant. We provide some examples of actual data as shown in Table 3. The probability that all of data blocks stored on the path $p(l)$ represented by a subtree's leaves (i.e. the shaded buckets in the Fig. 5) is very small. After a data block accessed, the redistribution of data enables the filled leaf nodes and its number to change randomly. Therefore our new method of reselecting fresh leaf nodes reveals no information to the server.

Table 3. The average number of reselecting l^*

N	Z,t	s	s'
$N = 2^{15}$	$Z = 4, t = 3$	$s = 4096$	$s' = 1.142852$
$N = 2^{15}$	$Z = 8, t = 4$	$s = 2048$	$s' = 1.066664$
$N = 2^{16}$	$Z = 4, t = 3$	$s = 8192$	$s' = 1.142855$
$N = 2^{16}$	$Z = 8, t = 4$	$s = 4096$	$s' = 1.066666$

4.2 Security Analysis

To prove the security of our scheme, let y be a data request sequence of size $M = \text{poly}(N)$. by the Definition 2, the server sees $A(y)$ which is a sequence

$$l = (l_1[a_1]); \ldots; l_{M-1}[a_{M-1}]; l_M[a_M]$$

where $l_j[a_j]$ is the address for the j-th load/store operation.

Notice that once $l_j[a_j]$ is revealed to the server, it is remapped to a fresh leaf node. $l_j[a_j]$ is statistically independent of $l_i[a_i]$ for $j < i$ with $a_j = a_i$ (a_i denotes the block identifier). $l_i[a_i]$ is also statistically independent of $l_i[a_i]$ for $j < i$ with $a_j \neq a_i$, therefore,

$$\Pr[l = A(y)] - \Pr[r \leftarrow_r \{0,1\}^{M(L+1)}] = \frac{1}{(N-s')^M} - \frac{1}{N^M} = negl(M(L+1))$$

Here $negl(\cdot)$ means negligible function.   ∎

This indicates that $A(y)$ is computationally indistinguishable from a random sequence of bit strings. Therefore our scheme is secure.

5   Cost Analysis

5.1   Client Storage

The storage on the client consists of two tables, including an address index table and a table that counting the number of real data block on the path p(l^*).

The client has N data and each block size is $B \geq c\log N$ bits. Thus, the client storage size is of $O(\frac{N \log N}{B})$.

We can adopt the Shi's scheme's recursion ORAM [27, 28] to reduce the client storage. Recursion technique reduces the client storage to $O(1)$.

5.2   Server Storage

We propose a modified way to reselect l^* and write data back to the tree, which reduces the bucket capacity Z to a constant level, thus reducing server storage and overhead of data access. The binary-tree has N leaf nodes, and the total number of nodes is $2N-1$. The server storage is $O(N)$.

5.3   The Cost of Data Access

For each load or store operation, the client reads a path of $Z\log N$ blocks from the server and then writes them back, resulting in a total of $2Z\log N$ blocks bandwidth used per access. Since Z is a constant, the bandwidth usage is $O(\log N)$ blocks. Considering the adopting of recursion ORAM, the bandwidth usage is $O(\log^2 N)$ blocks because of there being $\log_c N$ recursion ORAM.

Table 4 lists the cost of data access, client storage and server storage of five schemes based on the binary-tree ORAM. Note that all of these schemes have adopted recursion technique.

Table 4. comparison to other ORAM scheme

Scheme	Data access cost	Client storage	Server storage
Hierarchical Solution [26]	$O(N\log^2 N)$	$O(1)$	$O(N\log N)$
Tree ORAM [27]	$O(\log^3 N)$	$O(1)$	$O(N\log N)$
Path ORAM [29]	$O(\log^2 N)$	$O(\log N)$	$O(N)$
optimizing ORAM [1]	$O\left(\frac{k\log^2 N}{\log k}\right)$	$O(1)$	$O(N)$
Our scheme	$O(\log^2 N)$	$O(1)$	$O(N)$

6 Conclusion

We proposed a modified Tree-based ORAM scheme. In our work, we solved the problem of data overflow in traditional binary tree based ORAM. Moreover, compared with the former ones, our scheme has high performance with simplified algorithm and less data access necessity. Besides, our scheme maintains the security requirement of normal ORAM. ORAM is a cryptographic scheme that can be used to completely hide the data access pattern for IO operations, and has extensive application.

References

1. Gentry, C., Goldman, K.A., Halevi, S., Julta, C., Raykova, M., Wichs, D.: Optimizing ORAM and using it efficiently for secure computation. In: De Cristofaro, E., Wright, M. (eds.) PETS 2013. LNCS, vol. 7981, pp. 1–18. Springer, Heidelberg (2013)
2. Gentry, C., Halevi, S., Jutla, C., et al.: Private database access with he-over-oram architecture. Cryptology ePrint Archive, report 2014/345 (2014). http://eprint.iacr.org
3. Gordon, S.D., Katz, J., Kolesnikov, V., et al.: Secure two-party computation in sublinear (amortized) time. In: Proceedings of the 2012 ACM Conference on Computer and Communications Security, pp. 513–524. ACM (2012)
4. Keller, M., Scholl, P.: Efficient, oblivious data structures for MPC. In: Sarkar, P., Iwata, T. (eds.) ASIACRYPT 2014, Part II. LNCS, vol. 8874, pp. 506–525. Springer, Heidelberg (2014)
5. Liu, C., Huang, Y., Shi, E., et al.: Automating efficient RAM-model secure computation. In: 2014 IEEE Symposium on Security and Privacy (SP), pp. 623–638. IEEE (2014)
6. Wang, X.S., Huang, Y., Chan, T.H., et al.: Scoram: oblivious RAM for secure computation. In: Proceedings of the 2014 ACM SIGSAC Conference on Computer and Communications Security, pp. 191–202. ACM (2014)
7. Fletcher, C.W., Dijk, M., Devadas, S.: A secure processor architecture for encrypted computation on untrusted programs. In: Proceedings of the Seventh ACM Workshop on Scalable Trusted Computing, pp. 3–8. ACM (2012)
8. Fletchery, C.W., Ren, L., Yu, X., et al.: Suppressing the oblivious RAM timing channel while making information leakage and program efficiency trade-offs. In: 2014 IEEE 20th International Symposium on High Performance Computer Architecture (HPCA), pp. 213–224. IEEE (2014)

9. Maas, M., Love, E., Stefanov, E., et al.: Phantom: practical oblivious computation in a secure processor. In: Proceedings of the 2013 ACM SIGSAC Conference on Computer & Communications Security, pp. 311–324. ACM (2013)

10. Ren, L., Yu, X., Fletcher, C.W., et al.: Design space exploration and optimization of path oblivious RAM in secure processors. In: ACM SIGARCH Computer Architecture News, vol. 41(3), pp. 571-582. ACM (2013)

11. Stefanov, E., Shi, E.: Multi-cloud oblivious storage. In: Proceedings of the 2013 ACM SIGSAC Conference on Computer & Communications Security, pp. 247–258. ACM (2013)

12. Stefanov, E., Shi, E.: Oblivistore: high performance oblivious cloud storage. In: 2013 IEEE Symposium on Security and Privacy (SP), pp. 253–267. IEEE (2013)

13. Williams, P., Sion, R., Carbunar, B.: Building castles out of mud: practical access pattern privacy and correctness on untrusted storage. In: Proceedings of the 15th ACM Conference on Computer and Communications Security, pp. 139–148. ACM (2008)

14. Stefanov, E., Papamanthou, C., Shi, E.: Practical dynamic searchable encryption with small leakage. In: Network and Distributed System Security Symposium (NDSS 2014) (2014)

15. Curtmola, R., Garay, J., Kamara, S., et al.: Searchable symmetric encryption: improved definitions and efficient constructions. In: Proceedings of the 13th ACM Conference on Computer and Communications Security, pp. 79–88. ACM (2006)

16. Curtmola, R., Garay, J., Kamara, S., et al.: Searchable symmetric encryption: improved definitions and efficient constructions. J. Comput. Secur. 19(5), 895–934 (2011)

17. Islam, M.S., Kuzu, M., Kantarcioglu, M.: Poster: inference attacks against searchable encryption protocols. In: Proceedings of the 18th ACM Conference on Computer and Communications Security, pp. 845–448. ACM (2011)

18. Liu, C., Zhu, L., Wang, M., et al.: Search pattern leakage in searchable encryption: attacks and new construction. Inf. Sci. 265, 176–188 (2014)

19. Lindell, Y., Pinkas, B.: Secure two-party computation via cut-and-choose oblivious transfer. J. Cryptol. 25(4), 680–722 (2012)

20. Huang, Y., Evans, D., Katz, J., et al.: Faster secure two-party computation using garbled circuits. In: USENIX Security Symposium, vol. 201(1) (2011)

21. Lindell, Y.: Fast cut-and-choose based protocols for malicious and covert adversaries. In: Canetti, R., Garay, J.A. (eds.) CRYPTO 2013, Part II. LNCS, vol. 8043, pp. 1–17. Springer, Heidelberg (2013)

22. Gordon, S.D., Katz, J., Kolesnikov, V., et al.: Secure two-party computation in sublinear (amortized) time. In: Proceedings of the 2012 ACM Conference on Computer and Communications Security, pp. 513–524. ACM (2012)

23. Lomne, V., Dehaboui, A., Maurine, P., et al.: Side channel attacks. In: Badrignans, B., Danger, J.L., Fischer, V., Gogniat, G., Torres, L. (eds.) Security Trends for FPGAS, pp. 47–72. Springer, Netherlands (2011)

24. Kocher, P.C.: Timing attacks on implementations of Diffie-Hellman, RSA, DSS, and other systems. In: Koblitz, N. (ed.) CRYPTO 1996. LNCS, vol. 1109, pp. 104–113. Springer, Heidelberg (1996)

25. Dhem, J.F., Koeune, F., Leroux, P.A., Mestré, P., Quisquater, J.-J., Willems, J.-L.: A practical implementation of the timing attack. In: Schneier, B., Quisquater, J.-J. (eds.) CARDIS 1998. LNCS, vol. 1820, pp. 167–182. Springer, Heidelberg (2000)

26. Goldreich, O., Ostrovsky, R.: Software protection and simulation on oblivious RAMs. J. ACM (JACM) 43(3), 431–473 (1996)

27. Shi, E., Chan, T.H.H., Stefanov, E., Li, M.: Oblivious RAM with $O(\log N)^3$ worst-case cost. In: Lee, D.H., Wang, X. (eds.) ASIACRYPT 2011. LNCS, vol. 7073, pp. 197–214. Springer, Heidelberg (2011)

28. Stefanov, E., Shi, E., Song, D.: Towards practical oblivious RAM (2011). arXiv preprint arXiv:1106.3652
29. Stefanov, E., Van Dijk, M., Shi, E., et al.: Path ORAM: an extremely simple oblivious RAM protocol. In: Proceedings of the 2013 ACM SIGSAC Conference on Computer & Communications Security, pp. 299–310. ACM (2013)

A Novel WDM-PON Based on Quantum Key Distribution FPGA Controller

Yunlu Wang[1(✉)], Hao Wen[2], Zhihua Jian[1], and Zhendong Wu[1]

[1] College of Communication Engineering, Hangzhou Dianzi University Hangzhou,
Hangzhou, China
{wyl,zhjian,wzd}@hdu.edu.cn
[2] Chongqing Institute of Green and Intelligent Technology, Chinese Academy of Sciences,
Chongqing, China
wenhao@cigit.ac.cn

Abstract. A novel wavelength-division-multiplexed passive optical network base on quantum key distribution FPGA controller is presented here. QKD FPGA is responsible for 1.25 Gbps upstream PRBS source, clock regeneration, phase modulation control, key sifting, privacy amplification, and upstream time-divided-multiple-access control on quantum channels. An 8-user network experiment shows that over 20 km fiber, the mean secure exchange key rate can reach up to 500 bps in total, with the acceptable quantum bit error rate below safe limit and few impact on classical channels. This scheme can provide a promising way for the coexistence between quantum key distribution and classical data service.

Keywords: Quantum cryptography · WDM PON · Quantum key distribution · Quantum communication · Optical fiber communication

1 Introduction

Since huge information and privacy have been processed online, people's demand for security of broadband access network continues to rise. Though the data could be encrypted by the advanced encryption standard (AES), but the encryption/decryption keys exchange process is traditionally such as Diffie-Hellman algorithm [1], which is based on the discrete logarithm problem and has only limited computing complexity. QKD (Quantum Key Distribution) firstly presented in 1984 [2, 3] can share secure key between the two remote entities, and the process' security depends on the physical laws of quantum mechanics instead of computing complexity or other unproved methods [4–6]. Nowadays it is believed that QKD is the promising physical cryptography in the metropolitan area and can provide the enough exchange key rates for symmetric encryption communication [7, 8].

To support multiple users and save cost greatly, various QKD networks research [9, 10] and experiments were presented in recent years. British Townsend's group firstly published their work [11] in 1997, in which the central control end Alice is connected to three clients through a splitter in the several kilometers fiber. In 2006, BBN Technologies, Boston University and Harvard University announced the construction of a

© Springer Science+Business Media Singapore 2016
K. Li et al. (Eds.): ISICA 2015, CCIS 575, pp. 608–618, 2016.
DOI: 10.1007/978-981-10-0356-1_63

6-node QKD network [12]. Several Europe countries were also involved in the SECOQC project and have built a practical fiber quantum network covering Vienna city in 2009 [13], meanwhile Han's group announced a star type metropolitan network in China [14]. NEC, Mitsubishi and NTT have finished the field test of Tokyo QKD Network in 2011 [15]. Though plenty experiments have been made, none of above used the existing optical fibers or devices as quantum communication resources. Wholly new construction means too expensive to make QKD service available in practice.

Integrating QKD into existing fiber communication has to face the difficulty that the classical optical signals have tremendous noises on quantum optical signals when they are both transmitted in the same fiber. Efforts are made since 2005 by the quantum physicists and optical communication engineers [16–18]. In 2011, Townsend's group successfully continued to integrate QKD function into time-division-multiplexed PON (such as EPON) network [19], which definitely was a milestone. It showed the feasibility of co-existence between quantum cryptography and TDM PON [20].

On the other side, wavelength-division-multiplexed passive optical network (WDM-PON) has long been considered as an ultimate solution for the next-generation access network. In 2010, Townsend's group firstly gave an experiment of QKD in colorful WDM-PON, described in [21]. In this paper, we propose a novel scheme of WDM-PON integrated with QKD function. In our scheme, the WDM-PON is colorless, means in loopback configuration, in which the seed light provided from the optical line terminal (OLT) side is re-modulated with the upstream data at the optical network unit (ONU) side and sent back to the OLT utilizing reflective semiconductor optical amplifiers (RSOAs). The RSOA based WDM-PON implemented in re-modulation technique appears to have a great potential for practice [22–24], as shown in Fig. 1. Using the improved optical branch device and time-division multiple access method in upstream, quantum signals are well isolated or less infected by the classical signals.

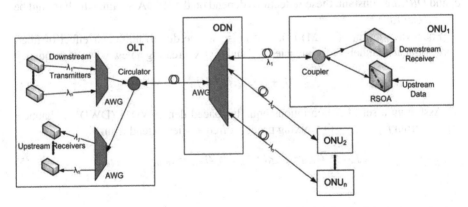

Fig. 1. RSOA based WDM-PON network demonstration utilizing re-modulation technique.

The rest of this paper is organized as follows. In Sect. 2, we firstly investigate the classical channel and quantum channel noises. In Sect. 3, we give the details of 8-user RSOA based WDM-PON with integrated QKD function. We experimentally evaluate the feasibility of our scheme, and give the measured results in Sect. 4. Finally, our work is summarized in Sect. 5.

2   Theoretical Analysis

2.1   Classical Channel Analysis

In the RSOA based WDM-PON, the bit error rate (BER) is mainly degraded by the reflected light both on upstream and downstream signals [25]. The back-reflected downstream signal (R_I) will interfere with the upstream signal. The re-modulated upstream signal reflected back to the ONU (R_{IIa}) will interfere with the downstream signal, and this reflected upstream signal will also be amplified again by the RSOA and sent back to the OLT (R_{IIb}) and interfere with the upstream signal. They are described as follows:

$$R_I \propto \frac{1}{c_1^2 G} t_{link}^2 \frac{1}{ORL_{OLT}} \tag{1}$$

$$R_{IIa} \propto c_1^2 G \frac{1}{ORL_{ONU}} \tag{2}$$

$$R_{IIb} \propto \frac{c_1^2 \Delta G^2}{G} \frac{1}{ORL_{ONU}} \tag{3}$$

Here G is the gain of RSOA, c_1 is the power transmittances of the coupler in the ONU the light sent to the RSOA, t_{link} is the total link loss, ORL_{OLT} or ORL_{ONU} is the optical return loss observed from the OLT or ONU side. For certain link setup, the t_{link}, c_1 and ORL are constant, these reflections depend on the RSOA's gain, which should be chosen carefully.

Four-wave mixing (FWM) is also a noise source due to nonlinear effect of fiber. Three optical channels at frequencies f_i, f_j and f_k mix, creating a new wave of frequency

$$f_{ijk} = f_i + f_j - f_k, k \neq i,j \tag{4}$$

Assuming a set of N contiguous equally spaced dense WDM (DWDM) channels ranging from f_{min} to f_{max}, the mixing products frequencies extend from

$$f_{low} = 2f_{min} - f_{max} \text{to } f_{high} = 2f_{max} - f_{min} \tag{5}$$

2.2   Quantum Channel Analysis

Raman scattering and FWM are the most serious limitations to quantum communication [26]. Noise photons will increase the QBER (quantum bit error rate) greatly, which limits the transmission distance and rate of the QKD, or even annihilates the quantum signal and makes QKD communication fail. As described in Eq. (5), DWDM system has the

FWM spectrum ranging from frequency f_{low} to f_{high}, if quantum signal wavelength is located below the flow or above the f_{high}, then it could be far from FWM effects, and the Raman scattering from classical signals will dominate the noise to quantum signal. For general, the equations that govern the interaction between the pump, P, and another optical signal, S, are as follows:

$$\begin{cases} \frac{dP}{dz} = -\alpha_P P + \beta_P S + \gamma_P PS \\ \frac{dS}{dz} = -\alpha_S S + \beta_S P + \gamma_S PS \end{cases} \tag{6}$$

where α, β and γ are the wavelength-dependent fiber attenuation, spontaneous Raman scattering coefficient and stimulated Raman scattering coefficient, respectively. By requiring that the initial pump and signal launch powers at $z = 0$ be $P(0) = P_0$ and $S(0) = 0$, respectively, the spontaneous Raman scattering power, $S(z)$ can be solved as:

$$S(z) = \begin{cases} P_0 \beta_S z e^{-\alpha_P z}, & \alpha_S = \alpha_P \\ P_0 \dfrac{\beta_S}{\alpha_S - \alpha_P} (e^{-\alpha_P z} - e^{-\alpha_S z}), & \alpha_S \neq \alpha_P \end{cases} \tag{7}$$

It has the maximum value at the Raman peak distance:

$$z_{max} = \begin{cases} \dfrac{1}{\alpha_P}, & \alpha_P = \alpha_S \\ \dfrac{1}{\alpha_S - \alpha_P} \ln\left(\dfrac{\alpha_S}{\alpha_P}\right), & \alpha_P \neq \alpha_S, \end{cases} \tag{8}$$

where α_P is the pump light attenuation coefficient, and α_S is the QKD light attenuation coefficient. At distances greater than z_{max}, the fiber attenuation diminishes the scattered noise signal more quickly than it can be replenished by the pump, at distances less than z_{max}, the pump scattered noise signal increases along with the distance. If the pump light is in the C-band with the attenuation 0.2 dB/km, the quantum light is in the O-band with the attenuation 0.4 dB/km, and then the Raman peak distance is approximately 15 km. Thus a reasonable system may be designed to keep the quantum signal and classical signal co-existing much less than z_{max}.

3 Network Architecture and Settings

3.1 WDM-PON Settings

The 8-user WDM-PON network architecture is shown in Fig. 2, which contains one OLT, one optical distribution network (ODN) and 8 ONUs. DWDM channels are assigned with frequency grid of 200 GHz (about 1.6 nm) from 192.6 THz (1557 nm) to 194 THz (1545 nm). DWDM wavelength multiplexers and de-multiplexers are all Arrayed Waveguide Gratings (AWGs). Classical communications are bi-directional as normal, while quantum communication is set to only upward direction to reduce the cost of quantum channel, since single photon detectors (SPD) is often extremely expensive. Thus only 2 InGaAs-APD SPDs (by BB84 scheme [2]) in total are needed inside OLT.

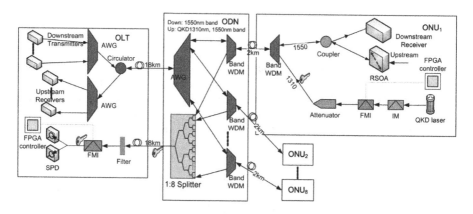

Fig. 2. 8-user QKD-integrated WDM-PON network setup.

The OLT has downstream transmitters and upstream receivers as follows. The downstream distributed feedback (DFB) lasers are modulated by phase modulators using 1.25 Gbps pre-coding non-return to zero (NRZ) pseudo-random bit stream (PRBS) to generate differential phase shift keying (DPSK) signals. Photo detector PIN receivers are used to receive the upstream signal in OLT. At the ONU, using 3 dB power coupler, portion of the DPSK signals are fed to DPSK balanced receivers, while other portion of the DPSK signals are re-modulated using 1.25 Gbps NRZ upstream PRBS by RSOAs.

The ODN contains 8 band WDMs (1310/1550 nm) connected to 8 distribution fibers, a PLC-type AWG de-multiplexer and a 1:8 power splitter. Two feeder fibers between the OLT and ODN are needed, one (primary) is for the classical channel, and the other (secondary) is for quantum channel, which usually acts as the backup fiber to provide network robustness.

3.2   QKD Systems

Quantum signal's wavelength is located at 1310 nm. This choice mainly has three advantages: (a) There is enough isolation (> 200 nm) between the classical signal (1550 nm) and quantum signal, according to (5):

$$f_{quantum} = 229\,(THz) >> f_{high} = 2f_{max} - f_{min} = 2 \times 194 - 192.6 = 195.4\,(THz) \qquad (9)$$

Therefore FWM noises of DWDM channels cannot have impact on quantum channel significantly; (b) This frequency is far beyond classical signals band (1550 nm), at anti-Stokes side, hence could effectively alleviate Raman-scattering noise; (c) The upstream selection of 1310 nm wavelength enables the client to use a lower-cost laser.

OLT includes a 50 GHz optical filter, an unbalanced Faraday-Michelson interferometer (FMI), two InGaAs-APD SPDs, and a field programmable gate array chip (FPGA) controller, while ONU includes a QKD laser source, an intensity modulator (IM), a FMI, a FPGA controller and an electronic-controllable attenuator. The DFB laser output is pulse-carved by an IM driven by the 1.25 GHz clock from the FPGA controller to generate a sequence of 400 ps duration, return-to-zero (RZ) pulses. Next, data are encoded by a FMI driven by 1.25 Gbps NRZ PRBS realized in FPGA controller to generate a sequence of 0 and π phase shifts in a phase shift keying (PSK) QKD scheme. Finally, the quantum signal is attenuated to the single-photon level, with the 0.2 photon per pulse mean photon numbers, then is multiplexed and de-multiplexed with classical signals by the band WDM (1310/1550) device (isolation > 60 dB). At the ODN, quantum signal is guided into a PLC-type 1:8 power splitter, then into the secondary fiber, and finally detected by the SPDs. Quantum transmission is synchronized with the upstream signal to adjust the detection window of SPDs, which are triggered by a 100 MHz clock recovered and divided from the 1.25 Gbps upstream data.

The quantum signal transmission uses time-division-multiple access (TDMA) method to realize multiple ONUs' sharing distribution. Each ONU is assigned a time slot to transmit the quantum signal without conflict, which is announced by the OLT's downstream signal. The accurate start and end time of a slot rely on the OLT's ranging mechanism (i.e., delay of each channel from OLT to ONU could be calculated by precise measurement), which is a dynamic real-time process.

4 Experimental Results

4.1 Experimental Settings

To transmit quantum signal in TDMA way, each ONU is assigned a time slot of 960 µs (1200000 symbols time), and additional 40 µs (50000 symbols time) is used as guard time to avoid the probable conflict, bringing 96 % effective transmission period. QKD provides the secure keys for AES symmetrical encryption application (i.e., IPsec protocols) both in downstream and upstream data, which separately needs one 128-bit key every four seconds for each ONU, bringing demand for 512 bps keys' bandwidth in total. The error correction, privacy amplification and other communications needed by QKD is performed by the classical channels.

The various parameters of the experimental network are shown in Table 1.

Table 1. Network settings and measurements

Parameters	Settings	
	Value	Unit
Upstream/downstream PRBS rate	1.25	Gb/s
Downstream light mean injection power	min:-15; max:3	dBm
Quantum pulse repetition rate f_{rep}	1.25	pulses/ns
Mean photon number	0.1	photons/ pulse
SPDs number n	2	–
SPD quantum detection efficiency	10	%
SPD dark count rate P_{dark}	1.0e-06	counts/gate
SPD detection gate width W	1	ns/gate
FMI Interference visibility V	98	%
FMI interference probability q	0.5	–
1546-1557 nm light fiber attentuation	0.185-0.22	dB/km
1310 nm light fiber attentuation	0.4	
1310/1550 WDM insertion loss	0.5	dB
Miscellaneous loss	1	
AWG insertion loss	0.6	
Coupler	3	
1:8 PLC Splitter insertion loss	10	

4.2   Results and Discussions

The RSOA's gain is determined by the optical injection power. Figure 3 shows the optical injection power against output gain characteristics of the RSOA. The downstream and upstream data BER at 193 THz are also figured in Fig. 3. According to Eqs. (1), (2), and (3), the downstream BER will slightly decrease along with the increase of RSOA's gain, while the upstream BER will increase. Hence the cross point of the two BER curves is an optimal point for both directions, which is about -15 dBm injection power. With our proposed scheme, low-penalty upstream and downstream transmission can be achieved simultaneously.

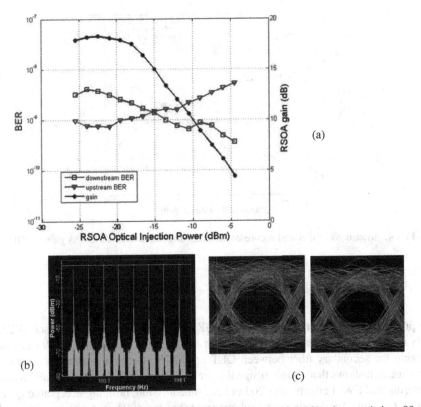

Fig. 3. Measured network performance for classical channels: (a) RSOA characteristic at 80 mA bias current and BER performances curves; (b) 8 DWDM wavelengths with spacing 200 GHz; (c) Downstream and upstream pulses eye diagrams of 193 THz.

To investigate the Raman-scattering impact on quantum signal, we changed the distribution fiber (from ONU to ODN) length and measured the corresponding quantum channel performance. Figure 4 is the measured mean QBER in 60 min along with the distribution fiber length changing at the RSOA's output gain 15 dB. It is shown that quantum channel will be affected slightly by the RSOA's reflection light when the co-existing in the same distribution fiber distance far below the peak distance mentioned in Eq. (8). If the distribution fiber length is less than 6 km, which is the most situations, QBER will be located under the theoretical safe limit about 11 % proved in [27], and QKD functions well. In Fig. 4, the total 8 ONUs' measured mean secure key (after error correction and privacy amplification) rate curve in 60 min is also given. More than 500 bps secure key exchanging bandwidth make the scheme meet most of the data security requirements.

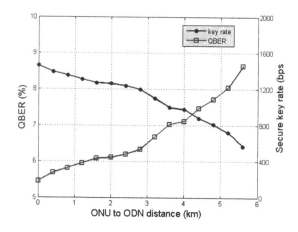

Fig. 4. Measured QBER and secure key exchanging rate in total at RSOA gain 15 dB.

5   Conclusions

We present a novel scheme integrated the QKD service into the RSOA based WDM-PON network architecture with some minor changes for optical branch devices and utilizing the secondary fiber between OLT and ODN. Given a 8-user PON network experiment, it shows that quantum signal is well isolated from classical signals' Raman scattering and FWM effects after 20 km fiber transmission, bringing acceptable QBER (below 11 % limit) and secure key exchange rate (beyond 500 bps) for encryption applications. Classical communications also function well by choosing optimal RSOA's gain and 8 200 GHz-spacing wavelengths. The compatibility between QKD and WDM-PON will greatly raise the access networks security level and facilitate the global distribution of quantum communication.

References

1. Diffie, W., Hellman, M.E.: New directions in cryptography. IEEE Trans. Inf. Theory **22**(6), 644–654 (1976)
2. Bennett, C.H., Brassard, G.: Quantum cryptography: public key distribution and coin tossing. In: India Proceeding of IEEE International Conference on Computers, Systems, and Signal Processing, Bangalore, vol. 175 (1984)
3. Gisin, N., Ribordy, G., Tittel, W., Zbinden, H.: Quantum cryptography. Rev. Mod. Phys. **74**(1), 145–195 (2002)
4. Li, J., Kim, K.: Hidden attribute-based signatures without anonymity revocation. Inf. Sci. **180**(9), 1681–1689 (2010). Elsevier
5. Li, J., Wang, Q., Wang, C., Ren, K.: Enhancing attribute-based encryption with attribute hierarchy. Mobile Networks and Applications (MONET) **16**(5), 553–561 (2011). Springer

6. Wu, Z., Liang, B., You, L., Jian, Z., Li, J.: High dimension space projection-based biometric encryption for fingerprint with fuzzy minutia. Soft Comput. (2015, in Press). doi:10.1007/s00500-015-1778-2

7. Wang, S., Chen, W., Guo, J.-F., Yin, Z.-Q., Li, H.-W., Zhou, Z., Guo, G.-C., Han, Z.-F.: 2 GHz clock quantum key distribution over 260 km of standard telecom fiber. Opt. Lett. **37**, 1008–1010 (2012)

8. Li, J., Chen, X., Li, M., Li, J., Lee, P., Lou, W.: Secure deduplication with efficient and reliable convergent key management. IEEE Trans. Parallel Distrib. Syst. **25**(6), 1615–1625 (2014)

9. Ciurana, A., Mart´ınez-Mateo, J., Peev, M., Poppe, A., Zbinden, N., Ciurana, H., Martin, V.: Quantum metropolitan optical network based on wavelength division multiplexing. Opt. Express **22**(2), 1576–1593 (2014)

10. Razavi, M.: Multiple-access quantum key distribution networks. IEEE Trans. Commun. **60**, 3071–3079 (2012)

11. Townsend, P.D.: Quantum cryptography on multiuser optical fibre networks. Nature **385**(6611), 47–49 (1997)

12. Elliott, C., Colvin, A., Pearson, D., Pikalo, O., Schlafer, J., Yeh, H.: Current status of the DRAPA quantum network. In: Proceeding of SPIE, Quantum Information and Computation III, 5815, p. 138 (2005)

13. Peev, M., et al.: The SECOQC quantum key distribution network in Vienna. New J. Phys. **11**, 075001 (2009)

14. Chen, W., Han, Z.F., et al.: Field experiment on a "star type" metropolitan quantum key distribution network. IEEE Photonics Technol. Lett. **21**(9), 575–577 (2009)

15. Sasaki, M., et al.: Field test of quantum key distribution in the Tokyo QKD network. Opt. Express **19**(11), 10387–10409 (2011)

16. Runser, R.J., Chapuran, T., et al.: Progress toward quantum communication networks: opportunities and challenges. In: Proceeding of SPIE, Optoelectronic Integrated Circuits IX, vol. 6476, p. 64760I (2007)

17. Bing, Q., Wen, Z., Li, Q., Hoi-Kwong, L.: Feasibility of quantum key distribution through a dense wavelength division multiplexing network. New J. Phys. **12**, 103042 (2010)

18. Patel, K.A., Dynes, J.F., Choi, I., Sharpe, A.W., Dixon, A.R., Yuan, Z.L., Penty, R.V., Shields, A.J.: Coexistence of high-bit-rate quantum key distribution and data on optical fiber. Phys. Rev. X **2**, 041010 (2012)

19. Choi, I., Young, R.J., Townsend, P.D.: Quantum information to the home. New J. Phys. **13**, 063039 (2011)

20. Hao, W., et al.: A GPON network architecture with integrated QKD service. Acta Photonica Sin. **43**, sup.1 (2014)

21. Choi, I., Young, R.J., Townsend, P.D.: Quantum key distribution on a 10 Gb/s WDM-PON. Opt. Express **18**(9), 9600–9612 (2010)

22. Cho, K.Y., Takushima, Y., Chung, Y.C.: 10-Gb/s operation of RSOA for WDM PON. IEEE Photonics Technol. Lett. **20**(18), 1533–1535 (2008)

23. Jin, N.Z., Xue, G.Y., Yue, G., Yong, Q.H., Ming, L.Z., Yan, G.Z.: A novel bidirectional RSOA based WDM-PON with downstream DPSK and upstream re-modulated OOK data. In: IEEE ICTON 2009

24. Guo, Q., Tran, A.V.: Reduction of backscattering noise in 2.5 and 10 Gbit/s RSOA-based WDM-PON. Electron. Lett. **47**(24), 1333–1335 (2011)

25. Keuo, Y.C., Yong, J.L., Hyeon, Y.C., Ayako, M., Akira, A., Yuichi, T., Yun, C.C.: Effects of reflection in RSOA-based WDM PON utilizing remodulation technique. IEEE J. Lightwave Technol. **27**(10), 1286–1295 (2009)

26. Peters, N.A., et al.: Dense wavelength multiplexing of 1550 nm QKD with strong classical channels in reconfigurable networking environments. New J. Phys. **11**, 045012 (2009)
27. Lutkenhaus, N.: Estimates for practical quantum cryptography. Phys. Rev. A **59**(5), 3301–3319 (1999)

New Security Challenges in the 5G Network

Seira Hidano[1]([✉]), Martin Pečovský[2], and Shinsaku Kiyomoto[1]

[1] KDDI R&D Laboratories, 2-1-15 Ohara, Fujimino-shi, Saitama 356-8502, Japan
se-hidano@kddilabs.jp
[2] Technical University of Košice, Košice, Slovakia

Abstract. Security is a fundamental aspect of the next generation mobile network. In this paper, features of the 5G network, such as IoT and D2D, are described along with the proposal of a new layered network model, which enables independence on multiple radio access technologies (RATs). Additionally, security requirements on the 5G network are stated to facilitate security examination. Security issues arising from new technologies to be used in the 5G network, for example, physical layer security, are investigated in the core section of the paper.

Keywords: 5G network · IoT · D2D · RATs · Physical layer security

1 Introduction

Mobile networks have dramatically changed our lives over the last few decades. From analogue voice networks in the 1st generation, through their digitalization in 2G and improvements on data access in 3G, and technologies currently being deployed in 4G, they have achieved capabilities which were unforeseeable a few decades ago. However, the development does not stop here. There were 3.6 billion unique mobile subscribers at the end of 2014, with 7.1 billion global SIM connections and 243 million machine-to-machine (M2M) connections. Half of the world's population has a mobile subscription now, and an additional one billion subscribers are predicted by 2020. Moreover, users still require the emergence of new services and new applications of M2M connections.

The concept of the 5G network is promising to satisfy the growing needs of mobile wireless communication. Along with increasing data rate, number of users, reliability and coverage of the mobile network, security is a matter of key importance that requires careful consideration. As with the upcoming spread of the Internet of Things (IoT) that the 5G network is going to propagate to almost all aspects of our lives, security will become even more crucial than it is now.

Many new technologies are emerging to be deployed in the 5G network and improve its performance. Their security issues should be examined so that appropriate countermeasures can be taken before new technologies are deployed into live operation. Novel approaches for security enhancement also have been proposed. In particular, physical layer security seems to offer reasonable solutions for many security requirements. It is important to recognize its possible risks and point out topics for further research.

K. Li et al. (Eds.): ISICA 2015, CCIS 575, pp. 619–630, 2016.
DOI: 10.1007/978-981-10-0356-1_64

1.1 Our Contribution

- We propose a new layered network model of the 5G network taking the collaborative multiple radio access technologies (RATs) into consideration from the security point of view (see Sect. 2). This section also summarizes basic features and technologies of the upcoming 5G network.
- We formulate security requirements on the 5G network as they arise from its features (see Sect. 3).
- We discuss new 5G technologies to clarify their possible security issues (see Sect. 4). Security issues on the 5G network are divided into two categories:
  - *Security issues inherited from previous generations of wireless networks*: Legacy networks with well-known security issues will be integrated into the heterogeneous architecture of the 5G network. Therefore, the new network will inherit their risks, and new security architecture must be well designed to be able to cope with them.
  - *Security issues new to the 5G network:* New technologies which were not widespread or were not available at all in previous wireless networks will introduce new issues into the 5G network. They need to be properly examined from a security perspective and countermeasures taken.

We especially focus on the new security issues in this paper and derive new security challenges that must be overcome.

2 Definition of the 5G Network

The so-called 5G network has not been standardized yet but after the 4G networks were deployed it has become a major focus of interest for researchers. ITU-R plans that final specifications of the 5G network (IMT-2020) will be issued in 2020 [9]. We point out its basic features and technologies followed by the proposal of a layered network model.

2.1 Features of the 5G Network

The most obvious and often compared feature of the new generation of mobile network is an increased data rate. It is widely agreed that the 5G network should be capable of a data rate of several Gbps per user. The main reason for increasing the data rates is growing demand on services with large-sized content, such as ultra-HD video and 3D video [15].

Apart from the data rate, the number of devices connected to the wireless network is also expected to increase dramatically. Not only are the numbers of human subscribers and devices per unique subscriber rising as mentioned above, but also an explosive growth in Machine-to-Machine (M2M) communication is predicted. The total number of mobile connections could become several hundred times the world population. In crowded areas, it could demand up to 1 million connections per square kilometer [8]. Latency in the 5G network for the load off the base station should be less than 1 ms round-trip for latency-sensitive applications with jitter of 20 μs. Low latency and high reliability are important for many services, e.g., self-driving cars, telemedicine and smart-grids, collectively called the Internet of Things (IoT), requires support for connecting large numbers of devices with different hardware and software capabilities and network resource requirements, which operate unattended by humans. For battery-operated devices, energy-efficient access to the network is required.

The two most significant differences between 5G and previous generations of mobile networks are device-to-device (D2D) communication and heterogeneous network architecture. D2D communication is a novel approach for communication in mobile wireless networks. Unlike the previous generations, devices should be allowed to communicate directly without passing data through the base station. Relaying using D2D should be enabled, resulting in taking the load off the base stations and saving power. Under the heterogeneous architecture, the integration of several wireless networks with various radio access technologies (RATs) is planned. Various RATs should be interconnected into the one network to achieve better coverage, flexibility in satisfying different service requirements and seamless handover between RATs. Moreover, high mobility up to speeds of 500 km/h should be achieved in the heterogeneous network.

2.2   Technologies of the 5G Network

Behind the expected features of the 5G network, a complex technical background should be established to fulfill the requirements.

To satisfy the requirements of high data rate and high user density, a heterogeneous network structure with densified small cells overlaying coverage is used. Small (micro-, nano-, pico-, femto-, atto-) cells offer high data rate and spectral efficiency due to the operation in high frequency bands with low power. On the other hand, macrocells operating at low frequencies provide wide coverage, mobility support and control. The small cells may be implemented as static WiFi, BlueTooth, millimeter wave, visible light communication (VLC), etc., and also as mobile femtocells deployed on vehicles. For the macrocells, the infrastructure of the previous generation's mobile networks can be used. Small base stations can be situated indoors, reducing the power loss caused by propagation through the walls and exploiting it for frequency reuse. A device will connect with a base station, which is the best way to satisfy its requirements and load-balance the network in the heterogeneous network. However if it is more efficient, data should be sent directly or through the relayed D2D link.

Another new network technology in 5G is software-defined networking (SDN). The main feature of SDN is decoupling the control plane and data plane by moving the control logic from underlying switches and routers to a centralized SDN controller in the control plane [5]. Switches simply follow the commands of the SDN controller, which has global knowledge of the network. This makes network reconfiguration easier because changes are made only in applications on top of the controller, resulting in enhanced network flexibility.

Deployment of new radio frequency (RF) technologies is expected in the 5G network. Massive MIMO with more than 10 antennas per data stream and 3D MIMO will enable accurate beamforming to save power and frequency resources [17]. Massive MIMO can be used in conjunction with spatial modulation. Spatial modulation is a recently developed transmission technique that uses multiple antennas. The basic idea is to map a block of information bits to two information-carrying units: a symbol chosen from a constellation diagram and a unique number of a transmit antenna chosen from a set of transmit antennas. The use of the transmit antenna number as an information-carrying unit increases the overall spectral efficiency [13].

Instead of looking for new frequency bands, it is possible to run the mobile communication service as a secondary service in bands assigned to other licensed primary radio services. This assumption is based on the fact that many radio communication services do not exploit all the frequencies assigned to them all the time. To enable dynamic spectrum allocation in the 5G network, the cognitive radio (CR) technology is a plausible method. It may be used together with software defined radio (SDR) to achieve hardware-independent compatibility with various RATs in an extremely wide frequency range.

A novel approach to security in the 5G network is physical layer security that exploits properties of local radio environments, which are complexity, dispersive and non-stationary [2]. Unlike conventional cryptography, physical layer security is based on randomness provided by the radio channel instead of computational hardness assumptions. Physical layer security has some advantages over the conventional key-based cryptography. It is independent of the device's computational capabilities. On a legitimate device, complicated computations are eliminated, allowing for power savings and low-cost hardware. On the eavesdropper's side, unlimited computational power, including quantum computing, will not allow the attacker to be successful. In the physical layer security concept, keys are neither needed nor exchanged. Because of that, no attacks against key exchange mechanisms are achieved and the link bandwidth necessary in conventional cryptography for key exchange can be saved for other services.

2.3 5G Layered Model

We propose a layered model for the 5G wireless network depicted in Fig. 1. The model is derived from ISO/OSI and TCP/IP basic network models. It consists of multiple layers, which are divided into the user plane and the control plane. The upper layers of the user plane protocol stack have the same functions as in the TCP/IP model.

The cooperation layer is the most specific part of the 5G network model. It enables compatibility among different RATs by taking control functions from the RAT-specific layers to the separated control plane, and it simultaneously abstracts the network access of higher layers. It can be implemented by SDN or a similar technique, which moves network control to a centralized controller with global knowledge over the network.

In the control plane, there is a controller layer that includes the centralized controller with its operating system and a robust network connection offering a solid background for control functions. The controller is implemented in a cloud

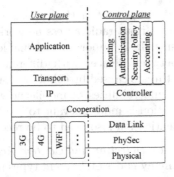

Fig. 1. Layered model of the 5G network.

that provides flexible computational resources and high reliability. The control modules are applications running on top of the controller, which are responsible for network and security functions such as routing, authentication, trust management, unified security policy, automated network management, accounting, etc. This architecture enables not only independence on RATs but also high flexibility and extensibility, as protocols can be changed only by updating the control module application running on the controller without needing to exchange or reconfigure multiple network devices.

The lower layers of the user plane are RAT dependent and are implemented as legacy wireless networks like 3G, 4G or WiFi or as new technologies designed for 5G based on millimeter waves, VLC or a similar model. The data link and physical layers are identical with the ISO/OSI protocol, except that they are controlled by the cooperation layer. Between those layers, the physical security (PhySec) layer is inserted to achieve additional security improvements based on the physical layer security principles. Along with security capabilities, other characteristics will also be different, e.g., available data rate, latency, required transmission power, etc. In this case, the cooperation layer should control which RAT to use for the transmission of each packet, according to the RAT's properties, QoS and security requirements of the packet.

3   Security Requirements on the 5G Network

ITU-T Recommendation X.805 [10] defined eight security dimensions as sets of security measures designed to address particular aspects of network security, which protect against all major security threats. When considering the security of the 5G network, these security dimensions can be taken as basic security requirements. In addition to these general requirements, we can formulate security requirements specifically for the upcoming 5G network. They originate in the new architecture, technologies and services, which will be deployed in the 5G network. These security requirements are:

Authentication and Trust Management in Heterogeneous Network: The 5G network is expected to include various radio access networks with formerly incompatible security mechanisms. It is important to enable secure authentication and trust management between devices with different capabilities and standards. Therefore, security threats caused by collaboration among multiple RATs as well as authentication vulnerabilities of mobile networks should be solved. Although device authentication is more relevant to the 5G network technology, some aspects of user authentication also need to be reconsidered.

Topology Discovery, Intrusion Detection and Content Access Control: 5G should be a flexible network with many nodes of different types moving in and out of the network and among different locations. Moreover, it can be expected that some base stations will be owned by the operator and others by private entities, which may change their settings unintentionally or try to perform an intentional attack. Therefore, these requirements are important for network flexibility and mitigating the damage from malicious nodes. In such a flexible and partly user-owned network enabled with relaying among multiple users, content access control will become a complex issue. New security techniques should prevent MitM attacks in both conventional and D2D scenarios as well as misconfiguration of access points by users or attackers.

Security Preservation and Enhancement: It is obvious that the level of security in the new network should be higher than in previous network generations. To satisfy this requirement, it is essential to be able to cope with the previous generation of security faults. Otherwise, security vulnerabilities of various network technologies which will be incorporated into the 5G network will be propagated into the new network. The concentration of inherited vulnerabilities can consequently lead to a decrease in overall network security. The security threats from multiple RATs must be mitigated, and the security of previous generations of wireless networks should be improved by additional new security technology, e.g., physical layer security. As physical layer security has both advantages and disadvantages, its issues must be analyzed.

Low Handover-Introduced Latency: Although the handover latency may not be considered as a security issue, handover is often accompanied by authentication procedures. High mobility among small cells results in frequent handover and authentication. The agreement is for security engineers to develop fast authentication algorithms for handover to avoid degrading the mobility performance of the network. Fast authentication procedures also help to prevent DoS attacks, as the bottlenecks created by demanding algorithms are mitigated. In the case of physical layer authentication, its vulnerabilities must be examined.

IoT Security: IoT is becoming a new phenomenon of the wireless world. It brings a huge number of simple low-cost devices with low computational capabilities to the network, operating unattended by humans and often located in distant places. Moreover, some IoT devices are employed in critical applications, such as smart grids, self-driving vehicles, smart cities, etc. IoT security is a strict requirement on the 5G as the number of IoT connections will soon exceed the number of personal mobile connections. Therefore, possible security issues arising out of IoT need to be considered and countermeasures taken.

Mobile Cloud Computing Security: Mobile cloud computing is expected to become widely spread in 5G networks. Although mobile cloud computing is mostly a matter of higher layers, its security should be taken into consideration when 5G protocols are designed. In addition to mobile access to a conventional cloud service, formation of a mobile cloud consisting of mobile devices co-located in a common geographical area is possible [6]. Malicious devices are a primary concern in the mobile cloud, where a mobile device shares work with other mobile devices, the device users are unknown and the mobile cloud is formed opportunistically with support for high mobility. Security issues with key management and authentication are therefore crucial in the multiuser mobile scenario [12].

Saving Resources: Security procedures in the 5G network should not degrade network performance. Not introducing too much computation and overhead is the key feature to mitigate latency, bandwidth occupation and power consumption. Limited energy resources of the mobile devices must be taken into consideration. Jamming and DoS attacks should be prevented to achieve efficient network operation. Moreover, CR and MIMO beamforming must be examined from the security point of view as the technologies enabling high spectrum reuse.

Flexibility, Inter-operability, Compatibility and Extensibility of Security Protocols: These general security requirements are becoming even more important in the 5G network. Extensibility of security protocols is required not only for their new applications but also for the ability to fix security vulnerabilities as they will appear after the 5G deployment. Moreover, backwards compatibility should be carefully considered to overcome as many inherited threats as possible and simultaneously not force expensive replacement of the previous components.

4 New Security Issues to 5G Networks

Besides the security requirements stated in Sect. 3, other topics which need to be taken into consideration when designing new security protocols are the security issues in the network. Although 5G security standards are not defined yet, possible vulnerabilities and threats are already emerging. As mentioned in Sect. 1, some of them originate in previous generations of wireless networks while the others arise from new technologies proposed for the 5G. In this paper, we discuss security challenges originating from novel technological concepts.

4.1 D2D Issues

D2D communication is a totally new technique in mobile wireless networks and its security issues should not be underestimated. As mobile end-user devices have lower computational capabilities than base stations, security protocols for D2D communication must be both simple and robust. Moreover, relaying in D2D manner needs to be supported.

Therefore, a relaying device with unknown ownership and configuration and uncertain location and mobility can be considered as the weakest point of the link in this scenario, possibly with the intention of eavesdropping on the communication. A variation of cooperative jamming is proposed as a possible solution in [14]. Assuming a half-duplex amplify-and-forward protocol, an effective countermeasure may be to have the destination jam the relay while it is receiving data from the source. This intentional interference can then be subtracted out by the destination from the signal it receives via the relay.

Location, relative mobility and power resources of the relaying device must be considered to ensure service availability during the required time period as a protection against sudden DoS. Correct content access control and billing is also an issue in the case where the services are not accessed through the operator's infrastructure. In this case, proper authentication is important to prevent masquerading attacks in which an attacker hidden in the relaying structure tries to access paid services under the identity of another user.

4.2 IoT Security

IoT is a novel approach to wireless networks as well, although it is already becoming widespread in 4G networks [1]. As the deployment of IoT devices is expected not only in low-priority applications but also critical ones, its security will become crucial. Similarly to D2D, low computational capabilities and battery life must be taken into consideration in IoT. In this case, attacks will include battery-draining of multiple devices and subsequent malfunction of the whole system. Moreover, IoT devices such as sensors and actuators may be located in distant areas, operating over an extremely long period of time without human oversight [1]. During this time, an intruder can gain physical access to the device and modify its function or launch side-channel attacks. IoT systems are expected to include a large number of devices with 5G connections. Being able to control multiple unattended devices in the network, an attacker may exploit them for distributed DoS attacks.

Authentication of IoT devices is important to prevent the attacker from introducing a false device into the system. The authentication protocol must be simultaneously simple and secure to achieve reliability and resource saving. Centralized protocols may be the solution for this task. In the centralized architecture, many low-cost devices with very low capabilities may be authenticated by a single authentication device that is able to carry out most of the computations, thereby saving power and implementation costs of the end devices. Moreover, by accessing the slave device with constrained functionalities, the intruder may gain less control over the network than in the case of autonomous devices in a peer-to-peer network. On the other hand, centralization has also disadvantages as described in Sect. 4.3. Decentralized protocols might be hard to implement, especially with incompatible devices from multiple vendors. Standardized security techniques should be used in the IoT segment among all vendors to prevent exploitation of incompatibilities by intruders.

4.3   Cognitive Radio Threats

In the CR architecture, special uplink and downlink channels are designed to support spectrum database downloading, spectral sensing and channel sounding procedures. These channels should be quickly recognizable, easily decodable and support accurate measurement. Their inherent properties make them very susceptible to both active and passive attacks. As the CR is an underlying structure of the wireless network, its failure will usually result in the DoS. However, in the opposite case, the CR infrastructure controlled by an intruder can be exploited to jam primary services in shared bands. Although such an attack could be easily detected, in case of emergency or military primary services, it should be considered seriously as it may become a threat to public security.

A primary user emulation attack is a well-known threat to CR. In the primary user emulation, a malicious node mimics the signal characteristics of licensed users in order to mislead the CR into vacating the spectrum [14].

4.4   Attacks Against Massive MIMO

Massive MIMO is generally considered to be a technology which enables an increase in security by proper beamforming and cooperation with the physical layer security. However, if the massive MIMO is successfully attacked, exactly the opposite may happen.

To enable accurate beamforming, the transmitter needs to perform channel estimation. During the channel estimation procedure, the mobile user is required to send his pilot sequence, which helps the MIMO device (e.g., a base station (BS)) to form the beam towards the user. In this phase, the eavesdropper can send a different pilot sequence to force the BS to form the beam towards the eavesdropper. This practice is called a pilot contamination attack. If successful, a significant drop in the secrecy capacity of the legitimate link occurs. As a result, the eavesdropper has better signal-to-noise ratio (SNR) than the legitimate user and is able to overhear the communication if solely physical layer secrecy is exploited. Moreover, this type of attack is hard to detect since there is a normal amount of pilot contamination already present in MIMO systems [11]. A security protocol to hinder channel state information (CSI) spoofing in MIMO systems by randomizing the channel sounding signals was proposed in [18] as a countermeasure against MIMO attacks.

4.5   Collaboration of Multiple RATs in 5G

Integration of multiple RATs into one network will be probably the hardest challenge for 5G, as today's technologies are incompatible. Incompatibility includes not only physical and MAC layer specifications but also security protocols. If not designed properly, merging the technologies may result in serious security faults. The most important challenges for 5G security are therefore unified authentication and the security policy. Software defined networking (SDN) may be an appropriate platform for unifying incompatible protocols. By separating the control and user plane, it moves control from the network devices to the centralized controller and makes it independent of the RATs. Apart from networking functions, security services like authentication and trust management could also be implemented in a similar centralized way. However, centralizing carries multiple risks. The centralized controller becomes the point of interest for possible attackers. By accessing the controller, an intruder gains full control over the network and access to confidential data about network traffic. In the event of controller failure or connection loss, the whole network collapses. Message exchange between the controller and devices can be attacked by an intruder. MitM and false controller attacks may occur. Therefore, robust authentication between the controller and devices is needed. Centralization creates a network bottleneck which may be exploited in DoS attacks.

Therefore, two possible solutions for cooperation of multiple RATs by unifying their networking and security protocols are emerging. Centralized architecture with global knowledge over the network which will be able to mitigate the risks mentioned above by proper countermeasures without degrading network performance. Decentralized protocols also mitigate some of the risks, but much harder to implement into the complex environment of heterogeneous network.

Another security threat that should be suppressed by unified security comes from the multi-RAT devices. As we have mentioned in previous sections, various RATs have different degrees of security and different vulnerabilities. In the overlaying coverage architecture, the multi-RAT device can be forced to operate in the least protected mode by a simple jamming procedure of the other modes. Consequently, attacks may be accomplished easily [4].

4.6 Physical Layer Security Challenges

In 5G networks, fast and secure authentication algorithms are needed to support high mobility. Physical layer security provides many ways to satisfy this requirement. Some methods include a secret key hidden in the modulation scheme, while others make use of non-ideal properties of the radio link or hardware equipment, so-called fingerprinting [14]. Various methods of physical layer authentication were described in [20]. Although its concept is a recommended solution for the 5G network, it has possible weaknesses and challenges for future research.

First, the secrecy capacity is discussed in terms of information-theoretic security. It can only ensure with high, but not certain, probability that a data block is secure, as the information-theoretic security is an average measure [3].

Second, practical technologies for physical layer security are still not widely available. Further research on secrecy coding and new security algorithms for practical channel models are required to implement all the theoretical knowledge into industrial standards. The secrecy coding schemes for both capacity and security maximization in realistic channel environments are still a challenge. An overview of research activities in this field was given in [7].

Also, a vulnerability of the physical layer security originates from its basic principles. It requires assumptions about radio channels, which might not be accurate. The CSI of a passive eavesdropper can only be guessed in practice.

This can lead to conservative assumptions and a significant drop in link capacity, or to secrecy capacity overestimation and leakage of confidential information. Moreover, channel estimation may be attacked in a similar way to a MIMO pilot contamination attack, resulting in information leakage. On the other hand, faked channel estimation leading to low secrecy capacity results can cause DoS.

All the assumptions above were made based upon a simple network, namely, the wiretap channel principle [19]. However, in a real mobile network, there are many users and possibly also many eavesdroppers. The following factors all lead to significant security degradation: allowing eavesdroppers to collude, increasing their number or the legitimate link length and decreasing SNR on the legitimate link [16]. In addition, the degradation must be correctly estimated by legitimate users to prevent information leakage. This will be a hard task, because the number and collusion ability of eavesdroppers is usually unknown, as well as the CSI of every single eavesdropper. On the other hand, the potential of multiple users distributed in different locations may be exploited as a security countermeasure in cooperative jamming of eavesdroppers as described in [14].

The vulnerabilities in state of the art technology for the physical layer security imply that it must be deployed carefully after sufficient research. However, coupling it with conventional cryptography can be an effective solution for security enhancements in the 5G network. Sometimes using a conventional encryption algorithm based on a secret key may be more efficient than exploiting information-theoretic security. In this case, physical layer security principles may be used to generate a common secret key and therefore overcome complex key-exchange mechanisms. Moreover, the physical layer may offer key generation at high rates. Under these circumstances, the key can be used as a one-time pad to enable ciphering with perfect secrecy. A secret key generation technique based on the most common principle of randomness sharing, reconciliation and privacy amplification was proposed in [3], and additional techniques can be found in [14].

5 Conclusion

In this paper, we gave a brief overview of the 5G network, its features and technologies. A new 5G layered network model aimed at cooperation among multiple RATs was proposed to assist network design and evaluation. Security requirements on the 5G network were stated to clarify goals of security designs for 5G. New security issues originating in the technologies to be deployed in 5G were examined to help to take appropriate countermeasures for the future network. There are many technologies proposed in related work which offer solutions for the 5G network, e.g., physical layer security, massive MIMO, centralized SDN architecture, etc. All of them have additional issues, which must be taken into consideration when designing the new network. Security research on 5G is a matter of key importance and should not be forgotten in the shadow of the race for ultra-high data rates. Otherwise, security flaws in the future network may downgrade its performance more seriously than robust security algorithms and cause disappointment to both network service provider and its users.

References

1. Americas, G.: 4G Americas' Recommendations on 5G Requirements and Solutions (2014)
2. Belfiore, J.C., Delaveau, F., Garrido, E., Ling, C., Sibille, A.: PHYSEC concepts for wireless public networks—introduction, state of the art and perspectives. In: The Proceedings of SDR-WinnComm 2013 (2013)
3. Bloch, M., Barros, J., Rodrigues, M.R.D., McLaughlin, S.V.: Wireless information-theoretic security. IEEE Trans. Inf. Theor. **54**(6), 2515–2534 (2008)
4. Delaveau, F., Evesti, A., Suomalainen, J., Savola, R., Saphira, N.: Active and passive eavesdropper threats within public and private civilian wireless networks—existing and potential future countermeasures—a brief overview. In: The Proceedings of SDR-WinnComm 2013 (2013)
5. Duan, X., Wang, X.: Authentication handover and privacy protection in 5G Het-Nets using software-defined networking. IEEE Commun. Mag. **53**(4), 28–35 (2015)
6. Fernando, N., Loke, S.W., Rahayu, W.: Mobile cloud computing: a survey. Future Gener. Comput. Syst. **29**(1), 84–106 (2013)
7. Harrison, W.K., Almeida, J., Bloch, M.R., McLaughlin, S.W., Barros, J.: Coding for secrecy: an overview of error-control coding techniques for physical-layer security. IEEE Signal Proc. Mag. **30**(5), 41–50 (2013)
8. IMT-2020 (5G) Promotion Group: 5G Visions and Requirements (2014)
9. ITU-R: Future technology trends of terrestrial IMT systems. Report M.2320 (2014)
10. ITU-T: X.805 Recommendation: Security architecture for systems providing end-to-end communications (2003)
11. Kapetanović, D., Zheng, G., Rusek, F.: Physical layer security for massive MIMO: an overview on passive eavesdropping and active attacks. IEEE Commun. Mag. **53**, 21 (2015)
12. Mane, Y.D., Devadkar, K.K.: Protection concern in mobile cloud computing–a survey. IOSR J. Comput. Eng. pp. 39–44 (2013)
13. Masleh, R.Y., Haas, H., Sinanović, S., Ahn, C.W., Yun, S.: Spatial modulation. IEEE Trans. Veh. Technol. **57**(4), 2228–2241 (2008)
14. Mukherjee, A., Fakoorian, A.A., Huang, J., Swindlehurst, A.L.: Principles of physical layer security in multiuser wireless networks: a survey. IEEE Commun. Surv. Tutorials **16**(3), 1550–1573 (2014)
15. Nokia Solutions and Networks: 5G use cases and requirements (2014)
16. Pinto, P.C., Barros, J., Win, M.Z.: Secure communication in stochastic wireless networks–part II maximum rate and collusion. IEEE Trans. Inf. Forensics Secur. **7**(1), 139–147 (2012)
17. Radio Access and Spectrum, FP7 Future Networks Cluster: 5G radio network architecture (2014)
18. Tung, Y.C., Han, S., Chen, D., Shin, K.G.: Vulnerability and protection of channel state information in multiuser MIMO networks. In: Proceedings of the 21st ACM SIGSAC Conference on Computer and Communications Security (CCS 2014), pp. 775–786 (2014)
19. Wyner, A.D.: The wire-tap channel. Bell Syst. Tech. J. **54**(8), 84–106 (1975)
20. Zeng, K., Govindan, K., Mohapatra, P.: Non-cryptographic authentication and identification in wireless networks. IEEE Wirel. Commun. **17**, 56–62 (2010)

A Method of Network Security Situation Assessment Based on Hidden Markov Model

Shuang Xiang[1]($\boxtimes$), Yanli Lv[1,2], Chunhe Xia[1], Yuanlong Li[1],
and Zhihuan Wang[3]

[1] Beijing Key Laboratory of Network Technology,
Beihang University, Beijing 100191, China
805051394@qq.com
[2] Information Center of Ministry of Science and Technology
of the People's Republic of China, Beijing 100862, China
[3] Yucheng Group, Beijing 100000, China

Abstract. In the network security situation assessment based on hidden Markov model, the establish of state transition matrix is the key to the accuracy of the impact assessment. The state transition matrix is often given based on experience. However, it often ignores the current status of the network. In this paper, based on the game process between the security incidents and protect measures, we improve the efficiency of the state transition matrix by considering the defense efficiency. Comparative experiments show the probability of the network state generated by improved algorithm is more reasonable in network security situation assessment.

Keywords: Network security situation assessment · Game matrix · Defense efficiency

1 Introduction

With the extensive application of network technology, network is expanding and opening, so the network will be affected by various security threats, such as an external attacker invasion, DDoS, worms, viruses, internal attacks etc. Network administrators need consider the impact of various factors on network security, it is difficult to discover network risk and take countermeasures effectively. If we can evaluate the risk of the network in real time, and discover the main problems in the network and improve solutions dynamically, in that way it will help administrators to find solutions to the risk quickly. So, real-time network security risk assessment have important theoretical and practical value.

2 Related Work

Network Situation was proposed by Tim bass [1,2] in 1999, which reflects the current state of the entire network. Chen [3,4] constructed hierarchical network system security threat posture, from top to bottom into network systems,

© Springer Science+Business Media Singapore 2016
K. Li et al. (Eds.): ISICA 2015, CCIS 575, pp. 631–639, 2016.
DOI: 10.1007/978-981-10-0356-1_65

hosts, services and attacks/vulnerabilities four levels to assess network situation. Ning [5,6] had realized the method of analyzing the threat of the network from the mass alert information through the alert correlation. Zhang [7] established a network security situation awareness method based on the Markov game model by analyzing the game relationship between the managers and the common users, but the model focused on the local threat assessment, the overall concept is not strong. Arnes [8,9] proposed using hidden Markov model (HMM) on network security state modeling, based on statistical model to solve the uncertainty problem of alarm information. But the state transition matrix cant reflect the change of the network security state objectively according to the empirical values. Masoud [10] approached yields scalability and integration of risk assessment and mitigation processes Using Bayesian decision networks. Xi [11] Based on the game process of security incidents and protective measures, proposed the method of determining the state transition matrix, but it ignores the current defense system of the network, which makes the method of network situation assessment is not accurate enough.

In this paper, we propose an improved method for the determination and correction of the state transition matrix, and the main work is to modify the transition matrix according to the real time state of the network, and to improve the validity of the transition matrix.

3   Risk Assessment Model

In the actual network, the network security is invisible, but the alert information is visible, and there is a specific probability relationship between the security state and the warning information. So we use Hidden Markov model to describe the transition of network security state.

3.1   Model and Terminology

Hidden Markov Model is represented formally by a 5-tuple $\lambda = \{S, V, P, Q, \pi\}$ [12], In order to apply it into the network security situation assessment, it will be extended to 7-tuple: $\lambda = \{S, V, P, Q, \pi, E, C\}$, where:

- S is a finite set of state, $S = \{s_1, s_2, ..., s_n\}$, where s_i represents an independent state, N represents the number of states. Different security event will lead to network into different security state, according to the division of security incidents, the network security state is divided into 4 states in this paper: Good state (G), Probed state (P), Attacked state (A) and Compromised state (C), $S = \{G, P, A, C\}$. Where:
  - Good(G): There is no attack behavior in the network, which is in a safe state,
  - Probed(P): There is behavior that someone is scanning the network in the network, and the attacker my be collecting the information,
  - Attacked(A): There is behavior that someone is attacking the authority of the system in the network,

- Compromised(C): The network has been compromised, the attacker may obtain system privileges.
- V is a finite set of observations, $V = \{v_1, v_2, ..., v_n\}$, where v_i represents the observation. Network security state is invisible, it can be only directly observed through security alert information generated by the equipment. If the mass alert is directly applied to the HMM model, it will lead to large scale of the model, which seriously affects the operation efficiency. To solve the above problems, the original alert should be classified. According to the security incident represented by the alert, the alert is divided into four categories, $V = \{g, p, a, c\}$. Where:
  - g: within the sampling period, not acquired any alert messages,
  - p: scanning alarm class information, such as ICMP ping, ICMP destination un-reachable,
  - a: Intrusion alarm class information, such as web cgi redirect access, bad traffic loopback,
  - c: Get root permissions alert information, such as services rsh root.
- P is state transition matrix, represents the probability of a transition from one state to another. $P = \{p_{ij}\}$, where $p_{ij} = P(\alpha_{(t+1)} = s_j | \alpha_t = s_i)$, which denote the probability that the network is in a state s_i at time t, and it transfer to state s_j at time t+1.
- Q is confusion matrix, $Q = \{q_i(v_k)\}$, where $q_i(v_k) = P(o = v_k | \alpha = s_i)$ denote the probability that the network is in a state s_i and the alert information v_k is observed.
- π, which is Initial state distribution matrix. $\pi = \{\pi_i\}$, where $\pi_i = P(\alpha_1 = s_1)$ denote probability that the network is in state s_1 at the initial moment.
- E, which is the efficiency of the current network defense, said the current network state of the defense capability of the efficiency, $0 \leq E \leq 1$, E is the higher and the network defense capabilities is the better.
- C is cost vector, $C = \{c_i\}$, where c_i denote When the network is in the state s_i, the system faces the risk value.

In this paper, we only discuss the generation of the state transition matric, the efficiency of the defense measures and the risk value calculation methods.

3.2 Generate State Transition Matrix

The state transition matrix is the key to evaluate the accuracy of the security situation. The traditional state transition matrix is often based on experience, so it is strong subjective. In this paper, the state transition matrix is determined based on game theory.

Event driven network security state transition process will be regarded as a state machine, security event is the basis of network state transition. Event driven network state transition model is shown in the following Fig. 1 [13]:

Where G, P, A, C represent the network state, E_i represent security incidents, D_i represent protect measures. If the network is in the state s_i at time t, when it a security incident E_p occurs, and the security measures in network is D_q, then it lead to the network enter the state s_j at time t+1. The process is expressed as:

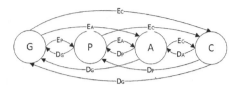

Fig. 1. Network security state transition process

$$s_i \xrightarrow{E_p \wedge D_q} s_j \qquad s_i, s_j \in S \tag{1}$$

Based on the game between security incidents and protective measures, probability of state can be calculate as follow:

$$p_{ij} = \sum P(s_i \xrightarrow{E_p \wedge D_q} s_j) = \sum_{s_k \in S} s_j / s_k \tag{2}$$

It can be obtained from the current state s_i of the safe transition of state transition vector.

We construct the state transition matrix of the current network state s_i, then we can obtain the initial transition matrix.

In real networks, when the attack occurs, the defensive measures will not necessarily be able to defend itself, so that the evaluation is more accurate.

3.3 Defense Efficiency and State Transition Matrix Correction

The defense equipment efficiency is mainly from the hardware capability and the actual load, considering whether the service is running healthy, all kinds of node in the network is able to work regularly and provide services to users in time. The basic operating index is used to reflect the safe state of the network equipment and services, and the calculation of the basic operation index is carried out mainly through the following evaluation factors [14]:

- Peak flow: Peak flow is the largest data stream received by a network device over a certain period of time, which is used to measure the security of the network equipment.
- Bandwidth utilization: High network bandwidth utilization of one or more hosts or all network devices. It will reduce the performance of the network and limit the normal activities in the network.
- CPU utilization: CPU utilization is very important to measure the performance of a host or device, the higher rate of CPU occupying, the worse it is used to accomplish other tasks. In this paper, the CPU utilization is the average utilization rate in a time period.
- Memory utilization: memory utilization is used to measure the real-time performance of a device. A number of denial of service attacks have been proven to be the target of the resource, so that it is important to measure the operating state of a host or device as important as the CPU utilization.

In the following steps, the quantitative analysis of the above situation assessment factors is carried out:

- The first step: measure the efficiency of each index in each time period to measure the severity. The efficiency is defined as follows:

$$e_i = \begin{cases} 1, & s_i \leq N \\ 1 - s_i/R, & s_i > N \end{cases} \tag{3}$$

Where $i = 1, 2, 3, 4$, denoting, the traffic peak, bandwidth utilization, CPU utilization and memory utilization. s_i represents the current state. R represents the total amount of resources in the system, and N indicates the threshold value of i.

- The second step: according to the target network state, determine the weight of the index. Then it can calculate comprehensive efficiency of defense equipment. The calculation formula is as follows:

$$E = \sum(e_i w_i) \tag{4}$$

- The third step: because of the different defense efficiency, the probability of the success of the attack is different, resulting in the transition of the transition matrix in different conditions, the introduction of the modified vector $\beta = \{\beta_t\}$,

$$\beta_t = \begin{cases} E, & o \in g \\ 1/E, & o \ inp, a, c \end{cases} \tag{5}$$

- The fourth step: the same as the Forward Algorithm [15] to update the T moments of the state.

$$\alpha_t(i) = \begin{cases} q_i(o_1)\pi_i, \\ q_i(o_t)\sum_{j=1}^{N}\alpha_{t-1}(j)p_{ij}\beta_t, \end{cases} \tag{6}$$

Where (α_{t-1}) probability that at time $t-1$ when the system is in state.
- Then the probability of the system in each state is then made by the Bayes formula:

$$\gamma_i(t) = \frac{P(s_i|o_i)}{\sum_{s_j \in S} P(s_j|o_i)} = \frac{\alpha_t(i)}{\sum_{s_j \in S} \alpha_t(j)} \tag{7}$$

3.4 Risk Assessment

The state of the network will correspond to a risk vector C, which indicates the potential risk of the system when the network is in this state. The risk value of the current network state can be calculated by the following formula:

$$R_t = \sum_{s_i \in S} \gamma_t(i)C(i) \tag{8}$$

where R_t represents the probability that the system is in state s_i at time T, c(i) represent When the network is in state s_i, the system faces the risk value.

3.5 Description of the Algorithm

When determined the observation sequence and parameter model.The pseudo-codes of algorithm are as follow:

Update state probability distribution

Input: O_t, γ the observation at time t, the hidden Markov model

Output: γ_t the security state probability at time t

if t = 1 then

   for i ← 1 to N do

   $_t(i) = q_i(O_1)_i$

   $\gamma_i(t) = \dfrac{\alpha_t(i)}{\sum_{s_j \in S} \alpha_t(j)}$

   repeat

else

   for i ← 1 to N do

   $\alpha_t(i) = q_i(o_t) \sum_{j=1}^{N} \alpha_{t-1}(j) p_{ij} \beta_t$

   $\gamma_i(t) = \dfrac{\alpha_t(i)}{\sum_{s_j \in S} \alpha_t(j)}$

   repeat

endif

return $\boldsymbol{\gamma(t)}$

---

4 Experiments

In order to verify the efficiency of the method, we use the classic data source LLDOS1.0 [16] Lincoln Laboratory provided for verification. LLDOS1.0 DDOS attack scenario is a test set that provides network traffic package which contains DDOS attacks. Attack scenario is divided into five stages, each stage of the whole incident led to network security in a different state.

According to the documentation, the various stages of the information as shown below. In order to reproduce the attack scenario, the use of flow replay technology, the original traffic to snort, by acting as collector of data for analysis. Documentation with respect to the attack, a slight difference in the number of alerts for each phase, mainly due to the different snort rule set, but the number of alerts the overall trend is the same, and all use the same data set algorithm does not affect the results.

Denoting the sampling period Δt is 5 min, in each sampling period, selected from snort flow generated the most alarm alerts as the observation vector. In order to have comparable, adoption and literature [11] same observation sequence /vecO, the state transition matrix /vecP, observation vector probability distribution matrix /vecQ, the initial state probability distribution matrix and the risk vectors /vecC.

Attack scenario is described in the following Fig. 2:

Figure 3(a) show improved algorithm to generate the safe state before 20 min and after 180 min, network security state between 20 min to 80 min is probing

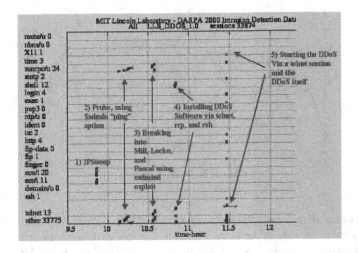

Fig. 2. Attack scenario

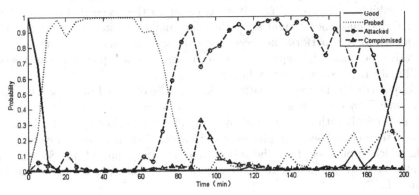

(a) Improved Algorithm

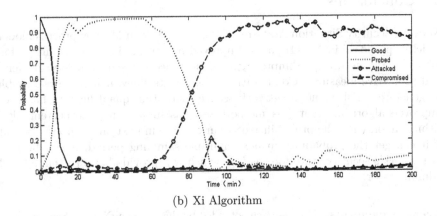

(b) Xi Algorithm

Fig. 3. Experiment of comparison

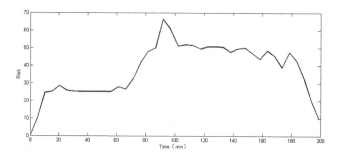

Fig. 4. Risk assessment

state, between 80 min to 180 min is intrusion state. Figure 3(b) show Xi Algorithm generate the safe state before 20 min, the network security state between 20 min to 80 min is probe state, then it always is attacked state. After the end of DDos attack, the original algorithm is still in the attack state, the state of the probability of improved algorithm is closer to the scene description. This is mainly because the original algorithm does not take into account when the attack end, system performance recovery, enhance the efficiency of the defense, protective measures and repair classes such measures will play a better role,so it will make the system recovery faster .

To quantify the network situation risk, it can be obtained network risk value R_t at time T by combining risk vector $C(i)$. As shown in Fig. 4:

Improved algorithm attack risk was significantly higher than the invasion phase detection and security phases, and in about 90 min there is a clear small peak, which is precisely that the system may be compromised, in 180 min, the threat decreased, indicating that the attack has ended. And scene description file match.

5   Conclusions

Network Security Situation Assessment Based on Hidden Markov process model, which is proposed in Xi [11] paper, improved by considering the current performance of the network, to improve the game process based on safety incidents and protective measures. Comparative experiments show, it is more reasonable to generate a risk value of network security situation quantification based on improved algorithm. But this method also has limitations, the current algorithm can only get the probability of security state in next one sampling period, if it can get the probability in next multiple sampling period, it can be used in network's security state forecasting, which will provide basis to the network administrator.

Acknowledgements. This research is funded by the National Natural Science Foundation of China (No. 61170295), the Project of National Ministries Foundation of China (A2120110006), the CoFunding Project of Beijing Municipal Education Commission (JD100060630) and the Research Project of Aviation Industry of China (CXY2011BH07).

References

1. Bass, T.: Multisensor data fusion for next generation distributed intrusion detection systems (1999)
2. Bass, T.: Intrusion detection systems and multisensor data fusion. Commun. ACM **43**(4), 99–105 (2000)
3. Xz, C., et al.: Quantitative hierarchical threat evaluation model for network security. J. Softw. **17**(4), 885–897 (2006)
4. Xiuzhen, X., et al.: Study on evaluation for security situation of networked system. J. Xi'An Jiaotong Univ. **38**(4), 404–408 (2004)
5. Ning, P., et al.: Techniques and tools for analyzing intrusion alerts. ACM Trans. Inf. Syst. Secur. (TISSEC) **7**(2), 274–318 (2004)
6. Xu, D., Ning, P.: Alert correlation through triggering events and common resources. In: 20th Annual Computer Security Applications Conference. IEEE (2004)
7. Yong, Z., et al.: Network security situation awareness approach based on Markov game model. J. Softw. **22**(3), 009 (2011)
8. Årnes, A., Valeur, F., Vigna, G., Kemmerer, R.A.: Using Hidden Markov Models to evaluate the risks of intrusions. In: Zamboni, D., Kruegel, C. (eds.) RAID 2006. LNCS, vol. 4219, pp. 145–164. Springer, Heidelberg (2006)
9. Årnes, A., Sallhammar, K., Haslum, K., Brekne, T., Moe, M.E.G., Knapskog, S.J.: Real-time risk assessment with network sensors and intrusion detection systems. In: Hao, Y., Liu, J., Wang, Y.-P., Cheung, Y., Yin, H., Jiao, L., Ma, J., Jiao, Y.-C. (eds.) CIS 2005. LNCS (LNAI), vol. 3802, pp. 388–397. Springer, Heidelberg (2005)
10. Khosravi-Farmad M, Rezaee R, Harati A, et al.: Network security risk mitigation using Bayesian decision networks. In: 2014 4th International eConference on Computer and Knowledge Engineering (ICCKE), pp. 267–272. IEEE (2014)
11. Rongrong, X., et al.: An improved quantitative evaluation method for network security. Chinese J. Comput. **38**(4), 749–758 (2015)
12. Rabiner, L.R.: A tutorial on hidden Markov models and selected applications in speech recognition. Proc. IEEE **77**(2), 257–286 (1989)
13. Han, R., Zhao, B., Xu, K.: Policy-based integrative network security management system. Comput. Eng. **8**, 069 (2009)
14. Jianfeng, Z.: Graduate School of National University of Defense Technology (2013)
15. Bishop, C.M.: Pattern Recognition and Machine Learning. Springer, New York (2006)
16. Lincoln Laboratory (2000). https://www.ll.mit.edu/ideval/data/2000data.html

Chaotic Secure Communication Based on Synchronization Control of Chaotic Pilot Signal

Honghui Lai[1](✉) and Ying Huang[2]

[1] School of Information Engineering, Gannan Medical University,
Ganzhou 341000, China
honghuilai@qq.com
[2] Faculty of Mathematics and Computer, Gannan Normal University,
Ganzhou 341000, China

Abstract. Nowadays, with the rapid development of social economy and technology, chaotic secure communication has been a hot issue of social development. This paper begins with describing the chaotic secure communication technology and then analyzes the design scheme for chaotic secure communication based on chaotic pilot signal for synchronization control. The experimental simulation shows that using chaotic pilot signal for synchronization control can help achieve chaotic synchronization between communication transmitting and receiving systems and further enhance security and confidentiality of exchange of information and transfer of data. The theoretical analysis and numerical simulation results show that the design approach is effective and universal.

Keywords: Chaotic pilot signal · Synchronous control · Chaotic secure communication

1 Introduction

Since the early 1990s when Pecora and Carroll realized synchronization between chaotic systems [1–3], the chaotic synchronization phenomenon has aroused more and more attention, and has been widely used in many areas, such as secure communication, chemical reactions and biological systems [4–7]. Chaotic signals feature broadband, complexity and orthogonality, and therefore been more widely used. In recent years, domestic research into chaotic secure communication has achieved excellent results. In general, the secure communication system is widely used in the chaos synchronization technology. At present, the chaotic secure communication scheme is divided into four parts: chaotic masking [8, 9], chaotic shift keying [10], chaotic spread spectrum [11, 12] and chaotic parameter modulation [13, 14].

H. Lai—Project supported by the Foundation of Jiangxi Nature Science Committee (No. 20142BAB217028).

K. Li et al. (Eds.): ISICA 2015, CCIS 575, pp. 640–647, 2016.
DOI: 10.1007/978-981-10-0356-1_66

Chaotic secure communication is mainly divided into three categories: analog-analog communication, analog-digital-analog communication and digital-digital communication [15]. The three security communication equipment developed as a foundation has also been divided into three types. The first is analog encryption devices, and the main mechanism is that all signals transmitted or received are analog signals rather than digital signals. The second is analog-digital-analog encryption devices, and the main mechanism is the transmission and receiver signals belong to analog signals. Signal processing and transmission belong to digital signals. The third is digital encryption devices, and the main mechanism is that is the transmission and receiver signals fall into digital signals. Such new digital encryption devices can fully adapt to and meet the needs of modern information transmission, and fully ensure the accuracy and validity of signal transmission. Today, the chaotic secure communication technology involves relatively broad fields and contains more abundant technical content. Moreover, further studies should also be strengthened.

In the chaotic secure communication technology, the core technology is chaotic hiding, which is also referred to as the chaotic masking [16], which mainly takes the chaotic signal as the carrier during the whole signal transmission activities and analyzes the secret signals in signal transmission for encryption processing. Because the chaotic signal itself contains broadband noise, the information signal can be hidden directly and is tricky to find, which is conductive to getting the optimal confidentiality effect and features strong secrecy.

Despite the chaotic secure communication technology has been commonly applied in contemporary society, it has extremely rapid development momentum. However, it is undeniable that it is still in a development stage. In addition, the development of chaotic secure communication can not only be achieved by using one single theory, but need the interdisciplinary interaction to help complete the final molding. At present, its development still has deficiencies to some extent, and there should be further explorations into its anti-interference ability and anti-decryption ability.

Chaotic masking communication is the core of the chaotic technology. As a result, when making innovative processing of the chaotic technology, it is necessary to be combined with chaotic masking communication synchronization of the neural networks and PCM coding communications, among others. When applying the synchronous digital processing technology, it is also essential to attach importance to comprehensively enhancing their communication flexibility and sensitivity and ensuring the dynamics of communication security. Further, it should also be noted that the future development pattern for the chaotic masking communication technology will gradually be timeless chaotic pattern, which can achieve excellent privacy and double the storage capacity compared with the conventional chaotic communication technology.

In this paper, the design scheme of chaotic secure communication based on the chaotic pilot signal for synchronization control is studied. The advantages of chaotic secure communication scheme are analyzed, and a chaotic communication system with chaotic pilot signal for synchronization control is constructed. The experimental simulation shows that using chaotic pilot signal for synchronization control can help achieve chaotic synchronization between communication transmitting and receiving systems and further enhance security and confidentiality of exchange of information and transfer of data. Experiment results also indicated that this scheme had universality and feasibility in actual applications.

2   Design Scheme of Chaotic Secure Communication Based on Chaotic Pilot Signal for Synchronization Control

2.1   Design Scheme of Chaotic Secure Communication

Chaotic masking communication is the first study of chaotic secure communication, the basic principle is based on the Pecora and Carroll driver response type self-synchronization principle, the use of a near Gauss white noise statistical characteristics of the chaotic signal as a carrier to hide the information to transmit, in the receiving end and then use the synchronization of chaotic signal to recover useful information. The useful information signal and the signal which is used to drive the receiving end chaotic synchronization, the size and the change of the information signal seriously affect the synchronization performance of the receiving end chaotic system.

Design scheme of chaotic secure communication is shown in Fig. 1, and the main idea is: chaotic synchronization and modulation are completed by two and more chaotic systems that are not necessarily the same and have a principal-subordinate relationship. And when there is only one chaotic system, chaotic synchronization and modulation are respectively performed by different signals. As a pilot signal, the chaotic signal for synchronization control is independent of the chaotic signal that has been modulated in the course of transmission.

Although the synchronization control signal is independent of the chaotic modulating signal in transmission, it does not mean that there should be two independent parallel carrier channels. Instead, signals that can be transmitted in only one carrier channel in whether TDD or FDD working mode can be employed.

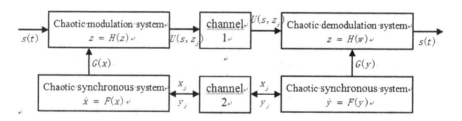

Fig. 1. Design scheme of chaotic secure communication

2.2   Advantages of Design Scheme of Chaotic Secure Communication

First, chaotic pilot signal for synchronization control may be used for independent transmission of synchronizing pilot signals, and both the transmitting and receiving ends are identical and synchronously control signals in coordination with the chaotic system, which can help improve the accuracy of control and synchronization of chaos, and help both the transmitting and receiving ends to realize more flexible control and synchronization of chaos. For example, we can use the forward chaotic pilot signal to synchronously control the chaotic pilot signal, or the reverse chaotic pilot signal to synchronously control the chaotic pilot signal [17]. Comparing the two control modes, the communication system of the latter is more secret.

Second, in the chaotic communication system, chaotic modulation is separated from synchronization control signal. The generation and synchronization of chaos are not affected by information and the relative size of the chaotic signal, and can help adopt a more flexible way of chaotic modulation.

Third, between the chaotic modulation system and the synchronization chaotic systems at both ends of transmitting and receiving ends, a more complicated synchronization control calculation method is accessible to help enhance the security of the whole chaotic communication.

3   Chaotic Communication System

3.1   Chaotic System

In order to scientifically verify the relevant conclusions, a chaotic communication system that synchronously control the chaotic pilot signals should be constructed. In 1994 American scholar J. C. Sprott used numerical methods to give nineteen three-dimensional chaotic system [18], these systems of equations or can be written with a quadratic nonlinear term of six polynomial, or contain two quadratic nonlinear term of five polynomial. These nineteen systems are known as the Sprott system, which is called Sprott-A system to Sprott-S system. The form of Sprott system is simple and has rich dynamic behavior. It is easy to realize the circuit. It can be obtained by the time scale transform. So the corresponding Sprott circuit in Sprott system has good application prospect in chaotic spread spectrum communication and chaotic secure communication.

In this paper, the Sprott-I system is selected as a chaotic model in the transmitting and receiving ends. The basic Sprott-I model is defined as follows:

$$\begin{cases} \dot{x} = -0.2y \\ \dot{y} = x + z \\ \dot{z} = x + y^2 - z \end{cases} \tag{1}$$

The system has a unique smooth point (0,0,0), which showed a single scroll chaotic attractor. The initial value $x_0 = [0.1, 0.2, 0.1]$, the use of MATLAB to draw the system phase diagram. Figure 2 shows Sprott-I system dynamic behavior of chaotic systems.

Suppose the chaotic system in the transmitting end of the chaotic communication system is:

$$\begin{cases} \dot{x}_1 = -0.2x_2 \\ \dot{x}_2 = x_1 + x_3 \\ \dot{x}_3 = x_1 + x_2^2 - x_3 \end{cases} \tag{2}$$

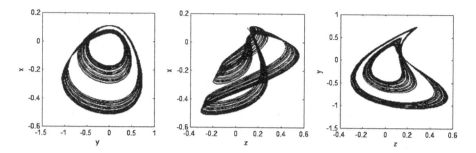

Fig. 2. Chaotic trajectories of Sprott-I system

where x_1, x_2 and x_3 all represent the three state variables for the chaotic communication system in the transmitting end, then the chaotic system in the receiving end is:

$$\begin{cases} \dot{y}_1 = -0.2y_2 \\ \dot{y}_2 = y_1 + y_3 \\ \dot{y}_3 = y_1 + y_2^2 - y_3 \end{cases} \tag{3}$$

where y_1, y_2 and y_3 represent the three state variables the chaotic communication system in the receiving end. In this paper, the encryption system is constructed by using chaotic encoding, which has large nonlinear property, strong security, and the modulation and demodulation of chaotic communication are realized.

3.2 Simulation System Construction

The simulation system mainly adopts the reverse synchronization control method, which is to realize synchronous control of y_1 in the transmitting chaotic system by using y_2 in

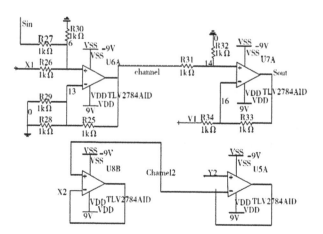

Fig. 3. Modulation and demodulation of chaotic communication and realization of reverse chaotic synchronization control

the receiving end chaotic system; x_1 is applied to the modulation of the transmitting end signal while y_1 is used for demodulation of the chaotic signal in the receiving end, see Fig. 3.

4 Simulation Data Analysis

Chaotic synchronization lays a basis in chaotic communications. The simulation results show that by using chaotic pilot signal for synchronization control, chaotic synchronization of the communication transmitting and receiving systems can be achieved, whilst some useful signals will not hinder or interfere in the chaotic synchronization activities between transmitting and receiving systems.

Because the separation between chaotic synchronization control signal and the chaotic modulating signal is used, the size of the relationship between the chaotic modulation and the modulated signal will not adversely affect chaotic synchronization activities. The simulation results show that the size of the relationship between the chaotic modulation and the modulated signal is only limited by circuit performance, especially that of some operational amplifiers [19].

Figure 4 shows that the receiving-end chaos demodulates the transmitting-end signals when the useful information signals peak at 38 mV, 3 kHz and the chaos signal amplitude is 580 mV.

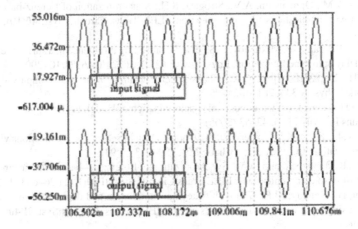

Fig. 4. Comparison between input and output signals in the transmitting-receiving ends of chaotic communication

The chaos system synchronization is realized by using the chaotic pilot signal to realize the chaotic system synchronization, which effectively solves the problems caused by the use of the signal and the signal which is used to drive the receiving end. The chaos modulation signal and chaotic synchronization control signal are separated from the two aspects of the research and development. It is good for the construction of the chaotic secure communication system with better performance.

5   Conclusion

Today's rapid development of social economy and technology enables modern information technology to be widely used in all sectors of society. In information technology development, safety is a top priority. Therefore, studies of the chaotic secure communication technology have also attracted more and more attention. The technology scheme of chaotic secure communication based on chaotic pilot signal for synchronization control is designed scientifically, in the hope of further enhancing the security and confidentiality of data transmission activities, and further enhancing the resistance to external thefts. A comprehensive exploration of the chaotic communication technology as well as design practice of new solutions can help update applications of information security, so that the safety and stability of our social economy, science and technology can receive comprehensive protection, thereby further promoting the sound and healthy development of our society.

References

1. Pecora, L.M., Carroll, T.L.: Synchronization in chaotic systems. Phys. Rev. Lett. **64**(8), 821–824 (1990)
2. Ott, E., Grebogi, C., Yorke, J.: Conntrolling chaos. Phys. Rev. Lett. A **64**, 1196–1199 (1990)
3. Lorenz, E.N.: Deterministic nonperiodic flow. J. Atmos. Sci. **20**, 130–141 (1963)
4. Guomo, K.M., Oppenheim, A.V., Strogatz, S.H.: Synchronization of Lorenz-based chaotic circuits with applications to communication. IEEE Trans. Circuits Syst. **40**(10), 626–633 (1993)
5. Beta, C., Bertram, M., Mikhailov, A.S., et al.: Controlling turbulence in a surface reaction by time-delay auto synchronization. Phys. Rev. E **67**(2), 46224/1–46224/10 (2003)
6. Lu, J.H., Yu, X.H., Chen, G.R.: Chaos synchronization of general complex dynamical networks. Phys. A **334**, 281–302 (2004)
7. Perez, O.C., Femat, R.: Unidirectional synchronization of Hodgk in Huxleg neurons. Chaos, Solutions Fractals **25**(1), 43–53 (2005)
8. Kocarev, L., Halle, K.S.: Experimental demonstration of secure communications via chaotic synchronization. Int. J. Bifurcat. Chaos **2**(3), 709–713 (1993)
9. Cuomo, K.M., Oppenheim, A.V.: Synchronization of Lorenzed—based chaotic circuits with applications to communications. IEEE Trans Circuits Syst. II **40**(10), 626–633 (1993)
10. Dedieu, H., Kenndy, M.P.: Chaos shift keying: modulation and demodulation of a chaotic carrier using self-synchronizing Chua's circuit. IEEE Trans. Circuits Syst.-II **40**(10), 634–642 (1993)
11. Abarbanel, H.D., Linsary, P.S.: Secure communication and unstable periodic orbit of strange attractors. IEEE Trans. Circuits Syst.-II **40**(1), 576–587 (1993)
12. Rulkov, N.F., Sushchik, M.M.: Digital communication using chaotic pulse-position modulation. IEEE Trans. Circuits Syst.-I **48**(12), 1436–1444 (2001)
13. Halle, K.S., Wu, C.W.: Spread spectrum communications through modulation of chaos. Int. J. Bifurcat. Chaos **3**(1), 469–477 (1993)
14. Itoh, M., Murakami, H.: New communication systems via chaotic synchronizations and modulations. IEICE Trans. Fund **E78-A**(3), 285–290 (1995)
15. Guohua, L.: Secure communication based on chaotic pilot signal controlling chaotic synchronization. Appl. Res. Comput. **31**(9), 2788–2790 (2014)

16. Lezhu, L., Jiqian, Z., Guixia, X., et al.: A chaotic secure communication method based on chaos systems partial series parameter estimation. Acta Phys. Sinica **63**(1), 010501 (2014)
17. Pang Jing, S., Shuangchen, L.J., et al.: Design of chaotic secure communication system. J. Hebei Univ. Technol. **40**(5), 17–21 (2011)
18. Sprott, J.C.: Some simple chaotic flows. Phys. Rev. E **50**(2), 647–653 (1994)
19. Ensheng, L., Shuangcheng, H.: Design of chaotic secure communication system based on wien bridge. Chin. J. Electron Devices **36**(3), 359–362 (2013)

The Privacy Protection Search of Spam Firewall

Kangshun Li and Zhichao Wen[✉]

College of Mathematics and Information,
South China Agricultural University,
Guangzhou 510642, China
283072731@qq.com

Abstract. Although most of the existing encryption system takes the privacy issues of storing data into consider, the reveal of user access pattern is inevitable during the e-mail filtering. Therefore, how to protect the private data in the process of spam filtering becomes one of the urgent problems to be solved. Combined with two filtering techniques which are based on keyword and blacklist respectively, this paper achieves the goal of sorting and filtering spams. Meanwhile, given the privacy issues in sorting and filtering the spams, the paper is based on an experimental project, the Pairing Based Cryptography, which is performed by Stanford University to achieve the e-mail encryption program. It adopts a searchable public key encryption in the process of sorting and filtering, which needs no decryption and can realize searching and matching operations. By this method, it fully protects the privacy and access patterns of the mail receiver from disclosing.

Keywords: Spam · Privacy-protection · Public-key-encryption · Searchable

1 Introduction

Contemporarily, with the advantages of simplicity, fastness, convenience and low-cost, e-mail has become the most widely used service of the Internet, changing the way of modern communication. However, since the first spam e-mails' appearance in the mid-1980s, the growing proliferation of spam e-mail inevitably became a widespread concern and a variety of spam filtering technology came into being naturally. But with the rapid development of network storage services, many enterprises and individuals use third-party servers to store large amount of mail data. Although the majority of existing encryption systems takes privacy issues of storing data into consider, yet in the process of e-mail filtering, there are more or less leakages of the user access mode. Therefore, how to protect the private data in the process of spam filtering becomes one of the urgent problems to be solved.

Spam Firewall privacy search is mainly to achieve two goals: spam filtering and protection of users' privacy. First of all, the basic designed objective is to realize message classification and filtering. In this project, the mail filters must be able to classify and filter the mails according to its content, so that the unexpected spam mails won't be received. Moreover, it is necessary to balance privacy concerns. In the process

© Springer Science+Business Media Singapore 2016
K. Li et al. (Eds.): ISICA 2015, CCIS 575, pp. 648–658, 2016.
DOI: 10.1007/978-981-10-0356-1_67

of sorting and filtering mail, it requires to achieve the searching and matching operation of the e-mail data without having to decrypt it in order to ensure the privacy and access mode of the mail recipient will not be let out.

2 Theoretical Foundation

2.1 Bilinear Pairing

The definition of Bilinear significance with cryptographical significance is as follow.

Set G1, G2 respectively as an additive group and a multiplicative group whose order are prime q and P is a generator of G1. Suppose that in the group G1, G2, the discrete logarithm problem is intractable. Bilinear mapping pair can be defined as $e : G1 \times Gw \rightarrow G2$.

And it meets the following properties:

(1) Bilinear mapping: $e(aP, bP') = e(P, P')^{ab}$, for all $P, P' \in G1$, it is approval for all the $a, b \in Z_q$.
(2) Non-degeneracy: if $e(P, P') = 1$, exists $P' \in G1$, then $P = O$.
(3) Computability: There exists efficient algorithms that for $P, P' \in G1$, can figure out $e(P, P')$.

2.2 Bilinear Diffe-Hellman Problem

2.2.1 BDH Parameter Generator

BDH parameter generator is an important concept of bilinear Diffe-Hellman problem. Input a security parameter k in BDH parameter generator, and output the prime p, description of G1 and G2 and the admissible bilinear map e().

2.2.2 Calculate BDH Problem

Footnotes Computing bilinear Diffe-Hellman (Computational Bilinear Diffe-Hellman, CBDH) can be defines as follow: Enter $g, ag, bg, cg \in G1$; Work out $e(g, g)^{abc} \in G2$.

2.2.3 Determinate BDH Problem

Determination of bilinear Diffe-Hellman (Decisional Bilinear Diffe-Hellman, DBDH) can be defined as follow: input $\{g, ag, bg, cg, abcg\}$ and $\{g, ag, bg, cg, kg\}$, among them, a, b, c, k, g are random parameters; if $\{g, ag, bg, cg, abcg\}$ and $\{g, ag, bg, cg, kg\}$ can be distinguished in polynomial time, then output the result YES; otherwise, output the result NO.

2.3 Public Key Encryption with Keyword Search

Public Key Encryption with keyword Search (PEKS) is a new type of cryptosystem, which allows us to go through a keyword search on the public key encrypted data. Thus not only does it protect the privacy and access mode of the receiver from leaking.

Meanwhile, it also offers a way that we will be able to match and search operation quickly and effectively without decrypting the data.

A public non-interactive scheme, with Public Key Encryption with keyword Search, includes the following four probabilistic polynomial time algorithms:

(4) Initialization algorithm KeyGen (s): Input a security parameter s, draw a key pair (public key A_{pub} and private key A_{priv}).

(5) Public Key Encryption with keyword Search algorithm PEKS (A_{pub}, W): Enter keyword W, use the acquired public key A_{pub} to calculate the ciphertext S of the keyword W used for searching.

(6) Construction algorithm Trapdoor (A_{priv}, W'): Enter keyword W', use the private A_{priv} to calculate the trapdoor T_w of keyword W '.

(7) Keyword search algorithm Test (S, T_w): Enter a searchable encryption ciphertext S of keyword W and a trapdoor Tw of keyword W', if $W = W'$, then the output the result YES, otherwise draw the result NO.

3   Transformation Plan of Mail System

3.1   Overall Design

As shown in the Fig. 1, the design consists of three parts:

(1) Mail encryption of sender.
(2) Setting of the recipient's e-mail trapdoor and black and white list.
(3) Classification and filtering of mail server.

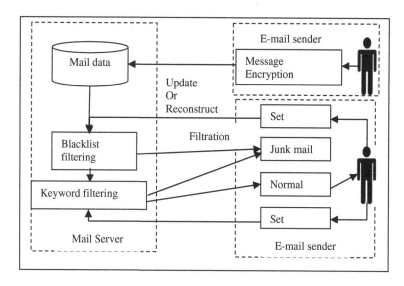

Fig. 1. The overall design

During the design, assuming that the message sender, message recipient and the third-party mail server have received their own key through the secure channel. The mail server has been stored black list defined by the users (Fig. 1).

First, the message sender uses the public key to encrypt the mail, and the message is sent to the mail server storage, waiting for the mail recipient to retrieve and receive it. Then the e-mail recipients use the private key to set black and white list and trapdoor for specific user name based on their own needs, and send the black and white lists and trapdoor to the mail server. Finally, the mail server uses the public key to match e-mail sender's user name with name on the list of black and white of the server, then match keyword list of the mail with the trapdoor set by the recipient, reaching to retrieve the filter effect, thus those don't meet the Mail filter condition will be set as spam.

Figure 2 shows the flowchart of achieving spam filtering and classification.

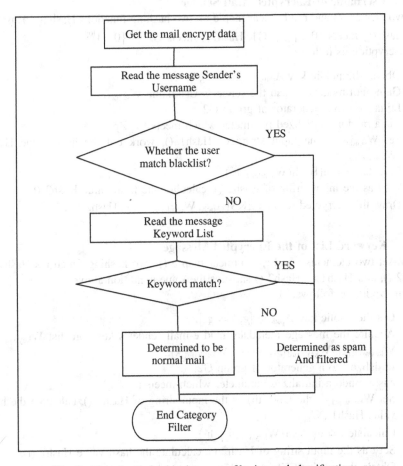

Fig. 2. Flowchart of achieving spam filtering and classification

3.2 Generation of the Key

Generate type A pairing based on A type parameter, define three elements g, h, α. Its steps are as follows:

(1) Based on the pairing, construct cyclic group $G1, G2, GT$ on elliptic curves.
(2) Initialize g to the generator of group $G1$.
(3) α is a random initialized parameter which meets $\alpha \in Z_p^*$.
(4) Calculate and draw $h = g^\alpha$.
(5) Output the public key $A_{pub} = [g, h]$, private key $A_{priv} = \alpha$.

3.3 Mail Encryption of Sender

3.3.1 Username of Encrypted Mail Sender

Set two elements t and r, a bilinear map e (), two hashing functions Hash1(), Hash2(), and Hash1() meets $\{0, 1\}^* \to G1$, Hash2() meets $G2 \to \{0, 1\}^{\log p}$.
Encryption as follows:

(1) Obtain the public key $A_{pub} = [g, h = g^a]$.
(2) Gain mail metadata, read the sender's Username W_{sender}.
(3) Initialize t to a generator of group $G2$.
(4) r is a random initialized parameter which meets $r \in Zp^*$.
(5) Set W_{sender} as the input string of Hash1 (), work out the hash value Hash1 (W_{sender}).
(6) Calculate $t = e(Hash(W_{sender}), h^r)$.
(7) Set t as the input string of Hash2 (), calculate the hash value Hash2 (t).
(8) Draw the encrypted result $PEKS(A_{pub}, W_{sender}) = [g^r, Hash2(t)]$.

3.3.2 Keyword List of the Encrypted Message

Again set two elements t and r, a bilinear map e (), two hashing functions Hash1 (), Hash2 (), and Hash1 (), Hash2 () satisfies the same condition above.
Encryption as follows:

(1) Get the public key $A_{pub} = [g, h = g^a]$.
(2) Acquire the message metadata, read e-mail sender's keyword list $W_{keyword}[1]$, $W_{keyword}[2]$, $W_{keyword}[3], \cdots \cdots, W_{keyword}[k],$.
(3) Initialize t to a generator of group $G2$.
(4) r is a random initialized parameter which meets $r \in Zp^*$.
(5) Set $W_{keyword}[i]$ (the key i-th) as the input string of Hash1 (), calculate the hash value Hash1 $(W_{keyword}[i])$.
(6) Calculate $t = e(Hash(W_{keyword}[i]), h^r)$.
(7) Set t as the input string of Hash2 (), calculate the hash value Hash2 (t).
(8) Obtain the result $PEKS(A_{pub}, W_{keyword}[i]) = [g^r, Hash2(t)]$.
(9) Repeat step (5) (6) (7) until all the keywords on the list have been computed.

(10) Obtain the result PEKS(A_{pub}, $W_{keyword}$[1])|| PEKS(A_{pub}, $W_{keyword}$[2])|| PEKS
(A_{pub}, $W_{keyword}$[3])......||PEKS(A_{pub}, $W_{keyword}$[k]).

3.4 The Settings of the Mail Recipients' Trapdoor and Black and White Lists

3.4.1 The Structure of Trapdoor

Set an element Tw, a bilinear mapping e (), a hash function Hash1 (), Where Hash1 () meets $\{0, 1\}^* \to G1$.

Construction as follows:

(1) Get private key $A_{priv} = \alpha$.
(2) Use the acquired the mail recipient for constructing trapdoor's keyword $W_{trapdoor}$.
(3) Initialize Tw to a generator of group G1.
(4) Set W' as the input string of Hash1 (), calculate the hash value Hash1 ($W_{trapdoor}$).
(5) Calculate Tw = Hash($W_{trapdoor}$)$^{\alpha}$.
(6) Output Trapdoor (A_{priv}, $W_{trapdoor}$) = Tw.

3.4.2 The Structure of Black and White List

Again set an element T_{blcak}, a bilinear mapping e (), a hash function Hash1 (), where Hash1 () meets.

Construction as follows:

(1) Get private key.
(2) Use the acquired the mail recipient for constructing username blacklist W_{sender}[1], W_{sender}[2], W_{sender}[3], ······, W_{sender}[k].
(3) Initialize T_{blcak} to a generator of group G1.
(4) Set W_{sender}[i] (the i-th username of blacklist) as the input string of Hash1 (), calculate the hash value Hash1 (W_{sender}[i]).
(5) Calculate.
(6) Repeat step (3) (4) (5), until all the usernames of blacklist have been calculated.
(7) Obtain result of structuring the blacklist Trapdoor(A_{priv}, W_{sender}[1] || Trapdoor (A_{priv}, W_{sender}[2])|| Trapdoor(A_{priv}, W_{sender}[3])|| ··· ··· Trapdoor(A_{priv}, W_{sender}[k]).

3.5 Sorting and Filtering the Mail Servers

3.5.1 Sorting and Filtering the Mail Based on the Black and White Lists

Set three elements A, B, S, a bilinear map e (), a hash function Hash2 () and Hash2 () meets $G2 \to \{0, 1\}^{\log p}$.

Mail sorting and filtering steps are as follows:

(1) Get the public key $A_{pub} = [g, h = g^{\alpha}]$.
(2) Get the encryption result of mail sender username $PEKS(A_{pub}, W_{sender}) = [g^r, Hash2(t)]$, set A=gr, B=Hash2(t).

(3) Read the processed blacklist stored in the server Trapdoor(A_{priv}, $W_{sender}[1]$)||
Trapdoor(A_{priv}, $W_{sender}[2]$)|| Trapdoor(A_{priv}, $W_{sender}[3]$)|| $\cdots$ $\cdots$ Trapdoor(A_{priv},
$W_{sender}[k]$). That is $T_{black}[1]$ || $T_{black}[2]$ || $T_{black}[3]$ || $\cdots \cdots$ $T_{black}[k]$ ||

(4) calculate S = e($T_{black}[i]$, A), and $T_{black}[i]$ is the i-th encrypted username of
blacklist.

(5) Set S as the input string of Hash2 (), calculate the hash value Hash2 (S).

(6) Match Hash2 (S) with B. If S = B, then the message sender is users in blacklist, so
determine it as spam and filter it; if S ≠ B, the message sender is in the whitelist,
therefore reserve the mail temporarily.

(7) Repeat step (4) (5) (6) until all the usernames in the blacklist have been matched.

(8) If the username does not exist in the blacklist, finally determined the message
sender is in the whitelist, reserve the mail temporarily.

3.5.2 Sorting and Filtering the Mails Based on the Keyword Matching

Set three elements A, B, S, a bilinear map e (), a hash function Hash2 () and Hash2 ()
meets $G2 \rightarrow \{0, 1\}^{\log p}$.

The sorting and filtering steps are as follows:

(1) Get public key $A_{pub} = [g, h = g^{\alpha}]$.

(2) Get trapdoor set by the message recipient Trapdoor (A_{priv}, $W_{trapdoor}$) = Tw.

(3) Get keyword list of the mail sender PEKS (A_{pub}, $W_{keyword}[1]$) || PEKS (A_{pub},
$W_{keyword}[2]$) || PEKS (A_{pub}, $W_{keyword}[3]$) $\cdots$ $\cdots$ || PEKS (A_{pub}, $W_{keyword}[k]$).

(4) Take the keyword |PEKS (A_{pub}, $W_{keyword}[i]$) = [g^r, $Hash2(t)$], Indicated the i-th
encrypted key, set $A = g^r$, B = Hash(t).

(5) Calculate S = e(Tw, A).

(6) Set S as the input string of Hash2 (), calculate the hash value Hash2 (S).

(7) Match Hash2 (S) With B. If S = B, then $W_{keyword}[i] = W_{trapdoor}$, this shows that the
recipient is reluctant to obtain the e-mail, so it is determined that the message is
spam mail and needs to be filtered; if S ≠ B, then $W_{keyword}[i] \neq W_{trapdoor}$, reserve
the mail temporarily.

(8) Repeat steps (4) (5) (6) (7) until all the keywords in the keyword list have been
matched.

(9) If trapdoor keywords set by the e-mail recipient does not exist in the keyword list,
then finally determine that the message is a normal mail, need not filtering.

4 Security and Efficiency Analysis

4.1 Security Analysis

(1) Provide verifiable encryption. When merely knows the ciphertext data of the mail,
the third-party server cannot know any of the information in plain text messages.
In this article, the random number used in the generation of the key and encrypted

key is unknown to the server. So when the keyword ciphertext is known only, the server cannot get any information about the keywords.

(2) Independent inquiry. In addition to search the matched results, the mail server cannot get any information about the plaintext message. In this article, we only use the received trapdoor, blacklists and the generated public key when the mail server is matching the ciphertext. There is no decryption during the searching and matching process, so the mail server cannot know anything about the plaintext message.

(3) The controlled query. Without the user's permission, the mail server and external attackers cannot search whether any message of the user contains certain keyword. In the article, the generated key is kept secret, servers and external attackers cannot know. When making a request for e-mail filtering, you need to enter the private key of the user. Therefore, we can ensure the server and external attacker cannot generate trapdoor and user blacklists keywords to retrieve whether the user's mail contains certain information.

(4) Supports implicit query. When the user sends filter conditions to the mail server, it can be achieved that any information of the filter conditions won't be leaked to the servers. In this paper, the mail server can only verify whether the sender of a message is in the blacklist or the e-mail contains the keyword during mail filtering. It is inaccessible to the filter conditions.

4.2 Efficiency Analysis

Analyze the efficiency of the design in its operating efficiency from the point of time. For example, in sending an e-mail, analysis the operational efficiency based on the filtering of blacklist and keywords.

The practical configurations in tests are as follows:

Table 1. Test the configuration parameters

Category	Parameters
Operating system	Windows 7
Programming language	C ++
CPU	Intel Core i5
RAM	2 GB

In the program, the main time consumed in encryption and matching process of operation of mapping pair e(), Hash functions and modular exponentiation, therefore we will focus on analyzing operational efficiency of these three operations. Suppose the message sender sends an e-mail with N keywords, and the current server has M names in the blacklist, the operation of various parts are shown in Tables 1, 2 and 3.

Table 2. Operation of the mail sender

Running function	Blacklist-based filtering	The keyword-based filtering
The number of calculations of mapping e ()	1	$1 \times N$
The number of calculations of Hash function	2	$2 \times N$
The number of calculations of modular exponentiation	2	$2 \times N$

Table 3. Operation of the mail recipient

Running function	Blacklist-based filtering	The keyword-based filtering
The number of calculations of mapping e ()	0	0
The number of calculations of Hash function	M	1
The number of calculations of modular exponentiation	M	1

Table 4. Analysis of operation of the mail server

Running function	Blacklist-based filtering	The keyword-based filtering
The number of calculations of mapping e ()	M	N
The number of calculations of Hash function	M	N
The number of calculations of modular exponentiation	0	0

In order to obtain a reasonable running time for time efficiency analysis, repeat the test for several times and the operating schedule is as shown in Table 4.

Table 5. Operational timetable

Times of operation	The mapping function e ()	Hash functions	Modular exponentiation
100	148.354 ms	2971.681 ms	0.015 ms
500	624.734 ms	14909.193 ms	0.105 ms
1000	1225.325 ms	29802.525 ms	0.192 ms
5000	6212.605 ms	148809.010 ms	0.981 ms

To facilitate the observation, assume that it has been run for 1000 times, and times required are:

(1) E-mail sender: 1225.325 ms + 29802.525 ms × 2 + 0.192 ms × 2 + 1225.325 ms × N + 29802.525 ms × 2 N + 0.192 ms × 2 N = 60830.759 ms × (N + 1)

(2) E-mail recipient: 29802.525 ms × M+0.192 ms × M + 29802.525 ms + 0.192 ms = 29802.717 ms × (M + 1)

(3) Mail Server: 1225.325 ms × M + 29802.525 ms × M + 1225.325 ms × N + 29802.525 ms × N = 31027.85 ms × (M + N)

The data above indicates that the time required in spam filtering is primarily related to the number of blacklist M and the number of keywords N and the running time is reasonable. It can be explained, the combination of searchable encryption scheme and spam filtering technology is feasible (Table 5).

5   Conclusion

In this article, we make spam mail as an object, design a spam filters which protect the personal privacy of e-mail users, achieving a searchable encryption scheme. In order to ensure the user's privacy when searching, a new type of encryption system is introduced - Public-key Searchable Encryption Technology. Based on the characteristics of Public-key Searchable Encryption Technology, the article achieve the goal of filtering respectively by black and white list filtering and keyword matching filtering.

Acknowledgements. This work is supported by the National Natural Science Foundation of China with the Grant No. 61573157, the Fund of Natural Science Foundation of Guangdong Province of China with the Grant No. 2014A030313454.

This work was jointly supported by Natural Science Foundation of Guangdong Province of China (#2015A030313408).

References

1. Guo-hua, C., Peng, X., Feng-yu, L.: Improved prototype scheme of PETKS and its expansion. Comput. Sci. **36**(3), 58–60 (2009)
2. Meng, X.: Analysis of privacy proetction of electornic mails. J. Northeast Agric. Univ. (Soc. Sci. Ed.) **1**, 37 (2009)
3. Hao-miao, Y., Shi-xin, S., Hong-wei, L.: Research on bilinear diffie-hellman problem. J. Sichuan Univ. (Eng. Sci. Ed.) **1**(38), 137–141 (2006)
4. Baek, J., Safavi-Naini, R., Susilo, W.: Public key encryption with keyword search revisited. In: Gervasi, O., Murgante, B., Laganà, A., Taniar, D., Mun, Y., Gavrilova, M.L. (eds.) ICCSA 2008, Part I. LNCS, vol. 5072, pp. 1249–1259. Springer, Heidelberg (2008)
5. Boneh, D., Di Crescenzo, G., Ostrovsky, R., Persiano, G.: Public key encryption with keyword search. In: Cachin, C., Camenisch, J.L. (eds.) EUROCRYPT 2004. LNCS, vol. 3027, pp. 506–522. Springer, Heidelberg (2004)

6. Baek, J., Safavi-Naini, R., Susilo, W.: Public key encryption with keyword search revisited. In: Gervasi, O., Murgante, B., Laganà, A., Taniar, D., Mun, Y., Gavrilova, M.L. (eds.) ICCSA 2008, Part I. LNCS, vol. 5072, pp. 1249–1259. Springer, Heidelberg (2008)
7. Boneh, D., Di Crescenzo, G., Ostrovsky, R., Persiano, G.: Public key encryption with keyword search. In: Cachin, C., Camenisch, J.L. (eds.) EUROCRYPT 2004. LNCS, vol. 3027, pp. 506–522. Springer, Heidelberg (2004)
8. Bei, Z., Xiao-ming, W.: Public key encryption schemes search with keyword. Comput. Eng. 36(06), 155–157 (2010). doi:10.3969/j.issn.1000-3428.2010.06.052
9. Dodis, Y., Katz, J., Xu, S., Yung, M.: Key-insulated public key cryptosystems. In: Knudsen, L.R. (ed.) EUROCRYPT 2002. LNCS, vol. 2332, pp. 65–82. Springer, Heidelberg (2002)
10. Boneh, D., Franklin, M.: Identity-based encryption from the weil pairing. SIAM J. Comput. 32(3), 586–615 (2003)
11. Ran, C., Halevi, S., Katz, J.: A forward-secure public-key encryption scheme. In: Biham, E. (ed.) EUROCRYPT 2003. LNCS, vol. 2656. Springer, Heidelberg (2003)
12. Li, J., Lu, Y.: A practical forward-secure public-key encryption scheme. J. Netw. 6(9), 1254–1261 (2011)

Study on Joint Procurement of Auto Parts Business Partner Selection

Bin Liu[1(✉)], Lengxi Wu[2], Xiaoyan Luo[1], and Youyuan Wang[2]

[1] Jiangxi Institute of Scientific and Technological Information,
Nanchang 330046, JiangXi, China
lboeing@qq.com
[2] Institute of Industrial Engineering, Nanchang Hangkong University,
Nanchang 330063, JiangXi, China
wulengxi@foxmail.com

Abstract. The automotive parts manufacturing companies choose partners in the joint procurement process, when it comes to multi-index heavy weight determination, proposed a combination of objective and subjective weighting method. It reflects both the decision-makers experience information, and reflects the actual reasonable and objective scientific and technical data on the comprehensive evaluation of the results.

Keywords: Joint procurement · The index weight · Subjective weight and objective empowerment

1 Introduction

The new economic era, the concept of competition in the market to break through the limitations of a single enterprise is defined between different companies set joint procurement is an effective way of modern production-oriented enterprises to gain competitive advantage. How were the characteristics of each alternative business information analysis, collation, evaluation, determine joint procurement partnership is the basis for successful operation of joint procurement. Auto parts enterprises joint first is the partner of choice, which involves the construction of evaluation criteria and selection evaluation methods. Currently, the implementation of joint procurement of specialized automotive parts manufacturers less relevant literature, for evaluation, in syndication direction, Wang Deliang [1] and so build quality control, production and organizational capacity, management capacity, market adaptability, information control several evaluation, the choice of method, domestic and foreign multi-AHP (analytic hierarchy process, AHP), packet seam analysis (data envelopment analysis, DEA), fuzzy comprehensive evaluation. Huang Ju, Yi Shuping [2] analyzed the significance of the selection of suppliers for modern manufacturing industry, pointed out the limitations of current theoretical research supplier selection method is proposed based manufacturing supplier comprehensive evaluation method introduced the basic principles and practical application of this method, and auto parts supplier

© Springer Science+Business Media Singapore 2016
K. Li et al. (Eds.): ISICA 2015, CCIS 575, pp. 659–664, 2016.
DOI: 10.1007/978-981-10-0356-1_68

selection example for analysis, the supplier comprehensive evaluation method be effective, simple and quick solution to manufacturers supplier selection problem. Chen Shi [3] to establish the use of improved AHP comprehensive evaluation model, combined with the specific business case analysis, to provide a scientific enterprise of China's auto parts suppliers selection method. Zou Yan et al. [4] multi-attribute decision making theory and evaluate the advantages of improved TOPSIS method combined entropy method, introduced to the auto parts supplier selection in the past, and in specific cases illustrate entropy method based on improved TOPSIS method is a reliable, efficient and automotive suppliers can continue to improve the evaluation model for enterprise vendor selection, management, monitoring, and other activities to improve the basis for more powerful. Zhou et al. [5] were established suppliers based on multi-level indicators selection method DEA-C2WH model assessment evaluation.

2 Joint Procurement Partnership to Determine the Principles of the Comprehensive Assessment System

(1) Comprehensiveness and integrity. As enterprises joint procurement factors involved are many and complex, so as to establish a comprehensive evaluation system. Especially the alternative enterprises for some of the major factors can't miss nor repeat, ensure alternative enterprises to conduct a comprehensive, comprehensive evaluation.

(2) Simple and scientific. The complexity of the evaluation system should be appropriate, should be based on science as a basic premise. Too complex evaluation system will evaluators in some tangled issues minutiae, thus ignoring the most crucial factor: while simplistic evaluation system can not fully reflect the true level of alternative enterprises.

(3) Flexibility and scalability. Due to changing market opportunities, alternative enterprises in joint procurement plays different roles, so the evaluation of alternative enterprises should also be different, the evaluation index system should have some flexibility and scalability.

(4) Quantitative and qualitative indicators combined. Evaluation Factors affecting the merits of alternative enterprises are mostly quantitative indicators can not be described, so the evaluation system and method using a combination of qualitative and quantitative methods to establish partner selection is very necessary.

3 Factors Partner Selection of Joint Procurement

(1) Comprehensive cost. Forming a purchasing alliance reflects the pursuit of low cost companies in the fierce competition in the market requirements, purchasing alliance in the choice of an alternative enterprise joint procurement, the overall cost to join an important indicator of choice for the enterprise.

(2) Core competencies. Joint procurement requires the participation of enterprises in close cooperation, and therefore the company's core capability of alternative to identify and evaluate the company's core capabilities meets the target joint procurement implementation.

(3) Agility. Joint procurement of fast response and organizational complexity of the joint venture must have a high degree of agility and flexibility, which is one of the important factors affecting the joint procurement partner of choice.

(4) Ability to resist risks. Purchasing Alliance selected partners to the greatest degree of avoidance and reduce the overall operational risk purchasing alliance, which is joint procurement partner selection is an important indicator.

(5) Target compatibility. Only co-operation in joint procurement in the business, under the circumstances in order to achieve coordinated to reduce procurement costs. Thus, compatibility is the joint procurement indispensable partner selection factor.

4   Joint Procurement Partner Selection Method

Joint procurement partner evaluation system, involving a lot of evaluation, some of them quantitative and others qualitative, qualitative indicators for accurate quantification is difficult to put them, so that the choice of partners can be considered as a multi-objective fuzzy comprehensive appraised process. In this paper, we used a combination of objective and subjective weighting method.

4.1   Analytic Hierarchy Process Principle and Weight Determining Step

1. For enterprises m, n evaluating indicators, judgment matrix. The United States proposed operations research Professor T.L.Saaty 1-9 scale method for evaluation of different pairwise comparison judgment matrix. This process will quantify thinking, about the 1-9 ratio scale method Scale and its contents are shown in Table 1.

2. Determine the index weight. Using the computer to solve the judgment matrix A's characteristic root. Find the maximum Eigen value $\lambda_{max}(\lambda_{max} = \sum_{i=1}^{n} \frac{(AW)_i}{nW_i})$ and its corresponding eigenvector W'. Then we get the weights ordering of each index which in the same level compared to the previous levels.

3. Consistency test. Random Consistency Index T.L.Saaty's average (see Table 2) to determine the consistency test matrix of A. According to the respective average consistency index, calculated judgment matrix consistency index $CI = (\lambda_{max} - n)/(n - 1)$ and random consistency ratio $CR = CI/CR$. If $CR < 0.1$, Is considered to have a satisfactory consistency judgment matrix, the matrix must be readjusted otherwise, until a satisfactory consistency.

Table 1. Saaty's 1-9 scale method

Scaling	Meaning
1	It represents two elements compared with the same importance
3	It represents two elements, the former is slightly more important than the latter
5	It represents two elements compared to the former than the latter obviously important
7	It represents two elements compared to the former than the latter strongly
9	Represent two elements compared to the former than the latter is extremely important
2,4,6,8	Represent the intermediate values of the neighboring judgment
Countdown values above	If the above value represents the importance of the elements compared, the elements and the comparison to the reciprocal of the value scale

Table 2. Mean random consistency index *RI*.

Order	2	3	4	5	6	7	8
RI	0.00	0.60	0.85	1.05	1.18	1.37	1.49

With the above method, we can obtain evaluation of our proposed weight vector

$$W' = (W_1, W_2, \ldots, W_n) \tag{1}$$

4.2 Objective Weighting Method

Determine the evaluation index weights, which had introduced the concept of entropy. Entropy is a measure of information using probability theory to a measure of uncertainty, it indicates that the data more dispersed, the greater the uncertainty. Decision of the available information for each indicator to represent its entropy of e_j.

First of all, we still construct the judgment matrix D. Since the dimension of the matrix is not uniform, it needs to be normalized. Normalized formula

$$r_{ij} = x_{ij} / \sum_{i=1}^{m} x_{ij} \tag{2}$$

$$e_j = -K \sum_{i=1}^{m} r_{ij} \ln r_{ij} \ (j = 1, 2, \ldots, n) \tag{3}$$

among them $K = 1/\ln m$. It is a constant on the candidate companies.

The dispersion index D_j of number j evaluation values can be expressed as $D_j = 1 - e_j$, $(j = 1, 2, \ldots, n)$. The distribution index value more dispersed, the corresponding value greater. This indicator shows that the higher the degree of importance. Conversely, if the value of the distribution of this indicator of concentration, showed the importance of lower. Consequently, the right to use this indicator to represent the entropy measure weight factor

$$W_j = \frac{D_j}{\sum_{j=1}^{n} D_j} = \frac{1 - e_j}{\sum_{j=1}^{n} (1 - e_j)} \tag{4}$$

Thus, we have come to the right desired target weight vector W''.

4.3 Subjective and Objective Weighting Method

Now, we will weight vector subjective and objective weight vector integrated, then let

$$W = \alpha W' + \beta W'' \tag{5}$$

W' is subjective weight vector, W'' is objective weight vector.
Finally, according to

$$\alpha = \sum_{i=1}^{m} \sum_{j=1}^{n} dij W_j' \bigg/ \sum_{i=1}^{m} \sum_{j=1}^{n} dij(W_j' + W_j'') \tag{6}$$

$$\beta = \sum_{i=1}^{m} \sum_{j=1}^{n} dij W_j'' \bigg/ \sum_{i=1}^{m} \sum_{j=1}^{n} dij(W_j' + W_j'') \tag{7}$$

Calculated weight vector W, that is what we ask of a comprehensive weight vector, the weight vector obtained by this method, an effective solution to the traditional method main, insufficient separation objective, not only to retain the subjective method of selection factors like policy makers, and effective integration of objective evaluation under amendments to the subjective method of choice.

5   Conclusion

This paper, auto parts manufacturers to conduct joint purchasing, involved in the selection process of partners in determining issues related to multi-attribute decision making weights are given an objective and subjective weighting method. This method subjective weights and objective weights combined. Contents of this paper makes up only using subjective weighting method or lack of objective weighting method, so that enterprises in the results when selecting partners in the joint procurement simultaneously reflects the degree of subjective and objective level. It should be noted, the study used herein Integrated Method of subjective and objective information has important theoretical and practical value, which is a need for in-depth study of the issue.

Acknowledgment. This work was financially supported by the National Science and Technology Support Program (No.2013BAF02B01), Scientific and Technological Support Projects of Jiangxi Province of China (No.20151BBE51064, No.20141BBE53005) and Scientific and Technological Projects of Nanchang City of China (No.2014HZZC005).

References

1. Deliang, W., Guanfei, D.: Application of fuzzy comprehensive evaluation method in the automotive parts supplier selection. Acad. Res. **1**, 77–78 (2007)
2. Huang, J., Shuping, Y.: Car companies supplier selection comprehensive evaluation method and its application. Chongqing Technol. Bus. Univ. Nat. Sci. **6**, 282–285 (2004)
3. Shi, C.: Evaluation and selection supply chain auto parts suppliers. Auto Ind. Res. **5**, 31–35 (2008)
4. Siyun, C., Yang, Z.: Automotive Components Supplier Selection Method applied research. Commod. Storage Maintenance **8**, 132–134 (2008)
5. Jiazhen, H., Xin, Z.: Study automotive suppliers contingency selection method based on DEA. Bus. Stud. **3**, 33–37 (2008)

Research on Ontology-Based Knowledge Modeling of Design for Complex Product

Xiaoyan Luo[1](✉), Yu Zhou[2], Bin Liu[1], and Youyuan Wang[2]

[1] Jiangxi Institute of Scientific and Technological Information,
Nanchang 330046, JiangXi, China
lboeing@qq.com
[2] Institute of Industrial Engineering, Nanchang Hangkong University,
Nanchang 330063, JiangXi, China
zhou012345yu@163.com

Abstract. To solve the problems of the design knowledge sharing and reuse difficulty in the process of complex product design, an ontology-based knowledge modeling method of complicated product design was proposed. A complex product design work model was established by analyzing the characteristics of complex product design firstly, and an ontology construction framework for complex product design was also put forward. At the research background of auto products design, the knowledge modeling process of complex product design based on ontology modeling theory was studied by utilizing the classification and description method of thinking. On this basis, a knowledge ontology model for complex products design was constructed. Finally, an application case was presented to illustrate the feasibility and validity of the knowledge modeling method.

Keywords: Complex product · Design knowledge · Knowledge modeling · Ontology

1 Introduction

Complex products means that research and development into large, complex product composition, product technical complexity of the manufacturing process is complicated, complex project management for a class of products, such as spacecraft, aircraft, automobiles, complex mechanical and electrical products, weapons systems [1]. Complex products usually involves a multidisciplinary field of mechanical, control, electronic, hydraulic, pneumatic and software, each of its components, subsystems are likely to be composed by the parts subject areas, these tens of thousands of components, subsystems interactions. Such a large-scale development of design knowledge sharing, interaction and management have put forward higher requirements. Complex product complex structure, involving many fields of knowledge, it is difficult exhaustive and express instances and objects, so it is hard to meet the needs of complex product design.

Ontology can describe the knowledge systems as a modeling tool [2], clear concepts and their relationships in the field of semantic and knowledge level, with a hierarchical

© Springer Science+Business Media Singapore 2016
K. Li et al. (Eds.): ISICA 2015, CCIS 575, pp. 665–671, 2016.
DOI: 10.1007/978-981-10-0356-1_69

structure to represent complex knowledge may have been designed to achieve a good knowledge of complex products common understanding reached accumulation of knowledge, sharing and reuse.

2   Complex Product Design Model

2.1   Design Knowledge of Complex Product

In the complex product design and manufacturing environments, including many types of design information, a variety of actors in collaborative design process [3]. Complex product design knowledge can be divided into two types of the body of knowledge and the auxiliary knowledge.

(1) Subject knowledge is the product designers and design experts in the field of common inductive principle, regular and typical models. Specifically include the principle of agency, design analysis and simulation methods, the typical structural model, field standards, as well as higher frequency reference experimental data, figures, formulas, and so on. This part of knowledge is more important, stability is the main enterprise knowledge accumulation.

(2) Auxiliary knowledge is abstract classification of information and common summary aided design groups involved. Information covered here include user information needs, critical information, suppliers supply the information, the user feedback information. Auxiliary knowledge of these aided design information with real-time and randomness, but the resulting sum up, for a modern product design also indispensable.

2.2   Decomposition Method of Complex Product Design Task

In the complex product design, which is often more complex product structure, product features and more, the amount of information corresponding to the design task is also large. Therefore, the task decomposition process uses the appropriate method for decomposing decomposition. Complex product design task division can be carried out according to the following method:

(1) By sector: responsibilities and capabilities of various departments generally different design tasks assigned to each department therefore different. By sector is mainly allocated appropriate responsibilities and the ability to design tasks according to the characteristics of products and sectors.

(2) By Function: The product is composed of a plurality of components, the function of each component are not the same, such as the car's gear shift system is mainly responsible for providing different speed, the hydraulic system is used to provide power and so on.

(3) According to the structure is divided: a product of a plurality of components, a member and composed of multiple parts, the design task decomposition can be divided according to the structure of the product.

(4) Divided by the design process: product design sequence, it can be divided according to product design order.

(5) According to the group division: to break down the product design tasks through organic combination of the above methods.

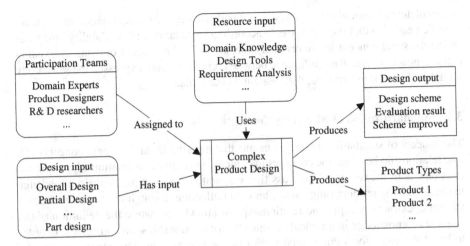

Fig. 1. Design model of complex product

Figure 1 is a complex product design model made after the division.

3 Modeling Method of Complex Product Design Knowledge Based on Ontology

Ontology is a conceptual description [4], is a concept and its relationship to the objective existence of description. One main purpose is to enable the introduction of design knowledge "standard." To achieve standardization in the field of knowledge and effective sharing of knowledge; second is to achieve knowledge of "reusable."

3.1 Data Collection and Analysis

According to the determined ontology field and scope, collect information modeling concepts and relations within the domain ontology related. It can access data network, vocabulary, stories, books; experts, documents, and existing ontology, in order after the data were analyzed to extract the body. Regardless of the collection methods, once to collect the required information, it can be traced. Ontology sharing and reuse characteristics, we should give priority to reuse existing ontology resources, select one of the standard terminology comprehensive definition, thereby reducing the workload of ontology.

3.2 Knowledge Structure Breakdown

Knowledge of product domain knowledge structural decomposition should be clear of the design process of design activities. After decomposition of each design activity, will each design activity is subdivided into sub-activities. Product design knowledge decomposition process should ensure that the body has: ① objectivity and clarity, and as complete; ② minimal resistance, defined body of knowledge acquisition just to meet specific needs, so that users can freely according to instantiate; ③ scalability, users can define the same concept in the original case, based on the concept of using an existing define a new concept; ④ no difference between representation, expressed the same body model with different symbols, the meaning is the same.

3.3   Standardization of Modeling Concepts

The concept of standardized concept means that a particular area is meta-concepts [5]. The relationship between the concepts may include synonymous relationship, antisense relations, is in the relationship, cross ties, disparate relationships. Domain knowledge to determine the product range and after data collection, through the work breakdown structure, extract concepts and relationships of knowledge, while the definition of the concept of knowledge in a logical design activity relationships, and attributes, in order to build the basis for a conceptual model of the entire body. This stage requires the following three aspects:

(1) Under the guidance of the concept of extraction of knowledge experts in the field, through the identification, analysis and statistics, extract concepts and relations from product design knowledge design activity /sub-activity, it is required that is capable of covering basic knowledge in the field.

(2) After define the class relationships and attributes, the concept of product knowledge extraction need to define the class relationships and attributes. This article is from the start of the most important concepts. The definition of the class restricted from the connotation and extension levels. Property is described ontology concept features an important part, is in the field with the properties of other entities entity distinct identity defined therefore required for this purpose.

(3) After creating an instance of the concept of the relationship between knowledge and attribute definitions, we need to create an instance. Examples are the corresponding class rank, first select a concept, and then add the specific instance of the class, and finally add specific attribute values.

3.4   Evaluation Standard

After the first few steps created a preliminary ontology. But this body meets the requirements of the field of knowledge organization, we need to be evaluated. More influential is TR Gruber [2] proposed five ontology rules as a standard to judge the completeness: clarity, consistency, scalability and compatibility. In addition, the need to contain the completeness, accuracy, etc. ontology evaluation. Test it according to the above standard. If the result meets the requirements, the completion of the final model body while forming the document; if not met, will return to the initial design stage, again and again, until finally constructed to meet the requirements of domain knowledge ontology model (Fig. 2).

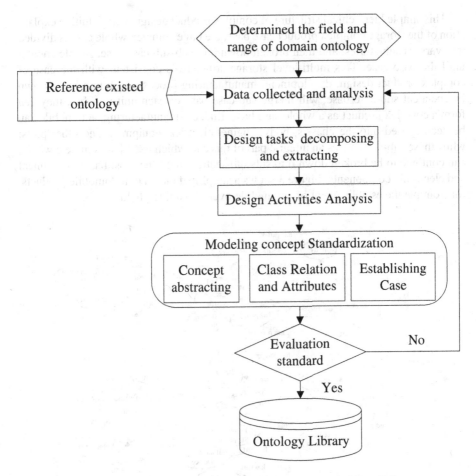

Fig. 2. Structure chart of complex product design knowledge modeling

4 Modeling Applications of Complex Product Design Knowledge Based on Ontology

Due to the complexity of complex products, multilevel nature and multi-module collaboration, this article will be applied to the case of complex ontology technology products expression. A complex product can be decomposed into a plurality of a collaborative development unit modules, and each module has a unit can be decomposed into a plurality of two unit module development, followed by decomposition continues until specific and detailed to each base unit module, a high level function module is based on low-level modules, each module with each other between the various elements of information and sharing of knowledge, collaborate with each other, work together to achieve high levels of functional modules;

This simple hierarchical structure of complex product design, very intuitive explanation of the storage complex product case base, the huge complex whole case is divided into various sub-stories, and then broken down into sub-sub-case case, simple enough until the base case. This multi-level storage structure is consistent with the case of complex product design, development, manufacturing process, each unit has its own independent sub-case base, will merge sub-case base of each unit together, they can form a complex product as a whole case base. Like car manufacturing, automobile can be decomposed into the chassis, body, engine, electrical equipment and other parts; while these sub-sections as an integral part of this car, which itself is a complete whole, can continue to be broken down into sub-sub-section, until the most basic mechanical and electronic components. Figure 3 shows a simplified model of automobile products, for example the establishment of automobile overall ontology field.

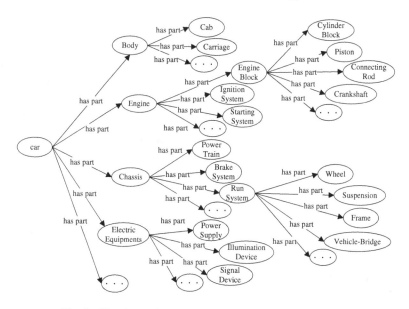

Fig. 3. Simple ontology concept tree of automobile products

5   Conclusion

Knowledge is a complex design and manufacturing process of knowledge management and the use of the main tools, knowledge modeling is the basis for building a knowledge base to solve the problem. This paper presents a new and complex product design knowledge representation, ontology theory is introduced, a detailed description of the steps of this method of knowledge modeling, and build a simple model for complex product design knowledge for the automotive design field knowledge representation and interaction within a unified system. Intelligent search, knowledge integration and other issues will continue to study in the future, and to further explore the impact of various factors on the knowledge modeled in the research process.

Acknowledgements. This work was financially supported by the National Science and Technology Support Program (No. 2013BAF02B01), Scientific and Technological Support Projects of Jiangxi Province of China (No. 20151BBE51064, No. 20141BBE53005) and Scientific and Technological Projects of Nanchang City of China (No. 2014HZZC005).

References

1. Hubelh, H., Colquhoungj, G.J.: A reference architecture for engineering data control (EDC) in capital plant manufacture. Comput. Ind. **46**(2), 149–165 (2001)
2. Kimky, Manleydg, Yangh. Ontology-based assembly design and information sharing for collaborative product development. Computer-Aided Design, **38**(12), pp. 1233–1250 (2006)
3. Guoquan, Z., et al.: Formalization of process model in mechanical product conceptual design. J. Comput. Aided Design Comput. Graph. **17**(2), 327–333 (2005)
4. Yi, W.W., Sun, Y., Zhang, S.K., et al.: The Application and Research of Ontology Construction Technology. Computer Society, pp. 618–623(2008)
5. Parmantier, J., Junqual, P.., Bertuol, S., et al.: Simplification method for the assessment of the EM response of acomplex cable harness. In: Proceedings of the 20th International Zurich Symposium on EMC, pp. 161–164, Zurich (2009)

Learning-Based Privacy-Preserving Location Sharing

Nan Shen[✉], Xuan Chen, Shuang Liang, Jun Yang, Tong Li,
and Chunfu Jia[✉]

College of Computer and Control Engineering, Nankai University, Tianjin, China
{shennan,cfjia}@nankai.edu.cn,
{chenxuan,liangshuang,junyang}@mail.nankai.edu.cn, rannicker@163.com

Abstract. With the improvement of mobile communication technology, mobile Online Social Networks (mOSNs) provide users with the corresponding location based services when compared with traditional social networks. Location sharing becomes a fundamental component of mOSNs now, and some practical methods and techniques have been proposed to protect user's privacy information. Some of these methods can accommodate privacy protection based on the input user profile and user's privacy preferences through personalization, but user may be unlikely to use them without easy operation and strong privacy guarantee. In this article, we make a further research on privacy-preserving location sharing in mOSNs and develop a framework to help user to choose his desired degree of the privacy protection based on context aware. An adaptive learning model is established to provide user privacy right decisions, based on analyzing a series of factors that influence the choice of user's privacy profile. This model will manage the different contexts of different user privacy preference with minimal user intervention and can achieve self-perfection gradually. So our proposed model can effectively protect users' privacy and motivate users to make use of privacy preferences available to them.

Keywords: Mobile online social networks · Privacy-preserving · Location sharing · Adaptive learning model

1 Introduction

In recent years, with great advances in mobile communication technology, the traditional social networks, as a new mode called mobile Online Social Networks (mOSNs), have gradually emerged. Compared with the traditional social networks, mOSNs can break the boundaries of time and space, so that no matter when and where people can communicate within their social relations [1].

Location sharing is an important factor which makes mOSNs more popular and brings people convenience greatly. There are two main types of location sharing: social-driven location sharing and purpose-driven location sharing [2]. Purpose-driven location sharing refers to that users obtain a specific query

© Springer Science+Business Media Singapore 2016
K. Li et al. (Eds.): ISICA 2015, CCIS 575, pp. 672–682, 2016.
DOI: 10.1007/978-981-10-0356-1_70

service by sharing their location. For example, a driver searches working gas station within 5 miles distance from his location. The push for more location sharing is largely driven by social media sites such like Twitter and Facebook, which is named social-driven location sharing. The mOSNs not only allow the user to share his location information by means of "check-in" [3], but also allow the user to use location sharing services to find and contact with their social relations nearby [4]. Also such social-driven sharing services provide a number of controls to users, to make their location disclosure decisions can be determined by users themselves. The popularization of mOSNs has profoundly influenced and changed the way of both social life and communication.

In this article, we focus more on the social-driven location sharing in mOSNs, which brings people great convenience. But at the same time, it also inevitably raises significant users' privacy concerns [5]. Since the current mOSNs are under centralized control, the users' privacy data will be collected when they use location sharing service every time. For example, the company that offers the service may be collecting and retaining users' detail records, and it is likely that these records may give away the sensitive private information, such as the personal data, living habits and health conditions [6]. Previous research has shown that users are likely to be hesitant to share their locations if privacy is not fully guaranteed [7]. However, in this type of location sharing, another key factor is the quality of service in addition to the privacy-preserving which often conflicts with the first one. Therefore, how to balance between the degree of them under different privacy requirements has become a more and more concerned problem of researchers.

To address this issue, there is a need for finer grained control of privacy by users, but most users are unwilling to set their personal privacy frequently. Location sharing service can sense the context in which they are being used and adapt their contents and presentation [8]. The sensing of context can analyze many important factors regarding a request for the service, which can automatically generate the corresponding protection degree for a specific user. In this article, we make a further research on privacy-preserving location sharing in mOSNs, and present an adaptive learning model to help users to choose their desired degree of the privacy protection in different queries.

This paper is organized as follows: in Sect. 2, the related work is discussed. A learning model and its corresponding framework are presented in Sect. 3. The learning model are analyzed and assessed in Sect. 4. Finally, we draw conclusion and future work in Sect. 5.

2 Related Work

Anonymity has been proved as an effective technique for privacy protection in mOSNs. The k-anonymity approach is the most commonly used anonymous technology, which is to obfuscate the users' real information by establishing cloaking regions covering k anonymous records. It is Sweeney who proposed k-anonymity technology in 2002 [9], and Gruteser et al. [10] who firstly used it to locate privacy protection.

In 2007, SmokeScreen [11] discussed location sharing presence with social networking and preserving user privacy by opaque identities. In 2013, Wei *et al.* proposed the Mobishare system [12] which provides flexible privacy-preserving location sharing in mOSNs. Similar to the Mobishare system, most popular solutions for location privacy have adopted the trusted third party anonymization service, which firstly processes query information through suitable methods to hide the users' true information, and then sends the processed query information to location sharing service provider.

Recently, personalization of privacy has attracted the attention of researchers. Most of the current solutions provide personalized k-anonymity for the social-driven location sharing. The personalized k-anonymity technique was first put forward in [13], which realized the users' willingness to share their location information may be affected by many factors. These factors are closely related to the users' contextual information, such as social context, task context and so on.

Many personalized k-anonymity approaches allow users to customize different k value to influence the quality of service depending on users' profile and different location privacy requirements. We need to make a trade off between the desired level of privacy protection and the quality of service from location sharing. If the k value is too low, it can imply lower guarantees of privacy protection. On the other hand, a higher k value can improve security. However, it may reduce the quality of service in mOSNs, such as taking a longer time to perform additional spatial and temporal cloaking or sending more than the required information back to users. What is more serious is that too high security requirements may result in inaccurate query results. As a result, our personalized approaches need to find the right balance between the quality of service and the degree of privacy protection.

Aiming at "personalized k-anonymity" for protecting location privacy, as the first time, Gedik *et al.* presented the CliqueCloak model [14] which supports k-anonymity for a wide range of users with context-sensitive privacy requirements. This model devises a spatio-temporal cloaking algorithms are proposed location privacy framework. Casper system introduced in [15], in which mobile users can entertain location based services without revealing their location information. Mobile users register with Casper specify their desired degree of privacy protection by a user-specified privacy profile. Knijnenberg *et al.* [16] discussed the effect of location sharing options on users' disclosure decisions when configuring a sharing profile in a location sharing service. In [17], Xu *et al.* studied the integration of various factors, and constructed a personalized anonymity model that included contrary factors capturing the delicate balance between privacy concerns, time costs and so on. After a field research of those factors influencing the privacy of purpose driven, a personalized k-anonymity approach is presented [8] to help users to choose and manage their privacy preferences. A feeling related privacy model is introduced [18], although it is difficult for one to scale their feeling by a number.

3   Learning-Based Model for Location Sharing

3.1   Influence Factors

Social-driven sharing services allow users to share their location in the mOSNs, and users make location disclosure decisions based on some contextual factors extracted from users' privacy profile. The social-driven location sharing rely on two types of servers, including the location based server and the social network server, which may be unreliable. The privacy is threaten by the location based server or the social network server who wants to get unauthorized access to target user's sensitive information. Therefore, the user's choice of k value should provide adequate the strength of the k-anonymity privacy protection.

There are many factors influence the user's privacy security in the query. For instance, when a user shares his own location information amidst his friend circle, "who he is" is inferred by analyzing the common friend relationship. Dealing with multiple upload location information, "where he is" is inferred that although it may be cloaked. With continuous tracking queries, the user's destination with reference to time factor is also inferred. Might as well assume that if the user's destination is a hospital, then "what he does" is inferred. Suppose if this hospital is a cancer hospital, his physical condition is likely very bad. Once an adversary has obtain this message, he will be able to spread negative information to the user's social relations. Through the above analysis, we can conclude that there are many factors that can influence the privacy profile decisions. Figure 1 shows four main contextual factors influencing a user's choice of k value in our learning-based model.

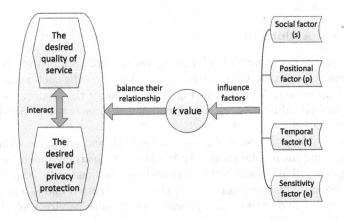

Fig. 1. Four factors influencing k value in learning-based model

Social Factor. The characteristics of social-driven sharing services, it is the user's location combined with social networks. The choices of privacy profiles are related to different identities in social networks. Users might feel more secure to share information in the core circle of friends than in circles of colleagues, and want to reduce anonymous in order to a better quality of service can be obtained. Also, users might want to be more anonymous when sharing information with strangers due to a fear of leakage of privacy.

Temporal Factor. Privacy is mainly associated with feeling, and the temporal factor is an important aspect that affects the user's feeling. For example, users are more concerned about privacy at night than during the day, and may want a better quality of service in leisure time than in busy time.

Positional Factor. Positional factor is another aspect relate to users' privacy feeling. Users often have different privacy feelings when they are in a different position. Some users may feel safe in their own place, such as home. Users will be obviously increased vigilance, when they arrive at a new place. Thus, users may prefer a higher k value in an unfamiliar position than in a familiar position for a safety reason.

Sensitivity Factor. The sensitivity is related to the purpose of the query, which influences the choice of anonymity required as users may have a set of parameters on query purpose of privacy. For instance, because a user does not want others to know about his physical condition when he is in the hospital, the results of the query should display his vague location information. In other situations, a user should provide his accurate location information for outdoor emergency query.

3.2   System Architecture

After we identified the factors influencing users' choice of privacy profile, we build a model that learns users' privacy profile based on the contextual effect on each query operation. As shown in Fig. 2, we consider the scenario of location sharing in mobile online network, which consists of three entities: Mobile Client, Anonymizer, Servers.

Users are able to access mobile social network and enjoy location sharing services by using his mobile client. We integrate the learning-based model with the mobile client. This model helps users and provides personalized privacy protection decisions. The Anonymizer is an absolutely trusted middleware between the mobile client and servers, and provides anonymous service using the given algorithm based on the registered privacy profile. Servers might be a server of any existing mOSNs that want to provide the location sharing service. They include two categories, one is the location based server, the other is the social network server. These two servers store users' anonymized data and exchange information with each other for users' query services.

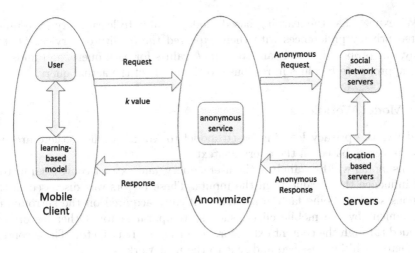

Fig. 2. System architecture of the learning-based model in mOSNs

3.3   Model Design

Before using the location sharing service, we should identify the contextual factors as shown in Table 1 for each user query.

Table 1. Summary of the notations and the possible value

Symbol	Description	Possible value
s	Social factor	friend, colleague, stranger
t	Temporal factor	leisure, busy
p	Positional factor	familiar, unfamiliar
e	Sensitivity factor	high, low

As the degree of privacy k value can be defined in accordance with these factors of the query, we derive that $k = f(s, t, p, e)$, where the k value should be within a given range of anonymous. It is clear that the k value is treated as an output of the learning-based model obtained as a result of choice of privacy profile and has a direct influence on the quality of service. At the same time, because the quality of service is also dependent on the feeling of the users, who are allowed to specify k value according to their own conditions, such as users' mood. Therefore, we define different symbols to express the different situations. k_i is the k value that the users specify while making the query. k_c is the k value that is extracted from the query context. The users may use the recommended the degree of privacy k_c by the associated contextual factors, or users may refuse recommendation and independent choose a different the degree of privacy k_i in delivering the desired quality of service.

We expect that the learning based model be able to learn every time users' desired privacy preferences with their expected the quality of service. That is to say, our proposed model can store the k values for each query, and finds the most appropriate k values in response to the needs of the actual query.

3.4   Model Workflows

We define the privacy learning based model to work, as follow, there are two main processes based on the query context:

This model will first analyze the user's query, and extract contextual factors that influence the k value from the inputs. These factors was discussed in the previous section. Some factors are automatically acquired on the surrounding environment by the mobile client, such as temporal factor. Other factors are provided through the content extractor, such as sensitivity factor. These contextual factors will be classified and sent to the next work.

Next, we will match the extracted set of parameters from the query and various sets of parameters which is stored in the model. Suppose there is no associated the preset k value for the given set of parameters from the query, the model will allow the user to input k value (k_i) independently. Then k_i value will be stored in the model and sent to the anonymizer as output. Suppose there is a match, the model then will compare the preset k value (k_c) for the extracted set of parameters with the user input k value (k_i). If the chosen k_c is less than k_i, the model will employ the k_i value to be anonymous and replace the k_c value with the k_i value in order to more secure protection. If it is the opposite, the user will receive a warning message, because the k_i value will reduce the degree of protection for the given context. The model will be waiting for the user to make a final judgment, choose the k_c value or the k_i value.

In addition, in order to prevent the extreme case that this query have no matching ($k_c = 0$) and no user input ($k_i = 0$), the model automatically records k_r value which registered with the anonymizer while making the query. Because the model cannot allow the output $k = 0$ to the anonymizer, if that happens, the model will output k_r value to protect the users' privacy.

3.5   Model Algorithms

The learning model contains two different modes according to actual situation. Let us consider all the possible scenarios when a query is being implemented on the different modes

Mode 1. $k_c = 0$

In this case, there is no associated preset k_c value for the extracted contextual factors in the model. At this moment, the k value used for the anonymizer is determined by comparison with k_i value and k_r value. There are three possible scenarios based on the different values of k_i and k_r. Following is the concrete process of algorithm for the model:

- $k_i = 0$: If the user does not input the k_i value in the given query, the user will receive a warning message that the pre-registered k_r value will be used for this query. Because $k_c = 0$, the k_r value as the basis to compare k_i value In order to prevent extreme case that the output k value is 0. If the user's response is "*yes*", the model will continue to use the k_r value to be anonymous. If the user's response is "*no*", or no response, the model will terminate this query and return the corresponding prompt information for the user.
- $k_i \geq k_r$: If the input k_i value in the given query is greater than or equal to the pre-registered k_r value, the model will use the k_i value to be anonymous without prompting the user for permission and set the k_i value to the preset k_c value for the contextual parameters on this query.
- $k_i < k_r$: If the input k_i value in the given query is less than the pre-registered k_r value, the user will receive a warning message that the degree of protection will be reduced. If the user's response is "*yes*", the model will use the k_i value to be anonymous and set the k_i value to the preset k_c value for the contextual parameters on this query. If the user's response is "*no*", or no response, the model will continue to use the k_r value to be anonymous.

Mode 2. $k_c \neq 0$

In this case, there is an associated preset k_c value for the extracted contextual factors in the model. At this moment, the k value used for the anonymizer is determined by comparison with k_i value and k_c value, and the model does not need to compare the pre-registered k_r value. There are also three possible scenarios based on the different values of k_i and k_c. Following is the concrete process of algorithm for the model:

- $k_i = 0$: If the user does not input the k_i value in the given query, this is the most common scenario, the model will use the k_c value to be anonymous without prompting the user.
- $k_i \geq k_c$: If the input k_i value in the given query is greater than or equal to the preset k_c value, the model will use the k_i value to be anonymous without prompting the user for permission and set the k_i value to the preset k_c value for the contextual parameters on this query.
- $k_i < k_c$: If the input k_i value in the given query is less than the preset k_c value, the user will receive a warning message that the degree of protection will be reduced. If the user's response is "*yes*", the model will use the k_i value to be anonymous and set the k_i value to the preset k_c value for the contextual parameters on this query. If the user's response is "*no*", or no response, the model will continue to use the k_c value to be anonymous.

4   Discussion and Assessment

4.1   Feasibility of Implementation

In this model, the contextual analysis of user queries can be done by using ontology based technology in pervasive computing. Ontology based technology analyzes complex context knowledge and provides a formal semantics to support sharing and integration of context knowledge, which involves context capturing and stores the captured context in web ontology language formats in context aware model. Web ontology language is a specially designed ontology language, which has many predefined classes and properties useful for expressing ontology information.

The context data can be collected and the context related factors could be identified in our learning model by constructing a context aware platform based on web ontology language. For example, Generic Context Management Model (GCoMM) [13] is an ontology based context aware model, which can provide proactive or reactive context aware services. There are a number of similar models based on the privacy ontology, which can be adapted into a single user data operation applied to the our model.

4.2   Characteristics of the Learning Model

Through detailed analysis of the social-driven sharing service, we take into account the four main factors influencing users' choice of privacy profile. We propose a model for learning these identified factors that address the actual threat to arrive at the desired quality of service.

The model can help users make decisions but not control users' decisions. It recommends the desired degree of privacy to users, but users can refuse to accept. Users are allowed to flexible choice a different value for their own actual situation. They may use the recommended the degree of privacy k_c value by the associated contextual factors, or input the degree of privacy k_i value if a better quality of service is expected.

The model will automatically learn the user's choice for each query and try to disturb the user as little as possible. In front of the two modes, all the possible analysis of the scenarios is shown in Tables 2 and 3.

Table 2. Scenario analysis of mode 1

$k_c = 0$	Degree of privacy	Response
$k_i = 0$	Flat	Warn user
$k_i \geq k_r$	Increase	No response
$k_i < k_r$	Decrease	Warn user

It can be found that the model will send a warning message waiting for the user to respond in the two situations by Tables 2 and 3. One situation is that the user reduces the degree of privacy of the model, the other is an extreme case that this query have no matching and no user input, and these two cases are only a very small part of all queries. As the learning model matures, the model has a matching ($k_c \neq 0$) for most of the queries the user generally makes. In addition to the user needs to reduce the existing degree of privacy, the model will automatically execute user's privacy profiles for each query in all other scenarios.

Table 3. Scenario analysis of mode 2

$k_c \neq 0$	Degree of privacy	Response
$k_i = 0$	Flat	No response
$k_i \geq k_c$	Increase	No response
$k_i < k_c$	Decrease	Warn user

5   Conclusion and Future Work

In this paper, we make a further research on privacy-preserving location sharing in mobile Online Social Networks and try to help users to consider their past choices of privacy preferences effectively when making privacy decisions. We identified and analyzed the contextual factors that influence the choice of user's privacy profile while making the query. Then, an adaptive learning model is constructed to help users to make right decisions based on the identified contextual factors and their past decisions. This model can be gradually self-perfection and only requires minimum users interference to register the suitable privacy profile in each query, unless users want to deliberately reduce the desired degree of protection. Therefore, our proposed model can effectively prevent users from privacy compromises as well as motivate them making use of privacy preferences available to them.

To implement the proposed model, further work needs to be done to discuss how to can more efficient and accurate to extract contextual factors extracted from user's query based on the ontology technology, and establish corresponding fault tolerant mechanism. Since the learning model based on k-anonymous proposed in social-driven sharing services is designed, other anonymous techniques should be investigated, such as encryption technology, offset technology and so on. Other privacy protection techniques, can also be extracted the contextual factors from the query that are converted into the corresponding parameters applied to their own. Moreover, a further research is needed on how to extend the learning model so that other personalized attributes can be considered for better quality of service.

References

1. Dinh, H.T., Lee, C., Niyato, D., et al.: A survey of mobile cloud computing: architecture, applications, and approaches. Wirel. Commun. Mob. Comput. **13**(18), 1587–1611 (2013)
2. Tang, K.P., Lin, J., Hong, J.I., et al.: Rethinking location sharing: exploring the implications of social-driven vs. purpose-driven location sharing. In: Proceedings of the 12th ACM International Conference on Ubiquitous Computing. pp. 85–94, ACM (2010)
3. Brooker, D., Carey, T., Warren, I.: Middleware for social networking on mobile devices. In: 2010 21st Australian Software Engineering Conference (ASWEC), pp. 202–211. IEEE (2010)

4. Xiao, X., Zheng, Y., Luo, Q., et al.: Inferring social ties between users with human location history. J. Ambient Intell. Humanized Comput. **5**(1), 3–19 (2014)
5. Gu, J., He, L., Yang, J., et al.: Location aware mobile cooperation-design and system[J]. Int. J. Sig. Proc. Image Proc. Pattern Recogn. **2**(4), 49–60 (2009)
6. Liu, L.: From data privacy to location privacy: models and algorithms. In: VLDB Endowment Proceedings of the 33rd International Conference on Very Large Data Bases, pp. 1429–1430 (2007)
7. Barkhuus, L., Dey, A.K.: Location-based services for mobile telephony: a study of users' privacy concerns. In: INTERACT, vol. 3, pp. 702–712 (2003)
8. Natesan, G., Liu, J.: An adaptive learning model for k-anonymity location privacy protection. In: IEEE 39th Annual Computer Software and Applications Conference (COMPSAC), vol. 3, pp. 10–16. IEEE (2015)
9. Sweeney, L.: k-anonymity: a model for protecting privacy. Int. J. Uncertainty Fuzziness Knowl. Based Syst. **10**(05), 557–570 (2002)
10. Gruteser, M., Grunwald, D.: Anonymous usage of location-based services through spatial and temporal cloaking. In: Proceedings of the 1st International Conference on Mobile Systems, Applications and Services, pp. 31–42. ACM (2003)
11. Cox, L.P., Dalton, A., Marupadi, V.: Smokescreen: flexible privacy controls for presence-sharing. In: Proceedings of the 5th International Conference on Mobile Systems, Applications and Services, pp. 233–245. ACM (2007)
12. Wei, W., Xu, F., Li, Q.: Mobishare: flexible privacy-preserving location sharing in mobile online social networks. In: 2012 Proceedings IEEE INFOCOM, pp. 2616–2620. IEEE (2012)
13. Ejigu, D., Scuturici, M., Brunie, L.: An ontology-based approach to context modeling and reasoning in pervasive computing. In: Fifth Annual IEEE International Conference on Pervasive Computing and Communications Workshops, PerCom Workshops 2007, pp. 14–19. IEEE (2007)
14. Gedik, B., Liu, L.: Protecting location privacy with personalized k-anonymity: architecture and algorithms. IEEE Trans. Mob. Comput. **7**(1), 1–18 (2008)
15. Mokbel, M.F., Chow, C.Y., Aref, W.G.: The new Casper: query processing for location services without compromising privacy. In: VLDB Endowment Proceedings of the 32nd International Conference on Very Large Data Bases, pp. 763–774 (2006)
16. Knijnenburg, B.P., Kobsa, A., Jin, H.: Preference-based location sharing: are more privacy options really better?. In: Proceedings of the SIGCHI Conference on Human Factors in Computing Systems, pp. 2667–2676. ACM (2013)
17. Xu, H., Gupta, S., Pan, S.: Balancing user privacy concerns in the adoption of location-based services: an empirical analysis across pull-based and push-based applications (2009)
18. Xu, T., Cai, Y.: Feeling-based location privacy protection for location-based services. In: Proceedings of the 16th ACM Conference on Computer and Communications Security, pp. 348-357. ACM (2009)

A Two-Lane Cellular Automata Traffic Model Under Three-Phase Traffic Theory

Yu Wang[⊠], Jianmin Xu, and Peiqun Lin

Institute of Civil Engineering and Transportation, South China University of Technology,
Guangzhou 510640, China
wangyu_0033@126.com

Abstract. In this paper, we propose a new two-lane cellular automata (CA) traffic model under the three-phase traffic theory framework. In the model, the velocity update and lane-changing rules are designed for the phase of wide moving jam, which fills the gap of two-lane CA model research in three-phase traffic theory. Then, a simulation system is designed and implemented. Numerical simulation results show that the model proposed in this paper can reproduce the typical phenomena of three-phase traffic flow. Finally, this model is used to explore the variation laws of average velocity under different phases. Our research results provide an useful reference for the management of two-lane system.

Keywords: Traffic model · Three-phase traffic theory · Traffic simulation system · Two-lane system

1 Introduction

To explain the complex nonlinear traffic flow phenomenon, Kerner [1] proposed three-phase traffic theory, which distinguished all the traffic states into three traffic phases, i.e. free flow, synchronized flow and wide moving jam. Traffic models based on this theory better conform to actual observations than the fundamental diagram approach in traffic simulation.

In 2002, Kerner et al. [2] proposed the Kerner-Klenov-Wolf model (referred to as KKW model), which was the first successfully established cellular automata (CA) model belonging to three-phase traffic theory. Since then, more CA models in three-phase traffic theory have emerged [3–6]. Nevertheless, the common shortcoming of these models is that they are all single-lane models without taking consideration of multi-lane interactions.

Traffic flow simulation systems based on CA model have the parallelization computational advantages; they can make full use of the up-to-date research results of IT [7–9]. For example, they can use communication and big data technologies to efficiently simulate the large scale road network.

Our contributions in this paper include these:

1. Based on the behavior characteristics of vehicles in wide moving jam, velocity update and lane-changing rules for the traffic phase of wide moving jam are put forward.

© Springer Science+Business Media Singapore 2016
K. Li et al. (Eds.): ISICA 2015, CCIS 575, pp. 683–688, 2016.
DOI: 10.1007/978-981-10-0356-1_71

2. We propose a new two-lane CA traffic model under the framework of three-phase traffic theory.

. 3. A simulation system based on the model is designed and implemented.

4. By means of numerical simulation, the fundamental diagram of two-lane CA traffic model clearly shows the three traffic phases.

5. We have analyzed the average speed of each lane in three traffic phases. The results provide certain references to the management of two-lane traffic.

The rest of this paper is organized as follows: In the second part, a two-lane CA traffic model applicable to the wide moving jam phase is brought up. The third part describes the design of the simulation system for our two-lane CA model. The fourth part analyzes the numerical simulation results of the model. The fifth part ends the paper with a summary.

2   Two-Lane CA Traffic Model in Wide Moving Jam

There are two parallel lanes with the same direction in a traffic system. We assume that the vehicles move from left to right. The lanes are called fast lane and slow lane respectively, and we mark them as lane 1 and lane 2. The lane is looked as a one-dimension discrete lattice chain. Each lattice can be either vacant or taken by one vehicle, as in Fig. 1.

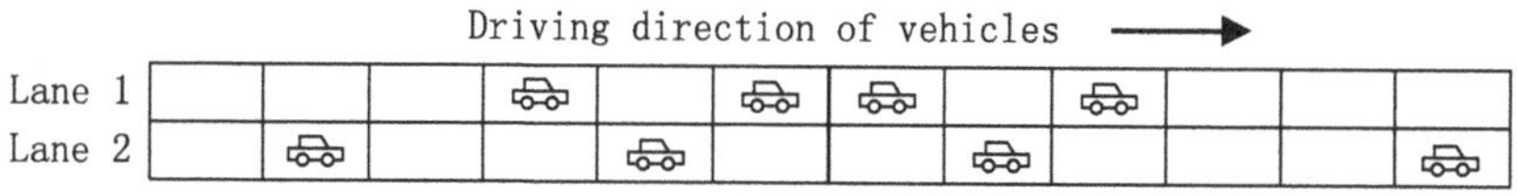

Fig. 1.   CA model diagram of two-lane system

During the evolution process of the model, each time step is divided into two stages of move and lane-changing, they respectively corresponding to velocity update rules and lane-changing rules.

In this context, the state of "wide moving jam" means: Vehicle i and at least three consecutive vehicles in front of vehicle i are stagnant or in slow moving. We denote V_{jam} and D_{jam} as the maximum velocity and maximum space headway in wide moving jam respectively. Slow moving means that the velocity of vehicle is not higher than V_{jam} and the space headway is not longer than D_{jam}.

In this model, the actual covered distance of each cellular is set to 5 m, and each car occupies 1 cellular. The time step $\triangle t$ is fixed as 1 s and the maximum speed V_{max} is set to 5 cells/s. We also defined V_{jam} as 1 cells/s and D_{jam} as 2 cells.

2.1   Velocity Update Rule

$$v_j^{(i)}(t+1) = \min\{V_{jam}, v_j^{(i)}(t) + 1, d_j^{+(i)}(t) + v_j^{(i-1)}(t)\}. \tag{1}$$

Here $v_j^{(i)}(t)$ represents the velocity of vehicle i in lane j during time step t, $d_j^{+(i)}(t)$ represents the gap between vehicle i and the adjacent vehicle ahead on lane j during time step t.

2.2 Lane-Changing Rules

In wide moving jams, vehicles will compete for the ahead gap, and based on this behavior, this paper presents lane-changing rules for the traffic phase of wide moving jam.

The Lane-Changing Rules of Lane 1. The vehicle i on lane 1 will change to lane 2 with probability $P_{1,2}$, if it satisfies the following condition:

$$d_1^{+(i)}(t) < \min\left(V_{\max}, v_1^{(i)}(t) + 1\right) \text{ AND } d_1^{+(i)}(t) < d_{1,2}^{+(i)}(t) \text{ AND}$$
$$d_{1,2}^{-(i)}(t) \geq \min\left(v_b(t) + a_b(t), V_{\max}\right) - \min\left(v_1^{(i)}(t) + a_i(t), V_{\max}\right) + 1. \quad (2)$$

In formula (2), $d_j^{x(i)}(t)$ represents the number of empty cells between vehicle i and the closest vehicle in some direction on lane j during time step t. $d_{j,k}^{x(i)}(t)$ represents the gap between vehicle i on lane j and the closest vehicle on lane k in some direction during time step t. x can be + or -, and + indicates the vehicle is in front of vehicle i, while - indicates the vehicle is behind it. $v_b(t)$ represents the velocity of vehicle b which is behind vehicle i on the other lane in time step t. $a_b(t)$ indicates the acceleration of vehicle b and $a_i(t)$ indicates the acceleration of vehicle i in time step t.

The Lane-Changing Rules of Lane 2. The vehicle i on lane 2 will change to lane 1 with probability $P_{2,1}$ if it satisfies the following condition:

$$d_2^{+(i)}(t) < \min\left(V_{\max}, v_2^{(i)}(t) + 1\right) \text{ AND } d_2^{+(i)}(t) < d_{2,1}^{+(i)}(t) \text{ AND}$$
$$d_{2,1}^{-(i)}(t) \geq \min\left(v_b(t) + a_b(t), V_{\max}\right) - \min\left(v_2^{(i)}(t) + a_i(t), V_{\max}\right) + 1. \quad (3)$$

The safety distance of this model is much less than $V_{\max}$ and this does not increase the risk of conflict occurrence.

3   Simulation System Design

Using the C# language, we design and implement a simulation system for our two-lane CA model. The simulation system integrates three models. When a vehicle is in the state of free flow or synchronized flow, the velocity update and lane-changing adopt the KKW model [2] and the right lane default lane-changing model [10] respectively. While a vehicle is in wide moving jam, the vehicle evolution adopts the rules of we proposed two-lane CA traffic model.

4   Simulations and Discussions

4.1   Fundamental Diagram

Figure 2 shows the results of simulations with two-lane CA traffic model. From Fig. 2, we can distinguish free flow, synchronized flow and wide moving jam clearly. The positive slope straight line in low-density area is corresponding to free flow, the scattered two-dimensional region is corresponding to synchronized flow, while the negative slope straight line in high-density area is corresponding to wide moving jam. The maximum density in free flow $\rho_{max}^{(free)}$ is 28.9 *vehicles/km* and the maximum density in stable free flow ρ_{min} is 19.1 *veh/km*. The maximum density in synchronized flow $\rho_{max}^{(syn)}$ is 92.3 *veh/km*. The stimulation results of our model and the measured results of Kerner [11] are consistent.

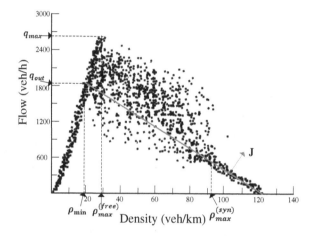

Fig. 2. Flow-density scatter plot

The line J is the characteristic line of wide moving jam [11] which is related to threshold states when the wide moving jam emergs. The wide moving jam spreading upstream velocity (the slope of the line J) v_g is 17.67 *km/h*, the flow rate outflowing from wide moving jam q_{out} is 1819 veh/h. Our results well correspond to results in the relevant literatures [12–14].

4.2   Average Velocity of Each Lane

Figure 3 shows the relationship between average velocity of each lane and density. F, S, J represent respectively free flow, synchronized flow and wide moving jam. From the diagram, it can be seen that the average of average velocity of lane 1 is higher than that of lane 2. In free flow, the average velocities of two lanes are in great disparity. In wide moving jam, the average velocities of different lanes are basically the same. The simulation results correspond to the empirical observations of the two-lane traffic.

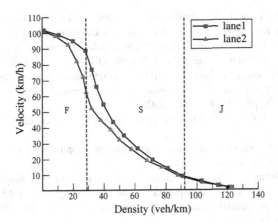

Fig. 3. Average velocity of each lane

5 Discussion and Conclusion

Based on the above analysis, two-lane system manifests typical asymmetry in free flow and synchronized flow. In wide moving jam, two-lane system manifests typical symmetry. Therefore, management strategies to two-lane system should elastically change when the traffic phase changes.

In conclusion, based on three-phase traffic theory, we have built our two-lane CA traffic model with new speed update and lane-changing rules. Through numerical simulation in our simulation system, our model can clearly display three traffic phases of two-lane traffic in fundamental diagram. We further studied the feature of average velocity of each lane in all phases. On the base of simulation results, we suggest different management strategies should be adopted for different traffic phases in two-lane road system.

References

1. Kerner, B.S.: Three-phase traffic theory and highway capacity. Phys. A **333**, 379–440 (2004)
2. Kerner, B.S., Klenov, S.L., Wolf, D.E.: Cellular automata approach to three-phase traffic theory. J. Phys. A. **35**, 9971–10013 (2002)
3. Schreckenberg, M., Lee, H.K., Barlovic, R.: Mechanical restriction versus human overreaction triggering congested traffic states. Phys. Rev. Lett. **92**, 238702 (2004)
4. Jiang, R., Wu, Q.S.: Spatial-temporal patterns at an isolated on-ramp in a new cellular automaton model based on three-phase traffic theory. J. Phys. A. **37**, 8197–8213 (2004)
5. Jiang, R., Wu, Q.S.: First order phase transition from free flow to synchronized flow in a cellular automata model. Euro. Phys. J. B. **46**, 581–584 (2005)
6. Gao, K., Jiang, R., Hu, S.X., Wang, B.H., Wu, Q.S.: Cellular-automaton model with velocity adaptation in the framework of Kerner. Phys. Rev. E. **76**, 026105 (2007)

7. Tao, M., Lu, D., Yang, J.: An adaptive energy-aware multi-path routing protocol with load balance for wireless sensor networks. Wireless Pers. Commun. **63**, 823–846 (2012)

8. Li, J., Chen, X., Li, M., Li, J., Lee, P., Lou, W.: Secure deduplication with efficient and reliable convergent key management. IEEE Trans. Parallel Distrib. Sys. **25**, 1615–1625 (2014)

9. Li, J., Au, M.H., Susilo, W., Xie, D., Ren, K.: Attribute-based signature and its applications. In: 5th ACM Symposium on Information, Computer and Communications Security (ASIACCS 2010), pp. 60–69. ACM (2010)

10. Pedersen, M.M., Ruhoff, P.T.: Entry ramps in the negel-schreckenberg model. Phys. Rev. E. **65**, 056705 (2002)

11. Kerner, B.S., Rehborn, H.: Experimental properties of complexity in traffic flow. Phys. Rev. E **53**, R4275–R4278 (1996)

12. Knospe, W., Santen, L., Schadschneider, A., Schreckenberg, M.: Single-vehicle data of highway traffic: microscopic description of traffic phases. Phys. Rev. E **65**, 056133 (2002)

13. Kerner, B.S.: Experimental features of self-organization in traffic flow. Phys. Rev. Lett. **81**, 3797–3800 (1998)

14. Kerner, B.S.: Experimental features of the emergence of moving jams in free traffic flow. J. Phys. A. **33**, 221–228 (2000)

Research on Knowledge Association and Reasoning of Product Design

Nan Jiang[1(✉)], Pingan Pan[1], Youyuan Wang[2], and Lu Zhao[2]

[1] College of Information Engineering, East China Jiaotong University, Nanchang 330013, JiangXi, People's Republic of China
jiangnan1018@gmail.com
[2] Institute of Industrial Engineering, Nanchang Hangkong University, Nanchang 330063, JiangXi, People's Republic of China
yywnc@sina.com

Abstract. The knowledge granularity is described, and the knowledge granularity model is constructed. With the help of granularity principle, design knowledge was classification, association and inference. The hierarchical structure of the domain knowledge was described by using the knowledge granularity, and the related knowledge was structured and formal. Through the analysis of a case, the method is proven to be effective to improve the relevance of knowledge and improve the efficiency of the knowledge service.

Keywords: Granularity principle · Product design · Knowledge correlation · Knowledge reasoning

1 Introduction

With the increasing complexity of product design, it involves domain knowledge more and more widely. In the knowledge organization service, with the increase of knowledge capacity, the discreteness, uncertainty and fuzziness of knowledge are more and more serious, and then the indiscernibility relationship between knowledge and knowledge is poor. All of this leads to the designers from the vast amounts of knowledge resources is more and more difficult to get the required knowledge. Meanwhile the efficiency of the knowledge service is becoming lower and lower.

Therefore, in order to solve such problems, domestic and foreign researchers have studied knowledge organization from different perspectives. Wang Youyuan [1] proposes product design knowledge model based on ontology knowledge model, including design concept classification and knowledge extraction. Hao jia [2] proposes non-geometric design knowledge organization method for user-oriented and establishes the unified product design knowledge model based on the product design knowledge navigation mode, the reference mode and automatic mode. According to the standard principle of knowledge organization, Estrada [3] proposes the standard general model adapted to the digital content and network through analyzing the concept of topic map. Mai [4] combines with user requirements from the cognitive perspective to describe and

© Springer Science+Business Media Singapore 2016
K. Li et al. (Eds.): ISICA 2015, CCIS 575, pp. 689–695, 2016.
DOI: 10.1007/978-981-10-0356-1_72

organize knowledge. Marcela Vegetti [5] uses ontology to comprehensive and consistent knowledge representation of product information.

More research on the product design knowledge organization have carried on the fruitful discussion, but there is still a problem, which is the effective classification and organization of knowledge, and distinguish indiscernibility relationship of knowledge. In this paper, it uses the method of rough granular structure to classify, associate and logical combination the indiscernibility knowledge, and describe the hierarchical structure of the domain knowledge by using the knowledge granularity. It improves the correlation of knowledge and resolving power, as well as the efficiency and accuracy of knowledge service.

2   Knowledge Granularity Description and Construction

2.1   Knowledge Granularity Description

Any granularity knowledge consists of three aspects: concept, object and relation attributes, then the concept is the description of the internal meaning of granularity knowledge, the description of the object is associated with other knowledge granularity knowledge, relationship attribute is internal and external knowledge granularity associated description [6, 7].

Through the structure and analysis method of the granule to describe the uncertain relationship between domain knowledge, the knowledge is standardized, structured and formalized. Schematic diagram of the domain knowledge granularity model as shown in Fig. 1.

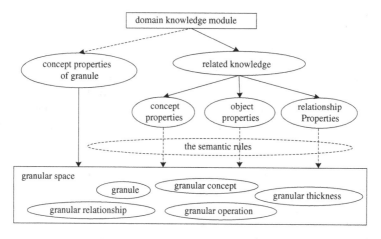

Fig. 1. Knowledge granularity model

The knowledge unit and knowledge point domain correlation is abstracted classification through the method based on granular computing, constructing knowledge granularity space from different angles. Knowledge granularity model can be described as a five-tuple (G_K, G_C, S, M, G).

Where G_K is a collection of domain knowledge unit; G_C is the attribute set of granular concept; S is grammar rules, which represents a binary relationship between knowledge concept and granular concept, $S = G\_K \times G\_C$; M is reflect the meaning of the rules of grammar between each KP and the attribute of granular concept C, $M(KP, C)$ represents a set of the significance of binary relations between all the knowledge and particles; G is granular set, where each granule of $Gr = (S (G\_K, G\_C), M (KP, C))$ corresponds to the domain knowledge to the transformation of the granular knowledge.

2.2 Knowledge Granularity Construction

In order to accurately express the multi-level granularity relationships in the Knowledge semantic relations, while according to the relationship or semantic, the conversion process of granule is called the granular layer, which reflects the problem of granular thickness in the diffident levels. For example, in the analysis of a mechanical product, it can be considered from diffident angles such as the structure, materials, parameters, etc., that is, from different granularity analysis of knowledge [8].

Definition 1 in granular collection $G = (G\_C, Gr, M)$, R is an equivalence relation on G_C. Let

$$[c_i]_R = \left\{ c_j \in G\_C \middle| (c_i, c_j) \in R \right\}$$ (1)

Then $[c_i]_R$ is called the c_i contains equivalence class.

Definition 2 in granular collection $G = (G\_C, Gr, M)$, B is an equivalence relation of R, and the granularity $GD (B)$ of B is defined by Eq. (2).

$$GD (B) = \frac{|B|}{|G\_C|^2}$$ (2)

In general, $0 \le GD(B) \le 1$; when the knowledge granularity is smaller, the resolving power is the stronger; On the other hand, the resolving power is weak.

Definition 3 Let any two equivalence relations: $R_1, R_2 \in R$, if for any $c_1, c_2 \in G\_C$ all have $c_1 R_1 c_2 \Rightarrow c_1 R_2 c_2$, then calls R_1 finer than R_2, writing for $R_1 \le R_2$.

By using method of granular structure analysis, a hierarchical structure of the granular space is constructed, which can reflect the size of the granularity of knowledge decomposition degree of thickness, through the uncertainty relation between knowledge concept and knowledge concept to classify association reasoning and logical combination, improve the ability to distinguish the particle knowledge, distinguish between different levels of granularity knowledge of the relevance of the design requirements.

3 Association and Reasoning of Product Design Knowledge

3.1 Knowledge Correlation Analysis

The relational analysis of knowledge is based on the semantic relations among different concepts, which can be identified and described quantitatively by similarity. The features of the concept are mainly embodied by the attributes, and the attributes are diverse, through the knowledge organization based on the principle of granularity, the concept of knowledge granularity, and the same level of attributes and attributes can be the intersection of the relationship. Each of the attributes in the hierarchy describes a concept of a particular granulation of view.

For any concept KC, its properties characteristics can be divided into m layer attributes, each property granule is represented by the $\{c_{i1}, c_{i2}, \ldots, c_{ik}, \ldots\}$, where i represents the i-layer granule. In the knowledge base, if a type automobile turbo product design concept has turbo axis, Turbo impeller, compressor impeller, etc., such as the concept of the turbine wheel can be divided leaf type, Leaf number, leaf material, leaf process, etc. of property characteristics. Then, we can get the semantic similarity between the concepts of KA and the concept of KC in the same layer granularity, this is to say

$$Sim_i\left(KA_i, KC_i\right) = \sum_{\substack{c\in G\_C \\ a\in G\_K}} \sum_{j=1}^{m} \left|c_j - a_j\right| \tag{3}$$

Where, M is the number of attribute characteristic granule in the same layer, as the different concept characteristics, the hierarchy and the number of granule are different. G_C is the semantic description of KA; c is the object of G_C; a is the object of G_K and c; and then the corresponding relation of object of G_K and G_C is the relative positions of each granular layer. Thus a set of similarity vector, $Sim = \{Sim_1 (KA_1, KC_1)$ and $Sim_2 (KA_2, KC_2)$ can be obtained,..., $Sim_i (KA_i, KC_i), \ldots\}$, can be obtained, where it is assumed that each requirement knowledge can be divided into N layer according to the relational structure of the concept, then the similarity between the demand knowledge and the retrieval knowledge can be calculated by the following formula (4).

$$Sim_T (KA, KC) = \frac{1}{n} \sum_{i=1}^{n} Sim_i\left(KA_i, KC_i\right) \tag{4}$$

Obviously, $0 \le Sim_T(KA, KC) \le 1$. When the $Sim_T(KA, KC) = 0$, completely different phase. The $Sim_T (KA, KC) = 1$, exactly the same. When $Sim_T (KA, KC) > 0.4$, both have relevance.

3.2 Knowledge Reasoning

A design task needs corresponds to one or more different design knowledge, the demand of multiple design task need one or more of the same design knowledge. Driven by the design requirements, homologous or classification relationship knowledge associated with design tasks is associated reasoning and abstracted into new knowledge by the granular operation, which is more clear and accurate describing the demand relation between the design task and knowledge.

According to the multi-level relevance between the granular space and domain knowledge, and the context classification relationship of the granular to the knowledge, the granular knowledge can be built into a structured and formal knowledge model of the granular to improve the effectiveness of the knowledge service and accuracy. From the perspective of semantic relations of knowledge, there is a logical relationship between the granule and the granule, granular property and granular concept property, as shown in Table 1.

Table 1. Object properties of granular knowledge for details

Type	Relation	Explanation
granular attribute	knowledge	It describes the concepts of the relevant fields and the classification of the corresponding ontology classes
father-granule	relationship of hierarchical	It represents the relationship of classification, cross and hierarchy between granular concept and granular concept, where the upper layer is rougher then the next layer, and the next layer is finer then the upper layer
son-granule	relationship of hierarchical	It represents the relationship of classification, cross and hierarchy between granular concept and granular concept, where the upper layer is rougher then the next layer, and the next layer is finer then the upper layer
new-granular	relationship of homology	It shows the hierarchical relationship between the granular concept and the granular concept, which belongs to the homologous relations or the concept of overlapping
the set of granular attribute	knowledge	It describes the multi-dimensional and multi-level relationship of the correlate ontology classes

4 Applications

In the process of automobile turbocharger product design involves multidisciplinary knowledge, such as design principle, process technology, the design manual, industry standard data number, product batch number, design knowledge. For example, In the process of turbine impeller design, the logical relationship between knowledge and knowledge is described by granularity principle, and using the structure and analysis method of the granular, the design knowledge will be granularity and realized the hierarchical reasoning between the knowledge and knowledge. The granularity knowledge correlation analysis of the impeller design is shown in Fig. 2.

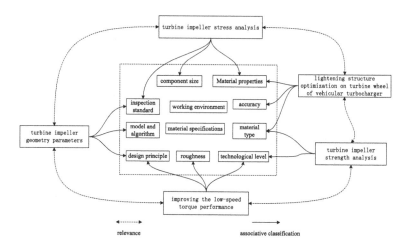

Fig. 2. The impeller design granularity knowledge correlation analysis

The different turbine impeller design knowledge related to the automobile industry is collected, which is granularity and relationship analysis using correlation analysis method in this paper. KC represents the concept of impeller design parameters, and the characteristics of its properties are divided into 5 layers including size dimension, material characteristics, accuracy, parts category and process level. Each layer contains many related property elements, for example, in the part category, c_{ik} = {impeller, roulette, moving blade, blade, rotor, gear, blade number}. And then the similarity calculation is shown as following:

Sim_T (turbine impeller structure parameters, the impeller design parameters) = Sim (turbine blade wheel geometry structure parameters, the impeller design parameters) + Sim(turbine impeller stress analysis, the impeller design parameters) + Sim(lightening structure optimization on turbine wheel of vehicular turbocharger, the impeller design parameters) + Sim(turbine impeller strength analysis, the impeller design parameters) + Sim(improving the low-speed torque performance, the impeller design parameters).

Where, $Sim_T > 0.6$. A design task needs one or more different design knowledge, the multiple design task needs one or more of the same design knowledge, the correlation analysis of turbine impeller design knowledge is shown in Fig. 3.

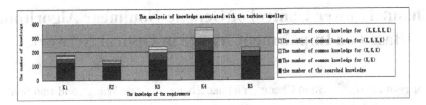

Fig. 3. The correlation analysis of turbine impeller design knowledge

5 Conclusions

Combining with the structured analysis method of the rough granular to construct granularity knowledge, the thickness of knowledge divided by the granularity will be structured and formal, which can solve the problem of uncertainty and discreteness of knowledge and improve the resolving power of knowledge to meet the design requirements. In the process of knowledge retrieval, it can reduce the repetitive search work, and same time improves the efficiency and accuracy of knowledge service.

Acknowledgements. This work is supported by National Natural Science Foundation of China under Grant No. 41402290, No. 61462028 and No. 81460275; Major Project for Natural Science Foundation of Jiangxi Province of China under Grant No. 20152ACB21011; Key Technology Research and Development Program of Jiangxi Province of China under Grant No. 20151BBE50068; External Science and Technology Cooperation Project of Jiangxi Province of China under Grant No. 20151BDH80010.

References

1. Wang, Y., Wang, F., Le, C., et al.: Ontology-based knowledge modeling of collaborative product design for multi-design teams. China Mech. Eng. **23**(22), 2720–2725 (2012)
2. Hao, J., Yan, Y., Wang, G.-X., Liu, J.-j.: User-oriented non-geometry design knowledge organization method. Comput. Integr. Manuf. Sys. **20**(6), 1300–1307 (2014)
3. Estrada, L.M.M.: Topic maps from a knowledge organization perspective. Knowl. Organ. **38**(1), 43–61 (2011)
4. Mai, T.E.: Actors, domains, and constraints in the design and construction of controlled vocabularies. Knowl. Organ. **35**(1), 16–30 (2008)
5. Vegetti, M., Leone, H., Henning, G.: An ontology for comprehensive and consistent representation of product information. Eng. Appl. Artif. Intell. **24**, 1305–1327 (2011)
6. Witold, P., Wei, L., Xiaodong, L., et al.: Human-centric analysis and interpretation of time series: a perspective of granular computing. Soft Compute. **18**, 2397–2411 (2014)
7. Huang, F., Zhang, S., et al.: Clustering web documents using hierarchical representation with multi-granularity. World Wide Web. **17**, 105–126 (2014)
8. Cao, J., Chen, J., et al.: Scene image semantic annotation method based on fuzzy theory research. J. Chongqing Normal Univ. Nat. Sci. **31**(2), 67–71 (2014)

Channel Power Control of Genetic-Nonlinear Algorithm Based on Impairment Aware in Optical Network

Dongyan Zhao[1,2(✉)], Shuo Cheng[3], Yichuan Zheng[2], Xiaoyu Wang[2], and Jian Sun[2,3]

[1] Beijing University of Posts and Telecommunications, Beijing, China
yellowduke@qq.com
[2] Institute of Advanced Network Technology and New Services,
University of Science and Technology Beijing, Beijing, China
[3] The University-Enterprise R & D Center of Measuring and Testing Technology & Instrument
and Meter Engineering, College of Electronic Science, Northeast Petroleum University,
Heilongjiang 163318, China

Abstract. The physical-layer impairments may weaken the signal quality in optical transmission link. And the signal quality degradation may substantially dent the performance of optical communications. This paper, taking into account the physical-layer impairments in performance optimization, proposes genetic-nonlinear algorithm to adjust channel power and optimize OSNR (Optical Signal Noise Ratio). Simulation results confirm the validity of the controlling strategy with the better dynamic and stable performance for adjustment of transmission power at the device nodes.

Keywords: Control strategy · OSNR · Optimization · Impairment · System performance

1 Introduction

Optical Wavelength-Division Multiplexing (WDM) communication networks are widely enabled by technological advances in optical devices such as optical add/drop MUXes (OADM), optical cross connects (OXC) and dynamic gain equalizer (DGE). Optical cables and fiber are spread out over vast surface areas in applications, so the signal attenuation cannot be avoided in fact. Nowadays, physical impairment is a hot topic due to the evolution of the optical networks to all-optical infrastructures and to the need to maintain or enhance Quality of Service (QoS). The tunable compensation technologies of the devices to optical impairments will be able to support dynamically established optical paths or links, which may traverse different routes and different optical fiber types [1, 2]. However, a proper control method is required to automatically adjust the devices parameters and optimize the QoS simultaneously according to the transmission requirement. The research on OSNR optimization and power impairment has been active, which is accomplished by maximizing the channel OSNR. A noncooperative game formulation for the OSNR optimization is presented by Pavel [3]. They suggested a link-level power control which adjusts the OSNR value of the signals toward channel OSNR optimization along with a game-theory-based control algorithm. Pan Y

© Springer Science+Business Media Singapore 2016
K. Li et al. (Eds.): ISICA 2015, CCIS 575, pp. 696–703, 2016.
DOI: 10.1007/978-981-10-0356-1_73

investigated the impact of amplifier gain ripple/tilt and fiber SRS on channel power excursions during wavelength reconfiguration events [4]. In recent years, the combination of genetic and nonlinear algorithm has become a new research direction for intelligent control and attracted many attentions from both the academia and industry. Salma Keskes proposed genetic algorithm (GA) optimization technique is applied to design power system stabilizer to search for optimal controller parameters [5]. Peng Qiong researched the nonlinear equations of the immune genetic algorithm method to quickly and easily obtain the effective solution of the production technology [6].

In this paper, we discuss the key physical-layer issues for performing impairment adjustment and compensation. A control-compensation strategy and system based on WSON standardization of the internet engineering task force (IETF) is provided. And this paper proposes the genetic-nonlinear control strategy to adjust channel power and optimize OSNR, which combines the advantages of the two algorithms by genetic and nonlinear algorithm in order to obtain the global and local optimal search.

1.1 Optical Link Model

The adjustable optical devices are comprised by functionalities of the pre-dispersion and dispersion compensators of the source node, the destination node and the WSS of the optical nodes. In the intermediate node, amplifiers can dynamically adjust the signal power level or OSNR. In order to solve the QoS-optimization problem, we refer to the schematic of an end-to-end light-path, as shown in Fig. 1. A set $M = \{1,..., m\}$ of channels are transmitted across the same optical fiber relying on wavelength-multiplexing. The link consists of N cascaded spans of optical fiber, with each followed by an OA(optical amplifier). It is assumed that all spans have an identical length. The channel optical power is assumed to be attenuated during the propagation through the optical fiber. OAs can be deployed along optical links to amplify the optical power of all channels simultaneously [7, 8].

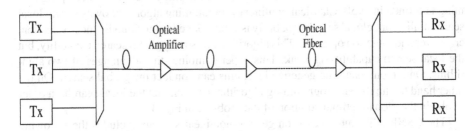

Fig. 1. Optical link model

We denote the vector of input powers by $u_i = [u_1,..., u_m]$ for all channels at the i-th node to the controller. The amplifier with the gain G_i is capable of simultaneously boosting the optical power of all channels.

It is important that channel transmission performance and quality of service (QoS) will be optimized and maintained. At the physical transmission level, channel performance and QoS are directly determined by the bit error rate (BER), which in turn depends on optical signal to noise ratio (OSNR), dispersion and nonlinear effects. Then, at the input of the node, through each intermediary optical span we have the following condition:

$$OSNR_i = u_i/(n_i^0 + G_iD_i) \tag{1}$$

n_i^o is the noise optical power for the ith channel. Here, in which P^0 is the same total power target. D_i is distributed coefficient. An optical span is composed of an optical amplifier (OA) with channel dependent gain G_i. Channel OSNR is affected by noise accumulation in optical amplifiers, which depends on all channel powers at the input.

Based on the above-mentioned model, an adaptive control model can be formulated towards the adjustment and optimization of the transmission power. Meanwhile, we should consider transmission performance limits. Since accumulating the minimum OSNR must be lower than OSNR, the total power target is required to be lower than the threshold of nonlinear effects. Furthermore, the system residual dispersion should locate between the minimum tolerate negative dispersion and the maximum tolerance positive dispersion, enabling the output power to be below the threshold of nonlinear effects. In addition, we assume that dispersion can be compensated in optical fiber transmission links and nodes, and only the OSNR-optimization problem where the OSNR of some channels is lower than the OSNR targets. Regulating the optical powers at the transmitter provides a proper way to achieve a better OSNR level for each channel at the receiver.

1.2 Genetic-Nonlinear Algorithm

The process of controlling operation in Fig. 2 includes input r_i and output y_i. We adjust the OA gain through control feedback system.

The control feedback system combines two controlling way. Nonlinear programming is formed in 1950. Classical nonlinear programming algorithm owns strong local search ability, but global search capability is weak. Genetic algorithm includes selection, crossover and mutation operator. This algorithm has strong global search capability, but the local search capability is weak. This paper combines the advantages of two algorithms. On the one hand the genetic algorithms can conduct the global search. On the other hand the nonlinear programming algorithm can conduct the local search, in order to obtain the global optimal solution of the problem in Fig. 3.

The OSNR adjustment based on genetic-nonlinear strategy includes the following steps:

(1) Initialize input value and fitness value, etc.
(2) Compute the nonlinear value.
(3) Adjust the parameters of amplifiers.
(4) Compute the output to nonlinear value stable.
(5) If the output doesn't meet the transmission requirement, then returns to step (2). Otherwise, the iteration terminates.

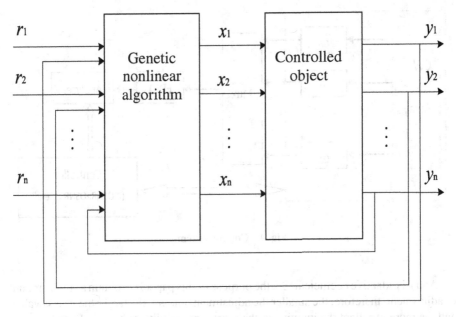

Fig. 2. Genetic nonlinear algorithm closed loop control feedback system

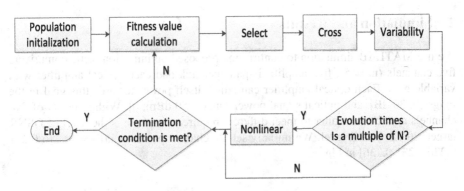

Fig. 3. Genetic-nonlinear algorithm

Genetic algorithm is based on the simulation biological genetic and evolutionary process. It is a self-adaption algorithm whose core is to make the fittest survive. Select a certain individual and self-adaption function to calculate, and then decide which to survive according to the probability statistics (Fig. 4).

In the paper, we set the fitness function:

$$F(f(x)) = 1/f(x) = 1/\rho Q \tag{2}$$

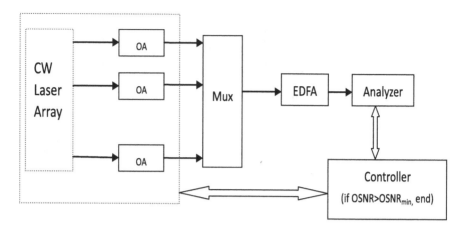

Fig. 4. Control system

And ρ is adaptive coefficient. So, the purpose of this paper is to obtain a small amount of adjustment, therefore, the smaller the adjustment amount, the larger the fitness value, and the more excellent the individual, the greater probability of being selected.

We set the fitness function as relevant portion with OSNR.

2 Simulation and Results

We use MATLAB simulation to confirm the proposed optimization tactic comprising five channels (m = 5), five amplified spans per link and each optical amplifier with variable gain. Each optical amplifier can adjust itself power towards this goal in the range [0, 30 dB], and optical signal power range [-50 dBm, 0]. Within the set of five channels and the controller to meet different requirement, there is the level of OSNR target, 20 dB OSNR level. We initially set the channel power as u(0) = [0.216 0.221 0.226 0.231 0.236] in Fig. 5.

Main	Frequency	**Power**	Polariza...	Simula...	Noise	Random...

Displ	Name	Value	Units	Mode
☐	Power[0]	0.216	dBm	Normal
☐	Power[1]	0.221	dBm	Normal
☐	Power[2]	0.226	dBm	Normal
☐	Power[3]	0.231	dBm	Normal
☐	Power[4]	0.236	dBm	Normal

Fig. 5. The channel power diagram

By using the genetic-nonlinear control algorithm to adjust all channel powers, the desired OSNR target is achieved through the G_i adjustment within range. The channel OSNRs after control tuning are shown in Fig. 6. It is shown that channel OSNR levels converge to new steady state values, and the OSNR is always greater than the original value and target value after adjustment and optimization. It satisfies the performance constraints and the simulation results show the effectiveness of the control strategy.

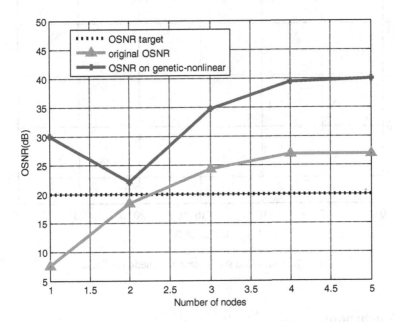

Fig. 6. OSNR in the optical system

Figure 7 exhibits the controlled variable about MATLAB control strategy. It shows that channel OSNR and channel power are capable of converging to the stable state during controlling process, making the final channel OSNRs meet the OSNR constraints. Consequently, the genetic-nonlinear control algorithm can availably make the transmission signal quality match the system requirements based on controlled robustness.

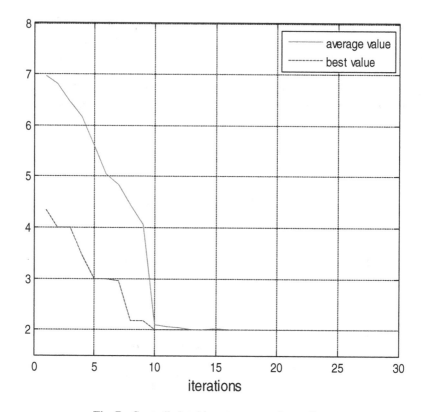

Fig. 7. Controlled stable state on genetic-nonlinear

3   Conclusions

This paper researches the performance optimization of the optical transmission links by improving OSNR. Simulation results exhibit that the control system has the better dynamic and stable performance for adjustment of transmission power at the OA nodes. It makes a broadband optical amplifier support individual channel power adjustment for large-scale network, despite that the control input may cause a slight jitter in the process of adjusting the controller. But the controlling strategy needs to consider the feedback time, which only is used as trial method. The inner running circulates to match the better parameter guiding delay-time in Genetic-nonlinear algorithm. In future, we will study some more stable and faster control strategies to improve the performance of the proposed scheme with Gain flatness or ASE noise.

Acknowledgments. This work was supported by the Research Foundation of China Mobile, and the Foundation of Beijing Engineering and Technology Center for Convergence Networks and Ubiquitous Services, the National Natural Science Foundation of China (No. 61302064), the China Postdoctoral Science Foundation (No. 2013M540862)

References

1. Yongli, Z., Ruiying, H., Haoran, C., Jie, Z., Yuefeng, J., Haomian, Z., Yi, L., Xinbo, W.: Experimental performance evaluation of software defined networking (SDN) based data communication networks for large scale flexi-grid optical networks. Opt. Expr. **22**(8), 9538–9547 (2014)
2. Lei, W., Jie, Z., Yongli, Z., Wanyi, G.: Study of optical control plane for translucent WDM networks. Photonic Netw. Commun. **20**(1), 64–74 (2010)
3. Pavel, L., et al.: A noncooperative game approach to OSNR optimization in optical networks. IEEE Trans. Autom. Control **51**(5), 848–852 (2006)
4. Pan, Y., Kilper, D., Morea, A., et al.: Channel power excursions in GMPLS end-to-end optical restoration with single step wavelength tuning. In: Proceedings of 2012 Optical Fiber Communication Conference and Exposition and the National Fiber Optic Engineers Conference, Los Angeles (2012)
5. Salma, K., Wissem, B., Mohamed, K.: Improvement of power system stability by static var compensator and tuning employing genetic algorithm. Int. J. Mod. Nonlinear Theory Appl. **3**(3), 11 (2014)
6. Peng, Q.: Nonlinear equations solving base on immune genetic algorithm. TELKOMNIKA Indonesian J. Electr. Eng. **13**(1), 174–179 (2014)
7. Pan, Y., Alpcan, T., Pavel, L.: A system performance approach to OSNR optimization in optical networks. Trans. Commun. **58**(4), 1193–1200 (2010)
8. Stefanovic, N., Pavel, L.: Robust power control of single sink optical networks with time-delays. In: 28th Chinese Control Conference, Decision and Control, pp. 2034–2039 (2009)

Optimal Low-Hit-Zone Frequency-Hopping Sequence Set via Cyclotomy

Haiyan Zhao[⊠], Xiangqian Dong, Changyuan Wang, and Wenfei Chen

Chengdu Neusoft University, Chengdu 611844, China
zhaohaiyan@nsu.edu.cn

Abstract. Frequency-hopping sequence set with low Hamming correlation within the fixed zone around the origin is called LHZ FHS set. In the quasi-synchronous frequency-hopping code division multiple access systems, the LHZ FHS set is often used to eliminate multiple-access interference. In this paper, basing on the Cyclostomes theory, we present a class of LHZ FHS set. It points out that the sequence set possesses excellent performance, and is optimal with respect to the Peng-Fan-Lee bound. It can be widely used in the frequency-hopping communication systems.

Keywords: Frequency-hopping sequence set · Low-hit-zone · Peng-Fan bound · Cyclotomy · Periodic Hamming correlation

1 Introduction

Frequency-hopping multiple accesses (FHMA) spread spectrum systems, with its anti-jamming, secure properties, have been found many applications in Bluetooth, military radio communications, mobile communications, and sonar echolocation systems [1–3]. In such systems each user is represented by a frequency-hopping sequence (FHS). Simultaneous transmission by more users over the identical frequency results in collisions. So, it is very desirable that such collisions over the identical frequency are minimized. The degree of such collisions is mainly related to the Hamming correlation properties of the frequency-hopping sequences (FHSs). In recent years, by combinatorial method or algebraic method, many FHSs meeting Lempel-Greenberger bound [3] have been presented [3–6]. And several optimal conventional FHSs meeting the Peng-Fan bounds [7] also have been reported.

In quasi-synchronous FHMA systems, different from the conventional FHSs, relative delays between different users are limited to a zone around the origin. So, frequency-hopping sequence set (FHS set) with low Hamming correlation within the fixed zone (LHZ) are more useful, this class of sequence set is called LHZ FHS set. The significance of LHZ FHS set is that the number of collisions will be kept at a very low level between different sequences as long as the relative delay does

H. Zhao—This work was supported partly by the general projects grant funded by the Education Department of Sichuan Province (No. 14ZB0350).

K. Li et al. (Eds.): ISICA 2015, CCIS 575, pp. 704–711, 2016.
DOI: 10.1007/978-981-10-0356-1_74

not exceed the LHZ, thus reducing the mutual interference (collisions). Now, several optimal LHZ FHS sets meeting the Peng-Fan-Lee bounds [8] have been covered in the literature. For example, by the Peng-Fan-Lee bounds [8], in 2010, optimal or near optimal LHZ FHS sets were constructed by Ma and Sun [9]. By interleaving techniques, in 2012 and in 2013, Niu [10,11] construct a new class of optimal LHZ FHS sets. And J. H. Chung [12,13] obtained several classes of LHZ FHS sets from the Cartesian product of some known FHS sets.

In this paper, our primary purpose is to construct a class of optimal LHZ FHS set, and analyze the maximum periodic Hamming correlation property within the LHZ. The rest of this paper is organized as follows: in Sects. 2 and 3, we introduce some notations, definitions and Cyclotomy theory; then present new generic construction for optimal LHZ FHS set by Cyclotomy techniques in Sect. 4; finally, we conclude the paper in Sect. 5.

2   Preliminaries

From now on, we will always suppose that (N, M, q, z, h) is a LHZ FHS set of length N over an alphabet of size q, whose LHZ is z and the maximum periodic Hamming correlation h within LHZ.

Let $\mathcal{F} = \{f_0, f_1, \cdots, f_{q-1}\}$ be a frequency slot set with size q, S a set of M FHSs of length N over $\mathcal{F}$. For any two FHSs $u = \{u_0, u_1, \cdots, u_{N-1}\}$ and $v = \{v_0, v_1, \cdots, v_{N-1}\}$ in S, any positive integer τ with $0 \leq \tau < N$, the periodic Hamming correlation $H_{u,v}(\tau)$ of u and v, at the time delay τ is defined as follow:

$$H_{u,v}(\tau) = \sum_{k=0}^{N-1} h(u_k, v_{k+\tau}), 0 \leq \tau < N,$$

where $h(u_k, v_{k+\tau}) = 1$ if $u_k = v_{k+\tau}$ and $h(u_k, v_{k+\tau}) = 0$ otherwise, and all the operations among the position indices are performed modulo N.

The maximum periodic Hamming autocorrelation $H_a(S)$, the maximum periodic Hamming cross-correlation $H_c(S)$ and The maximum periodic Hamming correlation $H_m(S)$ are defined as follows, respectively:

$$H_a(S) = \max_{1 \leq \tau < N} \{H_{u,u}(\tau) | u \in S\}$$

$$H_c(S) = \max_{0 \leq \tau < N} \{H_{u,v}(\tau) | u, v \in S, u \neq v\}$$

$$H_m(S) = \max\{H_a(S), H_c(S)\}.$$

In 1974, Lempel and Greenberger [3] established the following lower bound on $H_a(S)$.

Lemma 1. *For every FHS of length N over a given frequency slot set F with size of q, we have*

$$H_a(S) \geq \left\lceil \frac{(N-r)(N+r-q)}{q(N-1)} \right\rceil,$$

where r is the least nonnegative residue of N modulo q. Note that the Lempel-Greenberger bound is independent of the parameter M.

In 2004, Peng and Fan [7] developed the following lower bounds on $H_m(S)$ by incorporating the parameter M.

Lemma 2. *Let S be a set of FHS of family size M and length N over a given frequency slot set F with size q, then*

$$H_m(S) \geq \left\lceil \frac{(NM - q)N}{(NM - 1)q} \right\rceil. \tag{1}$$

and

$$H_m(S) \geq \left\lceil \frac{2INM - (I + 1)Iq}{(NM - 1)M} \right\rceil.$$

Let integers $l_a > 0, l_c > 0$. Then, the low-hit-zone L_{HZ} is defined as below:

$$L_{HZ} = \min\{\max_{1 \leq \tau \leq T, \forall u \in S}\{T | H_{u,u}(\tau) \leq l_a\}, \max_{0 \leq \tau \leq T, \forall u,v \in S, u \neq v}\{T | H_{u,v}(\tau) \leq l_c\}\}.$$

An FHS set S with $L_{HZ} > 0$ is called LHZ FHS set. For the LHZ FHS set, in 2006, Peng, Fan and Lee [8] obtained the following bound.

Lemma 3 (Peng-Fan-Lee bounds [8]). *Let S be an FHS set of size M and length N over a frequency slot set $\mathcal{F}$ with size q, and L_H the LHZ of S with respect to constants $H_m(S)$. Then, For any integer Z with $1 \leq Z \leq L_H$, we have*

$$H_m(S) \geq \left\lceil \frac{(MZ + M - q)N}{(MZ + M - 1)q} \right\rceil \tag{2}$$

and

$$H_m(S) \geq \frac{(Z + 1)(2INM + LM - Iq - I^2q) - (I + 1)Iq - ML^2}{(ZM + M - 1)LM}.$$

This bound include Peng-Fan bounds (1) as the special cases: $Z = N - 1$.

Definition 1. *Let S be a LHZ FHS set with parameters (N, M, q, L_H, h), S is said to be optimal if the equality in (2) is achieved.*

3   Cyclotomic Class and Cyclotomic Number

In this section, we give a very brief introduction of Cyclotomy [14] as we will employ it for the construction of LHZ FHS set in the sequel. Let $\mathcal{F}_{p^n}$ be the finite field with p^n elements, and let $ef = p^n - 1$, where e and f are positive integers. α is a primitive element of $\mathcal{F}_{p^n}$, $C_0 = <\alpha^e>$ is the multiplicative subgroup of $\mathcal{F}_{p^n}$ generated by α^e. We now define the cyclotomic class C_i of order e in $\mathcal{F}_{p^n}$:

$$C_i = \{\alpha^{i+ke} : i = 0, 1, \cdots, e-1; k = 0, 1, \cdots, f-1\}.$$

And note that the C_i's are pairwise disjoint, their union is $\mathcal{F}_{p^n}^*$ which is the multiplicative group of $\mathcal{F}_{p^n}$, if n' is any integer, then $C_{i+n'e} = C_i$. Furthermore, the cyclotomic number of order e with respect to $\mathcal{F}_{p^n}^*$ are defined as follow:

$$(i, j) = |(C_i + 1) \cap C_i|.$$

The following Lemma gives some basic properties of cyclotomic number which will be used in the sequel.

Lemma 4. *Let the symbols be defined as above. We have*
(1) If $\frac{1}{\theta} \in C_h$, then $|(C_i + \theta) \cap C_j)| = |\frac{1}{\theta}(C_i + 1) \cap \frac{1}{\theta}C_j)| = |(C_{i+h} + 1 \cap C_{j+h})|.$
(2) $\sum_{t=0}^{e-1}(i + k, i) = f - 1$ if $k = 0$; and $\sum_{t=0}^{e-1}(i + k, i) = f$ otherwise.

4   New Class of Optimal LHZ FHS Set

In this section, we will use cyclotomy techniques to construct a new class of optimal LHZ FHS set according to the Peng-Fan-Lee bound (2).

Construction A

Let $q = p^n$ be a prime or prime power, and $p^n = ef + 1$. C_i is the cyclotomic class of order e in $\mathcal{F}_q$. Define $\bar{G}_i$ as follow:

$$\bar{G}_i = C_i - 1.$$

For i, j with $0 \leq i \neq j \leq e - 1$, it is obvious that the equality holds.

$$\bar{G}_i \cap \bar{G}_j = \varnothing, \cup_{i=0}^{e-1}\bar{G}_i = Z_{q-1}.$$

Construct the desired FHS set $S = \{S_k(t), 0 \leq t \leq p^n - 1, 0 \leq k \leq e - 1\}$, such that
(1) $S_k(p^n - 1) = e.$
(2) If $t \in \bar{G}_{(i+k) \bmod e}$ and $t \neq le - 1, 1 \leq l \leq f - 1$

$$S_k(t) = i.$$

(3) If $t \in \bar{G}_{(i+k) \bmod e}$ and $t = le - 1, 1 \leq l \leq f - 1$

$$S_k(le - 1) = e.$$

Theorem 1. *S is an optimal LHZ FHS set with parameters $(p^n, e, e+1, e-1, f)$ if $e^2 f + 2e + 2 < e^3 + f$ according to the Peng-Fan-Lee bound (2).*

Proof. Let $\bar{G}_e = \{p^n - 1\}$, and for $1 \leq l \leq f - 1$, $\{le - 1\} \in \bar{G}_{i_l}$, that is $\{le\} \in C_{i_l}$. For any sequence S_i in S, by the definition of periodic Hamming autocorrelation, we have

$$H_{S_i,S_i}(\tau) = \sum_{i=0}^{e-1,i\neq i_l} |(\bar{G}_i + \tau) \cap \bar{G}_i| + \sum_{l=1}^{f-1} |\{p^n - 1 + \tau\} \cap \{le - 1\}|$$

$$+ |\{p^n - 1 + \tau\} \cap \{p^n - 1\}| + \sum_{l=1}^{f-1} |\{le - 1 + \tau\} \cap \{p^n - 1\}|$$

$$+ \sum_{l=1}^{f-1} |(\bar{G}_{i_l} \backslash \{le - 1\} + \tau) \cap (\bar{G}_{i_l} \backslash \{le - 1\})| + \Delta,$$

$$= f - 1 + \sum_{l=1}^{f-1} |\{\tau\} \cap \{lf\}| + \sum_{l=1}^{f-1} |\{le + \tau\} \cap \{0\}|$$

$$- \sum_{l=1}^{f-1} |(\{le\} + \tau) \cap C_{i_l}| - \sum_{l=1}^{f-1} |C_{i_l} \cap (\{le\} - \tau)| + \Delta',$$

where Δ and Δ' are given by

$$\Delta = \sum_{l=1}^{f-1} |\{e - 1 + \tau\} \cap \{le - 1\}|$$

$$+ \cdots + \sum_{l=1}^{f-1} |(\{(f-1)e - 1\} + \tau) \cap \{le - 1\}|,$$

$$\Delta' = \sum_{l=1}^{f-1} |\{e + \tau\} \cap \{le\}| + \cdots + \sum_{l=1}^{f-1} |(\{(f-1)e\} + \tau) \cap \{le\}|.$$

For $1 \leq \tau \leq p^n - 1$, it then follows that

$$\Delta' \leq f - 2.$$

So, when $0 < \tau < e$, the maximum periodic Hamming autocorrelation can be given by

$$H_a(S) = f - 1.$$

For any two distinct FHSs S_i, S_{i+u} in S, by the definition of periodic Hamming cross-correlation, we have

$$H_{S_i,S_{i+u}}(\tau) = \sum_{i=0}^{e-1,i\neq i_l} |(\bar{G}_i + \tau) \cap \bar{G}_{i+u}| + \sum_{l=1}^{f-1} |(\bar{G}_{e-u} + \tau) \cap \{p^n - 1\}|$$

$$+ |\{p^n - 1 + \tau\} \cap \bar{G}_{e+u}| + |\{p^n - 1 + \tau\} \cap \{p^n - 1\}|$$

$$+ \sum_{l=1}^{f-1} |\{p^n - 1 + \tau\} \cap \{le - 1\}| + \sum_{l=1}^{f-1} |\{le - 1 + \tau\} \cap \{p^n - 1\}|$$

$$+ \sum_{l=1}^{f-1} |(\bar{G}_{i_l-u} + \tau) \cap (\bar{G}_{i_l} \backslash \{le - 1\})|$$

$$+ \sum_{l=1}^{f-1} |(\bar{G}_{i_l} \setminus \{le - 1\} + \tau) \cap (\bar{G}_{i_l+u})| + \Delta,$$

$$= \sum_{i=0}^{e-1} |(C_i + \tau) \cap C_{i+u}| + |\{\tau\} \cap \{0\}| + \sum_{l=1}^{f-1} |\{\tau\} \cap \{le\}|$$

$$+ \sum_{l=1}^{f-1} |\{le + \tau\} \cap \{0\}| - \sum_{l=1}^{f-1} |(\{le\} + \tau) \cap C_{i_l+u}|$$

$$- \sum_{l=1}^{f-1} |C_{i_l-u} \cap (\{le\} - \tau)| + \Delta'.$$

For $0 \leq \tau \leq p^n - 1$, then
If $\tau \neq 0$

$$\Delta' \leq f - 2.$$

If $\tau = 0$

$$\Delta' = f - 1.$$

Thus, when $0 \leq \tau < e$, the maximum periodic Hamming cross-correlation can be given by

$$H_c(S) = f.$$

The above discussions have shown that for any $0 \leq \tau < e$, the maximum periodic Hamming correlation can be given by

$$H_m(S) = f.$$

Now, we check the optimality of the LHZ FHS set S according to the Peng-Fan-Lee bound (2). Putting the parameters of the FHS set S in (2), we get

$$H_m(S) \geq \left\lceil \frac{ef + 1}{e + 1} - \frac{fep^n}{((p^n - 1)e - f)(e + 1)} \right\rceil,$$

$$= f - \left\lceil \frac{e^2 f - f + e^2 + e + 1}{e^3 + e^2 - e - 1} \right\rceil.$$

Since $e^2 f + 2e + 2 < e^3 + f$, it then follows that

$$0 < \frac{e^2 f - f + e^2 + e + 1}{e^3 + e^2 - e - 1} < 1.$$

This leads to

$$H_m(S) \geq f.$$

It is obvious that the equality in Peng-Fan-Lee bound (2) is achieved. So that the conclusion follows.

The following example can illustrate our construction.

Example 1. Let $p = 37$, $e = 9$, $f = 4$. Applying the construction A, we obtain a LHZ FHS set S of length 37 as follows:

$$S = \{S_0 = \{0, 1, 8, 2, 5, 0, 5, 3, 9, 6, 3, 1, 2, 6, 4, 4, 7, 9, 8, 7, \cdots\};$$
$$S_1 = \{8, 0, 7, 1, 4, 8, 4, 2, 9, 5, 2, 0, 1, 5, 3, 3, 6, 9, 7, 6, 3 \cdots\};$$
$$\cdots$$
$$S_8 = \{1, 2, 0, 3, 6, 1, 6, 4, 9, 7, 4, 2, 3, 7, 5, 5, 8, 9, 0, 8, \cdots\}.\}.$$

The maximum periodic Hamming autocorrelation of S is shown in Fig. 1, the maximum periodic Hamming cross-correlation and the maximum periodic Hamming correlation of S are shown in Fig. 2. For $0 \leq \tau \leq 8$, it is easy to check that the periodic Hamming correlation of S satisfies

$$H_m(S) = 4.$$

Therefore, S is an optimal LHZ FHS set with parameters (37, 9, 10, 8, 4) with respect to the Peng-Fan-Lee bounds (2).

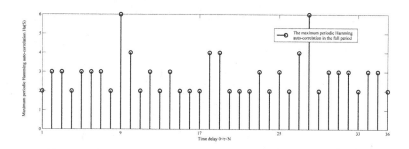

Fig. 1. The maximum periodic Hamming auto-correlation of S

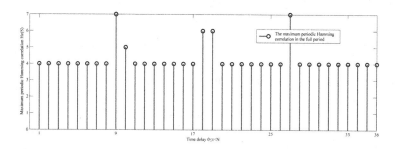

Fig. 2. The maximum periodic Hamming cross-correlation and correlation of S

5 Conclusions

In quasi-synchronous frequency-hopping code division multiple access systems, LHZ FHS set is often used to eliminate multiple-access interference. In this paper, basing on the Cyclostomes theory, we present a new class of optimal LHZ FHS set with respect to the Peng-fan-lee bound. It may be a challenging problem to find more classes of optimal LHZ FHS sets via Cyclotomy technique, which are not involved in this paper.

References

1. Fan, P.Z., Darnell, M.: Sequence Design for Communications Applications. Research Studies Press (RSP), Wiley, London (1996)
2. Golomb, S.W., Gong, G.: Signal Design for Good Correlation: For Wireless Communication, Cryptography and Radar. Cambridge University Press, Cambridge (2005)
3. Lempel, A., Greenberger, H.: Families of sequences with optimal Hamming correlation properties. IEEE Trans. Inf. Theory **20**(1), 90–94 (1974)
4. Chung, J.H., Han, Y.K., Yang, K.: New classes of optimal frequency-hopping sequences by interleaving technique. IEEE Trans. Inf. Theory **55**, 5783–5791 (2009)
5. Chung, J.H., Yang, K.: Optimal frequency-hopping sequences with new parameters. IEEE Trans. Inf. Theory **56**, 1685–1693 (2010)
6. Chung, J.H., Yang, K.: Optimal frequency-hopping sequences with new parameters. IEEE Trans. Inf. Theory **56**(4), 1685–1693 (2010)
7. Peng, D.Y., Fan, P.Z.: Low bounds on the Hamming auto and crosscorrelation of frequency hopping sequences. IEEE Trans. Inf. Theory **50**, 2149–2154 (2004)
8. Peng, D.Y., Fan, P.Z., Lee, M.H.: Lower bounds on the periodic Hamming correlations of frequency hopping sequences with low hit zone. Sci. China Ser. F Inf. Sci. **49**, 1–11 (2006)
9. Ma, W., Sun, S.: New designs of frequency hopping sequences with low hit zone. Des. Codes Crypt. **62**(2), 145–153 (2011)
10. Niu, X., Peng, D.Y., Zhou, Z.C.: New classes of optimal frequency hopping sequences with low hit zone. Adv. Math. Commun. **7**(2), 293–310 (2013)
11. Niu, X.H., Peng, D.Y., Zhou, Z.C.: New classes of optimal low hit zone frequency-hopping sequences with new parameters by interleaving technique. IEICE Trans. Fundam. **55**(11), 1835–1842 (2012)
12. Chung, J.-H., Yang, K.: Low-hit-zone frequency-hopping sequence sets with new parameters. In: Helleseth, T., Jedwab, J. (eds.) SETA 2012. LNCS, vol. 7280, pp. 202–211. Springer, Heidelberg (2012)
13. Chung, J.H., Yang, K.: New classes of optimal low-hit-zone frequency-hopping sequence sets by cartesian product. IEEE Trans. Inf. Theory **59**(1), 726–732 (2013)
14. Storer, T.: Cyclotomy and Difference Sets. Markham, Chicago (1967). Article in a Conference Proceedings

USPD Doubling or Declining in Next Decade Estimated by WASD Neuronet Using Data as of October 2013

Yunong Zhang[1,2,3](✉), Zhengli Xiao[1,2,3], Dongsheng Guo[1,2,3], Mingzhi Mao[1], and Hongzhou Tan[1,2]

[1] School of Information Science and Technology, Sun Yat-sen University, Guangzhou 510006, China
zhynong@mail.sysu.edu.cn, ynzhang@ieee.org, jallonzyn@sina.com
http://sist.sysu.edu.cn/~zhynong/indexe.htm
[2] SYSU-CMU Shunde International Joint Research Institute, Shunde 528300, China
[3] Key Laboratory of Autonomous Systems and Networked Control, Ministry of Education, Guangzhou 510640, China

Abstract. Recently, the total public debt outstanding (TPDO) of the United States has increased rapidly, and to more than \$17 trillion on October 18, 2013. It is important and necessary to conduct the TPDO projection for better policies making and more effective measurements taken. In this paper, we present the ten-year projection for the public debt of the United States (termed also the US public debt, USPD) via a 3-layer feed-forward neuronet. Specifically, using the calendar year data on the USPD from the Department of the Treasury, the neuronet is trained, and then is applied to projection. Via a series of numerical tests, we find that there are several possibilities of the change of the USPD in the future, which are classified into two categories in terms of projection trend: the continuous-increase trend and the increase-peak-decline trend. In the most possible situation, the neuronet indicates that the TPDO of the United States is projected to increase, and it will double in 2019 and double again in 2024.

Keywords: US public debt (USPD) · Ten-year projection · Total public debt outstanding (TPDO) · Feed-forward neuronet

1 Introduction

The US public debt is the amount owed by the Federal government of the United States [1,2]. It consists of two components: (1) the debt held by the public, and (2) the intragovernmental debt holdings [1–3]. Without doubt, the US public debt as a share of gross domestic product (GDP) plays an important role in the development of the United States. Because of the important role, many studies have been reported on the US public debt. It follows from the public debt reports provided by the Department of the Treasury [4] that the total

K. Li et al. (Eds.): ISICA 2015, CCIS 575, pp. 712–723, 2016.
DOI: 10.1007/978-981-10-0356-1_75

public debt outstanding (TPDO) of the United States on January 1, 1790 is $71.0605085 million. Since then, the TPDO of the United States has risen to more than \$17 trillion on October 18, 2013 (though, historically, it declined in some short time periods). Specially, the US public debt has recently increased with a rapid growth rate (mainly because of the financial crisis) [4]. With the rapid increase of the public debt, there may arise lots of problems that the US government has to face (e.g., the risks to economic growth and the potential existence of inflation) [1,2,5–9]. The US public debt has been a very appealing topic that influences the development of the United States.

Because the debt is closely related to the development of the United States, the projection of the US public debt, which can be used for planning and research (e.g., for the Federal budget), is becoming urgent and important. It is conducted by many organizations, including national and local governments and private companies [1,2,5,6,10–17], for better policies making and more effective measurements taken. For example, the Congressional Budget Office (CBO) of the United States releases its annual report *Updated Budget Projections* in 2013, which covers a ten-year window [5]. As shown in the report, for the 2014–2023 period, the debt held by the public is projected to remain above 70 percent of GDP. In addition to the above short-term projection, the CBO also projects the US public debt as part of its annual report *The 2013 Long-Term Budget Outlook* [6]. In such a report, the CBO forecasts that the debt held by the public will reach 100 percent of GDP in 2038. Another office, i.e., the Government Accountability Office (GAO) of the United States, also does some researches and provides the related reports on the US public debt projection [1,2]. According to the GAO, the debt-to-GDP ratio will be doubled in 2040, and doubled again in 2060, reaching 600 % in 2080.

As for the US public debt, the Department of the Treasury provides the fiscal year reports [4], from which the historical TPDO of the United States during 1790 ~ 2013 can be found and can also be viewed as the de facto standard. Note that the historical data contain the general regularity of change of the US public debt, and are also the comprehensive reflection of change of the US public debt under the influence of all factors (e.g., natural environment, policy, economy and culture). Based on the historical data, we can have the overall view on the trend of the US public debt from the past to the future. Therefore, in this paper, based on the calendar year data from the Department of the Treasury, we conduct the ten-year projection (termed also as the short-term projection) of the US public debt using a 3-layer feed-forward neuronet [18,19].

2   WASD Neuronet Method

Benefiting from parallel processing capability, inherent nonlinearity, distributed storage and adaptive learning capability, neuronet has been widely used in many areas [20–25]. In this section, the neuronet method for the projection of the USPD is developed and investigated. Specifically, based on the past 224-year data (i.e., from January 1, 1790 to October 18, 2013), a 3-layer feed-forward

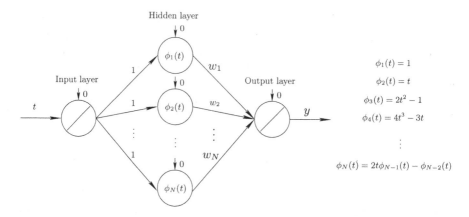

Fig. 1. The model structure and activation functions of the feed-forward neuronet for the projection of the USPD. As for the neuront, it has three layers: the input layer, the hidden layer and the output layer. Using the calendar year data on the USPD, we have the data set (of sample pairs) for the training of such a neuronet. Then, with the neuronet well trained, it is applied to the projection of the USPD.

neuronet is constructed and then applied to the projection of the USPD. Note that, in this paper, the US debt data are obtained from the Department of the Treasury.

2.1   Model Structure of Feed-Forward Neuronet

The 3-layer feed-forward neuronet used for the projection of the USPD is constructed in this subsection. The model structure of such a neuronet is detailed as follows.

Figure 1 intuitively shows the structure of the 3-layer feed-forward neuronet, which, in the hidden layer, has N neurons activated by a group of Chebyshev polynomials $\phi_j(\cdot)$ (with $j = 1, 2, \cdots, N$). In addition, the input layer or the output layer each has one neuron activated by the linear function. Moreover, the connecting weights from the input layer to the hidden layer are all fixed to be 1, whereas the connecting weights from the hidden layer to the output layer, denoted as w_j, are to be decided or adjusted. Besides, all neuronal thresholds are fixed to be 0. Note that, theoretically guaranteeing the efficacy of the neuronet, these settings can simplify the complexities of the neuronet structure design, analysis and computation.

Furthermore, as for the presented neuronet, we exploit the algorithm of weights and structure determination (WASD) that can determine effectively the weights connecting the hidden layer and the output layer, and obtain automatically the optimal structure of the neuronet. Please refer to the authors' previous work [18, 19] for more details about the WASD algorithm, of which the neuronet is termed WASD neuronet.

2.2 Normalization from $[1790, 2013]$ to $[-1, \alpha]$

In this subsection, to lay a basis for further discussion, we give some necessary explanations as follows. For the 3-layer feed-forward neuronet exploited in this paper, the input corresponds to date (e.g., January 1, 1790), and the output is the Treasury datum (i.e., the TPDO of the United States). In addition, the hidden-layer neurons of such a neuronet are activated by a group of Chebyshev polynomials of class 1 [18,19], which means that the domain of the input is $[-1, 1]$. It is thus necessary to normalize the data that are used as the neuronet input. Specifically, in this paper, the time interval $[1790, 2013]$ is normalized to the interval $[-1, \alpha]$ with the normalization factor $\alpha \in (-1, 0)$. Note that "1790" and "2013" here represent January 1, 1790 and October 18, 2013 (i.e., the day the TPDO of the United States reached \$17 trillion), respectively. Therefore, using the calendar year data on the USPD, based on the above data preprocessing, we have the data set (of sample pairs) for the training of the feed-forward neuronet. Then, with the neuronet well trained, it is applied to the projection of the USPD.

3 Approximation and Projection Tests

As mentioned above, the time interval $[1790, 2013]$ is normalized to the interval $[-1, \alpha]$ with the normalization factor $\alpha \in (-1, 0)$. In this section, different values of α are used to test the approximation and projection performances of the neuronet shown in Fig. 1.

To lay a basis for further discussion, we give $\{(x_i, \gamma_i)|_{i=1}^{Q}\}$ as the data set of sample pairs for the training of the neuronet, where, for the ith sample pair, $x_i \in \mathbb{R}$ denotes the input and $\gamma_i \in \mathbb{R}$ denotes the desired output (or to say, target output). As mentioned in the paper, the input corresponds to date (e.g., January 1, 1790), and the output is the Treasury datum (i.e., the TPDO of the United States). Then, the mean square error (MSE) for the neuronet is defined as follows:

$$E = \frac{1}{Q} \sum_{i=1}^{Q} \left(\gamma_i - \sum_{j=1}^{N} w_j \phi_j(x_i) \right)^2. \tag{1}$$

Note that, for the neuronet with a fixed number of hidden-layer neurons, we can obtain the related MSE (i.e., the training error) via (1). By increasing the number of hidden-layer neurons one by one, we can finally have the relationship between the MSE and the number of hidden-layer neurons. The WASD algorithm can thus be developed to determine the optimal structure of the neuronet (i.e., with the number of hidden-layer neurons determined corresponding to the smallest MSE). With the optimal structure obtained, the training procedure is finished, and the approximation performance of the resultant neuronet is evaluated.

Based on the well-trained neuronet, we can investigate its projection performance using the calendar year data on the USPD from 1790 to 2013. Specifically, the data corresponding to the time interval $[1790, 2010]$ are used to train the neuronet, and the other data are used for the projection test of the neuronet.

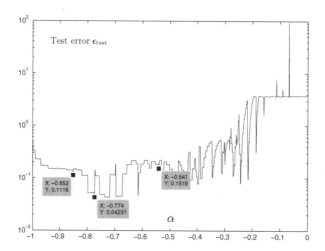

Fig. 2. The relationship between the normalization factor α and the test error of the neuronet with the optimal structure (i.e., the smallest MSE via (1)). Note that X corresponds to the abscissa and Y corresponds to the ordinate. As illustrated in the figure, different values of α lead to different test errors, showing different projection performances of the neuronet for the USPD.

Note that the time interval $[1790, 2010]$ also needs to be normalized to the interval $[-1, \alpha]$. Besides, the test error that is exploited to evaluate the projection performance of the neuronet is defined as follows:

$$\epsilon_{\text{test}} = \frac{1}{M} \sum_{m=1}^{M} \left| \frac{y_m - \gamma_m}{\gamma_m} \right|,$$

where M is the total number of the other data for test, and y_m and γ_m denote respectively the neuronet output and the public debt datum at the mth test point. Generally speaking, the smaller the test error ϵ_{test} is, the better the projection performance of the neuronet achieves.

Via a series of numerical tests, we find that different values of α lead to different (approximation and) projection results of the neuronet. For better understanding, Fig. 2 shows the relationship between the normalization factor α and the test error ϵ_{test} of the neuronet with the optimal structure. As shown in the figure, superior projection performance of the neuronet can be achieved by choosing a suitable value of α. More importantly, from the figure, we can observe that there exist several local minimum points and a global minimum point, such as the marked ones, which correspond to different values of α. Such local and global minimum points indicate the possible situations of the public debt projection via the neuronet, and the global minimum point further implies the most possible situation. Based on the results shown in Fig. 2, different α values are thus used for the neuronet to project the USPD in the next ten years, with the corresponding projection results presented in the ensuing section.

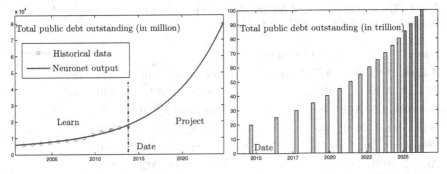

(a) The ten-year trend of the USPD   (b) The histogram of the selected data

Fig. 3. Projection results of the USPD via the neuronet using $\alpha = -0.774$ (being the global minimum point in Fig. 2). As shown in the figure, the USPD is projected to increase continuously in the next ten years.

4   Projection Results for USPD

As presented above, different values of α lead to different training performances of the feed-forward neuronet, thereby making the projection results of the USPD different. Thus, in this section, such different α values (or more specifically, the local and global minimum points marked in Fig. 2) are used for the neuronet to project the USPD in the next ten years (specifically, to September 30, 2024). The related projection results of the USPD via the neuronet simulations are presented as follows.

4.1   Projection of USPD with Most Possibility

$\alpha = -0.774$ is the global minimum point in Fig. 2, which indicates the situation with the most possibility for the projection of the USPD using the neuronet. The corresponding projection results via the neuronet simulation with $\alpha = -0.774$ are illustrated in Fig. 3 and Table 1. Specifically, Fig. 3(a) intuitively and visually shows the learning and projection results, which tells the rise of the USPD in the next ten years. In addition, as shown in Fig. 3(b) and Table 1, the TPDO of the United States is projected to increase to $20 trillion (actually, to $17.749172 trillion) on August 30, 2014, and it will double in 2019 and double again in 2024 (i.e., reaching $80 trillion on September 22, 2024). Moreover, from Fig. 3(b) and Table 1, we can observe that the growth rate of the USPD is increasing. With such a growth trend, as indicated by the neuronet simulation, the TPDO of the United States will increase further to $90 trillion on July 24, 2025, and then to $100 trillion on April 24, 2026 (see also Fig. 3 and Table 1).

4.2   More Possibilities for Projection of USPD

Differing from the global minimum point in Fig. 2, the local minimum point that indicates a possible situation of the debt projection is worth being investigated

Table 1. The actual datum of August 31, 2014 and the projection results that correspond to Fig. 3(b) for the USPD in the future.

Date	TPDO (in trillion)	Date	TPDO (in trillion)
August 31, 2014	$17.75 (actual)	August 30, 2014	$20
May 24, 2016	$25	October 2, 2017	$30
November 16, 2018	$35	November 1, 2019	$40
September 2, 2020	$45	June 1, 2021	$50
February 1, 2022	$55	September 13, 2022	$60
April 7, 2023	$65	October 14, 2023	$70
April 9, 2024	$75	September 22, 2024	$80
February 26, 2025	$85	July 24, 2025	$90
December 11, 2025	$95	April 24, 2026	$100

as well. Figure 4 in this paper shows the projection results of the USPD via the neuronet using $\alpha = -0.719$ (being the sub-global minimum point in Fig. 2), -0.852, -0.575 and -0.541. As seen from Fig. 4(a), the USPD is projected to increase in the next ten years. Note that such a projection result coincides with Fig. 3(a) that is related to the global minimum point (with $\alpha = -0.774$), except for the values being slightly different. Besides, Fig. 4(b) also shows the continually-increasing trend of the USPD in the next ten years, but with a gentle growth rate. As for Fig. 4(b), the corresponding neuronet simulation shows that the TPDO of the United States is projected to increase to $20 trillion on June 13, 2015 (actually, to $18.152852 trillion on May 31, 2015), to $30 trillion on September 17, 2019, to $40 trillion on September 10, 2022, and finally to $48.813838 trillion on September 30, 2024.

Here, it is worthwhile noting that the trends of the projection results shown in Fig. 4(c) and (d) are quite different from those shown in Fig. 4(a) and (b). That is, instead of increasing continuously, the projection of the US public debt is also found with possibilities to rise, peak and then decline, as shown in Fig. 4(c) and (d). More specifically, on the one hand, as for Fig. 4(c), the related neuronet simulation shows that the total public debt outstanding of the United States is projected to increase to $20 trillion on January 7, 2015 (actually, to $18.082294 trillion on January 31, 2015), reach the summit on June 3, 2020 with the value being $27.413432 trillion, and then decrease gradually. On the other hand, as for Fig. 4(d), the related neuronet simulation shows that the TPDO of the United States is projected to reach the summit on August 30, 2014 with the value being $17.51 trillion, and then decrease gradually and further to less than $10 trillion on August 4, 2017. Besides, according to the recent report provided by the Department of the Treasury [26], the TPDOs of the United States on October 31, 2013, November 30, 2013, December 31, 2013, January 31, 2014, and February 28, 2014 are $17.156117 trillion, $17.217152 trillion, $17.351971 trillion, 17.293020 trillion, and 17.463229 trillion, respectively. By contrast, the neuronet

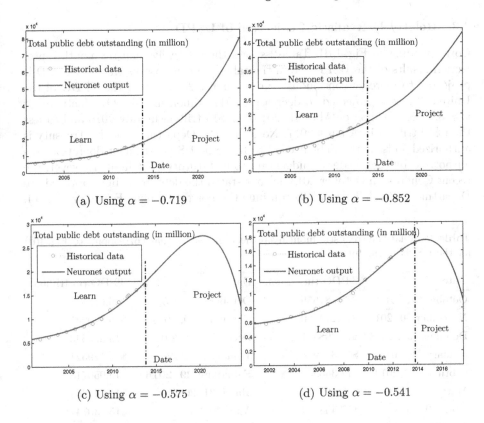

Fig. 4. Different projection results of the USPD via the neuronet using different values of α. Note that each of the subfigures which corresponds to a local minimum point given in Fig. 2 shows a possible situation of the debt projection. As shown in the figure, there are four different projection results for the USPD that can be classified into two categories in terms of projection trend: the continuous-increase trend and the increase-peak-decline trend. Besides, all of these projection results tell the rise of the USPD in the short future.

simulation corresponding to Fig. 4(d) shows that the TPDO of the United States is projected to increase to \$17.187655 trillion on October 31, 2013, to \$17.244600 trillion on November 30, 2013, to \$17.298451 trillion on December 31, 2013, then to \$17.346936 trillion on January 31, 2014, and further to \$17.385905 trillion on February 28, 2014. Evidently, the differences are tiny; specifically, \$0.031538 trillion (i.e., 0.183829 %), \$0.027448 trillion (i.e., 0.159422 %), \$0.053510 trillion (i.e., 0.308380 %), \$0.053916 trillion (i.e., 0.311779 %), and 0.077324 trillion (i.e., 0.442782 %), respectively. In view of this point, we thus pay more attention to Fig. 4(d) that may show the relatively realistic projection for the USPD.

4.3   Relatively Realistic Projection of USPD

Corresponding to Fig. 4(d), Table 2 shows the relatively realistic monthly projection results of the USPD. From the table, we can observe that the USPD is projected to increase continuously until August 2014. Then, the TPDO of the United States is projected to decrease to $17 trillion in July 2015 (actually, it is $18.152852 trillion on May 31, 2015), to $15 trillion in July 2016 and to less than $10 trillion in August 2017. Note that the Department of the Treasury is authorized to issue the debt needed to fund the US government operations (as authorized by each Federal budget) up to a statutory debt limit. However, the recent Congress in October 2013 has not stated the debt limit, but approved the Department of the Treasury to continue the borrowing until February 7, 2014.

Table 2. Relatively realistic monthly projection results that correspond to Fig. 4(d) for the USPD in the future.

Date	TPDO (in trillion)	Date	TPDO (in trillion)
October 31, 2013	$17.187655	October 31, 2015	$16.613578
November 30, 2013	$17.244600	November 30, 2015	$16.469235
December 31, 2013	$17.298451	December 31, 2015	$16.306543
January 31, 2014	$17.346936	January 31, 2016	$16.129521
February 28, 2014	$17.385905	**February 29, 2016**	$15.950718
March 31, 2014	$17.423112	March 31, 2016	$15.744864
April 30, 2014	$17.453283	April 30, 2016	$15.530741
May 31, 2014	$17.477971	May 31, 2016	$15.293400
June 30, 2014	$17.495326	June 30, 2016	$15.047667
July 31, 2014	$17.506394	**July 31, 2016**	$14.776580
August 31, 2014	**$17.509866**	August 31, 2016	$14.487729
September 30, 2014	$17.505957	September 30, 2016	$14.190557
October 31, 2014	$17.493863	**October 31, 2016**	$13.864486
November 30, 2014	$17.474147	November 30, 2016	$13.530212
December 31, 2014	$17.445095	December 31, 2016	$13.164693
January 31, 2015	$17.406600	**January 31, 2017**	$12.778280
February 28, 2015	$17.363987	February 28, 2017	$12.410525
March 31, 2015	$17.307264	**March 31, 2017**	$11.982011
April 30, 2015	$17.242524	April 30, 2017	$11.545547
May 31, 2015	$17.165411	May 31, 2017	$11.071140
June 30, 2015	$17.079894	**June 30, 2017**	$10.588983
July 31, 2015	$16.980503	July 31, 2017	$10.065911
August 31, 2015	$16.869273	**August 31, 2017**	$9.516749
September 30, 2015	$16.749625	**September 30, 2017**	$8.959872

In the day after such a Congress, the newly-increased TPDO of the United States is \$0.328 trillion. Thus, as for Table 2, the still-increasing projection of the USPD in the next few months (or at least, until February 2014) is acceptable and reliable. With the corresponding policies made (in or after February 2014), the USPD may be well under control. Then, it is hopeful that, in August 2014, the USPD reaches the summit, showing that the related measurements are taken effectively; and that, after the summit (or say, from September 2014), the USPD decreases gradually, as projected in Table 2. Besides, for comparison, the actual USPD from October 31, 2013 to May 31, 2015 is shown in Table 3.

Table 3. Actual USPD data from October 31, 2013 to May 31, 2015.

Date	TPDO (in trillion)	Date	TPDO (in trillion)
October 31, 2013	\$17.156117	**August 31, 2014**	\$17.749172
November 30, 2013	\$17.217152	September 30, 2014	\$17.824071
December 31, 2013	\$17.351971	October 31, 2014	\$17.937160
January 31, 2014	\$17.293020	November 30, 2014	\$18.005549
February 28, 2014	\$17.463229	December 31, 2014	\$18.141444
March 31, 2014	\$17.601227	**January 31, 2015**	\$18.082294
April 30, 2014	\$17.508437	February 28, 2015	\$18.155854
May 31, 2014	\$17.516958	March 31, 2015	\$18.152056
June 30, 2014	\$17.632606	April 30, 2015	\$18.152560
July 31, 2014	\$17.687137	**May 31, 2015**	\$18.152852

5   Further Discussion

In summary, the above numerical results synthesized by the neuronet show that there are two different trends of the projection of the USPD in the future, i.e., the continuous-increase trend and the increase-peak-decline trend. As mentioned previously, the TPDO of the United States increases year by year, and to more than \$17 trillion in October 2013 (provided by the Treasury report). Thus, the US government has made the policies for handling the rapid growth of the public debt. Note that these policies may not have much of an effect on such a huge TPDO within a short period. Therefore, as for Figs. 3 and 4 and Tables 1 and 2, it is projected that the USPD increases continually in the next ten years/months, which is reasonable and acceptable. On the one hand, as indicated by Fig. 3 and Table 1 [as well as Fig. 4(a) and (b)], for the huge TPDO that is still increasing, it may become more and more difficult for the US government to handle, but just keep it up (which can also be viewed as a vicious circle). In this sense, Fig. 3 and Table 1, showing the projection results of the USPD with the most possibility, imply that the current policies carried out by the government may be less effective on the debt-growth control. On the other hand, as indicated by Fig. 4(c) and (d) and Table 2, the USPD will reach its summit and then decrease gradually, thereby showing from one viewpoint the delayed effectiveness of those policies made by the US government on controlling the public debt growth. As noted from another viewpoint [5,6], these projection results may also possibly imply that some events including wars may happen, thereby resulting in the decrease of the USPD. Thus, by realizing these possibilities of the change of the future public debt, the US government together with other governments may need to make more effective policies to control the still-increasing TPDO and also take measures for handling the possibility of sudden events.

6   Conclusion

In conclusion, this paper has presented the neuronet projection for the US public debt (USPD), showing different possibilities of the change of the USPD in the future. Based on this work, we expect to do more and deeper researches on the USPD, such as the trend of the debt held by the public and the debt-to-GDP ratio of the United States. Besides, we anticipate this work to be a starting point for proposing and developing many more sophisticated projection models of the USPD.

Acknowledgments. This work is supported by the National Natural Science Foundation of China (with number 61473323), by the Foundation of Key Laboratory of Autonomous Systems and Networked Control, Ministry of Education, China (with number 2013A07), and also by the Science and Technology Program of Guangzhou, China (with number 2014J4100057).

References

1. US Government Accountability Office: Federal Debt Basics - How Large is the Federal Debt? Government Accountability Office, Washington, D.C. (2012)
2. US Government Accountability Office: The Nation's Long-Term Fiscal Outlook: September 2008 Update. Government Accountability Office, Washington, D.C. (2008)
3. Friedman, B.M.: Debt and Economic Activity in the United States. University of Chicago Press, Chicago (1982)
4. US Department of the Treasury: Public Debt Reports: Historical Debt Outstanding - Annual. Department of the Treasury, Washington, D.C. (2013)
5. US Congressional Budget Office: Updated Budget Projections: Fiscal Years 2013 to 2023. Congressional Budget Office, Washington, D.C. (2013)
6. US Congressional Budget Office: The 2013 Long-Term Budget Outlook. Congressional Budget Office, Washington, D.C. (2013)
7. Chernew, M.E., Baicker, K., Hsu, J.: The specter of financial armageddon-health care and federal debt in the united states. New Engl. J. Med. **362**(13), 1166–1168 (2010)

8. Young, J.A.: Technology and competitiveness: a key to the economic future of the united states. Science **241**(4863), 313–316 (1998)

9. Ooms, V.D.: Budget priorities of the nation. Science **258**(5089), 1742–1747 (1992)

10. Sarno, L.: The behavior of US public debt: a nonlinear perspective. Econ. Lett. **74**(1), 119–125 (2001)

11. Greiner, A., Kauermann, G.: Sustainability of US public debt: estimating smoothing spline regressions. Econ. Modell. **24**(2), 350–364 (2007)

12. Aizenman, J., Marion, N.: Using inflation to erode the US public debt. J. Macroecon. **33**(4), 524–541 (2011)

13. US Office of Management and Budget: The Budget for Fiscal Year 2013. Historical Tables. Washington, D.C. Office of Management and Budget (2013)

14. Celasun, O., Keim, G.: The U.S. Federal Debt Outlook: Reading the Tea Leaves. IMF Working Paper, International Monetary Fund, United States (2010)

15. Cecchetti, S.G., Mohanty, M.S., Zampolli, F.: The Future of Public Debt: Prospects and Implications. Bank for International Settlements, Monetary and Economic Department (2010)

16. Meuhleisen, M., Towe, C.M., Cardarelli, R.: United States Fiscal Policies and Priorities for Long-Run Sustainability. International Monetary Fund, United States (2004)

17. Bohn, H.: The behavior of U.S public debt and deficits. Q. J. Econ. **113**(3), 949–963 (1998)

18. Zhang, Y., Yu, X., Xiao, L., Li, W., Fan, Z.: Weights and Structure Determination of Artificial Neuronets (Chapter 5). Nova Science Publishers, New York (2013)

19. Zhang, Y., Chen, Y., Jiang, X., Zeng, Q., Zou, A.: Weights-directly-determined and structure-adaptively-tuned neural network based on chebyshev basis functions. Comput. Sci. **36**, 210–213 (2009)

20. Ferone, A., Maddalena, L.: Neural background subtraction for pan-tilt-zoom cameras. IEEE Trans. Syst. Man Cybern. Syst. **44**(5), 571–579 (2014)

21. Henriquez, P., Alonso, J.B., Ferrer, M.A., Travieso, C.M.: Review of automatic fault diagnosis systems using audio and vibration signals. IEEE Trans. Syst. Man Cybern. Syst. **44**(5), 642–652 (2014)

22. Ge, S.S., Yang, C., Lee, T.H.: Adaptive predictive control using neural network for a class of pure-feedback systems in discrete time. IEEE Trans. Neural Netw. **19**(9), 1599–1614 (2008)

23. Wang, Z., Liu, Y., Li, M., Liu, X.: Stability analysis for stochastic cohen-grossberg neural networks with mixed time delays. IEEE Trans. Neural Netw. **17**(3), 814–820 (2006)

24. Kabir, H., Wang, Y., Yu, M., Zhang, Q.: High-dimensional neural-network technique and applications to microwave filter modeling. IEEE Trans. Microw. Theory Tech. **58**(1), 145–156 (2010)

25. Liang, X., Wang, J.: A recurrent neural network for nonlinear optimization with a continuously differentiable objective function and bound constraints. IEEE Trans. Neural Netw. **11**(6), 1251–1262 (2000)

26. US Department of the Treasury: Public Debt Reports: Monthly Statement of the Public Debt (MSPD) and Downloadable Files. Department of the Treasury, Washington, D.C. (2013)

Prediction on Internet Safety Situation of Relevance Vector Machine about GP-RVM Kernel Function

Xiaolan Xie[1(✉)], Zhen Long[2], and Fahui Gu[3,4]

[1] College of Information Science and Engineering,
Guilin University of Technology, Guilin 541006, Guangxi, China
729020742@qq.com
[2] College of Mechanical and Control Engineering,
Guilin University of Technology,
Guilin 541006, Guangxi, China
[3] School of Mathematics and Informatics,
South China Agricultural University,
Guangzhou 510006, Guangdong, China
[4] School of Information Engineering,
Jiangxi Applied Technology Vocational College,
Ganzhou 341000, Jiangxi, China

Abstract. In prediction of network security situation, the prediction accuracy of traditional single kernel function vector machine is a little low. It can't describe the randomness and abruptness, and it has some limitation. A network security forecasting model was put forward which combined Gaussian kernel function and polynomial kernel to solve this problem. Proved by simulation experiment, this model can increase prediction accuracy and it has some practical meaning.

Keywords: Relevance vector machine · Compound kernel function · Situation prediction

1 Introduction

Informatization is the tendency of world economical society and important strength to promote the reformation and improvement of society. Human society has entered information era, enjoying achievements brought by informatization and facing more and more information security problems. As an important untraditional security factor, information security has become an important subject which determines the results of informatization, political stability, economic development and national security.

Network security situation awareness is a method for quantitative analysis on network security. It obtains, understands, evaluates and predicts factors which influence network security. It's also a fine measurement of network security, so it is of great meaning for the research of network security situation awareness. Some researchers put forward many forecasting methods such as time series method and evidence theory method and so on [1, 2]. But the results can just reflect the periodicity and similarity of network security, it

© Springer Science+Business Media Singapore 2016
K. Li et al. (Eds.): ISICA 2015, CCIS 575, pp. 724–733, 2016.
DOI: 10.1007/978-981-10-0356-1_76

can't describe the randomness and abruptness [3, 4]. In order to improve the accuracy of forecasting and solve the problem that the forecasting accuracy of traditional single kernel function is very low, this thesis put forward a network security forecasting model of a relevance vector machine, GP-RVM (Gaussian-Polynomial kernel function), which combined Gaussian kernel function and polynomial kernel.

2 RVM (Relevance Vector Machine)

Input target data set, Suppose target is a model with noise:

$$t_n = y(x_n; w) + \varepsilon_n. \tag{1}$$

ε_n is additive noise, and it matches: $\varepsilon_n \in N(0, \sigma^2)$. So $p(t_n|x) = N(t_n|y(x_n), \sigma^2)$, function $y(x_n)$ destination as follows:

$$y(x; w) = \sum_{i=1}^{N} \omega_i K(x, x_i) + \omega_0. \tag{2}$$

$K(x, x_i)$ is kernel function, ω_i is the weight of the model, and $\phi_i(x) \equiv K(x, x_i)$. Suppose t_n is independently distributed, the ratio of likehood estimation of the data group is:

$$p(t|w, \sigma^2) = (2\pi\sigma^2)^{-\frac{N}{2}} \exp\left\{ -\frac{1}{2\sigma^2} \|t - \Phi w\|^2 \right\}. \tag{3}$$

$t = (t_1, t_2, \cdots, t_n)^T$, $\omega = (\omega_1, \omega_2, \cdots, \omega_n)^T$ as vector. Φ is $N \times (N+1)$ high dimensional structure matrix. $\Phi = [\phi(x_1), \phi(x_2), \cdots, \phi(x_n)]^T$ and $\phi(x_n) = [1, K(x_n, x_1), K(x_n, x_2), \cdots, K(x_n, x_N)]^T$. Under Bayesian framework, weight w can be made out through maximum likelihood metho. In order to avoid over-fit phenomena, RVM defined Gaussian prior probability distributed parameter for every weight:

$$p(w|\alpha) = \prod_{i=0}^{N} N(\omega_i|0, \alpha_i^{-1}). \tag{4}$$

Conditional probability can be changed as follows according to probability prediction formula:

$$p(t_*|t) = \int p(t_*|w, \sigma^2) p(w, \alpha, \sigma^2|t) dw d\alpha d\sigma^2. \tag{5}$$

Formula (5) can be changed according to Bayesian theory:

$$p(w, \alpha, \sigma^2 | t) = \frac{p(t | w, \sigma^2) p(w | \alpha) p(\alpha, \sigma^2 | t)}{p(t | \alpha, \sigma^2)}$$

$$= (2\pi)^{-\frac{N+1}{2}} \left| (\sigma^{-2} \Phi^T \Phi + Z)^{-1} \right|^{-\frac{1}{2}} \exp \left\{ -\frac{1}{2} (w - L)^T (\sigma^{-2} \Phi^T \Phi + Z)(w - L) \right\}. \quad (6)$$

$L = \sigma^{-2} (\sigma^{-2} \Phi^T \Phi + Z)^{-1} \Phi^T t$, $\quad Z = diag(\alpha_0, \alpha_1, \cdots, \alpha_N)$. According to delta approximate function $p(\alpha, \sigma^2 | t) \approx \delta(\alpha_{mp}, \sigma_{mp}^2)$ (5) can be replaced by (7), predictive distribution of t_*, estimated value of x_* is (8).

$$p(t_* | t) = \int p(t_* | w, \alpha_{mp}, \sigma_{mp}^2) p(w | t, \alpha_{mp}, \sigma_{mp}^2) dw. \quad (7)$$

$$p(t_* | t, \alpha_{mp}, \sigma_{mp}^2) = N(t_* | y_*, \sigma_*^2). \quad (8)$$

$$y_* = \Phi(x_*) \sigma^{-2} (\sigma^{-2} \Phi^T \Phi + Z)^{-1} \Phi^T t, \sigma_*^2 = \sigma_{MP}^2 + \Phi(x_*)^T (\sigma^{-2} \Phi^T \Phi + Z)^{-1} \Phi(x_*).$$

3 Structure Hybrid Kernel Function

In practical use, the relevance vector structured by single kernel functiom has quite good effect in many areas. However, performance of kernel function varies a lot in different situation, because different kernel function has different features. Gaussian kernel function is a local kernel. It only has influence on few data around test point and slight influence on far test point. On the contrary, polynomial kernel has influence on both data around and far from test point, so polynomial kernel is considered global kernels. In document [5], it compared these two kernel functions. When there is information with different structure in data features [6, 7], or there is irregular distribution of multidimensional data caused by big sample size [10], or there is uneven data distribution in new space after data maps in high-dimensional feature space through kernel function [9], simple kernel function can't adapt well to the situation that sample data is various and complex. Therefore, we can combine the advantages of these two kernel functions to restructure a new function to make prediction model have a good learning ability and generalization ability under the effect of polynomial kernel.

3.1 Matrix of Gaussian Kernel Function

Gaussian kernel function:

$$K(x_i, x_j) = \exp \left\{ -\frac{(x_i - x_j)^2}{2\sigma^2} \right\}. \quad (9)$$

Kernel function can be structured according to the expression:

$$K(x_i, x_j) = \exp\left\{-\frac{\|x_i - x_j\|^2}{\sigma^2}\right\}. \tag{10}$$

According to the expression, kernel matrix $K(x_i, x_j)$ has following natures:

$$K(x_i, x_j) = \begin{cases} 0, & x_i = x_j \\ k(x_i, x_j) = k(x_j, x_i), x_i \neq x_j \end{cases}. \tag{11}$$

3.2   Kernel Matrix of Polynomial Kernel

Polynomial kernel:

$$K(x_i, x_j) = ((x_i, x_j) + r)^q. \tag{12}$$

Therefore, its kernel matrix can be structured as follows according to the expression:

$$K(x_i, x_j) = ((x_i \cdot x_j) + r)^q. \tag{13}$$

According to the expression, kernel matrix $K(x_i, x_j)$ has following natures:

$$K(x_i, x_j) = \begin{cases} (\|x_i\|^2 + r)^q, & x_i = x_j \\ k(x_i, x_j) = k(x_j, x_i), x_i \neq x_j \end{cases}. \tag{14}$$

3.3   Compound Kernel Function

Kernel matrix of polynomial kernel is as follows according to these two kernel matrixes:

$$K(x_i, x_j) = \lambda \cdot \exp\left[\frac{\|(x_i - x_j\|^2}{2\sigma^2}\right] + (1 - \lambda) \cdot [(x_i \cdot x_j) + r]^q. \tag{15}$$

4   Prediction Model of Network Security Situation

The process of prediction model of network security situation:

a. Collect data of network security situation and preprocess it, separate 70 % data as training set, 30 % as test set.

b. Normalize collected data within range from 0 to 1, and the method is as follows:

$$x_i' = \frac{x_i - x_{\min}}{x_{\max} - x_{\min}}. \tag{16}$$

x_i is the primitive value of network security situation. x_i' is the normalized situational value. $x_{\max}$ and $x_{\min}$ is the maximum and minimum value in network security situation.

c. Transferred one-dimensional data of network security situation into multidimensional data of network security situation by making sure of embedding dimension and time delay.

d. Train the training sample set repeatedly using the relevance vector of compound kernel function, choose reference parameter of kernel function according to cross-validation method, build network security situation model and predict test sample.

e. Predict network security situation using the prediction model. As shown in Fig. 1, the flow chart.

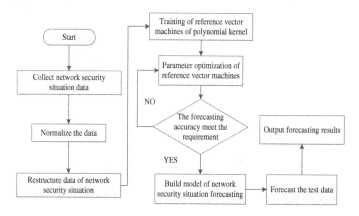

Fig. 1. Flow chart of network security situation prediction

5 Simulation Experiment

5.1 Experimental Data Sets

This thesis takes intrusion datas announced by Honeynet [8] as simulation object, 120 samples. Choose the last 50 samples as testing sample sets and normalize the samples. Figure 2 shows the normalized network security forecasting.

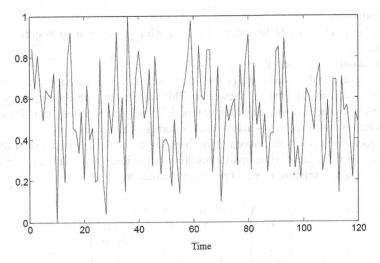

Fig. 2. The normalized network security data

5.2   Samples of Experiment Data for Study

This thesis chooses bacterial colony chemotaxis algorithms to select the best delay time and embedding dimension. Bacterial colony chemotaxis algorithms is discovered from species behavior nature and used to solve optimization problem. Foraging behavior of bacterial colony covers chemotaxis, swarming, reproduction, elimination and dispersal.

a. Chemotaxis

Chemotaxis of coli bacillus can be divided into 2 kinds, advancing and flipping. The position of bacteria $\theta^i(j+1, k, l)$ after chemotaxis behavior is:
$\phi(j)$ is random direction vector of unit length, $C(i)$ is step-size of chemotaxis behavior.

b. Swarming

The mathematics expression of swarming behavior of bacteria is:

$$
\begin{aligned}
J_{cc}(\theta, P(j, k, l)) &= \sum_{i=1}^{S} J_{cc}^i(\theta, \theta^i(j, k, l)) \\
&= \sum_{i=1}^{S} \{-d_{attract} \exp[-\omega_{attract} \sum_{m=1}^{\eta} (\theta_m - \theta_m^i)^2]\}. \\
&+ \sum_{i=1}^{S} \{-h_{repellant} \exp[-\omega_{repellant} \sum_{m=1}^{\eta} (\theta_m - \theta_m^i)^2]\}
\end{aligned}
\tag{17}
$$

θ_m is the position of the mth bacteria. S is the number of bacterial colony. $d_{attract}$ is the depth of the gravity. $\omega_{attract}$ is the width of the gravity; $h_{repellant}$ is the repulsion height. $\omega_{repllant}$ is the repulsion width.

c. Reproduction

After some chemotaxis behaviors, bacteria with good foraging results come to breed.

d. Elimination and Dispersal

Elimination and dispersal is the process that bacteria in actual environment killed by outside force or dispelled to a new area. This will destroy the chemotaxis process, but bacteria may find an area with more food. Therefore, in the long run, this kind of elimination and dispersal is also a kind of foraging behavior.

Use bacterial colony chemotaxis algorithms to gain data of network security situation, select best delay time and embedding dimension, gain results of 2 and 7, so study samples structure of network security situation is:

$$
\begin{pmatrix} X_1 \\ X_2 \\ \vdots \\ X_{106} \end{pmatrix}^T = \begin{pmatrix} x_1 & x_3 & \cdots & x_{13} \\ x_2 & x_4 & \cdots & x_{14} \\ \vdots & \vdots & \cdots & \vdots \\ x_{106} & x_{108} & \cdots & x_{120} \end{pmatrix}^T . \tag{18}
$$

6   Analysis of Experiment Results

6.1   Experiment Results

Use GP-RVM to fit training samples of network security situation, the fitting results gained by MATLAB show as Fig. 3. From the Figure, we can see the fitting accuracy of GP-RVM is quite high, and we can regard GP-RVM is a network security situation forecasting model with high fitting accuracy.

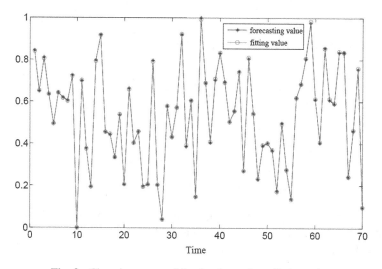

Fig. 3. Changing curve of fitted value and predictive value

In application of network security situation, fitting result can only describe the past and present situation of Internet. However, performance of a forecasting model is mainly valued by its generalization. Therefore, using GP-RVM to forecast the test sample, the results are shown as Fig. 4, and the changing curve of forecasting errors are shown as Fig. 5. As shown in Figs. 4 and 5, GP-RVM can accurately forecast the changing features of network security situation and the forecasting error is very slight. The results show that GP-RVM is a quite accurate network security situation prediction model with good generalization ability.

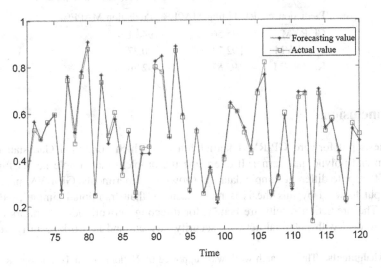

Fig. 4. Changing curve of normalized network security forecasting values and actual values

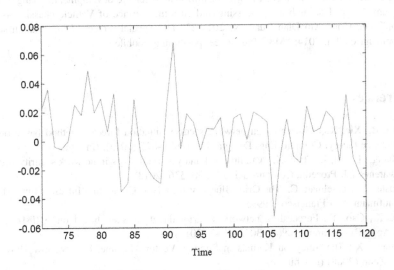

Fig. 5. Changing curve of forecasting error

6.2   Comparison Between GP-RVM and Other Forecasting Models of Network Security Situation

To show the advantages of GP-RVM, in GP-RVM, poly-RVM and RBF-RVM, compare MAPE with mean relative error, and the results show as Table 1. As shown in the table, compared with simple kernel function, GP-RVM can improve its accuracy of forecasting, and it has a good ability of study and generalization.

Table 1. Comparison between GP-RVM Model and other models

Prediction model	Fitting MAPE%	Prediction MAPE%
CF-RVM	95.56	94.21
Poly-RVM	92.74	90.37
RBF-RVM	93.87	92.76

7   Conclusions

This thesis put forward GP-RVM kernel function which combined Gaussian kernel function and polynomial kernel. It overcame the problem that simple kernel function couldn't adapt to different sample data. As shown by experiments, GP-RVM prediction model put forward by this thesis is more accurate than traditional simple prediction model. The prediction results are benefit for directing network administrators to take actions to deal with potential network security incident and network security trends.

Acknowledgments. This research work was supported by National High Technology Research and Development "Program 863" under Grant No.2013AA12A402. The scientific research and technology development plan of Guilin: "Four-wheel positioning mobile cloud platform", No. Gui Ke Gong 20150103-9. Special project of information service development, Guangxi, 2015: "Huge amount of data analysis, processing and customer service of Vehicle chassis based on Hadoop", Gui Gong Xin Dian Ruan, No.2015-239 and Special project of information service development, Guilin, 2014: "M4G four-wheel positioning mobile system".

References

1. Shi, B., Xie, X.Q.: Research on network security situation forecast method based on D-S evidence theory. Comput. Eng. Design. **34**(3), 821–825 (2013). (In Chinese)
2. Zhang, H., Shi, J., Chen, X.: A multi-level analysis framework in network security situation awareness. J. Procedia Comput. Sci. **17**, 530–536 (2013)
3. Foster, I., Kesselman, C.: The Grid: Blueprint for a New Computing Infrastructure. Morgan Kaufmann, San Francisco (1999)
4. Li, K., Cao, Y.: Forecasting network security threat situation based on ARIMA model. J. Appl. Res. Comput. **29**(8), 3042–3043 (2012)
5. Huang, X.: The Study on Kernels in Support Vector Machine. D. Soochow University, Su Zhou (2008) (in Chinese)

6. Foster, I., Kesselman, C., Nick, J., Tuecke, S.: The Physiology of the Grid: an Open Grid Services Architecture for Distributed Systems Integration. Technical report, Global Grid Forum (2002)
7. Bach, F.R.: Consistency of the group lasso and multiple kernel learning. J. Mach. Learn. Res. **9**(6), 1179–1222 (2008)
8. National Center for Biotechnology, Rakotomamonjy, A., Bach, F.R., Canu, S., et al.: Simple MKL. J. Mach. Learn. Res. **9**(11), 2491–2521 (2008)
9. Zheng, D.N., Wang, J.X., Zhao, Y.N.: Non-flat function estimation with a multi-scale support vector regression. J. Neurocomputing **70**(1–3), 420–429 (2006)
10. Liu, Y.-L., Feng, G.-D., Lian, Y.-F., et al.: Network situation prediction method based on spatial-time dimension analysis. J. Comput. Res. **51**(8), 1681–1694 (2014). (in Chinese)

Author Index

Printed in the United States
By Bookmasters